Advances in Intelligent Systems and Computing

Volume 905

The series "Advances in Intelligent Systems and Computing" contains publications on theory, applications, and design methods of Intelligent Systems and Intelligent Computing. Virtually all disciplines such as engineering, natural sciences, computer and information science, ICT, economics, business, e-commerce, environment, healthcare, life science are covered. The list of topics spans all the areas of modern intelligent systems and computing such as: computational intelligence, soft computing including neural networks, fuzzy systems, evolutionary computing and the fusion of these paradigms, social intelligence, ambient intelligence, computational neuroscience, artificial life, virtual worlds and society, cognitive science and systems, Perception and Vision, DNA and immune based systems, self-organizing and adaptive systems, e-Learning and teaching, human-centered and human-centric computing, recommender systems, intelligent control, robotics and mechatronics including human-machine teaming, knowledge-based paradigms, learning paradigms, machine ethics, intelligent data analysis, knowledge management, intelligent agents, intelligent decision making and support, intelligent network security, trust management, interactive entertainment, Web intelligence and multimedia.

The publications within "Advances in Intelligent Systems and Computing" are primarily proceedings of important conferences, symposia and congresses. They cover significant recent developments in the field, both of a foundational and applicable character. An important characteristic feature of the series is the short publication time and world-wide distribution. This permits a rapid and broad dissemination of research results.

** Indexing: The books of this series are submitted to ISI Proceedings, EI-Compendex, DBLP, SCOPUS, Google Scholar and Springerlink **

More information about this series at http://www.springer.com/series/11156

Qi Liu · Mustafa Mısır ·
Xin Wang · Weiping Liu
Editors

The 8th International Conference on Computer Engineering and Networks (CENet2018)

Springer

Editors
Qi Liu
School of Computing
Edinburgh Napier University
Edinburgh, UK

Xin Wang
School of Computer
Fudan University
Shanghai, China

Mustafa Mısır
College of Computer Science
and Technology
Nanjing University of Aeronautics
and Astronautics
Nanjing, China

Weiping Liu
College of Information Science
and Technology/College of Cyber Security
Jinan University
Guangzhou, China

ISSN 2194-5357 ISSN 2194-5365 (electronic)
Advances in Intelligent Systems and Computing
ISBN 978-3-030-14679-5 ISBN 978-3-030-14680-1 (eBook)
https://doi.org/10.1007/978-3-030-14680-1

Library of Congress Control Number: 2019933160

This Springer imprint is published by the registered company Springer Nature Switzerland AG
The registered company address is: Gewerbestrasse 11, 6330 Cham, Switzerland

Preface

This conference proceedings covers emerging topics for computer engineering and networking and their applications. The text covers state-of-the-art developments in areas such as artificial intelligence, machine learning, information analysis and communication.

This proceedings is a collection of the 8th International Conference on Computer Engineering and Networks (CENet2018) held on 17–19 August 2018 in Shanghai, China. It contains the five parts: Part I focuses on System Detection and Application (26 papers); Part II Machine Learning and Application (25 papers); Part III Information Analysis and Application (24 papers), Part IV Communication Analysis and Application (28 papers).

Each part can be used as an excellent reference by industry practitioners, university faculties, research fellows, and undergraduate as well as graduate students who need to build a knowledge base of the most current advances in the topics covered by this conference proceedings.

Thanks to the authors for their hard work and dedication as well as the reviewers for ensuring the selection of only the highest-quality papers; their efforts made the proceedings possible.

Contents

Communication Analysis and Application

List of Contributors

Pitri Bhakta Adhikari Tri - Chandra M. Campus, Tribhuvan University, Kathmandu, Nepal

Jiangwei Bai School of Information Science and Engineering, Lanzhou University, Lanzhou, China

Cheng Cai Power Dispatching Control Center, Guizhou Power Grid Limited Liability Company, Guiyang, Guizhou, China

Carlo Vittorio Cannistraci Biotechnology Center, Technische Universitat Dresden, Dresden, Germany

Haoshun Cao Institute of Information Security, Yunnan Police College, Kunming, China

Ning Cao School of Computer and Information, Hohai University, Nanjing, China

Xiawei Cao College of Computer Science and Technology, Hengyang Normal University, Hengyang, China

Yunpeng Cao School of Information Science and Engineering, Linyi University Institute of Linyi University of Shandong Provincial Key Laboratory of Network Based Intelligent Computing, Linyi, China

Weiling Chang CNCERT/CC, Beijing, China

Bingwei Chen State Key Laboratory of Networking and Switching Technology, Beijing University of Posts and Telecommunications, Beijing, China

Dengchi Chen Power Dispatching Control Center, Guizhou Power Grid Limited Liability Company, Guiyang, Guizhou, China

Dongcheng Chen Jiangsu Automation Research Institute, Lianyungang, China

Jia Chen School of Educational Information Technology, Central China Normal University, Wuhan, Hubei, China

Jiaying Chen Institution of Internet of Things, Nanjing University of Posts and Telecommunications, Nanjing, Jiangsu Province, China

Jinpeng Chen School of Software Engineering, Beijing University of Posts and Telecommunications, Beijing, China

Ming Chen Hainan Power Grid Co., Ltd., Haikou, China

Peng Chen CSG Power Dispatching Control Center, Guangzhou, China

Shuwen Chen School of Information and Communication of National University of Defense Technology, Xi'an, China

Si Chen Changchun Normal University, Changchun, Jilin, China

Tieming Chen Xi'an Research Institute of High-Technology, Xi'an, Shaanxi, China

Wenwei Chen State Grid Information and Telecommunication Group, Ltd., Beijing, China

Xi Chen State Grid Fujian Electric Power Co., Ltd., Fuzhou Power Supply Company, Fuzhou, China

Xingyu Chen Institute of Network Technology, Beijing University of Posts and Telecommunications, Beijing, China

Xue Chen School of College of Intelligence and Computing, Tianjin University, Tianjin, China

Yangling Chen State Grid Jiangsu Electric Power CO., Ltd., Nanjing Power Supply Company, Nanjing, China

Yanzhou Chen CSG Power Dispatching Control Center, Guangzhou, China

Yezhao Chen Power Grid Dispatching Control Center of Guangdong Power Grid Co., Ltd., Guangzhou, Guangdong Province, China

Yue Chen NARI Information & Communication Technology Co., Ltd., Nanjing, China

Zhangguo Chen NARI Information & Communication Technology Co., Ltd., Nanjing, China

Zhangguo Chen Electric Power Dispatching and Control Center of Guangdong Power Grid Co., Ltd., Guangzhou, China

Zhili Chen National Engineering Research Center for Information Technology in Agriculture, Beijing, China; College of Information and Electrical Engineering, China Agricultural University, Beijing, China

Zhixiong Chen North China Electric Power University, Baoding, Hebei Province, China

Zidi Chen State Key Laboratory of Networking and Switching Technology, Beijing University of Posts and Telecommunications, Beijing, China

Le Cheng Department of Computer Science and Engineering, Huaian Vocational College of Information Technology, Huaian, China; College of Computer and Information, Hohai University, Nanjing, China

Long Cheng State Key Laboratory of Networking and Switching Technology, Beijing University of Posts and Telecommunications, Beijing, China

Yihan Cheng School of Software Engineering, Beijing University of Posts and Telecommunications, Beijing, China

Jia Cui CNCERT/CC, Beijing, China

Xiao Cui China Academy of Railway Sciences, Beijing, China

Yifeng Cui School of Information Science and Engineering, Linyi University, Linyi City, China

Hongwei Deng School of Computer Science and Technology, Hengyang Normal University, Hengyang, Hunan, China; School of Information Science and Engineering, Central South University, Changsha, Hunan, China; Hunan Provincial Key Laboratory of Intelligent Information Processing and Application, Hengyang, China

Lin Deng The 29th Institute of CETC, Chengdu, China

Ming Deng The 29th Institute of CETC, Chengdu, China

Song Deng School of Software and Internet of Things Engineering, Jiangxi University of Finance and Economics, Nanchang, China

Wencheng Deng Hainan Power Grid Co., Ltd., Haikou, China

Feng Ding The Key Laboratory of Information System Engineering, The 28th Research Institute of China Electronics Technology Group Corporation, Nanjing, China

Ran Ding The Key Laboratory of Information System Engineering, The 28th Research Institute of China Electronics Technology Group Corporation, Nanjing, China

Gangsong Dong Information & Telecommunication Co. of State Grid Henan Electric Power Company, Zhengzhou, China

Wu Dong Power Dispatching Control Center, Guizhou Power Grid Limited Liability Company, Guiyang, Guizhou, China

Enbo Du School of Software Engineering, Beijing University of Posts and Telecommunications, Beijing, China

Ping Du Jiangxi Engineering Laboratory on Radioactive Geoscience and Big Data Technology, East China University of Technology, Nanchang, China

Yanhui Du Information Technology and Network Security Institute, People's Public Security University of China, Beijing, China; Collaborative Innovation Center of Security and Law for Cyberspace, People's Public Security University of China, Beijing, China

Xiao Feng State Grid Xintong Yili Technology Co., Ltd., Beijing, China

Guangyuan Fu Xi'an Research Institute of High-Technology, Xi'an, Shaanxi, China

Jianghui Fu East China University of Technology, Nan Chang, China

Wenyu Fu China Huadian Corporation LTD Sichuan Baozhusi Hydropower Plant, Guangyuan, Sichuan, China

Yingying Gai Institute of Oceanographic Instrumentation, Qilu University of Technology (Shandong Academy of Sciences), Shandong Provincial Key Laboratory of Marine Monitoring Instrument Equipment Technology, National Engineering and Technological Research Center of Marine Monitoring Equipment, Qingdao, China

Jihong Gao Department of Computer Science and Engineering, Huaian Vocational College of Information Technology, Huaian, China

Qiang Gao Information Center of Shandong Province People's Congress Standing Committee General Office, Jinan, China

Rui Gao School of Electronic and Information Engineering, Lanzhou Jiaotong University, Lanzhou, China; Key Laboratory of Opto-Technology and Intelligent Control, Ministry of Education, Lanzhou Jiaotong University, Lanzhou, China

Tilei Gao School of Software, Yunnan University, Kunming, China; School of Information, Yunnan University of Finance and Economics, Kunming, China

Zhipeng Gao State Key Laboratory of Networking and Switching Technology, Beijing University of Posts and Telecommunications, Beijing, China

Shuangquan Ge Institute of Computer Application, Chinese Academy of Engineering Physics, Mianyang, China

Songlin Ge School of Information Engineering, East China Jiaotong University, Nanchang, China

Qingtian Geng Department of Computer Science and Technology, Changchun Normal University, Changchun, China

Yi Gong The 29th Institute of CETC, Chengdu, China

Qixue Guan School of Information Science and Engineering, Shenyang Ligong University, Shenyang, China

Shijie Guan School of Information Science and Engineering, Shenyang Ligong University, Shenyang, China

Xinxin Guan Beijing Institute of Graphic Communication, Beijing, China

Qing Guo Institute of Network Technology, Beijing University of Posts and Telecommunications, Beijing, China

Wenming Guo Beijing University of Posts and Telecommunications, Beijing, China

Yihe Guo North China Electric Power University, Baoding, Hebei Province, China

Dongsheng Han North China Electric Power University, Baoding, Hebei Province, China

Xianbin Han Institute of Network Technology, Beijing University of Posts and Telecommunications, Beijing, China

Bo He China Southern Power Grid Co., Ltd., Guangzhou, China

Shangjun He State Grid Fujian Electric Power Co., Ltd., Fuzhou Power Supply Company, Fuzhou, China

Yueshun He Jiangxi Engineering Laboratory on Radioactive Geoscience and Big Data Technology, East China University of Technology, Nanchang, China

Jie Hong Hainan Power Grid Co., Ltd., Haikou, China

Gonghua Hou State Grid Fujian Electric Power Co., Ltd., Fuzhou Power Supply Company, Fuzhou, China

Yue Hou Beijing Guodiantong Network Technical Co. Ltd., Beijing, China

Bin Hu School of Technical College for the Deaf, Tianjin University of Technology, Tianjin, China

Hanjuan Huang Wuyi University, Wuyishan, China

Jiamin Huang Haylion Technologies Co., Ltd., Shenzhen, China

Jie Huang School of Software Engineering, Tongji University, Shanghai, China

Jinde Huang Guangxi College of Education, Nanning, China

Jun Steed Huang Southern University of Science and Technology, Shenzhen, China

Lin Huang Beijing University of Posts and Telecommunications, Beijing, China

Tianming Huang NARI Information & Communication Technology Co., Ltd., Nanjing, China

Yu Huang College of Informatics, Huazhong Agricultural University, Wuhan, China

Zengxi Huang School of Computer and Software Engineering, Xihua University, Chengdu, China

Zhipeng Huang School of Software Engineering, Beijing University of Posts and Telecommunications, Beijing, China

Chao Hu NARI Information and Communication Technology Co., Ltd., Nanjing, China

Congliang Hu 719 Research Institute, Wuhan, China

Feifei Hu Power Grid Dispatching Control Center of China Southern Power Grid, Guangzhou, China

Jingying Hu Jiangsu Automation Research Institute, Lianyungang, China

Peishan Hu School of Educational Information Technology, Central China Normal University, Wuhan, Hubei, China

Qinghua Hu School of Computer Science and Technology, Tianjin University, Tianjin, China

Shunbo Hu School of Information Science and Engineering, Linyi University, Linyi City, China

Yifei Hu Jinan University, Guangzhou, China

Zhaohui Hu Dingxin Information Technology Co., Ltd., Guangzhou, China

Ran Huo North China Electric Power University, Baoding, China

Ning Ji Air Force Command College, Beijing, China

Wei Jia NRGD Quantum CTek., Ltd., Nanjing, China

Xueming Jia Institute of Information Security, Yunnan Police College, Kunming, China

Meizhu Jiang East China University of Technology, Nan Chang, China

Rong Jiang School of Information, Yunnan University of Finance and Economics, Kunming, China

Xuesong Jiang Qilu University of Technology (Shandong Academy of Sciences), Jinan, China

Yingyan Jiang Power Grid Dispatching Control Center of Guangdong Power Grid Co., Ltd., Guangzhou, Guangdong Province, China

Yueqiu Jiang School of Information Science and Engineering, Shenyang Ligong University, Shenyang, China

Ge Jiao College of Computer Science and Technology, Hengyang Normal University, Hengyang, Hunan, China; Hunan Provincial Key Laboratory of Intelligent Information Processing and Application, Hengyang, Hunan, China

Pengfei Jiao School of Center of Biosafety Research and Strategy, Tianjin University, Tianjin, China

Di Jin College of Intelligence and Computing, Tianjin University, Tianjin, China

Houliang Kang College of Humanities and Art, Yunnan College of Business Management, Kunming, Yunnan, China

Zhongmiao Kang Power Grid Dispatching Control Center, Guangdong Power Grid Co., Ltd., Guangzhou, Guangdong Province, China

Huihong Lan Guangxi College of Education, Nanning, China

Shuwei Lei College of Information Science and Technology, Northwest University, Xi'an, China

Bo Li Hainan Power Grid Co., Ltd., Haikou, China

DongWang Li Qilu University of Technology (Shandong Academy of Sciences), Jinan, China

Dongchang Li Nari Group Corporation, Nanjing, China

Gao Li Southern University of Science and Technology, Shenzhen, China

Guoqiang Li School of Information Science and Engineering, Linyi University, Linyi City, China

Jie Li State Grid Xinjiang Information & Telecommunication Company, Urumqi, China

Juan Li School of Information and Communication of National University of Defense Technology, Xi'an, China

Lang Li College of Computer Science and Technology, Hengyang Normal University, Hengyang, China

Lian Li School of Information Science and Engineering, Lanzhou University, Lanzhou, China

Liang Li 719 Research Institute, Wuhan, China

Qingliang Li College of Computer Science and Technology, Changchun Normal University, Changchun, China

Tong Li Key Laboratory in Software Engineering of Yunnan Province, Kunming, China

Wei Li School of Information and Communication of National University of Defense Technology, Xi'an, China

Weijian Li Power Grid Dispatching Control Center, Guangdong Power Grid Co., Ltd., Guangzhou, Guangdong Province, China

Wencui Li Information & Telecommunication Co. of State Grid Henan Electric Power Company, Zhengzhou, China

Wenjing Li State Grid Information and Telecommunication Group Co., Ltd., Beijing, China

Xiong Li Information & Telecommunication Co. of State Grid Henan Electric Power Company, Zhengzhou, China

Yangqun Li College of Internet of Things, Nanjing University of Posts and Telecommunications, Nanjing, Jiangsu, China

Yeli Li Beijing Institute of Graphic Communication, Beijing, China

Yijin Li The 28th Research Institute of China Electronics Technology Group Corporation, Nanjing, China

Yuan Li Xi'an Research Institute of High-Technology, Xi'an, Shaanxi, China

Duan Liang Dingxin Information Technology Co., Ltd., Guangzhou, China

Lihong Liang China Special Equipment Inspection and Research Institute, Beijing, China

Xiaoman Liang School of Computer Science and Technology, Hengyang Normal University, Hengyang, Hunan, China

Lin Lin State Grid Fujian Electric Power Co., Ltd., Fuzhou Power Supply Company, Fuzhou, China

Mi Lin Hainan Power Grid Co., Ltd., Haikou, China

Peng Lin Beijing Vectinfo Technologies Co., Ltd., Beijing, China

Chaochao Liu College of Intelligence and Computing, Tianjin University, Tianjin, China; Tianjin Key Laboratory of Advanced Networking (TANK), Tianjin University, Tianjin, China

Enxiao Liu Institute of Oceanographic Instrumentation, Qilu University of Technology (Shandong Academy of Sciences), Shandong Provincial Key Laboratory of Marine Monitoring Instrument Equipment Technology, National Engineering and Technological Research Center of Marine Monitoring Equipment, Qingdao, China

Guangjie Liu Department of Computer Science and Technology, Changchun Normal University, Changchun, China

Jia Liu Key Lab of Networks and Information Security of PAP, Xi'an, China

Jianbo Liu Institute of Computer Application, Chinese Academy of Engineering Physics, Mianyang, China

Jin Liu College of Information Science and Technology, Northwest University, Xi'an, China

Jinsuo Liu State Grid Electric Power Research Institute, NARI Group Corporation, Nanjing, China; NRGD Quantum CTek., Ltd., Nanjing, China

Kai Liu College of Informatics, Huazhong Agricultural University, Wuhan, China

Li Liu School of Information Science and Engineering, Linyi University, Linyi City, China

Lu Liu Beijing University of Posts and Telecommunications, Beijing, China

Qifei Liu Information Technology and Network Security Institute, People's Public Security University of China, Beijing, China

Qing Liu Power Dispatching Control Center, Guizhou Power Grid Limited Liability Company, Guiyang, Guizhou, China

Ran Liu North China Electric Power University, Baoding, Hebei Province, China

Shengqing Liu Jiangxi Expressway Connection Management Centre, Nanchang, China

Song Liu CSG Power Dispatching Control Center, Guangzhou, China

Xu Liu Power Dispatching Control Center, Guizhou Power Grid Limited Liability Company, Guiyang, Guizhou, China

Yan Liu Changchun Normal University, Changchun, China

Yanchen Liu The 28th Research Institute of China Electronics Technology Group Corporation, Nanjing, China

Yang Liu School of Communications and Electronics, Jiangxi Science and Technology Normal University, Nanchang, China

Yingchu Liu North China Electric Power University, Baoding, Hebei Province, China

Zijian Liu Power Grid Dispatching Control Center, Guangdong Power Grid Co., Ltd., Guangzhou, Guangdong Province, China

Zhiyuan Long Hainan Power Grid Co., Ltd., Haikou, China

JiZhao Lu Information & Telecommunication Co. of State Grid Henan Electric Power Company, Zhengzhou, China

Qin Lu Qilu University of Technology (Shandong Academy of Sciences), Jinan, China

Tianliang Lu Information Technology and Network Security Institute, People's Public Security University of China, Beijing, China; Collaborative Innovation Center of Security and Law for Cyberspace, People's Public Security University of China, Beijing, China

Yiguang Lu School of Software Engineering, Beijing University of Posts and Telecommunications, Beijing, China

Wen Luo School of Software and Internet of Things Engineering, Jiangxi University of Finance and Economics, Nanchang, China

Jingya Ma Nari Group Corporation, Nanjing, China

Shaoping Ma Department of Computer Science and Technology, Tsinghua University, Beijing, China

Shaoping Ma Postdoctoral Research Station in Computer Science and Technology of Tsinghua University, Beijing, China

Tao Ma Nari Group Corporation, Nanjing, China

JiaYan Mao Nanjing University of Posts and Telecommunications, Nanjing, Jiangsu, China

Minghe Mao School of Computer and Information, Hohai University, Nanjing, China

Yanqin Mao Nanjing University of Posts and Telecommunications, Nanjing, Jiangsu, China

Dezhuang Meng Institute of Computer Application, Chinese Academy of Engineering Physics, Mianyang, China

Yisheng Miao Beijing Research Center for Information Technology in Agriculture, Beijing Academy of Agriculture and Forestry Sciences, Beijing, China

Dili Peng Power Dispatching Control Center, Guizhou Power Grid Limited Liability Company, Guiyang, Guizhou, China

Lin Peng Beijing Vectinfo Technologies Company, Ltd., Beijing, China

Ji Ping Hohai University Wentian College, Ma'anshan, China

Feng Qi Institute of Network Technology, Beijing University of Posts and Telecommunications, Beijing, China

Jinping Qi School of Electronic and Information Engineering, Lanzhou Jiaotong University, Lanzhou, China

Zhizhong Qiao NARI Information & Communication Technology Co., Ltd., Nanjing, China

Shengguang Qin Qingdao Leicent Limited Liability Company, Qingdao, China

Jinzhi Ran School of Information and Communication of National University of Defense Technology, Xi'an, China

Jiyang Ruan Beijing University of Posts and Telecommunications, Beijing, China

Lanlan Rui State Key Laboratory of Networking and Switching Technology, Beijing University of Posts and Telecommunications, Beijing, China

Mingchi Shao State Grid Jiangsu Electric Power CO., Ltd., Nanjing Power Supply Company, Nanjing, China

Subin Shen School of IOT, Nanjing University of Posts and Telecommunications, Nanjing, China

Yan Shen Wuhan Digital Engineering Institute, Hubei, China

Ziran Shen The 28th Research Institute of China Electronics Technology Group Corporation, Nanjing, China

Rui Shi State Grid Information and Telecommunication Branch, State Grid Corporation of China, Beijing, China

Wenchao Shi School of Software Engineering, Beijing University of Posts and Telecommunications, Beijing, China

Zhan Shi Electric Power Dispatch & Control Center, Guangdong Power Grid Co., Ltd., Guangzhou, Guangdong Province, China

Yanhong Song Department of Computer Science and Engineering, Huaian Vocational College of Information Technology, Huaian, China

Yang Su CSG Power Dispatching Control Center, Guangzhou, China

Zhou Su Electric Power Dispatch & Control Center, Guangdong Power Grid Co., Ltd., Guangzhou, Guangdong Province, China

Chengjie Sun Harbin Institute of Technology, Harbin, China

Meng Sun CSG Power Dispatching Control Center, Guangzhou, China

Mingyu Sun Department of Computer Science and Technology, Changchun Normal University, Changchun, China

Qiucheng Sun Department of Computer Science and Technology, Changchun Normal University, Changchun, China

Yueheng Sun School of College of Intelligence and Computing, Tianjin University, Tianjin, China; Tianjin Key Laboratory of Advanced Networking (TANK), Tianjin University, Tianjin, China

Han Tan Department of Mechanical Engineering, Wuhan Vocational College of Software and Engineering, Wuhan, China

Hao Tang Institute of Computer Application, Chinese Academy of Engineering Physics, Mianyang, China

Jianfeng Tang School of Software Engineering, Tongji University, Shanghai, China

Qi Tang Power Dispatching Control Center, Guizhou Power Grid Limited Liability Company, Guiyang, Guizhou, China

Wenwei Tao CSG Power Dispatching Control Center, Guangzhou, China

Dongbo Tian Nari Group Corporation, Nanjing, China

Fang Tian College of Informatics, Huazhong Agricultural University, Wuhan, China

Xincheng Tian State Grid Tangshan Power Corporation, Tangshan, China

Wentao Tong Wuhan Digital Engineering Institute, Hubei, China

Huaqing Wan 719 Research Institute, Wuhan, China

Bo Wang CNCERT/CC, Beijing, China

Changmei Wang Solar Energy Institute, Yunnan Normal University, Kunming, China

Haibo Wang Department of Computer Science and Engineering, Huaian Vocational College of Information Technology, Huaian, China

Haifeng Wang School of Information Science and Engineering, Linyi University Institute of Linyi University of Shandong Provincial Key Laboratory of Network Based Intelligent Computing, Linyi, China

Haizhu Wang Electric Power Dispatching and Control Center of Guangdong Power Grid Co., Ltd., Guangzhou, China

Hongling Wang East China University of Technology, Nanchang, China; Jiangxi Engineering Laboratory on Radioactive Geoscience and Big Data Technology, East China University of Technology, Nanchang, China; China University of Geosciences, Wuhan, China

Hongqiao Wang Xi'an Research Institute of High-Technology, Xi'an, Shaanxi, China

Jian Wang Shandong College of Information Technology, Weifang, China

Jiao Wang Nari Group Corporation, Nanjing, China

Jie Wang CSG Power Dispatching Control Center, Guangzhou, China

Jingyu Wang School of Information and Communication of National University of Defense Technology, Xi'an, China

Lihong Wang CNCERT/CC, Beijing, China

Qi Wang CSG Power Dispatching Control Center, Guangzhou, China

Shiwen Wang Information & Telecommunication Co. of State Grid Henan Electric Power Company, Zhengzhou, China

Wenjun Wang School of College of Intelligence and Computing, Tianjin University, Tianjin, China; Tianjin Key Laboratory of Advanced Networking (TANK), Tianjin University, Tianjin, China

Xiaoling Wang Institute of Computer Application, Chinese Academy of Engineering Physics, Mianyang, China

Yijie Wang School of Information Engineering, East China Jiaotong University, Nanchang, China

Ying Wang Electric Power Dispatch & Control Center, Guangdong Power Grid Co., Ltd., Guangzhou, Guangdong, China

Yixuan Wang School of Electronic and Information Engineering, Lanzhou Jiaotong University, Lanzhou, China

Zezhou Wang The 29th Institute of CETC, Chengdu, China

Zhili Wang Beijing University of Posts and Telecommunications, Beijing, China

Zhiqian Wang Beijing University of Posts and Telecommunications, Beijing, China

Han Wei School of Communications and Electronics, Jiangxi Science and Technology Normal University, Nanchang, China

Lei Wei State Grid Jiangsu Electric Power Company, Ltd., Nanjing, China

Nan Wei School of Electronic and Information Engineering, Lanzhou Jiaotong University, Lanzhou, China

Xiumei Wei Qilu University of Technology (Shandong Academy of Sciences), Jinan, China

Haojing Weng School of Software and Internet of Things Engineering, Jiangxi University of Finance and Economics, Nanchang, China

Huarui Wu Beijing Research Center for Information Technology in Agriculture, Beijing Academy of Agriculture and Forestry Sciences, Beijing, China

Jiang Wu College of Information Science and Technology, Northwest University, Xi'an, China

Jiawei Wu State Grid Electric Power Company Information and Communication Corporation, Shanghai, China

MaoYing Wu Qilu University of Technology (Shandong Academy of Sciences), Jinan, China

Shaobing Wu Institute of Information Security, Yunnan Police College, Kunming, China

Weiming Wu Hainan Power Grid Co., Ltd., Haikou, China

ZanHong Wu Electric Power Dispatch & Control Center, Guangdong Power Grid Co., Ltd., Guangzhou, Guangdong, China

Zhenyu Wu Institution of Internet of Things, Nanjing University of Posts and Telecommunications, Nanjing, Jiangsu Province, China

Nan Xiang State Grid Jiangsu Electric Power CO., Ltd., Nanjing Power Supply Company, Nanjing, China

Kai Xie Beijing Institute of Graphic Communication, Beijing, China

Xuanlan Xie College of Computer Science and Technology, Hengyang Normal University, Hengyang, China

Zhiyuan Xie North China Electric Power University, Baoding, China

Ao Xiong Institute of Network Technology, Beijing University of Posts and Telecommunications, Beijing, China

Guoquan Xiong East China University of Technology, Nan Chang, China

Fuxin Xu School of Electronic and Information Engineering, Lanzhou Jiaotong University, Lanzhou, China

Jie Xu CNCERT/CC, Beijing, China

Mengting Xu Southern University of Science and Technology, Shenzhen, China

Tong Xu Jingjiang College of Jiangsu University, Zhenjiang, China

Xueke Xu School of Software and Internet of Things Engineering, Jiangxi University of Finance and Economics, Nanchang, China

Yuanzi Xu Air Force Command College, Beijing, China

Jingzhi Xue Nari Group Corporation, Nanjing, China

Tianfeng Yan School of Electronic and Information Engineering, Lanzhou Jiaotong University, Lanzhou, Gansu, China

Fan Yang School of Communications and Electronics, Jiangxi Science and Technology Normal University, Nanchang, China

Hui Yang Beijing University of Posts and Telecommunications, Beijing, China

Jianhui Yang School of Electronic and Information Engineering, Lanzhou Jiaotong University, Lanzhou, China

Lifan Yang State Grid Fujian Electric Power Co., Ltd., Fuzhou Power Supply Company, Fuzhou, China

Liyuan Yang School of Information Science and Engineering, Shenyang Ligong University, Shenyang, China

Ming Yang School of Information, Yunnan University of Finance and Economics, Kunming, China

Tianqi Yang Jinan University, Guangzhou, China

Yang Yang State Key Laboratory of Networking and Switching Technology, Beijing University of Posts and Telecommunications, Beijing, China

Yi Yang School of Information Science and Engineering, Lanzhou University, Lanzhou, China; Silk Road Economic Belt Research Center, Lanzhou University, Lanzhou, China

Yifei Yang Taiyuan Works Section, Daqin Railway Co., Ltd., Taiyuan, China

Yuting Yang Culture and Tourism College, Yunnan Open University, Kunming, Yunnan, China

Zhifei Yang School of Electronic and Information Engineering, Lanzhou Jiaotong University, Lanzhou, China

Jianwei Yao Railway Science and Technology Research and Development Center, China Academy of Railway Sciences, Beijing, China

Shuaishuai Yao Qilu University of Technology (Shandong Academy of Sciences), Jinan, China

Leilei Yin The 28th Research Institute of China Electronics Technology Group Corporation, Nanjing, China

Shengyu You East China University of Technology, Nanchang, China

Fanhua Yu Department of Computer Science and Technology, Changchun Normal University, Changchun, China

Hecun Yuan State Key Laboratory of Networking and Switching Technology, Beijing University of Posts and Telecommunications, Beijing, China

Jinsha Yuan North China Electric Power University, Baoding, Hebei Province, China

Kungang Yuan Air Force Command College, Beijing, China

Qingtao Zeng Beijing Institute of Graphic Communication, Beijing, China; Department of Computer Science and Technology, Tsinghua University, Beijing, China; Postdoctoral Management Office of Personnel Division, Tsinghua University, Beijing, China

Guoyi Zhang Power Grid Dispatching Control Center of China Southern Power Grid, Guangzhou, China

Heming Zhang School of Instrumentation Science and Opto-Electronics Engineering, Beihang University, Beijing, China

Huanyu Zhang Hainan Power Grid Co., Ltd., Haikou, China

Minchao Zhang Institute of Network Technology, Beijing University of Posts and Telecommunications, Beijing, China

Ming Zhang School of Information Science and Engineering, Linyi University, Linyi, China

Wenzhe Zhang CSG Power Dispatching Control Center, Guangzhou, China

Xiaoyun Zhang China Electric Power Equipment and Technology Co. Ltd., Zhengzhou Electric Power Design Institute, Zhengzhou, China

Ying Zhang NRGD Quantum CTek., Ltd., Nanjing, China

Yingxin Zhang Unit 31002 of the PLA, Beijing, China

Yinuo Zhang School of IOT, Nanjing University of Posts and Telecommunications, Nanjing, China

Yu Zhang Department of Computer Science and Technology, Changchun Normal University, Changchun, China

Yu Zhang School of Electronic and Information Engineering, Lanzhou Jiaotong University, Lanzhou, Gansu, China

Zhiwei Zhang Air Force Command College, Beijing, China

Zhuo Zhang Xi'an High-Tech Institute, Xi'an, China; Key Lab of Networks and Information Security of PAP, Xi'an, China

Mengyuan Zhanghu Southern University of Science and Technology, Shenzhen, China

Dong Zhao Department of Computer Science and Technology, Changchun Normal University, Changchun, China

Gaofeng Zhao State Grid Electric Power Research Institute, NARI Group Corporation, Nanjing, China; NRGD Quantum CTek., Ltd., Nanjing, China

Kang Zhao State Key Laboratory of Networking and Switching Technology, Beijing University of Posts and Telecommunications, Beijing, China

Liang Zhao College of Informatics, Huazhong Agricultural University, Wuhan, China

Meili Zhao East China University of Technology, Nan Chang, China

Mingjun Zhao State Grid Xinjiang Information & Telecommunication Company, Urumqi, China

Qiling Zhao Wuyi University, Wuyishan, China

Xin Zhao The Key Laboratory of Information System Engineering, The 28th Research Institute of China Electronics Technology Group Corporation, Nanjing, China

Yanan Zhao School of Electronic and Information Engineering, Lanzhou Jiaotong University, Lanzhou, China

Liping Zheng Department of Computer Science and Engineering, Huaian Vocational College of Information Technology, Huaian, China

Zhiming Zhong CSG Power Dispatching Control Center, Guangzhou, China

An Zhou China Southern Power Grid Co., Ltd., Guangzhou, China

Bo Zhou NARI Information & Communication Technology Co., Ltd., Nanjing, China

Chufeng Zhou Beijing Institute of Graphic Communication, Beijing, China

Fang Zheng College of Informatics, Huazhong Agricultural University, Wuhan, China; Hubei Key Laboratory of Agricultural Bioinformatics, College of Informatics, Huazhong Agricultural University, Wuhan, China

Fan Zhou School of Software and Internet of Things Engineering, Jiangxi University of Finance and Economics, Nanchang, China

Fang Zhou The Key Laboratory of Information System Engineering, The 28th Research Institute of China Electronics Technology Group Corporation, Nanjing, China

Guiping Zhou State Grid Liaoning Electric Power Co., Ltd., Shenyang, Liaoning, China

Yan Zhou Institute of Oceanographic Instrumentation, Qilu University of Technology (Shandong Academy of Sciences), Shandong Provincial Key Laboratory of Marine Monitoring Instrument Equipment Technology, National Engineering and Technological Research Center of Marine Monitoring Equipment, Qingdao, China

Daohua Zhu Electric Power Research Institute, State Grid Jiangsu Electric Power Company, Ltd., Nanjing, China

Hong Zhu State Grid Jiangsu Electric Power CO., Ltd., Nanjing Power Supply Company, Nanjing, China

Huaji Zhu Beijing Research Center for Information Technology in Agriculture, Beijing Academy of Agriculture and Forestry Sciences, Beijing, China

Jinlong Zhu Department of Computer Science and Technology, Changchun Normal University, Changchun, China

Lida Zhu Hubei Key Laboratory of Agricultural Bioinformatics, College of Informatics, Huazhong Agricultural University, Wuhan, China

Luqing Zhu School of Instrumentation Science and Opto-Electronics Engineering, Beihang University, Beijing, China

Mingsi Zhu Hainan Power Grid Co., Ltd., Haikou, China

Rui Zhu School of Software, Yunnan University, Kunming, China

Yukun Zhu State Grid Information and Telecommunication Group, Ltd., Beijing, China

Yi Zou College of Computer Science and Technology, Hengyang Normal University, Hengyang, Hunan, China

System Detection and Application

This chapter aims to examine innovation in the fields of fault diagnosis, embedded, edge computing, electric field measurement, mobile storage, system control and scheduling optimization. This chapter contains 26 papers on emerging topics on system detection and application.

An Implementation of NIC Controller Integrated with Security Protection Module

Yan Shen and Wentao Tong[(✉)]

Wuhan Digital Engineering Institute, Hubei 430074, China
23862446@qq.com

Abstract. In order to solve the disadvantages of traditional measures for the network security. Network security software consumes a considerable amount of system resources, and dedicated security board or equipment has too much cost. This paper provides a new method of network security protection for personal users which is convenient and can overcome the disadvantages of common network security methods. A Gigabit Ethernet controller chip integrated with the security module is designed, and the security module is based on the state detection technology and misuse of intrusion detection technology. The test results on the FPGA show that the security network card controller designed in this paper occupies less system resources and has better security and protection performance. It can provide a low-cost hardware network security protect solution for individual and family users.

Keywords: NIC controller · Network security · FPGA · TCP · UDP · ICMP

1 Introduction

With the rapid development of information technology, the Internet has become the most important information channel and communication platform in the home and business. At the same time, Internet security problems such as hacker invasion, virus ravage, Trojan horse implantation, and botnets are also increasing, the harm is also increasing. At present, most networked computers adopt security software to deal with them. However, software products occupy computer resources, affect running speed of system, and the security software itself may also be attacked and destroyed, resulting in failure. Some companies use hardware firewalls and other specialized network security hardware devices, but it is high cost, not suitable for personal or home users. In this paper, a NIC with integrated network security protection is proposed, which has a small impact on system performance and a relatively low cost.

2 Research Status at Home and Abroad

Currently at home and abroad, a separate study or product has appeared both on network security protection technology and NIC design technology. Software, 360 security guard, Rising antivirus software etc., and dedicated board or equipment, H3C SecPath T5000-S3, RG-WALL 1600 Secure VPN, are common on the network

© Springer Nature Switzerland AG 2020
Q. Liu et al. (Eds.): CENet 2018, AISC 905, pp. 3–12, 2020.
https://doi.org/10.1007/978-3-030-14680-1_1

security. There are also many researches of security software and dedicated board in academic. Zhang and other implement Netfilter/Iptables firewall system on ARM embedded platform with Linux [1]. Zhang Studied and implemented a composite firewall based on the packet filtering principle [2]. The outstanding feature of this firewall is the combination of adaptive paths and functions, and also has the functions of common firewalls. Both of them realize part of the security defense function in the way of software, which has a certain influence on the running speed of the system.

Yu Shunzhi designed and implemented the hardware platform by using the protocol parsing technology and pattern matching technology. The network intrusion prevention system is achieved by hardware methods. But the hardware module exists independently, not with the NIC controller, the user use it inconveniently [3].

In this paper, by the use of EDA and microelectronics technology, the network security module and NIC controller are integrated into a single chip, which greatly improve the integration of the system and has little influence on the data transmission speed and system performance of the network, it is also easy to use.

3   Detailed Design

3.1   State Detection and Misuse Intrusion Detection Technology

At present, the firewall and intrusion detection technology is the mainstream of network security technology at home and aboard.

Processing speed of packet filtering firewall is fast, but its security is low [4]; security of application gateway firewall is high, but its processing speed is slow and it cost much. State detection firewall solves these issues. Efficiency, security, easy scalability, wide application and other advantages enable it to become the focus of firewall technology.

Misuse intrusion detection, also known as feature-based intrusion detection, assuming that a model can be used to represent activity of intrusion [5]. The target of system is to detect whether the state fit these modes, it is judged to be intrusion if the result is conformable. Therefore, matching technology is the core of misuse detection. The main disadvantage of misuse detection is that generally only the known attack patterns are monitored, and new patterns of attacks can only be detected by constantly updating the pattern library.

In this paper, state detection technology and misuse intrusion detection technology are combined to form a network security protection method which is more comprehensive. Based on this method, security protection module and Gigabit Ethernet controller has been integrated to achieve an NIC controller which is integrated with Security features.

3.2   System Design

The design of a Gigabit Ethernet card with network security features consists of two parts: A data transmission module and a security module. The data transmission module includes bus interface module, DMA module, TxFC module, RxFC module,

OMR module, MAC module and PHY interface module. Security module (Defender) is embedded in the MAC module of the NIC. A number of tests are carried out on the data received and sent by the NIC. The required data packets are released, and the threat packets are filtered out.

The structure is shown in Fig. 1.

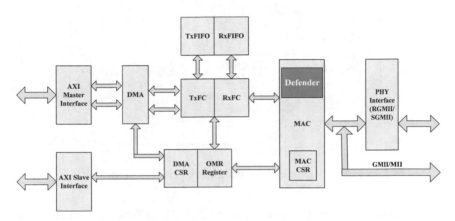

Fig. 1. Architecture of NIC controller.

3.3 Design of Data Transmission Module

The main functions of the data transmission module are:

1. The interface between microprocessor and NIC controller. This design uses a popular standard on-chip bus-the AXI interface. AXI is used to connect the NIC controller and ARM processor core.
2. Configuration and state register group.
3. DMA transmission channel. Achieve the rapid transfer between the main memory and Data channels which provide transmission and reception.
4. Data channels with transmission and reception. Channels provide the transmission interface between the transceiver queue and the PHY interface.
5. MAC module which is used to complete the functions by packaging, unpackaging, sending and receiving date frame. It is used to remove the preamble of receive packets and check Frame CRC and deal with receive error. It is also used to add preamble of receive packets and CRC, add padding and send error message when sending data packets.
6. RGMII interface to phy.

3.4 Design of Security Protection Module

As shown in Fig. 2, the security protection module includes a misuse intrusion detection module, a protocol preprocessing module, a TCP state detection module, a UDP state detection module, an ICMP state detection module, an audit module, a response module and a data storage module. They work together to complete the detection of data packets and protect local host security.

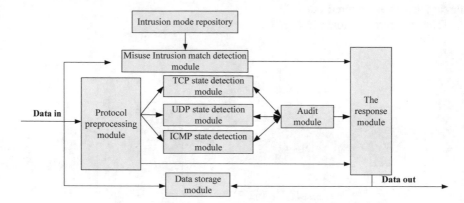

Fig. 2. Architecture of the security protection module.

The card must send and receive the data packets. The security protection module deals with these two kinds of situations different.

When sending packets, get them from TXFIFO and send to the security module for testing. Only when test ok, sent them to the MAC module for post-processing.

When receiving packets, send them to the MAC module in the NIC for processing, and then send to the security module for detection. Only after passing the detection, send the packets to the RXFIFO for transmission to the upper layer.

The specific functions of each module are as follows:

1. Misuse intrusion match detection module: provide misuse intrusion detection. Match the IP packet data transmitted over with the specific attack data which has been stored in intrusion mode repository. If the match is successful, it indicates that the packet is offensive. In this case, an error signal is sent to the response module. Intrusion mode repositories are implemented by multiple cams and can be updated regularly.
2. Protocol preprocessing module: the information of network layer header and transport layer header will be extracted according to the protocol. If it is a TCP, UDP or ICMP packet, pass it to the corresponding state transition detection module for detecting. If not, it will send the normal results to the response module directly.
3. TCP, UDP and ICMP state detection modules: provide firewall state transition detection. The module analyze whether the basic status and status changes of all types of packets are normal according to the normal conversion process of various types of protocol packets to ensure the normal packets passed and the abnormal packets are detected. If the packet is abnormal, an error signal is sent to the audit module.
4. Audit module: store all kinds of useful dates from state transition detection modules. The module keep communication with state transition detection modules, and pass the test results of the state transition detection module to the response module.

5. The response module: receive the decision information from the misuse match detection module, the audit module and protocol preprocessing module and decide whether the packet extract from the data storage module is discard or export.
6. Data storage module: store the data packet which is analyzed, and wait for analysis results from response module. The normal packet will be sent out and the abnormal packet will be abandoned.

Both the received and the transmitted data packets should be sent to the security protection module for detection. For the received network data packet, both the misuse intrusion and state transition detection should be detected. For the data packet to be sent, only the state transition detection is needed.

Design of Intrusion Matching Detection Module
In order to improve the speed of the model comparison, this design uses a special storage device- CAM to realize the function of data matching. CAM is a special storage device for parallel searches fast and massively. When searching, all the data in the storage device is compared with the search keyword at the same time, and the search result is the physical address of the matching item. The data table query can be done in the hardware. Each storage bit is compared by a dedicated comparison circuit. CAM pipeline structure is used for fast search. Each search can start every clock cycle, which can better complete the data matching function.

The pattern library is updated by downloading the latest virus database on the network to the CAM with the software tool.

Design of State Conversion Detection Module
Detection Principle
The safety NIC detects and tracks the state of TCP, UDP and ICMP packets. Different state detection sequences are used according to the different characteristics of the three protocols.

A complete TCP connection is divided into three state transition stages which are establishing connection, connection, disassembly connection. TCP establishes a connection by using a three-way handshake. A core part of this protocol is that the server does not enter the connection state immediately when receiving the connection request. It caches the request and waits for three responses before establishing the connection. In the connection process stage, whether the connection is normal is reflected in the relationship between confirmation number and serial number in packet header. Removing the connection also need the three-way handshake between both parties just like the establishment stage.

UDP (User Datagram Protocol) is a stateless protocol because it does not have any procedures when setup or shut down a connection, and most of the packet does not have a sequence number. Although there is no serial number, some information of the UDP packet can also be used to track the status of the UDP connection for state detection. The source, destination address and source, destination port in UDP packet can be retained as the status of UDP information. This connection, called "virtual connection", is built by artificial state information. The connection timeout time is set to detect state.

Therefore, a complete UDP connection is divided into three state transformation phases which are new, established 1, established 2.

UDP packets to establish a virtual connection, the state detection process and TCP packets similar. Once the UDP packet is received, first take out the source, destination address and source, port and other related information, check the status table to determine whether the connection has been established, as established, then determine whether the state transition requirements in line with, and if so change the status flag And then release; otherwise establish a "virtual connection", save the packet's source, destination address and source, port and other state attribute information, waiting for the next UDP packet connection to change the state.

ICMP is also a stateless protocol. There are request and response packets, but also perform other functions in ICMP. The request and reply packet types in ICMP packets are as follows: echo request and response, timestamp request and response, information request and response, address mask request and response. A complete ICMP connection is divided into two state transformation phases which are new, established. The state detection process is similar to UDP when a virtual connection is established by ICMP.

Based on the analysis of state transitions in the connection of TCP, UDP and ICMP, a state detection module for each packet is set up. TCP, UDP, and ICMP state connection information are recorded in the TCP status table, UDP Status table and ICMP status table in the module. The connection status is analyzed to ensure that normal packets are passed. Unusual packets that do not meet the normal conversion process are detected and error messages are sent to the response module. At the same time set the wait time for each state. The packet which comes within the wait time is valid. Otherwise, the connection is invalid because the delay is too high.

Key Technology Solutions

This design adopts special design methods to improve the speed of state detection and reduce the occupancy of the storage space of the system.

1. The storage and search methods of connection status

In order to realize the state detection, many connection state tables needed to be set up which needs to be recorded and searched at any time in the detection process. The performance of the state transition detection module is influenced greatly by the speed of state storage and search in the connection state table. Ideally, a definite relationship is built between the recorded storage location and its keywords. In the system, hash address mapping is used to solve this problem. The unique hash value corresponding to each connection is obtained according to the five values which are source and destination IP address (ipsa_reg, ipda_reg), the source and destination port number (sport_reg, dport_reg) and the protocol type (protocol_reg).

temp0_reg [103:0] <={ipsa_reg, ipda_reg, sport_reg, dport_reg, protocol_reg};
    hash0_reg<=temp0_reg[10:0]+temp0_reg[21:11]+temp0_reg[32:22]+temp0_reg[4
3:33]+temp0_reg[54:44]+temp0_reg[65:55]+temp0_reg[76:66]+temp0_reg[87:77]+te
mp0_reg[98:88];
    temp1_reg[103:0]<={ipda_reg,ipsa_reg,dport_reg,sport_reg,protocol_reg};
    hash1_reg<=temp1_reg[10:0]+temp1_reg[21:11]+temp1_reg[32:22]+temp1_reg[4
3:33]+temp1_reg[54:44]+temp1_reg[65:55]+temp1_reg[76:66]+temp1_reg[87:77]+te
mp1_reg[98:88];

As can be seen from the above formula, regardless of the package of the source direction or the package of opposite direction, the header information is relevant. if take the first package hash0 of package establishing connection as address which is used to store the connection information, calculate the hash1 value as hash value with package of opposite direction, or calculate hash0 value as hash value with the package of source direction. Therefore it can quickly find information stored in the address, and the searching speed will not be affected for the package is from different direction.

2. Timeout management of connection status

The different states of each connection have overtime. If the normal state is not transitioned in the overtime, the records in the table should be deleted. Otherwise, the extra data will take up a lot of storage space, and even affect the storage of the normal package. The system uses TDCMS [6] (time address cycle timeout management strategy) to solve the problem.

4 Test and Verification

In order to verify the effectiveness of the security methods described in this article, this paper builds a physical platform to test and verify. Combined with the simulation of the attack procedure, the card has been tested in two aspects in this platform.

1. without attack, the function and performance of the NIC in transmitting data.
2. under certain types of attacks, the monitoring and protection capabilities of the NIC in transmitting data.

4.1 Test Verification Platform

As shown in Fig. 3, the prototype design described in this article is achieved in the FPGA development board (Stratix IV GX FPGA Development Board, 530 Edition. Through the router, two ordinary pc, known as the host 1 and host 2, are connected respectively to development board to form a simulated external server and attack source. Development board and the host 1, host 2 are in the same network segment. FTP/TFTP server is started in host 1. The network attack programs are started in host 2.

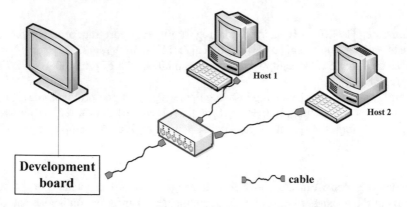

Fig. 3. Test and verification platform.

4.2   Test Process and Result

Normal Communication Ability Test

The FTP server and the TFTP server are started on the host 1. The development board downloads the data file from the host 1, calculates the downloaded file, checks the correctness of the downloaded file, tests the files with different lengths. The results are shown in Table 1.

Table 1. Test results of normal communication capability.

File size	Average download speed of FTP server	Average download speed of TFTP server	Correct rate
Less than 1 MB	3430 KB/s	1755 KB/s	100%
1 MB–100 MB	48485 KB/s	26493 KB/s	100%
100 MB–500 MB	72513 KB/s	49634 KB/s	100%
Mixing (each 1/3)	51168 KB/s	28594 KB/s	100%

Attack Protection Ability Test

The FTP server and the TFTP server are started on the host 1. Different attack programs are run in host 2 to initiate different types of attacks on the development board. At the same time, the development board downloads data files from host 1. Calculates the download rate, and compares the source files and downloaded files to check the correctness. The results are shown in Table 2.

Test Capability Test of Malicious Attack Feature

The FTP server and the TFTP server are started on the host 1. Using the network tester (model Smaribits 600B) to send 10000 packets with various attack features to the test platform. At the same time, the development board downloads data files from host 1. Test the download speed and file integrity of the fire. At the same time, check whether the NIC can filter out attack data packets through the capture software. The results are shown in Table 3.

Table 2. Test results of attack protection capability.

Attack type	Average download speed of FTP server	Average download speed of TFTP server	Correct rate
SynFlood	47265 KB/s	26698 KB/s	100%
UdpFlood	43562 KB/s	22393 KB/s	100%
Ping of Death	47656 KB/s	27473 KB/s	100%

Table 3. Attack detection capability test results.

File size	Average download speed of FTP server	Average download speed of TFTP server	Correct rate	Attack packet filtering ratio
Less than 1 MB	3246 KB/s	1668 KB/s	100%	100%
1 MB–100 MB	47189 KB/s	23853 KB/s	100%	100%
100 MB–500 MB	68596 KB/s	45697 KB/s	100%	100%
Mixing (each 1/3)	49618 KB/s	27691 KB/s	100%	100%

Analysis of Test Results

From the test results in Tables 1 and 2, with the attacks of SynFlood, UdpFlood, and death of Ping, the average download speed of FTP server can reach 92.4%, 75.1%, 93.1% respectively compared with non attack speed. The average download speed of the TFTP server can reach 93.4%, 78.3%, 96.1% respectively compared with non attack speed. The accuracy is 100% respectively. From the test results in Tables 1 and 3, with the various attacks and, the average download speed of FTP server can reach 97% compared with non attack speed when the various sizes of downloaded files are mixed. The average download speed of the TFTP server can reach 96.8% compared with non attack speed when the various sizes of downloaded files are mixed. The accuracy is 100% respectively. All the filtering ratio of attack data packets can reach 100%. So it can be considered that the security NIC controller has better state detection ability and misuse intrusion detection ability for several kinds of common network attacks, and has little effect on normal data communication speed.

5 Conclusion

In this paper, the state detection technology and misuse intrusion detection technology are combined to form a network security protection method which is more comprehensive. Security protection module based on this method has been integrated to gigabit Ethernet controller to achieve a NIC which is integrated with security controller. The FPGA verification is completed. Through the test we can see that the NIC controller can provide better security performance and have little impact on system performance. At the same time, the NIC is easy to use, cost low, can provide a low-cost hardware network security solutions for individuals or families.

References

1. Zhang, Q.S.: A design and implementation of embedded Linux firewall system. University of Electronic Science and Technology of China (2014). (in Chinese)
2. Zhang, Y.P.: A firewall design and implementation based on adaptive routing packet-filtering technology. Shandong University (2015). (in Chinese)
3. Yu, S.Z.: Research of the content processing in gigabit network intrusion prevention system. Beijing University of Posts and Telecommunications (2006). (in Chinese)
4. Lin, W.C., Ke, S.W., Tsai, C.F.: CANN: an intrusion detection system based on combining cluster centers and nearest neighbors. Knowl. Based Syst. **78**, 13–21 (2015)
5. Wu, Q.F.: Research and implementation of intelligent packet filtering firewall. Harbin University of Science and Technology (2003). (in Chinese)
6. Shen, Y., Fang, X.Y.: Design and implementation of TCP state detection based on the integrated circuit. In: 2nd International Conference on Data Storage and Data Engineering (2011). (in Chinese)

Optimization and Security Implementation of ITUbee

Xiawei Cao, Lang Li$^{(\boxtimes)}$, and Xuanlan Xie

College of Computer Science and Technology,
Hengyang Normal University, Hengyang 421002, China
lilang911@126.com

Abstract. ITUbee is the lightweight encryption algorithm that was proposed by the second International Symposium on lightweight encryption security and privacy in 2013. It is based on the Feistel network. We optimized S-box, round function and round constant addition. The optimized round constant is the variable which is converted from the related round number i. There is no need to allocate area resource for it. The experimental results show that the throughput of the optimized ITUbee algorithm reaches 364.695 Mb/s. The area is reduced to 10650 Slices. We studied and implemented the masking ITUbee algorithm to resist power analysis attack. The implemented performance is also compared. The area of the masking ITUbee is increased by about 4%. The clock frequency is raised from 100.291 MHz to 102.396 MHz, throughput is increased from 364.695 Mb/s to 372.349 Mb/s.

Keywords: Lightweight cipher · ITUbee · Optimization · Implementation

1 Introduction

The research of lightweight cryptography which is suitable for resource constrained has become the hot topic in the recent year [1]. It is very important how to implement the lightweight cryptographic algorithm effectively.

ITUbee was proposed in 2013 at the second International Conference on lightweight encryption security and privacy [2]. ITUbee has a shortage of hardware implementation area. In this paper, we optimized these problems. ITUbee has no key generation strategy in order to achieve the hardware implementation with less resource occupation. Therefore, ITUbee is easier to be attacked with power analysis attack compared with AES. It is necessary for ITUbee to be improved for resist power analysis attack in the higher security environment [3–5].

Masking is a common and effective method in various protection techniques against power analysis attack. The novel masking method is proposed [6]. We use the Boolean masking to resist power analysis attack in the paper [7, 8].

© Springer Nature Switzerland AG 2020
Q. Liu et al. (Eds.): CENet 2018, AISC 905, pp. 13–21, 2020.
https://doi.org/10.1007/978-3-030-14680-1_2

2 ITUbee Algorithm

As shown in the Fig. 1, round transformation includes the F function, addRoundkey, AddConstant, and L function. The round conversion was showed in Fig. 2, in which (1) and (2) modules were F functions, and module (2) multiplexes module (1).

Figure 1 is the operation diagram of the ITUbee algorithm.

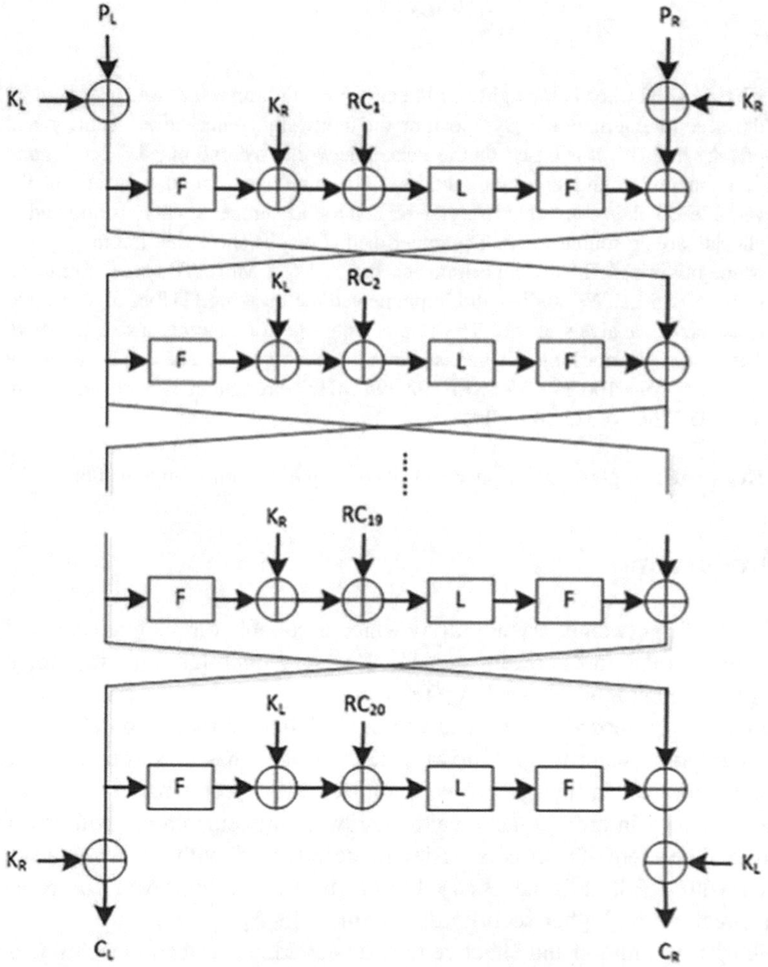

Fig. 1. ITUbee algorithm structure.

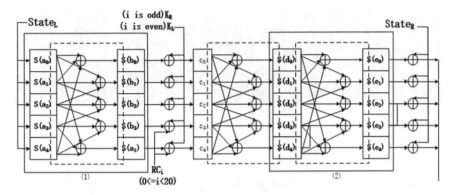

Fig. 2. The process of round transformation.

3 Optimization and Implementation of ITUbee

3.1 Hardware Implementation of AES

(1) We use assign statement to implement the ITUbee, at the same time to control the counter update with the clock signal. Repeating the 20 round of the same module will complete the encryption by this method, which can effectively save the implementation area.

(2) ITUbee algorithm reuses the AES S-box which is the 16×16 table. In this paper, the traditional case statements are not used in the optimization, but the register is used to save them. The great deal of encryption operation time can be saved by looking up the S-box table directly. The efficiency of directly lookup table is O(1).

Its optimized implementation code is described with Verilog HDL as follows.

```
begin
mem[0]=8'h63;
mem[1]=8'h7c;
......
mem[255]=8'h16;
end
always@(states) outs=mem[states];
```

Module correctness verification results are shown in Fig. 3.

(3) Optimization of round functions
As we all know, ITUbee round function calls the L function and S-box directly. Therefore, we can use the combination of L function and S-box to save the effect of calling the F function time in order to implement the algorithm without rewriting the F function. Its optimized implementation code is described with Verilog HDL as follows.

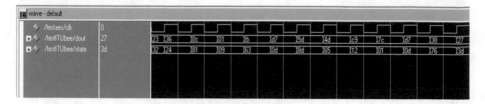

Fig. 3. S-box simulation waveform.

```
module ITUbeeRound(outR,stateR,key,r);
initial
begin
RC[0]=8'h14;
RC[1]=8'h28;
end
assign rkey=(r% 2==0)? key[39:0]:key[79:40];
Sbox O1(state1[39:32],stateR[39:32]);
......
Sbox O5(state1[7:0],stateR[7:0]);
L ll1(state2,state1);
Sbox P1(state3[39:32],state2[39:32]);
......
Sbox P5(state3[7:0],state2[7:0]);
assign state4=state3^rkey;
assign state5={state4[39:16],state4[15:8]^(RC[0]-r),state4[7:0]^(RC[1]-r)};
L ll2(state6,state5);
Sbox Q1(state7[39:32],state6[39:32]);
......
Sbox Q5(state7[7:0],state6[7:0]);
L ll3(state8,state7);
Sbox W1(outR[39:32],state8[39:32]);
......
Sbox W5(outR[7:0],state8[7:0]);
endmodule
```

Module correctness verification results are shown in Fig. 4.

(4) Optimization of round addconstant
   The result of using the subtraction of variables as an intermediate value is used in
   the Round Addconstant module. We do not use registers to save the round constant which can reduce the overhead of a large number of registers.

Fig. 4. Optimal verification of round function.

Its optimized implementation code is described with Verilog HDL as follows.

```
module ITUbeeRound(outR,stateR,key,r);
initial begin
RC[0]=8'h14;
RC[1]=8'h28;
end
assign state5={state4[39:16],state4[15:8]^(RC[0]-r),state4[7:0]^(RC[1]-r)};
endmodule
```

Module correctness verification results are shown in Fig. 5.

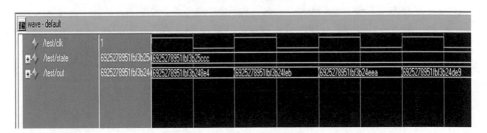

Fig. 5. Simulation waveforms of round constant addition.

3.2 Simulation Experiment

The correctness of the optimized ITUbee algorithm is verified by simulation. The simulation software is Modelsim 6.1f. The simulation results are shown in Fig. 6.

From the experiment results of Fig. 6, we can know that the optimized ITUbee algorithm is correct.

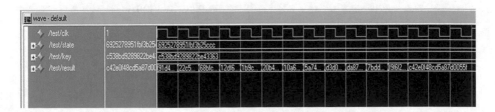

Fig. 6. ITUbee algorithm simulation screenshot.

3.3   FPGA Implementation

The optimized ITUbee is implemented in Xilinx Virtex-5LX50T. The performance analysis was done after the ISE13.2 composite Download. The results were shown in Figs. 7 and 8.

```
Slice Logic Utilization:
    Number of Slice Registers:              5,344 out of  28,800    18%
        Number used as Flip Flops:          5,327
        Number used as Latch-thrus:            17
    Number of Slice LUTs:                   5,306 out of  28,800    18%
        Number used as logic:               5,136 out of  28,800    17%
            Number using O6 output only:    4,881
            Number using O5 output only:       82
            Number using O5 and O6:           173
        Number used as Memory:                155 out of   7,680     2%
            Number used as Dual Port RAM:      64
                Number using O5 and O6:        64
            Number used as Shift Register:     91
                Number using O6 output only:   90
                Number using O5 output only:    1
        Number used as exclusive route-thru:   15
    Number of route-thrus:                     97
        Number using O6 output only:           92
        Number using O5 output only:            2
        Number using O5 and O6:                 3
```

Fig. 7. ITUbee area test data

```
Timing summary:
---------------

Timing errors: 0   Score: 0 (Setup/Max: 0, Hold: 0)

Constraints cover 398642 paths, 16 nets, and 29304 connections

Design statistics:
    Minimum period:   9.971ns (Maximum frequency: 100.291MHz)
    Maximum path delay from/to any node:   4.077ns
    Maximum net delay:   0.835ns
```

Fig. 8. ITUbee operating freguency.

From Fig. 7, the area of ITUbee hardware implementation is as follows: Size = 5306 + 5344 = 10650 Slices.

As shown in Fig. 8, the system clock cycle is 9.971 ns, the clock frequency is 100.291 MHz. The throughput rate of ITUbee is:

$$Throughout = \frac{100.291 * 80}{22} = 364.695 \, \text{Mb/s} \tag{3.1}$$

4   Masking ITUbee Against Power Analysis Attack

The basic idea is that the input data and every intermediate result are hidden by random numbers. The attacker cannot get the correct value. Usually, the random number masks the input. To recover, the correct ciphertext is removed masking at the end of the encryption. The implementation process of the masking technique is shown in Fig. 9.

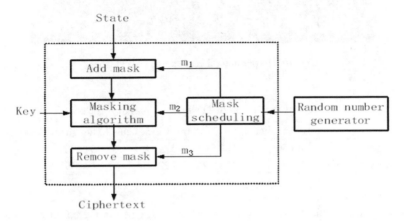

Fig. 9. The implementation process of mask technology.

S-box mask of ITUbee cipher is described with C language as follows:

```
Add(state,Rand);
encrypt(state,key,Rand);
Add(state,Rand);
```

The way S-box removes mask is described with C language:

```
void SubBytes(unchar *state,unchar *R)
{
int j;
for(j=0;j<5;j++){
state[j]=sbox[state[j] ^R[j]];
state[j]^=R[j];
}
}
```

The correctness of the ITUbee with the mask was verified. The result is showed in Fig. 10.

From the Fig. 10, the masking ITUbee is proved to be correct. In order to verify the area occupancy and encryption speed, the masking ITUbee is implemented with FPGA. The performance comparison between ITUbee and masking ITUbee was showed in Table 1.

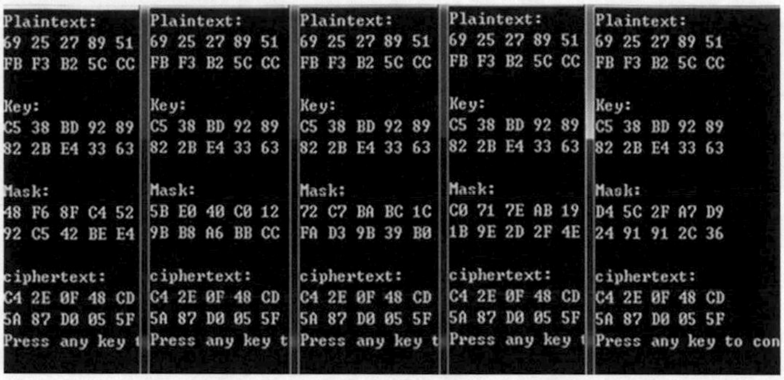

Fig. 10. Experimental results of mask ITUbee.

Table 1. Comparison of ITUbee and masking ITUbee.

	Area (slices)	Freq (MHz)	Clock count	Thoughput (Mb/s)
ITUbee	10650	100.291	22	364.695
Masking ITUbee	11091	102.396	22	372.349

As shown in Table 1, the area of the masking ITUbee is increased by about 4%. The clock frequency is raised from 100.291 MHz to 102.396 MHz; throughput is increased from 364.695 Mb/s to 372.349 Mb/s.

5 Conclusions

In this paper, we have studied the area and performance optimization of ITUbee. At the same time, lightweight cryptographic algorithms are vulnerable to power analysis attacks. The masking ITUbee is proposed to resist power analysis attack.

In the future, we will study the lightweight cryptographic algorithm how to implement in resource constrained smart cards.

Acknowledgement. This research is supported by the National Natural Science Foundation of China under Grant No. 61572174, Hunan Province Special Funds of Central Government for Guiding Local Science and Technology Development No. 2018CT5001, the Scientific Research fund of Hengyang Normal University with Grant No. 16CXYZ01, the Science and Technology Plan Project of Hunan Province No. 2016TP1020, Hunan Provincial Natural Science Foundation of China with Grant No. 2017JJ4001, Hengyang Normal University Experiment Programs of Learning and Innovation for Undergraduates of China No. CX1808, CX1839, Hunan Provincial Experiment Programs of Learning and Innovation for Undergraduates of China with Grant No. 2018749, Subject group construction project of Hengyang Normal University No. 18XKQ02.

References

1. Wang, G.L., Jiang, S.S., Shen, Y.Z., Yue, L.: Improved 3-dimensional meet-in-the-middle cryptanalysis of kTANTAN32. J. Sichuan Univ. (Eng. Sci. Edn.) **45**(6), 8–14 (2013)
2. Karakoc, F.: ITUbee: a software oriented lightweight block cipher. In: Procedings of Second International Workshop, Lightsec, Gebze, Turkey, pp. 17–27. Springer (2013)
3. Journault, A., Standaert, F.X.: Very high order masking: efficient implementation and security evaluation. In: International Conference on CHES, pp. 623–643. Springer (2017)
4. Guo, J., Peyrin, T., Poschmann, A., Robshaw, M.: The LED block cipher. In: International Conference on CHES, Nara, Japan, pp. 326–341. ACM (2011)
5. Liu, Z., Longa, P., Pereira, G., Reparaz, O., Seo, H.: FourQ on embedded devices with strong countermeasures against side-channel attacks. In: International Conference on CHES, pp. 665–686. Springer (2017)
6. Gross, H., Mangard, H., Korak, T.: An efficient side-channel protected AES implementation with arbitrary protection order. In: Cryptographers' Track at the RSA Conference, pp. 95–112. Springer (2017)
7. Kundi, D.E.S., Aziz, A., Ikram, N.: Resource efficient implementation of T-Boxes in AES on Virtex-5 FPGA. Inf. Process. Lett. **110**(10), 373–377 (2010)
8. Li, L., Zou, Y., Jiao, G.: FPGA implementation of AES algorithm resistant power analysis attacks. In: Proceedings of the 2017 the 7th International Conference on Computer Engineering and Networks, Shanghai, China, pp. 357–363 (2017)

General Parallel Execution Model for Large Matrix Workloads

Song Deng[✉], Xueke Xu, Fan Zhou, Haojing Weng, and Wen Luo

School of Software and Internet of Things Engineering,
Jiangxi University of Finance and Economics, Nanchang 330013, China
daonicool@sina.com

Abstract. Large scale statistical computing is crucial for extracting useful information from huge amount of data for both large companies and research scientists. The Solutions developed by high-performance communities have been more limited to clusters or high-end machines for decades. The cost of maintaining such dedicated clusters are prohibiting, people start to look at cloud computing where we can rent a cluster by time and pay-as-we-go. In a cloud setting, system features including fault tolerance and scalability become important. In this paper, we proposed a simple and universal parallel execution model for large matrix workloads. We implement the model in Hadoop MapReduce framework using map-only jobs. Because of the superiority of the model, experiments show that our Hadoop-based execution engine can reduce the execution time of matrix multiplication by half comparing with previous works.

Keywords: General parallel execution · Model · Large matrix workloads · Hodoop

1 Introduction

Meaningful statistical analysis of large datasets is extremely important in both academia and industry. Web companies are receiving huge amount of data every day and need to perform machine learning algorithms or statistical analysis to extract useful information. Scientists working in various fields also have the demand to analyze the vast amounts of data collected in the experiments.

Packages like R, LAPACK [1] and Matlab first assume that all data is suitable for memory. Works like SOLAR [2] can extend these in-memory packages to store data on disk, allowing it to handle larger data sizes, but these are still limited to a single machine. With the rapid increase in data size, the computing resources from a single machine are extremely limited, which is not enough to meet the data processing needs.

With the development of the times and technologies, various software packages have been successfully built in high-performance communities, such as Sca LAPACK [3] for parallel computation in a distributed memory setting. Typically, the CPU cores are arranged on a grid and the data are sent around using libraries like parallel virtual machine (PVM) and message passing interface (MPI). Therefore, the algorithm typically requires a large amount of interprocessor communication and/or coordination at each

Q. Liu et al. (Eds.): CENet 2018, AISC 905, pp. 22–28, 2020.
https://doi.org/10.1007/978-3-030-14680-1_3

step, and the level of parallelization is quite low. As a result, these packages are difficult to use in larger, and often heterogeneous, cluster settings. For instance, the speed of the entire cluster will decrease due to some nodes, which have lower speed CPUs or lower network bandwidth. Furthermore, the assumption that data has been partitioned and resides in the memory of each core still limits the size of the processed data. Consequently, these packages are more suitable for high-end clusters and tight coupling.

In some types of clusters, the data is assumed to be too large for the collective memory of all machines, when looking at this type of cluster, the Hadoop MapReduce [4] becomes promising because of its excellent scaling ability to leverage large clusters with fault tolerance, and because its design is based on a distributed file system, matrix data can be accessed and stored.

At the same time, there is great interest in building statistical engines in MapReduce. Because MapReduce has been successfully used in large-scale text analysis like Pig Latin and processing and shows its advantages, Hadoop cluster is already widely deployed and used. It will be nice if you can use the same platform for statistical workloads, no modifications or minor modifications.

As regards statistical workloads, there are many works based on MapReduce. In MapReduce, Mahout can be used to integrate machine learning algorithms by applying certain algorithms, but it does not provide a more adaptable solution. pR [5] dispatches single line of R program from an R runtime into a Hadoop cluster for parallel execution, but its limitations make it impossible to act as a complete, systematic, integrated parallelization solution. The method of integrating the Rhipe and MapReduce [6] program interfaces has been provided by RHIPE and Ricardo. However, MapReduce and hand-coded programs are what users really care about and need to understand. Users have decision-making power over algorithm selection, task partitioning, and data management. SystemML [7] provides an automatic conversion DML workflow solution that enables it to perfect limited optimizations and perform MapReduce jobs. Schelter [8] provides a technique for generating a wide variety of samples based on the sampling function given by the embedded user.

In this paper, we built a Hadoop-based execution engine that can handle matrix workloads more efficiently compared to previous works. Specifically, we have adopted a non-traditional but direct data access method that reduces the number of MapReduce jobs required for each matrix operation. Experiments show that this technique reduces the execution time dramatically compared to existing works.

2 Hadoop-Based Execution Engine

2.1 Matrix Blocking and Chunking

To map the matrix operations into Hadoop MapReduce jobs, the basic idea is to partition the matrix into chunks. Then each map/reduce task will deal with the computation of one chunk or chunk pairs. We observed that different the chunk sizes are preferred in different job and machine settings. It is desirable that the execution engine could allow flexible chunk sizes for different jobs. Then the best chunk partitioning factors could be specifically chosen for the job as job parameters in the optimization step.

For matrix storage, in one hand, if the matrix is stored in chunks, then we need to convert data chunk sizes frequently between jobs; in the other hand, the storage and data transfer overhead could be large if the data is too fine grained.

Considering the tradeoffs, we adopted a strategy where all matrices are saved in square blocks of a fixed size throughout the system. Each block is a 2048 by 2048 sub matrix. Only the blocks on the margin of the matrix could be of smaller size. A matrix chunk for a map/reduce task must consists of an integer number of blocks in both dimensions.

2.2   Physical Operators for Matrix Multiplication

Each physical operator will do certain computation on a matrix chunk (pair), and yield one chunk to its predecessor. All data transfer between operators is pull based like in DBMS. Each time the get Next function is called, the next block in the specified block order is returned. For operations like addition and transpose, the operator is trivial to implement assuming the block order is well defined.

For matrix multiplication, the strategies are more complicated. One physical operator we implemented is OpBufferOneMM. It reads in all the blocks in one of the chunks and holds all of them in memory. It will then read in the other input chunk block by block and do the multiplication and addition. Another column of blocks in memory is used to aggregate results and output the block one at a time. For instance, in the figure below (Fig. 1), the left chunk are chosen to be buffered. The blocks that reside in memory are shaded. The blocks in the right chunk are read in one at a time in column major order. The result blocks are buffered in a column of blocks and outputted also in column major order.

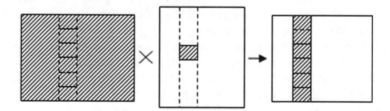

Fig. 1. Matrix multiplication.

The multiplication computation happens at block level. For different block format (sparse, dense or diagonal), different libraries for multiplication are called. For dense-dense multiplication, we used the JBLAS which is a Java wrapper for multithreaded ATLAS. In general, JBLAS prefer square sized and larger input matrices for higher CPU through put. For sparse MM, we provided a basic implementation in Java. One advantage of OpBufferOneMM is that by choosing the smaller input chunk to keep in memory, the memory requirement is low.

Note for dense-dense multiplication in OpBufferOneMM, JBLAS could only be called for block by block multiplication. The small input size limits the CPU utilization in JBLAS. To achieve higher CPU throughput, another strategy is to hold both chunks in memory and do the multiplication only once at the chunk level.

3 Experiments

3.1 Experiment Settings and Use Cases

We used the Elastic Computing Cloud (EC2) from Amazon Web Service as our test bed. Five machine types from Amazon EC2 are considered and the Table 1 below lists its basic information. We assume the machine is charged by seconds instead of hours, but we kept the ratio of prices same as EC2. We adopted a fixed matrix block size 2048 throughout the system. For dense blocks with entry of double precision, one block is 32 MB.

Table 1. Five machine types from Amazon.

Machine type	m1.smal	c1.medium	m1.large	c1.xlarge	m1.xlarge
Price per second	0.5	1	2	4	4
Number of virtual cores	1	2	2	8	4
Memory	1.7 G	1.7 G	7.5 G	7 G	15 G

Here are the two workloads we used in the experiments:

Gaussian Non-Negative Matrix Factorization, the workload used in SystemML. We start with an d × w document-word sparse matrix V, where w is the number of vocabularies and d is the number of documents. W is d by t and H is t by w, both are dense thin matrices. Typically, w > d > t. An iteration of GNMF can be expressed as

$$H = H.*(H^T V)./(W^T WH) \tag{3.1}$$

$$W = W.*(VH^T)./(WHH^T) \tag{3.2}$$

Same as in SystemML, we only considered the first expression because of symmetry. The input document-word matrix is reduced from the 6G wiki corpus from the Westbury Lab at http://www.psych.ualberta.ca/~westburylab.

A synthetic matrix multiplication benchmark: C = A * B. We included various and common seen matrix settings combinations for A and B, like dense or sparse, vector or square. Please refer to Sect. 3 for details.

We also use in the iteration of Gaussian Non-Negative Matrix Factorization to demonstrate the superiority of our execution engine. In SystemML, they convert the expression into five full MapReduce jobs in Table 2. While in our execution engine, four map-only jobs suffice. As our execution engine allow multiple blocks to appear in the same block position, job 3 is not needed. Below is the data we get using 10 machines of type c1.medium (Fig. 2).

Table 2. Five full MapReduce jobs.

Job ID	SystemML	Our approach
1	T1 = W' * W	T1 = W' * W
2	T2 = W' * V	T2 = W' * V
3	T3 = T1	
4	T4 = T3 * H	T4 = T1 * H
5	H = H. * (T2./T4)	H = H. * (T2./T4)

3.2 Impact of Different Choices and Parameters in the Job Plan Space

This is the multiplication of two 24B * 24B square dense matrices with different job parameters in 10 machines of type c1.xlarge. The first four columns are from the best a, b, c settings returned by our optimizer given the number of slots s. The last two columns are the job plan picked by SystemML where they fix b to be 1.

We record the timing of each part, including computation time and HDFS I/O time, within each map task. But even for the same work in a job, different job settings make the amount of work in a map task different, making the comparison and visualization hard. Thus we scale the average timing of each part in map tasks to the job execution time. As all map tasks are doing almost the same amount of work in a job, we roughly get the time spent on each part at the job time scale. Now we can easily compare the effect of job parameters on JBLAS computation time and HDFS I/O, etc.

As we can see in Fig. 3, more concurrent slots will enhance CPU utilization and thus less time spent on JBLAS computation. However, while the slot number increase, the memory per slots decrease and thus more tight constraint on a, b and c. Larger a, b, c value will result in more HDFS I/O.

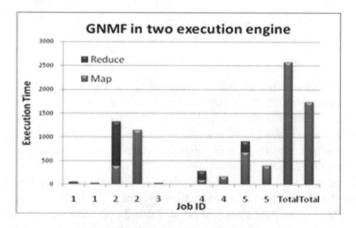

Fig. 2. GNMF in two execution engine.

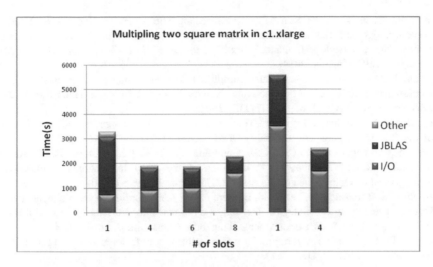

Fig. 3. Multipling two square matrix in c1.xlager.

SystemML limit their plan space to be either a = c = 1 or b = 1. Their settings for a, b, c is suboptimal for any slot number S, which leads to larger amount of HFDS I/O and worse job performance. Note that in C = A * B, each block in A is read c times, each block in B is read a times, and b blocks as partial result of each block in Care written back to HDFS.

4 Conclusion

In order to enhance the scalability and fault tolerance of the cloud system, we present a simple and generic parallel execution model for large matrix workloads. We implemented the model in the MapReduce Hadoop framework using the map-only jobs. The experiments show that our method can reduce the execution time of matrix multiplication by half comparing previous works. The next direction of our research is extending our method to the inverse operation or the pseudo-inverse operation of a matrix.

Acknowledgement. Our research was supported by the Natural Science Foundation of China under grant No: 61462037 and the Natural Science Foundation of Jiangxi under grant No: 20142BAB217014.

References

1. Castaldo, A.M., Whaley, R.C.: Scaling LAPACK panel operations using parallel cache assignment. In: ACM Sigplan Symposium on Principles and Practice of Parallel Programming, vol. 45, no. 5, pp. 223–232 (2010)
2. Toledo, S., Gustavson, F.G.: The design and implementation of SOLAR, a portable library for scalable out-of-core linear algebra computations. In: Proceedings of the Fourth Workshop on I/O in Parallel and Distributed Systems, pp. 28–40 (1996)

3. Wu, P., Chen, Z.Z.: FT-ScaLAPACK: correcting soft errors on-line for ScaLAPACK cholesky, QR and LU factorization routines. In: Proceedings of the 23rd International Symposium on High-performance Parallel and Distributed Computing (HPDC 2014), Vancouver, BC, Canada (2014)
4. Dean, J., Ghemawat, S.: MapReduce: simplified data processing on large clusters. In: 6th Conference on Symposium on Operating Systems Design & Implementation (OSDI 2004), Berkeley, CA, USA, vol. 6, pp. 107–113 (2004)
5. Li, J., Ma, X., Yoginath, S.B., Kora, G., Samatova, N.F.: Transparent runtime parallelization of the R scripting language. J. Parallel Distrib. Comput. **71**(2), 157–168 (2011)
6. Das, S., Sismanis, Y., Beyer, K.S., Gemulla, R., Haas, P.J., McPherson, J.: Ricardo: integrating R and Hadoop. In: Proceedings of the 2010 International Conference on Management of Data, Indianapolis, Indiana, USA, pp. 987–998 (2010)
7. Boehm, M., Tatikonda, S., Reinwald, B., Sen, P., Tian, Y., Burdick, D. R., Vaithyanathan, S.: Hybrid parallelization strategies for large-scale machine learning in SystemML. In: Proceedings of the VLDB Endowment, Hangzhou, China, vol. 7, no. 7 (2014)
8. Lee, D.D., Seung, H.S.: Algorithms for non-negative matrix factorization. In: Proceedings of the NIPS, Denver, CO, USA (2000)

The Improved Parallel Ray Casting Algorithm in Embedded Multi-core DSP System

Congliang Hu$^{(\boxtimes)}$, Huaqing Wan, and Liang Li

719 Research Institute, Wuhan, China
hucongliang0015@163.com

Abstract. An improved parallel ray casting algorithm in embedded multi-core DSP system is proposed in this paper. In order to speed up the process of intersection, the algorithm takes advantage of the improved bounding volume hierarchy (BVH) technology. By parallelizing and optimizing ray casting algorithm on TMS320C6678 of TI, the computing efficiency greatly increases. Master-slave mode is adopted to schedule the tasks and balance the load between 8 cores in parallel processing. Moreover, the code and Cache is optimized on the hardware platform. The experimental result shows that our method has good performance and exceeds the result of OpenMP (Open Multiprocessing). This paper proposes a new method to run rendering algorithm on embedded platform efficiently. The rendering algorithm usually runs on GPU, due to GPU's powerful computing capability.

Keyword: Ray casting · BVH · DSP · Parallel computing · Task scheduling · Load balancing

1 Introduction

3D scene rendering technology is a classic study in computer graphics. Ray casting is a method of direct volume rendering to generate high-quality images and preserve the details of the objects. Because every voxel needs to be processed, the computation of ray casting algorithm will be larger with the higher pixel resolution of imaging devices and larger data volume of the scene model. It leads to higher request to the efficiency of the algorithm.

With the further study of the rendering techniques, many kinds of methods have been proposed to accelerate ray casting algorithm, such as the software, hardware and parallel algorithms. Liu rasterized an octree to generate tight ray segments, which could avoid the traditional need of performing costly per-ray traversal. Goswami introduced a novel multi-way kd-trees based hierarchical data structure to simplify memory management and control the height of the tree. Kraus presented a GPU based volume rendering algorithm, adaptively sampling in three spatial directions to reduce the cost of computation. Kim used Coons patches to present a compact representation for the bounding volume. Each leaf node could be bounded very efficiently using the bilinear surface determined by the four corners.

© Springer Nature Switzerland AG 2020
Q. Liu et al. (Eds.): CENet 2018, AISC 905, pp. 29–37, 2020.
https://doi.org/10.1007/978-3-030-14680-1_4

2   Ray Casting Algorithm

The main idea behind ray casting is to trace rays from a viewpoint to each pixel on the imaging plane to find the closest object blocking the path of that ray. Using the material properties and the effect of the light in the scene, this algorithm can determine the shading of the objects, as shown in Fig. 1. Many approaches are taken to accelerate the algorithm, such as optimizing the sampling frequencies, reducing intersection computation and parallelizing the algorithm.

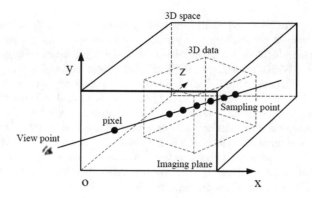

Fig. 1. The ray casting algorithm.

2.1   The Ray Casting Algorithm Flow

Flow diagram of the ray casting algorithm are presented in Fig. 2. Firstly, preprocess and extract the data from input. Secondly, classify the 3D data into different subsets without overlap and assign the color value and opacity value of the subsets. Thirdly, select multiple equidistance sampling points along the rays and use the color value and opacity value of eight voxels which are nearest a sample to calculate the sample's color and opacity value by means of three-line interpolation. Finally, combine the color and opacity of each sampling point on rays, mapping 3D image on the imaging plane. The computation of each pixel on the imaging plane is independent, which contributes to the parallel processing and acceleration of the ray casting algorithm.

3   Bounding Volume Hierarchy

BVH (Bounding Volume Hierarchies) is a tree structure on a set of geometric objects. All geometric objects are wrapped in bounding volumes. Performing collision tests on bounding volumes before testing the object geometry itself simplifies the intersection operation and results in significant improvement. Axis-Aligned Bounding Box (AABB) has a very simple shape and requires little memory space to be stored. Using AABB as bounding volume (BV) provides very fast collision detecting and helps to reduce the burden of performing testing. The tightness is another important property of BVs which needs to be considered.

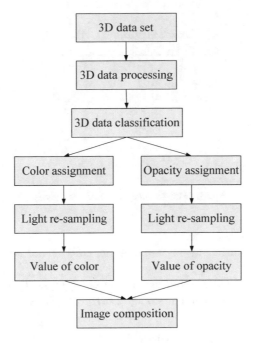

Fig. 2. The ray casting algorithm flow chart.

4   Hardware Acceleration of Multi-core DSP

4.1   Cache Optimization

The access speed of Cache is considerably faster than the external storage. The access time of L1 Cache is 1 clock cycle. The access time of L2 Cache is 2 clock cycles. However, the external storage needs at least 20 clock cycles to access data. Due to the limited memory space of Cache, it's very significant for CPU to efficiently read or write data between DDR and Cache.

When memory is cacheable access, the memory coherency problem should be taken into consideration as below:

(1) CPU and DMA are incoherent. DMA directly reads from DDR, while CPU reads from Cache.
(2) Data in Cache of each core is incoherent. When one core modifies the data in Cache, the data in external storage DDR at the same address doesn't change. When another core intends to access this data, the old data will be read from DDR to Cache.

The 3D data is stored in the external storage DDR and shared by 8 cores on DSP, because of its large amount. In order to resolve the problem of memory incoherency, this paper proposes a strategy to manually flush after writing and invalidate before reading. The solution includes the following steps, as shown in Fig. 3:

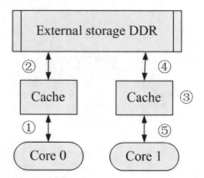

Fig. 3. The inconsistency of data access.

1. Core 0 writes data to Cache.
2. Cache of core 0 flushes to the external storage DDR at corresponding physical address.
3. Cache of core 1 invalids the old data at the same physical address.
4. Core 1 reads data from external storage DDR to Cache at corresponding physical address.

    Core 1 processes the data in Cache.

4.2   Parallel Processing in Multi-core DSP

The Communication Among Multicores
The communication between the master core and slave cores is based on the control variables and state variables. The control variables and the state variables are stored in non-cacheable shared memory segment. The control variables schedule CPUs when start to work, how to process data and summarize the results. The state variables represent the CPU is busy or idle, which contributes to the dispatch of tasks. In addition, the variable representing the progress of overall work should be shared and protected across multiple cores by using Semaphore. When one task is reading or writing the value of the variable, the other tasks can't get access to it, ensuring that every task is completed without repetition or omission.

Parallel Computing in Master-Slave Mode
In ray casting algorithm, the computation of each pixel on the imaging plane is independent which contributes to the parallelization of the algorithm on C6678. This paper uses master-slave model to process. The master-core is running to receive 3D model data in DDR shared by 8 cores, and it also should assign tasks to other cores.

After receiving start-up messages, eight cores operate their own computing task, send complete messages to the master-core, and wait for the next start-up order. Eventually the master-core summarizes the results from each core and outputs the final rendered image. As is shown in Fig. 4.

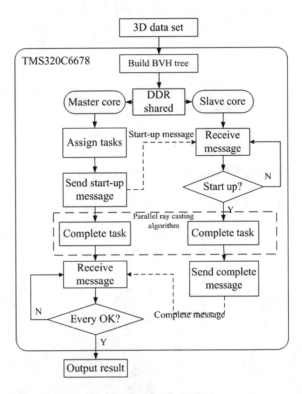

Fig. 4. System flow chart.

Task Scheduling and Load Balancing

There are 8 cores on C6678, so it's very significant to assign the tasks reasonably to ensure the execution efficiency of the system. Because pixels on the imaging plane are independent in ray casting algorithm, the whole imaging plane is divided into 8 parts which are assigned to 8 cores to calculate the color value.

Referring to the result in the next chapter, it can be found that the load is unbalanced, the maximum difference of operation time between is large, and the computing efficiency of the system is low. This paper proposes the increased density of assigned tasks and FCFS (First Come First Service) scheduling algorithm to improve load balancing of the system, two methods of task scheduling are shown in Fig. 5.

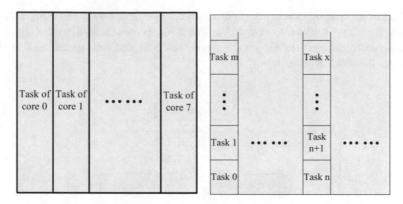

Fig. 5. Two methods of task scheduling.

5   The Experimental Results and Analysis

This paper implements the improved ray casting algorithm in two 3D scenes, and conducts the experiment on 8-core 1.4 GHz TMS320C6678. Scene 1 has 9000 vertex and 14283 facets. Scene 2 has 60083 vertex and 118031 facets. The resolution of the imaging plane is 1024 * 1024. As shown in Fig. 6.

Fig. 6. Rendering image of scene 1 and 2.

When the imaging plane is divided into 8 parts which are dispatched to 8 cores, the makespan of each core is shown in Table 1.

The maximum difference of operation time (MDM) between each core is used to measure the load balance. In scene 1, MDM is 1015.608 ms, the maximum different ratio of operation time (MDR) is 28.98%. In scene 2, MDM is 961.089 ms, MDR is 181.41%. We can conclude that dividing the imaging plane into 8 task parts ignores load balance, wastes CPU resource, and has low computation efficiency.

Table 1. CPU time consumption.

CPU	Scene 1 (ms)	Scene 2 (ms)
0	4438.156	529.793
1	4371.063	1244.220
2	4388.332	1317.647
3	3935.312	1289.964
4	3505.383	1235.327
5	3744.945	1384.709
6	4520.991	1490.882
7	3764.585	1213.636

This paper proposes the increased density of assigned tasks and FCFS (First Come First Service) scheduling algorithm to improve load balancing of the system and reduce MDM.

Table 2 indicates that increasing the number of tasks can significantly reduce MDR and improve the algorithm efficiency. But the increase of efficiency will grow at first, and gradually flatten. Eventually the efficiency doesn't increase any more.

Table 2. Data in scene 1 and 2.

Number of tasks	MDM(ms)	MDR	Increase of efficiency
64	417.152	10.22%	3.50%
128	225.342	5.55%	5.46%
256	123.521	3.05%	8.32%
512	58.009	1.43%	9.71%
1024	44.215	1.09%	10.01%
2048	13.793	0.34%	10.85%
4096	7.390	0.18%	10.88%
8192	4.962	0.12%	10.89%
Number of tasks	MDM (ms)	MDR	Increase of efficiency
64	47.927	3.45%	3.77%
128	43.287	3.16%	5.55%
256	25.042	1.83%	6.74%
512	9.260	0.67%	7.62%
1024	6.254	0.45%	7.96%
2048	5.931	0.43%	8.20%
4096	5.211	0.38%	8.22%
8192	2.981	0.22%	8.19%

As shown in Fig. 7, increasing the density of task parts will lead to the large cost of communication among cores and frequent Cache refresh in cores. The increasing density without limit will reduce efficiency and has negative impact on the algorithm. Like in scene 2, when the number of tasks increases to 8192, the algorithm efficiency decreases.

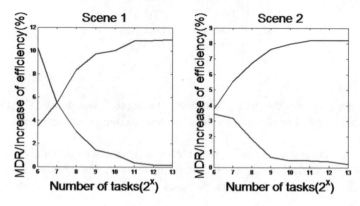

Fig. 7. The performance of our algorithm.

The TI compiler supports OpenMP (open multiprocessing), #pragma omp are used to optimize the computation of pixel color on the imaging plane. As shown in Table 3.

Table 3. Comparisons between OpenMP and our method

Scene	OpenMP speedups	Our method speedups
1	4.53	7.59
2	3.95	6.54

In order to ensure the consistency of data access, variables declared in OpenMP are shared and no-cacheable, so the optimization of loops in the algorithm is unsatisfactory. The speed-up ratio (SR) of OpenMP is 4.53 times in scene 1, and 3.95 times in scene 2. While the speed-up ratio of our method is up to 7.59 times in scene 1, and 6.54 times in scene 2, higher than OpenMP. The result shows our algorithm is superior to OpenMP.

6   Conclusion

Achieving the parallelization of ray casting algorithm on DSP system is an innovative combination of the embedded system and the rendering technology to adapt to the complex working environment. This paper proposes an improved parallel ray casting algorithm in embedded multi-core DSP system. BVH tree is built to speed up the process of intersection. The parallel ray casting algorithm in master-slave mode is adopted on TMS320C6678. And the increased density of assigned tasks and FCFS scheduling algorithm are proposed to improve load balancing of the system. The resource on DSP is in full use and the speed-up ratio is increased. The experimental results show that our algorithm has good performance and practical application value.

References

1. Bruckner, S., Groller, M.E.: VolumeShop: an interactive system for direct volume illustration. In: Proceedings of the IEEE Visualization, Minneapolis, pp. 671–678 (2005)
2. Goswami, P., Zhang, Y., Pajarola, R., Gobbetti, E.: High quality interactive rendering of massive point models using multi-way kd-trees. In: 18th Pacific Conference on Computer Graphics and Applications (PG), pp. 93–100. IEEE (2010)
3. Kraus, M., Strengert, M., Klein, T., Strengert, M.: Adaptive sampling in three dimensions for volume rendering on GPUs. In: 6th International Asia-Pacific Symposium on Visualization 2007, pp. 113–120. IEEE (2007)
4. Ling, T., Zhi, Y.Q.: An improved fast ray casting volume rendering algorithm of medical image. In: 4th International Conference on Biomedical Engineering and Informatics, pp. 109–112. IEEE (2011)

Measurement of Electric Field Due to Lightning Radiation

Pitri Bhakta Adhikari$^{(\boxtimes)}$

Tri - Chandra M. Campus, Tribhuvan University, Kathmandu, Nepal
pitribhakta_adhikari@hotmail.com

Abstract. Lightning phenomena is the electrical discharge from which the electromagnetic radiations of different wavelengths and different frequencies are produced. The electric and magnetic field signatures so produced are the basic physical parameters in understanding the mechanisms of lightning discharges. Two metallic plates, insulated from each other, are used that senses the electromagnetic radiations produced during the discharge process. Unusual lightning events by using this antenna system have been observed and recorded from hilly and subtropical mountainous country Nepal. This is a new innovation in the scientific community. The electric field signatures indicates just like a +CG stroke, but before that, the large opposite-polarity pulse occurs.

Keywords: Lightning phenomena · Electrical discharge · Unusual lightning · Opposite polarity pulse

1 Introduction

Lightning occurs when two different regions of the atmosphere gets sufficiently large electric charge, which produces electric field. The increased electric field converts insulating air into conductor. Lightning, one of the most common natural activities, is a transient and extremely complex electrical discharge during the course of which energy is released in the form of light, heat, sound and several other electromagnetic radiations. Consequently, the lightning channel acts as an effective transmitting antenna for electromagnetic waves over a wide range of frequencies. These electromagnetic waves so radiated propagate around the globe giving a number of features. The development in the understanding of lightning was initiated by the use of lightning photography on a moving film or plate. The development of magnetic links by measuring the amplitudes of lightning current is described [1, 2]. A third branch of research was concerned with the measurement of electric fields, the first result being reported by Wilson in 1916 and importance was given to lightning research in the second decade of last century [1, 3].

The lightning electromagnetic radiations travel from discharge channel in all possible direction, which is then sensed and captured by the parallel plate antenna. Different field signature sensed by this antenna connected with electronic circuit through the coaxial cable. Electric and magnetic field signatures produced due to lightning are

© Springer Nature Switzerland AG 2020
Q. Liu et al. (Eds.): CENet 2018, AISC 905, pp. 38–43, 2020.
https://doi.org/10.1007/978-3-030-14680-1_5

important in understanding the mechanisms of lightning discharges. The field signatures were captured by the parallel plate antenna system. The details of the antenna system and its calibration can be found [4].

2 Basic Theory of Antenna System

The electromagnetic wave produced due to lightning captured by parallel circular flat plate of the antenna as shown in Fig. 1. Consider a flat metallic plate employed to sense the time varying electric field over ground. Let Q be the charge that is induced on this parallel plate. Consider the size of metallic plate is smaller than the wavelength of electric field; the static electric field theory can be applied. From the Gauss's law, the total electric flux density over the surface gives the total magnitude of the charge:

$$Q = \int_{s} D.ds \qquad (1)$$

D is the electric flux density that can be expressed in terms of electric field intensity E. If the electric field intensity E is uniform over the metallic plate, Q = D.S where, S is the total surface area of the plate.

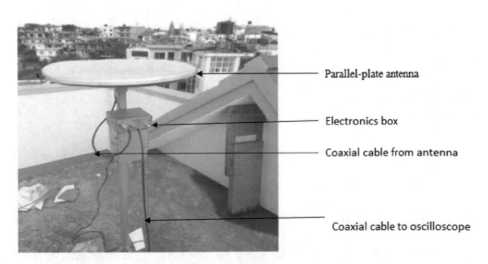

Parallel-plate antenna

Electronics box

Coaxial cable from antenna

Coaxial cable to oscilloscope

Fig. 1. The parallel-plate antenna set up on top of the building.

Then potential difference between two charged plate or the upper charged plate of the antenna and the ground is

$$V_g = E_n.d \tag{2}$$

If the voltage is measured across a resister 'R' connected between the upper plate and the ground, then the current through the resistor is $i = \frac{dQ}{dt}$, then the voltage is given by

$$V_g = Ri \tag{3}$$

On substituting the value of Q in terms of E_n, normal electric fields, then we get:

$$V_g = \varepsilon_0 \varepsilon_r R.S. \frac{dE_n}{dt} \tag{4}$$

Again, if the voltage is measured across a capacitor 'C', voltage across this capacitor is given by:

$$V_g = \frac{Q}{C} \tag{5}$$

On substituting the value of Q in terms of E_n, normal electric fields, then the voltage across the capacitor in terms of normal electric field is

$$V_g = \frac{\varepsilon_0 \varepsilon_r S}{C}.E_n \tag{6}$$

Hence, for practical purposes, the voltage is measured across the combination of capacitor and a resister, which are connected in parallel. R is chosen so large compared to C, that the effect of R on the voltage can be neglected (R = 100 MΩ, C = 15 pF); the effect of capacitor only can therefore be implemented. If C_g be the capacitance between upper plate and ground in absence of external loading (RC) circuit. The measured voltage across RC circuit is given by:

$$V_m = E_n.d.\frac{C_g}{C_g + C} = V_g \frac{C_g}{C_g + C} \tag{7}$$

3 Instrumentation

Two parallel metallic plates are used, which are insulated from each other. The lower plate is grounded and the upper plate is joined to the circuit via coaxial cable and they are separated by a metallic tube as shown in Fig. 1. This antenna has capacitance of 60 pF and the total output of the amplifier V_m is shown in Fig. 2. This output was connected to Pico-scope 6404D via a RG-58 coaxial cable.

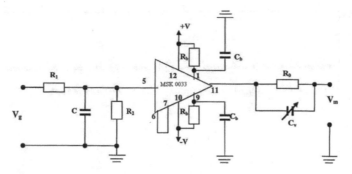

Fig. 2. The circuit used in this research.

The time constant of the circuit in the measuring system was made 13 ms, which is sufficient to record lightning electric field signatures. The instrumental decay affected the subsequent parts of longer waveforms. The measuring system which is described above has been used in many other experimental researches [5–10].

A digital storage oscilloscope (DSO) is an oscilloscope (or pico-scope) that digitizes analogue voltage and store the signal digitally. Nowadays, The DSO is the most widely used type of oscilloscope owing to its advanced trigger, data storage, measurement and display attributes. In this study, pico-scope (6404D) has been used to digitally store the electric field data. The scope has maximum sampling rate of 5 Gs/S with a time window of 500 ms.

4 Results and Discussion

Lightning electric fields were recorded in Kathmandu, Nepal in the year 2015 during April–June, at the height of 1300 m above sea level. In the process of lightning, the horizontal circular flat plate antenna sensed the vertical electric field and the signature passed through the RG-58 co-axial cable and pico-scope and was recorded in the storage device. The antenna of height 1.5 m was placed on the top of the building such that the total height of the antenna was 13.5 m above the ground. Electromagnetic radiations captured by parallel circular flat plate of the antenna helped in understanding some of the primary features of lightning activities, in Nepal. These include characteristic features of positive ground flashes, electric fields radiated by cloud flashes and unusual lightning field signatures.

We investigated the wave signature of +CG occurred and studied about their fine structures of high hills and rugged terrain of Nepal. This research was done on the regions of mountainous sub-tropical country. The result of this research was compared with temperate and tropical regions. During the pre-monsoon season of 2015, a total of 133 remarkable +CG were observed and the waveforms were single strokes. Some of them have two strokes and one waveform has four return strokes. Generally, multiplicity of the positive return strokes is unity, but we recorded four strokes in a single flash. The four stroke +CG flash is shown in the Fig. 3.

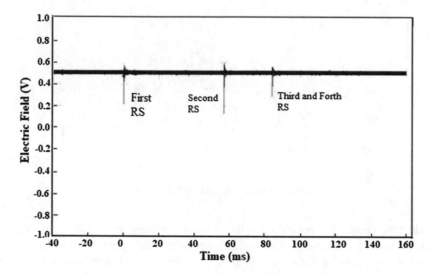

Fig. 3. The four stroke +CG flash.

One of the most important properties of lightning processes observed in Nepal using the parallel plate antenna system was the unusual lightning events. From all available studies and the analysis of electric field waveforms, the most regular pattern shown by the unusual lightning event is that they have the waveform of like +CG and opposite polarity leader pulses just before that. This type of unusual lightning phenomena was recorded in this pre-monsoon of the year 2015, which were about 38% of total events. Ratio of amplitudes of opposite leader pulse is 32.4% of the main waveform. The graphical representation of the unusual waveforms in Kathmandu is given in Fig. 4.

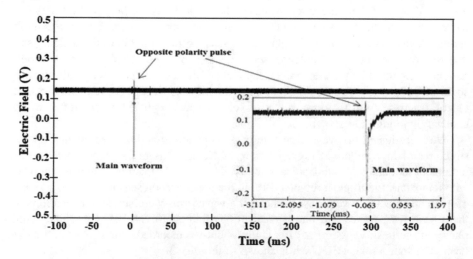

Fig. 4. Example of observed unusual lightning events.

5 Conclusion

Countless measurements have been conducted for natural lightning; yet, the phenomenon of lightning discharge has not been fully understood. Recently, from the mountainous country like Nepal, the unusual events were observed by using this antenna system. The four stroke +CG flash and the unusual lightning events were recorded and studied. The waveforms of unusual signature have the waveform of like +CG and opposite polarity leader pulses just before that event. In this study electric field measurements were conducted in hilly and subtropical country, Nepal with the help of the parallel plate antenna system which senses the electromagnetic radiations emitted during the phenomena, hence the unusual and +CG waveforms have been studied.

Acknowledgements. We would like to express our appreciation to the International Science Programme (ISP), Uppsala University, Sweden, for supplying the instruments used in this research. Also, we would like to thank the University Grant commission (UGC) to help for the conference and finally also thanks to Trichandra College, Amrit Science College, affiliated to Tribhuvan University for the facilities to do this research.

References

1. Berger, K.: The earth flash. In: Lightning, Physics of Lightning. Academic Press, New York (1977)
2. Uman, M.A.: The Lightning Discharge. Dover Publications, New York (2001)
3. Rakov, V.A., Uman, M.A.: Lightning: Physics and Effects. Cambridge University Press, Cambridge (2003)
4. Galvan, A., Fernando, M.: Operative characteristics of a parallel-plate antenna to measure vertical electric fields from lightning flashes. Institute for High Voltage, Uppsala University (2000)
5. Adhikari, P.B., Sharma, S.R., Baral, K.N., Rakov, V.A.: Unusual lightning electric field waveforms observed in Kathmandu, Nepal, and Uppsala, Sweden. J. Atmos. Solar Terr. Phys. **164**, 172–184 (2017)
6. Adhikari, P.B., Sharma, S.R., Baral, K.N.: Features of positive ground flashes observed in Kathmandu Nepal. J. Atmos. Solar Terr. Phys. **145**, 106–113 (2016)
7. Johari, D., Cooray, V., Rahman, M., Hettiarachchi, P., Ismail, M.: Characteristics of preliminary breakdown pulses in positive ground flashes during summer thunderstorms in Sweden. Electr. Power Syst. Res. **153**, 3–9 (2017)
8. Baharudin, Z.A., Fernando, M., Ahmad, N.A., Makela, J.S., Rahman, M., Cooray, V.: Electric field changes generated by the preliminary breakdown for the negative cloud-to-ground lightning flashes in Malaysia and Sweden. J. Atmos. Solar Terr. Phys. **84**, 15–24 (2012)
9. Sharma, S., Fernando, M., Cooray, V.: Narrow positive bipolar radiation from lightning observed in Sri Lanka. J. Atmos. Solar Terr. Phys. **70**, 1251–1260 (2008)
10. Sharma, S., Cooray, V., Fernando, M.: Isolated breakdown activity in Swedish lightning. J. Atmos. Solar Terr. Phys. **70**, 1213–1221 (2008)

Research on Variable-Lane Control Method on Urban Road

Yan Liu[✉]

Changchun Normal University, Changchun 130032, China
liuy78@126.com

Abstract. There is a kind of phenomenon of unbalanced road resources distribution in the city traffic congestion, called tidal traffic. In order to resolve the phenomenon of traffic congestion caused by the above reasons and improve the road capacity, a bi-level planning model is established based on the entire regional road network optimization. The total impedance of system, namely the control region is minimized in the upper level and Beckmann model is applied in the lower level and harmony ant colony algorithm is designed to solve the integer planning model. Finally, the proposed model is verified by practical examples. Results show that the new proposed lane adjustment method in this paper can solve the problem of "tidy traffic" effectively.

Keywords: Tide traffic · Bi-level programming model · Variable lane · HS-AC algorithm

1 Introduction

With the rapid development of economy, the ownership of private car is increasing and the requirements of the environment of people travel are also rising. Therefore, the condition of city traffic congestion exacerbates seriously. Along with a variety of policies and management programs have been introduced, congestion has eased, but the problem still exists. During the morning and evening peak hours, commuting trips lead to concentrate the centralized direction of traffic flow distribution, causing heavy traffic in one direction, while the other side of the road is idle and road resources are wasted. Recent studies of variable lane has gradually attracted more attention, but the actual variable lane programs are mostly given by the traffic police arbitrarily based on experience which is not able to give a more effective adjustment timely and accurately.

People determine empirically to implement the strategies of variable lane roads and driveways leading to some surrounding roads congestion severely. Therefore, the better strategy solving this kind of problems is the adjusted scheme of variable lane based on regional road network. An optimizing model for determining the steps of designing dynamic variable lane to improve the entire traffic capacity of road network and to decline the entire travel time of system is established in literature [1]. Literature involves the application condition of variable lane in the U.S.A [2, 3]. Based on the existing theories, a bi-level planning model considering the whole road network is proposed in this paper to resolve the traffic problem better.

© Springer Nature Switzerland AG 2020
Q. Liu et al. (Eds.): CENet 2018, AISC 905, pp. 44–50, 2020.
https://doi.org/10.1007/978-3-030-14680-1_6

2 Network Optimization

In order to increase the road capacity and resolve the problem of unbalanced supply and demand, the schemes of adjustment of variable lane need to take the regional road network into account, rather than only change a single lane, or is likely to cause road network congestion around variable lane more strength, losing the meaning of lane adjustment [4, 5]. Therefore, managers need to coordinate the various sections of the number of lanes in the region and achieve the best optimization results by reasonable adjustments [6].

For two-way traffic road, due to the uneven distribution of two-way traffic road, the lane congestion in one direction is significantly higher than the lane in the opposite direction, and even where congestion occurs in one direction, while in the opposite direction is smooth [7, 8]. In such a case, firstly, the utilization of road resource is low, cannot make full use of the existing road resources, even if the road network in the area is growing, and still cannot solve the problem of congestion [9]. Then, due to one or some directions of congestion, even though the area of road is increasing, which causes to produce delay and increase the travel cost largely, adding to the total impedance of road network system in terms of road network. Under fast pace of modern living conditions, people want to travel at a lower cost. Therefore, traffic travel costs have become a major indicator of the efficiency of transportation.

Bi-level planning model established in this paper considers the entire road network, in which the goal of the upper level is to minimize the total system impedance, namely, the shortest total travel time of road network. The lower level is the classic Beckmann model, namely, the optimization of user. The lower level planning reflects the feedback that the best capacity distribution scheme of each link to the upper level that improves the scheme of variable lane, which the two levels interact each other to obtain the best scheme. Specific model is as follows.

Take the form of BPR function as impedance function.

$$t = t_0[1 + \alpha(v/c)^\beta] \tag{2.1}$$

The upper level is as follows.

$$\min Z = \sum_a t_a(k_a) \cdot k_a + t'_a(k'_a) \cdot k'_a \tag{2.2}$$

The lower level is as follows.

$$\min Z(a) \tag{2.3}$$

Where,
a = any link in the network.
k_a = the traffic flow of link a in the direction of heavy traffic flow.
k'_a = the traffic flow of link a in the direction of light traffic flow.

3 Solving Algorithm

Ant colony algorithm using distributed parallel computer system is easy to combine with other algorithms and has strong robustness, but the disadvantage is that the complex issues are easy to fall into local optima, the advantages of harmony research algorithm are simple structure, small time complexity and easy to operate, etc. but the shortages are slow convergence speed and bad robustness [10–14]. This paper will combine the characteristics of two algorithms to propose harmony ant colony coupling algorithm to solve the problem that the ant colony algorithm is easy to fall into local optimal solution and that harmony search algorithm has slow convergence speed. Ant colony algorithm produces pheromone mechanism on each path of the optimum path node priorities, allowed to enter the harmony search algorithm memory and uses the optimal retention policy, thus making harmony algorithm to generate every new solution in a constantly updated optimum population. Proven, harmony ant colony algorithm compared to the ant colony algorithm and harmony algorithm, can get the same or even better the optimal solution in the same number of cycles [15], and optimize better. Algorithm steps are as follows:

Step 1: Initialize the value of the algorithm parameters. Population scales HMS, HMCR, PAR, etc.
Step 2: Generate harmony memory database. Each value selection probability of each independent variable is determined by the size of pheromone generating harmony memory database, and to calculate the objective function values.
Step 3: Produce new solution. New solution is generated by the three mechanisms of new solution vector based on harmony search algorithm, and then to calculate the objective function values.
Step 4: Compare the optimal solution from Step 2 with Step 3, and record the better one.
Step 5: Update the pheromone and recycle from Step 2 to Step 5 until arrive to the final condition.

4 Verify

This paper applies the data of network to prove, in which the largest traffic capacity of each link is 1800 pcu/h. The road network structure is shown in Fig. 1. The traffic capacities from four OD groups are 5-2, 1336 pcu/h, 6-9, 1887 pcu/h, 1-10, 5612 pcu/h, 10-11976 pcu/h. This paper applies the several approaches that is comparing different schemes of the total road impedance, the average delay in each link of the vehicle and the average queue length to prove the validity of approaches proposed in the paper, in which comparing to each other scheme one is that managers through experience give lane adjustment programs. Scheme two is adjustment program for the application of this paper method. Lane adjustment program is shown in Table 1, where scheme 0 is the original program of the lane without lane road network improvement.

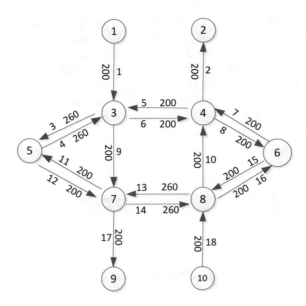

Fig. 1. The road network.

Table 1. Adjustment scheme in different programs.

Link	Program		
	Program0	Program1	Program2
1	2	2	2
2	2	2	2
3	4	2	1
4	4	6	7
5	4	2	1
6	4	6	7
7	2	1	1
8	2	7	7
9	2	2	2
10	2	2	2
11	4	5	6
12	4	3	2
13	4	5	6
14	4	3	2
15	1	1	1
16	1	7	7
17	2	2	2
18	2	2	2

The parameters of the new proposed algorithm are set as follows. The ant number m is set to 30. The parameter Alpha that represents the importance of pheromones is set to 1. Parameter Beta that represents the degree of heuristic factor importance is set to 5. Pheromone evaporation coefficient is set to 0.1. The maximum number of iterations is set to 200 and the intensity factor of pheromones increase is set to 100.

This paper verifies the road network under the environmental conditions of VISSIM simulation and the two-lane adjustment program is carried out microscopic simulation and data collection including average delay of each link and average queue length. Data analysis is shown below.

1. Total impedance in network.

The total impedance of the road network of two adjustments is calculated according to the proposed method in this paper respectively shown in Fig. 2.

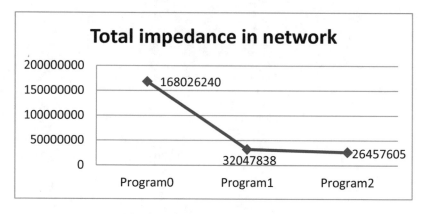

Fig. 2. Total impedance in network.

As can be seen from the above line chart, using this method to variable-lane setting, the total impedance of system is 26,457,605, compared to the scheme 0 and scheme 1 decreased by 84% and 12%. Overall, according to the proposed method, the effect of adjustment is obvious.

2. Delay and queue length.

Data show that the application of this method to set variable lane declines the average delay of road section improves the average queue length, and optimizes the traffic condition of the entire road network under ensuring the two-way traffic with normal prevailing circumstances. It proves the effectiveness of the proposed method for improving the "tide traffic" (Figs. 3 and 4).

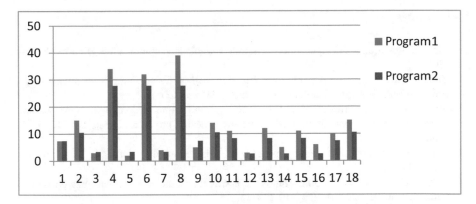

Fig. 3. The link of link.

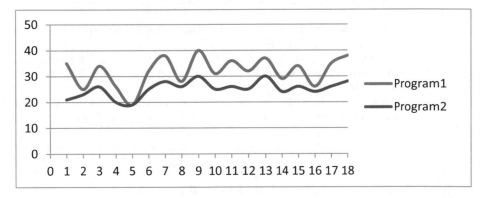

Fig. 4. The queue length.

5   Conclusion

This paper studies the issue of variable city road lane, establishes a bi-level planning model, designs AC-HS algorithm to solve the model, and finally verifies this model by instances. The result illustrates that this method proposed by this paper can improve the phenomenon of unbalanced distribution of city road resources effectively, optimizes traffic environment and increases the quality of traffic and capacity of road network in some extent. However, this paper cannot research the association between variable lane and signal timing and optimization yet, and needs further exploring.

Acknowledgment. This work is supported by Education Department of Jilin Province (JJKH20181181KJ).

References

1. Sun, Q.: Research about algorithm of dynamic variable lane optimization. Highway **8**, 45–50 (2009)
2. Zhang, H.Z., Gao, Z.Y.: A convex combination method for solving mixed- integer bi-level programming. J. Beijing Jiaotong Univ. **29**(6), 38–42 (2005)
3. Suwansirikul, C., Friesz, T.L., Tobin, R.L.: Equilibrium decomposed optimization: a heuristic for the continuous equilibrium net-work design problem. Transp. Sci **21**, 254–263 (1987)
4. Ma, Z.H.: Research on the setting methods of variable lane in urban roads. Jilin University, Changchun (2014)
5. Zhang, H.Z., Gao, Z.Y.: Optimization approach for traffic road network design problem. Chin. J. Manage. Sci. **15**(2), 86–91 (2007)
6. Brain, W., Laurence, L.: Reversible lane synthesis of practice. J. Transp. Eng. **9**, 933–944 (2006)
7. Brain, W., Laurence, L.: Planning and operational practice of reversible roadway. ITE J. **8**, 38–43 (2006)
8. Wang, X.: Bi-level programming model and algorithm of urban roads convertible lanes. Zhongnan University, Changsha (2013)
9. Gao, Z.Y., Zhang, H.Z., Sun, H.J.: Bi-level programming models, approaches and applications in urban transportation network design problems. J. Transp. Syst. Eng. Inf. Technol. **2**(4), 35–44 (2004)
10. Du, W.L., Zhang, H.L., Qian, F.: Chaotic particle swarm optimization algorithm with harmony search for industrial applications. J. Tsinghua Univ. (Science & Technology) **52**(3), 325–330 (2012)
11. Hao, B., Ren, X.H., Gao, Y.L.: Hybrid harmony search and estimation of distribution algorithm for multi-objective optimization problems. Appl. Res. Comput. **5**, 1659–1661 (2012)
12. Das, S., Mukhopadhyay, A., Roy, A., Abraham, A., Panigrahi, B.K.: Exploratory power of the harmony search algorithm: analysis and improvements for global numerical optimization. IEEE Trans. Syst. Man Cybern. B Cybern. **41**(1), 89–106 (2011)
13. Landa-Torres, I., Ortiz-Garcia, E.G., Salcedo-Sanz, S., Segovia-Vargas, M.J., Gil-Lopez, S., Miranda, M., Leiva-Murillo, J.M., Del Ser, J.: Evaluating the internationalization success of companies through a hybrid grouping harmony search-extreme learning machine approach. IEEE J. Sel. Top. Signal Process. **6**(4), 388–398 (2012)
14. Al-Betar, M.A., Khader, A.T., Zaman, M.: University course time tabling using a hybrid harmony search metaheuristic algorithm. IEEE Trans. Syst. Man Cybern. C Appl. Rev. **42**(5), 664–681 (2012)
15. Wei, L., Fu, H., Yin, Y.P.: Research and application of integer programming based on combinative harmony search and ant colony algorithm. Comput. Eng. Appl. **49**(20), 5–8 (2013)

Detecting Hot Spots Using the Data Field Method

Zhenyu Wu and Jiaying Chen[✉]

School of Internet of Things, Nanjing University of Posts
and Telecommunications, XinMoFan Road, Nanjing, Jiangsu Province, China
876179824@qq.com

Abstract. With the developments of the mobile devices and Internet of things, the location data have recorded amount of information about people activities. Mining the hot spots from the location-based data and studying the changing patterns of hot spots are useful to the early warnings of the disasters, traffic jams and crimes. Current researches on hot spots detections ignore the temporal factors. In this paper, the data field method is used to describe the interactions of spots, and the temporal factors are incorporated into the data field method. Furthermore, a hot spots detection method is proposed. Finally, the heat map is used to illustrate the effectiveness of the proposed method based on an open dataset.

Keywords: Data field · Hot spots · Location Based Service

1 Introduction

With the popularity of intelligent devices and GPS (Global Positioning System) sensors, the communications and connections among users are becoming more and more convenient. Nowadays, as the IOT (Internet of Things) technologies are being further studied, the location information could be used to identify the locations, statuses of users, so that the intelligent and personalized services could be provided. Therefore, there are large amounts of location big data which is usually dynamic.

The analysis and mining of location big data are meaningful. First, the location data has close relations with the daily life of users. Therefore, it includes the rich information such as places, time. The life patterns could be obtained by mining these location big data. For example, by using the different type of sensors located in the smart phones, Jeffery et al. proposed a data mining framework to analyze the daily behavior of users [1]. Second, with the development of urbanization, the population and its density are continuously changing. The movement patterns, such as the user activity patters, temporal and spatial patterns of the crowds, could be discovered. In this way, the city situations could be monitored, and the urbanization could be improved. In addition, the status of the communities could be effectively monitored, so that the whole society and city could be well managed [2].

With the development of the Internet technologies, users could share their information at any time and any places. Therefore, from the perspective of data, the source of location big data includes LBS (Location Based Services), and UGC (User

© Springer Nature Switzerland AG 2020
Q. Liu et al. (Eds.): CENet 2018, AISC 905, pp. 51–58, 2020.
https://doi.org/10.1007/978-3-030-14680-1_7

Generated Contents). The location big data could also be called the crowd sourcing geographic data, which means the geographic data collected by the crowds [3]. The typical data includes the trace data of GPS, the map data collectively annotated by users, the check-in data in social networks. These data have temporal, spatial, and social attribute, which record the trace of the life, and reflects the daily behavior of the crowds [4].

The time and the space are two important factors of the system. Therefore, it is much more likely to discover the behavior patterns and its changing trends by analyzing the location big data. In this paper, the hot spots detection methods are studied by mining the location big data. In general, the hot spots are the regions where an entity spends plenty of time [5]. By detecting the hot spots, the attention could be paid on the important area while the less important area would be ignored. Meanwhile, the studies of the trends of the hot spots could help find the changing patterns of the behavior. Therefore, the disasters could be warned, the road traffic could be controlled, and the crime behavior could be prevented [6].

In fact, the hot spots are continuously changing with time. However, the existing studies have ignored the temporal aspects. Therefore, spatial relations are modeled by the data field method, by which the interactions of different locations could be described. Moreover, the temporal information is introduced into the data field method, so that the changing patterns of the hot spots could be studied.

The remaining of the paper is organized as follows: the related works are introduced in the second part. In section three, the hot spots detection method using data field method is described. The forth section is about the experimental results of the proposed method. This paper is concluded in section five.

2   Related Works

There are two kinds of related works about the studies of the hot spots detection. The one is based on machine learning algorithms. For the given location big data, the machine learning algorithms are designed. For example, the cluster algorithms could be used to detect hot spots in location data. The other one is based on the anomaly detection methods, which is typically used to discover the data patterns that are different from the normal behavior. Anomaly detection is one of the important research directions in lots of domains. For hot spots detection studies, the hot spots could be regarded as the abnormal data points.

Detecting the hot spots by using the cluster based methods has been applied in lots of domains. Tony et al. proposed the machine learning algorithms to cluster the GIS (Geographic Information System) data. By means of the proposed method, the hot spots of the crime events could be detected [7]. Furthermore, a clustering algorithm based on the fuzzy c-means was proposed to identify the hot spots regions from the GPS data.

Hot spots detections are one of the meaningful solutions to effectively control the road traffic. In general, the places where are congested could be regarded as the hot spots. In order to effectively resolve the congested road traffic problem, Jitesh et al. pointed out that the detection of the congested regions on the road is necessary.

Moreover, the influence of the congested regions could help the administrators to formulate a better traffic management policy [8]. With the development of the wireless sensor networks, the hot spots detection is attacking more and more attention of the researcher. The distributed sensors could return the location information in near real time. In their works, Pedro et al. proposed a distributed method, in which a clustering algorithm is designed to analyze the data stream obtained from the sensor networks. Therefore, the correlations of the sensors could be identified [9].

Recently, the location data of the taxies are continuously generated with the rise of the taxi platforms. Generally, the location data are collected in the form of the GPS. The GPS data has reflected the intelligence of the taxi drivers, because the roads that the drivers are passing by are always the better road lines. Therefore, the GPS data provided by the taxi drivers reflects the crowd intelligence of the drivers. By mining the behavior patterns from these GPS data, the driving techniques of the skilled drivers could be concluded. Moreover, the discovered patterns could be used to optimize the planning of the cities. Daqing et al. studied the driving behavior of the taxi drivers by means of the matrix [10]. Furthermore, the hot spots and their changing patterns are studied. Hadi et al. proposed a hot spot detection method based on the tensor decomposition method [11].

In the studies of the anomaly detections, the abnormal trajectories or the gathering of the crowds in a specific place could be regarded as a special event, such as the holiday, the traffic accident, or the natural disasters. This kind of studies includes the detection of the important places [12, 13], the understanding of the activity patterns [14], and the prediction of the behavior of the crowds [12]. Apichon et al. constructed a framework to detect the abnormal events using the hidden Markov model. The spatio-temporal activity patterns of the crowds could be discovered by processing the GPS data [15]. Pawling et al. proposed the clustering algorithms to process the mobile phone data [16]. Therefore, the abnormal user behaviors could be identified. Liao et al. analyzed the mobile data points to detection anomaly [17]. The data was modeled in the time series, and the principal component analysis method is applied.

3 Hot Spots Detection Methods Using the Data Field

3.1 The Description Method of the Hot Spots

Recently, the data streaming mining has been becoming one of the popular research domains, because the knowledge could be discovered from the large scale, continuously generated big data [18]. Therefore, the location big data is modeled as the data streams in this paper.

The location data stream could be described as $pointStream = \{p_1, p_2, \ldots \ldots, p_n\}$. In $pointStream$, p_i is a data point, $p_i = \{lat_i, lon_i, ts_i\}$. For a given data point, lat_i is used to represent the latitude of the location data point i, and lon_i is used to represent the longitude of the location data point i. Moreover, ts_i is used to represent the timestamp of the location data point i. In this model, all the detected hot spots should be changed with time.

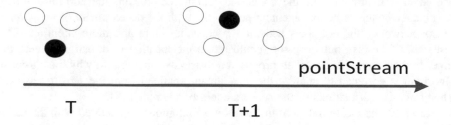

Fig. 1. The hot spots are changing with time.

As more and more location data are processed, the hot spots are changing. This is shown in Fig. 1.

The data field method has borrowed the idea of the field in the physics domain. The interactions among the particles and its description methods are introduced to the data space. Therefore, the interactions of the data points in data spaces could be formally described. Given the data space $\Omega \subseteq \mathbf{R}^p$, there are totally n data points constructing a dataset $D = \{x_1, x_2, \ldots, x_n\}$. The potential of the data point $x \in \Omega$ could be defined as follows [19]:

$$\varphi(X) = \varphi_D(X) = \sum_{i=1}^{n} \varphi_i(X) = \sum_{i=1}^{n} \left(m_i \times e^{-\left(\frac{\|x - x_i\|}{\sigma}\right)^2} \right)$$

Where, m_i is the mass of the data point i, σ is the influence range of the data point. $\|.\|$ is the distance between two data points. In this paper, the hot degree of a location data point could be defined as follows based on the data field method. The location data point x owns the hot degree $H(x, t)$ at time t:

$$H(X, t) = \sum_{i=1}^{n} \left(m_i(t) \times e^{-\left(\frac{\|x - x_i\|}{\sigma}\right)^2} \right)$$

Where, $m_i(t)$ represents the mass of the location data point i at time t, and its value is changing with time. The changing could be defined as follows:

$$m_i(t) = f(\Delta t) = t - t_0 + 1$$

Where, t_0 is the timestamp of the first appearance of the location data point i.

3.2 The Hot Spot Detection Algorithm

The proposed hot spot detection algorithm is shown in Algorithm 1. In this algorithm, the input data is the location data stream *pointStream*, the algorithm processes the location points one by one. When a location data point is processed, the hot spots are given. In the proposed algorithm, the hot spots map is used to show the hot spots.

Algorithm 1: Hot spots detection using data field
Input: location data stream *pointStream*;
Output: hot spots map
    Save the data stream into *existPoints*
    foreach point *dp* in *pointStream*:
        foreach point *ep* in *existPoints*:
            updateHeatValueofExistPoint(*ep*);
            calculateHeatValueofNewPoint(*dp*);
        end
    end
    print(*existPoints*)

The detailed steps of the proposed algorithm are shown in Algorithm 1. First, the inputted data stream is saved into *existPoints*. Second, read one data point from data stream. If this data point is a new one, it will influence other existing data points. Therefore, the hot degree values should be updated. Moreover, the hot degree value of the new data point should also be calculated. Finally, the hot spots map could be drawn. In this paper, the contours of the data field are used to describe the data points with the same hot degree value.

4 Experimental Results

Large amounts of open datasets have been released for the location based studies. In this paper, the dataset [20] released by Micheal et al. is used to verify the proposed method. This dataset is about the data of the taxi trajectory. Each item of the data is composed of four items, which is respectively latitude, longitude, free or not, and the timestamp. The latitude and the longitude data show the location of the taxi, and the timestamp shows the time when the taxi is at that place. In this paper, the data of latitude, longitude, and timestamp are used.

The experimental results are show in Figs. 2 and 3. In these two figures, the horizontal axis stands for the latitude, and the vertical axis stands for the longitude. Each location data is marked as a single black point in the figures. Moreover, the hot spots degree values are calculated using the proposed method. Furthermore, the contours of the hot spots degree values are drawn in the two figures.

Figure 2 shows the results of the first 100 records. The results demonstrate that the activity locations of the taxis are clustered together. The hot spot degree is higher in the clustering places than other places. Figure 3 shows the hot spots map of the 500 records. The results demonstrate that the hot spots are changing with the increase of the number of the new locations. New hot spots appear, and the old hot spots disappear.

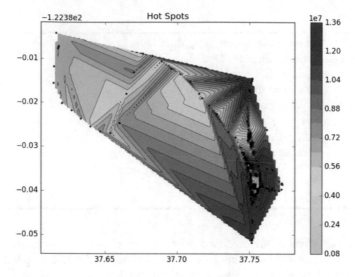

Fig. 2. Hot spots map of 100 records.

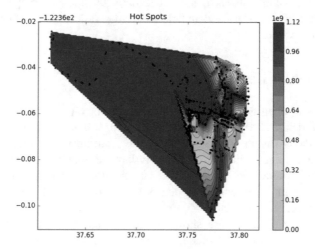

Fig. 3. Hot spots map of 500 records.

5 Conclusions

With the development of the mobile devices, users can fetch information at any places and at any time. Moreover, the IOT technologies would prompt this situation. Therefore, the location base service will become more and more popular.

In this paper, a data filed based method is proposed to discover the hot spots from the taxi activity data, which is meaningful for the studies, such as disaster preventions. Moreover, the hot spots map is used to verify the effectiveness of the proposed method. From the map, the hot spots and their changing patterns could be identified.

Acknowledgement. This work is supported by National Natural Science Foundation of China (No. 61502246), NUPTSF (No. NY215019).

References

1. Lockhart, J.W., Weiss, G.M., Xue, J.C., Gallagher, S.T., Grosner, A.B., Pulickal, T.T.: Design considerations for the WISDM smart phone-based sensor mining architecture. In: Proceedings of the Fifth International Workshop on Knowledge Discovery from Sensor Data, SensorKDD 2011, San Diego, CA, USA, pp. 25–33 (2011)
2. Guo, B., Wang, W., Yu, Z.W., Wang, Y., Yen, N.Y., Huang, R., Zhou, X.G.: Mobile crowd sensing and computing: the review of an emerging human-powered sensing paradigm. ACM Comput. Surv. **48**(1), 7–31 (2015)
3. Heipke, C.: Crowdsourcing geospatial data. J. Photogramm. Remote. Sens. **65**(6), 550–557 (2010)
4. Hu, Q.W., Wang, M.: Urban hotspot and commercial area exploration with check-in data. Acta Geod. Cartogr. Sin. **43**(3), 314–321 (2014)
5. Gudmundsson, J., Kreveld, M.V., Staals, F.: Algorithms for hotspot computation on trajectory data. In: Proceedings of the 21st ACM SIGSPATIAL International Conference on Advances in Geographic Information Systems, SIGSPATIAL 2013, Orlando, FL, USA, pp. 134–143 (2013)
6. Hwang, S.: Extending spatial hot spot detection techniques to temporal dimensions. In: Proceedings of the 4th ISPRS Workshop on Dynamic and Multi-dimensional GIS (2005)
7. Grubesic, T.H., Murray, A.T.: Detecting hot spots using cluster analysis and GIS. In: Proceedings of the 5th ISPRS Workshop (2001)
8. Tripathi, J.P.: Algorithm for detection of hot spots of traffic through analysis of GPS data. In: Computer Science and Engineering Department, THAPAR University (2010)
9. Rodrigues, P.P., Lopes, L.: Distributed clustering of streaming sensors: a general approach
10. Zhang, D., Sun, L., Li, B., Chen, C., Pan, G., Li, S.J., Wu, Z.H.: Understanding taxi service strategies from taxi GPS traces. In: IEEE Transactions on Intelligent Transportation Systems (2014)
11. Tork, H.F., Gama, J.: An eigenvector-based hotspot detection. In: Proceedings of 16th Portuguese Conference on Artificial Intelligence, Acores, Portugal, pp. 290–301 (2013)
12. Ashbrook, D., Starner, T.: Using GPS to learn significant locations and predict movement across multiple users. Pers. Ubiquitous Comput. **7**(5), 275–286 (2003)
13. Zhou, C.Q., Frankowski, D., Ludford, P., Shekhar, S., Terveen, L.: Discovering personally meaningful places: an interactive clustering approach. ACM Trans. Inf. Syst. **25**(3) (2007)
14. Liao, L., Patterson, D.J., Fox, D., Kautz, H.: Building personal map from GPS data. In: Progress in Convergence: Technologies for Human Wellbeing, vol. 1093, no. 1, pp. 249–265. Academy of Sciences, New York (2006)
15. Witayangkurn, A., Horanont, T., Sekimoto, Y., Shibasaki, R.: Anomalous event detection on large scale GPS data from mobile phones using hidden markov model and cloud platform. In: Adjunct Proceedings of the 2013 ACM Conference on Pervasive and Ubiquitous Computing Adjunct Publication, UbiComp 2013, Zurich, Switzerland, pp. 1219–1228 (2013)
16. Pawling, A., Yan, P., Candia, J.: Anomaly detection in streaming sensor data. Intell. Tech. Warehous. Min. Sens. Netw. Data, 99–117 (2008)
17. Liao, Z., Yang, S., Liang, J.: Detection of abnormal crowd distribution. In: IEEE/ACM International Conference on Green Computing and Communications, pp. 600–604 (2010)

18. Silva, J.A., Faria, E.R., Barros, R.C., Hruschka, E.R., de Carvalho, A.C.P.L.F., Gama, J.: Data stream clustering: a survey. ACM Comput. Surv. **46**(1) (2016)
19. Li, D.Y., Du, Y.: Artificial Intelligence with Uncertainty, 2nd edn. CRC Press, Boca Raton (2015)
20. Piorkowski, M., Sarafijanovic-Djukic, N., Grossglauser, M.: A parsimonious model of mobile partitioned networks with clustering. In: The First International Conference on Communication Systems and Networks, Bangalore, India (2009)

Denoising SOM: An Improved Fault Diagnosis Method for Quantum Power Communication Network

Xi Chen[1(✉)], Wenjing Li[2], Shangjun He[1], Gonghua Hou[1],
Lifan Yang[1], Lin Lin[1], and Xiao Feng[3]

[1] State Grid Fujian Electric Power Co., Ltd.,
Fuzhou Power Supply Company, Fuzhou, China
sxzyfx@163.com
[2] State Grid Information and Telecommunication Group Co., Ltd.,
Beijing, China
[3] State Grid Xintong Yili Technology Co., Ltd., Beijing, China

Abstract. With the continuous expansion of quantum power grids, big data analysis of the power grid has become an important issue in the management of smart grid operation and maintenance. Due to the complexity of the power grid system and the limitations of software and hardware environmental conditions, frequent faults in the power grid are very important for the processing of alarm data. This article analyzed the characteristics of power grid alarm data and proposes an improved fault diagnosis method, Denoising SOM. This method uses the self-organizing map (SOM) neural network as a basis, and proposes the idea of denoising the alarm data sample to improve the robustness of the model. In addition, this method also optimizes the Euclidean distance calculation and the neighbourhood function setting of the SOM neural network. The results of comparative experiments show that the fault diagnosis method proposed in this paper has higher accuracy in the task of multi-source fault diagnosis and reduces the time consumption.

Keywords: Quantum · SOM · Neural network · Fault diagnosis · Power communication network

1 Introduction

With the gradual construction of smart grids, the fault diagnosis of power grids has also received more and more attention. Therefore, the identification of the root cause fault is very important for the operation and maintenance of the power grid.

In recent years, fault diagnosis methods based on artificial intelligence have been studied by many scholars in order to accurately analyze and find faults in the system. Many researchers applied the SOM neural network model to multi-source fault diagnosis tasks and achieved good results. However, the traditional SOM model also faces many problems, such as noise problems. Grid data is also affected by noise during data collection and transmission. As a result, many abnormal data and irregular data are generated, leading to misdiagnosis.

© Springer Nature Switzerland AG 2020
Q. Liu et al. (Eds.): CENet 2018, AISC 905, pp. 59–66, 2020.
https://doi.org/10.1007/978-3-030-14680-1_8

To solve the above problem, we propose an improved fault detection model: Denoising SOM. The algorithm is based on the traditional SOM neural network model, and infers the failure type of the sample by comparing the similarity between the input sample and the cluster. The Denoising SOM model mainly considers the data noise problem. Our work was inspired by the idea of the denoising automatic encoder [1]. We added a random noise layer to the SOM model to simulate the noise and interference in the real environment, and designed a new Euclidean distance calculations process noise plus data. In addition, we optimized the design of the neighborhood function of the SOM neural network to solve the problem of unclear cluster boundaries. Finally, a comparative experiment was conducted in a simulated power grid environment. The experimental results show that Denoising SOM exhibits higher robustness and computational efficiency in fault diagnosis tasks.

2 Related Works

The SOM neural network algorithm model was first proposed by Kohonen [2] and has been validated for a variety of fault detection tasks. In recent years, Niva et al. [3] proposed a generic fault detection method based on SOM clustering for NFV. Li et al. [4] proposed a sensor fault diagnosis method based on wavelet packet and SOM neural network. Qiu et al. [5] proposed a fault diagnosis method based on wavelet singular entropy and SOM neural network for microgrid system. Li et al. [6] added the DAE (Denoising Automatic Encoder) model to the task and designed a multi-item condition identification method based on double DAE-SOM.

3 Denoising SOM Algorithm

3.1 The Model of Denoising SOM

The self-organizing map (SOM) neural network is a competitive neural network. It is characterized by its ability to preserve the high-dimensional features of the input vector and map it onto a low-dimensional neural network, thereby forming an ordered map in a topological sense.

The structure of the traditional SOM neural network [2] consists of two parts: the input layer and the competition layer. The input layer is responsible for receiving input signals. The number of input layer neurons is the number of dimensions of a data sample. The number of neurons in the competitive layer is often chosen to approximate the square of the dimension of the input vector, and it is arranged in a matrix as a plane. Each competitive layer neuron maintains a weight vector W_j, where j represents the serial number of the neuron. The dimension of the weight vector is the same as the dimension of the data sample. The input layer is fully connected to each competitive layer neuron, while the competitive layer neurons are not connected to each other.

$$d\left(W_{j^*}, X\right) = min_{j \in \{1,2,...,m\}} d\left(W_j, X\right) \qquad (3.1)$$

The operating mechanism of SOM neural network is competition learning rules (Winner-Take-All). First, the data sample vectors are normalized in the Euclidean space, and the weight vectors are initialized randomly in the Euclidean space. Then select the input vector and all the weight vectors to measure the similarity. The highest similarity is as the winning neuron, as shown in Eq. (3.1), d represents the similarity measure, X represents the input vector, and j^* represents the winning neuron. Finally, the winning neuron and its neighboring neurons are activated, and the neurons that are far away from the winning neuron are suppressed so that the competitive layer node maintains the topological characteristics of the input vector.

In our research on power grid operation and maintenance work, we found that the actual data is very different from the data in the standard experiment. The situation where the data is disturbed by noise is very common, causing the phenomenon of abnormal and missing data. In the traditional SOM neural network, the similarity measure used in the data classification process is the Euclidean distance. The absence and anomaly of some dimension attributes in the data sample vector will greatly affect the calculation of the Euclidean distance. It can lead to misclassification of data, that is, fault diagnosis. In order to solve the noise interference problem of fault diagnosis task, this paper introduces the idea of denoising [1] into SOM neural network model.

In order to train the noise reduction capability of the SOM model, it is necessary to add certain noise disturbances during the training of the model. Therefore, we add a random noise layer before the input layer of the SOM neural network, randomly adding Gaussian noise and Zero-set noise to the input data with a certain probability, forcing the competitive layer to learn more robust features in the data.

The structure of the Denoising SOM consists of three parts: the input layer, the random noise layer, and the competition layer. The input layer randomly acquires a data sample vector (which has been subjected to data normalization) as an input to the SOM. The random noise layer performs a random noise mapping q on the input data X to obtain a partially corrupted version $\tilde{X}$. The mapping process is as shown in formula (3.2). D represents a vector space:

$$\tilde{X} \sim q_D\left(\tilde{X}|X\right) \qquad (3.2)$$

For each data sample vector, the noise map randomly selects a fixed number of dimensions of data to add a Gaussian noise signal or a zero-set signal. In this experiment, the proportion of noise added is chosen to be 30%. When the data volume is large and the network model is more complex, the proportion of added noise data can be appropriately reduced. If there are few experimental data, in order to better mine the potential characteristics of the data and enhance the robustness of the model, the proportion of noise plus data should be increased.

The working modes of the random noise layer are mainly divided into two types: one is to set the selected dimensional data of the sample vector to 0 to simulate the absence of data; the other is to add Gaussian noise to the data to change its probability distribution. The Gaussian distribution is shown in Eq. (3.3):

$$F(x; \mu, \sigma) = \frac{1}{\sigma\sqrt{2\pi}} \int_{-\infty}^{x} \exp\left(-\frac{(x-\mu)^2}{2\sigma^2}\right) dx \tag{3.3}$$

Where σ is the variance of the data sample, μ is the average of the data samples, and x is an attribute of the sample vector X. The noise-add mapping can be expressed as Eq. (3.4), where d is a linear random number.

$$\tilde{x} = q_D(x) = x + \mu + \sigma * F(d) \tag{3.4}$$

In addition, the competitive layer of the traditional SOM neural network is to measure the similarity by calculating the Euclidean distance so as to select the winning neuron. For noise-added data, calculating the Euclidean distance will produce a large error. Therefore, we designed a fuzzy Euclidean distance calculation method to reduce noise effects and make the calculation process of the competition layer more robust:

$$d(W, X) = \sqrt{\sum_{i=1}^{n} (w_i - x_i)(w_i - \bar{x})} \tag{3.5}$$

Where $\bar{x}$ represents the mean of the sample vector X elements.

3.2   A New Neighborhood Function

In the algorithm, the class boundary is an important parameter that represents the range of influence of the winning neuron of the competitive layer. The neighborhood function is an important way for the SOM neural network to fit the boundary of the class.

The traditional SOM neural network uses a Gaussian function as a neighborhood function, which determines the degree of involvement of the neuron in the training process by calculating the Euclidean distance between the neuron and the current winning neuron. It makes the convergence speed of the SOM neural network algorithm faster than the general rectangular neighborhood function, but its excitation effect in the far-neighborhood range is not obvious. However, the Cauchy function is far less effective than the Gaussian function in the near-neighborhood excitation, but it has better output capability in the far-neighborhood range. This paper designs a compound neighborhood function that combines the advantages of the Gaussian function in the near domain and the advantages of the Cauchy function in the far-neighborhood.

The neighborhood function is shown in Eqs. (3.6) and (3.7), when $\sigma(t) > 3$:

$$h_i(t) = \begin{cases} \frac{1}{\pi\sigma(t)\left[1 + \left(\|r_c - r_i\|^2/\sigma(t)\right)\right]}, & \|r_c - r_i\|^2 < \sigma(t) \\ 0, & \|r_c - r_i\|^2 > \sigma(t) \end{cases} \tag{3.6}$$

When $\sigma(t) < 3$:

$$h_i(t) = \begin{cases} e^{\frac{\|r_c - r_i\|^2}{2\sigma^2(t)}}, & \|r_c - r_i\|^2 < \sigma(t) \\ 0, & \|r_c - r_i\|^2 > \sigma(t) \end{cases} \tag{3.7}$$

$\|r_c - r_i\|^2$ represents the Euclidean distance between winning node c and node i, $\sigma(t)$ represents the domain range. The neighborhood range is a decreasing function with respect to t and becomes smaller as t increases.

3.3 The Overall Flow of Denoising SOM Algorithm

Based on the above improvements, the training method for the Denoising SOM model designed for grid fault diagnosis tasks is as follows:

Algorithm : Denoising SOM

Parameters: Algorithm termination condition, η_{min}
        Training time limit, T
Input: Training set X
Initialization: $t = 0$
Weight vector W_j , $j = 1,2,\dots,m$
        Initial winning neighborhood, $h_{j,c}$
        Initial learning rate, $\eta(t) = \eta_0$

Start
While $\eta(t) > \eta_{min}$ **do** :
    Enter a data sample X_p , $p \in \{1,2,\dots,P\}$
    Add noise, $\widetilde{X_p} \sim q_D(\widetilde{X_p}|X_p)$
    Calculate winning neurons j^*, $d(W_{j^*},\widetilde{X}) = min_{j\in\{1,2,\dots,m\}} d(W_j,\widetilde{X})$
    Define winning neighborhood $h_{j^*}(t)$
    Decline in learning rate, $\eta(t) = \eta_0(1 - t/T)$
    Adjust weights, $W_j(t+1) = W_j(t) + \eta(t)h_{j^*}(t)[X_p - W_j(t)]$
    $t \leftarrow t + 1$
end while
Return Denoising SOM Model
End

---

4 Using Denoising SOM for Fault Diagnosis Experiments

We used the simulated power alarm data to conduct comparative experiments on the Denoising SOM model. The power alarm data is in the form of a numerical vector. The data itself does not contain the specific information of the root fault, but the data with the same root fault has certain similarities in the vector space. In the training phase, we input training data into the SOM neural network for clustering. After the clustering ends, each activated competitive layer neuron represents a clustering center. The failure type indicated by the clustering center is the root failure type of the training data of this cluster. In the test phase, if it is a single-source fault diagnosis task, the alarm data is input to the Denoising SOM neural network and a winning neuron is obtained. We believe that the type of root failure represented by the winning neuron is the root cause of the alarm data to be measured; if it is multi-source fault diagnosis tasks, fault types need to be located according to the similarity function, which will be described in detail in Sect. 4.1.

4.1   Multi-source Fault Diagnosis Comparison Experiment

Multiple-source fault diagnosis indicates that one alarm data may contain multiple root fault information and all related root fault information needs to be detected. We built a similarity function (4.1) to measure the degree of similarity between an input sample and a cluster. Therefore, in the diagnosis of multiple source faults, the main task is to select the root cause fault with high similarity and eliminate the root cause fault with low similarity.

$$s(\boldsymbol{W}, \boldsymbol{X}) = d(\boldsymbol{W}, \boldsymbol{X}) / \sum d(\boldsymbol{W}, \boldsymbol{X}) \tag{4.1}$$

Where $\sum d(\boldsymbol{W}, \boldsymbol{X})$ represents distance between all cluster centers and input data. We choose the k cluster centers that have the smallest distance, if $\sum_k s(\boldsymbol{W}, \boldsymbol{X}) < 1/n$, n represents the number of cluster centers, which means k-source fault.

We designed the simulation data set according to the characteristics of the power grid data, and simulated ten kinds of power grid source faults such as antenna anomaly, overload limit, sensor abnormality, and electromagnetic interference. The data set model is structured data. Each piece of data includes multi-dimensional attributes such as power consumption, sensor statistics, and traffic, which have been normalized into numerical vectors. Therefore, the data itself does not reflect the network failure, but the data with similar failures meets the clustering conditions.

In comparison with the choice of algorithm, we used the traditional SOM to compare with the improved model of this paper, and also selected the fault diagnosis framework in [7]. Smart Grid Fault Diagnosis Framework in [7] is able to autonomously learn the model of the nominal state using the respective data by means of hidden Markov models operating in the parameter space of linear time-invariant models. Subsequently, the framework is able to detect data not belonging to the nominal state and localize the potential fault at the cognitive level.

In the evaluation index of the experiment, we select two kinds of indicators that are more practical in fault diagnosis tasks: precision rate and recall rate. The precision rate indicates that the hypothesis algorithm classifies some data samples as A faults, and how many of them are true A faults. The recall shows that the hypothesis algorithm classifies some data samples as B faults, where the true B fault accounts for the proportion of all B faults in the data set. As shown in Table 1, the accuracy of the Denoising SOM algorithm is much higher than that of the other two algorithms in the multi-source fault diagnosis results, which are 9.0% and 6.3% higher than the traditional SOM algorithm

Table 1. Comparison of multiple source fault diagnosis.

Algorithm	Precision rate	Recall rate	Training time (s)
SOM	81.1%	87.7%	409.0
Algorithm in [7]	83.2%	**89.9%**	692.0
Denoising SOM	**88.4%**	89.0%	**377.4**

and the algorithm in [7]. In terms of recall, the Denoising SOM is 1.5% higher than the traditional SOM, but slightly lower than the algorithm in [7]. Due to the influence of the random noise layer, the Denoising SOM actually reduces the complexity of the input data, and it also significantly reduces the training time, which is 7.7% and 36.3% less than the traditional SOM and the algorithm in [7].

4.2   The Relationship Between the Number of Cascades and Accuracy of Fault Diagnosis

In the analysis of power alarm data, we guessed that there is a relationship between the accuracy of fault diagnosis and the number of cascaded alarms, that is, alarm data can trigger multi-level alarms. We have designed comparative experiments on the accuracy rate of root-source fault diagnosis under different cascaded alarm conditions.

As shown in Fig. 1, the precision of fault diagnosis increases with the increase of cascades. The reason is that when the cascade triggered by the root alarm is small, its anti-noise interference capability is weak. Since the Denoising SOM itself introduces a noise reduction mechanism, it can maintain a high accuracy rate even when the number of cascades is small.

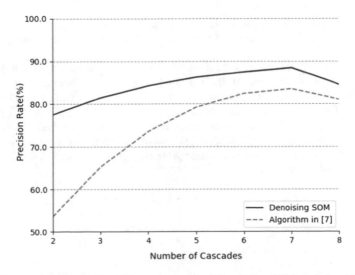

Fig. 1. The precision rate with different cascades of Denoising SOM and algorithm in [7].

5   Conclusions

For the problems of frequent faults and difficult management in power grid systems, this paper presents a fault diagnosis method based on SOM neural network, Denoising SOM. We mainly consider the data noise problem, add a random noise layer to the SOM model to simulate the noise and interference in the real environment, and designed a new Euclidean distance calculation method to process the noise-added data. In addition, this method also optimizes the design of the neighborhood function of the SOM neural network to solve the problem of unclear cluster boundaries. In the simulation experiment, we used clustering to divide the alarm data samples and determine the type of the root fault of the alarm data. The experimental results show that in the multi-source fault diagnosis task, the optimization algorithm of this paper, Denoising SOM, shows better accuracy and reduces the time consumption. In addition, the verification of the relationship between the number of cascades and the accuracy of fault diagnosis also shows that Denoising SOM has better robustness.

Acknowledgement. The project is supported by Science and Technology Project "Research and Development of Quantum Cryptography Equipment and Terminal Modules for Power Distribution Grid" in State Grid Corporation of China (536800170042).

References

1. Vincent, P., Larochelle, H., Bengio, Y., Manzagol, P.A.: Extracting and composing robust features with denoising autoencoders. In: International Conference on Machine Learning, pp. 1096–1103. ACM (2008)
2. Kohonen, T.: Self-Organizing Maps. Series in Information Science, vol. 30(4), pp. 266–270. Springer, Heidelberg (1997)
3. Niwa, T, Miyazawa, M, Hayashi, M., Stadler, R.: Universal fault detection for NFV using SOM-based clustering. In: Network Operations and Management Symposium, pp. 315–320. IEEE (2015)
4. Li, J., Meng, G., Xie, G.: Sensor fault diagnosis based on wavelet packet and SOM neural network. Chin. J. Sens. Actuators **30**(7), 1035–1039 (2017). (in Chinese)
5. Qiu, L., Ye, Y., Jiang, C.: Fault diagnostic method for micro-grid based on wavelet singularity entropy and SOM neural network. J. Shandong Univ. **47**(5), 118–122 (2017). (in Chinese)
6. Li, M., Du, W., Qian, F.: Performance recognition method based on multi-index and multi-layer DAE-SOM algorithm. CIESC J. **69**(2), 769–778 (2018). (in Chinese)
7. Ntalampiras, S.N.: Fault diagnosis for smart grids in pragmatic conditions. IEEE Trans. Smart Grid **9**(3), 1964–1971 (2018)

The Assessment Method of Network Security Situation Based on Improved BP Neural Network

Gangsong Dong[1], Wencui Li[1(✉)], Shiwen Wang[1], Xiaoyun Zhang[2], JiZhao Lu[1], and Xiong Li[1]

[1] Information & Telecommunication Co. of State Grid Henan Electric Power Company, Zhengzhou 450052, China
lliwencui@163.com
[2] China Electric Power Equipment and Technology Co. Ltd., Zhengzhou Electric Power Design Institute, Zhengzhou, China

Abstract. With the popularity of the Internet and the emergence of cloud computing, network security issues have become increasingly prominent. In view of the low efficiency and poor reliability of the existing network security situation assessment methods, this paper proposes a quantitative assessment method based on an improved BP neural network. Aiming at the disadvantages of slow convergence speed, easy oscillation, and local minimum in BP neural network, this paper optimized the algorithm by combining Cuckoo search algorithm, introducing momentum factor and adaptive learning rate. The simulation results show that the improved CS-BPNN algorithm in this paper has fast convergence rate and high evaluation accuracy, which provides a new method for network situation assessment.

Keywords: Network security situation assessment · BP neural network · Cuckoo search algorithm · Momentum factor

1 Introduction

With the rapid development of Internet technology, the security issues have become increasingly prominent and urgent. The Network Security Situation Assessment (NSSA) has gradually became the focus of next-generation network security technology research [1].

The network security situation assessment mainly focuses on the information fusion, comprehensive analysis and understanding of network security-related elements, so as to grasp the network security status. There are many assessment methods at present, including Bayesian technology, neural networks, fuzzy evaluation methods, DS evidence theory etc. These methods have guiding significance for our study, but also have many deficiencies such as single source of assessment information, large space-time overhead and low credibility. With the development of new technologies such as cloud computing and machine learning, network security assessment technology is moving in the direction of intelligence, comprehensiveness, and scale.

© Springer Nature Switzerland AG 2020
Q. Liu et al. (Eds.): CENet 2018, AISC 905, pp. 67–76, 2020.
https://doi.org/10.1007/978-3-030-14680-1_9

To make up for the determination of traditional network security assessment methods, this paper proposes a new quantitative assessment method based on the improved CS-BP Neural Network, which reduces the subjective impact of expert opinion, improves the efficiency and accuracy of assessments. The rest of this paper is organized as follows. Section 2 discusses related work. Section 3 put forward an improved CS-BPNN algorithm by combing the method of Cuckoo search algorithm, additional momentum term, and adaptive learning rate adjustment and describes detailed assessment process based on improved CS-BPNN. Simulation results are presented in Sect. 4 and we conclude in Sect. 5.

2 Related Works

Network security situation assessment is mainly divided into three categories. Firstly, assessment based on mathematical models includes Analytic hierarchy process, set-pair analysis and fuzzy comprehensive evaluation. Bian et al. [1] proposed a multi-level fuzzy comprehensive evaluation model (AHP_FCE) based on AHP and FCM for network security situation assessment. Han et al. [2] established a situation assessment model based on set pair analysis.

Secondly, assessment based on knowledge reasoning including Bayesian theory, DS evidence theory, probability theory, graph theory. For example, Liu et al. [3] proposed a hierarchical network threat situation assessment method based on D-S evidence theory. Szwed et al. [4] uses fuzzy cognitive maps to obtain the dependencies of important assets in the network and assess the degree of harm.

Finally, assessment based on pattern recognition which mainly draws on the concept of data mining algorithms such as SVM, neural network, rough set theory and Hidden Markov. For example: Xiao et al. [5] network security situation prediction method based on MEA-BP. Chen et al. [6] built an intelligent network security awareness system based on RF-SVM algorithm.

3 Network Security Situation Assessment Method Based on Improved CS-BPNN

3.1 Standard BP Neural Network

The BP neural network is a multi-layer feedforward network proposed in 1986. It contains one input layer, one output layer, and multiple hidden layers trained by forward and backward propagation process. Assume that the input vector is $X = (x_1, x_2, \cdots, x_n)$, the hidden layer output vector is $Y = (y_1, y_2, \cdots, y_m)$, the output layer vector is $O = (o_1, o_2, \cdots, o_l)$, the expected output vector is $D = (d_1, d_2, \cdots, d_l)$, the weight matrix between input layer and hidden layer is $V = (V_1, V_2, \cdots, V_j, \cdots, V_m)$, the weight value between the hidden layer and the output layer is $W = (W_1, W_2, \cdots, W_k, \cdots, W_l)$, the hidden layer threshold is θ_j, the output layer threshold is r_k. The output value of the j neuron of the hidden layer is y_j, and the output of the k neuron of the output layer is o_k the total error of the system is E.

$$y_j = f(\sum_{i=1}^{n} v_{ij} \times x_i - \theta_j) \tag{3.1}$$

$$o_k = f(\sum_{j=1}^{n} w_{jk} \times y_i - r_k) \tag{3.2}$$

In Eqs. (4.1) and (4.2), the transfer function is $f(x)$, as shown in Eq. (4.3).

$$f(x) = \frac{1}{1 + e^{-x}} \tag{3.3}$$

$$E = \frac{1}{2} \sum_{m=1}^{M} \sum_{n=1}^{N} [d_k - o_k]^2 \leq \varepsilon \tag{3.4}$$

In the backward propagation stage, the weight can be adjusted by the gradient descent method to reduce the error E. The calculation method is as follows, in which η is learning speed.

$$\Delta w_{ij} = -\eta \frac{\partial E}{\partial w_{jk}} \tag{3.5}$$

3.2   Improved Algorithm of BP Neural Network

Cuckoo Search Algorithm Optimization

The Cuckoo Search (CS) algorithm is an intelligence optimization algorithm proposed by the famous scholars Yang and Ded in 2009. There are numerous studies show that the SA algorithm is superior to genetic algorithm (GA), particle swarm algorithm (PSO) in solving optimization problems. The specific advantages are reflected in its strong global search capability, fast convergence, fewer parameters, and better versatility and robustness.

The initial weights and thresholds of the BP neural network are randomly assigned. So the training time is generally long, and the trained weights and thresholds may not be optimal. Therefore, the CS algorithm can be used to find the optimal weights and thresholds for BP neural network, so that to improve the training speed and training effect of BP.

(1) the path and position update formula

Set the following 3 ideal states: A. All cuckoos in the population only hatch one egg at once and place them randomly in a nest; B. In order to keep the population from degenerating, the best cuckoo individuals will be reserved for the next generation; C. The size of the population is constant, and the probability that the host bird has found an alien cuckoo egg is $p_a \in [0, 1]$. On the basis of these three ideal states, the path and position update formula for the cuckoo nest search is as follows:

$$x_i^{(t+1)} = x_i^{(t)} + \alpha \oplus L(\lambda) \quad i = 1, 2, \cdots, n \tag{3.6}$$

In which, $x_i^{(t)}$ represents the position of the i nest in the t generation. $\oplus$ is a point-to-point multiplication. $\alpha > 0$ is step length (usually take $\alpha = 1$). $L(\lambda)$ is a random search path, which obeys Lévy distribution, and determined by (3.7)

$$L - u = t^{-\lambda} \tag{3.7}$$

(2) Cuckoo coding method. The floating-point coding has the characteristics of easy control of code length, high coding accuracy, and large space search capability, which can reduce the computational complexity. According to the structure of the BP neural network, the weight and the threshold value are co-encoded into a floating-point cuckoo code:

$$W_{11}W_{21} \cdots W_{M1}V_{11}V_{21} \cdots V_{1J}\theta_1 \cdots W_{1I}W_{2I} \cdots W_{MI}V_{I1}V_{I2} \cdots V_{IJ}\theta_I r_1 \cdots r_J \tag{3.8}$$

(3) Fitness function design. Fitness is the basis for measuring the quality of individuals in a group. The higher fitness value indicates that the individual is closer to the optimal solution. In BP neural networks, the smaller the system error, the better the result. So we take the reciprocal of the total error as the fitness function:

$$F(w, v, \theta, r) = 10^{-3} / [\sum_{m=1}^{M} \sum_{n=1}^{N} (d_k - o_k)^2] \tag{3.9}$$

Momentum Factors Optimization

The standard BP algorithm is essentially a simple steepest descent static optimization method. In the correction of w(k), it is only corrected in the direction of the negative gradient of step k, and does not take into account the previous accumulated experience, which often causes the training process to oscillate, slows convergence. We can use the additional momentum method to optimize the weight adjustment formula (4.5). The weight adjustment formula with momentum items is designed as follows:

$$\Delta w(k+1) = (1 - \alpha)\eta[(1 - \beta)D(k) + \beta D(k-1)] \\ + [(1 - \beta)\Delta w(k) + \beta \Delta w(k-1)] \tag{3.10}$$

In which, $D(k) = -\frac{\partial E}{\partial w(k)}$, representing the negative gradient at time k. W is the network weight, Δw is the increment of the weight, k is the training frequency, and α is the momentum factor, $0 < \alpha < 1$, generally is 0.95. η is the learning rate. β is an adjustment factor which used to adjust the proportion of the influence of two weight corrections before. Since the weight change is mainly influenced by the previous weight adjustment trend, $0 < \beta < 0.5$.

The meaning of the above equation is to pass the influence of the previous two weight changes on the current weight adjustment trend through a momentum factor. When the network weight value enters the flat area at the bottom of the error surface, it prevents the appearance of $\Delta w_{ij} = 0$, and helps the network jump out of the local minimum of the error surface. The momentum factor added in this method is actually equivalent to the damping term, which reduces the oscillation tendency in the learning process and acts as a buffer smoothing, thereby improving the convergence.

Adaptive Learning Rate Optimization

In the standard BP algorithm, the learning rate η is constant. If η is chosen too small, it will lead to too long training time, too slow convergence; if η is too large, it may be overcorrected, causing oscillations and even diverging, thus making the system unstable. So it is difficult to determine a best learning rate that is suitable from beginning to end. In order to solve this problem, this paper adopts adaptive learning rate method in the network training process. The basic idea is heuristic adjustment is made to η based on the gradient information of the error change, so that making η increase and decrease according to people's expectations The expression of the mathematical model established in this paper is shown in formula (3.11):

$$\eta(k) = \begin{cases} m^{\gamma}\eta(k-1) & \Delta E < 0.2 \\ 0.95 & \Delta E \geq 0.2 \end{cases} \tag{3.11}$$

$$\gamma = sign[D(k)D(k-1)] \tag{3.12}$$

In which, $\eta(k)$ is the learning rate at time k, take $\eta(0) = 0.01$, $m > 0$ is an integer or decimal number. After verification, m = 2 in this paper. $f(x) = sign(x)$ is symbolic function, the value is either 1 or -1. When the gradient direction is the same for two consecutive iterations, it means that the drop is too slow, take $\gamma = 1$. Otherwise, take $\gamma = -1$.

3.3   Evaluation Process Based on Improved CS-BPNN Algorithm

The process of security situation assessment based on the improved CS-BP neural network algorithm are as follows:

> **Step 1** Collecting the situation-related data and extract indicators to generate input data of the CS-BPNN model.
>
> **Step 2** Initializing Population. According to the characteristics of neural network weights and thresholds, we produce n cuckoos randomly $x^{(0)} = (x_1^{(0)}, x_2^{(0)}, \cdots, x_n^{(0)})$, and encode them.
>
> **Step 3** Location update and calculation of fitness. Update the position of the cuckoo according to (3.6), $x^{(t)} = (x_1^{(t)}, x_2^{(t)}, \cdots, x_n^{(t)})$, Calculate the fitness of this generation cuckoo according to Eq. (3.9).

Step 4 Select, Replace and Delete. Find the optimal cuckoo $x_i^{(t)}$ which will be passed to the next generation. Generate a decimal r in the [0, 1] interval randomly, then compare r with the finding probability p_a. If $r > p_a$, then update the position according to the formula (3.6). The cuckoo with larger fitness is retained; Else if $r \leq p_a$, then keep the original cuckoo.

Step 5 Determine whether the optimal cuckoo meets the condition or whether the iterative algebra meets the requirements. If yes, then decode the optimal cuckoo to obtain the optimal weights and thresholds and assigns them to the BPNN part. Otherwise, Step 3 is performed.

Step 6 Training optimized BP neural network. (1) Initialize the remaining parameters, input and output vectors of the BP network. (2) Calculate the input and output of each neuron in hidden and output layer according to Eqs. (3.1), (3.2) and (3.3). (3) Calculate the total system error E according to Eq. (3.4). (4) Set the learning rates of the correction weight matrix W and V according to (3.11) and (3.12) respectively. Then adjust the weight of each neuron according to Eq. (3.10). (5) Iterative training continues until $E < \varepsilon$ or the number of iterations reaches N.

Step 7 The test situation data is input into the trained CSBPNN model with the evaluation capability. After the mapping, the network situation value SA is obtained.

4   Simulation

4.1   Experimental Environment and Parameters Setting

We set up a network experiment environment to test the performance of the assessment method proposed in this paper. Periodically collect network operation and attack information through dedicated software such as Nessus, Snort and Netflow, as the source of data for our experiment, then organize experts to conduct manual assessments of security risks and obtain the actual level and value of network security situation.

Extract eight indicators from the original sample, Number of security devices (x_1), Total number of open ports for key devices (x_2), Frequency of key devices accessing the mainstream secure website (x_3), Number of alarms (x_4), Usage rate of network bandwidth (x_5), Frequency of history security event (x_6), Change rate of subnet traffic (x_7), Average non-fault time of subnet (x_8). The eight evaluation indicators constitute a sample set as input for the evaluation model, and the expert evaluation results serve as expected output. We collected 1,500 samples, in which 1000 samples were randomly selected as the training set and the remaining 500 samples were used as test sets. We use Matlab software to realize the standard BP neural network algorithm (BPNN) [5], the optimized BP neural network by genetic algorithm (GA-BPNN) [7], and the improved CS-BP neural network algorithm (CS-BPNN) proposed in this paper. Respectively using these three algorithms to achieve security situation assessment, compare and analyze the assessment results.

In the cuckoo optimization section, we select the cuckoo population size n = 40, set the discovery probability $p_a = 0.1$, the maximum iteration algebra is 40; In the BP neural network part, set the enter node number is 8, the hidden node number is 6, the output node number is 1 (situation value, SA). The transfer function is the standard sigmoid function, the maximum number of iterations is set to 1000, the objective function error $\varepsilon = 0.05$, the initial learning rate $\eta(0) = 0.01$, and the momentum factor $\alpha = 0.95$.

4.2   Experimental Results and Analysis

We can analyze the experimental results from the following three stages:

Part 1: Weight Initialization Stage. Figure 1 shows the variation curve of the fitness value of individual cuckoos. After 18 iterations, the fitness rate converges to an optimal value. This shows the CS algorithm requires only a small amount of resources to achieve optimal. Therefore, using CS to find the initial optimal weights and thresholds of BP neural network is efficient and feasible, which can effectively reduces the later training time and energy consumption.

Part 2: The Training Stage. Figure 2 and Table 1 shows the training process of three algorithms. For iterations number and the improved CS-BPNN algorithm is 89, which is 431 times less than the BPNN and half of the GA-BPNN. The training time and training error are also greatly shortened. It can be seen in Fig. 2, the improved algorithm does not have extreme problem and has faster convergence speed and better

Table 1. Simulation results.

Algorithm	The iteration number	The training error	The execution time/s
Traditional BPNN	520	0.0408	3.734
GA-BPNN	182	0.0297	1.946
Improved CS-BPNN	89	0.0195	1.135

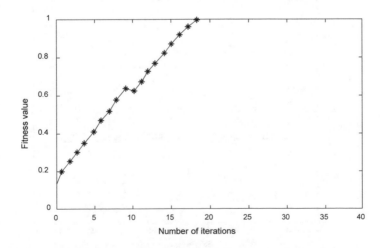

Fig. 1. The fitness value of individual cuckoos.

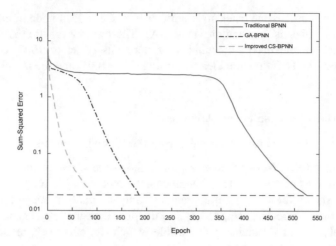

Fig. 2. Error curves.

convergence effect. In short, the new algorithm has significantly improved the training time and training effect.

Part 3: The Assessment Part. The final assessment results are the security posture SA and security level. We selects two indicators of relative evaluation error and evaluation accuracy to describe the evaluation effectiveness. As shown in Figs. 3 and 4, the CS-BPNN algorithm has smaller assessment errors and higher assessment accuracy. What is more, as the sample size increases, the error curve and accuracy curve of the new algorithm remains relatively stable, which shows that its assessment results are more stable.

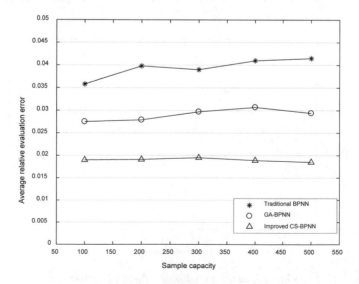

Fig. 3. Relative evaluation error curve.

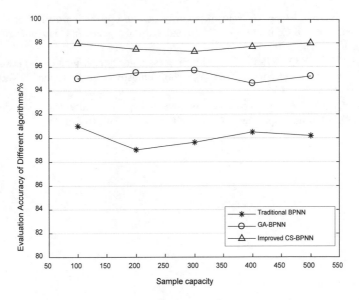

Fig. 4. Evaluation accuracy cure.

5   Conclusion

This paper presents a new quantitative assessment method of network security situation, based on improved BP neural network algorithm. In this paper, the momentum term and the adaptive learning rate is introduced into the BP algorithm. What's more, the cuckoo search algorithm is used to optimize the initial weights and thresholds of BP. These improvements avoid BP falling into local extremes and improve the training speed and assessment accuracy. Through experimental simulation, we can see that compared with the standard BPNN and GA-BPNN, the improved BP neural network in this paper has better evaluation results, fewer iterations, less error, and higher accuracy.

References

1. Bian, N., Wang, X., Mao, L.: Network security situational assessment model based on improved AHP_FCE. In: International Conference on Advanced Computational Intelligence, pp. 200–205. IEEE (2013)
2. Han, M.N., Liu, Y., Chen, Y.: Network security situational awareness model based on set pair analysis. Appl. Res. Comput. **29**(10), 3824–3827 (2012)
3. Liu, Z., Zhang, B., Zhu, N., Li, L.: Hierarchical network threat situation assessment method for DDoS based on D-S evidence theory. In: IEEE International Conference on Intelligence and Security Informatics, pp. 49–53. IEEE (2017)
4. Szwed, P., Skrzyński, P.: A new lightweight method for security risk assessment based on fuzzy cognitive maps. Int. J. Appl. Math. Comput. Sci. **24**(1), 213–225 (2014)

5. Xiao, P., Xian, M., Wang, H: Network security situation prediction method based on MEA-BP. In: International Conference on Computational Intelligence & Communication Technology, pp. 1–5. IEEE (2017)
6. Chen, G., Zhao, Y.Q.: RF-SVM based awareness algorithm in intelligent network security situation awareness system. In: The Workshop on Advanced Research & Technology in Industry (2017)
7. Chen, H.Y.: The research of computer complex network reliability evaluation method based on GABP algorithm. Appl. Mech. Mater. **556–562**, 6207–6210 (2014)

Research on Reconstruction Method of Random Missing Sensor Data Based on Fuzzy Logic Theory

Liang Zhao$^{(\boxtimes)}$ and Kai Liu

College of Informatics, Huazhong Agricultural University, Wuhan, China
zhaoliang323@mail.hzau.edu.cn

Abstract. Wireless sensor nodes are often deployed in the wild environment, and the data collected are often lost. It is very important to reconstruct the missing data for accurate scientific calculation or other applications. In this study, a random missing data reconstruction method based on fuzzy logic theory is presented. The method mainly studies how to combine the Euclidean distance between the sensor nodes and the correlation of the sensory data to construct a new method of determining neighbor nodes, while the weight calculation method of each neighbor node participating in reconstruction is studied, which is to solve the deficiencies of the neighbor node selection when there are obstacles between sensor nodes only rely on the Euclidean distance. The experimental results show that the accuracy of the proposed method is relatively high when the sensor data has a mutation or the acquisition time interval is large.

Keywords: Wireless sensor networks · Missing data reconstruction · Fuzzy logic theory · Spatial correlation

1 Introduction

Sensor nodes are often deployed in the wild environment, because of the disturbance of the natural environment, the obstacles in the transmission process or the damage of hardware, the data collected are often lost. It is very important to reconstruct the missing data for some applications that require very precise data.

Sensor data has obvious spatiotemporal correlation [1]. In many applications, the collected sensor data is generally continuously variable physical quantities, which has strong time correlation if the sensing data changes smoothly in a short time, also known as time stability. The collected sensor data on the physical location are often similar or have some functional relationship, which is called spatial correlation [2]. Generally, linear interpolation or piecewise linear function can be constructed to calibrate the abnormal data or to reconstruct the missing values. The linear approach usually uses the non-missing data of a node on the time point adjacent to the missing point, the collected sensor data usually considered as a time series in a certain time period, the linear regression [3–5] or multiple linear regression model [6, 7] is used to reconstruct the missing data.

© Springer Nature Switzerland AG 2020
Q. Liu et al. (Eds.): CENet 2018, AISC 905, pp. 77–87, 2020.
https://doi.org/10.1007/978-3-030-14680-1_10

In addition, there are also some nonlinear methods, such as cluster method [8], recurrent neural network (RNN) [9], Gaussian Process Regression [10] models or minimized similarity distortion (MSD) [11], Bayesian network [12] or Adams-Bashforth-Moulton algorithm [13] are used. Moreover, the method based on sparse dictionary or the matrix rank-minimization method is also the common method, it is necessary to calculate the possible correlations among multiple-attribute sensor data to estimate missing values [14, 15].

Another kind of reconstruction method is to draw on the idea of interpolation methods in geographical space, such as Kriging interpolation [16], inverse distance weighting (IDW) [17] interpolation, Gaussian Mixture Model [18] and so on. As general, the idea of geographic interpolation is calculating the sampled data in each grid which is usually divided into the same size. For every missing value, the main idea of these methods is how to find K preliminary predictive values and take each neighbor a weighted value as the correlation coefficient. However, the accuracy of some of these reconstruction methods is not very high when the time interval of the collected data is large or there are obstacles between nodes.

In this work, a random missing sensor data reconstruction method based on spatial correlation is presented, the main research contents are as follows: how to decide the neighbor nodes and how to assign different weights to each of them combining the spatial correlation and the similarity of sensor data between the reconstructed node and it is neighbors.

2   Spatial Correlation of Sensor Nodes

Suppose that the static sensor nodes are deployed in a certain area, if there is no obstacle, d is the Euclidean distance between any two nodes and R is the communication radius of the wireless sensor node. In theory, if $d(A, B) \leq R$, node B can receive messages sent by node A, called B as a single hop neighbor of A, show as Fig. 1.

If the node A has m neighbors, all neighbors is recorded as $N(A)$, then:

$$N(A) = \{A_i \in V : (A, A_i) \in E, i = 1, 2, \cdots, m\} \tag{1}$$

According to the spatial correlation, the sensor data of node A has a certain similarity with the neighbor node A_i, and the data at the time t are recorded as x_A^t and $x_{A_i}^t$, the difference between them is expressed as:

$$\left| x_A^t - x_{A_i}^t \right| \leq \delta_i^t \tag{2}$$

δ_i^t is the absolute difference of sensor data. If all the neighbor nodes are considered, then:

$$\left| x_A^t - x_{A_i}^t \right| \leq \delta_i^t, 1 \leq i \leq m \tag{3}$$

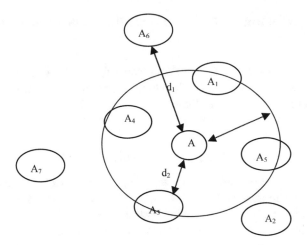

Fig. 1. Communication model.

It is obtained from the above methods:

$$x_A^t = \frac{1}{m} \sum_{i=1}^{m} (x_{A_i}^t + \delta_i^t) \tag{4}$$

The above equation shows that the data of a sensor node can be approximately calculated using the data of its m neighbor nodes. Different neighbors are assigned different correlation coefficients when the missing values are calculated, then:

$$x_A^t = f(\alpha_1 x_{A_1}^t, \alpha_2 x_{A_2}^t, \cdots, \alpha_m x_{A_m}^t) \tag{5}$$

$\alpha_i, i = 1, 2, \cdots, m$ is the correlation coefficient corresponding to each neighbor.

If there is an obstacle or interference between the nodes, the node A and B are in different spaces, while the Euclidean distance between B and C is far away, but they are in the same space, as shown in Fig. 2. In this case, we cannot simply use Euclidean distance as the only criterion to determine neighbor nodes.

So, how to find the neighbor nodes and to determine the different correlation coefficients for each neighbor are the key problems in the reconstruction of the missing sensor data based on spatial correlation.

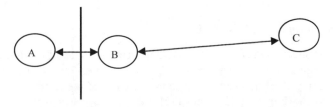

Fig. 2. Sensor node communication model with obstacles.

3 Random Missing Data Reconstruction Based on Fuzzy Theory

Fuzzy logic system (FLS) consists of four parts: fuzzer, fuzzy rule, fuzzy inference engine and defuzzification [19, 20]. The fuzzy function uses membership function to transform the input variables into the definite values in [0, 1]. A random missing data reconstruction method based on fuzzy logic is proposed.

3.1 Overall Framework of the Reconstruction Method

The reconstruction method adopts the fuzzy logic theory, which takes the Euclidean distance between different nodes as the input of the fuzzy inference system, and the correlation coefficient of each neighbor node is considered in the fuzzy calculation process. Especially when there are obstacles between two nodes, not only the Euclidean distance but the similarity coefficient of sensor data is considered to judge whether it is a neighbor node. The frame of the missing data reconstruction method is shown as Fig. 3:

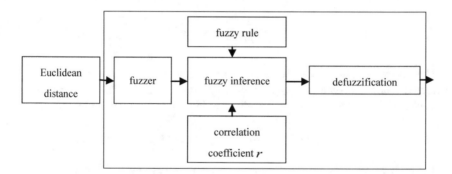

Fig. 3. The frame of the missing data reconstruction method.

The key is the definition of fuzzy rules and the calculation of the correlation coefficient of each neighbor node. In many of the existing research literature, the different correlation coefficients of different nodes are not considered, that is, all the nodes is set to 1. In practice, each node must be assigned a different correlation coefficient $r(0 < r < 1)$, representing the different correlation of the neighbor node A_i and the node A.

3.2 Calculation of Fuzzy Coefficient

Suppose the distance between a node (R) and its neighbor node (T) is divided into very near, near, medium, far and very far, which is represented by VC, C, M, F, and VF respectively. Using the Euclidean distance as the input of the fuzzy inference system, the methods in the literature [21] and [22] are used to select the Z function (VC), the Gauss function (C, M, F), S function (VF) and the combination of them as the membership functions. Suppose d_i is the Euclidean distance between the currently reconstruction node and its neighbor nodes, that is, the distance between a and f is d_i, as shown in Fig. 4.

a   b   c   d   e   f

Fig. 4. Correlation coefficient of Euclidean distance.

The very close membership function is:

$$\mu_{VC}(d_i) = \begin{cases} 1 & \text{when } d_i \leq a \\ 1 - \frac{2(d_i-a)^2}{(a-b)^2} & \text{when } a < d_i \leq \frac{a+b}{2} \\ \frac{2(b-d_i)^2}{(a-b)^2} & \text{when } \frac{a+b}{2} < d_i \leq b \\ 0 & \text{when } d_i > b \end{cases} \tag{6}$$

The membership functions of nearer, middle and far are:

$$\mu_C(d_i) = e^{-((d_i-b)^2/2\sigma^2)} \tag{7}$$

$$\mu_M(d_i) = e^{-((d_i-c)^2/2\sigma^2)} \tag{8}$$

$$\mu_F(d_i) = e^{-((d_i-d)^2/2\sigma^2)} \tag{9}$$

The very far membership function is:

$$\mu_{VF}(d_i) = \begin{cases} 0 & \text{when } d_i \leq d \\ \frac{2(d_i-d)^2}{(f-d)^2} & \text{when } a < d_i \leq e \\ 1 - \frac{2(d_i-d)^2}{(f-d)^2} & \text{when } e < d_i \leq f \\ 1 & \text{when } d_i > f \end{cases} \tag{10}$$

From the membership function, the distance d_i is separated into three sections. If $a < d_i < b$, the d_i is very near or nearer, if $b < d_i < d$, then d_i is medium, and if $d < d_i < f$, d_i is far or very far. The fuzzy coefficient is obtained as $\mu(d_i)$.

In order to modify the fuzzy coefficients produced by the member function, the regulatory factors r_{d_i} is introduced. If d_i is very close, r_{d_i} is 0.9, if d_i is near, r_{d_i} is 0.7, if d_i is medium, r_{d_i} is 0.5, if d_i is far away, r_{d_i} is 0.3, and if d_i is very far, r_{d_i} is 0.1. The formula for calculating the fuzzy coefficient is as follows:

$$\alpha_{x,y} = \mu(d_i) * r_{d_i} \tag{11}$$

$\alpha_{x,y}$ is the modified fuzzy coefficient.

3.3   Calculation of Similarity Coefficient

According to the spatial correlation model, the fuzzy coefficient can't be fully reflected the correlation if there are obstacles or interference between sensor nodes. The similarity of sensor data is used to modify fuzzy coefficient to decide the correlation coefficient comprehensively.

In this study, Pearson correlation coefficient is used to calculate the similarity of sensor data. The calculation formula is as follows:

$$\beta_{x,y} = \frac{1}{n-1} \sum_{i=1}^{n} \left(\frac{X_i - \bar{X}}{S_x}\right)\left(\frac{Y_i - \bar{Y}}{S_y}\right) \tag{12}$$

$$S_x = \frac{1}{n-1} \sum_{i=1}^{n} (X_i - \bar{X})^2 \tag{13}$$

$$S_y = \frac{1}{n-1} \sum_{i=1}^{n} (Y_i - \bar{Y})^2 \tag{14}$$

Where $\beta_{x,y}$ is the similarity coefficient. X_i and Y_i, $\bar{X}$ and $\bar{Y}$ represent the actual value and the average value respectively, and n is the sample size.

3.4   Calculation of Correlation Coefficient

Considering the Euclidean distance and the similarity correlation of sensor data, the correlation coefficient is calculated as follows:

$$r = \frac{f_1}{f_1 + f_2} \alpha_{x,y} + \frac{f_2}{f_1 + f_2} e^{\beta_{x,y}} \tag{15}$$

where $f_1 = \alpha_{x,y}$, $f_2 = \beta_{x,y}$, $\frac{f_1}{f_1+f_2}$, $\frac{f_2}{f_1+f_2}$ is the coefficients of $\alpha_{x,y}$ and $\beta_{x,y}$ respectively.

3.5   Implementation of Reconstruction Algorithm

According to the sensor data and the correlation coefficients of its neighbor nodes, the reconstructed values are:

$$T_A = \frac{\sum_{i=1}^{n} T_{A_i} r_{A_i}}{\sum_{i=1}^{n} r_{A_i}} (i = 1, 2, \cdots, n) \tag{16}$$

where, T_{A_i} is the sensor data of neighbor nodes, r_i is the correlation coefficient between neighbor nodes and the nodes to be reconstructed, T_A is the reconstructed value, and n is the number of total neighbor nodes.

Based on fuzzy distance, the algorithm of random missing sensor data reconstruction is described as follows:

The node A's neighbor nodes is set as:

$N(A) = \{A_i \in V : (A, A_i) \in E, i = 1, 2, \cdots, m\}$, the correlation coefficient threshold is set as ε, do:

① calculate the Euclidean distance d_i between node A and neighbor node A_i;

② divide the fuzzy interval and correspond d_i to the corresponding fuzzy interval.

③ According to the membership function, the membership value of each d_i in the fuzzy interval are calculated as $\mu(d_i)$.

④ According to the fuzzy modified coefficient r_{d_i} and the fuzzy coefficient $\mu(d_i)$, the fuzzy distance correlation coefficient $\alpha_{x,y} = \mu(d_i) * r_{d_i}$ is obtained.

⑤ Calculate the correlation coefficient $\beta_{x,y}$ between the reconstruction node and it's reference nodes.

⑥ According to $\alpha_{x,y}$ and $\beta_{x,y}$, the correlation coefficient r of each neighbor node is calculated.

⑦ If $r < \varepsilon$, A_i can't be used as a candidate neighbor node.

⑧ If $N(A)$ is not empty, a next candidate neighbor node is selected and go to ①, otherwise, go to ⑨.

⑨ Calculate the reconstructed value T_A.

⑩ End.

4   Experiment and Result Analysis

4.1   Experimental Data Set

This study used the Intel Berkeley Research Lab Data in the laboratory of Intel Berkeley University [23]. The data set has 32 million data generated by 54 sensor nodes in 36 days. Each data is the monitoring value of temperature, humidity, light and node voltage that the sensor nodes collect in about every 30 s.

Before using these data, it is necessary to preprocess the data due to the presence of some abnormal values or missing values in the data set. For a single node, a state identity is added to each node, 1 is not missing, and 0 is lack. In addition, because the data of partial node is abnormal or missing more than 95%, the data of the node is not available, so the 5th node and the 15th node are not used in this study. Meanwhile, for the sake of studying the influence of different time intervals on the reconstruction error, the temperature and humidity data of 52 nodes in March 7th and March 8th were divided into 6 subsets at different intervals.

4.2   Results and Analysis

In this study, when the time interval is 1 min, 5 min, 10 min, 15 min, 20 min and 30 min, and the number of neighbor nodes is 4, 8, 12, 16, 20 and 25 respectively, the random missing data reconstruction method based on fuzzy distance (RFUZZY), the K nearest neighbor (KNN) algorithm, the K nearest neighbor after the obstacle are considered (RKNN), the inverse distance weighting (IDW) and the fuzzy Euclidean distance method (FUZZY) are implemented.

The RFUZZY and FUZZY algorithms proposed in this study have realized the reconstruction with different number of neighbor nodes and different time intervals. The reconstruction error of node 1 with different neighbor nodes and different time interval of 1 min and 30 min is shown in Fig. 5.

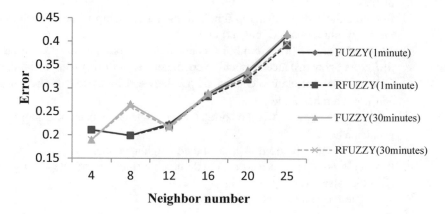

Fig. 5. Reconstruction error of different neighbors and different time interval.

It can be seen from the figure that when the number of neighbor nodes increases, the reconstruction error rate is also increasing. This shows that if the number of neighbor nodes increases, there may be multiple neighbor nodes added to the calculation which is not related to the reconstructed nodes, resulting in the increase of the error. Simultaneously, when the number of neighbors is 4, the error is slightly larger than the number of neighbor nodes is 8. This shows that when the missing values of the reconstructed nodes are estimated with less neighbor nodes, the neighbor nodes are too small, and the neighbor nodes with larger spatial correlation do not participate in the reconstruction calculation, the reconstruction error will be increased. When the number of neighbors is less than 10, the error is about 0.2, and the larger the time interval is, the greater the error. That is, using as few neighbors as possible to ensure that there is the strongest spatial correlation between sensor data and the highest reconstruction accuracy. When the number of nodes is more than 10, the situation is the opposite.

When there are 16 neighbors nodes, the reconstruction error of different algorithms at different time intervals is shown in Fig. 6, which includes FUZZY, RFUZZY, KNN, and IDW algorithm at different time intervals.

From the Figure, the reconstruction error varies slightly with the time interval, that is, the algorithm with spatial correlation as the main constraint is less influenced by the time correlation. The estimation error of the IDW algorithm is the largest, and the estimation error of the FUZZY proposed in this paper is lower. On this basis, the estimation error of the RFUZZY algorithm proposed by the elimination of the fuzzy coefficient and the similarity correlation is minimized, which shows the validity and accuracy of the algorithm.

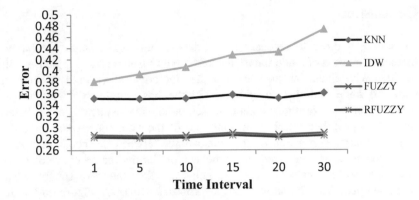

Fig. 6. Reconstruction error of different time interval when the neighbors is 16.

The comparison of reconstructed error of different neighbor nodes is shown in Fig. 7, and the data acquisition time interval is 30 min.

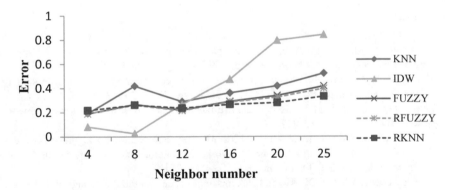

Fig. 7. Reconstruction error of different neighbors.

Figure 7 gives the error of 5 different algorithms to reconstruct the temperature of node 1. The reconstruction error of IDW algorithm increases sharply with the increase of neighbor nodes, and decreases sharply with the decrease of neighbor nodes. The maximum error of KNN algorithm is over 0.5, and the overall error of FUZZY algorithm proposed in this paper is the smallest and the maximum error is about 0.4. Moreover, when obstacles exist between nodes, the Euclidean distance and the similarity of sensor data between the reconstructed node and their neighbors are considered as the correlation coefficient. The blue dotted line in the figure shows the estimation error of the RFUZZY algorithm, which is less than the error only considered the Euclidean distance as the spatial correlation, the same as the RKNN.

5   Conclusion

In this study, in order to reconstruct the data of unstable change accurately, the reconstruction method of random missing sensor data based on fuzzy logic is studied in combination with Euclidean distance and the similarity of sensor data between the reconstructed node and their neighbors. On the basis of fuzzy logic theory, the Euclidean distance between the neighbor nodes and the reconstructed node is fuzzed by choosing the appropriate membership function. At the same time, when the obstacles exist, the similarity correlation of the sensor data between the neighbor nodes and reconstruction node is calculated, then the comprehensive correlation coefficient of each neighbor node is calculated, which helps to select the most suitable neighbor nodes. Experimental results show that the proposed RFUZZY algorithm has higher reconstruction accuracy than other conventional algorithms when there are obstacles between sensor nodes or the time interval between sensor data is longer.

Acknowledgement. This study is supported by the Natural Science Foundation of Hubei Province of China (Program No. 2016CKB705).

References

1. Vuran, M.C., Akan, Ö.B., Akyildiz, I.F.: Spatio-temporal correlation: theory and applications for wireless sensor networks. Comput. Netw. **45**(3), 245–259 (2004)
2. Vuran, M.C., Akyildiz, I.F.: Spatial correlation-based collaborative medium access control in wireless sensor networks. IEEE/ACM Trans. Netw. **14**(2), 316–329 (2006)
3. Zhen, Q.Q., Zhang, T.L.: A missing data estimation algorithm in wireless sensor networks. Boletín Técnico **55**(3), 212–217 (2017)
4. Xia, Y., Chen, J.W., Lei, J.J., Bae, H.Y.: Missing data estimation algorithm based on temporal correlation in wireless sensor networks. In: International Conference on Artificial Intelligence Science and Technology, pp. 309–314 (2017)
5. Gao, Z., Cheng, W., Qiu, X., Meng, L.: A missing sensor data estimation algorithm based on temporal and spatial correlation. Int. J. Distrib. Sens. Netw. **2**, 1–10 (2015)
6. Zhang, H., Yang, L.: An improved algorithm for missing data in wireless sensor networks. In: International Conference on Software Intelligence Technologies and Applications & International Conference on Frontiers of Internet of Things, pp. 346–350. IET (2015)
7. Pan, L., Gao, H., Gao, H., et al.: A spatial correlation based adaptive missing data estimation algorithm in wireless sensor networks. Int. J. Wirel. Inf. Netw. **21**(4), 280–289 (2014)
8. Zhou, Z., Fang, W., Niu, J., Shu, L., Mukherjee, M.: Energy-efficient event determination in underwater WSNs leveraging practical data prediction. IEEE Trans. Ind. Inform. **13**(3), 1238–1248 (2017)
9. Moustapha, A.I., Selmic, R.R.: Wireless sensor network modeling using modified recurrent neural networks: application to fault detection. IEEE Trans. Instrum. Meas. **57**(5), 981–988 (2008)
10. Karunaratne, P., Moshtaghi, M., Karunasekera, S., Harwood, A., Cohn, T.: Multi-step prediction with missing smart sensor data using multi-task Gaussian processes. In: IEEE International Conference on Big Data, pp. 1183–1192. IEEE (2017)

11. Niu, K., Zhao, F., Qiao, X.: A missing data imputation algorithm in wireless sensor network based on minimized similarity distortion. In: Sixth International Symposium on Computational Intelligence and Design, pp. 235–238. IEEE (2014)
12. Zhang, H., Liu, J., Pang, A.C., Li, R.: A data reconstruction model addressing loss and faults in medical body sensor networks. In: Global Communications Conference, pp. 1–6. IEEE (2017)
13. Islam, M., Al Nazi, Z., Hossain, A., Rana, M.: Data prediction in distributed sensor networks using adam bashforth moulton method. J. Sens. Technol. **8**, 48–57 (2018)
14. Zhao, L., Zheng, F.: Missing data reconstruction using adaptively updated dictionary in wireless sensor networks. In: 7th International Conference on Computer Engineering and Networks, p. 40 (2017)
15. Shao, Y., Chen, Z.: Reconstruction of missing big sensor data, pp. 1–13. CoRR, abs/1705.01402 (2017)
16. Mendez, D., Labrador, M., Ramachandran, K.: Data interpolation for participatory sensing systems. Pervasive Mob. Comput. **9**(1), 132–148 (2013)
17. Li, Y.Y., Parker, L.E.: Nearest neighbor imputation using spatial-temporal correlations in wireless sensor networks. Spec. Issue Resour. Constrained Netw. **15**(1), 64–79 (2014)
18. Yan, X.B., Xiong, W.Q., Hu, L., Wang, F., Zhao, K.: Missing value imputation based on gaussian mixture model for the internet of things. Math. Probl. Eng. **3**, 1–8 (2015)
19. Sugeno, M., Kang, G.T.: Fuzzy modelling and control of multilayer incinerator. Fuzzy Sets Syst. **18**(3), 329–345 (1986)
20. Takagi, T., Sugeno, M.: Fuzzy identification of systems and its applications to modeling and control. Read. Fuzzy Sets Intell. Syst. **15**(1), 387–403 (1993)
21. Khan, S.A., Daachi, B., Djouani, K.: Application of fuzzy inference systems to detection of faults in wireless sensor networks. Neurocomputing **94**(3), 111–120 (2012)
22. Zhao, L., He, L., Harry, W., Xing, J.: Intelligent agricultural forecasting system based on wireless sensor. J. Netw. **8**(8), 1817–1823 (2013)
23. Intel Berkeley Research Lab. http://db.csail.mit.edu/labdata/labdata.html

Power Quality Disturbances Detection Based on EMD

Ji Ping[(⊠)]

Hohai University Wentian College, Ma'anshan 243002, China
471374754@qq.com

Abstract. The power quality (PQ) disturbance signals have the characteristics of short duration and strong randomness, and often form complex disturbances, which make the disturbance signals difficult to detect and identify. In this paper, EMD algorithm is introduced to decompose the PQ disturbance signals and calculate the intrinsic mode function (IMF) of the disturbance signals. Then, Hibert transform are performed for each IMF to obtain the characteristic information of the disturbance signal. EMD transform is used to detect the type, duration, frequency and amplitude of PQ disturbances. To verify the effectiveness of the algorithm, several kinds of PQ disturbance signals are simulated with transient harmonic, voltage interruption, voltage drop and voltage surge and complex disturbances. Experimental results show that the algorithm can accurately detect power quality interferences. This paper provides a new method for the detection of PQ disturbances and a new idea for the power management.

Keywords: Power quality · Intrinsic mode function · Empirical mode decomposition · HHT

1 Introduction

The increasing use of non-linear load, impact load and unbalanced power systems seriously pollute the power grid, and influence the power quality (PQ). The transient disturbances such as voltage interruption, voltage surge and voltage sag cause serious distortion of power signals and reduce power quality. However, with the wide use of sensitive equipment in the high-tech fields such as computer, microelectronics and so on, power consumers have put forward higher requirements for power quality. Therefore, the power supply must provide a stable, efficient and high - quality electrical service. The accurate detection of power quality disturbances and the determination of the start time and the end time of disturbances can provide a basis for the power department to analyze the power quality problem, to make clear the obligations and responsibilities of both the power supply and the power user, and to provide the basis for the comprehensive management of the power grid.

The methods of detecting PQ disturbances mainly include the fast Fourier transform (FFT) and short-time Fourier transform (STFT). FFT is a whole or global transformation, so its limitation is that the analysis of the signal is either completely in the time domain or is completely in the frequency domain [1–3]. FFT cannot obtain the regularity of the frequency components of the signal with time. STFT is an improved

© Springer Nature Switzerland AG 2020
Q. Liu et al. (Eds.): CENet 2018, AISC 905, pp. 88–99, 2020.
https://doi.org/10.1007/978-3-030-14680-1_11

algorithm based on fast Fourier transform, and the local characteristic is better than the Fourier transform [4]. However, due to the limitation of the algorithm itself, the analysis of the non-stationary transient data has obvious disadvantages when extracting the time-frequency domain characteristics of the signal. Wavelet transform (WT) has the characteristic of localization analysis in time and frequency domain, but since the selection of wavelet basis is determined when the system design is tested, it cannot be changed at any time along with the type of disturbance signal [5, 6]. In order to accurately detect the transient PQ disturbances, the detection method has better local characteristics in time domain and frequency domain. Empirical mode decomposition (EMD) is a new adaptive time-frequency analysis method, and has higher time-frequency resolution and concentration. EMD can decompose any complicated signal into a finite number of intrinsic mode functions (IMFs) based on the local characteristic time scale of the signal, and then obtain the HHT spectrum to depict the parameter distribution with time-frequency. HHT is especially suitable for processing nonlinear and non-stationary signals. EMD is the core part of HHT, which is a method to deal with non-stationary signals and has achieved good results in many fields such as biomedical and geophysics [7]. This paper introduces the EMD into the power system for extracting the feature of the PQ disturbances. MATLAB simulation is used to verify the effect of this method on non-stationary disturbance signal detection of power system. The experimental results show that this method can effectively detect the PQ disturbances.

2 EMD

EMD is used to deal with nonlinear and non-stationary signals linearly and smoothly by decomposing the signal into multiple IMFs. By taking the Hilbert transform of IMF components, the instantaneous frequency, start-stop time and frequency of each vibration mode of complex data signals are obtained. Using EMD for signal analysis, the following two conditions need to be met:

(1) In all data vectors, the number of extremum points must be equal to the number of zero points, or the maximum difference cannot exceed one;

(2) At any time, the average value of the upper envelope formed by the local maximum point and the lower envelope formed by the local minimum point is zero, namely, the upper and lower envelops are locally symmetrical relative to the time axis. The EMD method can be used to decompose signal through the following steps.

(1) Firstly, the maximal points of the signal are obtained, and then the upper envelope of the original data sequence is calculated with cubic spline interpolation algorithm. In a similar way, the minimum point of the signal is obtained, and cubic spline interpolation algorithm is used as the lower envelope of the original data sequence. Therefore, the average of the upper envelope and the lower envelope of the original signal can be obtained.

$$m_1(t) = \frac{1}{2}[v_1(t) + v_2(t)] \tag{2.1}$$

Where, $v_1(t)$ is the upper envelope, $v_2(t)$ is the lower envelope, $m_1(t)$ the average of two envelops.

(2) The original data minus the average of the upper and lower envelope gets a new data sequence that removes the low frequency.

$$h_1(t) = x(t) - m_1(t) \tag{2.2}$$

Where, $x(t)$ is the original data. In the analysis of power quality disturbance signal, $x(t)$ is the original signal of power grid. If the $h_1(t)$ satisfies the IMF conditions, it will be the first IMF which is defined as $c_1(t)$:

$$c_1(t) = h_1(t) \tag{2.3}$$

Otherwise, the H1 will be the new raw data and repeat step (1) until the conditions are satisfied.

(3) And then to

$$r(t) = x(t) - c_1(t) \tag{2.4}$$

r(t) as new original data, and repeat the above steps (1) and (2), the successive IMF components can obtain until he selection is finished when the given termination conditions are satisfied.

(4) The original data can be represented by these IMF components and a trend term r(t), namely:

$$x(t) = \sum_{i=1}^{n} c_i(t) + r(t) \tag{2.5}$$

As can be seen from formula (2.5), the signal is decomposed into frequency from big to small, namely, the sum of n components and a trend term of the change from fast to slow. Since each component represents a data sequence of a characteristic scale, the selection process actually decomposes the original data sequence into the superposition of various characteristic wave sequences.

3   HHT

According to the time-scale analysis of the characteristic time scale of the decomposition signal, the detection signal is decomposed into multiple IMF, and then the instantaneous frequency and instantaneous amplitude of each IMF component are obtained.

According to formula (2.5), the Hilbert transform can be obtained from each c(t), i.e.

$$H_i(t) = \frac{1}{\pi} \int_{-\infty}^{+\infty} \frac{c_i(\tau)}{t - \tau} d\tau \qquad (3.0)$$

The constructor parsing function is as follows:

$$z_i(t) = c_i(t) + jH_i(t) = a_i(t)e^{j\phi_i(t)} \qquad (3.1)$$

Where, a_i is the amplitude function, φ_i is the phase function, i.e.

$$a_i(t) = \sqrt{c_i^2(t) + H_i^2(t)} \quad \varphi_i(t) = \arctan \frac{H_i(t)}{c_i(t)} \qquad (3.2)$$

Thus, the instantaneous frequency can be expressed as:

$$f_i(t) = \frac{1}{2\pi} w_i(t) = \frac{1}{2\pi} \frac{d\phi(t)}{dt} \qquad (3.3)$$

The Hilbert spectrum of IMF is as follows:

$$H(f, t) = Re \sum_{i=1}^{n} a_i(t)e^{j\phi_i(t)} = Re \sum_{i=1}^{n} a(t)e^{j2\pi \int f_i(t)dt} \qquad (3.4)$$

Where, Re represents the real part. Hilbert marginal spectrum is defined as:

$$h_i(f) = \int_{-\infty}^{+\infty} H(f, t)dt \qquad (3.5)$$

4 Experiments and Analysis

The PQ disturbances have the characteristics of strong randomness, short duration and great harm, and it has a variety of reasons, such as power network structure, load type and electric environment, etc. Therefore, The PQ disturbances detection is difficult and seriously affects the normal operation of power system. The power quality disturbances generation models and their parameters are shown in Table 1.

Table 1. Power quality disturbances modeling and its parameters.

	Transient harmonics			Voltage interruption	Voltage surge	Voltage sag
	Fundamental harmonic	Triple harmonic	Quintuple harmonics			
Amplitude	220	80	50	0	330	80
Frequency	50	150	250	50	50	50
Starting time	–	0.3	0.1	0.3	0.35	0.35
Ending time	–	0.45	0.24	0.5	0.55	0.55

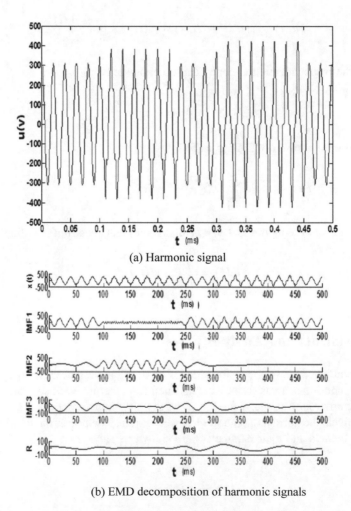

(a) Harmonic signal

(b) EMD decomposition of harmonic signals

Fig. 1. Harmonic signal and its analysis results.

4.1   Experimental Analysis of Single Disturbances

Figures 1, 2, 3 and 4 show EMD decompositions for detection of four typical power quality disturbances of transient harmonics, voltage interrupts, voltage surges and voltage signals respectively. From Figs. 1, 2, 3 and 4, Fig. a represents the waveform of the disturbance and Fig. b represents the graph of EMD analysis for the disturbance. Odd harmonic is the most serious damage to the power grid. Figure 1 shows the power grid signal with the third and fifth harmonics.

EMD can effectively and accurately detect the starting and ending time of the disturbance signal, and the detection error is within 0.003. The signal amplitude and frequency error are in the range of 0.04%. Only when the detection data is interrupted, the error is relatively large. Generally, the voltage is reduced to 0.1 times the rated voltage, which is considered as the voltage interruption.

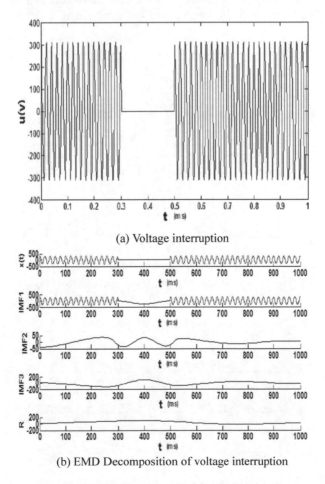

(a) Voltage interruption

(b) EMD Decomposition of voltage interruption

Fig. 2. Voltage interruption and its analysis results.

4.2   Experimental Analysis of Complex Disturbance

The disturbances of the power system may also appear as complex disturbances in addition to single disturbances. In this paper, EMD can also be used to detect complex disturbances. The complex harmonic signals of transient harmonic and inter-harmonic are simulated respectively.

The transient harmonic is based on the fundamental wave and three harmonics, among which the harmonic is 52 Hz. The complex interrupt signal consists of fundamental wave, three harmonics and interrupt signal. Figures 5, 6 and 7 respectively show the simulation data of the above-mentioned complex disturbances and the simulation results of EMD decomposition of each layer.

The detection error in the starting and ending time of the complex disturbances is similar to that of the single disturbances. It is kept in the error range of 0.003 and the detection accuracy is higher. In the amplitude detection of complex disturbances, the error of complex interrupt is the largest, and the error range of other disturbances is

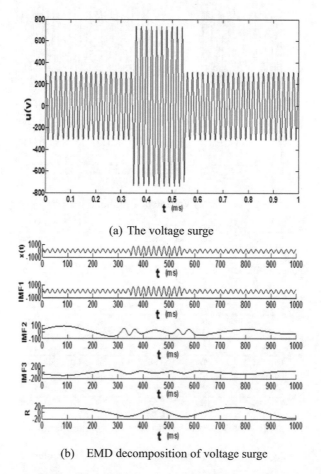

(a) The voltage surge

(b)   EMD decomposition of voltage surge

Fig. 3.   The voltage surge and its analysis results.

basically maintained in the range of 10-1. In the detection of voltage interruption, voltage surge and voltage sag, the accuracy of the detection data is basically the same regardless of the complex or single disturbances data processing. The detection effect of inter-harmonic is relatively poor, the error value is 0.15, the error rate is 0.05%, and the error rate of the relative other composite signals is basically kept at a quantitative level.

Through the simulation of single and complex disturbances and its data analysis, it can be seen that the EMD can be effective for power quality disturbances of the key data for testing, and testing data precision is higher. The accuracy of single distur-bances is slightly higher than that of complex disturbances, and the key data of dis-turbances can be detected by EMD. The maximum error of PQ disturbance detection using EMD and reference [8] is shown in Table 2. By comparing the data in Table 2, it can be seen that the accuracy of signal detection using EMD algorithm is relatively high. By comparing the data in Table 2, it can be seen that the accuracy of signal

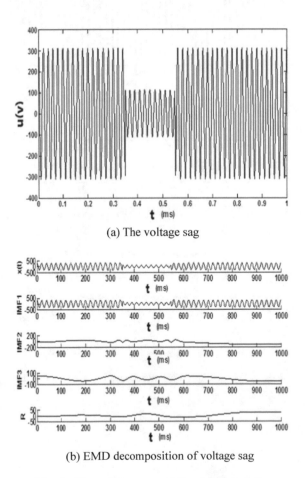

(a) The voltage sag

(b) EMD decomposition of voltage sag

Fig. 4. The voltage sag and its analysis results.

detection using EMD algorithm is relatively high. In the detection of starting and ending time of signal, it is difficult to detect the instantaneous value of signal, so the error of detection is larger than that of the other two terms.

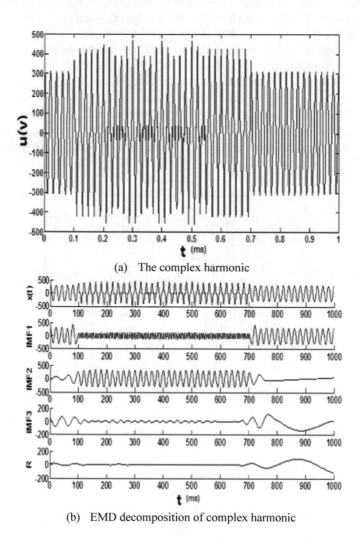

(a)   The complex harmonic

(b)   EMD decomposition of complex harmonic

Fig. 5. The complex harmonic and its analysis result.

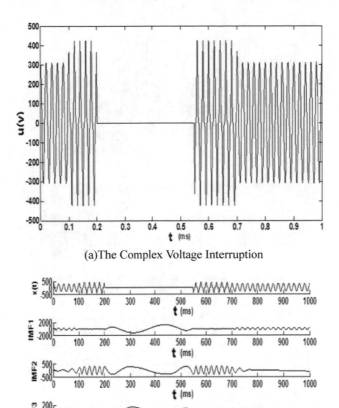

(a)The Complex Voltage Interruption

(b) EMD Decomposition of the complex voltage interruption

Fig. 6. The complex voltage interruption and its analysis results.

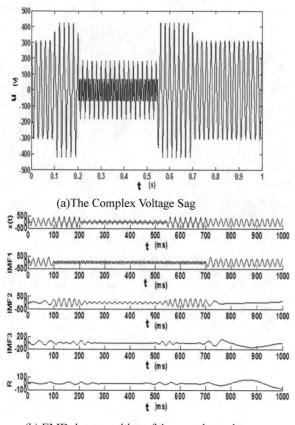

(a)The Complex Voltage Sag

(b) EMD decomposition of the complex voltage sag

Fig. 7. The complex voltage sag and its analysis results.

Table 2. The maximum error of PQ disturbance detection using EMD and DCT and WT.

	Amplitude	Frequency	Start time or end time
EMD	0.03%	0.036%	0.6%
DCT and WT	0.78%	1.4%	1%

5 Conclusion

In this paper, the establishment of a typical disturbance model of power quality is created. Using the EMD algorithm, the typical transient disturbances and its complex disturbances are simulated and tested respectively with the harmonic, voltage interruption, voltage surge and voltage sag. The EMD algorithm can more accurately detect the starting and ending time, frequency, and amplitude, and can achieve more satis-factory accuracy. The result of this paper provides a new way to detect the transient data of power quality, and opens up a new way for production practice and engineering application.

Acknowledgement. This study is supported by Provincial Natural Science Foundation of Anhui (KJ2018A0618).

References

1. Arrillaga, J.A., Watson, N.R., Chen, S.: Power System Quality Assessment. Wiley, New York (2000)
2. Fuchs, E., Trajanoska, B., Orhouzee, S., Renner, H.: Comparison of wavelet and Fourier analysis in power quality. In: Proceedings of the Electric Power Quality and Supply Reliability Conference, pp. 1–7 (2012)
3. Hao, Q., Zhao, R., Tong, C.: Interharmonics analysis based on interpolating windowed FFT algorithm. IEEE Trans. Power Delivery **22**, 1064–1069 (2007)
4. Robertson, D.C., Camps, O.I., Mayer, J.S., Gish, W.B.: Wavelets and electromagnetic power system transients. IEEE Trans. Power Delivery **11**, 1050–1058 (1996)
5. Poisson, O., Rioual, P., Meunier, M.: Detection and measurement of power quality disturbances using wavelet transform. IEEE Trans. Power Delivery **15**(3), 1039–1044 (2000)
6. Latran, M.B., Teke, A.: A novel wavelet transform based voltage sag/swell detection algorithm. Int. J. Electr. Power Energy Syst. **71**, 131–139 (2015)
7. Huang, N.E., Chern, C.C., Huang, K., Salvino, L.W., Long, S.R., Fan, K.L.: A new spectral representation of earthquake data: Hilbert spectral analysis of station TCU 129. Bull. Seismol. Soc. Am. **91**(5), 1310–1338 (2001)
8. Liu, A.D., Xiao, X.Y., Deng, W.J.: Detection and analysis of power quality disturbance signal based on discrete transform and wavelet transform. Power Syst. Technol. **29**(10), 70–74 (2005)

A Fault Injection Method for Resilient Information Service Cloud Environment Based on Association Model

Fang Zhou[✉], Feng Ding, Ran Ding, and Xin Zhao

The Key Laboratory of Information System Engineering, The 28th Research Institute of China Electronics Technology Group Corporation, Nanjing 210007, China
326zhoufang@163.com

Abstract. Considering the problem of how to estimate the influence on resilience of faults, and how to accurately inject faults, a fault injection method for resilient information service cloud environment based on association model is proposed. Firstly, five types of faults are presented based on cloud environment architecture, including the computing resources fault, communication networks fault, cloud platform fault, data fault and service fault, in order to implement the cross-layer mixed faults injection. Secondly, a fault injection description model is built through by multiple attribute group method, including fault type, injection point, injection mode, injection parameters and so on. Then, the "task-resource-failure" association model is proposed according to generate two relationships on "task-resource" and "resource-failure", in order to guidance the selection of fault injection objects. Finally, a large sample of resilient capability assessment experiments was repeated to summarize the fault injection rules, in order to support the selection of injected object and fault type setting.

Keywords: Fault injection · Resilience evaluation · Information service cloud environment · Task association model

1 Introduction

With the development of information systems and cloud computing, big data, and cyber confrontation technology, global integration and joint operations have put forward highest requirements on the information service cloud environment and should be completed with "sustained guarantee tasks". It is urgent to develop a resilient information service cloud environment that can demand diversified missions requirement.

Resilient information service cloud environment is introduced based on inheriting and developing common information infrastructure, introducing new ideas, models, and technical means, in order to make it more flexible, adaptable, and resilient [1–3]. At the same time, once suffering from partial failure, paralysis and damage, various anomalies that affect task execution can be identified and predicted, also quickly adopt active response control measures and strategies to ensure completion of core tasks. Therefore, in order to fully verify the capabilities of the resilient information service cloud environment in complex cyber attack environment, it is urgent to simulate

© Springer Nature Switzerland AG 2020
Q. Liu et al. (Eds.): CENet 2018, AISC 905, pp. 100–109, 2020.
https://doi.org/10.1007/978-3-030-14680-1_12

possible internal failures about the cloud environment, including the incidents and external threat scenarios, software, hardware, and platform mixed failures. Then the failure, overload, and traffic anomalies are generated, in order to verify the resilient capacity on cloud environment under conditions of resource damage.

Considering the problem of how to verify the resilient capabilities of cloud environment under cyber attack. From the perspective of IaaS layer, PaaS layer, DaaS layer, and SaaS layer of cloud environment, computing resource fault, networks fault, platform fault, data fault, and service fault are proposed. The fault injection description model is established to implement cross-layer hybrid fault injection. At last, a three-level association model of "task-resource-fault" is established to solve problems of how to select the fault injection object and set the fault type in the resilient information service cloud environment.

2 Related Research

Research on fault injection of the cloud platform, data center, and virtualization system has been extensively studied at home and abroad, in order to support fault location, abnormal analysis and prediction, fault tolerance mechanism, and resilience capability assessment.

Arzani [4] introduced a fault injection and location method based on data center, through by test beds to simulate all possible faults. The fault injection objects contain servers, clients, communication networks, and so on. The fault type includes high CPU load, high memory load, high I/O load, network packet loss, delays and bandwidth limitations. The real-time simulated failures are injected into the training nodes of data center. Monitoring agents deployed on the training nodes are used to collect various monitoring indicators in real time, the models are built through machine learning method. Through the symptoms of actual monitoring indicators, the approximate location of failure sources is inferred. The fault location is performed by using fault fingerprinting, random forest, logistic regression, Markov chain methods.

Le et al. [4] proposed a fault injection method to evaluate the reliability of virtualized systems and taking the Xen virtualization system as examples, the code faults, memory faults, and register faults are injected into the VMMs and VMs. A memory fault injection method by modifying the page table in VMM is introduced. The fault injection tool is mainly aimed at the virtualized system object in order to test the robustness of the virtual machine itself. In paper [5], considering the problem of the toughness resilience ability assessment of hardware fault tolerant mechanism, an toughness capability assessment technique based on fault injection is proposed to improve the credibility and coverage of toughness resilience capability assessment results. The fault injection technique proposed in this paper is intended to evaluate hardware fault tolerance mechanisms. In paper [6], Considering the cloud platform, a fault injection testing framework FATE (failure testing service) for the cloud platform is designed. The system's ability can be tested by FATE in order to recover in the face of various failures. In paper [7], a reliability model is proposed to test the high availability of the cloud platform, using the fault injection method to crash the cloud component, recording the time of cloud platform collapsed, and building the model about expected impact. In the application of different complex scenes, the relationship between functions and components is regarded as a priori knowledge.

At present, research on fault injection methods in domestic is mainly focused on cloud computing platforms and virtualized systems. In paper [8, 9], a virtual machine fault design method for cloud computing platforms is proposed, and a fault injection platform is developed for Xen virtualization systems, in order to test and evaluate the fault tolerance performance of cloud computing platforms and virtualized systems. The platform integrates a variety of fault injection tools, including fault injection tools for the underlying system and virtualized management layers. It can inject CPU, memory, and file system failures, and event channels, virtual machine migration, access control, and memory management failures. In paper [10], a virtual machine system fault insertion tool design method based on software simulation is proposed. The faults are designed to evaluate the fault recovery capability, fault tolerance mechanism, and reliability of the virtual machine system. In paper [11], a software vulnerability test framework based on fault injection technology is designed, the injected fault target location is located in software and Xen virtualization. The paper [12] designs and implements a message-driven fault injection platform that addresses the limitations of existing single-fault approaches. It integrates the original decentralized fault injection tools into a single architecture and provides them to testers.

3 The Description Model of Fault Injection

3.1 The Concept of Resilience Information Service Cloud Environment

Resilience refers to a capability or quality characteristic that the system presents in response to various disturbances and changes, that is, the ability of system to predict, resist, absorb, react, adapt, and recover from natural disturbances or human events. Resilient information service cloud environment can realize the unified convergence, organization and integrated of computing, storage, software, information resources, and also it can support system users to obtain and share information services on-demand. On the other hand, once the information service cloud environment is subject to cyber attack, physical damage, and node failure, it can analyze and identify abnormal events and adopt response control strategies to eliminate or mitigate the impact of attacks or failures on system operations tasks.

3.2 The Fault Categories

The physical hardware and software failures that occur in physical infrastructure resources, cloud platforms, services of resilient information service cloud environment are analyzed. In order to meet the mixed fault injection requirements including IaaS (Infrastructure as a Service Layer-by-layer) layer, PaaS (Platform-as-a-Service) layer, DaaS (Data as a Service) layer, and SaaS (Software as a Service) layer, from the perspective of the components of resilience information service cloud environment, five layers of fault types are proposes, including computing resources fault, communication network fault, platform services fault, data services fault, and services fault, as shown in Fig. 1.

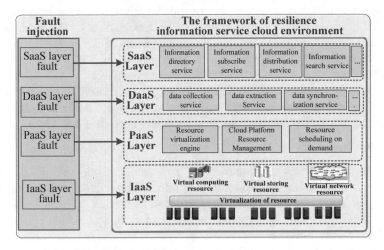

Fig. 1. The fault injection concept for resilience information service cloud environment.

(1) Computing resource fault

From physical paralysis, resource overload, and network attack, suffered by computing resources, five failures are proposed including server/virtual machine damage, high CPU load, high memory load, high I/O load, malicious failure.

(2) Communication network fault

From the perspective of physical damage to the communication network and external attacks, there are five failures including communication network node damaged, link damaged, network continuously lost and network high-latency are proposed.

(3) Platform layer fault

Focusing on the physical damage and external attack of the service container, it proposes three types of failures: service container overload, service container downtime, and service container malicious failure.

Among them, the service load failure is means that service container has high throughput of CPU and memory, resulted in high throughput of service container, which cannot meet the requirements for delay and capacity required by the service processed.

The service container damaged failure refers to an abnormal shutdown of service container due to a service environment changed or abnormal deployment of software.

The malicious failure means that the service container fails to run, since service container cannot respond to the request under external malicious code attack conditions.

(4) Data layer fault

From the perspective of database access performance degradation and database failure, this paper proposes two types of typical failures, including database access overload and database abnormal damaged.

Among them, the database access overload fault refers to a large number of unintended database access requests, resulted in a persistently high database access request processing, which cannot meet the requirements of the application system for database query, real-time data storage, and other processing requirements.

Database abnormal damaged fault is produced by physical reasons such as power outages, operating systems, and CPU hardware failures, caused all transactions in the database are terminated abnormally, application services and business systems unable to obtain the data in the database, such as air intelligence data and so on.

(5) Service layer fault

The service layer covers common services such as information transmission, information access, information subscription, information release, and general-purpose business processing services. From the perspective of service degradation and failure, there are three types of failures are proposed, including unexpected service request overload, service non-response.

An unexpected service request overload fault refers that a large number of illegal users submit concurrent service requests within a certain period of time.

Service unresponsive failure refers to services that cannot service external services for a long period of time due to service crashes or communication failures, other service-dependent service failures, and service-dependent data failures.

Service paralysis failure is produced by the external network attack conditions and semi-legal service data is injected.

3.3 The Fault Injection Description Model

Considering the typical failures of the server, virtual machine and service resources faced by the proposed ductile information service cloud environment. The fault injection description model based on eight attribute group is proposed as follows.

$$FaultM ::= \ <Type, Point, Level, Param, Mode, StartT, DurT, Tool>$$

① *Type* refers the types of injection faults, including physical damage, overloading, failure, and network precision attack.

② *Point* refers the fault injection point, that is, the object or node injected into the fault at present, such as physical CPU, memory, disk, virtual machine, service, database, etc.

③ *Level* refers the fault injection level, including computing resource layer, communication network layer, platform service layer, data service layer, common/general service layer.

④ *Param* refers the behavior parameters of fault injection are used to describe the characteristic parameters of the fault injection execution of various categories.

⑤ *Mode* refers the failure injection mode, including the trigger fault injection, the periodic fault, the fault injection which conforms to the distribution law.

⑥ *StartT* refers the occurrence time of the fault injection is expressed.

⑦ *DurT* refers the duration of fault injection is used to describe the duration of fault injection which is consistent with a specific distribution law.

⑧ *Tool* refers to tools for fault injection, including CPU, memory, I/O and other resources overload, high consumption requests for services, denial of service attacks and other tools.

(1) Computing resource fault injection description model

Taking the virtual machine high memory load failure as an example, the fault injection detailed description model established as follows:

$$FaultM(VM_{cpu}) ::= <Type, Point, Level, Param, Mode, StartT, DurT, Tool>$$

In the formula, $Param ::= <LeakM, StackS, Intel, HandS>$, *LeakM* specifies the size of leaked memory, *StackS* specifies the stack size, *Intel* specifies the time interval for each push, *HandS* specifies the handle size.

(2) Communication network resource fault injection description model

Taking the network high delay fault as an example, the fault injection description model established is as follows:

$$FaultM(Net_{delay}) ::= <Type, Point, Level, Param, Mode, StartT, DurT, Tool>$$

In the formula, $Param ::= <Type, IP, Port, Count, PacketN, PacketR>$, *IP, Port* specify IP address and port number of the attacked node, *Count* specifies the number of attacks, *PacketN* specifies the number of attack packets per attack, *PacketR* specifies the attack packet sending rate.

(3) Platform layer fault description model

The fault of the platform layer covers the overload and damaged of the service container. Taking the service container CPU resource overload fault as an example, the fault injection description model established is as follows:

$$FaultM(Contain_{over}) ::= <Type, Point, Level, Param, Mode, StartT, DurT, Tool>$$

In the formula, $Param ::= <Id, IP, ThreadNum, Activity>$, ID refers the attacked service container, *IP* refers the IP address of virtual machine or server deployed, *ThreadNum* refers the number of threads that perform CPU overload attacks. *Activity* refers the priority of thread running.

(4) Data layer fault injection description model

Data layer faults cover database failures and database access overloads. The fault injection description model established for database access overload faults is as follows:

$$FaultM(Data_{over}) ::= <Type, Point, Level, Param, Mode, StartT, DurT, Tool>$$

In the formula, $Param ::= <url, userCount, reqCount, reqRate>$, among them, *url* refers the protocol and data source identifier, *userCount* refers the number of users accessing database, and *reqRate* refers the rate at which the user sends a database access request.

(5) Service fault description model

The service layer fault mainly involves three types of failures: unexpected service request overload, service non-response, and service damaged. Taking the unexpected service request overload as an example, the fault description model is shown in the following formula:

$$FaultM(Ser_{over}) ::= <Type, Point, Level, Param, Mode, StartT, DurT, Tool>$$

$$Param ::= <serName, serIP, userCount,$$
$$serreqNum, serreqRate, subCon>$$

In the formula, *serName* refers service type; *IP* refers the virtual machine/server IP address; *userCount* refers the number of batch users requested for submitting situation subscription service, *serreqN* refers the rate for user to submit situation subscription service request, *subCon* refers the condition parameter.

4   Fault Injection Method Based on "Task-Resource-Fault" Correlation Model

The purpose of fault injection is to verify the capability of resilient information service environment in complex cyber attack condition, through by injecting cross-layer hybrid faults into computing, network, and service resources in the cloud environment. The fault injected will triggers resource overloading and traffic abnormal, in order to verify and evaluate the adaptive tuning and response recovery capabilities of resilience information service cloud environment.

The difficulties on the fault injection of resilience information services cloud environment are mainly reflected in the following aspects: Firstly, how to select the fault injection object/injection point, in order to meet the resilience test demand. The second is that how to set the fault injection category for the selected object.

Considering the above problems, a fault injection method of "task-resource-fault" three-level association model is established, as shown in Fig. 2. Firstly, analyzing the core supported tasks of cloud environment, the relationship between "task-resources"

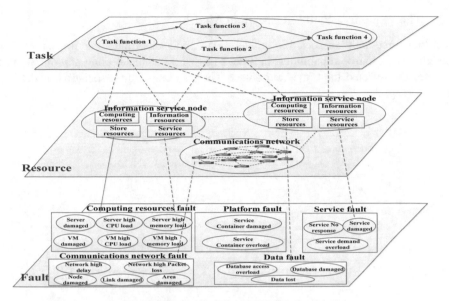

Fig. 2. "Task-resource-fault" association model

are established, in order to guidance the selection of fault injection objects. Secondly, the possible failures of resources occurred in the process of operation are analyzed, a correlation model of "resources-faults" is established. Finally, based on the correlation model between "task-resources" and the "resource-failure" correlation model, a large sample of toughness capability assessment experiments was repeated to summarize the fault injection rules, in order to support the selection of injected object and fault type setting.

4.1 Task-Resource Association Model

The "task-resources" association mapping model is used to describe the relationship between the protection and support of resources to tasks, including the relationship between tasks and tasks, and the relationship between tasks and resources.

Among them, the task-task functional relationship is to describe the timing or logical relationship between task functions decomposed by the task, including serial, parallel and conditional relationships.

The function-resource relationship is to describe the protection or support relationship of a resource to a task function, involving multiple resources at the same time to guarantee one or more task functions, and one resource to guarantee single and multiple task functions.

4.2   The "Resource-Fault" Association Model

The "resource-fault" association model is used to determine the impact of fault injection on resource capabilities, in order to assist tester in selecting appropriate fault type for the selected fault injection object. The resource- fault association model is as follows (Fig. 3):

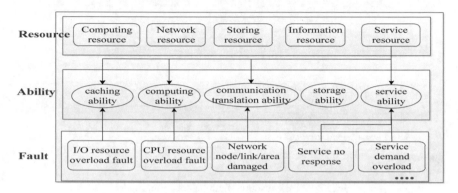

Fig. 3. The resource-fault association model.

Firstly, establish a mapping relationship between "resources-resource capabilities". From the core task function to its resource scheduling strategy, to analyze the differences in resource capabilities required to ensure the completion of core tasks, and dividing the ability of resource into five categories: cache ability, computing ability, communication ability.

Secondly, a mapping relationship between resource capabilities and failures is established. Analyzing the influence of faults at each level of computing resources, communication networks, platforms, and data services on resource capabilities.

Finally, according to the mapping relationship between resource and resource capabilities, resource capabilities and failures, an association model between "resources-faults" is established as a basis for assisting the selection of fault injection objects.

5   Conclusion

The verification and testing for resilience information service capability in complex cyberspace threat environment involving how to inject mixed failures, and how to select fault injection objects. In order to realize the physical damage, failure of the node and other abnormalities in the cloud environment, a system-wide type of faults, a description model for fault injection and a three-level model of "task-resource-fault" are established, in order to solve the problem of how to inject fault objects and how to select fault types.

References

1. Goerger, S.R., Madni, A.M., Eslinger, O.J.: Engineered resilient systems: a DoD perspective. Procedia Comput. Sci. **28**, 865–872 (2014)
2. Pflanza, M.: An approach to evaluating resilience in command and control architectures. Procedia Comput. Sci. **1**(31), 141–146 (2012)
3. Arzani, B., Ciraci, S., Loo, B.T., Schuster, A., Outhred, G.: Taking the blame game out of data centers operation with NetPoirot. In: Proceedings of the 2016 ACM SIGCOMM Conference, pp. 440–453 (2016)
4. Le, M., Gallagher, A., Tamir, Y.: Challenges and opportunities with fault injection in virtualized systems, April 2008
5. Hoang, M.L., Herdt, V.L., Grobe, D.: Injection on intermediate code. In: Design, Automation & Test in Europe Conference & Exhibition, pp. 19–25 (2018)
6. Haryadi, G.H.S., Do, T., Joshi, P., et al.: FATE and DESTINI: a framework for cloud recovery testing. In: Proceedings of 8th USENIX Symposium on Networked Systems Design and Implementation, p. 239 (2011)
7. Benz, K., Bohnert, T.: Dependability modeling framework: a test procedure for high availability in cloud operating systems, in vehicular technology. In: IEEE 78th Vehicular Technology Conference (VTC Fall), pp. 1–8 (2013)
8. Ma, Y.D.: Research and design of fault injection. Harbin Institute of Technology (2015)
9. Feng, G.: Research and design of fault injectors for virtual machine in cloud computing. Harbin Institute of Technology (2013)
10. Che, J.H., He, Q.M., Chen, J.H., Wang, B.: Sofware simulation-based fault injection tool of virtual machine system. J. Zhejiang Univ. (Eng. Sci.) **45**(4), 614–621 (2011)
11. Zeng, F., Li, J., Li, L., et al.: Fault injection technology for software vulnerability testing based on Xen. In: WRI World Congress on Software Engineering, pp. 206–210. IEEE (2009)
12. Liu, W.N., Yang, H.T., Zhang, Z.: Fault injection automation test platform based on message driven. Comput. Eng. 1–6 (2011)

Task Scheduling of GPU Cluster
for Large-Scale Data Process
with Temperature Constraint

Haifeng Wang and Yunpeng Cao[✉]

School of Information Science and Engineering, Linyi University Institute
of Linyi University of Shandong Provincial Key Laboratory of Network Based
Intelligent Computing, Linyi 276000, China
caoyunpeng@lyu.edu.cn

Abstract. With the development of GPU general-purpose computing, GPU heterogeneous cluster has become a widely used parallel processing solution for Large-scale data. Considering temperature management and controlling becomes a new research topic in high-performance computing field. A novel task scheduling model for GPU cluster with temperature limitation was built to balance the heat distribution and prevent the temperature hotspots occur. The scheduling index was introduced by combining the utilization of GPU and temperature. And the state matrix was designed to monitor the GPU cluster and provided status information for scheduler. When the temperature exceeds specific threshold value, the scheduler can improve the speed of fans to reduce the temperature. The experimental results show that the proposed scheduler can balance the heat distribution and prevent the temperature hotspots. Compared with the benchmark scheduling model, the loss of scheduling performance is in the acceptable range.

Keywords: GPU cluster · Task scheduling model · Temperature limitation · Large-scale data

1 Introduction

MapReduce is a widely applied computing model in large-scale data processing. This model divides the computing job into multiple tasks over distributed nodes in heterogeneous cluster. Graphic Processing Units (GPU) general computing is very suitable for data-intensive computing due to the tremendous computational power, such as massive parallel threads and high memory bandwidth as compared to CPU [1]. So with the development of GPU device, CUDA and OpenCL, GPU is gradually integrated into the MapReduce model. GPU clusters, such as Hadoop, Spark, Flink. GPU clusters are more suitable for data mining and machine learning which require many iterative operations. A GPU cluster to deal with large-scale data is a continuous process, which inevitably requires high workloads. So the temperature and power consumption of GPU and CPU may continue to rise. If temperatures of GPU nodes excess a certain threshold, the computing reliability will be depredated. The high temperature of device incurs hardware errors. To solve this issue, the task scheduling scheme is

© Springer Nature Switzerland AG 2020
Q. Liu et al. (Eds.): CENet 2018, AISC 905, pp. 110–117, 2020.
https://doi.org/10.1007/978-3-030-14680-1_13

designed to control the increasing temperature in the continuous process. The scheduling model can prevent the nodes from running for a long time that leads to high temperature and forming hotspots in hardware device. The mainly trend of Large-scale data computing is power consumption conversation [2–4]. Here we focus on the task scheduling model on GPU cluster to process large-scale data. And the main advantage is that the proposed scheduler is designed based on the heat distribution to reduce the GPU temperature and prevent the hotspots emerge. This schedule model can improve the reliability of GPU cluster.

2 Related Works

The GPU cluster is used to process the Large-scale data and emerge many MapReduce designed in GPU cluster. The task scheduling is always the hot research topic. However the Large-scale data processing job is a continuous computing process. So the computing reliability and energy consumption should be considered. Here we discussed the task scheduling about GPU computing. Currently, the scheduler for MapReduce model is mainly to reduce power consumption in continuous processing. A novel job aware power consumption scheduler was designed by integer programming, which was used to divide the map and reduce phase into different sub-phases [2]. To reduce the power consumption of short jobs in Hadoop cluster, a resource scheduler was proposed to adjust the computing resource configuration by heartbeat mechanism [3]. Aiming at the heterogeneous environment, the scheduler model was built with economic cost properties, such as the price of electricity, in order to reduce the peak power consumption [4]. To control the energy consumption of Web server clusters, the scheduler was built by combining the feedback control and mixed integer programming [4]. The model predictive controlled theory was introduced to build the scheduler and this model can optimize the energy consumption from the global perspective [5]. To reduce the energy consumption of nodes at idle state, a novel scheduling model was proposed. This approach can improve the resource utilization by combining DVFS and task partition [6]. In brief, the current researchers focus on task scheduling of GPU cluster by adjusting CPU/GPU core voltage, frequency, hardware statistics information, et al. However, there are few works to consider the temperature constrains. When temperature is too high, energy consumption will increase and reliability decline. The hardware error may occur. Therefore the temperature should be controlled in a reasonable range to ensure computing reliability. In this work, the task scheduling model was built to limit the temperature of GPU node and prevent the temperature Hotspot emerges.

3 Scheduling Model

The scheduling model is to dispatch the map tasks of large-scale data processing job into the GPU nodes. It determines that the nodes assignment scheme to complete the concurrent computing.

3.1   Scheduling Index

High temperature can lead to hardware errors and reduce the computing reliability. We should consider the temperature of GPU device. Additionally, the utilization of GPU is an important factor in computing process. So the comprehensive index P_g is defined by Eq. (1).

$$P_g = w_1 U_g + w_2 T_g \tag{1}$$

In Eq. 1, U_g denotes the utilization of GPU; T_g represents the current temperature of GPU; The weights to balance the importance of utilization of GPU and temperature are w_1 and w_2. The temperature T_g is normalized into [0, 1] by Eq. (2).

$$T_g = T/T_{\max} \tag{2}$$

In Eq. (2), T is the current temperature of GPU device. And $T_{\max}$ denotes the maximum temperature of GPU, such as $T_{\max} = 100$ °C. The experimental cluster is installed CUDA 7.0 on Ubuntu 16.0. The utilization and temperature of GPU is monitored and provided by Nvidia-Sim. The Python script was designed to log the reports of Nvidia-Sim and to extract the key data.

3.2   Scheduling State Matrix

The scheduling model is to assign specific sub-node set to jobs. So the status of nodes in cluster is very important to scheduling model. The status of node is described by the comprehensive index P_g. And the state of cluster can be described by state matrix. Assume that the cluster is consisted of m racks, which includes n nodes. Then the state of cluster M^p is defined by Eq. 3.

$$M^p = \begin{bmatrix} P_{g11} & P_{g12} & \cdots & P_{g1n} \\ P_{g21} & P_{g12} & \cdots & P_{g2n} \\ P_{gi1} & P_{gi2} & \cdots & P_{gin} \\ P_{gm1} & P_{gm2} & \cdots & P_{gmn} \end{bmatrix}, \tag{3}$$

P_{gij} represents the status of jth node in the ith rack. M^p is continuous monitored and updated by the scheduling model.

3.3   Scheduling Model

As shown in Fig. 1, the scheduling model is selected a job from the job queue. A job is assigned to some nodes in GPU cluster. Assume that job queue is $J = <j_1, j_2, \ldots, j_m>$, node set is $N = <n_1, n_2, \ldots, n_m>$. There are four kinds of nodes: Unassigned node, Assigned node, cooled node and unCooled node. This scheduling model is to determine the assigned of nodes that are labeled with grey color. And this model is responsible for lowering the temperature of nodes that are labeled with black color. These black nodes called cooled node are the hottest nodes, whose temperatures exceed specific threshold values. On the contrary, the unCooled node should be lowered the speed of fans due to its temperature under the threshold. So the main idea is to improve the utilization of GPU with the temperature limitation. The scheduling algorithm is as follows.

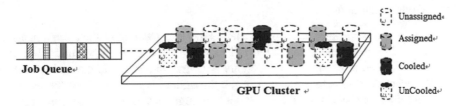

Fig. 1. Scheduling model.

Algorithm: Scheduling algorithm of GPU cluster.

Input: Job queue J, the state matrix M^P;

Output: the subset of nodes $\overline{N}=<n_1,n_2,...,n_k>$.

1) **While** J is not Null Do
2) j=**deQueue**(J);
3) m=**preScheduling**(j);
4)   **Traverse**(M^P);
5) $\overline{N}$=**topSmallNode**(M^P);
6)   Output $\overline{N}$ and execute job j;
7) $t_{i+1}=t_i+\Delta t$;
8)   Updata(M^P) in t_{i+1};
9)   **Traverse**(M^P)
10)   If each $t>\delta$ in N Then **SpeedFan**(n)
11)   If each $t<\gamma$ in N Then **LowerFan**(n)

12) EndWhile

In scheduling algorithm, the main loop is controlled by the job queue. If the job queue is not null, the scheduling process continuous to work (line 1–12). The procedure preScheduling is the original scheduling scheme, which is used to obtain the number of assigned nodes for one job j (line 3). Then the state matrix M^P is scanned to sort the comprehensive index P_g of all nodes in cluster (line 4). And the assigned node set $\overline{N}$ is generated by the procedure topSmallNode, which is to select the first m smallest nodes (line 5). So the job j is distributed into $\overline{N}$ and executed on every node in $\overline{N}$ (line 6). In the next scheduling period t_{i+1}, the state matrix M^P is updated and scanned to find the hottest nodes (line 8–9). When the nodes exceed the specific threshold δ, the cooling device will be launched (line 10). When the nodes are below the specific threshold γ, the node can be reduced the fan speed. Note that we control the fan speed to cool the processors. This algorithm mainly traverses and updates the state matrix. So the algorithm complexity is $O(n)$ and n is the number of cluster node. Note that when the GPU cluster has more than one thousand nodes, this algorithm complexity should be reduced to guarantee scheduling limitation.

3.4   Model Parameters

This section introduces the model parameters. Firstly, the weight of the comprehensive index P_g is to control the tradeoff the utilization of GPU and temperature. Here the experiment is used to determine the weights. The first values of weight are both set to 0.5. We selected 10 jobs to execute. The job throughout is used to adjust the weights. We repeat the experiments to optimize the job throughout with different weight values. The step of adjusting weight is step 0.1. Finally, to obtain the optimal solution of job throughout, the weight of utilization GPU is 0.7 and the weight of temperature is 0.3. The threshold of temperature δ and γ is set to 75 °C and 55 °C. Those two parameters are determined by empirical data from the production environment. Additionally, the step of adjusting fan speed is 20%.

4   Experiment and Analysis

The testbed of GPU cluster is consisted of 32 nodes, which is configured as follows: Intel i7-3930k/16G/NVIDIA GTX660i, Ubuntu12.04/JDK1.7, Hadoop2.2.0. The benchmark scheduling scheme denoted as MR is the original scheduler in Hadoop. The proposed scheduling model is denoted as ATCTS (Aware Temperature Constraint Task Scheduling). GPU-Z and Nvidia-Sim were used to monitor the temperature of GPU node and utilization of GPU. And the python script was designed to control the speed of fans based on the interface of Nvidia-Sim. The large-scale data is selected GPS trajectory of taxi with 2 TB. The simulation jobs are mainly trajectory similarity matching with different data size.

4.1   Effect of Temperature Control

The main goal of scheduling scheme is to limit the GPU temperature. So we Firstly compared the temperature distribution of two scheduling schemes. As shown in Fig. 2, the temperatures of GPU are increasing during processing the large-scale. With the number of tasks increasing, the temperatures of GPU improve quickly and the hotspots are formed. The difference of the minimum and maximum is about 30 °C.

As shown in Fig. 3, the proposed scheduling model was used in the same experimental environment. The results show that the difference of temperature of each node is significant lower. Note that the temperature of GPU increased, the Hotspot didn't occur. And the temperature of each node does not increase continuously when temperature exceeds specific threshold value. This means that our proposed scheduling model can balance the temperature distribution and prevent temperature Hotspot from occurring in GPU cluster.

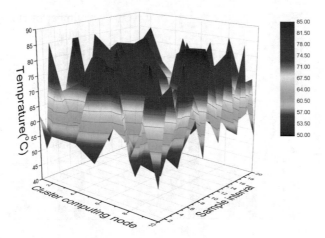

Fig. 2. Temperature distribution of original scheduler.

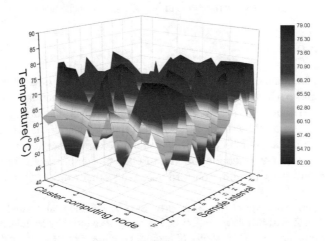

Fig. 3. Temperature distribution of proposed scheduler.

4.2 Job Throughout of Scheduling Model

The job throughout is very important to the scheduling model. In this section we compared the job throughout of these two scheduling models. The simulation jobs are trajectory matching computing with different data size, such as 200 MB, 300 MB, 500 MB, etc. The simulation job queue contains 1000 jobs. And GPU cluster executes with different nodes, such as 8 nodes, 16 nodes, 32 nodes.

As shown in Fig. 4, Our proposed model ATCTS considers the temperature limitation. So this may lead to scheduling performance loss. Compared with the original scheduler, the scheduling performance loss is in the acceptable range.

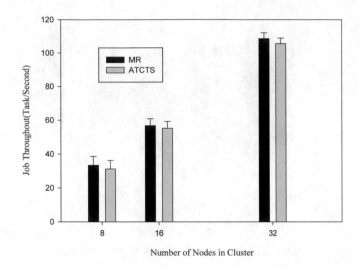

Fig. 4. Comparison job throughout of two schedulers.

5 Conclusion

The proposed scheduling model of GPU cluster is to control the hotspot of temperature for each node and balance the heat distribution by scheduling the tasks of large-scale data processing. Though this scheme may lead to the loss of scheduling performance, the experimental results show the loss within acceptable range. The scheduling model can guarantee computing reliability. To the best of our knowledge, few research focuses on the aware temperature scheduling model of GPU cluster. And our future work is to study and compare the scheduling performance with other scheduling models.

Acknowledgement. This project is supported by Shandong Provincial Natural Science Foundation, China (No. ZR2017MF050), Project of Shandong Province Higher Educational Science and technology program (No. J17KA049), Shandong Province Key Research and Development Program of China (No. 2018GGX101005, 2017CXGC0701, 2016GGX109001) Shandong Province Independent Innovation and Achievement Transformation, China (No. 2014ZZCX02702).

References

1. Kaur, T., Chana, I.: Energy efficiency techniques in cloud computing: a survey and taxonomy. ACM Comput. Surv. **48**(2), 22–54 (2016)
2. Lena, M., Mahyar, M.N., Zhang, Q., Shi, W.: Energy-aware scheduling of MapReduce jobs for big data applications. IEEE Trans. Parallel Distrib. Syst. **26**(10), 2720–2733 (2015)
3. Neetesh, K., Deo, P.: An energy aware cost effective scheduling framework for heterogeneous cluster system. Future Gener. Comput. Syst. **71**, 73–88 (2017)
4. Shi, Y.L., Zhang, K.H., Cui, L.Z., Liu, L., Zheng, Y.Q., Zhang, S.D., Yu, H.: MapReduce short jobs optimization based on resource reuse. Microprocess. Microsyst. **47**, 178–187 (2016)

5. Wang, H.F., Cao, Y.P.: GPU power consumption optimization control model of GPU clusters. Acta Electronica Sin. **43**(10), 1904–1910 (2015). (in Chinese)
6. Huo, H.P., Hu, X.M., Sheng, C.C., Wu, B.F.: An energy efficient task scheduling scheme for node-layer heterogeneous GPU clusters. Comput. Appl. Softw. **30**(3), 283–286 (2013). (in Chinese)
7. Lee, Y., Kulkarni, I., Pompili, D., Parashar, M.: Proactive thermal management in green data centers. J. Supercomputing **60**(2), 165–195 (2012)
8. Zhang, S., Chatha, K.S.: Approximation algorithm for the temperature-aware scheduling problem. In: IEEE/ACM ACM International Conference on Computer-Aided Design, San Jose, USA, pp. 281–288 (2007)
9. Li, X., Jiang, X.H., Wu, Z.H., Ye, K.J.: Research of thermal management methods for green data centers. Chin. J. Comput. **37**(5), 1–21 (2014). (in Chinese)
10. Vanderster, D.C., Baniasadi, A., Dimopoulos, N.J.: Exploiting task temperature profiling in temperature-aware task scheduling for computational clusters. In: Choi, L., Paek, Y., Cho, S. (eds.) Advances in Computer Systems Architecture, pp. 175–185. Springer, Heidelberg (2007)
11. Liu, H., Wang, J.G., Ge, Z.Z., Gu, Q., Chen, Q., Du, J.C.: Self-learning load balancing scheduling algorithm for GPU heterogeneous cluster. J. Xi'an Shiyou Univ. (Nat. Sci. Ed.) **30**(3), 105–111 (2015). (in Chinese)

A Proactive Heuristic for Task-Resource Allocation with Resource Uncertainty

Yuanzi Xu[1], Yingxin Zhang[2($\boxtimes$)], Kungang Yuan[1], Zhiwei Zhang[1], and Ning Ji[1]

[1] Air Force Command College, Beijing 100097, China
[2] Unit 31002 of the PLA, Beijing 100094, China
zhyingxin@163.com

Abstract. Course of Action (COA) planning is a complex problem which involves allocating limited resources to a given set of tasks. A rule-based schedule heuristic is proposed to solve the COA planning problem. In the heuristic, a resource buffering strategy is adopted to resolve the resource uncertainty, i.e. extra resources are adopted to absorb the unexpected resource breakdowns. To decide where and how much resource slacks to insert in the schedule, a resource uncertainty metric namely reliable resource capability is introduced. For illustration, a joint-task-force test scenario is utilized to show the feasibility of the heuristic in solving the COA planning problem. Empirical results validated that the proposed heuristic for the COA planning problem is available, and the resource buffering strategy can effectively absorb the resource uncertainty breakdowns.

Keywords: Course of Action · Task-resource allocation · Proactive heuristic

1 Introduction

The Course of Action (COA) planning problem is an essential component of the military mission planning which involves time and space synchronization as well as resource and capability allocations. The output schedule determines the starting time of each task and consequently also the amount of resource units of each resource type required in each time period. In the past few years, many researchers have concentrated on its resolving algorithms. [1] The current studies includes multi-dimensional dynamic list scheduling (MDLS) algorithm [2, 3], Tabu Search based algorithm [4], evolutionary algorithm (EA) [5], multi-objective optimization evolutionary algorithm [6], Fuzzy-genetic decision optimization [7], Role-based approaches [8–10]. A proactive heuristic adopting resource buffering strategy is proposed in this paper, which aims to strength the schedule robustness by inserting resource slacks.

The rest of this paper is organized as follows: Sect. 2 discusses the mathematical formulation of the COA planning problem; Sect. 3 outlines the proposed method; Sect. 4 provides a case study on how the scheduling process can be exploited to enhance the solution robustness; and Sect. 5 draws the conclusion.

© Springer Nature Switzerland AG 2020
Q. Liu et al. (Eds.): CENet 2018, AISC 905, pp. 118–125, 2020.
https://doi.org/10.1007/978-3-030-14680-1_14

2 Mathematical Description of the Problem

The problem of COA planning can be formally described as follows: given a set T of J tasks $T = (t_1, t_2, \cdots, t_J)$ related by precedence constraints and a set P of I platforms $P = (p_1, p_2, \cdots, p_I)$. Each task $t_j(j \in J)$ is specified by its duration d_j, resource requirement $R_j = (r_{j1}, r_{j2}, \cdots, r_{jK})$, and location $L_j = (x_j, y_j)$. Each platform $p_i(i \in I)$ is specified by its function capability $C_i = (c_{i1}, c_{i2}, \cdots, c_{iK})$ and velocity v_i. The purpose of COA planning is to assign starting time and performing platforms for each task with respect to the precedence and resource constraints, as well as the given objective. The mathematical formulation of the COA planning problem can be described as follows:

$$s_j + d_j \leq s_{j'}, \forall t_{j'} \text{ and } t_j \in Prec_j \tag{1}$$

$$\sum_{p_i \in M_j} c_{ik} \geq r_{jk}, \forall j \in J \tag{2}$$

$$f = \min s_J \tag{3}$$

Equation (1) describes the precedence constraint, which specifies that t_j cannot start until all its predecessors Pre_j are completed. Equation (2) describes the resource constraint, which specifies that each task $t_j(j \in J)$ must be executed by being assigned a group of platforms, named task mode M_j, wherein the aggregated function capabilities are greater than or equal to its resource requirement. Equation (3) describes the objective function, which demands the schedule being finished as soon as possible.

3 Resource Buffering Scheduling Heuristic

When resource uncertainties come into play, the available resource capability of a platform becomes a random variable instead of a deterministic value. The idea of resource buffering is to construct the COA baseline schedules using an available resource capability that is lower than the deterministic given resource capability. In this case, the baseline schedules can be protected against disruptions by including resource slacks.

3.1 Resource Buffering Algorithm

The key of assigning resource slacks is to calculate the resource uncertainty. In the following, a concept of reliable resource capability of platform is introduced to do this, which is modeled based on the following assumptions: (1) The platform being assigned for each task is fixed, and cannot change during the schedule execution; (2) The failure time X_{ik} and repair time Y_{ik} for each unit resource c_{ik} (the resource type k of platform p_i) are identical and independent; (3) X_{ik} and Y_{ik} both obey exponential distribution with parameters λ_{ik} and μ_{ik} respectively; (4) Resource capacity can only break down when they are being used.

The stationary availability of a unit resource c_{ik} can be calculated as follows: [11]

$$EA_{ik} = \frac{E(X_{ik})}{E(X_{ik}) + E(Y_{ik})} = \frac{1/\lambda_{ik}}{1/\lambda_{ik} + 1/\mu_{ik}} = \frac{1}{1 + \varphi_{ik}} \tag{4}$$

where $\varphi_{ik} = \mu_{ik}/\lambda_{ik}$.

To define the reliable resource capability, the dynamic availability of a unit resource c_{ik} is defined as follows:

$$PA_{ik} = \frac{1}{1 + \theta \cdot \varphi_{ik}} \tag{5}$$

where $\theta (0 \le \theta \le 1)$ is a tunable variable namely resource buffering parameter.

The available resource capability $AC(c_{ik})$ can be calculated as follows:

$$AC(c_{ik}) = \sum_{a=0}^{c_{ik}} a \cdot Pr(AC(c_{ik}) = a) = \binom{c_{ik}}{a} \cdot PA_{ik}^{a} \cdot (1 - PA_{ik})^{(c_{ik} - a)} \tag{6}$$

where $Pr(AC(c_{ik}) = a)$ denotes the probability of $AC(c_{ik}) = a$.

3.2   Schedule Generation Algorithm

In the following, a priority rule-based heuristic is proposed to construct the initial COA schedules based on the available resource capability. Firstly, a solution representation is proposed. Then, the potential task modes for each non-dummy task are calculated. Finally, feasible schedules are constructed using the serial schedule generation scheme.

Platform Allocation

As mentioned earlier, a task mode must satisfy the following two conditions: First, it must satisfy the resource requirement, that is, the aggregated resource capability vector of a platform group should be greater than, or equal to the task resource requirement vector. Second, none of the platforms within the execution mode is redundant. In other words, the absence of any platform within a task mode will make the platform group not satisfy the resource requirement. The calculation of the task mode set M_j for non-dummy tasks can be formally stated as follows:

Step 1: Let j = 2.
Step 2: Let $l = 1, M_j = \emptyset$.
Step 3: Calculate the candidate task mode m_j that is constituted by l platforms.
Step 4: Let $M_j = M_j \cup m_j$, if $|M_j|$ (the number of execution modes including in M_j) is equal to the predefined maximum number of task modes, repeat step 3, otherwise, go to step 6.
Step 5: Let l = l + 1, if $l \le K$ (K is the number of resource types), repeat step 3, otherwise, go to step 6.
Step 6: Let j = j + 1, if $j \le J - 1$, repeat step 2, otherwise, go to step 7.
Step 7: End.

Schedule Representation

An appropriate solution representation for the problem should reflect both the scheduling problem (which assigns starting times to tasks) and the mode assignment problem (which assigns platforms to tasks). Moreover, both of the problems should be dealt with simultaneously because they interact with each other. With this in mind, an extended task list representation is proposed. The proposed solution representation contains two lists, namely, a precedence feasible task list (TL), which determines the sequence in which the tasks are executed, and a mode list (ML), which determines the execution platforms for each task. Figure 1 shows an illustrative example of the proposed solution representation containing eight tasks. As can be seen in Fig. 1, the planning order of the eight tasks is given by $t_1, t_4, t_6, t_7, t_3, t_5, t_2, t_8$, and the corresponding task modes are $m_{11}, m_{23}, m_{31}, m_{42}, m_{52}, m_{62}, m_{73}, m_{81}$ respectively.

TL:	1	4	6	7	3	5	2	8
ML:	1	2	2	3	1	2	3	1

Fig. 1. Solution representation example.

Schedule Generation Scheme

Schedule generation scheme is a basic component of heuristics based on priority rules. It iteratively generates a feasible schedule by extending a partial schedule. The task-oriented serial Schedule generation scheme and the time-oriented parallel Schedule generation scheme are commonly used in literature. In the serial Schedule generation scheme, each stage is associated with two disjoint task sets, namely, the set of planned tasks and the set of all eligible tasks. Eligible tasks are the tasks in which all predecessors have been planned. In each stage, an eligible task is chosen and planned at the earliest precedence and resource feasible time. In the parallel Schedule generation scheme, each stage is associated with one time point tp and three disjoint task sets, including the set of finished tasks, the set of active tasks, and the set of eligible tasks. Active tasks are the tasks that have been planned before time tp, but will finish after the time tp. The eligible tasks correspond to the tasks in which all predecessors have completed up to time tp and sufficient platforms are available when they start in time tp. In this study, we adopt the serial schedule generation scheme.

4   Case Study

4.1   Test Scenario

A test joint-task-force scenario being modified by the command and control (A2C2) experiment seven [2–5] is introduced to validate the proposed heuristic, which contains 20 tasks and 20 platforms. A directed acyclic active-on-node graph $G\ (V,\ E)$ is introduced as shown in Fig. 2 to represent the precedence constraints between tasks. The nodes represent the tasks, and the arcs represent the zero-lag finish-start precedence constraints. The precedence constraints specify that any task cannot be performed until all its predecessors are completed. The first and the last nodes respectively denote the dummy start and end tasks.

The task attributes are described in Table 1. Each non-dummy task is specified by its duration, resource requirement $(r_1, r_2, \cdots, r_8)$, and location. Table 2 depicts the platform attributes. Each platform is specified by its velocity and resource capability $(c_1, c_2, \cdots, c_8)$. The vectors $(r_1, r_2, \cdots, r_8)$ and $(c_1, c_2, \cdots, c_8)$ have the same length and each entry corresponds the same type of function capability.

Table 1. Task parameters.

Task	Duration	Resource requirement								Location
		r_1	r_2	r_3	r_4	r_5	r_6	r_7	r_8	
1	-	-	-	-	-	-	-	-	-	-
2	30	5	3	10	0	0	8	0	6	70,15
3	30	5	3	10	0	0	8	0	6	64,75
4	10	0	3	0	0	0	0	0	0	15,40
5	10	0	3	0	0	0	0	0	0	30,95
6	10	0	3	0	0	0	0	10	0	28,73
7	10	0	0	0	10	14	12	0	0	24,60
8	10	0	0	0	10	14	12	0	0	28,73
9	10	0	0	0	10	14	12	0	0	28,83
10	10	5	0	0	0	0	5	0	0	28,73
11	10	5	0	0	0	0	5	0	0	28,83
12	10	0	0	0	0	0	10	5	0	25,45
13	10	0	0	0	0	0	10	5	0	5,95
14	20	0	0	0	0	0	8	0	6	25,45
15	20	0	0	0	0	0	8	0	6	5,95
16	15	0	0	0	20	10	4	0	0	25,45
17	15	0	0	0	20	10	4	0	0	5,95
18	10	0	0	0	0	0	8	0	4	5,60
19	20	0	0	0	8	6	0	4	10	5,60
20	-	-	-	-	-	-	-	-	-	-

4.2　Experiments Results

To validate the proposed heuristic, we adopt the Monte Carlos simulation method to analyze the planning results. The simulation schemes are supposed as follows. The schedule implement is according to the railway time table scheme, i.e. the task can not start before its planned starting time; The task breakdown go on executing at the breakdown time after the resource repairment. The simulation parameters are set as follows. The expected resource failure times are drawn from a uniform distribution between the minimal makespan MK_{min} and its twice amount; The lower bound MK_{min} is obtained by solving the problem without resource slacks, i.e. $\theta = 0$; The expected resource repair times are drawn from a uniform distribution between 1 and 10.

Table 2. Platform parameters.

Platform	Velocity	Resource capability							
		c_1	c_2	c_3	c_4	c_5	c_6	c_7	c_8
1	2	10	12	6	0	9	5	0	0
2	2	1	4	12	0	4	3	0	0
3	2	10	12	6	0	9	2	0	0
4	4	0	0	0	4	0	0	8	0
5	1.35	1	0	0	10	2	2	2	0
6	4	5	0	0	0	0	0	0	0
7	4	3	4	0	0	6	10	2	0
8	4	1	3	0	0	10	8	2	0
9	4	1	3	0	0	10	8	2	0
10	4	1	3	0	0	10	8	2	0
11	4.5	6	2	0	0	1	1	0	0
12	4.5	6	2	0	0	1	1	0	0
13	4.5	6	2	0	0	1	1	0	0
14	2	0	0	0	0	0	0	10	0
15	5	0	0	0	0	0	0	0	8
16	7	0	0	0	0	0	0	0	8
17	2.5	0	0	0	8	6	0	2	14
18	1.35	2	0	0	12	2	2	2	0
19	1.35	2	0	0	12	2	2	2	0
20	1.35	2	0	0	12	2	2	2	0

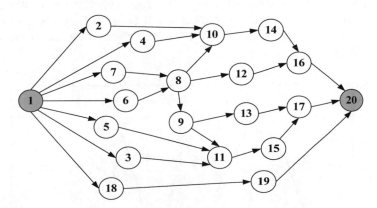

Fig. 2. Task graph.

The result schedule of $\theta = 0$ (135.1 s) is depicted as the Fig. 3. According to the task and platform attributes description in Tables 1 and 2, task 2 and 3 both need platform 2 (at least 90.1 s). Furthermore, the sum duration of the succeeding tasks of task 2 (10, 14, 16) and task 3 (11, 15, 17) are both not less than 45 s. Hence, the makespan of Fig. 3 is the minimal makespan.

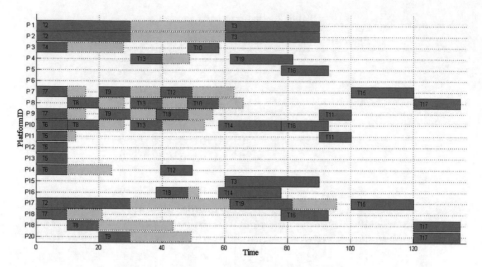

Fig. 3. Schedule result without resource slacks of $\theta = 0$.

In order to analyze the effects of resource buffer to the schedule robustness, we respectively evaluate the resource buffering parameter $\theta = 0 : 0.1 : 1$ to gain the planning schedules. And for each schedule, we execute it 100 times with the above Monte Carlos simulation parameters. Finally, we calculate the standard division σ of the 100 schedule real starting times as depicted in Fig. 4. We can see that the best resource buffering parameter value is not 1, but 0.1. To analyze the result, we can see that when the resource buffers are very little such as $\theta = 0$, the number of the resource slacks is very little, and the schedule is very vulnerable to break down. However, as the θ value increase, the resources (or platforms) will be more busy and easier to fail. Hence, for each special scenario, the best resource buffering parameter value must be between 0 and 1.

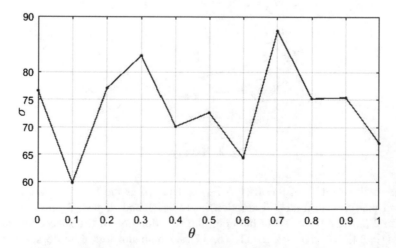

Fig. 4. The standard deviation of Monte Carlos simulation results.

5   Conclusion

Resource uncertainty is a common problem in Course of Action planning problem. In this paper, a resource buffering strategy is adopted to enhance the robustness of the schedule results. In order to do this, a resource uncertainty metric namely reliable resource capability is introduced and calculated. Then, a rule-based COA schedule algorithm is proposed to obtain the schedule results. Finally, a test scenario is introduced to validate the proposed heuristic algorithm.

References

1. Boukhtouta, A., Bouak, F., Berger, J.: Description and analysis of military planning systems. Defence Research and Development Canada (2006)
2. Levchuk, G.M., Levchuk, Y.N., Luo, J., Pattipati, K.R., Kleinman, D.L.: Normative design of organizations. I. Mission planning. IEEE Trans. Syst. Man Cybern. 32(3), 346–359 (2002)
3. Cheng, K., Zhang, H., Zhang, R.: A task-resource allocation method based on effectiveness. Knowl.-Based Syst. 37, 196–202 (2013)
4. Belfares, L., Klibi, W., Lo, N.: Multi-objectives Tabu search based algorithm for progressive resource allocation. Eur. J. Oper. Res. 177(3), 1779–1799 (2007)
5. Yu, F., Tu, F., Pattipati, K.R.: Integration of a holonic organizational control architecture and multiobjective evolutionary algorithm for flexible distributed scheduling. IEEE Trans. Syst. Man Cybern. Part A Syst. Hum. 38(5), 1001–1017 (2008)
6. Bui, L.T., Michalewicz, Z., Parkinson, E., Abello, M.B.: Adaptation in dynamic environments: a case study in mission planning. IEEE Trans. Evol. Comput. 16(2), 190–209 (2012)
7. Kewley, R.H., Embrechts, M.J.: Computational military tactical planning system. IEEE Trans. Syst. Man Cybern. Part C Appl. Rev. 32(2), 161–171 (2002)
8. Zou, Z., Che, W., Wang, S., Bai, Y., Fan, C.: Role-based approaches for operational tasks modelling and flexible decomposition. J. Syst. Eng. Electron. 27(6), 1191–1206 (2016)
9. Li, N., Huai, W., Wang, S.: The solution of target assignment problem in command and control decision-making behaviour simulation. Enterp. Inf. Syst. 11(7), 1059–1077 (2017)
10. Bao, Y.W., Zhang, W.X., Zhang, S.M.: An improved task and role-based access control model with multi-constraint. Appl. Mech. Mater. 32(6), 713–715 (2015)
11. Lambrechts, O., Demeulemeester, E., Herroelen, W.: Proactive and reactive strategies for resource-constrained project scheduling with uncertain resource availabilities. J. Sched. 11(2), 121–136 (2008)

Integrating Latent Feature Model and Kernel Function for Link Prediction in Bipartite Networks

Xue Chen[1], Wenjun Wang[1], Yueheng Sun[1(✉)], Bin Hu[2],
and Pengfei Jiao[3]

[1] School of College of Intelligence and Computing,
Tianjin University, Tianjin, China
yhs@tju.edu.cn
[2] School of Technical College for the Deaf,
Tianjin University of Technology, Tianjin, China
[3] School of Center of Biosafety Research and Strategy,
Tianjin University, Tianjin, China

Abstract. Link prediction aims to infer missing links or predict future links from existing network structure. In recent years, most studies of link prediction mainly focus on monopartite networks. However, a class of complex systems can be represented by bipartite networks, which containing two different types of nodes and the no links exist in the same type. In this paper, we propose Kernel-based Latent Feature Models (KLFM) framework which can extract nonlinear high-order information in the existing network through kernel-based mappings. Then a kernel-based iterative rule has been developed. Extensive experiments on eight disparate real-world bipartite networks demonstrate that the KLFM framework achieves a more robust and explicable performance than other methods.

Keywords: Link prediction · Bipartite network · Latent feature model · Kernel function

1 Introduction

Link prediction used to infer the missing links or predict future links [1]. The key issue of link prediction is to estimate the likelihood of potential links in networks. Many real-world networks display a natural bipartite structure, i.e. nodes can be separated into two different sets and no links exist in the same type. Therefore, link prediction in bipartite networks has significant importance in various fields such as biology, e-commerce and others.

Many algorithms have been proposed for link prediction in bipartite networks. Roughly, the algorithms can be classified into three categories, e.g. projection-based methods [2–4], similarity-based methods [5–7] and latent feature models [8–10]. Projection-based methods map bipartite networks into monopartite networks by unweighted or weighted manner. However, the projection-based methods are always less informative than the original bipartite networks. Similarity-based methods can be

© Springer Nature Switzerland AG 2020
Q. Liu et al. (Eds.): CENet 2018, AISC 905, pp. 126–134, 2020.
https://doi.org/10.1007/978-3-030-14680-1_15

defined by a variety of local and global indices. However, the calculations of most similarity indices only use the information of the network topology. Latent feature models have been widely used in link prediction of bipartite networks. Compared with other methods, latent feature models can learn nonlinear representations from network structures to obtain satisfactory prediction performance. However, the choice of similarity function in the latent space is usually linear.

Inspired by the idea of kernel function [11], which map the data into higher dimensional spaces so that in this higher-dimensional space the data could be linear modeling. In this paper, we propose Kernel-based Latent Feature Models (KLFM) which can extract nonlinear high-order information in the existing network through kernel-based mappings. Then a kernel-based iterative rule has been developed. Extensive experiments on several disparate real-world bipartite networks demonstrate that the KLFM has a more robust and explicable performance than the other methods.

2 Methods

2.1 Overview Notation

A bipartite network can be represented by $G(U, V, E)$, where U and V are two different parts of the nodes and E is the set of links. The given network can be represented by $m \times n (m = |U|, n = |V|)$ bi-adjacency matrix B, where the element $B_{ij} = 1$ if nodes $i \in U$ and $j \in V$ are connected; otherwise $B_{ij} = 0$. To test the algorithm's accuracy, in each dataset, all existing links E are randomly split into two parts: a training set E^T and a test set E^P.

2.2 Kernel-based Latent Feature Models (KLFM)

Latent feature models predict links based on their latent features or latent groups, which can be extracted through low rank decomposition of network's adjacency matrix. Let B denotes the adjacency matrix of the bipartite network. $U \in \mathbb{R}^{m \times k}$ and $V \in \mathbb{R}^{n \times k}$ denote a matrix where each row is a k-dimensional vector of latent features, respectively.

To estimate U and V in the latent feature model, a reasonable approach is to minimize the squared error. It can lead to the following objective:

$$\min O(U, V) = \|B - F(U, V)\|_F^2 + \lambda \|U\|_F^2 + \lambda \|V\|_F^2 \tag{2.1}$$

$F(\cdot)$ is usually expressed by a linear combination of U and V, e.g., UV^T. However, it could extract higher-order features hidden in the existing network through kernel-based nonlinear mappings. The latent feature model in kernel space defined as follows:

$$\min O(\phi(U), \phi(U)) = \left\|B - \phi(U)\phi(V)^T\right\|_F^2 + \lambda \|\phi(U)\|_F^2 + \lambda \|\phi(V)\|_F^2 \tag{2.2}$$

Where $\phi(\cdot)$ is a nonlinear mapping function, and we do not need to know its specific form.

It is impractical to directly solve (2.2) since $\phi(\cdot)$ is unknown. In the following, this problem can be solved by deriving a novel method. Firstly, we randomly select k rows $[c_1, c_2, \cdots, c_k]^T$ from adjacency matrix of bipartite networks. Here, k is the dimension of the latent space $(k \ll \min(m, n))$. The dimension of latent space can be determined by cross validation method. Then we assume that the feature matrix $\phi(U)$ can be represented as a linear combination of $\{c_i\}_{i=1}^k$ in kernel space as follows:

$$\phi(U) = \begin{bmatrix} a_{11} & \cdots & a_{1k} \\ \vdots & \ddots & \vdots \\ a_{m1} & \cdots & a_{mk} \end{bmatrix} \begin{bmatrix} \phi(c_1) \\ \vdots \\ \phi(c_k) \end{bmatrix} = A\Phi \tag{2.3}$$

Similarity, we also assume that $\phi(V)$ can be represented as:

$$\phi(V) = \begin{bmatrix} d_{11} & \cdots & d_{1k} \\ \vdots & \ddots & \vdots \\ d_{n1} & \cdots & a_{nk} \end{bmatrix} \begin{bmatrix} \phi(c_1) \\ \vdots \\ \phi(c_k) \end{bmatrix} = D\Phi \tag{2.4}$$

Combining (2.2), (2.3) and (2.4), the mathematical expressions of our proposed KLFM model is as follows:

$$\min O(A, D, \Phi) = \left\| B - A\Phi\Phi^T D^T \right\|_F^2 + \lambda tr(A\Phi\Phi^T A^T) + \lambda tr(D\Phi\Phi^T D^T) \tag{2.5}$$

The kernel function denotes an inner product in feature space and is usually denoted as: $K = \Phi\Phi^T$. So we can rewrite (2.5) as

$$\min O(A, D) = \left\| B - AKD^T \right\|_F^2 + \lambda tr(AKA^T) + \lambda tr(DKD^T) \tag{2.6}$$

Using the kernel function, the algorithm can be carried into a higher-dimension space without explicitly mapping the input points into this space. In this paper, we use three simple kernels: linear kernel function $K_1(c_i, c_j) = c_i c_j^T$, polynomial kernel function $K_2(c_i, c_j) = (1 + c_i c_j^T)^d$, Gaussian kernel function $K_3(c_i, c_j) = \exp(-\|c_i - c_j\|^2/\sigma^2)$.

The partial derivatives of O with respect to A and D are

$$\partial O/\partial A = -2BDK^T + 2AKD^T DK^T + 2\lambda AK^T \tag{2.7}$$

$$\partial O/\partial D = -2B^T AK + 2DK^T A^T AK + 2\lambda AK^T \tag{2.8}$$

Setting $\partial O/\partial A = 0$ and $\partial O/\partial D = 0$, we can obtain the optimal solution of A and D as

$$A = BDK^T(KD^TDK^T + \lambda K^T)^{-1} \tag{2.9}$$

$$D = B^TAK(D^TA^TAK + \lambda K^T)^{-1} \tag{2.10}$$

At last, we rank all non-observed links in a descending order according to their corresponding values in matrix AKD^T, with the top L links constituting the predicted results.

3 Experiment

To examine the performance of the KLMF method, this section describes an extensive experimental study. We detail the experimental setup, including datasets, comparison algorithms and evaluation metric.

3.1 Datasets

We perform the KLFM framework on eight bipartite networks, which are (1) G-protein coupled receptors (GPC) [12], (2) Ionchannels [12], (3) Enzymes [12], (4) Southern Women (referred here as "SW") [13], (5) Malaria [14], (6) Drug-target [15], (7) Country-organization [11], (8) Movielens [11]. The description of the networks can be found in references there.

3.2 Comparison Algorithms

For comparison, we introduce fourteen benchmark algorithms. The basic algorithmic expressions are shown in Table 1.

Table 1. Overview of the compared methods in bipartite networks.

Methods	Formula or description
CN [3]	$S_{\{xy\}}^{\{CN\}} = |(N(x) \cap N(N(y))) \cup (N(y) \cap N(N(x)))|$
JC [3]	$S_{\{xy\}}^{\{JC\}} = S_{\{xy\}}^{\{CN\}} \big/ |N(x) \cup N(y)|$
AA [3]	$S_{\{xy\}}^{\{AA\}} = \sum_{z \in (N(x) \cap N(N(y))) \cup (N(y) \cap N(N(x)))} \frac{1}{\log_2 |N(z)|}$
RA [3]	$S_{\{xy\}}^{\{RA\}} = \sum_{z \in (N(x) \cap N(N(y))) \cup (N(y) \cap N(N(x)))} \frac{1}{|N(z)|}$
PA [3]	$S_{\{xy\}}^{\{PA\}} = |N(x)| \cdot |N(y)|$
CAR [3]	$S_{\{xy\}}^{\{CAR\}} = S_{\{xy\}}^{\{CN\}} \cdot S_{\{xy\}}^{\{LCL\}}$
CJC [3]	$S_{\{xy\}}^{\{CJC\}} = S_{\{xy\}}^{\{CAR\}} \big/ |N(x) \cup N(y)|$

(*continued*)

Table 1. (*continued*)

Methods	Formula or description
CAA [3]	$S_{\{xy\}}^{\{CAA\}} = \sum_{z \in (N(x) \cap N(N(y))) \cup (N(y) \cap N(N(x)))} \frac{|\gamma(z)|}{\log_2 |N(z)|}$
CRA [3]	$S_{\{xy\}}^{\{CRA\}} = \sum_{z \in (N(x) \cap N(N(y))) \cup (N(y) \cap N(N(x)))} \frac{|\gamma(z)|}{|N(z)|}$
CPA [17]	$S_{\{xy\}}^{\{CPA\}} = e(x) \cdot e(y) + e(x) \cdot S_{\{xy\}}^{\{CAR\}} + e(y) \cdot S_{\{xy\}}^{\{CAR\}} + (S_{\{xy\}}^{\{CAR\}})^2$
BPR [2]	Random walk
Jac [2]	Jac similarity
Euc [2]	Euclidean distance
Cos [2]	Cosine similarity
Pea [2]	Pearson correlation
NMF [4]	Nonnegative Matrix Factorization

3.3   Evaluation Metric

We use two standard metrics to quantify the accuracy of algorithms: area under the receiver operating characteristic curve (AUC) [16] and precision [17]. AUC value reflects the overall ranking performance of the algorithm while precision value concentrates on the performance of the top-L links.

3.4   Analysis on the Experimental Results

For each of the eight bipartite networks, the links E are divided into training set E^T (90%) and probe set E^P (10%). The precision values and AUC values are shown in Tables 2 and 3, respectively. For each bipartite network, the best result on each network is boldface. Our KLFM method with different kernels method are better than other algorithms.

Table 2. Precision values of different methods on 8 bipartite networks.

Precision	C2O	Drug	Enzymes	GPC	Ionchannel	Malaria	Movie	SW
KLFM1(linear)	0.9002	0.7065	**0.7215**	0.3734	0.6723	0.3033	0.1983	0.1889
KLFM2(poly)	**0.9011**	**0.7163**	0.7155	0.4078	**0.6811**	0.3027	0.1973	0.2278
KLFM3 (Gaussian)	0.8999	0.7087	0.7142	**0.4125**	0.6713	**0.3034**	**0.1985**	**0.2556**
CN	0.8733	0.6087	0.3741	0.3063	0.1905	0.1886	0.1384	0.1444
RA	0.8857	0:6935	0.2935	0.3500	0.1905	0.2212	0.1035	0.1778
AA	0.8733	0.6435	0.2941	0.3469	0.2132	0.2158	0.1321	0.1667
JC	0.5985	0.3848	0.0321	0.0063	0.0122	0:2505	0.0010	0.0222
PA	0.8730	0.3065	0.0232	0.0969	0.0365	0.0182	0.1453	0.1222
NMF	0.0056	0.0176	0.0049	0.0009	0.0048	0.0000	0.0000	0.0324

(*continued*)

Table 2. (*continued*)

Precision	C2O	Drug	Enzymes	GPC	Ionchannel	Malaria	Movie	SW
Cos	0.6649	0.4877	0.3307	0.2023	0.3493	0.1354	0.1200	0.1630
Euc	0.6210	0.1467	0.0244	0.0421	0.0500	0.0071	0.0451	0.1190
BPR	0:8947	0.6844	0.4950	0.2714	0.4377	0.2403	0.1849	0.1630
CAR	0.8733	0.5978	0.5078	0.2906	0.4324	0.1879	0.1776	0.1889
CRA	0.8781	0.6304	0.6416	0.3594	0.5486	0.2532	0.1848	0.2111
CAA	0.8733	0.5913	0.5037	0.3281	0.5378	0.1896	0.1913	0.1222
CJC	0.8731	0.6130	0.4962	0.3125	0.4946	0.2256	0.1848	0.1889
CPA	0.8733	0.5935	0.5181	0.2906	0.4324	0.1852	0.1776	0.1778

Table 3. AUC values of different methods on 8 bipartite networks.

AUC	C2O	Drug	Enzymes	GPC	Ionchannel	Malaria	Movie	SW
KLFM1(linear)	**1.0000**	0.9250	0.9130	0.8740	0.9545	**0.9385**	0.9000	0.8330
KLFM2(poly)	**1.0000**	**0.9300**	0.9180	0.8713	**0.9590**	0.9330	**0.9160**	0.8153
KLFM3(Gauss)	**1.0000**	0.9270	**0.9250**	**0.8848**	0.9503	0.9220	0.9000	**0.8445**
CN	0.9910	0.9170	0.8532	0.8140	0.9200	0.8985	0.8730	0.7305
RA	1.0000	0.9270	0.8570	0.8400	0.9280	0.9170	0.8880	0.7660
AA	0.9880	0.9250	0.8693	0.8480	0.9340	0.9090	0.8780	0.7225
JC	0.9520	0.9105	0.8815	0.8200	0.8550	0.9010	0.7850	0.6640
PA	0.8980	0.8795	0.7884	0.7060	0.8230	0.5885	0.8820	0.6480
NMF	0.9899	0.8876	0.7612	0.7025	0.8538	0.8598	0.8880	0.6921
Cos	0.9600	0.8680	0.8360	0.8010	0.8350	0.8240	0.8310	0.6894
Euc	0.8710	0.7920	0.6900	0.7280	0.6800	0.5990	0.8000	0.7610
BPR	0.9980	0.9170	0.8880	0.8410	0.9130	0.9040	0.9000	0.7400
CAR	0.9900	0.9045	0.8670	0.7960	0.9160	0.9060	0.9120	0.7265
CRA	1.0000	0.9280	0.8920	0.8200	0.9360	0.9060	0.9200	0.7740
CAA	1.0000	0.9060	0.8532	0.8300	0.9220	0.9235	0.9100	0.7590
CJC	0.9860	0.9115	0.8671	0.8280	0.9250	0.9125	0.8820	0.7620
CPA	0.9700	0.9040	0.8703	0.8210	0.9080	0.8985	0.9160	0.7420

Furthermore, we analyze the experimental results on four networks with different fraction of training set from 0.4 to 0.9. As reported in Figs. 1 and 2, we show the results of drug target, GPC, Ionchannel and malaria based on precision and AUC, respectively. The black lines represent the performance of the proposed KLFM methods with different kernels. Our proposed framework is superior to other methods.

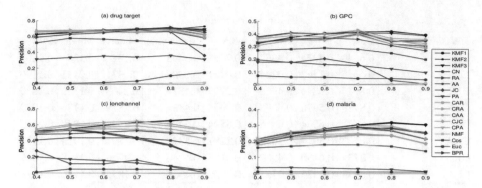

Fig. 1. The precision on the four bipartite networks for different sizes of the probe sets.

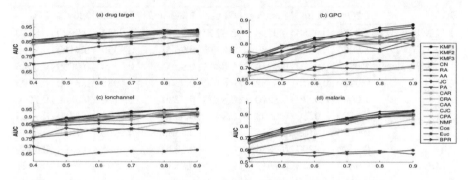

Fig. 2. The AUC on the four bipartite networks for different sizes of the probe sets.

We study the dependence of the prediction accuracies on the regularization parameter λ. Figure 3 shows the KLFM with Gaussian kernel methods' sensitivity of parameter λ, which stands for the regularization parameter. Other datasets show the similarity pattern.

Finally, the complexity analysis of our KLFM framework is given. The time cost of $BDK^T(KD^TDK^T + \lambda K^T)^{-1}$ is $nmk + nk^2 + k^3$. The time cost of $B^TAK(D^TA^TAK + \lambda K^T)^{-1}$ is $nmk + mk^2 + k^3$, so the total time cost of the algorithm is $O(nmk + nk^2 + mk^2 + k^3)$.

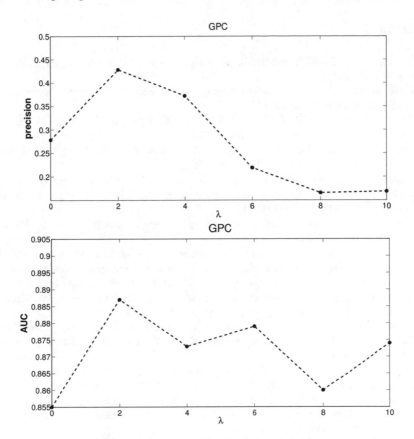

Fig. 3. Comparison of different values of λ (regularization parameter).

4   Conclusion

In the paper, a bipartite link prediction algorithm is given by using kernel-based latent feature model. The framework is called Kernel-based Latent Feature Models (KLFM) which can extract nonlinear high-order features hidden in the original data through some kernel-based mappings. Then a kernel-based iterative rule has been developed to solve the objective function. The experiments show that our KLFM framework can achieve a more robust and explicable performance compared with other methods.

Acknowledgment. The National Key R&D Program of China (2018YFC0809800, 2016QY15Z2502-02, 2018YFC0831000).

References

1. Lü, L., Zhou, T.: Link prediction in complex networks: a survey. Phys. A **390**(6), 1150–1170 (2011)
2. Yildirim, M.A., Coscia, M.: Using random walks to generate associations between objects. PLoS ONE **9**(8), e104813 (2014)
3. Gao, M., Chen, L., Li, B., Li, Y., Liu, W., Xu, Y.C.: Projection-based link prediction in a bipartite network. Inf. Sci. **376**, 158–171 (2017)
4. Zhou, T., Ren, J., Medo, M., Zhang, Y.C.: Bipartite network projection and personal recommendation. Phys. Rev. E **76**(4), 046115 (2007)
5. Daminelli, S., Thomas, J.M., Durán, C., Cannistraci, C.V.: Common neighbours and the local-community-paradigm for topological link prediction in bipartite networks. New J. Phys. **17**(11), 113037 (2015)
6. Newman, M.E.: Clustering and preferential attachment in growing networks. Phys. Rev. E **64**(2), 025102 (2001)
7. Durán, C., Daminelli, S., Thomas, J.M., Haupt, V.J., Schroeder, M., Cannistraci, C.V.: Pioneering topological methods for network-based drug-target prediction by exploiting a brain-network self-organization theory. Brief. Bioinform. **19**(6), 1183–1202 (2017)
8. Lee, D.D., Seung, H.S.: Algorithms for non-negative matrix factorization. In: Advances in Neural Information Processing Systems, pp. 556–562 (2001)
9. Menon, A.K., Elkan, C.: Link prediction via matrix factorization. In: Joint European Conference on Machine Learning and Knowledge Discovery in Databases, pp. 437–452. Springer, Heidelberg (2011)
10. Wang, W., Chen, X., Jiao, P., Jin, D.: Similarity-based regularized latent feature model for link prediction in bipartite networks. Sci. Rep. **7**(1), 16996 (2017)
11. Yamanishi, Y., Araki, M., Gutteridge, A., Honda, W., Kanehisa, M.: Prediction of drug-target interaction networks from the integration of chemical and genomic spaces. Bioinformatics **24**(13), 232–240 (2008)
12. Davis, A., Gardner, B.B., Gardner, M.R.: Deep South: A Social Anthropological Study of Caste and Class. University of South Carolina Press, Columbia (2009)
13. Larremore, D.B., Clauset, A., Buckee, C.O.: A network approach to analyzing highly recombinant malaria parasite genes. PLoS Comput. Biol. **9**(10), e1003268 (2013)
14. Yamanishi, Y., Kotera, M., Moriya, Y., Sawada, R., Kanehisa, M., Goto, S.: DINIES: drug-target interaction network inference engine based on supervised analysis. Nucleic Acids Res. **42**(W1), 39–45 (2014)
15. Coscia, M., Hausmann, R., Hidalgo, C.A.: The structure and dynamics of international development assistance. J. Globalization Dev. **3**(2), 1–42 (2013)
16. Herlocker, J.L., Konstan, J.A., Terveen, L.G., Riedl, J.T.: Evaluating collaborative filtering recommender systems. ACM Trans. Inf. Syst. **22**(1), 5–53 (2004)
17. Hanley, J.A., McNeil, B.J.: The meaning and use of the area under a receiver operating characteristic (ROC) curve. Radiology **143**(1), 29–36 (1982)

Study on Dynamic Wheel/Rail Interaction Caused by Polygonal Wheels

Xiao Cui[1(✉)] and Jianwei Yao[2]

[1] China Academy of Railway Sciences, Beijing 100081, China
cx890626@163.com
[2] Railway Science and Technology Research and Development Center,
China Academy of Railway Sciences, Beijing 100081, China

Abstract. In order to study the influences of polygonal wheel on wheel/rail dynamic characteristics, a transient finite element model of a high-speed train with 20 orders polygonal wheel abrasion is established. The parallel computing is used to improve the calculation speed. The work conditions with the passing speeds of 200 km/h, 240 km/h, 280 km/h and 320 km/h are chosen. The vertical forces and the longitudinal forces of wheel/rail contact are calculated to analyze the dynamics performance in the time domain and the frequency domain. The peak-peak value of the maximum vertical contact force at the fastening increases linearly by about 3 kN per 10 km/h when the passing speed is above 200 km/h and then grows slowly with the passing speed above 280 km/h. The vertical contact force dominant frequencies are almost the same as passing frequencies be caused by the polygonal wheels at different passing speeds and the energy are concentrated at the only dominant frequencies. While the energy distributions of longitudinal contact force are scattered due to the nonlinear contact system that will affect the development of the wheel polygon.

Keywords: Wheel polygon · High-speed railway · Wheel/rail interaction · Frequency domain analysis

1 Introduction

Polygonal wheel abrasion is a phenomenon that wheel radial profile differs in different position due to operational wear. It will cause the severe vibration of the wheel/rail system and the bolt fracture of the wheel axle-box. With the increase of railway operation speed and the axle load, the damage of wheel polygon becomes serious, which leads to the higher railway operation cost and directly affects the safety of railway transportation. Figure 1 shows the different orders of the wheel polygon [1].

Many researchers have taken efforts on the problems. Cui detected the non-circularization of high-speed wheels and investigated the mechanism of the phenomenon [2]. Li studied the vibration behavior and vibration propagation relationship of the vehicle/bogie with high-order polygonal wheel [3]. Song studied the evolution of wheel polygon in EMUs and the effects of wheel polygon on dynamic performance during operation [4]. Liu investigated the effects of wheelset elasticity on wheel vibration with the polygon [5]. Yin simulated the influence of train speed, wheel

© Springer Nature Switzerland AG 2020
Q. Liu et al. (Eds.): CENet 2018, AISC 905, pp. 135–143, 2020.
https://doi.org/10.1007/978-3-030-14680-1_16

polygon amplitude and its order on the wheel/rail vertical force [6]. Wu revealed that the impact load generated by the polygonal wear can lead to the resonance of the wheelsets [7]. Almost all the researches are based on the multibody dynamic method rather than the element finite method. There are two reasons of this situation. The first reason is that element finite method is time consuming compared with the multibody dynamic method. Another reason is that researchers pay more attention to whole structure than flexibility of each component. The disadvantage is obvious that many high frequencies components and their effects are ignored and the transient contact effect is also neglected.

Fig. 1. The different orders of wheel polygon: (a) High-order polygon; (b) Low-order polygon.

In order to study the dynamic characteristics of wheel polygon, a nonlinear finite element model to simulate the physical mechanisms of wheel/rail rolling contact is built. The parallel computing is used to improve the calculation speed of the element finite method. The finite element method is a numerical method derived from variational principle and widely used by people. All the model is based on the Newton's second law in the discrete form. This model accounts for the vehicle/track system coupling dynamics and the finite element dynamic analysis code ABAQUS is used. The high frequencies components and the transient contact effects are taken into account. The study would discuss the effects of wheel polygon with different speeds on the vehicle/track system dynamic response.

2   Dynamic Finite Element Model of Polygonal Wheel

2.1   Finite Element Model

There is a polygon phenomenon in some wheels of the high-speed railway as shown in Fig. 2. It is an irregularity along the circumferential direction of the wheel. According to the statistics of the fault wheelset, the wheel polygon is usually about 20 orders and the roughness level of the high order polygon is generally more than 25 dB. As can be seen from Fig. 2, the polygon roughness of the wheel presents a simple harmonic vibration, and therefore a sinusoidal curve with the amplitude of 30 dB is selected in this paper to fit the 20-order wheel polygon.

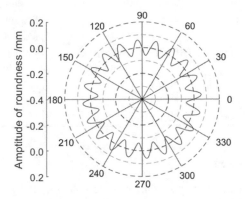

Fig. 2. 20-order wheel polygon.

For the high-speed railway track/vehicle systems, one three-dimensional wheelset with polygon abrasion and the ballastless track are selected as shown in Fig. 3. The wheel rolling diameter is 860 mm and the rail profile is CN60 with a 1/40 inclination. In order to facilitate the establishment of the wheel polygon model, the wheel rim is simplified as a cylinder. The distance between the two sleepers is 0.65 m. The system structure and material property parameters selected in the study are as shown in Table 1. The bottom of the mortar layer is fixed. The 8-node reduced linear integral element is used during the calculation of ABAQUS software. The minimum element with the size of 1.25 mm is used in the initial wheel/rail contact surface. The total number of finite element model nodes are approximately 2.0×10^6. The Cartesian coordinate system is used for the global coordinates. The origin of the coordinates is the initial position of the wheel/rail contact, the X-axis is the longitudinal direction which is along the rolling direction, the Y-axis is the longitudinal direction and the Z-axis is the vertical direction.

Fig. 3. Finite element model: (a) Vehicle/track model; (b) Wheel/rail contact zone.

Table 1. Parameters of the model.

Component	Parameter	Value	Component	Parameter	Value
Friction coefficient	f	0.5	Ballast material	Young's modulus	34.5 GPa
Lumped sprung mass	Mc	8000 kg		Poisson's ratio	0.25
				Mass density	2400 kg/m³
Primary suspension	Stiffness	0.88 MN/m	Wheel/rail material	Young's modulus	205.9 GPa
	Damping	4 kN s/m		Poisson's ratio	0.3
Rail fastening	Stiffness	22 MN/m		Mass density	7790 kg/m³
	Damping	200 kN s/m			

2.2   The Passing Speed

When the wheel is a 20-order polygon, we can calculate the passing frequencies by Eq. (2.1) based on different passing speeds of the vehicle.

$$f = \frac{nv}{3.6\pi d} \tag{2.1}$$

where f is the passing frequency, and v is passing speed. The diameter of wheel is d and n is the order of polygonal wheel. Then we work out the frequencies corresponding to the speeds of 200 km/h, 240 km/h, 280 km/h and 320 km/h, respectively, as shown in Table 2.

Table 2. Passing frequency at different speeds.

Speed [km/h]	200	240	280	320
Passing frequency [Hz]	411.25	493.50	575.75	658.00

2.3   Parallel Computing

The degrees of freedom of established model reach up to 600000, so we need to speed up the solve process. One of the effective methods is to adapt the parallelization algorithm. Parallelization in Abaqus/Explicit is implemented in two different ways: the domain-level method and the loop-level method [8]. The domain-level method divides the model into lots of domains and assigns each domain to one processor. The loop-level method parallelizes at a more basic level loops that are responsible for the most of the computational cost. The majority of the low-level parallelized routines are made up of the element, node, and contact pair operations. We establish the computer clusters and choose the second method.

The domain-level method breaks the model up into a number of domains. These domains are regarded as the parallel domains to discern them from other domains during the analysis. The domains are submitted evenly among all the processors.

The analysis is then implemented independently in every domain and the all information passes between the domains in each increment by MPI-based parallelization modes.

3 Calculation Result Analysis

3.1 Vertical Contact Force

From the calculation, we can obtain the vertical contact force at different speeds in Fig. 4(a). The location of the fastening is identified in the figure. The peak-peak value is another value we care about which represents the difference between the maximum and the minimum values in one period. It can also characterize the intensity of vibration. The maximum peak-peak values are shown in Fig. 4(b).

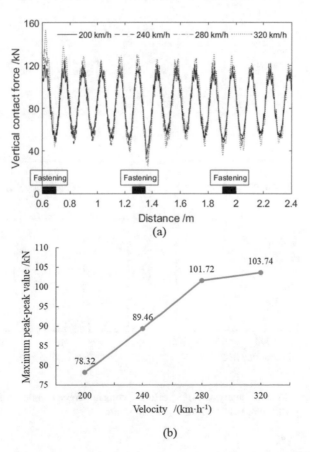

(a)

(b)

Fig. 4. Vertical contact force under different speeds: (a) The time-domain data; (b) Maximum peak-peak value.

The wheel/rail vertical contact force appears periodic vibration with the 20-order wheel polygon and the vibration period is 1/20 of the wheel circumference. The vibration increases with the increase of the speed. The peak-peak values of vertical contact force near the fastenings are larger, because the rail is discretely supported and the vertical stiffness near the fastening is greater. The maximum peak-peak values are shown in Fig. 4(b) at different speeds. We can divide the value in the graph into two parts. The first part is from 200 km/h to 280 km/h which maximum peak-peak values are almost proportional to passing speed and increases linearly by about 3 kN per 10 km/h. The second part starts with 280 km/h. The growth rate of maximum peak-peak values begin to slow down, which indicates that the amplitudes increase of the vibration and the influence of polygonal wheels are gradually reduced.

The power spectral density analysis is carried out to explore the vertical contact force in frequency domain as shown in Fig. 5 and Table 3. The power spectral density indicates the distribution of energy in the frequency domain.

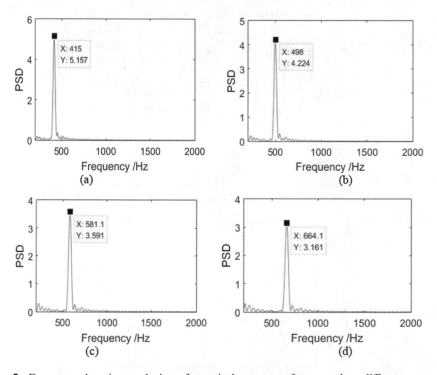

Fig. 5. Frequency-domain analysis of vertical contact force under different speeds: (a) 200 km/h; (b) 240 km/h; (c) 280 km/h; (d) 320 km/h.

Table 3. Main frequency of vertical contact force under different speeds.

Speed [km/h]	200	240	280	320
Main frequency [Hz]	415.0	498.0	581.1	664.1

From Fig. 5, we can see that the most of the energy is concentrated in a single dominant frequency for each passing speed and the energy at other frequencies is much smaller than the energy at the dominant frequencies. These frequencies can correspond to the polygonal wheels passing frequencies at different passing speeds. The reason for this phenomenon is that the polygonal wheel provides a vertical excitation resulting in a vertical force.

3.2 Longitudinal Contact Force

The longitudinal contact forces at different speeds in time domain are as shown in Fig. 6. The location of the fastening is identified in the figure.

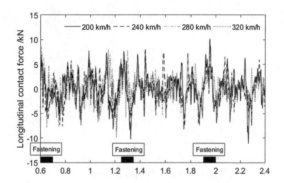

Fig. 6. Longitudinal contact force under different speeds.

The wheel/rail longitudinal contact force is more complex than the vertical. The vibration of the longitudinal contact forces near the fastenings are serious. We also performed power spectral density analysis on the longitudinal forces and obtained different results from the vertical force as shown in Fig. 7 and Table 4.

From the Fig. 7, we can see that there are several main frequencies in the longitudinal contact force. It means that the energy of the corresponding passing frequency of the polygonal wheels are still the highest, but they have no dominant position. This indicates that the energy of the longitudinal force is not concentrated on a single dominant frequency. The probable cause is that the contact model is not a linear model and vertical forces at a single frequency can lead to longitudinal forces at different frequencies and redistribute energy at different frequencies. Many components of longitudinal forces at high frequencies are exited which is neglected by the previous method. As the speed increases, the components of the vibration energy of the other main frequency increases and will affect the development of the wheel polygon.

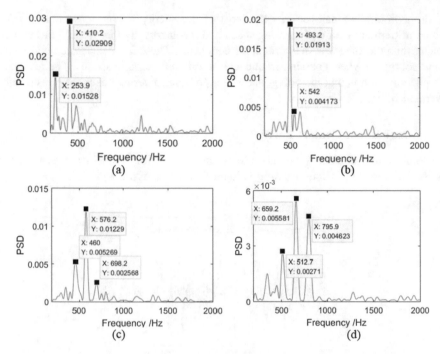

Fig. 7. Frequency-domain analysis of longitudinal contact force under different speeds: (a) 200 km/h; (b) 240 km/h; (c) 280 km/h; (d) 320 km/h.

Table 4. Main frequency of longitudinal contact force under different speeds.

Speed [km/h]	200	240	280	320
Main frequency with peak PSD [Hz]	410.2	493.2	576.2	659.2
The other main frequency [Hz]	253.9	542.0	460.0 and 698.2	512.7 and 795.9

4 Conclusion

In order to study the influences of the wheel polygon on the vehicle/track system coupling dynamics, a three-dimensional finite element model is built. Compared the physical mechanisms differences of high-speed wheel/rail interaction with different speeds, the following conclusions are drawn:

(1) The vertical contact force vibration increases with the increase of passing speed. Due to the track vertical stiffness irregularity, the peak-peak value of the vertical contact force near the fastening is larger. It increases linearly by about 3 kN per 10 km/h when the passing speed is above 200 km/h and then grows slowly with the passing speed above 280 km/h.

(2) The power spectral density analysis results of vertical contact force reveal that dominant frequencies are almost the same as passing frequencies at different passing speeds. Most of the energy are concentrated at the dominant frequencies

and other frequencies share very little energy which prove that the influence of the wheel polygon on the vertical force is relatively straightforward.

(3) The longitudinal contact force vibration is more complex than the vertical in time domain. The power spectral density analysis results demonstrate that the energy distributions are scattered in spite of the most of the energy still concentrated at the dominant frequencies. The reason of the phenomenon is that rail/wheel contact model is not a linear system. The single frequency normal force will stimulate multi-frequency longitudinal force and affect the development of the wheel polygon.

Acknowledgement. This study is supported by CHINA RAILWAY Scientific and Technological Research and Development Project (Contract No. 2017J003-B).

References

1. Tao, G., Wang, L., Wen, Z., Guan, Q., Jin, X.: Measurement and assessment of out-of-round electric locomotive wheels. J. Rail Rapid Transit **232**(1), 275–287 (2018)
2. Cui, D.B., Lin, L., Song, C.Y.: Out of round high-speed wheel and its influence on wheel/rail behavior. J. Mech. Eng. **49**, 8–16 (2013)
3. Li, D.D., Dai, H.Y.: Research on wheel polygonization frequencies based on modal analysis of rail. Railw. Locomot. Car **37**(4), 6–11 (2017)
4. Song, D.L., Wu, H.L., Zhang, W.H., Xu, T.Y., Jiang, Y.N., Qi, X.Y.: Study on evolvement rule of the polygonal wear of EMU wheels and its effect on the dynamic behavior of a vehicle system. In: Proceedings of the 25th Symposium of the International Association of Vehicle System Dynamics, pp. 1095–1101. CRC Press, London (2018)
5. Liu, W., Ma, W., Luo, S., Li, X.L.: Research on influence of wheel vibration and wheel polygonization on wheel-rail force in consideration of wheelset elasticity. J. China Railw. Soc. **35**(6), 28–34 (2013)
6. Yin, Z.K., Wu, Y., Han, J.: Effect of polygon wear high-speed train wheels on vertical force between wheel and rail. J. China Railw. Soc. **39**(10), 26–32 (2017)
7. Wu, X., Chi, M., Wu, P.: Influence of polygonal wear of railway wheels on the wheel set axle stress. Veh. Syst. Dyn. **53**(11), 1535–1554 (2015)
8. Yamileva, M., Yuldashev, A.V., Nasibullayev, I.S.: Comparison of the parallelization efficiency of a thermo-structural problem simulated in SIMULIA Abaqus and ANSYS mechanical. J. Eng. Sci. Technol. Rev. **5**(3), 39–43 (2012)

Sea Land Air Selabot Swarm Design Considerations

Tong Xu[1], Jiamin Huang[2(✉)], Mengyuan Zhanghu[3], Mengting Xu[3], Gao Li[3], and Jun Steed Huang[3(✉)]

[1] Jingjiang College of Jiangsu University, 301 Xuefu, Zhenjiang 212000, China
[2] Haylion Technologies Co., Ltd., Shenzhen 518000, China
[3] Southern University of Science and Technology,
1088 Xueyuan, Shenzhen 518055, China
huangj@mail.sustc.edu.cn

Abstract. To provide a platform that can change shape, status, and form for the rapid development of Artificial Intelligent, we describe a design concept of aeroamphibious unmanned boat-car-plane combination robot and its development plan. Considering the facts of the design contradictions among aerodynamics, hydromechanics, and powertrain, this design focuses on the land collaborative algorithm as well. We divided the design tasks into the aeroamphibious robot with sensor, action and communication parts. For the sensor module and communication, we prepare basic sensor and keep other sensor communication interfaces available. As the robot action, we choose roller wheel and proposed a three-mode transportation platform with only one engine. We follow open source strategy of both software and hardware to make the robot develop quick, upgradeable, expandable. We introduced a new multiple-party formula that was used to coordinate more than two moving platforms, which enable it meets both academic research and market development needs.

Kerwords: Cyclocopter · Swarm · Artificial · Intelligence · Amphibian

1 Introduction

An adjustable sensor carrier platform on both body shape and hardware function is never done in the history, whose function is very important for testing of swarm intelligence algorithm and environment interacting ability. Meanwhile, applications like volcanic gas collection, glacier polar research, central air conditioner cleaning, and infectious disease object transportation, coal mine flood and fire controlling or even intelligent educational toy, it has a board vertical markets.

The platforms we had now all have their limits, e.g. e-puck is a bit outdated for software development and debug; Kilobot is newer, but has limited movement and sensors which cannot be expanded; Lego has much more sensors, the development tools evolved constantly, but based on the traditional closed source industrial computer LabView framework, which is against the popular ROS/LINUX structure, finally Turtlebot has the whole package of development and LINUX structure but lack of

© Springer Nature Switzerland AG 2020
Q. Liu et al. (Eds.): CENet 2018, AISC 905, pp. 144–153, 2020.
https://doi.org/10.1007/978-3-030-14680-1_17

acting ability, cannot get into water or fly in the sky. The drone of DaJiang can fly but cannot swim, and does not have open source hardware.

These platforms cannot support the sensor adding all kinds of SI (Swarm Intelligent) group cooperation algorithm and action ability [1]. As such we hope to mimic the develop mode of Raspberry Pi, use the way of open source on both software and hardware, to contribute to AI development. Not only the department of academic research can do more exploration and at temptation based on ours platform but also the relevant industries in the market can sell products based on this platform.

We proposed a three mode transportation platform with only one engine that has never been done on the recorded history of human being. And we introduced a new multiple party formula that is used to coordinate more than two moving platforms, that is also never done in the past.

2   Feasibility Analysis

We design an aero amphibious robot with sensor, action and communication parts separately, meanwhile, keep the upgradeable of hardware so that everyone can explore it in their own favorite way.

For the sensor module, we use the hearing sensor, smelling sensor, and vision sensor and keep other sensor interfaces like the speaker, indicator available. We've chosen three options for communication as well, there are ZigBee for science research, Z-Wave for toy market and Wi-Fi for indoor transportation. And we will keep other communication interfaces open as well.

We choose roller wheel to realize robot action, the main power is from the direct current motor. Instead of regular reaction lift force, roller blade plane using the upper wing to provide pressure lift force [2]. It is similar to the normal fixed-wing airplane, different from helicopter rotor drone and multiple rotor drones, it only has the problem of balance adjustment in a right-left way and doesn't need the complex algorithm to balance the front-back way. The fixed-wing drone and helicopter are both flat type spin, while roller blade wheel is a stereoscopic type called cyclocopter [3]. Lift force of the former is vertical to its action surface, while the latter is parallel to the action surface, so the cyclocopter is a stable gyroscope itself. At the same time, the wings below gives low gravity center force, makes it much stable than the flat types in a strong wind. Just keep the axial of cyclocopter against the wind direction to make sure it flies or hovers at sky for stability. There is an animation about the theory of cyclocopter:

https://en.wikipedia.org/wiki/Cyclogyro#/media/File:Cyclogyro-Mechanics.gif.

Roller wheel cyclocopter with round fences can be used as wheels for robot to move on the ground, it saved the additional landing gear and wheels to reduce the robot weight, not just which, if we change the up looking angle of the blade, it can just replace the modern ship propeller as classic ship power paddler. And we get three usages with one wheel and one motor, just change the switchgear then you will get a robot that can fly, run and swim at one shot.

Selabot uses GPS/DR (calculate with information of chassis like wheel speed difference) combination to make up the lack of GPS indoors. DR is useful because it is good at tracking the course of the robot (the turning angle difference between the left and right wheels as the track angle of the robot itself) [4]. When GPS signal remains bad for a long time, the locating error would accumulate, we use collaborative average in statistics to correct it. Every single robot report position of itself and the neighbor robot it catches, and then calculate the new position, and iterative the steps above until each location is most accurate [5]. We make robots cooperate using the function of Finsler Geometry dynamic window approach:

$$G(v, w) = total(a * heading(v, w) + b * dist(v, w) + c * velocity(v, w) + d * cola(v, w))$$

$$dist = (x^p + y^p)^{\frac{1}{p}}$$

$$p = 1.5 + \frac{1}{1 + \frac{1}{e^{2*(DE*ds)}}}$$

$$DE = (x^2 + y^2)^{\frac{1}{2}}$$

$$cola(v, w) = \left| \frac{\left(\sum_{i=1}^{n} (v_i - \bar{v})^{\frac{3}{4}} (w_i - \bar{w})^{\frac{3}{4}} \right)^{\frac{4}{3}}}{\left(\sum_{i=1}^{n} (v_i - \bar{v})^{\frac{3}{2}} \sum_{i=1}^{n} (w_i - \bar{w})^{\frac{3}{2}} \right)^{\frac{2}{3}}} \right|$$

In above formula, dist stands for Finsler Geometry Distance variable, dimension p said everything from zero to infinite, and DE is Euclidean distance. Cola is collaboration measure theory put forward by Jun Steed Huang in 2015. Where v is linear velocity, w is angular velocity, x and y is the distance, a, b, c, d are the weight, between 0 and 1.

a, b, c are weight coefficient and the best value of a, b, c are 0.05, 0.07, 0.1. The best values are obtained by a huge amount of simulations done by many researchers in the field of robot path planning. According to the geometric relations on a, b, c, we set a series value of d (range from 0.05 to 1), the table below shows the above swarm calculation converges time and parameters used in the simulation, for different d.

In the following simulation, we generate a series of robots, destinations, obstacles. Through cooperation robots can find the best way to destination, wasting the shortest time, avoiding all the obstacles. The five small squares represent dynamic obstacles, the small circle represent destination, the green line represents the robot way. Robots start from the origin move to the end of the destination with constant displacement of the square, constantly adjust the direction in the process.

As we can see from pictures the most harmonic coordination is achieved when d is 0.28 in this 5 robot case. In the process, some of the robots change the goal, by coordinating with neighbor robot (Figs. 1, 2 and Table 1).

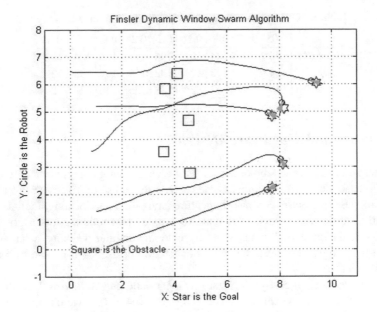

Fig. 1. The simulation when d = 0.05.

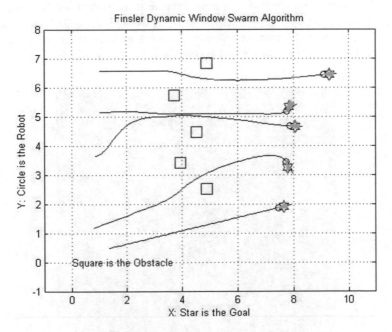

Fig. 2. The simulation when d = 0.28.

Table 1. Convergence time of swarm *vs.* the coordination weight.

d	0.05	0.07	0.1	0.14	0.2	0.28	0.4	0.58	0.83
First (S)	98	98	104	101	106	97	101	104	102
Last (S)	134	129	132	127	132	122	137	132	131
Delta (S)	36	31	28	26	26	25	36	28	29

3 Software and Hardware Design

The smell sensor uses the smoke sensor MQ-2 as poison gas represent TVOC sensor for sea volcano, outdoor coal mine, home renovation and so on [6]. We use SEC290 as 9 axles full gesture electric compass in gesture navigation and locating, includes 3 axles acceleration to do line velocity calculation, 3 axles magnetic sensor to do magnetic azimuth calculation and 3 axles gyroscope to do angular velocity calculation, we combine them to get the trace. The chip has EKF filter itself and is able to satisfy our accuracy demand.

The platform designed by University of Maryland can fly and roll on the ground [7]. Their 900 g roller wheel drone is sized $0.3 \times 0.4 \times 0.5$; we change it to around $0.2 \times 0.4 \times 0.7$ and 1000 g, and focusing more on skidding capability on water. It could only fly when the volume is in proportion to its weight or we can say it as effective density can't bigger than 10 times of the air density, or it will take bigger than 10 times of energy to overcome the air friction, this would bring too much speed to the upper wings and design pressure of eccentric connecting string arm, it would square the noise and the efficiency would get down directly [8] (Figs. 3 and 4).

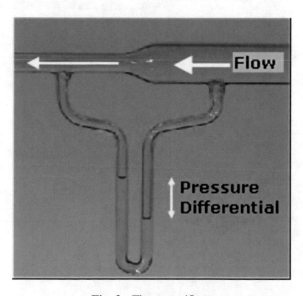

Fig. 3. The venturiflow.

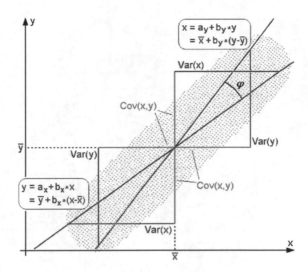

Fig. 4. Regression lines.

The skeleton uses carbon fiber completely, with slides and sticks, the connect parts use resin completely and the wheels are 3D-printed using nylon, to reduce the weight and keep the strength [9]. All the waterproof cover use heat conduction silica gel slices entirely to ensure the cooling, waterproof, shockproof and dustproof. The power motor use 30 A/5 A DC Sky Walker and the battery use 3 S 11.1 V 2.2 h 30 C lithium battery.

The landing gear on water uses stable trapezoidal to steady the square master frame, like the hovercraft, just we install the float balls on the left and right side and using the trapezoidal to maintain the balance of front and back. Using float balls instead of barrels to add the best surface area to volume ratio and the total of the fluid balance pivot, reduce the weight and lower the resistance of flying. We drop the idea of movable landing support for now because the steering engine would add lift burden while the water float balls already had some weight. We can make them roll to the special landing gear seat on the ground and flies one by one. We can add moving the part on square master frame later on to adjust the ball position as the lifter (Fig. 5).

For the CPUs of the robot, we choose to design them separately, there is an open source fly controller AVR which can run on 8-bit CPU named ATMEGA2560 as the epencephala to control the motor PWM, the 64-bit Raspberry Pi serial as the main brain to support ROS sensor and communication algorithm. The table below compared between several CPUs.

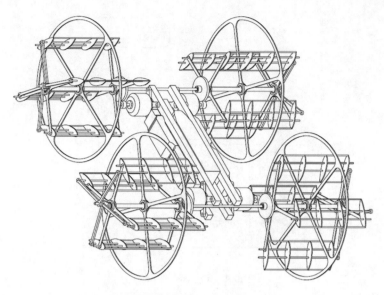

Fig. 5. Selabot structure.

For the next version, we will consider designing our own PCB board, the main idea of it is to reduce the weight and unnecessary plug-ins, make it highly integrated to ensure the swim and take off ability and energy efficiency.

The first hardware version is 3D-printed and handmade. The software has three modules: sensor, controller, and communication. Sensor module reads two sensors reading, compares their strength and order to reckon the source of sound, smoke or light. To control the speed difference between the left and right wheels to make selabots get close to or far away from the sensed signal source. Software structure uses the simple open source ROS to dock with the cloud simulation platform, to connect the mirror bots, virtual bots or self-mixed bot team (Figs. 6, 7 and Table 2).

Table 2. Comparison of different CPUs.

CPU	BCM Raspberry Pi	TI Cortex	Intel Edison	Arduino serial
Bit	64	32	32	8
Interface	HDMI	Ethernet	FPGA	GPIO
Wireless	WiFi	none	none	none
Power consumption	360 mW	300 mW	625 mW	75 mW
Software	ROS	Android	Ubuntu	LINUX

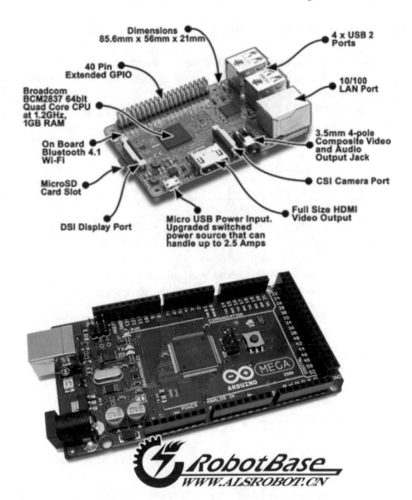

Fig. 6. Selabot main brain and epencephala.

Fig. 7. Selabot prototype.

4 Market Research and Conclusion

We have proposed an open hardware and software for aeroamphibious unmanned robot platform with just one engine, this design has never been done on the recorded history of human being, which focusing on swarm intelligent study. The platform is good for skidding on water, collecting sea volcano gas and similar applications. We will increase the number of wings to make it bulletproof (Table 3).

Table 3. The list of some similar products.

Name	Company website
e-puck	http://www.e-puck.org/
Kilobot	https://www.k-team.com/mobile-robotics-products/kilobot
Lego EV3	https://www.lego.com/zh-cn/mindstorms/about-ev3
Turtlebot3	http://www.turtlebot.com/
Matrice100	https://www.dji.com/cn/matrice100
mBot	http://store.makeblock.com/product/mbot-robot-kit
Root	http://www.codewithroot.com/root

Acknowledgments. We would like to thank Prof. Yuhui Shi for offering C-BSO code, Prof. Hongwei Shi, Haiqin's manager of Mr. Qian for funding, Mr. Fan Yang, Mr. Jian Sun, Mr. William Chen for 3D print drawing, Prof. Bin Chen's support, Miss Lijun Sun, Mr. Dabu Zhang for offering Market research chart, Miss Elena Shrestha and Mr. Vikram Hrishikeshavan for helps on dimensioning.

References

1. Cheng, S., Chen, J.F., Lei, X.J., Shi, Y.H.: Locating multiple optima via brain storm optimization algorithms. IEEE Access **99**, 1 (2018)
2. Tang, J.W., Hu, Y., Song, B.F.: A detailed aerodynamic analysis of the cyclorotor using dynamic mesh in CFD. In: 52nd Aerospace Sciences Meeting, AIAA SciTech Forum, (AIAA 2014-0406) (2014)
3. Runco, C., Benedict, M., Coleman, D.: Design and development of a Meso-scale cyclocopter. In: AIAA SciTech Forum (2015)
4. Pei, D., Qin, D.G.: Research on non-linear fault-tolerant filtering for GPS/DR integrated navigation system. In: IEEE International Conference on Communication Software and Networks, pp. 231–235, IEEE, Beijing, China (2016)
5. Eiben, A.E., Smith, J.E.: Introduction to Evolutionary Computing. Springer, Heidelberg (2015)
6. Chen, C.J., Chen, B., Li, Z., Wang, Z.Q.: Important role of magma mixing in generating the Mesozoic monzodioritic-granodioritic intrusions related to Cu mineralization, Tongling, East China: evidence from petrological and *in situ* Sr-Hf isotopic data. Lithos **248–251**, 80–93 (2016)

7. Shrestha, E., Williams, B., Hrishikeshavan, V., Chopra, I.: Development of MAV-scale quad-cyclocopter capable of aerial and terrestrial locomotion. In: AHS International Technical Meeting 7th VTOL Unmanned Aircraft Systems Program, Mesa, AZ, vol. 7 (2017)
8. Monteiro, J.A.L., Páscoa, J., Xisto, C.: Aerodynamic optimization of cyclorotors. Aircr. Eng. Aerosp. Technol. **88**(2), 232–245 (2016)
9. Li, Y.L., Wang, S.J., Wang, Q., Xing, M.: A comparison study on mechanical properties of polymer composites reinforced by carbon nanotubes and graphene sheet. Compos. B Eng. **133**, 35–41 (2018)

TDOA Time Delay Estimation Algorithm Based on Cubic Spline Interpolation

Tianfeng Yan and Yu Zhang[✉]

School of Electronic and Information Engineering, Lanzhou Jiaotong University,
Anning District, Lanzhou 730070, Gansu, China
zhangyu_lzjt@163.com

Abstract. In order to solve the fence effect caused by the fixed number of sampling points in the passive time difference positioning system in practical applications, a TDOA time delay estimation algorithm based on cubic spline interpolation is proposed. The algorithm uses cubic spline interpolation to interpolate the spectral peak curve of the cross-correlation function. While ensuring the stability and convergence, the spectral peak curve is smoother, the accuracy of the peak value is improved, and the accurate time delay estimation of the signal source is further obtained. The measured results show that the TDOA time delay estimation method based on cubic spline interpolation can solve the fence effect well and obtain the time delay estimation value accurately. This method can use the algorithm to improve the time delay estimation accuracy under the condition that the hardware sampling rate is fixed, thus reducing the dependence of passive time difference positioning on the system hardware.

Keywords: TDOA · Fence effect · Cubic spline interpolation · Sampling rate

1 Introduction

Because of its wide application in the fields of electronics and aerospace, passive time difference positioning technology has the characteristics of small size and high concealment [1], which has become a hot research field. Passive time difference positioning technology can be applied to civil signals such as FM, AM [2], mobile phone signals, Wifi [3] and so on. The methods of passive time difference positioning mainly include: Received Signal Strength (RSS) [4], Arrival of Angle (AOA) [5, 6], Time of Arrival (TOA) [7], Time Difference of Arrival (TDOA) [8], Frequency Difference of Arrival (FDOA) [9] and so on. In this paper, relying on the existing TDOA system, aiming at the fence effect caused by the fixed number of sampling points in the real environment, the cubic spline interpolation method is used to improve the accuracy of the spectrum peak value, thereby improving the time delay estimation accuracy of the signal target.

© Springer Nature Switzerland AG 2020
Q. Liu et al. (Eds.): CENet 2018, AISC 905, pp. 154–162, 2020.
https://doi.org/10.1007/978-3-030-14680-1_18

2 Principle of Generalized Cross-Correlation

2.1 TDOA Signal Model

Assume that the receiving signals of TDOA delay estimation for two antennas is

$$\begin{cases} x_1(n) = y(n) + n_1(n) \\ x_2(n) = Sy(n-d) + n_2(n) \end{cases} \tag{1}$$

where $y(n)$ is the radiation source signal; $x_1(n)$ and $x_2(n)$ are the signals received by two monitoring stations; $n_1(n)$ and $n_2(n)$ are the ideal mean Gaussian white noise, and assume that the noise and the radiation source are not related to each other; S is the attenuation factor; d is the time delay.

2.2 Generalized Cross-Correlation

The basic algorithm for finding TDOA delay is to find the cross-correlation function for $x_1(n)$ and $x_2(n)$. The cross-correlation function $R_{12}(\tau)$ of the two signals can be expressed as

$$\begin{aligned} R_{12}(\tau) &= E[x_1(n)x_2(n-\tau)] = SE[y(n)y(n-D-\tau)] + E[y(n)n_2(n-\tau)] \\ &+ SE[s(n-d-\tau)n_1(n)] + E[n_1(n)n_2(n-\tau)] \end{aligned} \tag{2}$$

Since the radiation source signal and the noise are not related to each other, besides the noises are mutually opposed, Eq. (2) can be simplified as

$$R_{12}(\tau) = SE[y(n)y(n-d-\tau)] = SR_{yy}(\tau - d) \tag{3}$$

$R_{yy}(\tau)$ represents the degree of correlation between the two signals. Since the signal received by the receiving station is the same signal at different times, the characteristics of the signal do not change. According to the nature of the cross-correlation function, the two signals do cross-correlation and have the greatest correlation at $R_{yy}(0)$, so the maximum value of $R_{12}(\tau)$ can be obtained when $\tau - d = 0$ and the time difference d can be obtained as the abscissa point corresponding to the maximum value of $R_{12}(\tau)$ [10, 11].

3 Principle of Cubic Spline Interpolation

The cubic spline interpolation has good stability and convergence [12], assuming that $f(x)$ is a continuously differentiable function in interval $[m, n]$ and given in this interval:

$$m = x_0 < x_1 < x_2 < \cdots < x_n = n \tag{4}$$

Make function $y(x)$ meet the following conditions:

(1) In interval $[x_i, x_{i+1}]$ $(i = 0, 1, 2, \cdots, n-1)$, the maximum number of polynomials $y(x)$ is q;
(2) $y(x)$ has $q - 1$ continuous derivatives in $[m, n]$. $y(x)$ is define as a q-spline function in $[m, n]$, where $x_0, x_1, x_2 \cdots$ is a spline node. The boundary nodes are x_0, x_n and the remaining $x_1, x_2, \cdots, x_{n-1}$ are the inner nodes. When $p = 3$, $y(x)$ is called a cubic spline interpolation function.

Assume that the cubic spline function $f(x)$ in subinterval $[x_{i-1}, x_i]$ is:

$$f(x) = f_i(x) = a_i x^3 + b_i x^2 + c_i x + d_i, x \in (x_{i-1}, x_i), i = 1, 2, \cdots, n \qquad (5)$$

Where a_i, b_i, c_i, d_i is the undetermined coefficient and reaches the insertion condition:

$$f(x_i) = s(x_i), (i = 0, 1, \cdots, n) \qquad (6)$$

According to the definition of cubic spline interpolation, $f(x)$ can be found after solving the four undetermined coefficients in subinterval $[x_{i-1}, x_i]$, and the number of subintervals is n, so $4n$ coefficients need to be determined. Since the second derivative of the function is continuous in interval $[m, n]$, the following conditions must be satisfied at the spline junction:

$$\begin{cases} f(x_i - 0) = f(x_i + 0) \\ f'(x_i - 0) = f'(x_i + 0), (i = 1, 2, \cdots, n-1) \\ f''(x_i - 0) = f''(x_i + 0) \end{cases} \qquad (7)$$

From Eqs. (6) and (7), it can be seen that the existing condition number is $4n - 2$, and two conditions are also needed to determine the function $f(x)$.

A boundary condition can be added at the endpoint $m = x_0, n = x_n$ of the interval $[m, n]$:

(1) The first derivative when $m = x_0, n = x_n$:

$$\begin{cases} f'(x_0) = s'(x_0) = m_0 \\ f'(x_n) = s'(x_n) = m_n \end{cases} \qquad (8)$$

(2) The second derivative when $m = x_0, n = x_n$:

$$\begin{cases} f''(x_0) = s''(x_0) = M_0 \\ f''(x_n) = s''(x_n) = M_n \end{cases} \qquad (9)$$

a natural boundary condition when $M_0 = M_n = 0$:

$$f''(x_n) = s''(x_n) = 0 \tag{10}$$

(3) If $s(x)$ is a function whose period is $x_n - x_0$, then $f(x)$ is also a periodic function, which satisfies:

$$\begin{cases} f(x_0 + 0) = f(x_n - 0) \\ f'(x_0 + 0) = f'(x_n - 0) \\ f''(x_0 + 0) = f''(x_n - 0) \end{cases} \tag{11}$$

From Eq. (6), we can determine that the function is a periodic spline function of $f(x)$, so $f''(x)$ is a linear function.

Make $l_k = f''(x_k)$ $(k = 0, 1, 2, \cdots, n)$, and set $x \in [x_k, x_{k+1}]$, then the linear function of the points of (x_k, l_k) and (x_{k+1}, l_{k+1}) can be expressed as:

$$f''(x) = l_k \frac{x_{k+1} - x}{h_k} + l_{k+1} \frac{x - x_k}{h_k} \tag{12}$$

where $h_k = x_{k+1} - x_k$.

In the interval $[x_k, x_{k+1}]$, two ends of Eq. (12) are integrated two consecutive times:

$$\begin{aligned} f(x) = &l_k \frac{(x_{k+1} - x)^3}{6h_k} + l_{k+1} \frac{(x - x_k)^3}{6h_k} [\frac{x_{k+1} - x}{h_k} - \frac{h_k}{6}(l_{k+1} - l_k)](x - x_k) \\ &+ x_k - l_k \frac{h_k^2}{6} \end{aligned} \tag{13}$$

Therefore, it determines l_k, and also determines the cubic spline interpolation function for each interval.

The first derivative of $f(x)$ in $[m, n]$ is continuous:

$$f(x - 0) = f(x + 0) \tag{14}$$

Therefore:

$$\frac{6}{h_{k-1} + h_k} [\frac{x_{k+1} - x}{h_k} - \frac{x_k - x_{k-1}}{h_{k-1}}] = 2l_k + \frac{h_{k-1}}{h_{k-1} + h_k} l_{k-1} + \frac{h_k}{h_{k-1} + h_k} l_{k+1} \tag{15}$$

The number of equations is $n - 1$, and the number of l_k is $n + 1$. To solve this problem, two equations are still needed. In this paper, we choose the natural boundary condition of Eq. (10), solving the $n + 1$ equations jointly to obtain l_k, and determining the spline function $f(x)$.

In summary, the steps for the cubic spline interpolation calculation are:

a. determine the boundary conditions;
b. select the appropriate boundary conditions and calculate the boundary values to obtain the equations;
c. solve equations to get l_k $(k = 0, 1, 2, \cdots, n)$;
d. substitute l_k from $f(x)$ to get an approximation of the function point in interval $[m, n]$.

4   Experiments and Analysis

4.1   Experimental Equipment and Experimental Design

The main experimental equipment is: TRS-3600T small radio monitoring station, BDS100 MHz time synchronization system module, QF1480C RF signal generator.

The experimental site marked by three monitoring sites and a radiation signal source point shown in Fig. 1.

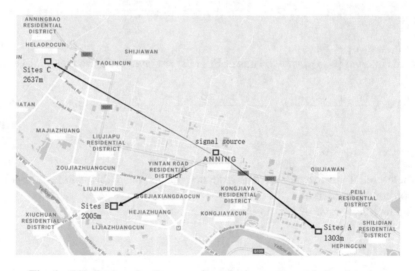

Fig. 1. Distribution site diagram of monitoring station and signal source.

4.2   Experimental Results and Analysis

The distance difference between the signal source and the monitoring stations A and B is 703 m. Generally, we think approximately that the propagation speed of the radio signal in the air is equal to the speed of light, and the distance difference is converted into a time difference of 2.343 μs. The signal sampling rate of the monitoring station is 2.5 MHz, so the delay per point is 0.4 μs. A total of 110 packets of data were measured in this experiment, including 2048 sampling points for each packet of I/Q data.

Single Package Data Analysis

Select one of the packages and use the generalized cross-correlation method to obtain the cross-correlation function which is shown in Fig. 2.

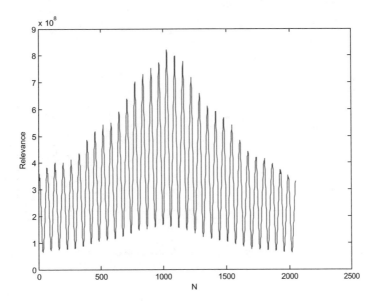

Fig. 2. Cross-correlation function of single package data.

As can be seen from Fig. 3, the peak point is located at 1031 points, which translates into a delay of 2.4 µs. Due to the apparent jitter of the peaks, three points on the right and left of the spectrum peak were selected to perform the cubic spline interpolation. The number of interpolation points was 60 points. The obtained results are shown in Fig. 4.

It can be seen from Fig. 4 that the actual number of delay points after cubic spline interpolation is $1030 - (N/2 + 1) + (-0.2) = 5.8$, and the time delay after the conversion is 2.32 µs. From the figure we can see that neither 0 nor −1 corresponds to the peak point of the function. If we calculate according to the original cross-correlation method, we will mistake the point corresponding to 0 as the peak point, which obviously causes errors. The improved algorithm, through interpolation linear prediction, better restores the peak point of the cross-correlation function, which is closer to the actual delay.

Multi-packet Data Analysis

The time delay distribution statistics are made for the 110 packet data, and the results of cross-correlation delay estimation and cross-correlation delay estimation based on the cubic spline interpolation are shown in Fig. 5.

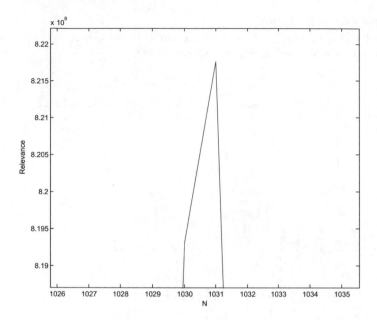

Fig. 3. Cross-correlation function of single package data (peak amplification).

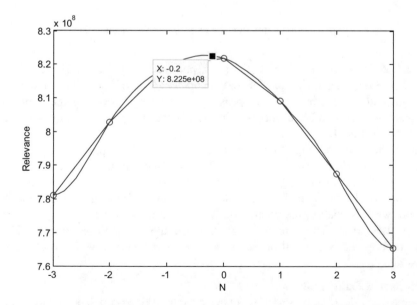

Fig. 4. Cross-correlation function diagram after cubic spline interpolation.

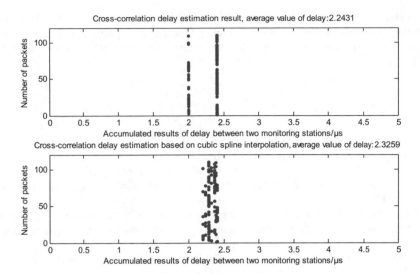

Fig. 5. Distribution of delay values for 110 packet data.

From Fig. 5 we can see that the cross-correlation delay estimation results based on cubic spline interpolation are more concentrated near the actual delay value. From the overall experimental situation, the improved algorithm can improve the accuracy of delay estimation. Compared to the cross-correlation delay estimation results, the error percentage of the time delay average of increases from 4.26% to 0.73%.

4.3 Comparison of Two Algorithm

When the sampling rate is fixed at 2.5 MHz, due to the problem of sampling point and the barrier effect, the peaks of traditional cross-correlation function are stable at 1030 points and 1031 points. The actual delay points are $0.4\,\mu s \cdot [1030 - (N/2 + 1)] = 2.0\,\mu s$ and $0.4\,\mu s \cdot [1031 - (N/2 + 1)] = 2.4\,\mu s$. Uncertainties in delay values at low sample rates can lead to a decrease in passive time difference positioning accuracy.

The improved algorithm uses cubic spline interpolation to make a linear prediction of the time delay value of the fence effect, and gets the time delay value closer to the real environment, so that it is concentrated in the range of 2 µs to 2.4 µs, which improves the time delay estimation accuracy better.

5   Conclusions

The TDOA time delay estimation algorithm based on cubic spline interpolation further enhances the accuracy of the spectrum peak by linearly approximating the value of the function near the peak, which largely solves the problem of time delay fluctuation due to the barrier effect, thereby improving the capability of time delay estimation of TDOA target positioning.

Acknowledgement. This study is supported by Youth Science Foundation of Lanzhou Jiaotong University under Grant No. 2018003, Scientific Research plan projects of Gansu Education Department under Grant (2017C-09), Lanzhou Science and Technology Bureau under Grant (2018-1-51).

References

1. Farina, A., Kuschel, H.: Guest editorial special issue on passive radar (Part I). Aerosp. Electron. Syst. Mag. IEEE **27**(10), 5 (2012)
2. Colone, F., Bongioanni, C., Lombardo, P.: Multifrequency integration in FM radio-based passive bistatic radar. Part I: target detection. Aerosp. Electron. Syst. Mag. IEEE **28**(4), 28–39 (2013)
3. Wu, J., Lu, Y., Dai, W.: Off-grid compressed sensing for WiFi-based passive radar. In: IEEE International Symposium on Signal Processing and Information Technology, pp. 258–262 (2017)
4. Weiss, A.J.: On the accuracy of a cellular location system based on RSS measurements. IEEE Trans. Veh. Technol. **52**(6), 1508–1518 (2003)
5. Nonsakhoo, W., Sirisawat, P., Saiyod, S., Benjamas, N.: Angle of arrival estimation by using stereo ultrasonic technique for local positioning system. In: IEEE International Colloquium on Signal Processing & ITS Applications, pp. 112–117 (2017)
6. Patzold, M., Gutierrez, C. A.: Modelling and analysis of non-stationary multipath fading channels with time-variant angles of arrival. In: Vehicular Technology Conference, pp. 1–6. IEEE (2017)
7. Nguyen, T.T.T., Denney, K.A., Syuhei, O., Nagao, Y., Kurosaki, M., Ochi, H.: High-accuracy positioning system based on ToA for industrial wireless LAN. In: NAFOSTED Conference on Information and Computer Science, pp. 37–41 (2017)
8. Meyer, F., Tesei, A., Win, M.Z.: Localization of multiple sources using time-difference of arrival measurements. In: IEEE International Conference on Acoustics, Speech and Signal Processing, pp. 3151–3155. IEEE (2017)
9. Adelipour, S., Hamdollahzadeh, M., Behnia, F.: Constrained optimization of sensors trajectories for moving source localization using TDOA and FDOA measurements. In: RSI International Conference on Robotics and Mechatronics, pp. 200–204. IEEE (2015)
10. Jing, S.Y., Feng, X., Zhang, Y.H.: Study of a generalized cross-correlation time delay estimation based acoustic positioning algorithm. Tech. Acoust. **33**(05), 464–468 (2014). (in Chinese)
11. Jin, Z.W., Jiang, M.S., Sui, Q.M., Yaozhang, S., Shizeng, L.U., Cao, Y., Cao, Y.Q., Zhang, F.Y., Jia, L.: Acoustic emission localization technique based on generalized cross-correlation time difference estimation algorithm. Chin. J. Sens. Actuators **26**(11), 1513–1518 (2013). (in Chinese)
12. Liu, J.X., Tang, J.N., Lang, Y.X.: Approach for time delay location of UHF signals of GIS partial discharge based on Interpolating cross-correlation function. High Volt. Appar. **54**(02), 62–67 (2018). (in Chinese)

Modeling and Simulation of Self-organized Criticality of Intelligent Optical Network Based on Sand Pile Model

Jingyu Wang[✉], Wei Li, Juan Li, Shuwen Chen, and Jinzhi Ran

School of Information and Communication of National University
of Defense Technology, Xi'an, China
junxiao2010@163.com

Abstract. With the continuous development of optical communication technology, the structure of optical network is becoming more and more complex, and complex network theory can analyze the nature of the actual system. From the angle of self-organizing criticality, the "sand pile" model of intelligent optical network is constructed, the model parameters are designed, and the self-organized critical model of intelligent optical network is established. Finally, the computer algorithm is designed according to the model, and the critical value of the whole network load is solved. The research shows that there is a quantitative relationship between the critical value of network load and the number of the betweenness, which has a certain guiding significance for improving the processing ability of key nodes and optimizing the network security.

Keywords: Self-organized criticality · Sand pile model ·
Intelligent optical network

1 Introduction

Self-organization Criticality (SOC) theory is an important branch of complexity science, which explains the organizational principle of the extended dissipative power system [1]. When the system reaches the self-organized critical state, the intensity or scale distribution of the energy dissipation event obeys the power law relation (non-characteristic scale phenomenon). From the perspective of application, the theory has successfully explained the behavior characteristics of the complex spatiotemporal complex systems that contain thousands of short interactions. It has been widely used in forest fire, geotechnical mechanics, cosmic origin, social economy, seismic science, solar flare and so on, and has made breakthrough progress [2–4]. The proponents of the theory even think that from the point of view of the evolution patterns of complex systems, self-organized criticality is the only theoretical concept so far the only general mechanism that can explain how complexity arises. In recent years, self organized criticality theory has been applied in power network, Internet and urban road network.

The intelligent optical network has the advantages of dynamic flexible scheduling of resources and self-healing protection recovery, but the intelligent optical network has been put into operation and operation and maintenance management for a period of

Q. Liu et al. (Eds.): CENet 2018, AISC 905, pp. 163–172, 2020.
https://doi.org/10.1007/978-3-030-14680-1_19

time. The current work practice has found that the optical network originally planned topology and bandwidth allocation. There is a gradual mismatch between nodes, and the capacity allocation among the links of the nodes is gradually unbalanced, and the network capacity is also increasingly unable to meet user requirements [5]. The long-term interaction between this increasing demand for user bandwidth and the growing scale of optical networks has the following: "Energy injection is continuous, slow, and uniform, and energy dissipation is instantaneous relative to energy injection. "Avalanche" features. Therefore, it may cause the optical network system to enter the self-organization critical state. When the system reaches the self-organized critical state, there is a greater probability that a large-scale cascading failure will result in a partial or large area of network congestion and collapse. At present, there are many researches on the management and diagnosis of multi-layer faults in optical networks, but the critical phenomenon of the occurrence of interlocking faults has not attracted the attention of the industry. The theoretical study of self-organized criticality for optical networks has not been reported in the literature. Therefore, from the perspective of organizational criticality, this paper constructs an evaluation model and obtains the self-organizing critical value of the intelligent optical network through simulation. The steps of modeling are: First, through research and analysis, determine the parameters used when the sand model is used to study the self-organized criticality, map these parameters to the intelligent optical network, and find the corresponding intelligent optical network parameters; then, the mapped parameters are further analyzed and related parameters are calculated by the formula. Finally, in combination with the evolution rules of the sand model, a self-organized critical model of the intelligent optical network is established to solve the critical point of the entire network load. The following steps are introduced.

2   Sand Model Parameters Mapping of Intelligent Optical Network

The parameter map of the sand model of the intelligent optical network is shown in Table 1.

Table 1. Parameter mapping.

Sand pile model	Backbone router IP carrying network
The position of sand in the sand pile	The location of nodes in the network
The rate of adding sand to the sand pile	The rate of data packets sent by a client node
Direction of "sand avalanche"	Node i forwards the packet direction, that is the neighbor node set
Sand number	Node i the number of packets in the current queue, that is node load
Critical value of "sand avalanche"	The critical value of node i load

(continued)

Table 1. (*continued*)

Sand pile model	Backbone router IP carrying network
"Avalanche" after the rest of the number of grains of sand	The lower limit of the remaining quantity after node i forwards the packet
The number of sand grains reduced after "sand avalanche"	Total number of packets forwarded to neighbor nodes by node i
Number of sand grains that slipped into the surroundings	Number of packets forwarded to each neighbor node

Through the parameter mapping relationship in Table 1, the intelligent optical network has been modeled as a sand pile model.

3   Analysis of Evolution Rules Based on Sand Pile Model

From the principle of the sand-pile model, the impact of a packet on the local or global intelligence optical network is similar to the impact of sand collapse on the entire sand pile. When the load of the customer layer of the smart optical network reaches a certain critical value, the nodes in the network will collapse, and the data packets inside the nodes transfer to the neighbourhood nodes. This process is basically the same as the "sand collapse" process of the sand pile model. But we need to pay attention to the difference between the two, that is, the direction of the neighbor nodes in the intelligent optical network is uncertain. Because the degrees of its nodes are different, when the request packet is forwarded, it is no longer as simple as the sand model, simply forwarding the packet to the top, bottom, left, and right directions in the regular grid. It needs to be forwarded according to the connection of the actual network nodes. When all the nodes in the network that are affected by the initial collapse no longer transmit the packet, the network reenters a new equilibrium state. This is the specific evolution process of intelligent optical network combined with the core idea of sand pile model.

In addition to the analysis of the evolution rules when the intelligent optical network is "collapsed", the calculation of the critical load value of the node itself needs to be further analyzed. The test of the sand pile model was done on the regular square paper, which set the rule of the test environment, thus setting a critical height of 4 for each square. However, the actual intelligent optical network studied in this paper has been proved to be a complex network with scale-free characteristics [5]. There are some large Hub nodes in the network and a large number of small degrees of nodes in the network. Therefore, the critical threshold setting for node load will be different from the sand pile model. In the network, the larger nodes have strong data processing ability, and the characteristic parameters in the complex network theory, the number of nodes, can better reflect this characteristic. Therefore, in modeling, the critical load values of each node in the intelligent optical network are calculated by combining the number of nodes betweenness, so that the unique properties of the network can be more accurately displayed.

At the same time, in order to avoid the collapse of the network, a node, as a neighbor node of many different nodes, reaches a high load in a moment, which leads to the long queuing time and the instantaneous reduction of network performance. It also needs to consider the load balance, that is to set the maximum lower limit for each node. Then the process of the network in the event of "sand avalanche" is that if the node 1 load exceeds the critical value, the load is transferred to the neighbor node, and the lower limit of the load of the rest of the neighbor nodes is also needed. Assuming that neighbor node 2 has already shown excessive load, it collapses and begins to forward packets to 3 of neighbor node 2 neighbors. If the neighbor node 3 exceeds its critical load value, it will continue to transfer the load to the next neighbor node. By analogy, until the load of all nodes is below its own critical value, the network reaches a stable state again.

Based on the above analysis, the evolution rules of the sand pile model are mapped to the intelligent optical network, and the critical load of the intelligent optical network with scale-free characteristics is studied by the introduction of the node number, and the characteristics of the network are more accurately reflected.

4   Parametric Analysis

4.1   The Process of "Sand Avalanche" in Simulated Sand Pile Model

At the current moment, the total number of data packets stored in the node's queue is expressed as the load of the node. Then the number of data packets that enter the node during the time from t − 1 to t is the number of packets that leave the node during this time. The number of packets. In queuing theory, it is generally considered that the service rate of the system is equal to the rate of arrival of the customer to just let the system be in the state of non queuing. Therefore, the rate of arrival of packets in the network is slightly less than the rate of the node processing packets, then the queue changes in the time segment can be expressed as the following differential:

$$\frac{dL_i(t)}{dt} = Lin_i(t) - Lout_i(t) \tag{1}$$

At this point, the number of packets and the number of packets in the node $Lin_i(t)$ and $Lout_i(t)$ can be obtained from the nodes. Therefore, the load of nodes at t + 1 is:

$$L_i(t+1) = L_i(t) + Lin_i(t) - Lout_i(t) \tag{2}$$

In the two-dimensional sand pile model, when the sand avalanche occurs, the sand grains of the collapse nodes only slide to the four nodes around [6]. In real networks, due to the different importance of nodes, the neighbor nodes connected by each node are not fixed values. At this time, it is necessary to find the neighbor nodes of the node according to the actual connection condition of the node i in the network. The set of the neighbor nodes is B, and there are b_i neighbors. When the load of node i exceeds its critical value, it becomes unstable, collapses, and begins to forward packets to the neighbor node j. In order to make the network load unbalance as much as possible, the load lower le of the node is introduced, and the number of packets forwarded to the neighbor node j is determined by calculating this parameter. The limit of the load of the node is defined as follows:

$$LE_i = \left| \frac{L_i(t) + \sum_{j=1}^{b_i} L_i(t)}{1 + b_i} \right| \tag{3}$$

The lower bound of the node load calculated by (3) is actually the average value of the load of the collapsed node and its neighbor node load. The reason for calculating the mean is to ensure network load balancing. Then the total number of packets sent by the node is:

$$S_i(t) = LC_i(t) - LE_i(t) \tag{4}$$

The number of packets that each neighbor node receives from node i is calculated as follows. j is the number of the neighbor node.

$$F_j(t) = \begin{cases} LE_i(t) - L_j(t), & LE_i(t) > L_j(t) \\ 0, & LE_i(t) < L_j(t) \end{cases} \tag{5}$$

The formula shows that when the load of the neighbor node is greater than the load lower limit of the sending node, the collapse node i does not consider the neighbor node when the packet is forwarded. When the load is less than or equal to the load lower limit of the sending node, the packet is forwarded to the neighbor node.

The sum of the data packets received by each neighbor node is equal to the total number of data packets forwarded by the node i that has collapsed. It is expressed as:

$$S_i(t) = \sum_{j=1}^{b_i} F_j(t) \tag{6}$$

When a collapse ends, it is assumed that there are a nodes in the network that are affected, these nodes are saturated because of the overload and can not continue to work normally. The nodes affected by the next collapse are not as good as the normal, so these nodes are removed from the network, then a new round of data transmission is carried out, and the analogies are carried out in turn. It is equivalent to every time a collapse occurs, the routing table of the network is updated once, and the whole process is a chain reaction process.

At this point, a sand pile evolution model based on QoS is built, and then the critical load of the network in the critical state is solved.

4.2   Optical Network Load Threshold Calculation

First define the network load: network load = current throughput/network bandwidth. This value reflects the current load of the network, which is usually a percentage. The higher the percentage, the greater the network load. In the model established in this paper, the percentage f of nodes that send data packets per unit time in the network is regarded as the network load. Therefore, the critical load calculated in this paper refers to the full network load threshold when the whole network packet loss rate is equal to 0.5%, and the intelligent optical network will cause a large area of fault and even paralysis because of a slight disturbance. The following is a description of the critical solution process:

According to Little Theorem, the average number of users (number of customers) = [average arrival rate of users (customers)] × [service delay of users (customers)] in a stable state. Applying this theorem to the study of this paper, it means that the number of data packets entering the network per unit time is equal to the number of data request packets processed at the current time when the intelligent optical network does not experience congestion or failure. as follows:

$$\frac{dP(t)}{dt} = Nq\omega - \frac{P(t)}{\tau(t)} \tag{7}$$

In the formula, $Nq\omega$ represents the average rate of arrival of packets per unit time and network, and $P(t) = \sum_{i=1}^{N} L_i(t)$ represents the number of packets in the network at present time. $\tau(t)$ represents the time that a request packet queues in the queue plus the time used to process the packet, and $P(t)/\tau(t)$ represents the rate of packet transmission in the network per unit time.

In the initial transmission, the load in the network is low, and the length of each node's queue is short. When calculating the low load, the average delay of the network can be approximated by the average path length $\overline{l}$, that is $\tau \approx \overline{l}$; However, when the network is running for a period of time, the number of data packets increases, and the network gradually becomes in a state of high load. After the data packets enter the node, they need to be queued. In this case, the average path length cannot be used to indicate the delay. Instead, it should add the time the request packet was queued in the node queue, expressed as:

$$\tau = \overline{l} + t(P(t), \overline{l}) \tag{8}$$

In order to further solve the critical load of the network, it is assumed that the time used by the request packet from the source node to the destination node is approximate to the time spent in the queue, which is expressed as:

$$\tau^*(s, d) \approx \sum_{i=1}^{N} LC_i = \sum_{i=1}^{N} \frac{\overline{Q}}{i} = \sum_{v \in R(s,d)} \overline{C_B}(v)P^* \tag{9}$$

The $R(s,d)$ in the formula indicates the path between the source node s and the destination node d through node v. Formula (9) indicates that the node number is a parameter reflecting the characteristic of the scale-free network from the point of time delay, that is, the length of the average queue length is long, so the time spent on the node is longer.

Calculated by the above formula, combined with the time delay expression at low load and high load, the total delay of the network is approximately

$$\tau(t) \approx \overline{T} + \frac{1}{S}\sum_{s\in V}\sum_{d\neq s\in V}\sum_{v\in R(s,d)} \overline{C_B}(v)P(t) \tag{10}$$

Where S represents the size of the set of source nodes and destination nodes. Simplified formula, Let

$$T = \frac{1}{S}\sum_{s\in V}\sum_{d\neq s\in V}\sum_{v\in R(s,d)} \overline{C}_B(v) \tag{11}$$

At this point, the number of betweenness of the entire network is defined as the mean value of the betweenness of all source node pairs in the network. Then

$$\tau(t) = \overline{T} + TP(t) \tag{12}$$

When the network is in a stable state, simultaneous (3.5), (3.11) are available

$$P(t) = \frac{\overline{T} Nq\omega}{1 - TNq\omega} \tag{13}$$

At the critical load point, the number of packets gradually diverges, that is $P(t) \rightarrow \infty$. At this point, $1 - TNq\omega = 0$ can get the critical load of the network.

$$q_c = \frac{1}{TN\omega} \tag{14}$$

The formula (14) solves the critical load of the intelligent optical network after running for a period of time. The parameter T is expressed as the mean value of the betweenness of the nodes in the network, and can be obtained by formula (13). According to the above formula relationship, there is a quantitative relationship between the critical load of the whole network and the intervening node.

5 Computer Simulation Test

Because the object of this paper is intelligent optical network, compared with the traditional optical network, the intelligent optical network increases the control plane and has the function of automatically finding neighbors. As a result, the self-organizing nature of traffic flows in the network. Based on the related theories of queuing theory, in the simulation of this paper, it is assumed that the arrival rate of data services in the network has Poisson distribution characteristics, i.e., obeys Poisson distribution. Where $\lambda = Nq\omega$, and in general, the service rate of the node processing data has an exponential characteristic. Therefore, assuming that the service rate obeys the exponential distribution, set μ equals λ to ensure that the nodes can process the request packets in a timely manner. Since each simulation is a random selection of the number of sending nodes, it is not the only result of each simulation. For the sake of the simulation results, the simulation results can be better verified, and the simulation operation has been carried out 10 times, and the most accurate results are obtained and analyzed.

First, using visio2013 to abstract the actual intelligent optical network into an undirected graph with 52 nodes and 95 edges. The topology of this intelligent optical network is shown in Fig. 1.

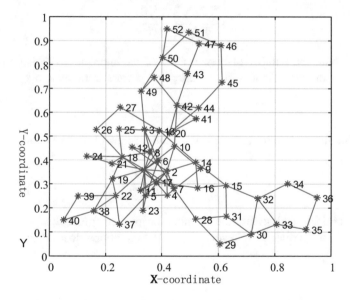

Fig. 1. The actual intelligent optical network topology.

Because this model is based on the assumption of nodes betweenness and network critical load has a direct relationship, Therefore, in the calculation of type (14), according to the different values of ω, different critical loads corresponding to different medials can be calculated, and the critical load q_c change diagrams of different scale networks are drawn as shown in Fig. 2.

In Fig. 2, the two curves expressed in different arrival rate under the condition of network critical load relationship between the average and all betweenness values, map position marked hollow curve is approximately equal to the critical value of intelligent optical network load is studied in this paper. Comparing the two curves show that when

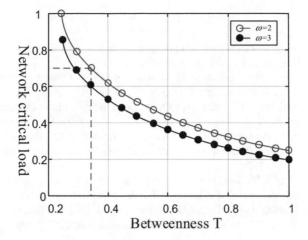

Fig. 2. The relationship between critical load and node betweenness.

the network node of the average value between the same, namely when T = 0.4, q_c hollow = 0.6231, q_c solid = 0.5679, q_c hollow > q_c solid, available in the intelligent optical network scale under the condition of the same packet arrival rate is high, the network critical load is relatively small; In the case of the same packet arrival rate, that is, when ω = 2, q_c hollow 0.4 = 0.6231, q_c solid 0.8 = 0.3472, q_c hollow 0.4 > q_c solid 0.8, this shows that the larger the average network betweenness, the larger the size of the intelligent optical network, the more complex the connection, the smaller the critical load value of the network. With reference to the critical network load we have studied, we can see from Fig. 2 that for some of the networks that we study more, that is, a network with more complex connections, the critical value of the network load will gradually decrease. Rather than for this research network betweenness of small network, the critical load value is relatively small.

6 Conclusion

Based on the sand pile model, the self-organized criticality model of intelligent optical network is constructed. The study shows that there is a quantitative relationship between the critical value of network load and the betweenness. To improve the processing ability of the key nodes, which has a certain guiding significance to optimize the network security capability, which can start from the betweenness of nodes, some key nodes in the network, also is the betweenness of nodes taken to increase the capacity, to enhance the reliability of network security capabilities.

Acknowledgement. This work was supported by National Natural Science Foundation of China (61605247) and Research Foundation of National University of Defense Technology (ZK17-03-26).

References

1. Yin, Z.Z.: Electrical investigations of self-organized behavior of steel sphere packing. Harbin University of Science and Technology, Harbin (2007)
2. Rao, B.: Self-organized criticality in complex systems. Graduate School of National University of Defense Technology, Changsha (2005)
3. Tyler, H.S., Fabrizio, L.C., John, L.: On submodularity and controllability in complex dynamical networks. IEEE Trans. Control Netw. Syst. **3**(1), 91–101 (2006)
4. Djeundam, S.R.D., Yamapi, R.E., Filatrella, G., Kofané, T.C.: Dynamics of disordered network of coupled hind-marsh-rose neuronal models. Int. J. Bifurcat. Chaos **26**(3), 1–18 (2016)
5. Zhu, E.L., Zhang, Y.F., Liu, Q.: Empirical study of complex network characteristic of intelligent optical network. Opt. Commun. Technol. **42**(1), 9–12 (2018)
6. Zhang, J., Jin, Z., Sun, G.Q., Sun, X.D., Wang, Y.M., Huang, B.: Determination of original infection source of H7N9 avian influenza by dynamical model. Sci. Rep. **4**, 1–16 (2014)

Incentive Cooperation Enforcement Based on Overlapping Coalition Formation Game Framework for Ad Hoc Networks

Bo Wang[✉], Lihong Wang, Weiling Chang, Jie Xu, and Jia Cui

CNCERT/CC, Beijing 100029, China
wbxyz@163.com

Abstract. Cooperation enhancement in ad hoc networks becomes a research hotspot to encourage cooperative forwarding among selfish nodes. In this paper, we model an overlapping coalition formation game to solve the problem. In this game, each node can freely make its decision to join multiple coalitions simultaneously, in order to maximize its utility for its cooperation. Then, we propose a distributed algorithm (called OCF) using three move rules for the different choices for each node to find a stable coalition structure. At last, we verify the performance of OCF algorithm through extensive simulation. The analysis results demonstrate that the OCF algorithm has a better performance than the classical algorithms for coalition games with disjoint coalition formation (called DCF) and the AODV algorithm with non-cooperative scheme.

Keywords: Coalition game · Game theory · Coalition formation ·
Incentive cooperation · Ad hoc networks

1 Introduction

The ad hoc network is an autonomous system, where each node freely communicates with other nodes by wireless channels and cooperative forwarding [1, 2]. So far, ad hoc networks have attracted a lot of applications, which mainly focus on emergency and military situations. In such applications, it is usually assumed that all nodes in the network are willing to cooperate. However, as the network is deployed in emerging civilian applications, some nodes shows their selfish behaviors to save their limited resources. Hence, stimulating cooperative forwarding among selfish nodes is critical to ensure the proper functionalities of ad hoc networks.

Recently, coalitional game theory is powerful analytical tool for modeling players' cooperation in wireless networks. Some works in [3–10] have focused on studying the players' behaviors and strategies for incentive cooperation in wireless networks. These works aim at modeling a non-overlapping coalition formation (called DCF) game and finding a stable non-overlapping coalitional structure [3, 4]. However, the DCF game restricts the cooperation of each node and affects the improvement of network performance. Moreover, overlapping coalition formation game is a branch of coalitional game, where a player can freely join multiple coalitions simultaneously [5, 6]. Therefore, in this paper, we address the above issue as an overlapping coalition

© Springer Nature Switzerland AG 2020
Q. Liu et al. (Eds.): CENet 2018, AISC 905, pp. 173–181, 2020.
https://doi.org/10.1007/978-3-030-14680-1_20

formation game instead of DCF game. Then, we propose a distributed coalition formation algorithm to find a stable coalition structure using three move rules, so as to enhance cooperation for each node within the same coalition further.

The rest of this paper is organized as follows: We discuss related work in Sect. 2. In Sect. 3, we propose an overlapping coalition formation game framework in ad hoc networks. In Sect. 4, we present and analyze the simulation results. In Sect. 5, we summarize the paper.

2 Related Work

Here, we briefly review the related works on cooperation-incentive schemes using coalitional game theory in [6–10]. In [6], the authors propose an overlapping coalition game to collaborative spectrum sensing in cognitive radio networks. They also propose a distributed coalition formation algorithm for coalition formations so that it can reach a stable overlapping coalitional structure. In [7], the authors propose a coalition game scheme using the selective decode-and-forward (SDF) framework, and analyze the newly formed coalitional structure's properties and stability. In [8], the authors propose a dynamic Bayesian coalitional game model to motivate each node to cooperate with other mobile nodes for packet forwarding in a coalition, and then propose an individual preference-based algorithm to solve this coalition game to find the Nash-stable coalitional structures. In [9], the authors devise a coalition formation mechanism to stimulate cooperative forwarding in a coalition, and formulate a Markov chain model to obtain the payoff for each node when the node stays in a coalition. In [10], the authors model the players' cooperation through a game theoretical algorithm based on coalitional game theory, so as to maximize their utilities.

3 Overlapping Coalition Formation Game

In this section, we address the problem as an overlapping coalition formation (OCF) game for cooperation stimulation among nodes in ad hoc networks. First, we present the following definitions of an OCF games for further modeling and analyzing. Then, we propose a distributed coalition formation algorithm based on the three choice decisions to solve the proposed game.

A. Overlapping Coalition Formation Concepts

Definition 1: An OCF game $G(IN,v)$ is described as a non-transferable utility (NTU) game if the players in a given coalition cannot divide the worth of coalition arbitrarily.

Definition 2: An OCF game $G(IN,v)$ is specified by a finite set of rational players IN and a characteristic function v mapping a real value for each coalition in the game.

Here, we note that the empty set in v is 0, namely, $v(\phi) = 0$. A coalition denoted by S is defined to be a nonempty subset of players over $IN = \{1,\ldots,N\}$, where $S \subseteq IN$. Each player $i \in IN$ can freely join multiple coalitions simultaneously.

The characteristic function $v(S)$ of coalition S is represented as below:

$$v(S) = \sum_{i \in S} u_i(S) \tag{1}$$

Note that $u_i(S)$ is the allocated utility of node i in coalition S. The allocated utility $u_i(S)$ of node i consists of the cost $c_i(S)$ due to its receiving or forwarding packets for helping other nodes and the $r_i(S)$ of receiving the payoff due to its cooperative forwarding. The allocated utility $u_i(S)$ can be denoted as:

$$u_i(S) = \alpha \cdot r_i(S) - \beta \cdot c_i(S) \tag{2}$$

Where α and β denote the positive weight of $r_i(S)$ and $c_i(S)$, respectively. The solution to the proposed game is to find a stable coalition structure. Next, we give the definition of an overlapping coalitional structure as below:

Definition 3: In OCF game $G(IN,v)$, an overlapping coalitional structure over IN is denoted as $\eta = \{S_1, \ldots, S_M\}$, where $\cup_{i=1}^{M} S_i = N$ and M is the number of coalitions.

Where any two coalitions S_i, S_j, $S_i \cap S_j \neq \phi$ for $i \neq j$. That is, any node may belong to multiple coalitions simultaneously.

Let S_j denote one of coalition in η. The individual utility of node i over η is represented as follows:

$$u_i(\eta) = \sum_{S_j \in \eta} u_i(S_j) \tag{3}$$

In OCF game $G(IN,v)$, the value of a coalition structure η can also be defined as the system utility over η. It can be denoted as follows:

$$u(\eta) = \sum_{i \in IN} u_i(\eta) \tag{4}$$

We note that the system utility over η is the sum of the total utility $u_i(\eta)$ of each node. That is, the $u(\eta)$ can indirectly reflects the stability of a coalition structure η using the system utility.

Therefore, using the above two utility values, we can compare two or more coalition structures and choose the one which has more system utility and individual utility simultaneously. That is, we define preference order of coalition structures for comparison as:

Definition 4: In OCF game $G(IN,v)$, given two coalition structures η_x and η_y over IN. The preference order is represented as:

$\eta_x \succ_i \eta_y \Leftrightarrow u_i(\eta_x) > u_i(\eta_y)$ and $u(\eta_x) > u(\eta_y)$, where $i \in IN$.

Note that: $\eta_x \succ_i \eta_y$ also means η_x is i-preferred over η_y, which is equivalent to $u_i(\eta_x) > u_i(\eta_y)$ and $u(\eta_x) > u(\eta_y)$.

According to Definition 4, a coalition structure is preferred by some node to the other if satisfying the following condition: the system utility and the individual utility are both increased from one to the other. This preference order on one hand guarantees the increasing of system utility for some coalition structure, on the other hand, it also ensures the improvement of individual utility for some node.

In the game, each node can make its decision on joining or leaving a coalition with the above preference order. To find a stable coalitional structure, we define three move rules for the different choices for each player: (1) The player can join another new coalition, which it does not contain the player, (2) The player can leave its current coalitions for next step choice, and (3) The node can switch from its current coalitions to other coalitions. So, we give the definitions of three choice rules as:

Definition 5 (Join Rule): In OCF game $G(IN,v)$, given a coalition structure η_x, a coalition $S_j \in \eta_x$ and some node $i \in IN \backslash S_j$. If $\eta_x \succ_i \eta_y$, node i can decide to join coalition S_j and η_x can evolve into a new coalition structure η_y. That is, η_y is replaced with η_x as: $\eta_y = \{\eta_x \backslash S_j\} \cup \{S_j \cup \{i\}\}$.

Definition 6 (Leave Rule): In OCF game $G(IN,v)$, given a coalition structure η_x, a coalition $S_j \in \eta_x$ and some node $i \in IN \cap S_j$. If $\eta_x \succ_i \eta_y$, node i can decide to leave coalition S_j and η_x can evolve into a new coalition structure η_y. That is, η_y is replaced with η_x as: $\eta_y = \{\eta_x \backslash S_j\} \cup \{S_j \backslash \{i\}\}$.

Definition 7 (Switch Rule): In OCF game $G(IN,v)$, given a coalition structure η_x, two coalitions $S_j, S_k \in \eta_x$, some node $i \in IN$ satisfying $i \in S_j$ and $i \notin S_k$. If $\eta_x \succ_i \eta_y$, node i decide to switch from coalition S_j to coalition S_k and η_x can evolve into a new coalition structure η_y. That is, η_y is replaced with η_x as:

$$\eta_y = \{\eta_x \backslash \{S_j, S_k\}\} \cup \{S_j \backslash \{i\}\} \cup \{S_k \cup \{i\}\}.$$

At last, we give the definition of overlapping stable coalitional structure as:

Definition 8: In OCF game $G(IN,v)$, an overlapping coalitional structure η is stable if any node $i \in IN$ satisfying $i \in S_j$ and $i \notin S_k$ for coalitions $S_j, S_k \in \eta$ and $j \neq k$, node i does not have a profitable from coalition S_j or join another coalition S_k.

According to Definition 8, if any node $i \in IN$ stays in a stable coalitional structure, it will not have incentive to make any other choices for its leaving or joining decision. Therefore, all the nodes would not deviate from their current coalitions and do not make any other choices.

B. Distributed Coalition Formation Algorithm

In order to describe the coalition formation process, we propose a distributed OCF algorithm using the three rules to find a stable coalition structure. As each node executing the proposed algorithm, it seeks to improve its individual utility while increasing the system utility of the newly formed coalitional structure. The three different rules of nodes' choice can lead to form a new coalition structure at each iteration. Thus, each node can obtain a larger individual utility when the coalition structure η evolves into a new one after each iteration. The OCF algorithm is presented in Table 1.

Table 1. The proposed OCF algorithm using the three rules.

Algorithm1 The proposed algorithm using the three rules
1:Intialization: $t=0$ and $\eta(t) = \{S_1(t),...,S_s(t)\}$
2:Repeat
3:   For each node i calculates its individual utility $u_i(\eta(t))$ and the system utility $u(\eta(t))$
4:   For each node i randomly chooses a coalition structure $\eta'(t)$ after joining
5:   If $\eta'(t) \succ_i \eta(t)$ then
6:      Node i makes its decisions in $\eta'(t)$ using the joining rule
7:      $\eta'(t)$ can be replaced with $\{\eta(t) \setminus S_j(t)\} \cup \{S_j(t) \cup \{i\}\}$
8:      $\eta(t+1)$ can be replaced with $\eta'(t)$
9:   End if
10:   $t=t+1$
11:   Node i randomly chooses a coalition structure $\eta'(t)$ after leaving
12:   If $\eta'(t) \succ_i \eta(t)$ then
13:      Node i makes its decisions in $\eta'(t)$ using the leaving rule
14:      $\eta'(t)$ can be replaced with $\{\eta(t) \setminus S_j(t)\} \cup \{S_j(t) \setminus \{i\}\}$
15:      $\eta(t+1)$ can be replaced with $\eta'(t)$
16:   End if
17:   $t=t+1$
18:   Node i randomly chooses a coalition structure $\eta'(t)$ after switching
19:   If $\eta'(t) \succ_i \eta(t)$ then
20:      Node i makes its decisions in $\eta'(t)$ using the switching rule
21:      $\eta'(t)$ can be replaced with $\{\eta(t) \setminus \{S_j(t), S_k(t)\}\} \cup \{S_j(t) \setminus \{i\}\} \cup \{S_k(t) \cup \{i\}\}$
22:      $\eta(t+1)$ can be replaced with $\eta'(t)$
23:   End if
24:   $t=t+1$
25: Until a stable coalition structure is obtained

4   Simulation Results

A. Simulation Environment

We simulate the proposed algorithm using NS2 [11]. We simulate an ad hoc network scenario including 50 mobile nodes, which are randomly deployed in a 1500 m × 1500 m square area. The mobile speed of each node can be randomly selected between 0 m/s and 20 m/s from a random source point to a random destination point. We use the IEEE 802.11b protocol of wireless network as the MAC layer of our simulation. The transport protocol of our simulation is set to User Datagram Protocol (UDP). We choose Constant-Bit-Rate (CBR) as the traffic sources of our simulation.

The rate of the packet generation is set to 4 CBR. The number of source nodes and destination nodes are both set to 10. The bandwidth of wireless link and transmission radius of each node are 2 Mbps and 250 m, respectively. The initial battery energy of each node is set to 100 joules. The transmission cost is 1.4 W power, the receiving cost is 1 W power, the idle and the sleeping power consumption is 0.83 W power and 0.13 W power, respectively. Table 2 summarizes the other simulation parameters.

Table 2. Simulation parameters.

Parameter	Meaning	Value
T	Maximum simulation time	200 s
P	Data payload size	512 bytes/packet
α	The weight coefficient of $r_i(S)$	0.6
β	The weight coefficient of $c_i(S)$	0.4

B. Performance Metrics

To verify the effectiveness of the proposed OCF algorithm using simulations, we also compare its performance with the other classic algorithms (e.g. DCF algorithm with coalitional cooperation and AODV without coalition and cooperation) by the following three metrics:

(1) Packet delivery ratio: The ratio of the data packets received by destinations to those sent by sources. It denotes the successful probability of sending data packets from sources to destinations.

(2) Average end-to-end delay: The average delay experienced by the data packets from sources to destinations. It includes the queuing delays at the intermediate nodes, propagation delays along the links, buffer delays due to routing establishment, et al.

(3) Total payoff of all the nodes: The sum of individual payoff for each node during performing overlapping coalition formation.

C. Simulation Results

Figure 1 gives packet delivery ratio comparisons when the simulation time increases from 0–200 (s). As shown in the figure, both OCF and DCF algorithms have a better packet delivery ratio than AODV algorithm. This is because in OCF and DCF algorithms, each node can join multiple coalitions, which increases its chance to cooperation for packet forwarding. However, each node can make its decision for joining only one coalition, and cooperate with other nodes in this coalition in the DCF algorithm. Therefore, its chance of packet forwarding for cooperation in the whole network is reduced relatively. Furthermore, the packet delivery ratio of AODV algorithm is a litter lower than that of the other two algorithms. This is because in the AODV algorithm, each node only performs AODV routing establishing procedure and it does not introduce the cooperation mechanism, so it shows a lower performance.

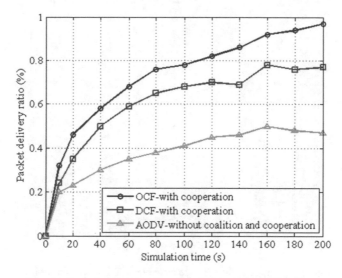

Fig. 1. Packet delivery ratio with different simulation time.

Figure 2 shows average end-to-end delay comparison as the simulation time increases from 0–200 (s). We find that OCF algorithm obtains a lower delay overhead than the other two algorithms. This is due to the fact that the OCF algorithm incorporates the overlapping coalition game mechanism to improve the cooperation between nodes in multiple coalitions. Therefore, in this algorithm each node can maximize its individual utility, and the coalition formation game can has the optimal system utility until the coalition structure reaches stable. That is, in each coalition each node can cooperate with each other for packet forwarding to reduce the packets delay overhead. However, the DCF algorithm also enhances the cooperation of the whole network, but it does not take into account the characteristic of overlapping coalition formations for improvement performance due to the broadcasting characteristic in wireless network. Furthermore, due to the lack of a cooperation mechanism of the AODV algorithm, its delay overhead is a litter higher than the other two algorithms.

Figure 3 shows total payoff of all the nodes as the simulation time increases. The total payoff of the three algorithms increases gradually when the simulation time increases from 0–200 (s). This is because in the three algorithms, due to the coalition game mechanism, cooperative node receives more payoffs for its cooperative packet forwarding behaviors so as to stimulate selfish nodes' forwarding initiative. So the two algorithms achieve a slightly higher total payoff than AODV algorithm. However, the performance difference between OCF and DCF algorithms is that each node can choose multiple coalitions simultaneously for pursuing more payoffs, in order to make up for its cost overheads in OCF.

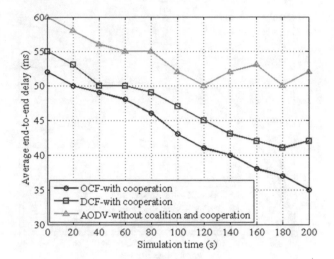

Fig. 2. Average end-to-end delay with different simulation time.

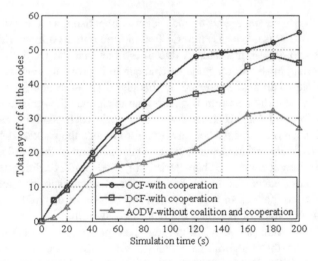

Fig. 3. Total payoff of all the nodes with different simulation time.

5   Conclusions

In this paper, for the sake of address the incentive cooperation problem for ad hoc networks, we have modeled an overlapping coalitional formation game to motivate the players' cooperation in a coalition. Then, we have devised a distributed OCF algorithm using three move rules to make the formed coalitional structure convergence to a stable coalition structure. At last, we have evaluated the performance of OCF algorithm through extensive simulations, and have compared its performance with the other classic algorithms using three performance metrics. Comparison results have showed that the OCF algorithm achieve a better performance than the other two algorithms.

In our future work, we will extend it into other applications such as secure routing and trusted routing in multi-hop network for further analysis.

Acknowledgement. The work is supported by National key Research and Development Program of China (Project No. 2016QY03D0504, 2016YFB0801304) and the National Natural Science Foundation of China (Project No. 61300206). The authors wish to thank many referees for their suggestions on this paper.

References

1. Murthy, C.S.R., Manoj, B.: Ad Hoc Wireless Networks: Architectures and Protocols. Prentice Hall, Upper Saddle River (2004)
2. Zhang, C., Zhou, M.C., Yu, M.: Ad hoc network security: a review. Int. J. Commun. Syst. **20**(8), 909–925 (2007)
3. Ray, D.: A Game-Theoretic Perspective on Coalition Formation. Oxford University Press, New York (2007)
4. Saad, W., Zhu, H., Debbah, M., Hjorungnes, A., Basar, T.: Coalitional game theory for communication networks. IEEE Signal Process. Mag. **26**(5), 7797 (2009)
5. Chalkiadakis, G., Elkind, E., Markakis, E., Polukarov, M., Jennings, N.R.: Cooperative games with overlapping coalitions. J. Artif. Intell. Res. **39**(1), 179–216 (2010)
6. Wang, T., Song, L., Han, Z., Saad, W.: Overlapping coalitional games for collaborative sensing in cognitive radio networks. In: Proceedings of IEEE Wireless Communication and Networking Conference, Shanghai, China, pp. 4118–4123 (2013)
7. Pu, J.W., Li, C.P., Yu, C.S., Wang, T.Y., Li, H.J.: A coalitional game analysis for selfish packet-forwarding networks. In: Proceedings of WCNC 2012, Shanghai, China (2012)
8. Akkarajitsakul, K., Hossain, E., Niyato, D.: Coalition-based cooperative packet delivery under uncertainty: a dynamic Bayesian coalitional game. IEEE Trans. Mob. Comput. **12**(2), 371–385 (2013)
9. Akkarajitsakul, K., Hossain, E., Niyato, D.: Cooperative packet delivery in hybrid wireless mobile networks: a coalitional game approach. IEEE Trans. Mob. Comput. **12**(5), 840–854 (2013)
10. Saad, W., Han, Z., Debbah, M., Hjorungnes, A.: A distributed coalition formation framework for fair user cooperation in wireless networks. IEEE Trans. Mob. Comput. **8**(9), 4580–4593 (2009)
11. The network simulator-ns-2 (2017). http://www.isi.edu/nsnam/ns/
12. Perkins, C.E., Royer, E.M.: Ad-hoc on-demand distance vector routing. In: Proceedings of the 2nd IEEE Workshop on Mobile Computing Systems and Applications, New Orleans, LA, USA, pp. 90–100 (1999)

Research on Simulation System for Ship Transport Operation Based on HLA

Dezhuang Meng[✉], Jianbo Liu, Shuangquan Ge, Xiaoling Wang,
and Hao Tang

Institute of Computer Application,
Chinese Academy of Engineering Physics, Mianyang 621900, China
295547318@qq.com

Abstract. The cabin of a certain type of ship is loaded with various types of weapons and its supporting equipment. Therefore, various factors such as space between equipment and environmental constraints must be considered in the design of transport operation scheme. Aiming at solving the problem of shortage of computing resources, Multi-angle battle positions and joined semi-physical simulation, a distributed simulation method based on HLA (High Level Architecture) is proposed. The federal members (nodes) are divided according to calculation and functions. Method to test delay time of communication between nodes is studied, and communication mechanism between nodes is developed. The results showed that the method can meet design requirements of the scheme and improve the expansibility and maintainability of the system. The system has been successfully applied to the design of a certain type of ship. Compared with the traditional method, the system can improve design efficiency of transport operation scheme and reusability of the simulation resources.

Keywords: Transport operation · High level architecture · Federal members · Communication mechanism · Simulation tests

1 Introduction

Transport operation is about all operations related to position's changes of operations entities in ship, including vehicle running, towing, and cargo transfer. It is widely used in various tasks such as replenishment, mission support, and cargo transportation. Due to a large number of types of weapons and equipment in the cabin, designing of operation schemes involves path planning and time sequence of each equipment and its support equipment. Therefore, designing of operation schemes is extremely complicated. Raised with complication and diversification of ship operations, transport operation is faced with a huge increase in environmental constraints and complexity of work plan, therefore problems such as space occupation by other operation entities and conflict between operation processes are more likely to occur. So it puts forward higher requirements for work plan's design of transport operation.

Using simulation methods to study transport operation can fully verify dynamic changes in various operation processes, so as to obtain a more accurate assessment of design. A distributed simulation system is established in this paper. On one hand,

© Springer Nature Switzerland AG 2020
Q. Liu et al. (Eds.): CENet 2018, AISC 905, pp. 182–189, 2020.
https://doi.org/10.1007/978-3-030-14680-1_21

compute load can be distributed to multiple hardware nodes to ensure real-time performance of various calculations. On the other hand, a system composed of several computers also supports multi-person to observe in multi-point, Then we can control and evaluate transport operation simulation in different aspects.

HLA is a general technology framework dedicated to distributed systems [1]. It was proposed by US Department of Defense in 1995 and was accepted as an official IEEE standard in 2000 [2]. The article [3] analyzes the development of US military's simulation system structure, and introduces the phenomenon of a variety of simulation systems coexisting and a large amount of HLA usage. The article [4] discusses real-time capabilities among federates in the intelligent vehicle infrastructure cooperative systems simulation. The article [5] analyzes the implementation method of multi-operator training simulation system based on HLA. The article [6] studies the principle of the combat system. Federation member model, simulation object model and the interface of federation member are designed in the simulation system to solve visualization problem. The article [7] establishes a HLA-based simulation system framework to improve the expansibility and maintainability of the distributed visual simulation system, Article [8] developed an effective testing tool for federal based on HLA, but could not test communication performance. There is no public report on the application of HLA in transportation operation system of ship. This paper integrates characteristics of transport operation, establishes a distributed simulation system based on HLA, designs a coordination mechanism for nodes, and verifies high efficiency of the system.

2 Design of Simulation System

This paper analyzes simulation system on a stand-alone machine based on the following principles. Each part is divided into different federates and deployed to different hardware nodes of distributed system.

(1) Computational load balancing

When simulation system performs work plan of transport operations, module of interference checking and visual simulation have a large amount of calculation. In order to balance calculation, it is divided into different members of federation, namely, collision federate and visual federate.

(2) Separation of different functional positions

Taking into account that control information of simulation can't be affected by other nodes, functions such as simulation controlling are integrated in simulation control federate; However, functions such as operation-driven and path planning are divided into mathematical federate; Almost all battle positions can be divided into three categories: pilots, commanders, and drivers. Therefore, several visual federals are set up to meet the needs of different battle positions.

(3) Independence for driving hardware

Driving console node is used to implement human-in-the-loop simulation. The integrated display, which simulates driver's view, is same as visual federates and based on Vega Prime (VP) for visual driving. Drivers can simulate driving vehicles or tanks at this node. In this paper, it is developed based on MAK-RTI which implements through VR-Link interface.

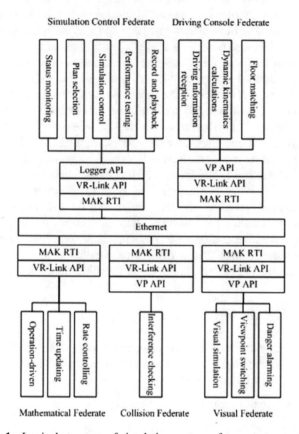

Fig. 1. Logical structure of simulation system of transport operation.

In summary, the logical structure of this system is shown in Fig. 1. Each federate corresponds to a node and is a single running application program.

2.1   The Information Interaction Between Nodes

During federation running, HLA object model template is used to illustrate information that federation and federates need to exchange, and is one of important mechanisms for interoperability and reuse [9]. Federate publishing object class or interaction class provides other federates with updated attribute values, and federate ordering corresponding object class or interaction class receives data from other federates, which will be localized for use [10]. The design of object classes in simulation system is shown in Table 1, and the design of interaction class is shown in Table 2.

Table 1. Design of object classes.

Class name	Meaning	Node name				
		Simulation control	Mathematical	Collision	Driving console	Visual
uDtInter Time	Simulation time	subscribe	publish	subscribe	subscribe	subscribe
uDtInter Main Status	Simulation running status	publish	subscribe	subscribe	subscribe	subscribe
uDtInter Entity Loc and Pos	Well planned work kind and entities position	–	publish	–	subscribe	–
uDtInter DynCal Loc and Pos	Actual work kind and entities position	–	–	publish	publish	publish

Table 2. Design of interaction classes

Class name	Meaning	Node name				
		Simulation control	Mathematical	Collision	Driving console	Visual
uDtInter Plan	Plan number	publish	subscribe	subscribe	subscribe	subscribe
uDtInter Time Test	Communication time in performance test	publish	subscribe	subscribe	subscribe	subscribe
uDtInter Collision	Collision information	–	–	publish	–	subscribe
uDtInter Heart Beat	Heartbeat information	subscribe	publish	publish	publish	publish

2.2   Application Framework

Each node of simulation system adopts a dual-threaded structure. One is a thread of window interface. It is mainly used to deal with the interaction between users and window. The other is a federation thread of simulation and is mainly used to execute simulation model and information exchange between federates.

3   Node Control Mechanism

The system begins with user's startup. Mathematical node performs time deduction, operation-driven and path planning according to operation schemes, and send time information to other nodes through collaborative network. When driving console node receives work kind and has a driving input, it sends real information such as entities' position to collision node and visual node which is based on the vehicle dynamics model. Collision node performs interference checking which is transmitted to visual node. Visual node receives collision information, work kind, entities' positions and so on. Finally visual node renders virtual three dimensions scene. As simulation time advances, all nodes will work together to repeat the above steps. The information interaction between nodes is shown in Fig. 2.

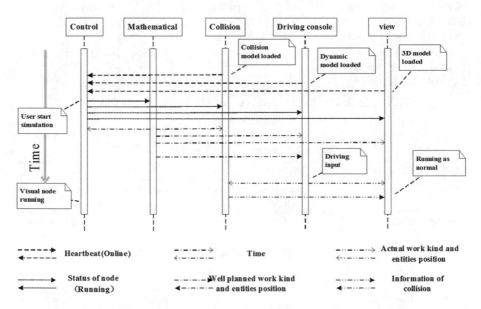

Fig. 2. Mechanism of information interaction between nodes in running state.

4   Performance Test of Time

Time is an important factor in distributed simulation and determines the level of system performance [11]. In a collaborative network, delay of communication between nodes is mainly composed of interaction time between nodes and calculating time within a node. After performance test starts, simulation control federate creates a clock of high-precision, sends heartbeat information to the target federates, and records current clock value when receiving returned message from the target federates. According to formula (1), the delay between simulation control federate and the target one can be calculated.

$$Time = \frac{ReturnTime - StartTime}{Frequency * 2} \tag{1}$$

Run each node's application program, and start the performance test without operation scheme of simulation system. So the calculation time in a node can be neglected. Delay time of communication between nodes is interaction time between nodes; Run the application program of each node, then choose a operation scheme, and start performance test again. Record delay time of communication between nodes, eliminate interaction time between nodes to calculate the calculation time within a node.

5   Simulation Tests

5.1   Simulation Experiment

Run all applications of simulation system, and select one of operation schemes in simulation control federate. Through simulation of operation system, it was found that the total response time is far less than 0.1 s, which starts from simulation running to cabin environment searching, to physical constrained transmission, to work flow deduction and three dimensions displaying. So we can conclude that it satisfies the realistic, fluent and functional requirements of the simulation.

5.2   Analysis of Test Data

The maximum value of interaction time between nodes is shown in Fig. 4, and the comparison result of computation time is shown in Fig. 5. If a single computer runs all simulation system modules, the sum of system operation time is about 5 (Visual node) + 3 (Collision node) + 2 (Mathematical node) + 1 (Driving console node) = 10 > 8.1 ms, and the balance is visible. After calculating the load, delay time of simulation system is shortened. In an actual simulation, it is limited by computer resources, and computing time of a single computer is much longer than 10 ms.

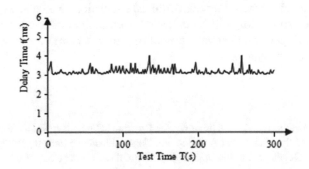

Fig. 4. Maximum value of interaction time between nodes.

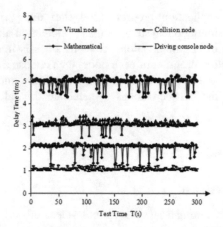

Fig. 5. Comparison of computation time in nodes.

6 Conclusion

In this paper, the federation and federates of simulation system are established for transport operation system of ship. Object classes and interaction classes in object model template are formulated, working mechanism of nodes is analyzed, time management strategy is studied, communication performance of nodes are tested, and finally a distributed simulation system is built based on HLA.

(1) Based on HLA, this paper builds a distributed simulation system for ship transport operation, which satisfies the needs of operation schemes for transport operation in terms of function and performance, and provides a simulation verification platform for transport operation. At the same Time, the system has good extensibility [12].

(2) The system is stable and delay between nodes is within an acceptable range. In the future, research on recording of distributed simulation data will be carried out, and raw data of simulation will be collected from each node to prepare for the next step that will perform the evaluation of operation schemes for transport operation system in the simulation.

References

1. Kuhl, F., Weatherly, R., Dahmann, J.: Creating Computer Simulation Systems: An Introduction to the High Level Architecture. National Defence Industry Press, Beijing (2003)
2. Wang, Z.C.: Development and application of simulation technology. Eng. Sci. **5**(2), 40–43 (2003)
3. Chen, X.X., Luo, X., Qu, K., Feng, J.J.: Research on the development status and trend of simulation architecture. J. Comput. Eng. Appl. **50**(9), 32–36 (2014)
4. Li, S.H., Shangguan, W., Cai, B.J., Wang, J.: Cooperative vehicle-infrastructure system simulation strategy and method. J. Central South Univ. (Sci. Technol.) **46**(10), 3944–3953 (2015). (In Chinese)

5. Tan, B., Guo, S.B., Fu, B.: Design of equipment operation training simulation system based on HLA. J. Syst. Simul. **23**(S1), 177–179 (2011). (In Chinese)
6. Peng, L., Huang, X.H.: Missile combat virtual simulation system based on HLA and Vega Prime. J. Central South Univ. (Sci. Technol.) **42**(04), 1015–1020 (2011). (In Chinese)
7. Bao, J.S., Feng, H.K., Wu, D.L., Jin, Y.: An high extendable distributed hardware in the loop real time simulation system. J. Dong Hua Univ. (Nat. Sci.) **37**(04), 446–448 (2011). (In Chinese)
8. Wang, Y.H., Bian, X.Q., Wang, H.J., Xia, J.Q.: Design of ship operation integrated simulation platform based on HLA. J. Syst. Simul. **21**(2), 385–388 (2009). (In Chinese)
9. Technologies, M.A.K.: VR-Link Developers Guide. MAK Technologies, Cambridge (2005)
10. Huang, X.X., Long, Y., Zhang, Z.L., Gao, Q.H., Yuan, J., Guan, W.L.: Summarization of distributed visual simulation technology. J. Syst. Simul. **22**(11), 2742–2747 (2010). (In Chinese)
11. Wang, Q., Cai, X.B., Du, C.L., Li, G.: A study of distributed virtual test supporting environment. Comput. Simul. **25**(5), 16–19 (2008). (In Chinese)
12. Chen, V.P., An, S.L., Dong, Y.Q., Yao, Y.P.: Distributed simulation architecture with new time management method. J. Syst. Simul. **21**(18), 5754–5759 (2009). (In Chinese)

Research and Design of Mobile Storage Monitoring Based on File Filter Driver

Bo He[1(✉)], An Zhou[1], Chao Hu[2], Duan Liang[3], Zhizhong Qiao[2], and Zhangguo Chen[2]

[1] China Southern Power Grid Co., Ltd., Guangzhou 510670, China
576705701@qq.com
[2] NARI Information & Communication Technology Co., Ltd.,
Nanjing 210033, China
[3] Dingxin Information Technology Co., Ltd., Guangzhou 510627, China

Abstract. In order to deal with the problem that the confidential information in the intranet is easily leaked, the method based on file filter driver is used for mobile storage monitoring to solve such problem. Through the close relationship between the file filter driver and other components of the operating system, we expound the principle of file filter driver. Except that, we discuss about the reason why the file filter driver is applied to the control of mobile storage device. Besides, the file filter technology related to this is elaborated. On this basis, we probably design a mobile storage monitoring system based on file filter driver.

Keywords: Mobile storage monitoring · File filter driver · Intranet

1 Introduction

Currently, mobile storage media have become popular due to its flexibility and convenience, and have become an indispensable file storage, transmission, and exchange tool. At the same time, more and more sensitive information and confidential data are stored in mobile storage. If the management is confusing and improper use [1], it will lead to the leakage of confidential information, which will bring huge security risk to users.

Given that mobile storage poses a great threat to important data from enterprises, governments, and other organizations, there have been some mobile storage security management products on the market to control the use of mobile storage in the internal network. These products complete the security management of the media by registering, authorizing, and auditing the removable media. However, these products generally have some defects. Most of them are developed based on the application layer. They use thread timing to detect whether media is inserted and the type of media and take corresponding measures. Such illegal media can bypass the system detection thread and use it illegally. With the development of technology, some products use kernel driver to solve problems. These products control the mobile storage from the bottom layer of the system, manage the use of mobile storage, and some use USB filter drivers, and some use the combination of the file filter drivers and USB filter drivers, some directly use the file filter driver.

Q. Liu et al. (Eds.): CENet 2018, AISC 905, pp. 190–199, 2020.
https://doi.org/10.1007/978-3-030-14680-1_22

Some people use the filter driver to implement mobile storage access control. Someone uses the WDM framework [2, 3], or uses the WDF framework [4], but does not consider the volume-based hierarchy based on the file-level hierarchy to control the device, also does not consider whether the device is infected. So others begin to use file filter drivers for device control. Somebody uses USB filter drivers and file filter drivers for device control [5]. Besides, it may use file filter drivers to control storage devices [6], but none of them considered multiple volumes. With the uniqueness of volume identification, a mobile storage device may generate multiple volumes at the file system level.

Based on the above situation, this paper proposes a mobile storage access control method based on file filter driver. This method can uniquely identify each volume stored on the move and perform access control based on the volume access privileges. According to the policy, transparently control the access rights of the specified volume or certain types of volumes to ensure safe use during the process of reading the mobile storage files to prevent the host from being damaged by malicious viruses.

In this paper, we expound the concept of volume, file system and related components and file filter driver about mobile storage detailedly in Sect. 2. In Sect. 3, based on the previous research on file filter driver, we design the file filter driver process. In Sect. 4, we make a simple comparison with other methods. Finally, we summarize the paper in Sect. 5.

2   Primaliary Knowledge

2.1   Volume Control Technology of Mobile Storage

Volume is concept at the file system level. On Windows system, mobile storage from a plug-in USB interface to a file on the last-accessed storage medium contains a series of actions [7], one of which is to mount the removable storage volume before it can access the file in the removable storage.

Before the volume is mounted, the file system is traversed first, and the device object of the volume is generated. If the generation of the device object of the volume fails, then there is no device object for the next mount operation [8]. Only if the volume is successfully generated, the volume may be hung. Volume objects are generated by the file system. Therefore, from the perspective of the file filter driver hierarchy, if you want to prohibit the loading of volumes, you can filter the process of generating volume device objects from the file system to filter unauthorized use of removable storage devices [9]. This feature can be used to achieve unauthorized mobile access denied access.

For the rejection function, there are two *read* and *write* operations. From these two perspectives, *read* and *write* operations are prohibited for unauthorized mobile storage, and denial of access can also be implemented. From the point of view of flexibility and ease of integration of the system, in terms of response speed, this solution is more flexible than blocking the generation of volumes.

At the file filter driver level, read-only and write-only, it is achieved by determining the two *IRP* main functions *IRP_MJ_WRITE* and *IRP_MJ_READ*. This point of contact is exactly as same as using the *read* and *write* operations to achieve the volume's denial of access function [10]. At this level, the speed of response is very high once the permissions of the volume are queried.

2.2   Windows File System and Related Components

The design principle and method logic of the Windows operating system come from the UNIX and OPENVMS operating systems and is also influenced by MS-DOS and OS/2. The design of the Windows NT operating system includes designing a minimal kernel and implementing a C/S structure to facilitate the transfer of messages between each kernel module [11]. The operating system presents a hierarchical structure, and communication between each layer and other layers is performed on a well-defined interface. The overall structure of the simplified Windows is shown in Fig. 1 below.

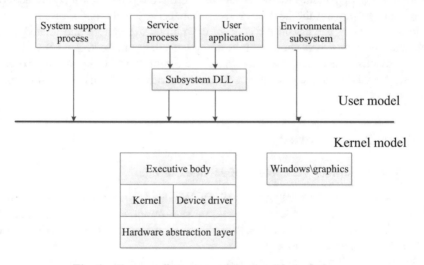

Fig. 1. The overall structure of the simplified windows.

We can see from Fig. 1, similar to most multi-user operating systems, Windows system application and the operating system itself are divided into two different levels, one is user model and the other is kernel model. In kernel model, the model includes the executive body, kernel, device driver and hardware abstraction layer, and the window and graphics system. The user model includes environmental subsystem, subsystem DLL, user application, service process, and so on.

The executive body provides a large number of system service APIs for the subsystem. In addition, executive body widely supports software such as third-party drivers [12], installable file system drivers, and filter drivers. The executable windows mainly includes the following kernel components: I/O manager, virtual memory manager, cache manager, configuration manager, process manager, security reference monitor, and plug and play manager (Pnp). The I/O manager, virtual memory manager, cache manager, and object manager are closely related to file system drivers.

2.3   Principles of File Filter Driver

Windows is a well-defined hierarchical structure array. The I/O manager manages the positivity of all operating devices. The kernel software of the hardware device driver, intermediate driver, filter driver, and file system driver completes the operations through mutual cooperation. The system model of the operation is shown in Fig. 2.

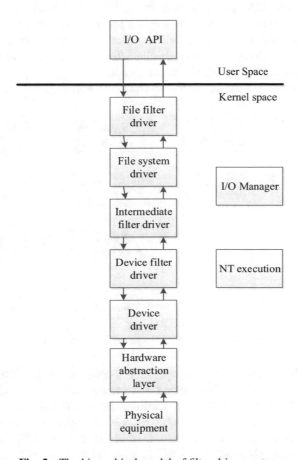

Fig. 2. The hierarchical model of filter driver system.

From Fig. 2, we can see that the physical equipment is at the lowest level of the hierarchy and interacts with the hardware abstraction layer to directly handle the hardware [6]. The intermediate driver accepts the IRP of the file system driver, processes it [13], or sends it to the device driver. The file system driver provides kernel services on various operations of the file, such as file opening, creating, reading and writing, and manages various data structures on the file, such as BIOS parameter blocks, logical partition numbers, and so on. The filter driver filters the upper IRP operations and filters the data or functions that need to be taken care of. The user reads and writes the file [14], firstly passes the request to the I/O manager, and the file path is parsed by the I/O manager, and the file request is sent to the file system driver that meets the function. The I/O manager starts from the internal registration list traverses to find each file system driver, so if a file system driver wants to have the right to control the operation file, the file system must first register with the I/O manager and notify the I/O manager.

The file filter driver creates Control Device Objects (CDO) and Filter Device Objects (FiDO). The CDO uniquely represents a driver. For the operating system, CDO is the driver that creates its object [15]. FiDO binds the device object VDO of each loaded volume to complete the actual filtering work. The interception is performed for each volume's *read* and *write* operations. Specifically, the I/O manager first sends the IRP to the volume device object bound with FiDO.

The file filter driver loading process is shown in Fig. 3.

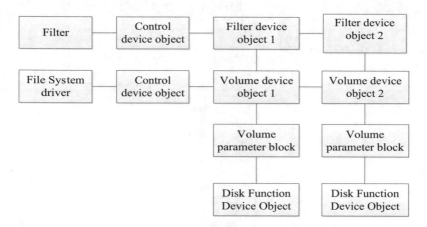

Fig. 3. The loading process of file filter driver.

As shown in Fig. 3, when the operating system starts up, the bus driver is responsible for scanning the disk device and then calling the disk device driver to generate the PDO, and then the bus driver informs the function driver to create the FDO bound to the PDO. At this point, FDO notifies the operating system to load the corresponding file system. The file system driver creates the CDO to represent the driver and creates the corresponding VDO (Volume Device Object) [7]. Then the VPB (Volume Parameter Block) connects the VDO and FiDO. This completes the file system loading process. The file filter driver is loading together with the file system load. After the file system creates the CDO, the operating system scans the registry and loads the corresponding filter driver. At this time, the filter driver creates a filter-driven CDO to represent the filter driver and binds the CDO to the file system-driven CDO. In this way, each file system creates a device object through its CDO notification filter driver CDO. After the filter-driven CDO receives the VDO information from the file system, it creates the corresponding FiDO binding to the VDO. In this way, all IPR packets that are sent to the VDO will be sent to FiDO first. The filter driver intercepts the user's request for accessing the file, which means that the entire process of loading the file filter driver is completed.

3 Design of Mobile Storage Monitoring Based on File Filter Driver

The above context describes the concept and principle of file filter driver. The following context will describe how to implement the file filter driver. Generally, the following four steps are required to design a filter driver.

Step 1. The filter driver creates a device object, attaches it to the target device object, and intercepts all requests to the device object.

Step 2. Filter driver intercepts IRP and deals with it. The filter driver can directly pass the IRP to the lower filter driver, or it can give up the IRP, create a new IRP, and then send it to the specified target device object.

Step 3. Creating a completion routine for the attached device object to call when it completes the IRP. Some actions must wait for the completion of the routine. The obtained data is meaningful, such as the reading of files. It must wait for the IRP finished before reading the data.

Step 4. Creating a deletion routine and releasing the attachment from the attached driver when appropriate. This is generally not required, and sometimes the file filter driver does not need to be attached from the file system driver.

3.1 Attach to Target Device Object

In order to successfully complete the additional operations, the following steps are required:

(1) Get the target pointer of the target device object.
(2) The filter driver creates its own device object.
(3) The filter driver ensures that various fields of the device object can be set correctly, such as device object type, device object attributes, and processing routines corresponding to the device object.
(4) After all preparations have been completed, the filter driver uses manager service function and attaches to the target device object.

If the attachment is successful, the manager will send all the objects destined for the target device to the filter driver first. After the filter driver is processed, it will be transmitted to the file system driver. The process is shown in Fig. 4.

The I/O manager sets a domain *Attached Device* for each device object structure. The I/O manager uses this field to record all the device objects attached to a device object. This design allows multiple filter drivers to be added to the filter hierarchy and have the opportunity to process the IRP sent to the target device object.

We use *IoAttachedDeviceByPointer* to complete the functionality attached to the target device object. When we attach a device object to another device object with *IoAttachDeviceByPointer*, the masculine manager operates in the following order:

(1) The I/O manager gets a pointer to a device object pointing to the top of the target device's attached queue. The I/O Manager calls the function *IoGetAttachedDevice()* to get the top-pointing device object *ReturnedDeviceObject*.

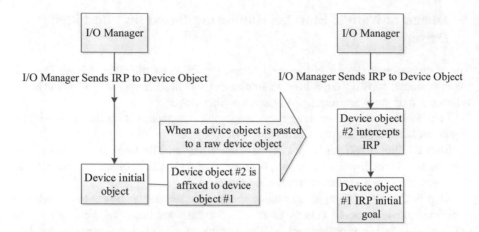

Fig. 4. The change process of the IRP sequence sent by the I/O manager before and after the additional operation.

(2) Actually completing additional operations. The I/O manager completes the additional operations as follows:

ReturnedDeviceObject->AttachedDevice domain points to the source device object.

The source device object domain *StackSize* is set to (*ReturnedDeviceObject->Stacksize+1*).

3.2 Intercept "Create or Open" Request

The file with "*create or open*" request is first passed to the I/O manager, and then the I/O manager decides to send the "*create or open*" request to the target file system driver.

(1) The I/O manager determines the target driver by creating the relevant file object specified in the open request.
(2) For all other "*create or open*" requests, the I/O manager either sends the IRP to the highest-level device object, or send to the device object that represents the target physical virtual logical device.

If the I/O manager gets the IRP that directly opens the target device object, the I/O manager uses the passed target device object as the target device object and passes the IRP to the device object, so that the filter devices attached to the target device object are unable to intercept this "*create or open*" request. This usage is for the recursive operation of special files, so as to avoid repeated requests for file requests to be intercepted by filter drivers.

If the target device object is not directly opened, the I/O manager queries whether the target physical storage has an accessible volume. This volume is created by the file system. If none filter driver is attached to the device object, the request is sent directly to the file system. If filter driver is attached to the device object, the I/O manager obtains the highest-level device object of the attached filter-driven queue and sends the request.

If we want to access the files on the disk, the volume corresponding to the disk must be mounted. If the filter driver needs to intercept all requests for a volume, we must attach itself to the volume device object after the file filter driver finishes mounting the volume. So that, all IRPs sent to the volume device object will be intercepted by the filter driver.

3.3   Complete Routine

The purpose of the completing routine is to call *IoCompleteRequest()* to complete the IRP. If the acquisition of some data must wait for the completion of the driver at the lowest level, a completion of routine can be set up to obtain data within or after completion of the routine. The I/O manager provides the function *IoSetCompletionRoutine()* to complete routine.

3.4   Relieve from the Target Device Object

The function *IodetachDevice()* can be used to relieve from the target device object. The function parameter is the target device object pointer. The condition for successful use of this function is that there is only one filter driver on the target device object. If there are multiple filter drivers on the target device object, the function will fail to call. One possible solution is to iterate through the list of additional device objects and then relieve them from the top device object.

4   Comparison and Analysis

Because USB mobile storage devices play an important role in intranet security, researchers have studied the use of kernel drive technology to control the use of devices. Currently, there are several methods, such as using USB filter driver [2–4], combining USB filter driver and file filter driver [5], using the file filter driver directly [6].

We will compare the following aspects. One is the comparison of control function test, including setting the no-access mode, read-only access mode, and write-only access mode for all volume comparisons, we will see whether these methods achieve authorized access to any volume of mobile storage. There is also a comparison of anti-virus capabilities, including searching and killing feature code, searching and killing Autorun viruses, and searching and killing other executable files. Then we can check whether there is a corresponding searching and killing plan. The results are shown in Table 1.

The references [2–4] use USB filter driver to implement, do not take into account the volume-based hierarchy, and do not examine the file-based hierarchy to control the device, and whether the device is infected with a virus Trojan. The references [5, 6] do not consider the problem of unique identification of multi-volume and volume.

In this section, we compare with other methods in the aspect like control function test and anti-virus capabilities. Then, tests have shown that the our method is effective and have advance in relative work done by other authors.

Table 1. The comparison of control function tests and anti-virus ability.

	References [2–4]	Reference [5]	Reference [6]	Our method
No-access mode	Yes	Yes	Yes	Yes
Read-only access mode	No	Yes	Yes	Yes
Write-only access mode	No	Yes	Yes	Yes
Searching and killing feature code	No	No	No	Yes
Searching and killing Autorun virus	No	No	No	Yes
Searching and killing Other executable files	No	No	Yes	Yes

5 Conclusion

In this paper, we design a monitoring system based on file filter driver through volume control technology of mobile storage, file system with its components and file filter driver technology, including attaching to target device object, intercepting the object of target devices, setting the completion routine, and releasing the completion routine. In addition, this article implements the authorized access of mobile storage of any number of volumes, and has certain anti-virus capabilities from the function, which has certain advantages comparing to some other methods.

References

1. Sandhu, R.S., Coyne, E.J., Feinstein, H.L.: Role-based access control models. IEEE Comput. **29**(2), 38–47 (1996)
2. Yu, J., Yang, Y.J., Zhao, X.Y.: Based filter driver design and implementation. Microcomputer **23**(9), 97–98 (2007). (In Chinese)
3. Qin, J., Wang, B.L.: Drive-based mobile media management system. Comput. Digit. Eng. **38**(4), 113–114 (2010). (In Chinese)
4. Zhou, J.X., Cai, W.D.: USB storage device monitoring system based on WDF filter driver. Comput. Digit. Eng. **32**(3), 42–44 (2010). (In Chinese)
5. Li, J.S., Shu, H., Dong, W.Y.: Driveline based storage device security monitoring technology. Comput. Eng. **34**(8), 255–257 (2008). (In Chinese)
6. Zhang, J.W., Luo, H., Qiao, X.D.: Design and implementation of real-time monitoring system for mobile storage devices based on file filtering. Commun. Technol. **42**(2), 283–285 (2009). (In Chinese)
7. Frincke, D.: Balancing cooperation and risk in intrusion detection. ACM Trans. Inf. Syst. Secur. **3**(1), 1–29 (2000)
8. Liu, H.X., Xiao, S.: Designing WDM device driver I/O Synchronized sampler time-code generator. Chin. J. Sens. Actuators **17**(4), 700–703 (2004)
9. Jandura, P., Cernohorsky, J., Richter, A.: Electric drive and energy storage system for industry modular mobile container platform, feasibility study. IFAC Papers On Line **49**(25), 448–453 (2016)

10. Wang, A., Li, Z., Yang, X.: New attacks and security model of the secure flash disk. Math. Comput. Model. **57**(11–12), 2605–2612 (2013)
11. Gorman, M.: Mobile storage set up helps these A/C techs in the field. Prof. Tool Equipment News **27**(5), 54 (2016)
12. He, D., Kumar, N., Lee, J.H.: Enhanced three-factor security protocol for consumer USB mass storage devices. IEEE Trans. Consum. Electron. **60**(1), 30–37 (2014)
13. Aminzadeh, N., Sanaei, Z., Ab Hamid, S.H.: Mobile storage augmentation in mobile cloud computing: Taxonomy, approaches, and open issues. Simul. Model. Pract. Theory **50**, 96–108 (2015)
14. Wenge, C., Guo, H., Roehrig, C.: Measurement-based harmonic current modeling of mobile storage for power quality study in the distribution system. Arch. Electr. Eng. **66**(4), 801–814 (2015)
15. Vasiliev, L.L., Kanonchik, L.E., Tsitovich, A.P.: Adsorption system with heat pipe thermal control for mobile storage of gaseous fuel. Int. J. Therm. Sci. **120**, 252–262 (2017)

Research on Multistage Forwarding of Media Stream

Meili Zhao(✉), Guoquan Xiong, Jianghui Fu, Meizhu Jiang,
and Hongling Wang

East China University of Technology, Nan Chang 330013, China
714656976@qq.com

Abstract. In order to make full use of existing video monitoring systems, one urgent task of the network operation and management is to establish the provincial networking video monitoring system currently. According to the actual demand of the expressway management project in Sichuan Province as the background, based on the Video network monitoring platform, we design a parallel to the administrative structure of the media flow in multistage forwarding mechanism, multi-level and multi-domain resource access, and design the media stream and control flow between stages. With the design idea to realize the centralized monitoring, distributed management, realize the sharing of video resources in different jurisdictions. The results of this research have been successfully applied in Sichuan Province, and have achieved remarkable results.

Keywords: Media stream management · Dispatch mechanism ·
Multistage forwarding

1 Introduction

The traditional analog closed-circuit television monitoring system has limited monitoring ability and only supported local monitoring, the video load is heavy, the user can only get or replace the new video tape from the video recorder, and the content of the video tape is easily lost, stolen or deleted. The quality of video will continue to decrease as the number of copies increases. Circuit system video switching and control are independent. As the transmission distance is far away, the wiring is complex, the operation is complex, it is difficult to realize multiple center control, and the system capacity is relatively small, and the expansion is difficult, so it is difficult to achieve regional networking. The urgent task of the road network operation management is to make full use of the original video monitoring system to establish the video network monitoring system of the provincial highway.

Video network monitoring structure

This system mainly controls all road network information in Sichuan Province, so it is necessary to organize effectively the management organizations in the province. The traffic monitoring domain refers to the traffic monitoring resources in a certain area and the logical structure within the same network nodes. The traffic monitoring resources

© Springer Nature Switzerland AG 2020
Q. Liu et al. (Eds.): CENet 2018, AISC 905, pp. 200–206, 2020.
https://doi.org/10.1007/978-3-030-14680-1_23

and the corresponding road infrastructure are controlled according to the same network rules. After determining the concept of the traffic monitoring domain, we can recombine the resources and infrastructure in accordance with the requirements of the network monitoring, so that the basis of the network monitoring can be better.

The concept of multilevel is mainly for the administrative structure within the scope of the province. The Sichuan Provincial Department of communications belongs to the primary monitoring domain. It is the information summary organization of the road network monitoring and control system. It is responsible for coordinating the monitoring and control of all road networks. It is responsible for the overall coordination, scheduling and command of the traffic monitoring system within the entire road network, all of which come from its subordinate departments (such as service bureaus and companies). The all original data and video images are all put together at the Network Monitoring Center of the Communications Department. Therefore, the subordinate service bureaus and companies are the secondary monitoring domains. The secondary monitoring domain can be further subdivided.

This system takes the Department of communications and the monitoring sub centers as the central monitoring center. They need to complete the access to the media stream, select routing, multi-level forwarding forwarding and decoding, playing and so on. So in the tree structure of the entire domain, they are called forwarding nodes. In each forwarding node, a video network monitoring platform is configured, which includes streaming media server, database server, and tuned server.

2 Key Technology of Multilevel Forwarding

2.1 Management Mode Based on Monitoring Domain

In order to facilitate the comprehensive management of video network monitoring resources, this paper reclassifies the entire monitoring network system according to the domain model, and proposes the concept of intra-domain and extra-territorial information resources. In general, intra-domain resources can only be used in this domain. At the same level, extraterritorial users of the same level cannot view the domain resources, only the superior domain users can view the resources of their lower domain. When users in a peer domain perform resource interworking, they must pass the permission of the upper domain. In this system, resource management follows the principle of "intra-domain management and common use." Each working domain is only responsible for managing all resources within the domain, following the principle of terrestrial management. The resources of the domain can only be controlled by the domain itself.

2.2 Media Stream Acquisition

By connecting the encoder, the media stream can get the video stream from the encoder, and then open the video source. For the acquired format of the video coding and media stream obtained should be decoded simply, and the important information in the media packet can be obtained. Finally, you can send the media streaming server to other domains directly or to the calling server as required.

To achieve this function, the media streaming server communicates with the underlying encoder through the network, control it to start or stop sending images. The operation process is shown in Fig. 1.

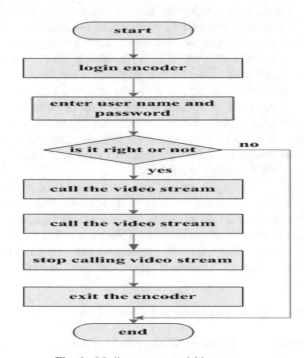

Fig. 1. Media stream acquisition process.

2.3   Implementation of Multilevel for Warding Technology

The system must solve the problem of resource access between different domains. For monitoring resources that do not belong to this domain, it is necessary to find its neighboring media stream server and issue an access request to it. In this way, a recursive resource access plan can be designed, and the media stream server can subordinate to it. The server and the local user provide a uniform recursive algorithm so that access or addressing of the monitoring resources can be achieved. This algorithm flow is shown in Fig. 2.

Through this recursive resource access method, users can access all monitoring devices in the monitoring system across multi-media streaming servers. This recursive resource access method is relatively simple and effective, and can save bandwidth when multiple upper-level servers access the same monitoring device, which is very important for the transmission of video images.

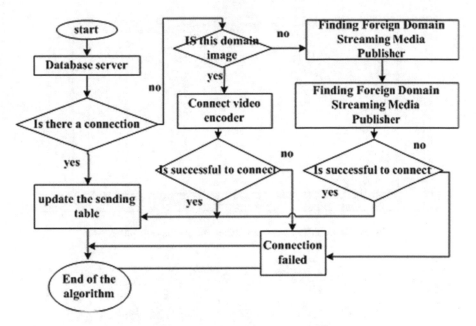

Fig. 2. Cross-level resource access algorithm flow chart.

3 Performance Analysis

In order to fully understand the robustness and compression of the video network monitoring platform under multilevel forwarding, the maximum user concurrency of the media stream server and the video resources are tested respectively.

3.1 User Maximum Concurrency Test

Login and logout are the basic functions of the video networking monitoring platform system. The purpose of this test case is to test the maximum number of concurrent users that the system can withstand.

In the testing process, the maximum number of concurrent users was set to 15. Gradual pressurization mode was used during the test. One concurrent user (Vuser) was started every 8 s, a total of 15 Vu were pressurized, all users started running for 4 min, and 2 users were reduced every 4 s when finished.

The data in Fig. 3 reflects that the response time of transactions increases gradually as the number of concurrent users increases from 0 to 15. When the operation reached 2 min and 45 s, the maximum concurrent number is 15. After that, the average response time of the system was stable at about 12 s, and there was no failed user.

From Fig. 4, it can be seen that as the number of concurrent users increases from 0 to 15, the system throughput also gradually increases. After reaching 15 concurrent users, the system throughput reaches the highest and remains relatively stable.

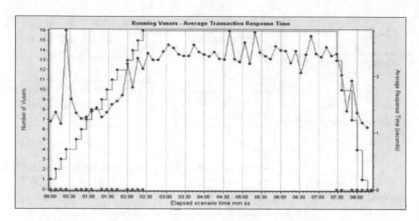

Fig. 3. Relationship between Vuser and average response time.

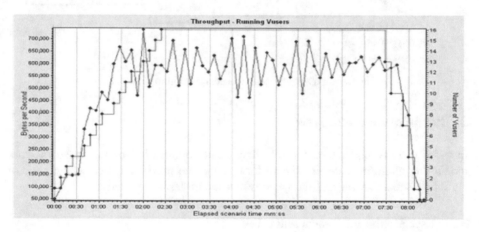

Fig. 4. Vuser and throughput diagram.

3.2 Video Resource Viewing Test

Viewing video is the most important function in the video network monitoring platform system. As the quality of video and speed of transmission are greatly affected by the outside world.

During the testing process, the maximum number of concurrent users was set to 15 and the operating conditions of the system under this pressure were observed. The test was gradually pressurized. A concurrent user (Vuser) is started every 8 s, and a total of 15 Vu are pressed. All users start to run for 4 min, and 3 users are reduced every 10 s when finished. In the testing process, a general request such as repeated user login, viewing video, and exiting the system is initiated according to the set time interval.

From Fig. 5, it can be seen that the average utilization rate of the streaming media server CPU is about 50%. Under the premise of 15 concurrent users, the system does not occupy too much system resources. In addition, no failed users were found during

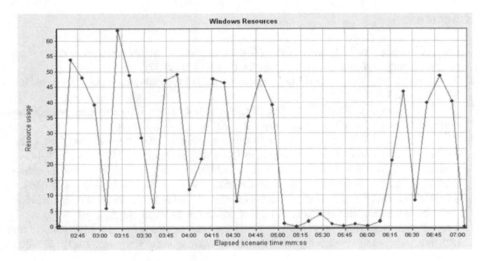

Fig. 5. Streaming server resource utilization.

the testing process, indicating that with the increase of bandwidth (from 20 M to 30 M), there has been a lot of improvement in viewing video performance.

3.3   Application of Results

The research results have been successfully applied to video surveillance of road network in Sichuan province. The multi-stage forwarding function of streaming media server video stream has been successfully completed, and the design requirements are achieved. At the same time, the call and transmission of streaming media in the monitoring system are solved in cross domain and concurrent access. It has achieved the expected effect and can be used for reference in other industries.

4   Conclusion

The management and scheduling mechanism of streaming media server enables the functions of the monitoring platform system to be implemented according to the initial planning, and the image effect is satisfactory. When multiple users are transdomain, and when the number of concurrent users reaches the maximum, each user sends a request to respond in a short time. At the same time, the system can reasonably allocate and schedule permissions, link resources, etc. The overhead of the system will not suddenly increase due to the increase of the number of users. It is controlled in the expected plan, and the system has a more perfect function of fault and exception handling. It has achieved the design requirements, and has a certain advanced nature and good application value.

Acknowledgement. This study is supported by Open Fund of Jiangxi Engineering Laboratory on Radioactive Geoscience and Big data Technology (No. JELRGBDT201708).

References

1. Cao, X., Wang, Y.C., Huang, H.P.: Wireless multimedia sensor network video stream multipath route algorithm. Softw. Report **23**(1), 108–121 (2012)
2. Tao, X.L., Zheng, Y.B.: A Level task allocation method for multiple agents. J. Comput. Inf. Syst. **9**(2), 813–820 (2013)
3. Bai, G.W., Tao, J.J., Shen, H.: A link-lifetime based dynamic source routing protocol for multimedia over MANETs. J. Chin. Inst. Eng. **33**(5), 761–768 (2010)
4. Wu, D.P., Lou, Y.W., Fan, S.L.: Intermittently connected wireless network data forwarding mechanism based on code node dynamic management. J. Commun. **35**(2), 25–32 (2014)
5. Aguilar, I.M., Carrascal, F.V.: Self-configured multipath routing using path lifetime for video-streaming services over ad hoc networks. Comput. Commun. **33**(15), 1879–1891 (2010)
6. Sha, C., Sun, L.J., Wang, X.C., Huang, H.P.: Energy-efficient sampling and transmission methods in wireless multimedia sensor networks. Commun. Stud. **32**(2), 1–10 (2011)
7. Tao, J.J., Bai, G.W., Shen, H., Cao, L.: ECBRP: an efficient cluster-based routing protocol for real-time multi-media streaming in MANETs. Wireless Pers. Commun. **61**(2), 283–302 (2011)

Dynamic Hybrid Timing Switching and Power Splitting SWIPT for Multi-hop AF Relaying Systems

Mingjun Zhao[1], Minghe Mao[2(✉)], Ning Cao[2], Le Cheng[3], Jie Li[1], and Rui Shi[4]

[1] State Grid Xinjiang Information & Telecommunication Company, Urumqi 830002, China
[2] School of Computer and Information, Hohai University, Nanjing 210098, China
mmh_1988@126.com
[3] Department of Computer Science and Engineering, Huaian Vocational College of Information Technology, Huaian 223003, China
[4] State Grid Information and Telecommunication Branch, State Grid Corporation of China, Beijing 100761, China

Abstract. In this paper, a dynamic hybrid time switching (TS) and power splitting (PS) simultaneous wireless information and power transfer (SWIPT) algorithm in a multi-hop amplify-and-forward (AF) relaying system is proposed. The trade-off between the power transfer efficiency and the information transmission performance under the constraint of total transmitted power and throughput of each relay is evaluated. Numerical results show that, there are two arguments can be dynamically adjusted along with the different relaying hops to achieve optimal SWIPT performance, which is a optimization problem that can be overcome.

Keywords: Emergency communications · SWIPT · Multi-hop AF relaying · Dynamic hybrid TS and PS

1 Introduction

Emergency communications systems can guarantee the patency operation of the command and dispatch management system after encountering emergencies such as: accident hazards, natural disasters, social security events, public health incidents, and so on. Thus, it is competent to improve the efficiency of post-disaster reconstructions and power facilities restoration. However, the most basic requirement for the construction of emergency communication systems is rapid action due to the emergency of these events. It is necessary to construct a wireless emergency communication and power supplying system which can be complementary with existing power facilities and communication networks.

For a common scenario, the emergency communication operation usually takes place in the longest distance from a fixed regular base station and thus a single-hop relaying nearly does not make any sense. For the sake of further expanding the

Q. Liu et al. (Eds.): CENet 2018, AISC 905, pp. 207–215, 2020.
https://doi.org/10.1007/978-3-030-14680-1_24

transmission range, multi-hop wireless relaying [1] can be employed. Amplify-and-forward (AF) is a commonly used protocol for multi-hop wireless relaying [2], where the relaying node receives signals from the source node or their previous-hop relaying node and before forwarding them to the next relaying node or the destination node, just amplifies them simply without any processing. The works of references [3] studied the wireless channel parameters performance for multi-hop relaying systems and showed its simplicity just by carrying out amplification at the relay node. Therefore, the multi-hop AF relaying is considered in this paper.

However, there is a serious problem with the traditional AF relaying protocol that all relay nodes must be powered from the wired power facilities or the solar battery, which is obviously not desirable in certain applications such as: those relay nodes in the same network that are peer users and depend on batteries to operate. The relaying tasks would drain the batteries faster. Therefore, the relay node itself will be subject to more energy constraints. References [4] adopted wireless energy harvesting (WEH) to solve this issue. In the wireless relay nodes, there are two parts of WEH relaying implementation. They are an energy harvester which harvests the energy from the transmitted signals not only from the source nodes but from other adjacent sources and a repeater which carry out the AF operations. Therefore, the relay node can ignore the limitation of the battery issue any more. The seminal paper [4] proposed some kinds of continuous-time energy harvesting relaying protocols and analyzed their performances. The two important protocols of them are time switching (TS) and power splitting (PS). Where the TS receiver switches between WEH and wireless information transmission (WIT) on the basis of the TS ratio [5], and in consequence, the received radio frequency (RF) signal is switched to energy harvester firstly and then to the signal receiver. While the PS receiver splits the received signal into two parts on the basis of the PS ratio [6], one for the energy harvester and the other for the signal receiver.

Back to the discussion of emergency communications system, in order to complement the communication-damaged area or the communication coverage blind area, WIT may be utilized, while to make the power-damaged area or power coverage blind area be complemented, wireless power transmission (WPT) should be employed. The wireless nodes of the simultaneous wireless information and power transfer (SWIPT) [6] system collect energy from the RF signals they receive to provide energy for data transmission. This system exactly meets the above two conditions. Using RF energy harvesting (EH) for remote power charging together with wireless relaying to forward the received signals [4], SWIPT is an effective way to expand the network coverage or enlarge the signal transmission range for a mobile handheld terminal equipment or a mobile base station without much additional charging infrastructure on emergency rescue areas, especially for the so-called last kilometer communication problem [7] in emergency rescue operations.

For a SWIPT multi-hop system, only in the perspective of extending the transmission rage or increasing the network coverage for a base station, the analyzation results of [8] showed that TS outperforms PS. However, for a one-relaying system, in the perspective of improving communication performance of signal-to-noise ratio and bit error rate, [9] showed that PS outperforms TS on the contrary. Therefore, based on philosophy mind, there is a optimization problem on the trade-off of choosing the TS protocol or the PS protocol. Some heuristic methods have been proposed to illustrate the advantage of hybrid TS and PS protocol on improving the whole jointed SWIPT performance. In [10], the authors proposed a beam-domain hybrid TS and PS SWIPT protocol for full-duplex massive MIMO system to show the superiority in terms of spectral efficiency compared to conventional protocols. And [11] presented an artificial-noise aided hybrid TS and PS protocol for OFDM systems to secure data transmission with energy transferring. However, the hybrid protocols in references [10] and [11] did not involve the multi-hop relaying scenario, which limits the application range of the hybrid protocol especially on the emergency communication area. The aforementioned works focus on hybrid TS and PS protocol for dual-hop relaying. There have been very few works on multi-hop relaying for hybrid TS and PS SWIPT.

Thus, motivated by these observations and based on the scenario of [8], the dynamic hybrid TS and PS SWIPT for multi-hop AF relaying is proposed in this paper. Optimized relaying performance can be achieved via dynamically adjusting the factors of TS and PS.

Numerical results have revealed that the optimal SWIPT performance can be achieved by adjusting different time and power portion of EH in the hybrid TS and PS protocol in case of a given initial source energy and relay distances in a single multi-hop AF relaying down-link SWIPT system. And that the maximum number of achieved hops depends on several important system parameters such as: initial source energy and the relay distance. On the other hand, the energy conversion efficiency is not so important, and in practical condition it cannot be improved ideally due to the hardware device limitation. Although, it is found that PS outperforms TS [9], only using PS protocol the hop number cannot be improved significantly and in fact it can only support single relay, which limits the extension of communication range. While the dynamic hybrid TS and PS protocol can not only ensure the communication performance but also achieve the same maximum number of hops as the TS protocol.

2   System Model and Derivations

Similar to [8], a downlink relaying system with a relaying link that passes several hops from the source node to the destination node is considered in this paper. It is assumed that each node has a single antenna and is half-duplex, which is quite easy to expand the results to multiple antennas or multiple links. In order to increase the network coverage, assuming that there is no direct link between the source and the destination, or there is no direct link between nodes that more than one hop apart. The signal carries both power and information from the source node and is relayed through multi-hop relay nodes, one node per hop, until it reaches the destination node. In each hop, the transmitter of the previous node transmits the packaged signals to the next node and the receiver of the next node harvests the energy and then uses it to relay the repackaged signal to the next hop. Where $m = 1, 2 \cdots$ represents the number of hops, d_m is the node distance of the m-th hop, g_m is the amplification coefficient of the m-th relay R_m, $r_m[k]$ is the received signal of the m-th hop. For $s[k] = -1$ and $s[k] = 1$, $r_0[k] = s[k]$ is a binary phase shift keying signal with equal priori probabilities assumed to be transmitted. According to [4], $n_{ma}[k]$ is the additive white Gaussian noise (AWGN) introduced by the RF front of the m-th relay, and $n_{mc}[k]$ is the AWGN introduced in the RF to baseband conversing operation at the m-th relay. Also, $n_{ma}[k]$ and $n_{mc}[k]$ are both assumed as mean-zero Gaussian random variables with variances σ_{mc}^2 and σ_{ma}^2, respectively.

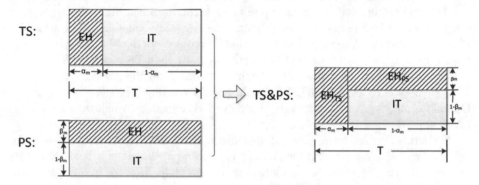

Fig. 1. Frame structure of the novel hybrid TS and PS SWIPT protocol.

Further assume that it takes T seconds to transmit the signal in each hop. The frame structure construction process of the hybrid TS and PS SWIPT protocol is designed in Fig. 1. Firstly, the relay spends a part of T to collect energy from the source or the previous hop signal. The part at the m-th relay in the m-th hop is denoted as α_m, where $m = 1, 2 \cdots$ and $0 \leq \alpha_m < 1$. Secondly, the relay splits a part of the received signal power for the $1 - \alpha_m$ part of the entire relaying time T as collected energy. The splitting part coefficient of the m-th relay is denoted as β_m, where $m = 1, 2 \cdots$ and $0 \leq \beta_m < 1$. One should be noted that, when $\beta_m = 0$ the hybrid protocol degenerates into the pure TS protocol, otherwise, when $\alpha_m = 0$ the hybrid protocol degenerates into the pure PS protocol.

Therefore, for the multi-hop AF Relaying, the received signal from the previous hop is just amplified and then forwarded directly to the next hop using its switched and splitted energy. And thus, the received signal at the m-th relay can be given by

$$r_m[k] = \sqrt{(1 - \beta_m)P_{m-1}}g_{m-1}u_m r_{m-1}[k]/\sqrt{d_m^v} + \sqrt{(1 - \beta_m)}n_{ma}[k] + n_{mc}[k], \quad (1)$$

where P_{m-1} is the transmission power from the $(m - 1)$-th relay node and P_0 denotes the source transmission power, u_m is the fading factor of the m-th hop and it is assumed that each block is invariable, v is the path loss exponent. Furthermore, in order to simplify expressions, assume that u_m is known and that the distance d_m is set to 1. When explicit of the distance is required, one can easily replace u_m with $u_m/\sqrt{d_m^v}$ in the following results. Then one further has

$$r_m[k] = \prod_{i=1}^{m} u_i \prod_{i=1}^{m-1} g_i \prod_{i=0}^{m-1} \sqrt{(1 - \beta_{i+1})P_i} s[k]$$
$$+ \sum_{i=1}^{m} \left\{ \prod_{j=i+1}^{m} u_j \prod_{j=i}^{m-1} g_j \prod_{j=i}^{m-1} \sqrt{(1 - \beta_{i+1})P_i} [(1 - \beta_i)n_{ia}[k] + n_{ic}[k]] \right\}. \tag{2}$$

Where the amplification factor g_i corresponds to two kinds of AF relaying gain: fixed-gain and variable-gain with $g_i = 1/\sqrt{P_{i-1}\Delta_i + \sigma_{ic}^2 + \sigma_{ia}^2}$, where $\Delta_i = E\{|u_i^2|\}$, and $g_i = 1/\sqrt{P_{i-1}|u_i^2| + \sigma_{ic}^2 + \sigma_{ia}^2}$, respectively. Also, it should be noted that the product notations of $\prod$ in (2) are not well defined mathematically when the lower limit is greater than the upper limit. And if this happens, for the sake of avoiding this notational problem, the product result is set to 1. For example, when $m = 1$, $\prod_{i=1}^{0} g_i = 1$.

Based on (1), the harvested energy at the m-th relay is

$$E_{hm} = \eta \prod_{i=1}^{m} u_i^2 \prod_{i=1}^{m-1} g_i^2 \prod_{i=0}^{m-1} (1 - \beta_{i+1})P_i \alpha_m T$$
$$+ \eta \prod_{i=1}^{m} u_i^2 \prod_{i=1}^{m-1} g_i^2 \prod_{i=0}^{m-1} (1 - \beta_{i+1})P_i(1 - \alpha_m)\beta_m T, \tag{3}$$

where η is the energy conversion efficiency. Since the entire transmission time per hop is T, the transmission power at the m-th node is

$$P_m = E_{hm}/T = \eta \prod_{i=1}^{m} u_i^2 \prod_{i=1}^{m-1} g_i^2 \prod_{i=0}^{m-1} (1 - \beta_{i+1})P_i[\alpha_m + (1 - \alpha_m)\beta_m]. \tag{4}$$

One should note that (4) is a normalization form of T, because T determines the transmission power.

Then, the throughput of the m-th relay is $C_0 = (1 - \alpha_m)\log_2(1 + \gamma_m)$, where γ_m is given by $\gamma_m = U/V$, where

$$U = \prod_{i=1}^{m} u_i^2 \prod_{i=1}^{m-1} g_i^2 \prod_{i=0}^{m-1} (1 - \beta_{i+1})P_i(1 - \alpha_m)T, \tag{5}$$

and

$$V = \sum_{i=1}^{m} \left\{ \prod_{j=i+1}^{m} u_j^2 \prod_{j=i}^{m-1} g_j^2 \prod_{j=i}^{m-1} (1 - \beta_{j+1})P_i[(1 - \beta_i)\sigma_{ia}^2 + \sigma_{ic}^2] \right\}. \tag{6}$$

As m increases, the harvested energy per hop decreases at an accelerated rate. Therefore, there is a critical value of m, based on a certain standard, the harvested energy will not be sufficient for transmission. Thus, it is very meaningful to find the critical value of $(1 - \alpha_m)\log_2(1 + \gamma_m) < C_0$. Before this happens, the m hops has the throughput of C_0 such that the optimal coefficient α_m and β_m can be found to maximize the relaying transmission range under the initial power condition P_0. Although there is not a closed-form expression of α_m and β_m to make $(1 - \alpha_m)\log_2(1 + \gamma_m) < C_0$ achieve the critical value, it can be easily solved by using math software.

3   Numerical Results and Discussion

In this section, numerical results will be given to examine the maximum relaying range for different harvesting factors and different system parameters. In the examination, without loss of generality, the parameters are set as $\sigma_{ma}^2 = \sigma_{mc}^2 = 0.01$, $u_m^2 = 0.1$ and $\Delta_m = 0.1$ for $m = 1, 2, \cdots, v = 3$, $T = 1$ and $C_0 = 2$. And let $\gamma_0 = u_1^2 P_0 T / (\sigma_{1c}^2 + \sigma_{1a}^2)$ be the initial signal-to-noise ratio (SNR) of the first relay. This initial SNR is directly related to the initial amount of energy $P_0 T$ of the source node.

Figures 2, 3 and 4 illustrate the maximum number of achieved hops under a given node throughput for different values of γ_0, η and d when different pair of hybrid factors are implemented. As expected, larger initial SNR γ_0, higher conversion efficiency η and smaller value of d lead to larger number of hops. However, in practical cases, the initial SNR and conversion efficiency are usually limited, and the network coverage is related with the product of relay distance and hop number, which cannot be improved by only increasing d as the maximum number of achieved hops decreases when d increases. Compared with the three parameters, one can see that the biggest impact on network coverage extension are the hybrid protocol factors α_m and β_m.

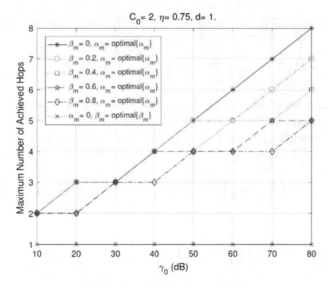

Fig. 2. SNR *vs.* hop number.

On the other hand, Figs. 2, 3 and 4 show that the smaller β_m is the larger number of hops will be achieved after α_m being optimized, but small β_m will reduce the power used for strengthening the relaying signals especially in the long distance case of the end several hops. That is because smaller β_m makes larger portion of TS ingredient in the hybrid protocol and thus the number of hops will be extended as larger as possible, however at the end relaying hop the WIT signal will become very weak where the larger β_m will be required to strengthen the weak WIT signal. This is a trade-off

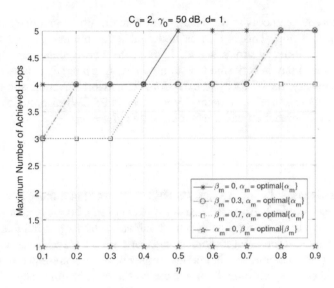

Fig. 3. η *vs.* hop number.

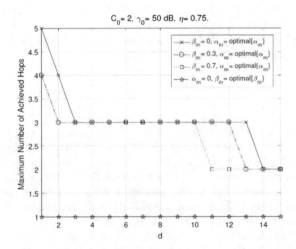

Fig. 4. d *vs.* hop number.

problem that can be dynamically optimized based on specific issues. One should note that, when α_m is 0 the hybrid protocol degrades into a PS protocol which only achieves single hop under the throughput constraint for all the three cases. This can be explained as follows. The powers of harvesting and relaying are basically different for the PS protocol in AF relaying (if and only if $\beta_m = 0.5$, the two parts have the same power). As time (T) is fixed, even though the first relay splits a small part of power for relaying and thus harvests very little energy, in order to transfer most of the energy for later use by the subsequent relays, the first relay actually only uses a large transmission power for transmitting a weak signal because of the small portion of power splitted. In which case, the most source energy is wasted and therefore the WIT cannot be extended more than one relaying hop. This is also as undesirable as splitting a small transmission power for transmitting a strong signal. In both the above situations, the harvested power of the next relay is very small, resulting in wasting a large quantity of conserved energy. In summary, the best way of extending the network coverage and guaranteeing energy efficiency is to use the dynamic hybrid TS and PS SWIPT protocol. One should avoid using the degradation cases when $\alpha_m = 0$ or $\beta_m = 0$.

4   Conclusion

In this paper, dynamic hybrid TS and PS SWIPT protocol in a multi-hop AF relaying system has been studied. An un-closed-form expression of double hybrid parameters have been derived. Using the mathematical software, the largest number of relaying hops under a given throughput limitation together with the harvested energy and BER of each hop have been obtained. Numerical results have been presented in this paper to indicate that there is a trade-off between the power transfer efficiency and the information transmission performance under the constraint of total transmit power and throughput of each relay. The trade-off can be made via dynamically choosing the two hybrid factors α_m and β_m to achieve optimal SWIPT relaying performance.

Acknowledgement. This study is supported by the National Natural Science Foundation of China for Young Scholars under grant No. 61701167.

References

1. Laneman, J.N., Tse, D.N.C., Wornell, G.W.: Cooperative diversity in wireless networks: efficient protocols and outage behavior. IEEE Trans. Inf. Theory **50**, 3062–3080 (2004)
2. Mao, M., Cao, N., Chen, Y., Chu, H.: Novel noncoherent detection for multi-hop amplify-and-forward relaying systems. Int. J. Commun. Syst. **29**, 1293–1304 (2016)
3. Yu, M., Li, J.: Is amplify-and-forward practically better than decode-and-forward or vice versa? In: Proceedings of IEEE International Conference on Acoustics, Speech, and Signal Processing, pp. 365–368, Philadelphia, PA, USA (2005)
4. Nasir, A.A., Zhou, X., Durrani, S., Kennedy, R.A.: Relaying protocols for wireless energy harvesting and information processing. IEEE Trans. Wireless Commun. **12**, 3622–3636 (2013)
5. Lu, X., Wang, P., Niyato, D., Kim, D.I., Han, Z.: Wireless networks with RF energy harvesting: a contenmporary survey. IEEE Commun. Surv. Tutorials **17**, 757–789 (2015)
6. Pan, G., Lei, H., Yuan, Y., Ding, Z.: Performance analysis and optimization for SWIPT wireless sensor networks. IEEE Trans. Commun. **65**, 2291–2302 (2017)
7. Sbeiti, M., Tran, T., Subik, S., Wolff, A., Wietfeld, C.: MuSE: novel efficient multi-tier communication security model for emergency and rescue operations. In: 2011 IEEE Eighth International Conference on Mobile Ad-Hoc and Sensor Systems, pp. 929–934, Valencia, Spain (2011)
8. Mao, M., Cao, N., Chen, Y., Zhou, Y.: Multi-hop relaying using energy harvesting. IEEE Wirel. Commun. Lett. **4**, 565–568 (2015)

9. Li, L., Mao, M., Cao, N., Li, J.: LMMSE channel estimation for wireless energy harvesting AF relaying. Phys. Commun. **27**, 133–142 (2018)
10. Xu, K., Shen, Z., Xia, X., Wang, Y., Zhang, D.: Hybrid time-switching and power splitting SWIPT for full-duplex massive MIMO systems: a beam-domain approach. IEEE Trans. Veh. Technol. **67**(8), 7257–7274 (2018)
11. Shafie, A.E., Tourki, K., Al-Dhahir, N.: An artificial-noise-aided hybrid TS/PS scheme for OFDM-based SWIPT systems. IEEE Commun. Lett. **21**, 632–635 (2017)

A Data Acquisition and Processing Scheme Based on Edge Calculation

Yanqin Mao, JiaYan Mao, Zhenyu Wu[✉], and Subin Shen

Nanjing University of Posts and Telecommunications, Nanjing, Jiangsu, China
zhenyu.wu@njupt.edu.cn

Abstract. In order to reduce network load and network latency of centralized server in smart home system, optimization method of data acquisition and application framework based on environment intelligence are proposed. Data acquisition and processing scheme based on edge calculation is designed and implemented. Empirical results show that optimizing the data acquisition and application framework will help improve the efficiency of data acquisition. Data acquisition and processing scheme based on edge calculation can effectively reduce network communications.

Keywords: Smart home · Data acquisition · Edge calculation · ZigBee

1 Introduction

Comparing with traditional home, smart home [1] is an application of pervasive computing that provides context awareness automation or ancillary services to occupants through environmental intelligence, remote control, or home automation and provides a safer, more convenient and more comfortable living environment for human.

In the smart home system, different types of sensing devices will collect various sensing information through Bluetooth [2], Wi-Fi, ZigBee [3, 4] and other communication and networking technologies [5, 6] and then send the data to the controller, such as location information, displacement information, energy consumption information, etc. After that the controller sends the data to the remote server. With the continuous growth of sensor data in the smart home environment, centralized storage and analysis of data through the central server [7] cannot meet the real-time requirements of the application due to limited computing capability. Besides, transferring unprocessed data increases network load and causes network latency.

In order to resolve the above problems, a data acquisition and application framework based on environmental intelligence is proposed. A data acquisition and processing scheme based on edge calculation is designed and implemented. The edge sensor nodes and edge gateways are used to collect the sensor data, Data filtering, outlier handling, feature extraction and other processing, thus effectively reduce network load, reduce network latency, and meet real-time application requirements.

Q. Liu et al. (Eds.): CENet 2018, AISC 905, pp. 216–222, 2020.
https://doi.org/10.1007/978-3-030-14680-1_25

2 A Data Acquisition and Application Framework Based on Environmental Intelligence

A data acquisition and application framework based on environmental intelligence is shown in Fig. 1. This framework adopts the data collection method of environmental intelligence [8, 9]. Sensing data is collected in real-time using sensing devices deployed throughout the home environment. Considering short-range wireless communication, ZigBee network has features such as low power consumption, low cost, large network capacity, self-organizing, etc. Star topology of ZigBee network is used to realize communication between sensor nodes. Interconnection between ZigBee and Internet and data preprocessing are implemented through intelligent gateways.

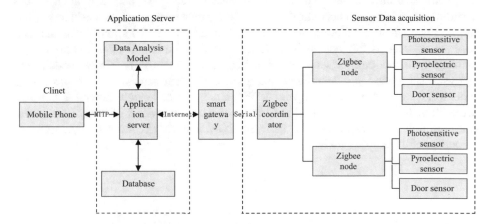

Fig. 1. A data acquisition and application framework based on environmental intelligence.

3 Design of a Data Acquisition and Processing Scheme Based on Edge Computing

At present, data acquisition and processing for smart homes mainly use centralized analysis and processing of server. This method cannot meet the increasing demand for data processing. In the data transmission process, the network load increases and the network delay increases. Considering that ZigBee nodes, intelligent gateways, and application servers have certain data analysis and processing capabilities, it is considered that part of the source data is filtered at the ZigBee node; timestamp addition, processing of missing value of outliers, and feature extraction are performed on the sensor data at the smart gateway. The application server can use a fixed-time sliding window method and a behavior recognition model to output daily behavior labels in real time and provide remote access monitoring functions.

3.1   Sensing Data Acquisition

Sensing data acquisition provides raw data for the system, including ZigBee network foundation, wireless sensor data acquisition, and data communication. Among them, ZigBee network is divided into network initialization and node joining network. Wireless sensor data collection is divided into photosensitive sensor data acquisition function, door sensor data acquisition function and pyro electric sensor data acquisition function; data communication is divided into coordinator data communication function with the terminal node and data communication function between the coordinator and the smart gateway.

The ZigBee node belongs to the edge device of the ZigBee network in the smart home data acquisition and processing application which is responsible for collecting the local environment data and the status data in real time and sending it to the coordinator. Taking into account periodic sensory data such as photosensitive sensors send a photosensitive value to the ZigBee coordinator every 10 s, but the actual data application value is low and increase ZigBee network communication overhead, and increase ZigBee node energy consumption. To make full use of the ZigBee node's data analysis and processing capabilities, a ZigBee node data acquisition and processing scheme is designed as shown in Fig. 2. When the data collection method is periodic acquisition, threshold data is used to filter the source data, which not only reduces the communication overhead and the network delay, but also reduces the energy consumption of the ZigBee node.

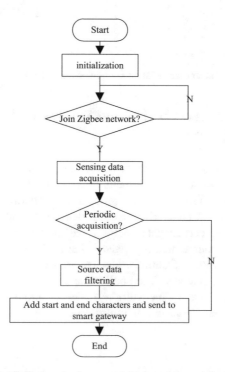

Fig. 2. A ZigBee node data acquisition and processing scheme.

3.2 Smart Gateway Data Preprocessing

The smart gateway includes data communication functions and feature extraction functions to implement interworking between the ZigBee network and the Internet and provide pre-processed data for the server. Among them, data communication is divided into intelligent gateway and coordinator serial communication function and intelligent gateway and remote server HTTP communication function; feature extraction mainly to achieve the original sensor data feature vector extraction.

The smart gateway is a bridge connecting the ZigBee network and the Internet. It has the functions of transmitting data, providing simple services, and storing and processing data. According to the data analysis capability of the smart gateway, a data processing scheme for the smart gateway is designed as shown in Fig. 3 which places the data pre-processing part of the application system at the smart gateway to prevent the growing sensor data from increasing the number of application servers and improve system response time and real-time performance.

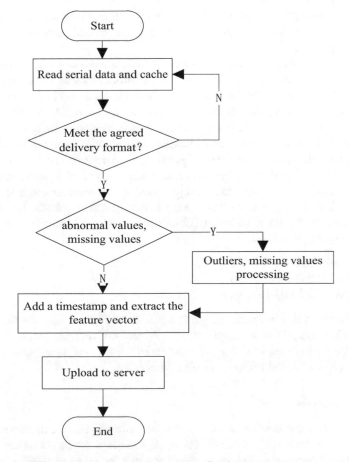

Fig. 3. A data processing scheme for the smart gateway.

4 Solution Implementation and Testing

4.1 Solution Implementation

Sensing data acquisition adopts the integrated development environment developed by IAR Systems of Sweden. The sensor node adopts CC2530 produced by Texas Instruments TI, the photosensitive sensor adopts Risym 5549, the pyro electric sensor adopts HC-SR501, and the intelligent gateway adopts S3C2440 development board.

Pyro electric sensors and gate sensors adopt the triggering method that requires defining a corresponding external interrupt function and processing the sensing event. For the door sensor, the signal changes from 0 to 1 when the door is opened, and an ON message is sent to the coordinator. When the door is closed, the signal changes from 1 to 0, and an OFF message is sent to the coordinator. For the pyro electric sensor, when the person enters the pyro electric sensor detection range, the signal changes from 0 to 1, and sends an ON message to the coordinator; when the person leaves the pyro electric sensor detection range, the signal changes from 1 to 0, and send an OFF message to the coordinator.

The smart gateway mainly realizes the feature extraction of acquired data and the communication with the application server. It is developed using the Python programming language.

The feature extraction function is to extract the feature data of the sensor data and define four feature vectors as date, time, sensor identifier and sensor message. Wherein, the date and time, that is, the date and time of sensing data acquisition, provides time series; the sensor identifier identifies the sensor type, for example, M indicates that the position sensor L represents the photosensitive sensor, and since the sensor is deployed in a deterministic manner, the sensor identifier also identifies the sensor deployment. Location information; the sensor message refers to the perceived local environment state. The message has discrete data and continuous data such as position sensor and gate sensor message as discrete data, and photosensitive sensor message as continuous data. The following is the extracted feature vector, the first column is the date, the second column is the time, the third column is the sensor identifier, and the fourth column is the sensor message.

2018-01-16 10:03:15 M03 ON
2018-01-16 10:04:42 L04 00
2018-01-16 10:19:09 D01 ON

Interaction function between the smart gateway and server uses the urllib module and the urllib2 module to establish an HTTP long connection, sends the extracted feature vector through the POST request and waits for the server to respond. When the response code is 200, the response result is processed.

4.2 Test Analysis

Wireless sensor data collection is divided into photosensitive data acquisition, gate data acquisition and pyro electric data collection; all test cases are based on the success of ZigBee network formation. The specific test settings are shown in Table 1.

Table 1. Raw data of different sensors.

Type of sensor	Setting	Raw data
Photo sensor	1. Photosensitive sensors connect the pin of terminal node 2. USB data cable connection coordinator serial port module and computer serial port 3. Open the serial debugging assistant, set the baud rate as 9600, set serial as COM1 4. The light illuminating sensor is turned off after 10 s	The serial port debugging assistant displays: $L04 XX# The specific value increases from small to large
Pyro electric sensor	1. Pyro electric sensors connect the pin of terminal node 2. People close to the sensor and leave	The serial port debugging assistant displays: People close $M03 ON# People leave $M03 OFF#
Door sensor	1. Door ensures connect the pin of terminal node 2. Close the door after opening the door for 10 s	The serial port debugging assistant displays: Door open $D01 ON# Door close $D01 OFF#

Use the manual test method to input three kinds of raw data to be tested to the serial port. Start the remote server and switch to Debug mode, and set a breakpoint at the getParameter code of the request method corresponding to the doPost method to view the received feature vector from smart gateway. Test results are shown in Table 2.

Table 2. Test results.

NO.	Input data	Test results
1	$M03 ON#	2018-01-16 10:03:15 M03 ON
2	$L04 00#	2018-01-16 10:04:42 L04 00
3	$D01 ON#	2018-01-16 10:19:09 D01 ON

The test results show that when the serial port inputs the pyro electric sensing data type, the door sensing data type and the photosensitive sensing data type, the server side checks the test result in the Debug mode to be consistent with the prediction result, which proves that the functions of the intelligent gateway module work well.

5   Conclusion

For the continuous growth of sensor data in the smart home environment, the server cannot meet the data processing requirements by adopting centralized analysis storage, and the increase of the network load during the data transmission leads to a long network delay problem. To solve these problems, the innovation of this paper is to propose an edge computing application framework. With the computation and storage capabilities of ZigBee nodes, smart gateways and the application server, a data collection and preprocessing scheme for smart homes is proposed using edge calculation methods. However, the performance of the solution has not been quantitatively analyzed and compared. In the future, relevant work in conjunction with actual application will be carried out.

Acknowledgement. This study is supported by National Natural Science Foundation of China (61502246) and Innovative Research Joint Funding Project of Jiangsu Province (BY2013095-108).

References

1. Alam, M.R., Reaz, M.B.I., Ali, M.A.M.: A review of smart homes-past, present, and future. IEEE Trans. Syst. Man Cybern. Part C **42**(6), 1190–1203 (2012)
2. Piyare, R., Tazil, M.: Bluetooth based home automation system using cell phone. In: International Symposium on Consumer Electronics, pp. 192–195. IEEE, New York (2011)
3. Byun, J., Jeon, B., Noh, J., Kim, Y., Park, S.: An intelligent self-adjusting sensor for smart home services based on ZigBee communications. IEEE Trans. Consum. Electron. **58**(3), 794–802 (2012)
4. Liu, W., Yan, Y.: Application of ZigBee wireless sensor network in smart home system. Int. J. Adv. Comput. Technol. **3**(5), 154–160 (2011)
5. Lee, J.S., Su, Y.W., Shen, C.C.: A comparative study of wireless protocols: Bluetooth, UWB, ZigBee, and Wi-Fi. In: Conference of the IEEE Industrial Electronics Society, pp. 46–51. IEEE, New York (2008)
6. Chandane, M.M., Bhirud, G.S., Bonde, V.S.: Performance analysis of IEEE 802.15.4. Int. J. Comput. Appl. **40**(5), 23–29 (2012)
7. Banerjee, A., Sufyanf, F., Nayel, M. S., Sagar, S.: Centralized framework for controlling heterogeneous appliances in a smart home environment. In: 2018 International Conference on Information and Computer Technologies, pp. 78–82. IEEE (2018)
8. Acampora, G., Cook, D.J., Rashidi, P.: A survey on ambient intelligence in healthcare. Proc. IEEE Inst. Electr. Electron. Eng. **101**(12), 2470 (2013)
9. Che-Bin, F., Hang-See, O.: A review on smart home based on ambient intelligence, contextual awareness and Internet of Things (IoT) in constructing thermally comfortable and energy aware house. Res. J. Appl. Sci. Eng. Technol. **12**(7), 764–781 (2016)

Research on Controllability
in Failure-Recovery Process
of Dynamic Networks

Ming Deng, Yi Gong, Lin Deng[✉], and Zezhou Wang[✉]

The 29th Institute of CETC, Chengdu 610036, China
1302191330@qq.com, 350276267@qq.com

Abstract. The recovery model of complex networks is an important method to research the robust performance of the network. When a complex network dynamic failure occurs, the stability of its controllability measure is a prerequisite for ensuring full control of the network and failure network reconfiguration. In order to explore the trend of the controllability measure of the dynamic failure network model and the correlation between the parameters and the network controllability. This paper introduces the recovery model proposed by Majdandzic. Based on this model, the simulation model of spontaneous failure-recovery dynamic for the node was proposed by using the dynamic recovery mechanism. And the dynamic change process of the structure controllability in the dynamic recovery model under different parameters is analyzed. The simulation results show that when the network node has a spontaneous failure-recovery dynamic, its controllability measure has a significant phase transition phenomenon, and the position of the phase transition point shows a different trend with the probability of recovery, the activity threshold and the adjustment of network average degree.

Keywords: Complex networks · Controllability · Dynamic failure ·
Recovery model

1 Introduction

In the past 20 years, researchers have conducted extensive and in-depth research on the structure and parameter changes of dynamic networks in recovery models [1–9]. Among them, the cascade recovery model [10] proposed by Motter et al. is the most concerned. The model focuses on the dynamic impact of fault propagation on the network caused by the node exceeding the load limit and it's mainly used to analyze the impact of fault propagation process on network topology parameters, including network connectivity, network community, network controllability etc. under different conditions [11–16].

The cascade recovery model provides an effective model framework for studying the dynamic failure process of complex networks, analyzing the changes of network topology characteristics and formulating network defense strategies. However, there is still a type of system in the real world. When a node fails in the network, it will affect the neighboring node with a certain probability, but at the same time, it will return to

© Springer Nature Switzerland AG 2020
Q. Liu et al. (Eds.): CENet 2018, AISC 905, pp. 223–237, 2020.
https://doi.org/10.1007/978-3-030-14680-1_26

the normal state within a certain time interval. The above-mentioned dynamic failure process can be called a spontaneous failure-recovery model, which often appears in the network of financial systems and nervous systems [17]. But so far, there are few studies been found on the impact of such dynamic failure processes on network controllability. Therefore, researching on the spontaneous failure-recovery model can provide a new idea for studying the trend of network controllability in the dynamic failure process.

In order to analyze the controllability of complex networks based on spontaneous failure-recovery dynamic model of the node, this paper introduces the network recovery model proposed by Majdandzic [17] and Xiao [18], and the improved recovery model is proposed via introducing the external recovery probability r. By adjusting the recovery probability, activity threshold and network average degree in the model, the dynamic change process of the network controllability measure under the recovery model and the correlation with the above three model parameters are analyzed. In addition, the numerical simulation in this paper also consider the trend of the controllability measure in the recovery model under the directed BA scare-free network, undirected BA scare-free network 1 and ER network 2. And finally, we analyzed the law of phase transition point migration for controllability measure in the recovery model under different type of networks.

2   The Failure-Recovery Process of Dynamic Network

2.1   Spontaneous Failure-Recovery Model

The spontaneous failure-recovery model was first proposed by Majdandzic [17] to study the spontaneous recovery capability of dynamic networks in failure states. In reference, based on the average activity of the network $\langle z \rangle$, the dynamic repair capabilities of the regular network, ER random network and scale-free network are studied. The internal relationship between the internal failure probability and the external failure probability is analyzed. The critical value of the phase change of parameter $\langle z \rangle$ is obtained. Here, we give a brief description of the spontaneous failure-repair model analysis framework.

There are two states (active and inactive) to define the status of the nodes in the failure-recovery model of the dynamic networks. First, assume that the network node generates internal failures with probability within the same time interval, and the failed nodes do not directly cause the state changes of the remaining normal nodes. Then, the node external failure policy is established, that is, when the total number of active nodes of the neighbor nodes of the node is less than a fixed integer, the neighbor node will cause the node itself to fail by probability, and otherwise remain active. Finally, the network itself has a recovery process, the internal failed node will automatically recover within the set time period, and the external failed node will automatically recover within the time period. Compared to the failure process in the Watts model [19], an increased spontaneous recovery strategy will cause the network to spontaneously switch between crash and recovery states, a feature that is common in transportation networks.

2.2 Improved Spontaneous Failure-Recovery Model

In Sect. 2.1, Majdandzic attributes the failure process to the internal failure and external failure of the node, and gives the failure probability respectively. However, when discussing the recovery process of the network, only the failed node is defined to recover automatically within a certain time interval. This modeling method obviously weakens the practical problems faced by network failure nodes in the recovery process. In a real network, there are usually two recovery states after a node fails [18]. One is internal recovery. For example, when an airport node in an aviation network encounters extreme weather, the node is functionally in a failed state, but as the weather improves, the failure state is reversible, and the process of its failure can be determined by setting the time interval and the recovery probability. The other is the recovery of the failed node caused by external factors. Similarly, in the case of an aviation network, when the airport node fails due to the number of flights exceeding the load, it cooperates with other nearby airports (transfer or change the landing place of the proposed landing flight). The mode is effective to reduce the overload state of the failed node so that the failed node returns to normal.

Based on the above analysis, Xia [18] established an improved spontaneous failure-recovery model by adding the internal and external factors recovery rules of the failed nodes. In literature [18], the concept of repairing energy constraint is introduced, and the repair cost function is established which analyze the change of the robustness index of the network arrival steady state and the cost change curve of the network repair in the spontaneous failure-recovery model. The following is a description of the dynamic recovery process of the improved spontaneous failure-recovery model.

In the model, the network is a single network without coupling edges, and statistical models are used to establish ER networks and scale-free networks that meet the requirements. There are two states of active and inactive nodes in the network. The definition of the failure process of the node refers to Sect. 2.1. There are internal recovery and external recovery in the recovery process of the node. The former assumes that the node fails at the moment, and it must be restored at the moment. The latter considers the synergy effect of the neighbor nodes in the failure process and sets the activity evaluation of the adjacent nodes. Standard, determine whether the failed node meets the recovery conditions. If the condition is met, the failed node recovers with probability at the moment.

3 Numerical Simulation and Analysis

In this section, numerical simulation studies the controllability of the failure-recovery process in dynamic networks from two aspects. On the one hand, consider the impact of improved network recovery mechanism on network controllability. On the other hand, consider the influence of important parameters of the network in the spontaneous failure-recovery process on the structure controllability of the network. Therefore, this paper uses the proposed model and the model in the literature [17] for comparative simulation. The specific simulation settings are as follows. The typical directed ER network and Scale-Free network are used in this simulation. During the simulation, the internal failure probability of the network is adjusted from 0 to 1. The activity threshold of the node due to the failure of the neighbor node is the external failure probability, the time interval of the spontaneous recovery of the internal failure node is considered, and the time interval of the spontaneous recovery of the external failure node set to 1. At the same time, the simulation will also analyze the recovery model parameters and the influence of network topology parameters on structural controllability from three aspects, including (1) external failure probability; (2) activity threshold; (3) network average degree; simulation will be repeated 50 times and the average value is taken as the final result.

3.1   Spontaneous Failure-Recovery Model

(1) External failure probability r

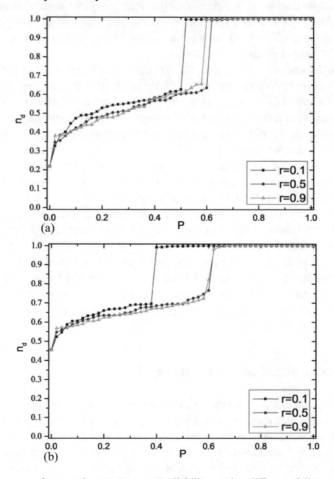

Fig. 1. The curve of network structure controllability under different failure probability r. Figs. (a), (b) and (c), (d) are simulation results of directed and undirected ER networks and SF networks respectively. In the simulation process, The activity threshold is set to $k = 0.5$, the network scale $N = 1000$, the average degree $< k > = 2$, and the power rate coefficient of the scale-free network $\gamma = 2.5$.

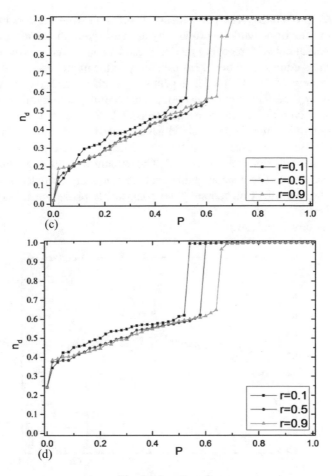

Fig. 1. (*continued*)

Figure 1 shows the trend of the network structure controllable under three failure probability r. It can be seen from the figure that the controllability n_d in the four networks all have controllable phase transitions within the range $p \in [0.5, 0.7]$, and the phase transition points of the curves under the three failure probabilities are consistent with the phase transition points of the network average activity index $<z>$ in the literature [17]. It shows that in the spontaneous failure-recovery model, the network controllability is significantly related to the $<z>$. When the network is in the high-active area, the controllability measure changes with the increase of the network node failure probability, and the corresponding controllability decreases. When the network is transformed from a high-active area to a low-active area, the network is directly controlled from theoretical controllable to uncontrollable. In addition, in different network models, the offset law of controllable phase transition points also differs.

Observing the undirected network in Figs. 1(c), (d), the theoretically controllable phase transition point shifts to the right with the increase of the external failure probability, and the robustness of the controllability increases with the increase of the external failure probability. The phenomenon is different from the conclusion that the robust performance of network controllability decreases with the increase of the failure ratio in the general network recovery model. It can be considered that the enhancement of this robustness comes from the repair capability of the model itself, and the recovery control of the network in the model depends on With internal recovery time τ and external recovery time. The trend of network controllability at that time is equivalent to the conclusions given in the general network model. In the directed networks of Figs. 1 (a), (b), the structure controllable phase transition point tends to shift to the right as the probability of failure increases, but as the probability of failure increases to a certain extent, the phase transition point is biased. The amount of shift is gradually reduced. This shows that in the spontaneous failure-recovery model, the external failure probability has a limited effect on the structure controllable phase transition point of the directed network.

(2) The activity threshold k

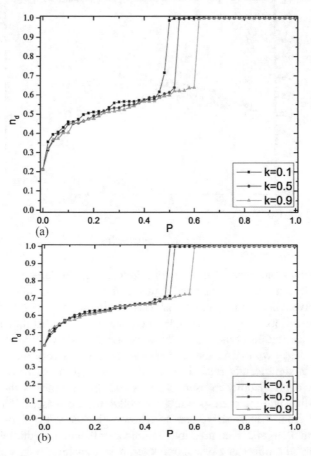

Fig. 2. The curve of the structure controllable under different activity thresholds. Figs. (a), (b) and (c), (d) are simulation results of directed and undirected ER networks and SF networks, respectively. In the simulation process, the external failure probability is set to $r = 0.5$, the network scale $N = 1000$, the average degree $<k> = 2$, and the power rate coefficient of the scale-free network $\gamma = 2.5$.

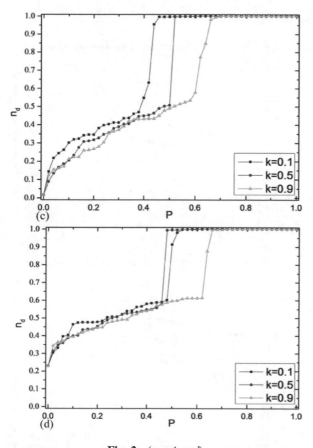

Fig. 2. (*continued*)

In Fig. 2, as the activity threshold increases, there is a significant difference in the structure controllable phase-change point migration trend between the ER network and the SF network. In the ER network, the greater the activity threshold, the stronger the structure controllable persistence of the network, and the more the phase change point is shifted to the right. In the SF network, the activity threshold has a tendency to decrease first and then increase the position of the structure controllable phase change point. At $k = 0.5$, the SF network controllable phase change point is shifted to the left relative to the position with $k = 0.1$, and when $k = 0.9$ the controllable phase change point is the right to the position with $k = 0.1$. The results indicate that the SF network structure controllable phase transition point has a quadratic dependence on the activity threshold. In addition, from the perspective of the minimally required drive nodes which can control the whole network, under the same failure conditions, the ER network is theoretically easier to control than the SF network, and the undirected network is theoretically easier to control than the directed network.

(3) The average degree $< k >$

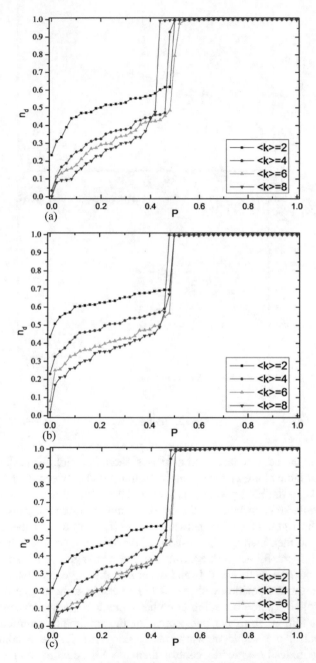

Fig. 3. The curve of the structure controllability under the different average degree. Figs. (a), (b) and (c), (d) are simulation results of directed and undirected ER networks and SF networks, respectively. In the simulation process, the external failure probability is set to $r = 0.5$, the network scale $N = 1000$, the activity threshold $k = 2$, and the power rate coefficient of the scale-free network $\gamma = 2.5$.

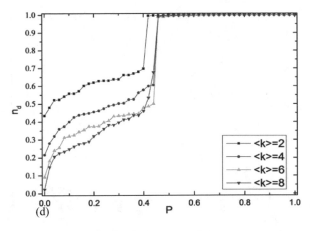

Fig. 3. (*continued*)

In Fig. 3, due to the different density of the network, the initial network controllability are different. The structure controllability of different networks is consistent with Figs. 1 and 2. In the ER network, the position of the structure controllable phase change point shows a tendency to shift to the left and then to the right with the increase of the average degree. The position of the structure controllable phase change point in the undirected SF network monotonically shifts to the right with the average degree $< k >$, while the phase transition points of different averaging degrees in the directed SF network are in the same position. Overall, the intensity of the network has a weaker impact on the offset of the theoretically controllable phase transition point. In contrast, the external failure probability and the active threshold have a more pronounced effect on the phase transition point.

Based on the above three sets of numerical simulation results, the structure controllability measure has obvious phase transition processes in the spontaneous failure-recovery model. The structure controllability of the network always appears in the high-active area of the node of the model, and the phase transition point from the high-active area to the low-active area is also the phase change point that the network can be theoretically controlled. The external failure probability, active threshold and average degree of the network have different degrees of influence on the offset trend of the phase transition point. The structure controllability measure shows weak linearity, linearity and quadratic dependence under the above three influencing factors. In addition, the ER network is easier to control during network dynamic failure, while the number of driver nodes required for the three-group numerical simulation of the undirected ER network is lower than other networks.

3.2   Improved Spontaneous Failure-Recovery Model

(1) The average degree $< k >$

In Fig. 4, which can be seen from the analysis of the curve of the structure controllability that its controllability also presents two stages. when the network is in the

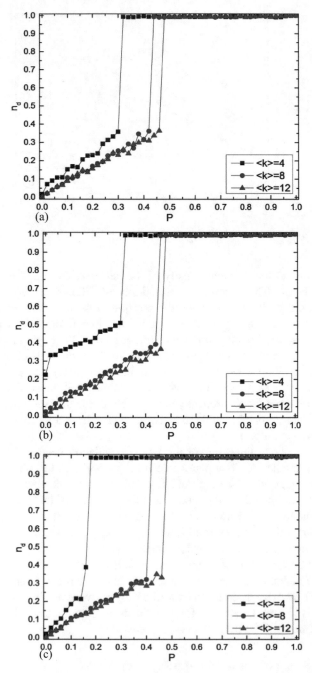

Fig. 4. The curve of the structure controllability under the different average degree $<k>$. Figs. (a), (b) and (c), (d) are simulation results of directed and undirected ER networks and SF networks Parameter setting: Model external recovery probability $r = 1$, activity threshold $k = 0.5$, internal recovery time interval $\tau = 5$, network scale $N = 1000$, $\gamma = 2.5$.

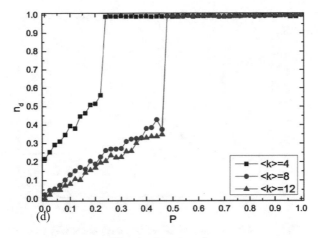

(d)

Fig. 4. (*continued*)

high-active area, the corresponding network controllability has a linear dependence on the failure probability; when the network is in the low-active area, the corresponding network directly changes to an uncontrollable state. The exponent has the same critical probability of phase transition as the structure controllable measure, and the value gradually shifts to the right as the value rises. In addition, in the undirected network, when the exponential decreases linearly with the probability of failure, the structure controllable measure still transitions to an uncontrollable state under the critical probability. It is indicated that in such dynamic recovery models, the controllability measure always has a critical phase transition probability.

(2) External recovery probability r

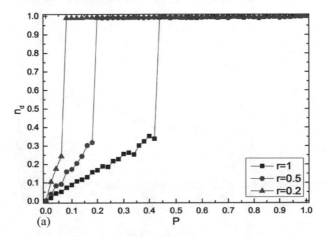

(a)

Fig. 5. The curve of the structure controllability under different external recovery probability. Figs. (a), (b) and (c), (d) are simulation results of directed and undirected ER networks and SF networks Parameter setting: activity threshold $k = 0.5$, the average degree $< k > = 8$, internal recovery time interval $\tau = 5$, network scale $N = 1000$, $\gamma = 2.5$.

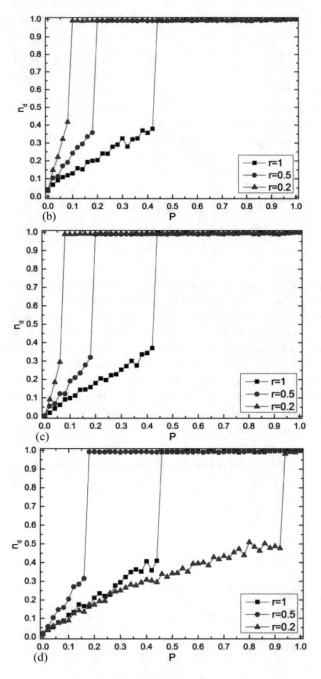

Fig. 5. (*continued*)

As shown in Fig. 5, the network structure controllable measure curve has a significant dependence on the external recovery probability r. On the basis of the improved repair strategy, the robustness of the network controllability is improved. Also in an undirected SF network, when the external recovery probability is exceeded, the network controllability maintains its value under the failure probability changes in a long range. As can be seen from Figs. 4 and 5, the external recovery strategy can improve the average activity index of the network while ensuring the robustness of the network controllability. Compared with the original model in Sect. 3.1, the recovery ability of the model in the dynamic failure process is improved (Fig. 6).

(3) The activity threshold k

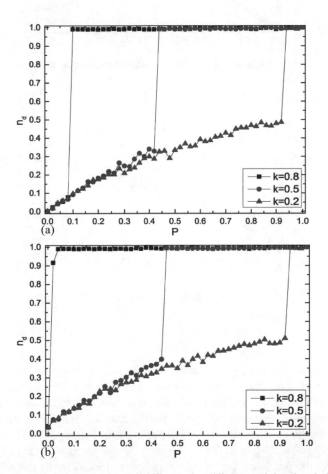

Fig. 6. The curve of the structure controllability under different activity threshold k. Figs. (a), (b) and (c), (d) are simulation results of directed and undirected ER networks and SF networks Parameter setting: Model external recovery probability $r = 1$, internal recovery time interval $\tau = 5$, network scale $N = 1000$, $\gamma = 2.5$.

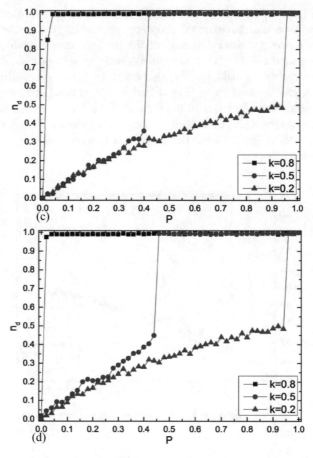

Fig. 6. (*continued*)

4   Conclusion

The research work in this paper mainly focuses on the trend of the controllability measure of complex networks in the dynamic failure process. The numerical simulation results show that when the network node has spontaneous failure-recovery dynamics, as the internal failure probability of the node increases, there is a phase transition phenomenon in the change curve of the network controllability. At the same time, the external failure probability and the model activity threshold will cause the controllability measures the positive offset of the critical phase transition point, while the change of the network average degree $<k>$ is not significantly related to the phase transition point. After improving the node recovery mechanism, the phase transition points of the controllability measure under the same conditions have shifted to the right, which significantly improves the robustness of the network controllability. In the improved model, the external failure probability, the model activity threshold, and the network average degree will cause the controllability measure to change the position of the phase transition point. In addition, by comparing the trend of the controllability measure of ER network and BA scale-free network in numerical simulation, it can be seen that there is no obvious correlation between the two types of network models of the same scale and the change of controllability measure in the process of node recovery.

Acknowledgement. This study is supported by technical basis scientific research plan channel of State Administration of Science, Technology and Industry for National Defense (JSZL201610B001).

References

1. Barabási, A.L.: Linked: The New Science of Networks. Perseus, Cambridge (2002)
2. Watts, D.J.: Six Degrees: The Science of a Connected Age. Norton, New York (2003)
3. Sachtjen, M.L., Carreras, B.A., Lynch, V.E.: Disturbances in a power transmission system. Phys. Rev. E: Stat. Phys. Plasmas Fluids Relat. Interdisc. Top. **61**(5A), 4877 (2000)
4. Glanz, J., Perez-Pena, R.: 90 seconds that left tens of millions of people in the dark. New York Times (2003)
5. Moreno, Y., Gómez, J.B., Pacheco, A.F.: Instability of scale-free networks under, node-breaking avalanches. J. Complex Netw. **58**(4), 630–636 (2007)
6. Herrmann, H.J., Roux, S.: Statistical models for the fracture of disordered media. Cryst. Res. Technol. **26**(8), 1076 (1990)
7. Moreno, Y., Gomez, J.B., Pacheco, A.F.: Fracture and second-order phase transitions. Phys. Rev. Lett. **85**(14), 2865–2868 (2000)
8. Moreno, Y., Pastor-Satorras, R., Vázquez, A., Vespignani, A.: Critical load and congestion instabilities in scale-free networks. Europhys. Lett. **62**(2), 292–298 (2003)
9. Motter, A.E., Lai, Y.C.: Cascade-based attacks on complex networks. Phys. Rev. E: Stat., Nonlin, Soft Matter Phys. **66**(2), 065102 (2002)
10. Motter, A.E.: Cascade control and defense in complex networks. Phys. Rev. Lett. **93**(9), 098701 (2004)
11. Crucitti, P., Latora, V., Marchiori, M.: Model for cascading failures in complex networks. Phys. Rev. E: Stat. Nonlin. Soft Matter Physics **69**(4 Pt 2), 045104 (2004)
12. Kinney, R., Albert, R.: Modeling cascading failures in the north american power grid. Eur. Phys. J. B: Conden. Matter Complex Syst. **46**(1), 101–107 (2005)
13. Overbye, T.J.: Reengineering the electric grid. Am. Sci. **88**(3), 220–229 (2000)
14. Electricity Technology Roadmap, 1999 Summary and Synthesis. http://www.epri.com/corporate/discover_epri/roadmap/
15. Liu, Y.Y., Slotine, J.J., Barabási, A.L.: Controllability of complex networks. Nature **473**(7346), 167–173 (2011)
16. Pósfai, M., Liu, Y.Y., Slotine, J.J.: Effect of correlations on network controllability. Sci. Rep. **3**(3), 1067 (2013)
17. Majdandzic, A., Podobnik, B., Buldyrev, S.V., Kenett, D.Y., Havlin, S., Stanley, H.E.: Spontaneous recovery in dynamical networks. Nat. Phys. **10**(1), 34–38 (2013)
18. Xiao, W., Yang, C., Yang, Y.P., Chen, Y.G.: Phase transition in recovery process of complex networks. Chin. Phys. Lett. **34**(5), 132–136 (2017)
19. Watts, D.: A simple model of global cascades on random networks. Proc. Nat. Acad. Sci. U. S.A. **99**(9), 5766–5771 (2002)

Machine Learning and Application

This chapter aims to examine innovation in the fields of deep learning, algorithmic design, similarity analysis and clustering analysis. This chapter contains 25 papers on emerging topics on machine learning and application.

Learning High Level Features with Deep Neural Network for Click Prediction in Search and Real-Time Bidding Advertising

Qiang Gao[1(✉)] and Chengjie Sun[2]

[1] Information Center of Shandong Province People's Congress Standing
Committee General Office, Jinan, China
172663066@qq.com
[2] Harbin Institute of Technology, 92 West Da-zhi Street, Harbin, China

Abstract. Here you can write the abstract for your paper. Sponsored search advertising and real-time bidding (RTB) advertising have been growing rapidly in recently years. For both of them, one of the key technologies is to estimate the click-through rate (CTR) accurately. Most of current methods utilize shallow features, such as user attributes, statistical data. As in sponsored search advertising and RTB advertising, all parties are connected because of the interests from users, hence the user features may contain richer latent factors or abstract information on higher levels which are helpful to improve the accuracy of click prediction. Based on this assumption, the object of this paper is to use high level features learned from basic features, specially user features, to improve the performance of CTR. A deep neural network framework is proposed to learn the high level features in this work. The proposed framework consists of two different deep neural network model in order to process different types of user features respectively. Experimental results on sponsored search advertising dataset and RTB advertising dataset show that the learned high level features can improve the accuracy of click prediction.

Keywords: Click-through rate prediction · Deep neural network ·
Real-time bidding advertising

1 Introduction

With the rapid development of the Internet, online advertising plays an increasingly important role in the IT industry. Its revenue has increased from $6.0 billion in 2002 to $36.6 in 2012 [1]. Facing to such a rapid and scaled growth, Broder proposed the concept of computational advertising [2], and prepared to use computational methods to improve the efficiency of online advertising.

Nowadays there are many forms of online advertising [3] including sponsored search advertising [4, 5], real-time bidding auctions, display advertising, contextual advertising and so on [6, 7]. No matter what the advertising it is, the important key is the click-through rate (CTR). In recent years, click prediction about the on-line advertising has received increasing attention from the industry and research communities. Most works in the ads are either new features or new models development.

© Springer Nature Switzerland AG 2020
Q. Liu et al. (Eds.): CENet 2018, AISC 905, pp. 241–252, 2020.
https://doi.org/10.1007/978-3-030-14680-1_27

In terms of model development, decision trees [8] and maximum entropy [9, 10] are common models used for click prediction in on-line advertising. Models based on test hypothesis, such as a dynamic Bayesian network [11] model, click chain model and so on, also have achieved good results. In addition, the matrix factorization [12] is utilized in CTR predicting. But traditional matrix factorization method is two-dimension, in advertisement the dimensions is more, not only user and ad, but also media, advertiser and so on.

In terms of feature development, Chakrabarti et al. [13] used click feedback features that are based on the historical click data to improve the click prediction result. Liu et al. [14] put forward to utilize syntactic features to model the relevance between advertisement and query for sponsored search by viewing the ad's text as a document. Cheng et al. [9] took the statistical information about the attribute and the number of user to calculate CTR. There are some methods using recommendation or topic extraction technology to do query expansion in order to select more elect features.

In deep learning field, extensive studies have been done by Hinton and his co-workers [15, 16], who initially propose the deep belief nets (DBN). Now deep learning and unsupervised feature learning have shown great promising in many practical applications, such as speech recognition [17], visual object recognition [18] and text processing [19]. Many of them are using deep network to reduce feature dimensions or to learn new representations of features from data to get better results.

For CTR prediction problem, the traditional methods often involve huge amount of surface features. This work proposes to use deep learning methods to obtain new representation of features in order to get better performance. We explore the importance of the basic (low level) feature, then use the stable prediction models to select some features and input them to different deep designed neural network structure. Finally, we combine the output of the deep neural network with the other features and put them into the logistic regression prediction model. No matter on sponsored search advertising dataset or RTB dataset, the results of experiments show that our approach improved the AUC in different degree.

2   Method

In our method, we first extract some basic feature (low level feature) and do feature selection. After that, deep network framework is utilized to learn the high level features for CTR estimating from the selected low level features. The whole framework is shown in Fig. 1.

2.1   Basic Features

As we know, features are important for supervised learning methods. In the CTR prediction problem, many features come from users, ads and publishers. On the publisher's side, there are some display urls, formats and so on. On the user's side, some demographic information, such as gender, age and nationality, are common features that are used for click prediction and other user behavior information. On advertisement's side, there are many attributes about the ad such as the size of ad slot and the

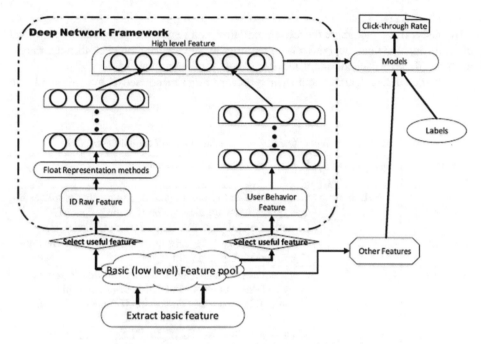

Fig. 1. The framework of the proposed method.

visibility of the ad. The features we extracted are as follows. And we call them low-level features.

Users' Behavior Features
We observe that different users use different browsers while surfing on the Internet. Meanwhile, we find the top browsers maybe not generate more click actions.

Besides the user-agent, there is another important field in the logs, user-tags, that can more accurately indicate the user's behavior. A 44-dimension vector is used to represent the tags, and the value of each dimension is binary value.

Raw ID Features
Each ID is taken as a feature with binary value. Some pruning rules are used to reduce the feature dimension. Taking the user ID as an example, totally we have 23,907,634 user IDs in one dataset, but just 3,291,677 ids make click actions.

ID Value Features
Apart from the above raw ID feature, in KDD Data set, transformed ids are provided by the organizer. We find the value of the transformed id also contains useful information.

Similarity Features
We take the query, keyword, title and description as documents, and calculate the tokens' TF-IDF values, then use the TF-IDF vector to get the cosine similarity of each other.

Historical CTR

Obviously, if we recognize the CTR prediction as a regression problem, the historical click-through rate can be a referenced feature. So we calculate the click-through rate of each ad id, advertiser id, query id, user id, url, user-agent and so on.

All the surface features and their representations mentioned above are shown in Table 1.

Table 1. Features types, features representations and detailed descriptions.

Features type	Feature Rep.	Feature description and representation
User behavior	f_userBehvior	user tags (f_userTag); user-agent (f_ua); number of tokens in query, title, keyword, description (f_numQ, f_numT, f_numK, f_numD); gender (f_gender); age (f_age)
Raw ID	f_rawID	user id (f_uid); ad id (f_aid); query id (f_qid); keyword id (f_kid); position (f_pos); url (f_url)
Similarity	f_sim	query and keyword (f_simQK); query and title (f_simQT); query and description (f_simQD); keyword and title (f_simKT); keyword and description (f_simKD); title and description (f_simTD)
Historical CTR	f_ctr	ad ctr (f_adCTR); user ctr (f_userCTR); advertiser ctr (f_aderCTR); query ctr (f_queryCTR); url ctr (f_urlCTR); user-agent ctr (f_uaCTR); creative ctr (f_creCTR)
ID value	f_valID	user id value (f_valU); query id value (f_valQ)

2.2 Deep Network for High Level Feature

In this work, we consider using deep network to learn latent feature representation (called high-level features henceforth) embedded in the surface features. As Fig. 1 shown, after selecting the useful feature, we split the features into three types, raw ID feature related to user, user's behavior feature, and other features. Then, we take two deep networks with different hidden layers to different types of features to learn high-level features. The basic building block of the two deep networks is the restricted Boltzmann Machine (RBM). After getting the high level features, we join them with the third types feature to form the input of our logistic regression prediction model.

Restricted Boltzmann Machines (RBM). A RBM is a generative stochastic neural network that can learn a probability distribution over its set of inputs [20]. It is a two-layer network [21], which consists of m visible units $\mathbf{V} = (V_1, \cdots, V_m)$ representing observable data, also named input layer, and n hidden units $\mathbf{H} = (H_1, \cdots, H_n)$ named hidden layer. The RBM is a undirected graph and the weight between input layer and hidden layer is $\mathbf{W}$.

For visible layer $\mathbf{V}$, we use V_i denote the i th state of the visible unit, H_j represent the jth state of the hidden unit, and w_{ij} is the weight associated with the edge between units V_i and H_j. For a_i and b_j, they are respectively express the real valued bias terms associated with the ith visible and the jth hidden variable.

So given a state $(\mathbf{V}, \mathbf{H})$ takes values (v, h), the energy function of the RBM is formula (1).

$$E(v, h) = -\sum_{i \in visible} a_i v_i - \sum_{j \in hidden} b_j h_j - \sum_{i,j} v_i h_j w_{ij} \tag{1}$$

With the parameters of the model, it is joint distribution is

$$p(v, h) = \frac{1}{Z} \exp(-E(v, h)) \tag{2}$$

where $Z = \sum_{v,h} \exp(-E(v, h))$.

According to the special structure of the RBM (the connections is just between the layers not between two variables of the same layer), the hidden variables are independent given the visible variables and vice versa. So the probability of the jth hidden unit is

$$p(h_j = 1|v) = \sigma\left(b_j + \sum_{i=1}^{m} w_{ij} v_i\right) \tag{3}$$

and when given the hidden variables, the probability of the ith visible unit is formula (4), where $\sigma(x) = \frac{1}{1+e^{-x}}$ denotes the sigmoid function. The parameters is updated by Contrastive Divergence [22] algorithm.

$$p(v_i = 1|h) = \sigma\left(a_i + \sum_{j=1}^{n} w_{ij} h_j\right) \tag{4}$$

Training the Deep Network. We use two deep networks (DN) to process the raw id features the user behavior features respectively. The details of our deep networks are illustrated in Fig. 2. There are four layers in DN for raw ID features. The function of the first layer is to re-construct the raw feature. The rest of layers are used to reduce the dimension of the features and make them to be dense real value rather than sparse values. The DN for user behavior features has one hidden layer to re-construct the general feature and mine the latent factors under the surface features.

Here we take the DN for user behavior features as an example to explain the training process. First, the user information and the behaviors are fetched as the primitive features and formed the input vector. Then the bottom layer generates the hidden layers, based on the formula (3). The hidden vector calculate the visible layer vector, used formula (4). We update the parameters using 1-step CD algorithm.

Estimate the CTR. After the two deep networks establish the high level features, we will add the features of the other types in the feature pool. Then, logistic regression model is involved to make the final prediction.

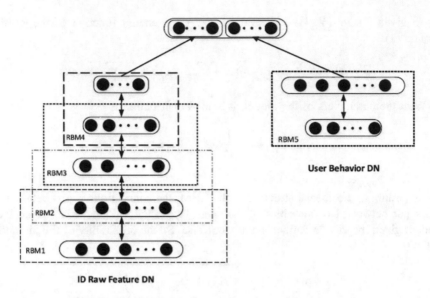

Fig. 2. The detailed design of the proposed deep network.

3 Dataset

In this work, we use two public dataset. The first dataset (DataSet_KDD) is about sponsored search advertising, which derived from the track 2 of KDD Cup 2012 provided by Tencent. The second dataset (DataSet_iPinyou) is about the RTB advertising that provided by iPinYou company [23], which consists of a set of processed iPinYou DSP bidding, impression, click, and conversion logs from some advertising campaigns among seven days.

In DataSet_KDD, the training set contains 149,639,105 instances that comes from log messages of search sessions, where a search session refers to an interaction between a search engine and a user. It contains the following ingredients: the user, the query issued by the user, some ads returned by the search engine to the user, and zero or more ads that were clicked by the user. Every session is divided into multiple instances, where each instance describes an impressed advertisement under a certain setting (i.e. with certain depth and position values). Each instance has 12 columns, which can be viewed as a vector (Click, Impression, DisplayURL, AdID, AdvertiserID, Depth, Position, QueryID, KeywordID, TitleID, DescriptionID, UserID). In addition to every instance, the dataset also contains token lists of keyword, query, description and title, where a token is represented by its hash value. Apart from these, there are also some gender and age information of each user in the dataset. The test set share the same format as the training set, except for the counts of ad impressions and ad clicks that are needed for computing the empirical CTR. The test set is derived from the log messages that come from sessions which are latter than those in the training set.

In DataSet_iPinyou, iPinYou provides a set of processed iPinYou DSP bidding, click, impression and conversion logs from several advertising campaigns among seven days for training, and a set of iPinYou DSP bidding logs chosen from the succedent three days from the same set of advertising campaigns for testing, which is not visible. So we choose the first six days logs as the training data, and the next one as the test data. For each bidding log, it contains the user, ad slot, and bidding information. IPinYou ID column is the user cookie set by iPinYou. Timestamp column is when the bid request arrives at the DSP server. IP address is the first three bytes of user IP address. Domain is the name of domain where the ad impression will be appear. The ad information is described with the ad slot ID, width, height, visibility and the format. The column of creative ID is the ad creative of the advertiser for which the DSP bid on the exchange. User tags are the user's tags IDs in iPinYou's database. User agent column is the user's browser. For the impression and click log, most columns are the same as those of bidding log, except some additional columns that as follows. Timestamp is the time when the event (including impression, click, or conversion) happens. Log type is 1 (for impression), 2 (for click), or 3 (for conversion). Key page URL is the link that the user visits if the user clicks the ad for the log type 2, or the link where the user makes the conversion for type 3. Paying price is what the winning DSP actually pays on the exchange. There are totally 3,625,087 impression records and 2235 click records in our training dataset.

4 Experiments

4.1 Select Useful Features

In the feature selection process, we mainly use the sponsored search advertising dataset, and for proving the validity of our framework we test on both of the dataset. In DataSet KDD, we split the training dataset into two parts randomly. The details of the data statistics are shown in Table 2.

Table 2. Data statistics of DataSet KDD.

	#instance	#impression	#click
Sub-training	213,616,893	213,616,893	7,488,134
Validation	21,965,986	21,965,986	729,499

Select features utilizing Logistic Regression. Due to work condition limit, we don't have enough memory to calculate the total sub-training data with the logistic regression. So we randomly sample 10% of the sub-training data. In Logistic Regression model, we use L1 regularization, our feature dimension is 5798979, and we choose below features following the work of the winner in KDD Cup 2012 Track2: user id (Numbers: 3,291,677), ad id (Numbers: 630,868), query id (Numbers: 1,722,825), token id (Numbers: 153,548), user id value (after regularization), query id value (after regularization), cosine similarity (numbers: 6), ctr of user, ad, advertiser and query, the depth and position, numbers of tokens of query, title, description and keyword, the user's gender and age. And the result of these features is considered as the baseline of our experiment.

Then we test the other four types of feature with raw ID feature, and the result is shown in Table 3.

Table 3. Test validity of some features with LR model in validation data.

Features	AUC
f_rawID + f_valID	72.91%
f_rawID + f_ctr	75.66%
f_rawID + f_sim	76.03%
f_rawID + f_userBehavior	78.74%
All features	81.36%

Further, we analyze the weights of id features calculated by logistic regression model, finding that the user id has a large fluctuation, which indicates that this category feature contains more complex features and may have some in fluence for the CTR prediction. And the absolute values of user behavior features' weights are larger than the other types. So we also choose the user behavior feature as another input for our deep network.

4.2   Estimate CTR

Based on our framework of the deep network, we use user ID feature as the input of the left deep network in Fig. 1. And we call the network as the user id network. Considering the cost of time and machine's memory, we cannot put all user id into the deep network. So we utilize the weights of the user id calculated by logistic regression model and select ids with the absolute value greater than a threshold. Finally, we choose 4,768 user ids from total 3,291,677 user ids that have clicked some ads.

Also, we do not use the binary value in the visible layer, instead the frequency of the user id appeared in the training data is used because frequency is more consistent with the true situation. The unit numbers of other 3 layers are 2000, 500 and 50. So we refer to the settings that Hinton proposed in [15]. The reason why we utilize this setting is that we think reducing the dimension excessively fast may lost useful information.

The second deep network (called user behavior network) is used to enlarge the dimension of the raw user behavior feature vector. We choose the raw user behavior feature as the input vector of this deep network.

Table 4. Total results on DataSet_KDD.

Model	Features	AUC
SVR	All basic (low level)	77.54%
	High level (user id) + High level (user behavior) + Part basic	77.80%
LR	All basic (low level)	77.90%
	High level (user id) + Part basic	78.17%
	High level (user behavior) + Part basic	78.04%
	High level (user id) + High level (user behavior) + Part basic	78.27%

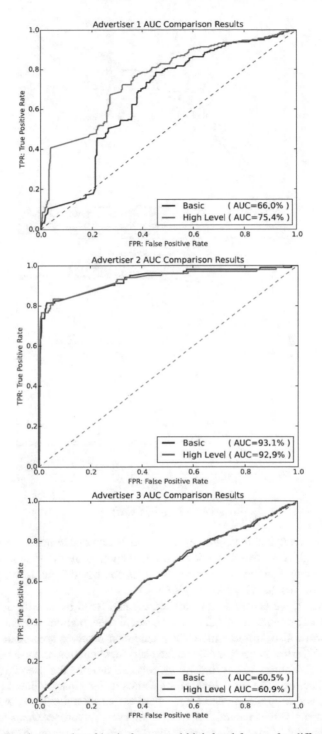

Fig. 3. The comparison results of basic feature and high level feature for different advertisers.

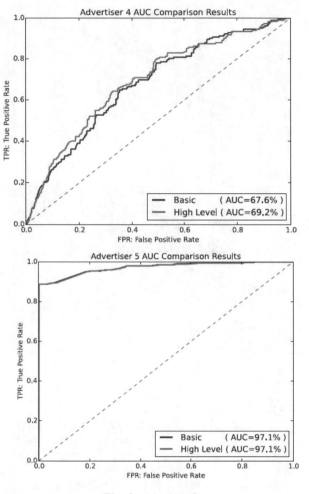

Fig. 3. (*continued*)

Finally, we combine the high level user id features outputted by the first deep network, the high level user behavior features outputted by the second deep network and the other features in the feature pool. With the combined features, we get the finally CRT predication results as shown in Table 4.

From Table 4, we can find that, just represents 0.15% users with user id network can bring an improvement of 0.27% AUC score. Meanwhile, we can see the user behavior network also promote the CTR predication result to some extent.

For RTB advertisements, we choose the user-Tags features, the user-agent features, and the position features in our logistic models and process the user-Tags in our deep network architecture. The input layer has 44 units using binary value representing the different tags, and only one hidden layer where we set 88 units. Here, we test different advertisers respectively since in RTB advertising every advertiser wants to optimize his own ads. In DataSet_iPinyou, there are 5 advertisers in all. We show the results in Fig. 3.

We can see that the performance has been improved for three advertisers, especially for the advertisers which have collected more user's tag and interests, such as the first advertiser.

Generally, in our proposed approach, we first choose efficient features, then put them into different deep network structure to learning high level features. And last, use logistic regression model or other models to predict the CTR. The proposed framework can make a good performance both in sponsored search advertising and real-time bidding advertising. Meanwhile, we can see that most selected features used for deep network learning correlate to the users, which may indicate the users linked all the aspects in the advertisement. We conjecture that there are three mainly reasons for the phenomena: First, users are the connecting bonds of the multi-aspects in CTR prediction, which may contain rich information. Second, features in different levels need proper representation, and the deep network can achieve it. Third, high level features may be easy for models to fit, just like the kernel function.

5 Conclusion

In this paper, we have provided a new perspective to predict click-through rate: learning high level features through different deep network structures instead of only using the basic features. Specifically, we proposed a deep network framework and simple feature selection methods to obtain the high level features. We test the high level features with the other basic features with common models that widely used in industries. Experimental results have demonstrated that the proposed approach can capture the latent factors and complementarity between the user and other elements, which is useful to improve the CTR prediction results compared to the baseline model with basic features.

References

1. Yuan, S., Wang, J., Zhao, X.: Real-time bidding for online advertising: measurement and analysis. In: Proceedings of the Seventh International Workshop on Data Mining for Online Advertising, p. 3. ACM (2013)
2. Broder, A.Z.: Computational advertising. In: Proceedings of the Nineteenth Annual ACM-SIAM Symposium on Discrete Algorithms, SODA 2008, San Francisco, California, USA (2008)
3. McMahan, H.B., Holt, G., Sculley, D., Young, M., Ebner, D., Grady, J., Nie, L., Phillips, T., Davydov, E., Golovin, D., Chikkerur, S., Liu, D., Wattenberg, M., Mar Hrafnkelsson, A., Boulos, T., Kubica, J.: Ad click prediction: a view from the trenches. In: Proceedings of the 19th ACM SIGKDD International Conference on Knowledge Discovery and Data Mining, pp. 1222–1230. ACM (2013)
4. Graepel, T., Candela, J.Q., Borchert, T., Herbrich, R.: Web-scale bayesian click through rate prediction for sponsored search advertising in microsoft's bing search engine. In: Proceedings of the 27th International Conference on Machine Learning, pp. 13–20 (2010)

5. Hillard, D., Schroedl, S., Manavoglu, E., Raghavan, H., Leggetter, C.: Improving ad relevance in sponsored search. In: Proceedings of the Third ACM International Conference on Web Search and Data mining, pp. 361–370. ACM (2010)
6. Richardson, M., Dominowska, E., Ragno, R.: Predicting clicks: estimating the click through rate for new ads. In: Proceedings of the 16th International Conference on World Wide Web, pp. 521–530. ACM (2007)
7. Zhu, Z.A., Chen, W., Minka, T., Zhu, C., Chen, Z.: A novel click model and its applications to online advertising. In: Proceedings of the Third ACM International Conference on Web Search and Data Mining, pp. 321–330. ACM (2010)
8. Dave, K.S., Varma, V.: Learning the click-through rate for rare/new ads from similar ads. In: Proceedings of the 33rd International ACM SIGIR Conference on Research and Development in Information Retrieval, pp. 897–898. ACM (2010)
9. Cheng, H.: Personalized click prediction in sponsored search. In: Proceedings of the Third ACM International Conference on Web Search and Data Mining, pp. 351–360. ACM (2010)
10. Berger, A.L., Pietra, V.J.D., Pietra, S.A.D.: A maximum entropy approach to natural language processing. Comput. Linguist. 22(1), 39–71 (1996)
11. Chapelle, O., Zhang, Y.: A dynamic bayesian network click model for web search ranking. In: Proceedings of the 18th International Conference on World Wide Web, pp. 1–10. ACM (2009)
12. Koren, Y., Bell, R., Volinsky, C.: Matrix factorization techniques for recommender systems. Computer 42(8), 30–37 (2009)
13. Chakrabarti, D., Agarwal, D., Josifovski, V.: Contextual advertising by combining relevance with click feedback. In: Proceedings of the 17th International Conference on World Wide Web, pp. 417–426. ACM (2008)
14. Liu, C., Wang, H., Mcclean, S., Liu, J., Wu, S.: Syntactic information retrieval. In: IEEE International Conference on Granular Computing, p. 703. IEEE (2007)
15. Hinton, G.E., Salakhutdinov, R.R.: Reducing the dimensionality of data with neural networks. Science 313(5786), 504–507 (2006)
16. Hinton, G.E., Osindero, S., Teh, Y.W.: A fast learning algorithm for deep belief nets. Neural Comput. 18(7), 1527–1554 (2006)
17. Dahl, G.E., Yu, D., Deng, L., Acero, A.: Context-dependent pre-trained deep neural networks for large-vocabulary speech recognition. IEEE Trans. Audio Speech Lang. Process. 20(1), 30–42 (2012)
18. Coates, A., Ng, A.Y., Lee, H.: An analysis of single-layer networks in unsupervised feature learning. In: International Conference on Artificial Intelligence and Statistics, pp. 215–223 (2011)
19. Bengio, Y., Schwenk, H., Senecal, J.S., Morin, F., Gauvain, J.L.: Neural probabilistic language models. In: Innovations in Machine Learning, pp. 137–186. Springer (2006)
20. Smolensky, P.: Information processing in dynamical systems: foundations of harmony theory. In: Parallel Distributed Processing: Explorations in the Microstructure of Cognition, pp. 194–281. MIT Press (1986)
21. Fischer, A., Igel, C.: An introduction to restricted Boltzmann machines. In: Progress in Pattern Recognition, Image Analysis, Computer Vision, and Applications, pp. 14–36. Springer (2012)
22. Hinton, G.E.: Training products of experts by minimizing contrastive divergence. Neural Comput. 14(8), 1771–1800 (2002)
23. iPinYou: Global Bidding Algorithm Competition. http://contest.ipinyou.com/index.shtm

Research on Influence of Image Preprocessing on Handwritten Number Recognition Accuracy

Tieming Chen[✉], Guangyuan Fu, Hongqiao Wang, and Yuan Li

Xi'an Research Institute of High-Technology, Xi'an 710025, Shaanxi, China
chentieming1995@163.com

Abstract. In the process of handwritten number recognition, image pretreatment is a key step that has a great influence on the recognition accuracy. By unifying the standard, handwritten digital images are normalized, which can improve the adaptability of handwritten digital recognition algorithms to different writing habits. This article mainly considers the four characteristics of the angle, position, size and strength when writing characters, and how these factors influence four classical handwritten recognition algorithms. According to the four characteristics, tilt correction, offset correction, size normalization and thinning preprocessing were performed one by one to observe the changes of recognition accuracy in four classical algorithms. Through experiments, it is found that the recognition accuracy of the original data set and the scrambling data set are both greatly improved after preprocessing operation. In conclusion, it is resultful to increase the recognition accuracy by image preprocessing in handwritten digital recognition.

Keywords: Image preprocessing · Handwritten number recognition · CNN · Pretreatment

1 Introduction

As a branch of Optical Character Recognition (OCR), handwritten digital recognition has been a very hot research topic in the field of image recognition [1–3]. The essence of recognition is the classification of pictures, that is, the algorithm matches the largest similarity number as the final classification result. Although there are only ten numbers, and the strokes are relatively simple, it is extremely difficult to identify handwritten digits due to the different handwriting habits. To solve this problem, it is critical to find a universal processing method to improve the accuracy of recognition.

Up to now, there have been plenty of studies on image preprocessing, most of which are brief summary methods [4–6]. This article explains in detail how to perform image preprocessing in handwritten digit recognition and also simplifies the complexity and difficulty of method, when meeting the accuracy requirement. In addition, this paper proposes a preprocessing solution especially for four kinds of handwriting habits from different perspectives of angles, positions, sizes, and strengths when people writing characters. In order to verify the influence of preprocessing from various aspects, this

© Springer Nature Switzerland AG 2020
Q. Liu et al. (Eds.): CENet 2018, AISC 905, pp. 253–260, 2020.
https://doi.org/10.1007/978-3-030-14680-1_28

paper uses four classical algorithms. By comparing the experimental results with or without preprocessing, the influence of image preprocessing can be seen clearly.

2   Image Preprocessing

2.1   Character Segmentation

The character segmentation is the basis of image pretreatment, which identifies the area needed to be processed. The image is first grayed and binarized. The binarization process makes every pixel pure black (pixel value is 0) or pure white (pixel value is 255). These grayscale images can help speed up calculations and improve recognition accuracy.

Fig. 1. The character zone.

Fig. 2. The character segmentation image.

The core of the character segmentation algorithm is to find the coordinates of the four vertices on the character border. The process is as follows:

First, input the image that has been grayed and binarized;

In the second step, scan the input image progressively, and compare pixel value of each point with the size of 255 (when the pixel value is 255, it represents white, that is, the foreground color of the handwritten digital image). Then record the coordinates of points where the pixel value is equal to 255.

Third, repeat the second operation until all the pixels are traversed. Then the minimum horizontal (vertical) coordinate of all recording points is left margin (bottom), which is denoted by $x_{min}(y_{min})$. Similarly, the maximum horizontal (vertical) coordinate founded is the right margin (top), which is denoted by $x_{max}(y_{max})$.

Forth, the segmented character area can be shown in Figs. 1 and 2, which can be described as $S = \{(a, b)|x_{min} \ll x \ll x_{max}, y_{min} \ll y \ll y_{max}\}$.

2.2 Tilt Correction

When a person writes numbers with different habits, the characters often have a certain angle of tilt. Therefore, the angle of characters need to be corrected. At the same time, the tilt-corrected character image also facilitates the segmentation of the character and improves the recognition accuracy.

In general, the character image after character segmentation should have similar height of left and right pixels. If the average height of pixels on both sides has a large difference, it indicates that the image is tilted and needs to be adjusted. So adjust the image by using the average height of the left and right white pixels of the image, the main process of tilt correction is shown in Fig. 3.

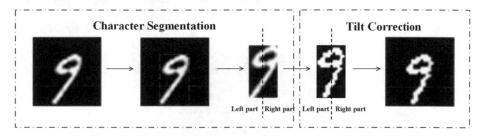

Fig. 3. The tilt correction after character segementation.

First, calculate the average weighted height of the white pixels in the left part and right part of image. The rule for the weight setting is that the weight is higher when the pixel is closer to two edges. Then calculate the slope of inclination and reorganize the image:

$$S = \frac{H_l - H_r}{W/2} \tag{2.1}$$

In the above equation, S denotes the slope of inclination, H_l and H_r represent the average weighted height in the left and right half image respectively, W represents the width of a character image. Then, the coordinates of corrected point can be calculated as:

$$\dot{x}_i = \lfloor x_i - (y_i - W/2) \times S \rfloor, \dot{y}_i = y_i \tag{2.2}$$

where $(\dot{x}_i, \dot{y}_i)$ denotes the coordinate of the point after the inclination correction of the point (x_i, y_i), and $\lfloor x \rfloor$ indicates rounding down to x.

2.3 Offset Correction

In most cases of writing numbers, the center of character will deviate from the center of the image, which can easily affect the recognition accuracy. And the center-of-gravity normalization method is usually used. First, calculate the center of gravity A in the image, which should have coincided with the center of the image B. Point A and point B can make up an offset vector $\boldsymbol{AB}$. Later, the handwritten digit offset can be corrected by moving the red frame in Fig. 4 in accordance with the direction and distance of the offset vector. The formula for calculating the center of gravity (x_h, y_h) of the image is:

$$
\begin{cases}
x_h = \dfrac{\sum_{i=1}^{n} x_i}{n} \\
y_h = \dfrac{\sum_{j=1}^{n} y_j}{n}
\end{cases}
\tag{2.3}
$$

where n is the number of white points, $\sum_{i=1}^{n} x_i$ represents the sum of the abscissas of white points, and $\sum_{j=1}^{n} y_j$ represents the sum of the vertical ordinates of white points.

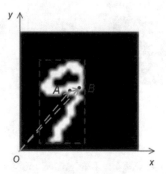

Fig. 4. The center of gravity and center of the image.

As shown in Fig. 4, the center of gravity of image is $A(x_h, y_h)$. Construct a direction vector $AB = (x_H - x_h, y_H - y_h)$ with its center $B(x_H, y_H)$.

$$
\begin{cases}
\dot{x} = x + (x_H - x_h) \\
\dot{y} = y + (y_H - y_h)
\end{cases}
\tag{2.4}
$$

In the above equation, $\dot{P}(\dot{x}, \dot{y})$ is a new coordinate point obtained after offset correction from point $P(x, y)$, and $x_{min} \ll x \ll x_{max}, y_{min} \ll y \ll y_{max}$. After moving all the white pixels in the red box by the vector AB, the center of gravity can finally coincide with the center of image, and the corrected image is shown in Fig. 5.

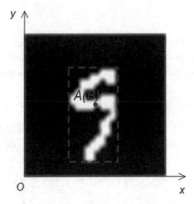

Fig. 5. The center of gravity and center of the image coincide.

2.4 Size Normalization

The purpose of size normalization is to make the length and width ratio of the segmented character images basically similar, so as to eliminate the interference caused by the inconsistent handwritten digits. The advantage is to improve the accuracy of image recognition after unified size. Common image size normalization methods include linear normalization and nonlinear normalization. The nonlinear normalization method is based on the linear normalization method and it adds nonlinear change factors which can better solve the problem of image deformation after scaling. In the actual writing process, the handwritten digit size difference will be small, i.e., the influence of the deformation of the image before and after the normalization process will not cause too much influence on the recognition accuracy.

And compared with the nonlinear normalization method, the linear normalization method has a simpler method, a smaller amount of calculation, so use the linear normalization method which can also meet the image scaling requirements.

The linear normalization method is an image scaling method that linearly zooms in or zooms out the original image to a predetermined size in a certain proportion. The linear normalization formula is given now:

$$\begin{cases} m = \frac{i \times M}{I} \\ n = \frac{j \times N}{J} \end{cases} \tag{2.5}$$

The point (m, n) indicates the normalized coordinate point corresponding to original point (i, j).

Taking into account the characteristics of the digital shape, it is specified that the image of number one after segmentation is normalized to 20×6, and the rest is normalized to 20×14. The algorithm is to obtain the height and width of the segmented character image firstly. Then the height and width are compared to obtain the transform coefficient. Later, the points in the new image are mapped to the original image according to the interpolation method.

2.5   Image Thinning

Different writing strength leads to different thicknesses of digital strokes, which will have a great impact on extracting feature values in the process of digital recognition. In order to enhance the adaptability of digital recognition, the process of image thinning is necessary.

A good image thinning algorithm has the following requirements:

1. The skeleton image must maintain the connectivity of the original image.
2. The skeleton image should be the centerline of the original image as much as possible.
3. The thinning skeleton should be thin for a pixel-wide line image as much as possible.
4. The algorithm should use as few iterations as possible.

There are already many kinds of image thinning algorithms such as Classical Image Thinning Algorithm, Pavlidis Asynchronous Image Thinning Algorithm, Deutseh Algorithm, Zhang Fast Parallel Image Thinning Algorithm and so on. Among them, Zhang's fast parallel image thinning algorithm is a practical thinning method. It is an 8-adjacent and parallel image thinning algorithm, which has the advantage of high speed and ability to maintain the connectivity of curves. Due to the disadvantage of missing local information and the presence of redundant pixels after thinning, this paper uses the improved Zhang fast parallel image thinning algorithm to thin handwritten characters [7].

Because of ink brushing, thick strokes, or overly random writing, handwritten numbers are prone to black blocks, especially numbers 0, 6, 8, and 9, as shown in Fig. 6. And Fig. 7 is the image after binarization, size normalization, tilt, and offset correction. If use the algorithm proposed by Zhang directly, as shown in Fig. 8, discrete and irregular lines appear in image, which losts the original structure and information of the character. The use of the improved algorithm proposed by Zhang Cuifang and others for such problems can effectively maintain the shape of the characters after thinning, as shown in Fig. 9.

Fig. 6. The original image

Fig. 7. The binary image

Fig. 8. The thinning corrected image using Zhang's algorithm

Fig. 9. The thinning corrected image using improved Zhang's algorithm

3   Result and Conclusion

The data used in the experiment comes from the MNIST Handwritten Numeric Character Database, which contains 70,000 handwritten digital samples from 0 to 9 divided into 60,000 training data sets and 10,000 test data sets. The shape of all pictures are 28×28. And the experimental environment is: 64-bit Windows 10 operating system, programmed by Python, the four classical algorithms used are k-Nearest-Neighbor (KNN), Support Vector Machine (SVM), Back Propagation Neural Network (BP) and Convolutional Neural Network (CNN). The experimental steps are as follows:

In the first step, the MNIST raw data set is grayed and binarized, and four classical algorithms are used for training and testing. Then, the original data is preprocessed, with parameters still unchanged, and four algorithms are reused. Identify and obtain experimental results as shown in Table 1.

In the second step, in order to explore the influence of different writing habits on the recognition accuracy of handwritten digits, the MNIST data set is scrambled. The method is that 25% of samples are randomly selected from the training set and the test set, then perform position offset, angular offset, size scaling, and dilation erosion operations respectively, which simulates the writing habits of different people. Keep the same algorithm parameters, train and test new data sets, and the result is shown in Table 2.

The third step is to preprocess the scrambled data set obtained in the second step, that is, perform the tilt correction, offset correction, size normalization and thin images operations. Then use the four classical algorithms and the result is shown in Table 2.

The fourth step is to do experiments for multiple times, compare, analyze data, and make a conclusion.

It can be seen from Table 1 that the four algorithms all have a good representation on MNIST original data set. After the data set is preprocessed, only the recognition rate of KNN algorithm has a slight decrease. The recognition rate of the other three algorithms has a little increased, and CNN algorithm can improve even up to 99.31%. Through experiments, the recognition of number 1 in four algorithms after preprocessing is generally improved a lot. What's more, the recognition rate of number 7 is also improved which has a similar shape and simple structure with number 1.

Observe the result in Table 2 and it is found that compared with the original data set, the recognition rates of the first three algorithms with scrambled data set are all reduced, while the recognition rate of the CNN algorithm has been improved, still lower than the original data set after preprocessing. The possible reason is that the diversity of the sample is increased after scrambling, and from the side it also reflects the strong learning ability of the deep learning algorithm.

Table 1. Handwritten digit recognition accuracy results before and after preprocessing on raw data set.

Algorithm	KNN		SVM		BP		CNN	
Condition	Before	After	Before	After	Before	After	Before	After
Average accuracy rate (%)	96.85	96.52	94.39	95.54	97.37	97.47	97.56	99.31

Table 2. Handwritten digit recognition accuracy results before and after preprocessing on scrambled data set.

Algorithm	KNN		SVM		BP		CNN	
Condition	Before	After	Before	After	Before	After	Before	After
Average accuracy rate (%)	88.53	91.82	89.56	91.22	95.77	97.43	98.61	99.43

The above data shows that the preprocessing of handwritten digital images is very necessary for the traditional machine learning algorithms like KNN and SVM algorithm. Meanwhile, the neural network algorithm has higher adaptability to the recognition of handwritten digits with different characteristics. Because of the strong self-adaptive and self-learning ability of deep learning, the scrambled data set to some extent increases the diversity of input samples and further improves the recognition accuracy of CNN. In conclusion, whether it is machine learning or deep learning method, the preprocessing of handwritten images can achieve improved recognition accuracy generally.

References

1. Zhang, M., Yu, Z.Q., Yao, S.W.: Image pretreatment research in recognition of handwritten numerals. Microcomput. Inf. **22**(16), 256–258 (2006). (in Chinese)
2. Chen, H., Guo, H., Liu, D.Q., Zhang, J.Q.: Handwriting digital recognition system based on tensorflow. Inf. Commun. **3** (2018). (in Chinese)
3. Zhang, Y.S.: License plate recognition key technology related algorithms research and implementation. North Minzu University (2017). (in Chinese)
4. Zhang, C.F., Yang, G.W.: Research on image pretreatment technique in recognition of handwritten number. Comput. Sci. Appl. **6**(6), 329–332 (2016). (in Chinese)
5. Huang, Q.Q.: Research on handwritten numeral recognition system based on BP neural network. Central China Normal University (2009). (in Chinese)
6. Zhang, S.H.: Study and realization of algorithms for chinese characters image's preprocessing. Microcomput. Dev. **13**(4), 53–55 (2003). (in Chinese)
7. Zhang, C.F., Yang, G.W., Yue, M.M.: Improving of Zhang parallel thinning algorithm. Inf. Technol. Informatiz. **6**, 69–71 (2016). (in Chinese)

Generative Information Hiding Method Based on Adversarial Networks

Zhuo Zhang[1,2](✉), Guangyuan Fu[1], Jia Liu[2], and Wenyu Fu[3]

[1] Xi'an High-Tech Institute, Xi'an 710025, China
adam_zz01@163.com
[2] Key Lab of Networks and Information Security of PAP, Xi'an 710086, China
[3] China Huadian Corporation LTD Sichuan Baozhusi Hydropower Plant, Guangyuan 628000, Sichuan, China

Abstract. Traditional Steganography need to modify the carrier image to hide information, which will leave traces of rewriting, then eventually be perceived by the enemy. In this paper, an information hiding scheme based on Auxiliary Classifier Generative Adversarial Networks (AC-GANs) model is proposed for Steganography. This method designs and trains the networks model based on AC-GANs by constructing a dedicated dictionary and image database. The sender can map the secret information into the category labels through the dictionary, and then use the labels generate the real looking images to be sent through the model. On the contrary, the receiver can identify the image label through the model and obtain the secret information. Through experiments, the feasibility of this method is verified and the reliability of the algorithm is analyzed. This method transmits secret messages by generating images without overwriting the carrier images. It can effectively solve the problem of modification of carrier images in traditional information hiding.

Keywords: Steganography · Information hiding · AC-GANs · GAN

1 Introduction

Steganography is the science of communicating secret information in a hidden manner [1]. Traditional image information hiding methods are generally studied in the image Spatial domain or frequency domain. There are many algorithms, such as LSB-based methods in the spatial domain [2–4], DCT-based methods in the transform domain [5], DFT-based methods [6], and DWT-based methods [7]. Although these algorithms are ingenious, they always need to modify the carrier image to hide information, which will inevitably leave traces of rewriting, then eventually be perceived by the enemy and lose the significance of information hiding. In recent years, some new information hiding methods have gradually attracted people's attention. For example, there were carrier-free information hiding [8] and coverless information hiding [9]. The sender selects an image in the normal image library according to the secret information, and the receiver obtains the secret information by calculating the image Hash. These new methods try to avoid the drawbacks of traditional information hiding methods to modify images, and

Q. Liu et al. (Eds.): CENet 2018, AISC 905, pp. 261–270, 2020.
https://doi.org/10.1007/978-3-030-14680-1_29

expand the connotation and extension of information hiding from a new perspective, which is more in line with the characteristics of information forms in the current big data network environment.

Inspired by these new methods [8, 9], we propose a generative information hiding method based on Auxiliary Classifier Generative Adversarial Networks (AC-GANs) model. Our method constructs a dedicated dictionary and corresponding image database, then designs and trains a generative networks model based on ACGANs for steganographic (stego-ACGANs for short) on the database. The model can learn the data distributions of various image samples from the database and generate the real looking images according to the labels from the dictionary. Moreover, the model's discriminator network can quickly and accurately identified the image label. Applying the model to secret communication can achieve the purpose of information hiding. Compared with other methods, our method is relatively novel and highly secure, and the generated images are various and difficult to be found by the enemy. Theoretically, it cannot be detected by traditional statistical steganalysis tools based on statistics.

2 The Basic Theory of AC-GANs

The AC-GANs model is a model proposed by Odena et al. [10] in 2017. The basic idea of the model is basically the same as that of the standard Generative Adversarial Networks (GANs) model [11]. Compared with ordinary GANs, AC-GANs adds sample class label information in addition to random noise at the input of the generator network, and an auxiliary classifier is added to the judgment network to determine the value of the sample label. In this way, the constraints of the model are increased, the training of the model is more stable and the convergence is faster, and the final generated image quality is much higher than that of standard GANs. The AC-GANs model structure is shown in Fig. 1.

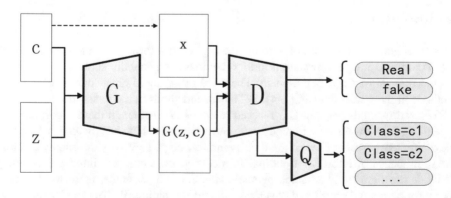

Fig. 1. AC-GANs model structure.

c represents the sample class label, z represents random noise; G represents the generator network used to generate the image, G(z, c) represents the generated image after c and z input G; D represents the discriminator network used to determine the input probability that the sample x and the generated sample G(z, c) are "true"; and an auxiliary classifier Q (Q can be considered as part of D) is attached to D to determine the label's class to which the sample belongs.

Loss function of the model is defined as:

$$L_s = E[\log P(S = real|X_{real})] + E[\log P(S = fake|X_{fake})] \tag{2.1}$$

$$L_c = E[\log P(C = c|X_{real})] + E[\log P(C = c|X_{fake})] \tag{2.2}$$

where X_{real} represents a real image, X_{fake} represents an generated image, C represents an image label, and S represents whether the image is judged to be a real image or generated image. The goal of D training is the maximum $L_S + L_C$ and the goal of G training is the maximum $L_C - L_S$. The gradient descent method can be used to optimize the loss function $L_S + L_C$ and $L_C - L_S$.

3   The Proposed Steganography Framework

The AC-GANs model can generate real looking image according to the given label, and the generated image can be effectively recognized by the model as a label. Using these characteristics, we propose a generative information hiding method based on AC-GANs. The method is mainly composed of three parts:

(1)   Formulate word segmentation dictionary and image database;
(2)   Design Generative Information Hiding Model: Stego-ACGANs;
(3)   Design information hiding algorithm and extraction algorithm.

3.1   Word Segmentation Dictionary and Database Construction

Dictionaries need to be able to cover commonly used characters and phrases in secret information, it can be flexibly constructed according to the actual needs of use. Binary numbers can be used to encode these participles to facilitate the query of the codebook.

Image database is built for the label items in the dictionary, and the image set and the label set are stored in the database, as shown in Fig. 2. The image set can use the current Internet image data to collect various types of images. Each image corresponds

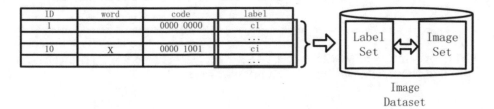

Image
Dataset

Fig. 2. Word segmentation dictionary and database structure diagram.

to a label and is stored in the label set. Requires various types of images to be stored in K, K >= 1000. At the same time, each image in the library should be normalized to a uniform size, which is convenient for neural network training.

3.2   The Stego-ACGANs Model Design

In order to effectively generate the label information as an image and can be identified, we designs Stego-ACGANs model based on the AC-GANs. The structure follows the basic model structure of AC-GANs in Fig. 1. The Stego-ACGANs model is composed of three neural networks. The structure of each neural networks is designed as follows:

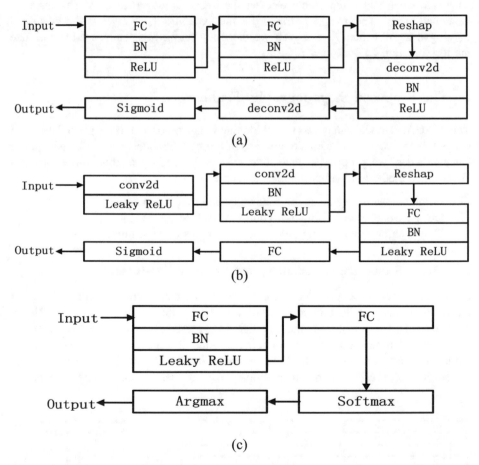

Fig. 3. Stego-ACGANs Model Structure: (a) is generator network structure; (b) is discriminator network structure; (c) is auxiliary classifier network structure.

In Fig. 3, FC (fully connected layers) represents a fully connected layer; conv2d (convolution2d) represents a 2D convolution layer; BN (Batch Normalization) represents a batch normalization operation; deconv2d (Deconvolution2d) represents a 2D Deconvolution layer. The Loss function of the model is defined as follow:

$$L_{D,Q}^{Stego-ACGANs} = E[log(D(x))] + E[log(1 - D(G(z)))] \\ + E[log(P(class = c|x))] + E[log(P(class = c|G(z)))] \tag{3.1}$$

$$L_{G}^{Stego-ACGANs} = E[log(D(G(z)))] + E[log(P(class = c|G(z)))] \tag{3.2}$$

where x represents a real image, z represents a noise signal, c represents an image label, $G(.)$ represents the generator network, $D(.)$ represents the discriminator network. At the same time, three neural networks are trained. The goal of D, Q training is the maximum $L_{D,Q}^{steog-ACGANs}$, and the goal of G training is the maximum $L_{G}^{steog-ACGANs}$.

Model training uses a gradient descent method to update network weights according to the following rules.

For G networks: Keeping D, Q fixed, update G by

$$W_G \leftarrow W_G + \gamma_G \nabla_G L_G^{stego-ACGANs} \tag{3.3}$$

For D, Q networks: Keeping G fixed, update D, Q by

$$W_{D,Q} \leftarrow W_{D,Q} + \gamma_{D,Q} \nabla_{D,Q} L_{D,Q}^{stego-ACGANs} \tag{3.4}$$

γ represents the learning rate (It is an important hyperparameter, and this settings often require experience as a judgement.) and needs to be set manually.

3.3   Hiding and Extracting Algorithms

After the Stego-ACGANs model has trained, the network structure and parameters of the generation network and the auxiliary classifier in the model are the encryption and decryption keys of the steganographic method. Since the training of GANs cannot be completely converged, we need one of the correspondents to construct a dictionary and image database locally and train the model before steganography. The information of the dictionary and model parameters is passed to the other party through the secret channel to ensure that both parties use the same dictionary and model weights to hide and extract the secret information.

The sender uses the message and the random noise as the driver, generates the image sequence through the generation network G in the trained Stego-ACGANs model, and then transmits it. The hiding algorithm is described as follows:

Step 1: Input the secret information S and combine the private dictionary to segment the secret S information to obtain the binary code sequence $S = \{s_1, s_2, \ldots s_n\}$ which is the n partial words in S, and its corresponding label sequence is $C = \{c_1, c_2, \ldots c_n\}$.

Step 2: Divide each i code into a segment $S^{(k)} = \{s_1^{(k)}, s_2^{(k)}, \ldots, s_i^{(k)}\}$, k represents the present number of the segment information. A dictionary is used to determine the label of each code, and finally the label sequence $C^{(k)} = \{c_1^{(k)}, c_2^{(k)}, \ldots, c_i^{(k)}\}$ of the k-th segment secret information is obtained.

Step 3: Input $C^{(k)}$ as a driver, input to the generator network of the Stego-ACGANs model, and combine the input noise to generate an image sequence $I^{(k)} = \{i_1^{(k)}, i_2^{(k)}, \ldots, i_i^{(k)}\}$.

Step 4: Combine this image sequence into a single image for transmission or publish it publicly on a web platform.

After the receiver acquires the image, the image label is identified by the auxiliary classifier Q in the model, and then the secret message is extracted through the dictionary. The extracting algorithm is described as follows:

Step 1: After receiving the image, the image is first normalized to ensure that the image is in the correct orientation. The image is scaled to the standard size and divided into image sequences.

Step 2: Input the processed image sequence into the model one by one, and use the auxiliary classifier of the Stego-ACGANs model to determine the label of the image sequence to obtain the label sequence $\hat{C}^{(k)} = \{\hat{c}_1^{(k)}, \hat{c}_2^{(k)}, \ldots, \hat{c}_i^{(k)}\}$.

Step 3: Use the dictionary to get the corresponding binary code sequence $\hat{S}^{(k)} = \{\hat{s}_1^{(k)}, \hat{s}_2^{(k)}, \ldots, \hat{s}_i^{(k)}\}$.

Step 4: Search the dictionary in combination with the code $\hat{S}^{(k)}$ to obtain a segment in the secret information. Finally, the pieces of information are combined to obtain the secret information $\hat{S}$.

4   Experiments

The experimental platform is python3.5+Tensorfolw1.1.0. The database uses the MNIST dataset [12]. Select a segment of one piece of secret information. After a dictionary query, a word segmentation and encoding are represented as S = {0000 0001, 0000 0110, 0000 0010, 0000 1001, 0000 0010, 0000 0011, 0000 0111, 0000 0010, 0000 0010}, and the corresponding label is C = [1, 2, 2, 2, 2, 3, 6, 7, 9]. This label sequence is used as the driver for generating the image in the experiment to verify the feasibility of the model.

We build the Stego-ACGANs model according to the structure designed in Sect. 3.2, and train the model on the MNSIT data set, and set the training parameters as follows:

Batchsize (the size of the training sample batch) is 100; the dimension of the random noise z is set to 10; the number of trainings is set to 50000; the optimization method is Adam, and the learning rate is set to 1e−3 (Here, we set it to 1e−3 based on experience). Noise z is randomly initialized to a 100×10 random matrix with values in the $(-1,1)$ interval. Each layer network weight w is initialized with the Xavier method, and the offset b is initialized to zero.

In the experiment, for each training of the model, the two loss function values DC_loss and GC_loss are counted, where $DC\_Loss = -L_{D,Q}^{Stego-ACGANs}$, $GC\_Loss = -L_G^{Stego-ACGANs}$; every 100 times, the label data C of secret information is input into the model, the quality of the image generated by the model is viewed, and the time used to generate each image is counted; for each 1000 trainings times of the model, the accuracy of the auxiliary classifier is viewed and counted.

4.1 Checking the Convergence of the Model

During the training process, the two loss function values of the Stego-ACGANs model were shown in Fig. 4.

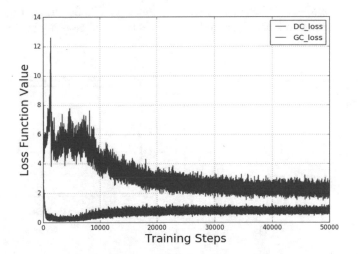

Fig. 4. Two loss function values in model training.

The red line (lower curve) represents the loss function DC_loss of the model. The blue line (upper curve) represents the loss function GC_loss of the model. As can be seen from the figure, the two loss function values in the training process are a process that gradually tends to balance through the game. Although the individual loss function values oscillate greatly, the overall training trend of the entire model tends to be stable. Especially after 30,000 trainings, the model tends to be stable.

4.2 Checking the Generated Image

When the model is trained 100 times, the label data C is input into the generator network once to generate a picture sequence. In the training, the generated image sequences were extracted for viewing to verify the quality of image generation in the generator network.

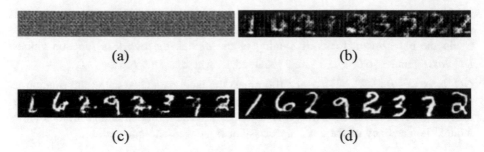

(a)   (b)

(c)   (d)

Fig. 5. Picture generated in the model training.

In Figs. 5(a), (b), (c), and (d) respectively show the generated images of the model training 1, training 10,000, training 20,000, and training 30,000 steps. As we can be seen from the figure, as the number of training steps increases, and the image quality is significantly improved. When the model is trained to 30,000 times, digital images generated by the network are clear, and visual effects cannot be distinguished from real images.

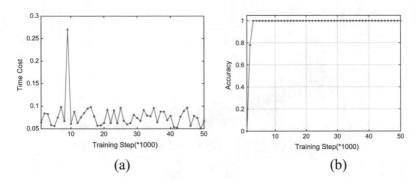

(a)   (b)

Fig. 6. (a) The time cost to generate each image during training; (b) Pediction accuracy curve of auxiliary classifier.

Randomly count the time for 50 times when images were generated, as shown in Fig. 6. As can be seen from the figure, the time consumed by the model to generate the image is basically stable between 0.05 s and 0.1 s, which show that the model generates images at a very fast speed.

4.3 Checking Model Information Extraction

For every 1000 times of training, the generated image sequence is passed to the auxiliary classifier to determine the label, and then the judged label sequence is compared with the input tag sequence C. The forecast accuracy rate of the auxiliary classifiers in the training process is shown in Fig. 6.

In summary, The Stego-ACGANs model through training can generate real looking images of the secret information, and can accurately identify the images.

5 Discussion

Experiments show that our model can perform covert communication. In addition, this method has a strong security and capacity too. The analyses are as follows:

5.1 Security Analysis

The training of GANs model is a complex problem. When the network structure, hyperparameter setting and initialization method are different, the results of network training are different. There is no unique convergence optimal solution. The quality of GANs training often requires human judgment, so there is a great deal of uncertainty in the method and model training steps. In addition, neural networks use specific datasets as the basis for training, and network parameters trained by different data sets are quite different. This increases the difficulty of inversely deriving neural network parameters. Even if the enemy accidentally acquired the generated the image label, it is still difficult to restore the original secret information without a dedicated word segmentation dictionary.

5.2 Capacity Analysis

This method is to directly map the word in the dictionary into a picture. Each picture carries one item of information in the dictionary. So the average hidden capacity of each picture, which is related to the number of words contained in each item in the dictionary, can be expressed as:

$$\bar{C} = \frac{\sum_{i=1}^{n} C_i}{n} \tag{5.1}$$

$\bar{C}$ represents the average hidden capacity of one generated image, the unit is words; C_i represents the number of vocabularies of the term i in the dictionary, the unit is words; n is the length of items in the dictionary. It can be seen that the more words that each dictionary item contains, the more information that each image carries.

6 Conclusions and Future Work

The contributions of this letter are as follows:

(1) We propose a method for information hiding using the AC-GANs model. Compared with the traditional information hiding method, our method does not modify the carrier image, but combines the special dictionary and the generative adversarial networks model to generate the image from the secret information, and the

image can be effectively recognized by the receiver. This approach is relatively novel and safe.

(2) The structure of the dictionary and Stego-ACGANs model are designed, and the information hiding algorithm based on the model is proposed. The feasibility of the algorithm is verified by experiments.

In the future, we plan to improve the model structure and experiment with larger datasets.

References

1. Cheddad, A., Condell, J., Curran, K., Mc Kevitt, P.: Digital image steganography: Survey and analysis of current methods. Sig. Process. **90**(3), 727–752 (2010)
2. Wu, H.C., Wu, N., Tsai, C.S., Hwang, M.S.: Image steganographic scheme based onpixel-value differencing and LSB replacement methods. Proc. Vis. Image Signal Process. **152**(5), 611–615 (2005)
3. Mielikainen, J.: LSB matching revisited. IEEE Signal Process. Lett. **13**(5), 285–287 (2006)
4. Yang, C.H., Weng, C.Y., Wang, S.J.: Adaptive data hiding in edge areas of images with spatial LSB domain systems. IEEE Trans. Inf. Forensics Secur. **3**(3), 488–497 (2008)
5. Cox, I.J., Kilian, J., Leighton, F.T.: Secure spread spectrum watermarking for multimedia. IEEE Trans. Image Process. **6**(12), 1673–1687 (2010)
6. Ruanaidh, J.J.K.O., Dowling, W.J., Boland, F.M.: Phase watermarking of digital images. In: International Conference on Image Processing, pp. 239–242 (1996)
7. Sun, Q.D., Guan, P., Qiu, Y., Yan, W.Y.: DWT domain information hiding approach using detail sub-band feature adjustment. Telkomnika Indones. J. Electr. Eng. **11**(7) (2013)
8. Zhou, Z.L., Sun, H.Y., Harit, R.: Coverless image steganography without embedding. In: ICCCS 2015. LNCS, vol. 9483, pp. 123–132 (2015)
9. Fridrich, J.: Steganography in Digital Media: Principles, Algorithms, and Applications. Cambridge University Press, Cambridge (2010)
10. Odena, A., Olah, C., Shlens, J.: Conditional image synthesis with auxiliary classifier GANs. In: Proceedings of the 34th International Conference on Machine Learning, Sydney, Australia, PMLR, vol. 70 (2017)
11. Ian, G., Pouget-Abadie, J., Mirza, M., Xu, B., Warde-Farley, D., Ozair, S., Courville, A., Bengio, Y.: Generative adversarial nets. In: Ghahramani, Z., Welling, M., Cortes, C., Lawrence, N.D., Weinberger, K.Q. (eds.) Advances in Neural Information Processing Systems, vol. 27, pp. 2672–2680 (2014)
12. The MNIST Database of Handwritten Digits. http://yann.lecun.com/exdb/mnist/

Computer Image Processing Technology Based on Quantum Algorithm

Si Chen[✉]

Changchun Normal University, Changchun 130032, Jilin, China
47116227@qq.com

Abstract. As the most concerned new model, quantum computing is always a hot topic in the field of information processing. In particular, after the quantum algorithms of Shor large number factor decomposition and Grover search are proposed, people gradually realize that quantum computing is expected to break through the classical computing limit, thus bringing subversive upheaval to the whole information processing field. However, at present, the research on quantum computing theory for image processing is still in its infancy. People still linger on how to use quantum information to characterize images. It is difficult to design a quantum image processing algorithm that can solve practical problems. On the basis of this problem, a quantum image processing system is set up. It is gradually progressive from image storage to image preprocessing, and then to image classification, and then the construction of the whole system is realized, so as to expand the research and analysis. It is found that the processing of computer images by quantum algorithm is accurate and real-time, and it can effectively deal with the problems in traditional image processing. On the one hand, the construction of the quantum image processing system solves the problem of the performance of traditional image processing. On the other hand, the bottom-up research method from the information storage angle can also provide guidance on how to use the online algorithm in other information fields.

Keywords: Image processing · Quantum algorithm · Information field

1 Introduction

Quantum computers provide services for quantum computation, so the research theory of quantum computers has been emerging one after another. At present, there are more and more related design schemes and system models for realizing quantum computers. The relevant studies show that, in 2020, human beings are possible to produce the quantum computers. But in this context, the related research on quantum algorithms is relatively few and relatively shallow. Therefore, the quantum algorithms in practical applications are mainly studied.

Image processing is an important part in the information field, mainly responsible for data collection, sorting, analysis and processing. Generally, there are more applications in weapons, radar monitoring, medicine and so on. The construction of computer quantum image processing system is divided into three parts, including: image storage, mainly responsible for the storage of image information and simple quantum

© Springer Nature Switzerland AG 2020
Q. Liu et al. (Eds.): CENet 2018, AISC 905, pp. 271–278, 2020.
https://doi.org/10.1007/978-3-030-14680-1_30

image conversion; image preprocessing, mainly responsible for the collection and preprocessing of the characteristics of quantum images; image classification, responsible for the design of contrast algorithms similar to quantum images. The quantum image processing system will be used in the quantum computer, which can effectively and quickly collect the whole image information, and then analyze and process the content of the image.

Graph kernel function is an important measure of measuring similarity in the field of machine learning. It advocates mapping the graph in high dimensional space, so that the similarity degree of the graph is defined by using the distance of space. Through the experimental comparison, it is found that the quantum graph kernel designed has higher classification accuracy than some of the most popular graph kernels at present. At the same time, for some special graphs, such as the residual and regular graphs, the general graph kernel cannot distinguish their differences, but the method proposed here can accurately depict the small differences between different graphs.

2 State of the Art

The study of Caraiman and Manta showed that in classical computation, bit is the most basic unit of computation. Every bit has a state in a time, 0 or 1. In quantum computation and quantum information processing, the concept of qubits, commonly used in qubit, is also presented. Similar to bit, qubit also has a state at every moment [1]. In the study of Jiang and Wang, the current research on quantum algorithms is based on three quantum algorithms that have core functions to be further applied and improved. For example, aiming at the failure of Grover search algorithm, the algorithm is improved, and Grover technology is used to complete more complex statistical algorithm design [2]. In the research of Jiang et al., many application layer quantum algorithms based on core quantum algorithm extension are proposed, such as Graph search algorithm, computational geometry quantum algorithm and dynamic programming quantum algorithm [3]. In the study, Zhou and Sun pointed out that, in order to use quantum mechanism to complete information processing, the classical information is first needed to be stored in a quantum state in a certain way. Through several quantum unitary transformations, the probability of the ground state in the quantum superposition state can be adjusted so that the algorithm can read the result from the quantum state with a large probability in the collapse measurement of the quantum state [4]. Youssry et al. proposed in the study that the new quantum image model used the base vector of the qubit column to express the color information of each pixel, which made the image processing operations related to the color of the image unable to be completed in the FRQI model can be conveniently completed under the new quantum image model [5].

3 Methodology

3.1 Quantum Image Storage Model

The advantage of the FRQI quantum image is that it uses the form of a qubit column to store the location information of the image pixels so that the pixels in the various positions of the digital image can be operated simultaneously. The drawback of FRQI quantum image is that it only uses the quantum amplitude of 1 bit qubits to represent the color information of pixels. Therefore, a new quantum image representation method NEQR designed inherits and improves the FRQI quantum image model. It uses two qubit columns to store the pixel location information and color information, respectively, and simultaneously superposes the two entanglements, so that all the pixels of the image can be stored simultaneously.

For a 2n * 2n size image, the expression of the new quantum image NEQR is as follows:

$$|I> = \frac{1}{2^n}\sum_{Y-0}^{2^n-1}\sum_{X-0}^{2^n-1}|f(Y,X)>|YX> = \frac{1}{2^n}\sum_{Y-0}^{2^n-1}\sum_{X-0}^{2^n-1}|C^i>|YX> \qquad (1)$$

Supposing a 2 * 2 pixels' image is represented by a new quantum image model, if the pixel color range is [0, 255], a 8-bit qubit column is needed to represent the color information of the pixel. Therefore, for a 2n * 2n image with 2q color range, the new quantum image model NEQR needs a total of 2n + q quantum bits to preserve the image information (Fig. 1).

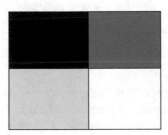

Fig. 1. 2*2 Graphical example.

Quantum measurement is the only way to get classical information from quantum states. Unfortunately, the existing quantum image models preserve the color information of the image pixels in the probability amplitude of the qubits, which represent the value of the color by using the amplitude α and β in the quantum bits, as shown in the following formula:

$$|C> = \partial|0> + \beta|1> \qquad (2)$$

According to the principle of quantum measurement, it is impossible to obtain accurate information of α and β in quantum states by quantum measurement. Therefore, the current quantum image model can only reconstruct the classical image probabilistically. This also affects the efficiency of quantum model used for image storage and image transmission, and also limits the application of quantum image processing to a certain extent.

In the new quantum image model NEQR, the pixel color information is stored by the base vector of the qubit column, so that the quantum states of the different color values are orthogonal to each other, and each color value corresponds to a fundamental vector of the qubit column. Therefore, when measuring each pixel of the NEQR image, the color information of the pixel can be obtained accurately, so that the classical image can be accurately reconstructed from the NEQR image model.

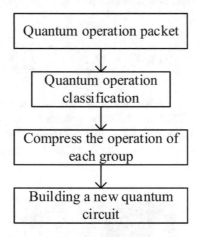

Fig. 2. The compression process of quantum image model.

Figure 2 shows the flow of NEQR quantum image compression. At the same time, the process is explained through the example below. In this example, q = 8 (the quantum sequence of quantum images consists of q qubits). According to the different types of quantum operations, each set of quantum manipulations will be divided into two parts: the blue part and the red part, which represent the quantum unit operation and the quantum control gate operation, respectively. It is found by experiments that the whole circuit consists of 8 quantum gates before compression. Next, the Espresso algorithm is used to compress this part of the circuit. The control information of the quantum gate circuit of the red part is input into the algorithm, and then the algorithm outputs X_0, which means that a quantum gate is only needed to complete the function of the whole circuit. Therefore, this process is used to compress the quantum image construction process.

Image compression for NEQR quantum image model is more effective than FRQI model. In the FRQI quantum image model, it can only be compressed with the smallest Boolean expression only when the adjacent pixels in a certain region have the same

0	0	1	1
0	0	1	1
0	0	1	1
0	0	1	1

Fig. 3. Graphical example.

color values, because the FRQI model only uses one bit to save the color information of the image. In the NEQR model, the color values of pixels are stored in the ground state of a quantum sequence. Each of the qubits in this sequence is independent of each other, so the compression algorithm can compress the quantum operations of each of the quantum bits in this sequence. Therefore, qualitatively, compared to the FRQI model, the NEQR model can obtain a higher compression rate for the image construction process using the minimum Boolean expression method.

3.2 Quantum Image Preprocessing

Sobel algorithm is currently one of the most representative algorithms for classical image edge extraction. Many studies based on Sobel algorithm or improved research has been widely applied. In this section, first of all, the traditional algorithm of image edge extraction based on Sobel algorithm is introduced.

The whole translation operation of the quantum circuit will consist of 1 single qubit flip gate, 1 two-qubit flip gate and 3 qubit turnaround gates. The number of 3 qubit turnaround doors is calculated as follows:

$$1 + \sum_{i=3}^{n-1} number(i - cnot) = 1 + \sum_{i=3}^{n-1} (4i - 8) = n^2 - 5n + 7 \tag{3}$$

According to the problem of image registration, it can be seen that at least one image in the image database should be measured with the same image. Next, the Grover search algorithm is used to perform quantum image retrieval. An auxiliary qubit is needed to initialize it as follows:

$$| - > = \frac{1}{\sqrt{2}}(|0 > - |1 >) \tag{4}$$

The whole process of feature extraction algorithm based on NEQR quantum image model is shown in the following figure (Fig. 4):

The quantum image registration algorithm flow is as follows:

Step 1: Quantum image construction. In this step, the reference image and the measured image will first be stored in the quantum state based on the QUALPI model.

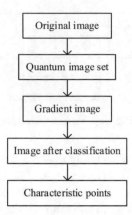

Fig. 4. The process of feature extraction of quantum images.

Step 2: Quantum image expansion. A quantum image expansion algorithm based on Fourier transform is used to transform the reference image into a set of quantum images, each of which represents a quantum image that is obtained by rotating a reference image in a different angle.

Step 3: Quantum image retrieval. The quantum image retrieval algorithm based on Grover algorithm is used to find the same quantum image in the image set obtained in the previous step.

Step 4: Quantum measurement. In order to be able to output classical information, quantum measurement will extract the information needed from the evolutionary quantum state, that is, the sequence number of the quantum image found in Step 3.

3.3 Quantum Image Classification

MCS has been extensively studied in the past decades, and there are two kinds of solutions: exact algorithm and inexact algorithm. In many exact MCS algorithms, pruning and backtracking are used, such as Ullmann, VF2, Mc Gregor, and so on. Another way is to get the result by constructing the association graph and finding the largest complete subgraph in the correlation graph. Quantum computation is a new computing theory that uses quantum superposition and quantum entanglement as special quantum mechanisms to calculate. Different from classical computation, the information in quantum system will be stored in the unit vector of complex Hilbert space. A common quantum state is the superposition of all base vectors, as shown in the following formula:

$$| \varphi > = \sum_{i=1}^{n} \alpha_i | \varphi_i > \tag{5}$$

In quantum computers, the quantum states are nondeterministic, and all the fundamental vector states are preserved simultaneously in the quantum superposition state. It is possible to observe the ground state in the quantum state from the quantum state through quantum measurement, and the quantum state will collapse to the quantum ground state after observation. Topological molecular structure of algorithm effectiveness is shown in Fig. 3.

First, for MUTAG data sets, only 1 pair of molecules of the same size fail in the Qwalk algorithm, and there are n pairs of number of tests that can run successfully in the whole data set. Figure 5 gives the topological structure of this pair of invalidation molecules. For the image data in the database, only 3 groups of tests in the different tests will fail in the Qwalk algorithm for the comparison of all the samples of each target. Therefore, in a real database, the problem of failure is basically impossible to appear, so it is believed that in the actual operation, the problem can be ignored.

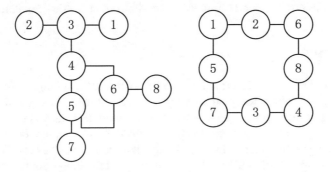

Fig. 5. Topological molecular structure of algorithm effectiveness.

The acceleration effect of this algorithm comes from the introduction of the largest isomorphic neighborhood pair. By using ING pairs to get the largest common subgraph, the search space in the algorithm is greatly reduced. However, when all the nodes in the two input graphs have different degrees, it is very difficult to find an effective ING pair that matches the point to point matching. If this happens, the performance of the Qwalk algorithm will be minimized, and the algorithm will fail.

4 Results and Discussion

Quantum image storage model: As the foundation of the whole system, the quality of the quantum image storage model will directly determine the complexity and performance of the image processing task that the whole image processing system can accomplish. Through the research on the existing quantum image storage model, it is found that the existing model has many shortcomings in the storage of image pixel color information. As a result, it causes the existing models to have many shortcomings and directly hinders the algorithm design based on the quantum image model.

The algorithm of quantum image preprocessing: through analysis, the algorithm can achieve an exponentially improved performance compared with the classical image edge extraction algorithm. This method can improve the performance of square magnitude compared with the traditional histogram rendering algorithm. The algorithm can achieve the performance improvement about the index magnitude compared with the traditional feature point extraction algorithm in the same way. The algorithm achieves a performance improvement of about four times compared with the simple image registration algorithm.

Quantum image classification: compared with the existing isomorphic inexact algorithms of the largest subgraphs, the algorithm shows better versatility, accuracy and noise immunity. At the same time, the size of the maximum isomorphic subgraph is applied to judge the similarity of graph. For some real databases, the algorithm also shows good classification ability and clustering ability. In addition, due to the sampling of quantum mechanism, the computational complexity of the algorithm is about $O(NE^{1.5})$ (of which N is the number of nodes and E is the number of edges of the graph). It is the fastest general imprecise algorithm to solve the problem.

5 Conclusion

From the classical image information, the research on the storage level of quantum image saves information in a quantum state in a specific way. However, the level of the preprocessing of quantum image is based on the quantum images in the quantum state for information mining, thus extracting the abstract information of the edges and features of the quantum image. Finally, the quantum image classification layer uses the abstract feature information to classify and understand the image content and return the result. As a whole, the quantum image processing system takes the classical image information as input, and returns the result of image processing by calculation, and the work completed is consistent with the classical image processing. But by using the quantum mechanism to design the process, in terms of theory and experiment evaluation, every level of the system has obvious performance improvement. Therefore, applying the system to image processing on future quantum computers can solve the performance problems of classical image processing.

References

1. Caraiman, S., Manta, V.I.: Image segmentation on a quantum computer. Quantum Inf. Process. **14**(5), 1693–1715 (2015)
2. Jiang, N., Wang, L.: Quantum image scaling using nearest neighbor interpolation. Quantum Inf. Process. **14**(5), 1559–1571 (2015)
3. Jiang, N., Wang, J., Mu, Y.: Quantum image scaling up based on nearest-neighbor interpolation with integer scaling ratio. Quantum Inf. Process. **14**(11), 1–26 (2015)
4. Zhou, R.G., Sun, Y.J.: Quantum multidimensional color images similarity comparison. Quantum Inf. Process. **14**(5), 1605–1624 (2015)
5. Youssry, A., El-Rafei, A., Elramly, S.: A quantum mechanics-based algorithm for vessel segmentation in retinal images. Quantum Inf. Process. **15**(6), 1–21 (2016)

The Cross-Language Query Expansion Algorithm Based on Hybrid Clustering

Huihong Lan and Jinde Huang[✉]

Guangxi College of Education, Nanning 530023, China
lanlandoll@163.com

Abstract. To propose a cross-language query expansion algorithm based on hybrid clustering aiming at problems such as theme shift, word dismatch, translation ambiguity and polysemy etc. in the cross-language information retrieval. The HPH-CLQE for cross-language query expansion is put forward by combining advantages of partition clustering and hierarchical clustering. This algorithm can achieve a good clustering effect by avoiding selection of the initial cluster center and cluster number. Then, the CLQE-HPH based on hybrid clustering is presented. This algorithm applies HPH-CLQE to conduct clustering analysis on top documents selected at the initial cross-language retrieval aiming to obtain the expansion word that is highly relevant with user query to achieve the cross-language query post-translation expansion. The experimental results based on NTCIR-5 CLIR data set indicate that MAP, P@5 and P@10 related to the algorithm proposed in this paper get improved greatly with the maximal increase of 20.52% by comparing with three standard algorithms. This algorithm can reduce problems such as theme shift and word dismatch etc. in the cross-language information retrieval effectively and improve the cross-language information retrieval performance.

Keywords: Text clustering · Partition clustering · Hierarchical clustering · Cross-language information retrieval · Query expansion

1 Introduction

The cross-language query expansion is one the key technologies to solve problems such as theme shift, word dismatch, translation ambiguity and polysemy etc. in the cross-language information retrieval. In the recent ten years, scholars at home and abroad have carried out deep exploration and researches on the cross-language query expansion models and algorithms, from which rich theoretical achievements are gained including the cross-language query expansion based on relevance feedback, latent semantic and association pattern mining etc. [1]. However, those achievements fail to solve the recall ratio and precision ratio in the cross-language information retrieval completely. The quality of extended source is the premise for effective query expansion. How to improve the quality of extended source in the cross-language query has become a pressing problem. In recent years, Bi Jianting attempt to introduce the text clustering technology into the cross-language query expansion so as to improve the information retrieval performance. Therein, the *K*-means algorithm is used to conduct

© Springer Nature Switzerland AG 2020
Q. Liu et al. (Eds.): CENet 2018, AISC 905, pp. 279–286, 2020.
https://doi.org/10.1007/978-3-030-14680-1_31

clustering on the bilingual documents to improve relevancy of the extended document set. Then, the latent semantic analysis technology is applied to find out the target word with the highest relevancy to the source query from the extended bilingual text set as the expansion word to achieve Chinese-English cross-language query expansion, which has gained excellent experimental effects [2].

The text clustering, as an unsupervised machine learning method, can gather documents with similar themes and relevant contents together, while the dissimilar ones will be separated. It is an effective method to gain rather high-quality documents. The partition clustering and hierarchical clustering are the most commonly used text clustering methods, of which the partition clustering is represented by K-means algorithm. This algorithm firstly selects K objects randomly as the initial cluster centers, and then includes the rest of objects into the closest cluster in accordance with its distance from various cluster centers, further re-calculate every cluster center. This process will be repeated continuously until the criterion function converges or data objects are invariable. The K-means algorithm can classify large data set efficiently with rather high execution efficiency. However, its cluster number shall be firstly determined due to the random selection of the initial cluster center, which will lead to unstable clustering results [3]. The hierarchical clustering can be divided into cohesive hierarchical clustering and divisive hierarchical clustering, of which the cohesive hierarchical clustering method firstly takes every object as a cluster, then merge two clusters with the closest distance until all clusters are merged into one or reach to a specified threshold value, while the divisive hierarchical clustering do not need to specify expected cluster numbers in advance, which can avoid selection of cluster number and the initial point, but once division or merging is executed, then it cannot be amended any more. The restricted clustering quality and rather complex algorithm time make it not suitable for large text clustering [4].

In view of advantages and disadvantages of partition clustering and hierarchical clustering, this paper firstly proposes a hybrid clustering algorithm based on partition and hierarchical method for cross-language query expansion (HPH-CLQE), and then puts forward a cross-language query expansion algorithm based on hybrid clustering. This algorithm translates the query word in the source language into the target language, and then conducts initial retrieval on the target document set. Later, the HPH-CLQE algorithm is adopted to conduct clustering analysis on m initially top reviewed documents to select a document cluster with the highest relevancy to user query as the target document set, which improves quality of expansion sources and enables to extract the expansion word with the highest relevancy to user query in the clustering results so as to achieve cross-language post-translation expansion. Compare the algorithm proposed in this paper with three standard algorithms, the experimental results based on NTCIR-5 CLIR data set indicate that MAP, P@5 and P@10 related to the algorithm presented in this paper get improved greatly with the maximal increase of 20.52%, which means it can improve cross-language information retrieval performance effectively.

2   HPH-CLQE Algorithm

2.1   Related Definition

Definition 1. Overall similarity: A measure of similarity of two objects. The agglomeration in the cluster is used to measure the clustering quality. The high-quality clustering is characterized by uniform and close distribution. Various objects in the cluster are very close to the cluster center with relatively high closeness. Let presume C_i has a data set of $\{x_1, x_2, \ldots , x_n\}$, then the overall similarity of this cluster $OS(C_i)$ is defined as follows:

$$OS(C_i) = \frac{1}{\frac{1}{n}\sum_{i=1}^{n}(x_i - c_i)^2} \tag{2.1}$$

Of which, x_i is the ith object of cluster C_i, while c_i is the center of cluster C_i.

Definition 2. Relative closeness: the relative closeness $RC(C_i, C_j)$ of cluster C_i and cluster C_j is defined as follows:

$$RC(C_i, C_j) = \frac{1}{dist_{min}(C_i, C_j)} \tag{2.2}$$

Of which, $dist_{min}(C_i, C_j)$ defines the smallest distance between cluster C_i and cluster C_j, $dist_{min}(C_i, C_j) = min_{x_i \in C_i, x_j \in C_j}\{|x_i - x_j|\}$.

Definition 3. Relative interconnection: If there exist m data points in any two cluster C_i and C_j to make distance between each data point reach $dist(x_i, y_i) \leq \alpha$ (specified threshold value), $x_i \in C_i$, $y_i \in C_j$. The relative interconnection $RI(C_i, C_j)$ of cluster C_i and C_j is defined as follows:

$$RI(C_i, C_j) = \alpha m \tag{2.3}$$

2.2   Basic Thoughts of HPH-CLQE Algorithm

The basic thought of HPH-CLQE algorithm is to divide the clustering algorithm into two stages including division and merging. First of all, divide the text set into two clusters by using K-means method based on partition clustering, and then calculate overall similarity of each cluster. If it is less than the specified threshold value, then repeat the above procedure until the overall similarity of all sub clusters is larger than the specified threshold value; then, the hierarchical clustering is adopted for further clustering; calculate relative similarity and relative interconnection of any two clusters, merge two very close clusters with high interconnection. Then, it comes to the final clustering result.

2.3   Description of HPH-CLQE Algorithm

Significance of symbols in the HPH-CLQE algorithm: γ is division stop parameter; α and β are merging stop parameter; then the description of HPH-CLQE algorithm is as follows:

> Algorithm 1: HPH-CLQE
> Input: γ, α and β
> Output: clustering result
> Begin
> 1) Select two documents in the current cluster as the initial cluster centers;
> 2) In regards to the rest of documents, calculate its distance from various cluster centers and then include into the document cluster with the smallest distance;
> 3) Recalculate every cluster center; the center is the arithmetic mean value of all points in the document cluster;
> 4) Repeat Step 2) and Step 3) until various clusters are invariale;
> 5) Calculate the $OS(C_i)$ of every cluster C_i , if $OS(C_i) \leq \gamma$ then go back to Step 1).
> 6) Calculate $RC(C_i,C_j)$ and $RI(C_i,C_j)$;
> 7) If $RC(C_i,C_j) \geq 1/\alpha$, and $RI(C_i,C_j) \geq \beta$, then cluster C_i and C_j will merge into one cluster;
> 8) Repeat Step 6) and Step 7).
> End.

The overall similarity is used to evaluate whether to divide this cluster in the HPH-CLQE algorithm; while the relative closeness and relative interconnection are applied to judge whether to merge two clusters. Therefore, HPH-CLQE algorithm can deal with clustering with different shapes, sizes and even those with noise, and can achieve better clustering effect.

3   The Cross-Language Query Expansion Based on Hybrid Clustering

3.1   Basic Thoughts of Cross-Language Query Expansion Based on Hybrid Clustering

The basic thought of cross-language query expansion based on hybrid clustering is to translate the query word in the source language into the target language through machines, and then conduct initial retrieval on the target document set to extract m documents from the initially reviewed results. Further, the HPH-CLQE algorithm is adopted to conduct clustering analysis on m documents from the initially reviewed results and then select a document cluster that is the most relevant to the user query to analyze user's query intention. That is, take the document cluster with the highest relevancy with user query as the target document cluster. Then, select n_{ET} feature words with the highest relevancy with the query word in the target document cluster as the cross-language post-translation expansion word, add it into the post-translation original query to combine into a new query; conduct retrieval on the target document set again to obtain the target document set of the final retrieval results.

3.2 Cross-Language Query Expansion Algorithm Based on Hybrid Clustering

The cross-language query expansion algorithm based on hybrid clustering can be formalized into cross language query expansion base on hybrid partition and hierarchical method (CLQE-HPH). The significance of symbols in the algorithm is as follows: O_{sour} is the query word in the source language; O_{tar} is the query word in the post-translation target language; m is the number of top documents from the initially reviewed results; Doc_{CLIR} is the top document set of the initially reviewed results; n_{ET} is the number of final expansion word; Doc_{tar} is the target document set of the final retrieval results. The specific description is as follows:

Algorithm 2: CLQE-HPH
Input: O_{sour}, m, n_{ET}.
Output: target document set Doc_{tar} from the final retrieval result
Begin
1) Translate the query word O_{sour} in the source language into the query word O_{tar} in the target language through machines;
2) Conduct initial retrieval on the target document set by applying the query word O_{tar} in the target language; extract top m documents as the initial document set Doc_{CLIR};
3) Conduct pre-treatment on Doc_{CLIR}; that is, word segmentation, stop word deletion, feature word extraction and weight calculation etc.;
4) Call HPH-CLQE algorithm to conduct clustering on document set Doc_{CLIR};
5) Select the clustering cluster with the largest similarity with O_{tar} as the target document cluster, and then extract the candidate expansion word in the target document cluster.
6) Calculate similarity of every candidate expansion word with O_{tar} then descending order; select top n_{ET} candidate expansion words with highest similarity as the final cross-language post-translation expansion word, and then combine it with O_{tar} to obtain the new query O_{new}.
7) Conduct retrieval on the target document set by using the new query O_{new} again to obtain the target document set Doc_{tar} of the final retrieval results.
End.

The weight $w_{i,j}$ of the feature word in Step (3) can be calculated as below [5]. The related formula is as follows:

$$w_{i,j} = \frac{tf_{ij}}{max_j tf_{ij}} \times log\left(\frac{m}{n_i}\right) \tag{3.1}$$

Of which, n_i is the number of documents with feature word t_i in the initially top reviewed document sets; tf_{ij} is the frequency of the feature word t_i shown up in the document d_j.

4   Experiment and Results Analysis

4.1   Experimental Data and Pre-treatment

This paper took English as the source language and Chinese as the target language to conduct English-Chinese cross-language query expansion experiment. The experimental data is the Chinese corpus test set provided by NTCIR-5 CLIR (http://research. nii.ac.jp/ntcir/permission/ntcir-5/perm-en-CLIR.html), which comes from Economic Daily News, News Media and News Test in 2000, totally 79,380 Chinese text information. This corpus includes the document test set, the result set and the query set. Thereinto, the query set has 50 topics. Each query topic can be classified into TITLE, DESC, NARR and CONC. The TITLE query is used in the experiment of this paper.

This paper adopts ICTCLAS2014 to conduct pre-treatment on Chinese corpus. That is, conduct word segmentation, wipe off stop word and extract feature words etc. The Microsoft Translator API is used as the interface of the machine translation system.

4.2   Experimental Evaluation Index and Standard Algorithms

The experimental evaluation indexes refer to mean average precision (MAP), the precision ratio of top five and ten results P@5 and P@10.

The experimental evaluating standards are as follows; (1) Cross-language baseline (CLB): Translate 50 English words into Chinese, and then conduct retrieval on Chinese documents to obtain results; (2) Cross-language query expansion based on K-means (CLQE-KM): Conduct clustering on m top documents from the initially reviewed results by using K-means algorithm, and then select the document cluster with the largest relevancy to user query as the target cluster; extract top n_{ET} feature words with the largest similarity to user query as expansion words; combine the expansion word and original query as a new query; conduct retrieval again to obtain results. The experimental parameters are set as k = 6, m = 500, n_{ET} = 25, of which k is the cluster number, m is the number of documents and n_{ET} is the number of expansion words; (3) Cross-language query expansion based on cohesive hierarchical clustering (CLQE-CHC): Conduct clustering on m top documents from the initially reviewed results by using cohesive hierarchical clustering algorithm; select the document cluster with the largest relevancy to user query as the target cluster; extract n_{ET} feature words with the largest similarity to user query as expansion words; combine the expansion word with the original query as a new query; conduct retrieval again to obtain results. The experimental parameters are set as m = 500 and n_{ET} = 25.

4.3   Experimental Result and Analysis

The experimental results of the algorithm in this paper and standard algorithms are shown in Table 1, and their retrieval performance is compared as shown in Table 2.

It is found through the experiment that for given data sets, if γ is larger, than it will generate more sub clusters in the division process, which will further affect the execution efficiency, while if γ is too small, then it cannot separate those two types, which will then affect the clustering precision. The smaller α between any two sub clusters

Table 1. Experimental results of algorithm in this paper and standard algorithms.

Evaluation index	CLB	CLQE-KM	CLQE-CHC	CLQE-HPH
MAP	0.3124	0.3318	0.3201	0.3703
P@5	0.4371	0.4615	0.4504	0.5268
P@10	0.4249	0.4482	0.4361	0.5052

Table 2. Comparison on retrieval performance of CLQE-HPH and standard algorithms.

Evaluation index	impr. over CLB (%)	impr. over CLQE-KM (%)	impr. over CLQE-CHC (%)
MAP	18.53	11.60	15.68
P@5	20.52	14.15	16.96
P@10	18.90	12.72	15.84

indicates closer distance. The larger β indicates high interconnection of two clusters. At this time, those two sub clusters are more likely to be of the same kind. The experiment indicates that when $\gamma = 10/OS_{total}$ (of which OS_{total} is the overall similarity of data sets), α is between 1 and 4 and β is between 5 and 20, then the clustering effect is rather ideal. When m and n_{ET} are relatively small, some lexical items with large relevancy to query contents will be left out; when m and n_{ET} are relatively large, some expansion words that are not relevant to query theme will also be included, both of which will decrease the retrieval performance. After repeated experiments, the parameters of this paper are set as $\gamma = 1/200$, $\alpha = 2$, $\beta = 15$, m = 500 and $n_{ET} = 25$.

The experimental results in Table 2 indicate that compared with standard algorithms CLB, CLQE-KM, CLQE-CHC, the MAP, P@5 and P@10 of CLQE-HPH presented in this paper get improved greatly with the increasing amplitude between 11.6% and 20.52%. Those results indicate that the algorithm proposed in this paper is effective, which can improve the cross-language information retrieval performance. The time complexity of CLQE-HPH is $O(n^2)$, which is lower than that of CLQE-CHC, but higher than those of CLB and CLQE-KM.

5 Conclusion

The cross-language query expansion is one of key technologies to improve the cross-language information retrieval performance. This paper proposes a cross-language query expansion algorithm based on hybrid clustering by introducing the text clustering technology. This algorithm firstly put forward HPH-CLQE by combining advantages of partition clustering and hierarchical clustering, which can avoid selection of the initial cluster center and cluster number to achieve better clustering effects and improve quality of the source document set for cross-language query expansion. Then, the CLQE-HPH is put forward to conduct clustering analysis on top documents selected from the initially retrieval results to obtain the expansion word that is highly relevant to the user query to realize cross-language query post-translation expansion. The experimental results based on NTCIR-5 CLIR indicate that the CLQE-HPH algorithm proposed in this paper can reduce problems such as theme shift and word dismatch etc. in the cross-language information retrieval effectively compared with CLB, CLQE-KM and CLQE-CHC. More specifically, its MAP, P@5 and P@10 get improved greatly with the maximal increase of 20.52%. The further research will focus on discussion on application of CLQE-HPH algorithm into practical cross-language information retrieval system to improve the corresponding performance. Besides, researches on influencing rules of parameters on cross-language query expansion will be further carried out.

Acknowledgement. This work is supported by National Natural Science Foundation of China (No. 61262028), and the Scientific Research Project of Education Department of Guangxi Province (No. KY2015YB337), and the Education Science fund of the scientific research projects of Guangxi College of Education (No. YB2014587).

References

1. Huang, M.X.: Vietnamese-English cross language query expansion based on weighted association patterns mining. J. China Soc. Sci. Tech. Inf. **36**(3), 307–318 (2017). (in Chinese)
2. Bi, J.T., Su, Y.D.: Expansion method for language-crossed query based on latent semantic analysis. Comput. Eng. **35**(10), 49–53 (2009). (in Chinese)
3. Zhang, C.C., Zhang, H.Y., Luo, J.C., He, F.: Massive data analysis of power utilization based on improved *K*-means algorithm and cloud computing. J. Comput. Appl. **38**(1), 159–164 (2018). (in Chinese)
4. Zhang, F.L., Zhou, H.C., Zhang, J.J., Liu, Y., Zhang, C.R.: A protocol classification algorithm based on improved AGNES. Comput. Eng. Sci. **39**(4), 796–803 (2017). (in Chinese)
5. Wu, D., Qi, H.Q.: Development of information retrieval model and its application in cross-language information retrieval. J. Mod. Inf. **29**(7), 215–221 (2009). (in Chinese)

Layer Clustering-Enhanced Stochastic Block Model for Community Detection in Multiplex Networks

Chaochao Liu[1,2], Wenjun Wang[1,2], Carlo Vittorio Cannistraci[3], Di Jin[1], and Yueheng Sun[1,2(✉)]

[1] College of Intelligence and Computing, Tianjin University,
Tianjin 300350, China
yhs@tju.edu.cn
[2] Tianjin Key Laboratory of Advanced Networking (TANK),
Tianjin University, Tianjin, China
[3] Biotechnology Center, Technische Universitat Dresden,
Tatzberg, Dresden, Germany

Abstract. Nowadays, multiplex data are often collected, and the study of multiplex-network (MN)s' community detection is a cutting-edge topic. multiplex-network (MN) layers can be grouped by clustering, and there are correlations between network layers that are assigned to the same cluster. Although the differences between network layers entail that the node community membership can differ across the layers, Stochastic-Block-Models (SBM)-based MN-community-detection methods current available are theoretically constrained to assume the same node community membership across the layers. Here, we propose a new SBM-based MN-community-detection algorithm, which surpasses this theoretical constraint by exploiting a two-stage procedure. Numerical experiments show that the proposed algorithm can be more accurate and robust than multilayer-Louvain algorithm, and may help to contain some inference issues of classical monolayer SBM. Finally, results on two real-world datasets suggest that our algorithm can mine meaningful relationships between network layers.

Keywords: Multilayer networks · Stochastic block model · Community detection · Network layer clustering

1 Introduction

In the last decades network science made exciting progresses, however traditional complex network approaches have mostly concentrated on a simple single-snapshot (therefore static) network. This may not fully capture the intrinsic dynamic or multiform organization of many real-life systems [1]. A neat example is given by biological systems, whose nodes can be proteins that are simultaneously involved in different non-overlapping pathways [2, 3], and the pathways can coincide with different layers of a multiplex network (MN). The need to merge in a holistic representation this multifaceted network organization, led network scientists to recently propose the new paradigm of MNs. Interestingly, recent studies observed that multilayer networks can

© Springer Nature Switzerland AG 2020
Q. Liu et al. (Eds.): CENet 2018, AISC 905, pp. 287–297, 2020.
https://doi.org/10.1007/978-3-030-14680-1_32

also present redundant layers [5], and that the detectability of community structure in multilayer networks can be enhanced through layer aggregation [6]. The fundamental rationale that justifies layer aggregation is that each layer of multilayer networks contains a hidden structure, which is the similar across some layers and different across others. For instance, friends and families may tend to communicate through WhatsApp, WeChat and QQ, while researchers and colleagues may tend to communicate through email, research gate and twitter.

Numerous methods for community detection in multilayer networks have been proposed. Mucha et al. [4] developed a generalized framework of network quality functions to study the community structure of arbitrary multislice networks. The stochastic block model (SBM) is a generative model of random graphs that can help to understand the mechanisms of node and link organization that bring to community formation in complex networks. The extension of traditional SBM to the multilayer framework is receiving growing attentions [9–11]. To be specific, Han et al. [7] proposed a multi-graph stochastic block model applied to dynamic and multi-layer networks, which assumes that nodes share the same block structure over the multiple layers, but the class connection probabilities may vary across layers. Barbillon et al. [8] proposed a multiplex SBM, which restricts the block structure and community membership of the nodes to be the same. Stanley et al. [9] described a strata multilayer SBM, which also maintains the community membership of the nodes the same across the strata layers, while computes different community memberships of nodes in each layer when clustering the network layers. This procedure may face curse of dimensionality problem when clustering network layers composed of large nodes, and has to deal with model selection problem for the number of network layer clusters. The multiplex networks may have common hidden structure in some layers, while the connection strength between hidden structures may vary through some layers. And because of the individual difference of the nodes in each layer, the community memberships of the nodes should not be restricted to be the same between the common hidden structure layers. For such a reason, in this paper we propose a new algorithm namely multiplex network stochastic block model community detection framework (MNSBM), aimed to solve this limitation of current methods.

The rationale behind our algorithm is the following. Since network layers can be partitioned into clusters, and the layers inside the same cluster share a common hidden structure, we thought to design our proposed community detection algorithm according to an iterative procedure based on two stages. In the first stage, we compute node community memberships. In the second stage, we partition network layers into clusters. Since our assumption is that the community memberships of the nodes should not be restricted to be the same between the common hidden structure layers, as a consequence we had the necessity to propose a new SBM approach that was redesigned to be used in the first stage. Hence, in the second stage-to avoid the curse of dimensionality problem that occurs when network layers composed of large nodes are clustered-we calculated pairwise similarities between layers by the information theory method proposed by Iacovacci et al. [10], and then we used the Louvain algorithm [11] to aggregate the network layers into clusters, which can avoid model selection. The network layer clustering result of the Louvain algorithm obtained in phase two, can be used in phase one to iterate the process. This framework will continue until the clustering result of Louvain algorithm in phase two, after some iterations, converges to the

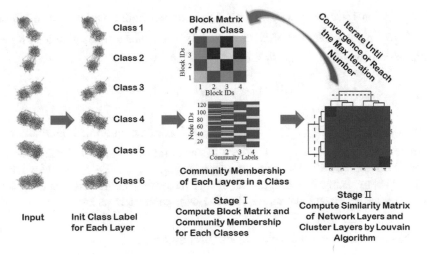

Fig. 1. An illustrative example of the MNSBM flowchart.

same result, or the maximum iteration number is reached. An example of the MNSBM flowchart is reported in Fig. 1.

2   Methods

2.1   Multiplex Network Community Detection in the First Stage

We now introduce a Bayesian degree-corrected stochastic block model for multiplex networks. For an undirected multiplex network G, with number of network layers Ω, and number of nodes N in each layer. $A_{ij}^{(\omega)}$ represents the number of links between nodes i and j in layer ω, and a positive number of self-links $A_{ij}^{(\omega)}$ is allowed in our model definition. We assume that the links between nodes i and j in layer ω follow a Poisson distribution, for analytical convenience the model assumes a particular parametrization of the self-links $A_{ij}^{(\omega)}$.

$$A_{ij}^{(\omega)} \sim Poisson(\theta_i^{(\omega)} \eta_{z_i^{(\omega)} z_j^{(\omega)}} \theta_j^{(\omega)}) \ for \ i<j \ (linkweight)$$

$$A_{ii}^{(\omega)} \sim Poisson(\frac{1}{2}(\theta_i^{(\omega)})^2 \eta_{z_i^{(\omega)} z_i^{(\omega)}}) \ for \ i=j \tag{1}$$

The parameter η_{fg} controls the probability of links between nodes in groups f and g in all the layers, $z_i^{(\omega)} = f$ indicates that node i is assigned to group f, and $\theta_i^{(\omega)}$ is a node-specific parameter that regulates this link probability in each layer, and thus accounts for heterogenous node degrees. The model is subject to the constraint that $\sum_i \delta_{z_i^{(\omega)} f} \theta_i^{(\omega)} = 1$ for all groups in all layers, i.e. the sum of the $\theta_i^{(\omega)}$ within each group in a layer ω is 1. For each group f in a layer ω containing $n_f^{(\omega)}$ nodes, we introduce an

$n_f^{(\omega)}$-dimensional vector of weights $(\phi_i^{(\omega)})_{z_i^{(\omega)}=f}$ drawn from a Dirichlet distribution in (2), where γ is the parameter of layer ω, and $\mathbf{1}_k$ is a vector of ones with length k. The relationship between $\phi_i^{(\omega)}$ and $\theta_i^{(\omega)}$ is $\theta_i^{(\omega)} = n_{z_i^{(\omega)}}^{(\omega)} \phi_i^{(\omega)}$.

$$(\phi_i^{(\omega)})_{z_i^{(\omega)}=f} \sim Dirichlet(\gamma \mathbf{1}_{(n_f)}) \tag{2}$$

We draw $\mathbf{z}^{(\omega)} = \{z_1^{(\omega)}, z_2^{(\omega)}, \ldots, z_i^{(\omega)}, \ldots, z_N^{(\omega)}\}$ from a multinormal distribution with parameter $\boldsymbol{\rho}^{(\omega)} = \{\rho_1^{(\omega)}, \ldots, \rho_F^{(\omega)}\}$, where F stands for the number of groups, and $\boldsymbol{\rho}^{(\omega)}$ follows Dirichlet distribution with parameter $\alpha_1, \alpha_2, \ldots, \alpha_F$. We follow Nowicki and Snijders [18] by fixing the components of this vector to a single value, and by default $\alpha = 1$, and get η_{fg} from Gamma distribution with parameters κ, and λ.

2.2 Network Layers Clustering in the Second Stage

Since structure correlations are ubiquitous in multilayer networks, Iacovacci et al. [10] proposed an information theory measure $\tilde{\Theta}^S$, which can be used to define similarities between the layers of a multiplex network with respect to their mesoscopic structures. In this paper, we used this similarity measure to build network of network layers that can avoid the curse of dimensionality problem. Then we choose the Louvain algorithm [11] to cluster the network layers, which can avoid model selection that is to determine the number of network clusters.

2.3 Optimization Algorithm

In order to infer $\mathbf{z}^{(\omega)}$ in each layer in the first stage, we use Gibbs sampling method. And because our algorithm is based on the Bayesian framework, we can treat the hyperparameters α, κ, λ, and γ as random variables.

$$(N_{fg}^+)^{(\omega)} = \begin{cases} \sum\limits_{\substack{i:z_i^{(\omega)}=f \\ j:z_j^{(\omega)}=g}} A_{ij}^{(\omega)}, & f \neq g \\ \sum\limits_{\substack{i<j: \\ z_i^{(\omega)}=z_j^{(\omega)}=f}} A_{ij}^{(\omega)}, & f = g \end{cases}, \quad N_{fg}^{(\omega)} = \begin{cases} n_f^{(\omega)} n_g^{(\omega)}, & f \neq g \\ \frac{n_f^{(\omega)} n_f^{(\omega)}}{2}, & f = g \end{cases} \tag{3}$$

We introduce the shorthand notation for between- and within-group link counts in (3), as well as node degrees $k_i^{(\omega)} = \sum_j A_{ij}^{(\omega)}$ and $\hat{k}_i^{(\omega)} = k_i^{(\omega)} + A_{ii}^{(\omega)}$ so we can get (4) which can be used in the Gibbs sweeping.

$P(\{\mathbf{A}^{(\omega)}\}, \{\mathbf{z}^{(\omega)}\} | \kappa, \lambda, \{\gamma\})$

$$= \frac{N!}{B(\alpha \mathbf{1}_F) \Gamma(N + F\alpha) \prod\limits_{\omega \in \Omega, i<j} A_{ij}^{(\omega)} \prod\limits_{\omega \in \Omega, i} 2^{A_{ii}^{(\omega)}}} \left[\prod\limits_{\omega \in \Omega, f} \frac{B(\gamma \mathbf{1}_{(n_f^{(\omega)})} + (\hat{k}_i^{(\omega)})_{i:z_i^{(\omega)}=f}) (n_f^{(\omega)})^{\hat{k}_f^{(\omega)}} \Gamma(n_f^{(\omega)} + \alpha)}{B(\gamma \mathbf{1}_{(n_f^{(\omega)})}) n_f^{(\omega)}!} \right] \frac{\prod\limits_{f \leqslant g} G(\sum\limits_{\omega \in |\Omega|} (N_{fg}^+)^{(\omega)} + \kappa, \sum\limits_{\omega \in |\Omega|} N_{fg}^{(\omega)} + \lambda)}{G(\kappa, \lambda)}$$

$$\tag{4}$$

The pseudo-code of the optimization algorithm for the MNSBM is showed in Algorithm 1. Since each layers' cluster is uncorrelated with each other, the first stage can run synchronously for each cluster. The most time-consuming part is the processing of sample $Z^{(cl)}$ according to (4). Thus, the most time-consuming part is calculating the Beta function values for communities at each Gibbs sweep. Then the time complexity of the whole algorithm is $O(MG * C * N * K * K * nf)$, where MG is the maximum Gibbs iteration number, C the maximum layers number of clusters, N the number of nodes in each layers, K the dimension of block matrix, and nf the maximum number of nodes in all the communities.

Algorithm 1. MNSBM(Multiplex network stochastic block model community detection framework)

Input: $G = A^{(1)}, A^{(2)}, ..., A^{(\Omega)}$ which is the multiplex network; N which is the number of nodes in each layers; K which is the initial dimension of block matrix; α, κ, and γ which are the hyper-parameters; M which is the maximum iteration number of the framework; MG which is the maximum Gibbs iteration number; $Z = z^{(1)}, z^{(2)}, ..., z^{(\Omega)}$ which is the community membership for all the layers; and $C = c^{(1)}, c^{(2)}, ..., c^{(\Omega)}$ which is the layers' cluster labels

Output: Z which is the Community membership of network layers; and C which is the partition of network layers

1: initialize $m = 0$;
2: **while** True **do**
3:   **for** c in C_{last} **do** // can run synchronously for each cluster
4:     **for** $mg = 1$; $mg \leqslant MG$; $mg ++$ **do**
5:       **for** network each layer cl in c **do**
6:         sample $Z^{(cl)}$ according to (4);
7:       **end for**
8:     **end for**
9:   **end for**
10: **for** $i = 1$; $i \leqslant \Omega$; $i ++$ **do**
11:    **for** $j = 1$; $j \leqslant i$; $j ++$ **do**
12:      compute similarities between layers $\tilde{\Theta}_{ij}^{s}$ according to symmetric measure defined in [10];
13:    **end for**
14: **end for**
15: layers' partition $C_{current} \leftarrow Louvain(\tilde{\Theta}^{s})$;
16: $m = m + 1$
17: **if** $C_{current} = C_{last}$ or $m = M$ **then**
18:    Break;
19: **end if**
20: **end while**

---

3   Results

3.1   Artificial Datasets Example

To demonstrate the effectiveness of MNSBM, we generated three synthetic multiplex networks using the graph_tool public software, which is a network tool proposed by Peixoto. We used two different block matrices (see Fig. 2) to generate 6 layers, and set the dimension to 4 for all the synthetic multiplex networks. The block matrix dimension represents the hidden structure of the multiplex network. We set the number of dimensions to 4 because this will make it the same as the number of communities desired in the synthetic networks. Each block matrix generated 3 layers, and each layer contained the same node set. The synthetic multiplex networks contain respectively 128 nodes, 256 nodes, and 512 nodes in a network layer. Besides, we selected the option to make the node-community-memberships diverse across different network layers.

Multilayer Louvain (ML) [4] is one of the best state-of-the-arts for community detection in non-overlapping multilayer networks, hence we compared the proposed MNSBM algorithm with ML. We conjectured that the multiplex version of an algorithm, if well-designed, should outperform the respective monolayer version. Therefore, we considered for comparison also a recently proposed monolayer stochastic block model algorithm (SBM) [12] and the classical monolayer Louvain algorithm (Louvain) [11]. These comparisons can help to investigate and numerically quantify the advantage offered by multilayer algorithms for community detection. To compare the accuracy and robustness of these algorithms, we generated an ensemble of 20 synthetic multiplex networks, and considered as performance indicator the mean of the normalized mutual information (NMI) [13] values across the 20 networks calculated independently for each layer. All parameters in MNSBM were set to a default value 1 during the tests in our experiments. Interestingly, MNSBM partitions the 6 network layers into two clusters, which match the ground truth exactly (result not shown).

Figure 3 shows the mean and standard error values in NMI obtained by the algorithms compared for each layer. Monolayer SBM offered a poor performance in comparison to monolayer Louvain. This result is not surprising since a recent study of Vallès-Català et al. [14] suggested that SBM, when the single most plausible model is selected, might display overfitting issues in learning and predicting the intrinsic topological organization of complex networks. The same study clarifies that this issue is contained when-to prevent overfitting-the average over collections of models that are individually less plausible is used. However, recently Muscoloni et al. [15] showed that also when the average over collections of less-plausible models is used, the overfitting behaviour of SBM is still clearly affecting its performance. Although the study of Vallès-Català et al. [14] and Muscoloni et al. [15] were performed considering evaluations in link prediction, our results suggest that some further studies should also investigate the inference limitations of SBM in community detection.

The results suggest that MNSBM is in general-regardless of network size - much more accurate and robust than the monolayer SBM, because the mean NMI curve is higher across the layers and displays a smaller error. This significant performance improvement of MNSBM in respect to monolayer SBM might be related with an

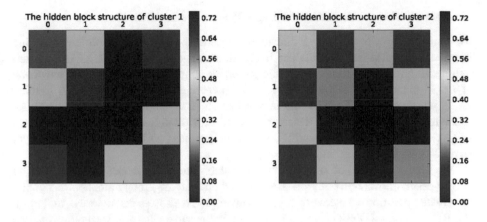

(a) Block Structure of Cluster 1   **(b) Block Structure of Cluster 2**

Fig. 2. Two block structures to generate multiplex networks.

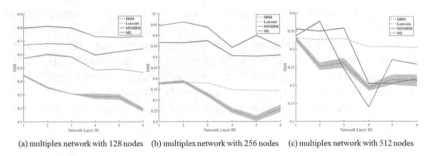

(a) multiplex network with 128 nodes   (b) multiplex network with 256 nodes   (c) multiplex network with 512 nodes

Fig. 3. The mean and standard error value of NMI for each layer using the monolayer stochastic block model (SBM), monolayer Louvain algorithm (Louvain), multilayer Louvain algorithm (ML), and the proposed algorithm (MNSBM).

important overfitting reduction, which can be triggered by the multilayer design strategy adopted in our algorithm. Interestingly, MNSBM performs better than ML in the multiplex networks with 128 nodes, but worse in the multiplex networks with 256 nodes. While, in the multiplex networks with 512 nodes MNSBM seems again to slightly outperform ML. However, considering networks of different sizes, both MNSBM and ML displays a similar trend along the x-axis, and this fact suggests that adopting a multilayer design philosophy can in general positively increase and homogenize the performance even for very different algorithms. Finally, as far as concern our results, we can conclude that multilayer algorithms for community detection-if well-designed-can provide improved performance in respect to their respective monolayer versions.

3.2   Roethlisberger and Dickson Bank Wiring Room Dataset

We applied the MNSBM algorithm to the Roethlisberger & Dickson Bank Wiring Room dataset of UCINET [16]. The interaction categories include: RDGAM, participation in horseplay; RDCON, participation in arguments about open windows; RDPOS, friendship; RDNEG, antagonistic (negative) behaviour; RDHLP, helping others with work; and RDJOB, the number of times workers traded job assignments. Since the RDHLP and RDJOB are non-symmetric matrixes, we only consider the RDGAM, RDCON, RDPOS, and RDNEG four matrixes, which construct a multilayer network with four layers, in this paper.

We set γ, κ, λ and α of multiplex network stochastic block model to be 1, and using the Bayesian information criterion (BIC) to find out that the group number is 2. Finally, the cluster partition of network layers converged to the results reported in Fig. 4. The RDCON and RDNEG layers are included in the same cluster, this suggests that the participation in arguments about open windows behaviour and antagonistic (negative) behaviour influence each other; the RDGAM and RDPOS layers are in the same cluster, indicating that the participation in horseplay behaviour and friendship relation always influence each other. All the network layers and community partitions are showed in Fig. 5. The colour and shape of each node indicate the community membership of the node. The nodes with larger and lower degrees tend to be respectively in the same community. We speculate that human personality may drive different connections (network links) and community partitions, i.e. extroversive people tend to horseplay and make friendship with each other, while introversive people assume an opposite trend. The nodes of lower degrees are in the same community in the RDCON network layer, while not all of these nodes are still in the same class in the RDNEG network layer.

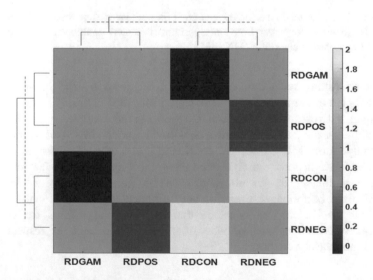

Fig. 4. Roethlisberger & Dickson Bank Wiring Room multiplex network layers clustering result.

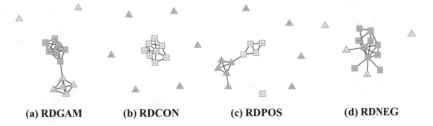

(a) RDGAM   (b) RDCON   (c) RDPOS   (d) RDNEG

Fig. 5. Roethlisberger & Dickson Bank Wiring Room network.

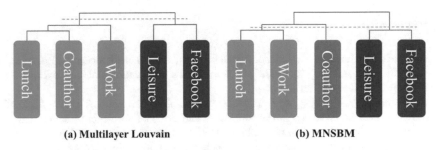

(a) Multilayer Louvain   (b) MNSBM

Fig. 6. Network layer clustering results on the multiplex social network dataset.

3.3 A Multiplex Social Network Dataset CS-AARHUS

We run MNSBM and ML on the multiplex social network dataset CS-AARHUS, which is proposed by Magnani et al. [17]. We set all the parameters of MNSBM and ML to be 1, and using the Bayesian information criterion (BIC) to find out the number of groups. The network layers clustering results are reported in Fig. 6. Figure 6(a) shows the network layers clustering result by using ML, and Fig. 6(b) shows the network layers clustering result by using MNSBM. The colour of rectangles represents layer clusters. Our algorithm detected Coauthor layer as a single cluster, while ML located Lunch, Coauthor and Work layers into the same cluster. As Magnani et al. [17] noticed, the "current working relationships" network is quite distinct from the "co-authorship relation" network, hence the layer clustering result of MNSBM seems to be more meaningful.

4 Discussion

Multiplex network layers can be clustered, and there are correlations between network layers that are included in the same cluster. And because of the differences between network layers, the community memberships of nodes through layers can be different. Current multilayer network community detection methods constrain the community memberships of the nodes to be the same between the common hidden structure layers. The proposed MNSBM algorithm relaxes this constraint to allow community memberships of the nodes to be different between the common hidden structure layers.

The results provided in the numerical experiments showed that MNSBM performs much better than the multilayer Louvain algorithm. In particular, the two analysed real world datasets show that our algorithm can successfully detect the relationships between the network layers.

Acknowledgement. This work was supported by the National Key R&D Program of China (2018YFC0809800, 2016QY15Z2502-02, 2018YFC0831000), the Project of National Social Science Fund 15BTQ056, and the National Natural Science Foundation of China (91746205, 91746107, 51438009); The Klaus Tschira Stiftung (KTS) gGmbH, Germany (Grant number: 00.285.2016).

References

1. Boccaletti, S., Bianconi, G., Criado, R., del Genio, C.I., Gómez-Gardeñes, J., Romance, M., Sendiña-Nadal, I., Wang, Z., Zanin, M.: The structure and dynamics of multilayer networks. Phys. Rep. **544**, 1–122 (2014)
2. Cannistraci, C.V., Ogorevc, J., Zorc, M., Ravasi, T., Dovc, P., Kunej, T.: Pivotal role of the muscle-contraction pathway in cryptorchidism and evidence for genomic connections with cardiomyopathy pathways in rasopathies. BMC Med. Genomics **6**, 5 (2013)
3. Durán, C., Daminelli, S., Thomas, J.M., Joachim Haupt, V., Schroeder, M., Cannistraci, C.V.: Pioneering topological methods for network-based drug-target prediction by exploiting a brain-network self-organization theory. Brief. Bioinform. **19**, 1183–1202 (2017)
4. Mucha, P.J., Richardson, T., Macon, K., Porter, M.A., Onnela, J.P.: Community structure in time-dependent, multiscale, and multiplex networks. Science **328**, 876–878 (2010)
5. De Domenico, M., Nicosia, V., Arenas, A., Latora, V.: Structural reducibility of multilayer networks. Nat. Commun. **6**, 6864 (2015)
6. Taylor, D., Shai, S., Stanley, N., Mucha, P.J.: Enhanced detectability of community structure in multilayer networks through layer aggregation. Phys. Rev. Lett. **116**, 228301 (2016)
7. Han, Q., Xu, K., Airoldi, E.: Consistent estimation of dynamic and multi-layer block models. In: Proceedings of the 32nd International Conference on Machine Learning, pp. 1511–1520 (2015)
8. Barbillon, P., Donnet, S., Lazega, E., Bar-Hen, A.: Stochastic block models for multiplex networks: an application to a multilevel network of researchers. J. R. Stat. Soc. Ser. (Stat. Soc.) (2016)
9. Stanley, N., Shai, S., Taylor, D., Mucha, P.J.: Clustering network layers with the strata multilayer stochastic block model. IEEE Trans. Netw. Sci. Eng. **3**, 95–105 (2016)
10. Iacovacci, J., Wu, Z., Bianconi, G.: Mesoscopic structures reveal the network between the layers of multiplex data sets. Phys. Rev. E **92**, 042806 (2015)
11. Blondel, V.D., Guillaume, J.L., Lambiotte, R., Lefebvre, E.: Fast unfolding of communities in large networks. J. Stat. Mech. Theory Exp. (2008)
12. Peixoto, T.P.: Efficient monte carlo and greedy heuristic for the inference of stochastic block models. Phys. Rev. E **89**, 012804 (2014)
13. Meilă, M.: Comparing clusterings-an information based distance. J. Multivar. Anal. **98**(5), 873–895 (2007)
14. Vallès-Català, T., Peixoto, T.P., Guimerà, R., Sales-Pardo, M.: On the consistency between model selection and link prediction in networks. Mach. Learn. (2017)
15. Muscoloni, A., Cannistraci, C.V.: Local-ring network automata and the impact of hyperbolic geometry in complex network link-prediction. Phys. Soc. (2018)

16. Roethlisberger, F.J., Dickson, W.J.: Management and the Worker: Social Versus Technical Organization in Industry. Harvard University Press, Cambridge (1939)
17. Magnani, M., Micenkova, B., Rossi, L.: Combinatorial analysis of multiple networks. Soc. Inf. Netw. (2013)
18. Nowicki, K., Snijders, T.A.B.: Estimation and prediction for stochastic blockstructures. J. Am. Stat. Assoc. **96**, 1077–1087 (2001)

Clustering Ensemble for Categorical Geological Text Based on Diversity and Quality

Hongling Wang[1,2,3(✉)], Yueshun He[1,2], and Ping Du[1,2]

[1] East China University of Technology, Nanchang 330013, China
whl9win@163.com
[2] Jiangxi Engineering Laboratory on Radioactive Geoscience and Big Data Technology, East China University of Technology, Nanchang 330013, China
[3] China University of Geosciences, Wuhan 430074, China

Abstract. Clustering analysis for geological text makes the navigation, retrieval or extraction of geological text more effectively. Clustering ensemble can be employed to obtain more robust clustering results. However, most generation approaches focus on the diversity of clustering members rather than their quality. Too much emphasis on the diversity of clustering members reduces the accuracy of clustering results. In order to solve the problem, a new generation method of clustering members is proposed in this paper. Hierarchical clustering algorithm and k-means algorithm alternately combined with random projection method are employed to generate diverse base members and a new selection strategy for the number of clusters is presented to improve the quality of clustering members. Furthermore, a clustering ensemble framework for geological text is constructed. The framework involves geological text preprocessing, geological text feature representation, clustering members generation and ensemble integration. Experimental results on two UCI datasets and one real-world geological text demonstrate that the clustering ensemble based on diversity and quality is superior to those clustering ensemble algorithms that only consider the diversity of clustering members.

Keywords: Clustering ensemble · Diversity · Geological text · Quality

1 Introduction

In recent decades, with the continuous improvement in the service level of geological information and in-depth study of geological science, the exponential growth of harvested geological texts has been accumulated. Therefore, it is increasingly challenging to efficiently obtain and analyze these geological texts and then transform them into geological knowledge. Data mining technology undoubtedly becomes the most alternative means to analyze the massive geological texts and discover useful knowledge that can be used for decision making to meet the requirements of different application levels. As a main task in data mining, clustering analysis aims to divide objects with more similarity into same group. Clustering analysis for geological text makes the geological text navigation, retrieval or information extraction more effectively.

© Springer Nature Switzerland AG 2020
Q. Liu et al. (Eds.): CENet 2018, AISC 905, pp. 298–306, 2020.
https://doi.org/10.1007/978-3-030-14680-1_33

The conventional clustering algorithms such as k-means and hierarchical clustering all have their own dependence on data distribution or data structure that result in unsatisfactory clustering results. Clustering ensemble, as the improvement and extension of clustering algorithm, is to integrate different clustering results and obtain a superior final clustering result. Research shows that the stability, applicability and scalability of clustering ensemble are superior to those of single clustering algorithm [1, 2]. In clustering ensemble, the main tasks include generating clustering members on the given dataset, and integrating these clustering members. Therefore, in recent years, the research has mainly focused on two aspects, how to produce effective clustering members and how to design consensus functions to merge clustering members. It has been recognized that diversity between clustering members contribute to improve the accuracy of clustering ensemble by correcting classification errors between different clustering. Many literatures present the ways to generate diverse clustering members, which include different clustering algorithm, different parameters of the same algorithm, data sampling and feature selection. Recent research contributions in clustering ensemble indicate that many existing methods pay more attention to how to produce diverse clustering members and give less consideration to the quality of the clustering members themselves. In fact, some studies have demonstrated that the quality of the clustering member is low, when the diversity is the highest. Inspired by the observations above, a new clustering ensemble algorithm based on quality and diversity is proposed in this paper. Also, some experiments are performed on geological text to establish the superiority of the proposed methods. The main contributions of our proposed approach can be summarized as follows:

(1) In the process of clustering member generation, quality and diversity as guidance are utilized to generate base clustering members. As a result, the base clustering members with high diversity and quality provide a reliable source for the clustering ensemble process.
(2) An efficient clustering ensemble framework for geological text is constructed. The framework involves geological text preprocessing, geological text feature representation, base clustering members generation and ensemble integration. To the best of our knowledge, this is the first study in geological text analysis.

The remainder of the paper is organized as follows. Section 2 reviews previous works on clustering ensemble. Section 3 presents our proposed clustering ensemble approach based on diversity and quality. In Sect. 4, experimental results are presented. Finally, conclusions are given in Sect. 5.

2 Related Works

It has been generally recognized that clustering ensemble can improve the clustering performance and computational efficiency. Strehl and Ghosh proposed the concept of clustering ensemble for the first time. In their opinion, due to the limitations and differences of traditional clustering algorithms, the performance of clustering can be strengthened and improved if the advantages of different algorithms are integrated. They also creatively put forward three ensemble algorithms based on hyper graph,

CSPA, HGPA and MCLA. Once the concept is proposed, it has become a new hot spot in the field of clustering [3]. Fred et al. propose EA algorithm. First, it needs to obtain the clustering members, and then the interconnected matrix is calculated. The final clustering results are obtained by using the hierarchical clustering algorithm based on the minimum spanning tree. Topchy presents a probability model, which uses the polynomial distribution in clustering space, and then the maximum likelihood parameter estimation algorithm is used to get the final clustering results. The influence of incomplete information and missing class labels on the ensemble effect is also analyzed. Clustering ensemble based on voting is to share the information of the data objects among the clustering members as much as possible. The final clustering result is obtained by calculating the voting ratio of a data object belongs to a cluster [4]. The base idea of the ensemble algorithm based on neural network is to use a large number of initial clustering members as the input of the neural network, and the final clustering results are obtained by using the neural network integration algorithm.

Strchl and Ghosh propose that the clustering members can be generated by using different clustering algorithms. Minaei-Bidgoli et al. use randomly sampling to generate the subset of data. The k-means algorithm is then employed to produce the clustering members on this dataset. Fred and Jain use k-means algorithm with different initial parameter to produce clustering members [5–7]. Zhou et al. propose a cluster generation method based on k-means algorithm with fixed k value and random cluster centers. Dudoit and Fridlyand use Bootstrap method to get clustering members. Topchy et al. study the effect of the quality of a single clustering member on the clustering results [8–10]. Fern et al. present an ensemble generation method based on the random projection method. They use random project, random project combined with PCA and PCA combined with random sampling to generate the base clustering members. AI-Razgan et al. use LAC algorithm with different initial parameters to create the base clustering members. Luo proposes CEAN and ICEAN methods to create ensemble members by adding artificial noise data. Jia et al. put forward a new method to create clustering members required by spectral clustering ensemble based on randomly zooming parameters [11, 12]. Kuncheva et al. study the diversity among clustering members in the clustering ensemble, and propose a variant method for clustering generation method, which randomly selects the clustering member from the surplus members. The diversity between clustering members can be evaluated by Rand Index, Adjusted Rand Index, Jaccard Index and Mutual Information. Hadjitodorov et al. demonstrate the relationship between diversity and quality of the clustering members is not monotonic. In some cases, the moderate diversity of clustering member can get a more accurate clustering result [10, 13].

Most of the previous works tend to emphasize the importance of generating clustering members with more diversity and give less consideration of the quality. Furthermore, there are few studies exploring the predictive performance of clustering ensemble method on geological text clustering analysis.

3 Proposed Clustering Ensemble Approach Based on Diversity and Quality

3.1 Clustering Ensemble

The clustering ensemble can be described as follows. Let $X = (x_1, x_2, \cdots, x_n)$ represent a dataset of n objects. Also, let $P = (p_1, p_2, \cdots, p_m)$ be a set of clustering members with m clustering members. $p_i = (c_1^i, c_2^i, \cdots, c_{ki}^i)(i = 1, 2, \cdots, m$ denotes the ith clustering member, where k is the cluster number of clustering member p_i, such that $U_k^{j=1} c_j^i = X$. And then, the final integrated clustering result will be produced from the available m clustering members by a consensus function F.

3.2 Generating Clustering Members

As mentioned before, many existing literatures focus more on how to produce clustering members with great diversity, and pay less attention to both diversity and quality. This paper mainly discusses the strategy of generating clustering members with better quality and diversity simultaneously. The most commonly used method to generate clustering members is k-means algorithm by randomly selecting different initial points and number of clusters. The main advantages of this method are low complexity and easy implementation. However, it is not effective for non-spherical data. Random projects can transform non-spherical clusters into spherical clusters and maintain the distance information between the original data points at a high probability. Regarding the algorithms to be used in the process of base clustering members generation, k-means, hierarchical clustering algorithm and random projection method are used. We name this new algorithm as RKHE. Random projection method is used to generate multiple instances of the original dataset, and then k-means and hierarchical clustering algorithms are used alternately to generate clustering members on these instances. k-means algorithm is initialized with a random cluster center and different number of clusters k. So, diverse clustering members are produced. In order to enhance the quality of clustering members, a selection strategy of k is proposed as follows.

(1) Optimal-k: The optimal number of clusters is obtained by using majority voting strategy on the results calculated by Hartigan, Dunn and DB indices to the dataset $X = (x_1, x_2, \cdots, x_n)$. In the clustering generation stage, the optimal k has a large probability PK_o to be employed as the number of clusters to produce clustering members.

(2) Medium-k: A clustering is created by randomly selecting the parameter k between $\{2, \cdots, k-2, k-1, k+1, k+2\}$, and the medium k has a lower probability to be selected.

(3) Perturbed-k: A few perturbed members are created by randomly selecting the parameter k between $\{min\_k, \cdots, max\_k\}$, where max_k is the maximum range of k.

The aim of optimal k and medium k is to ensure the quality of clustering results with the stable clustering results. The purpose of perturbed k is to increase the diversity of clustering results by the random perturbation.

3.3   Clustering Analysis for Geological Text

The classification of geological text mainly includes three phases: text preprocessing, vector representation of text and clustering analysis based on clustering ensemble.

Text Preprocessing
The main tasks of text preprocessing include Chinese word segmentation and stop words removal. Word segmentation is carried out by using the Jieba toolkit. In order to enhance the segmentation accuracy, a dictionary of geological terms is constructed. Then, some words with high frequency but no practical meanings are removed by introducing the stop words list.

Vector Representation of Text
The key words that can represent the text feature are extracted as text feature words, and the feature words vector are constructed to transform unstructured geological text into structured vectors. Each geological text d is represented as a t dimensional vector composed of t feature words, which can be represented as formula (3.1), where t_i is a feature term in the text d, and w_i is the weight value of the feature term. TF-IDF method is used to represent the term weight w_i in this paper, which can be described as formula (3.2).

$$d = (t_1, w_2, t_2, w_2, \cdots, t_i, w_i, \cdots, t_n, w_n) \tag{3.1}$$

$$w_i = \frac{N_{t_i}}{N} * \log\left(\frac{D}{D_{t_i}}\right) \tag{3.2}$$

where N_{t_i} represents the number of times that feature term t_i occurs in text d, N represents the total number of all feature terms in text d, D denotes the total number of texts, and D_{t_i} is the number of documents contains the feature item t_i.

Clustering Analysis Based on Clustering Ensemble
After generating the clustering members, these clustering members are integrated based on voting strategy. The number of times that each object belongs to a class in each clustering member is recorded in a co-association matrix, and then, the object is divided into the cluster with the greatest number.

4   Experiments

This section presents the datasets utilized in the experiment, the experimental procedure and the experimental results on UCI datasets and real-world dataset from CNKI.

4.1   Data Sets

The datasets used in the experiment include two standard UCI datasets and one real-world geological patent text extracted from cnki.net. The detailed description of these datasets is shown in Table 1.

Table 1. Descriptions of experimental datasets.

No	Dataset	# objects	# of classes
1	Iris	150	3
2	Breast Cancer	683	2
3	Real-world Geological Text	462	3

4.2   Experimental Evaluation

The clustering performance is evaluated by Classification accuracy (CA) and Normalized mutual information (NMI). In the following, CA and NMI are briefly described.

(1) CA

CA measures the correct percentage of clustering result, the better the clustering effect, the greater the CA value. Let $P = \{p_1, p_2, \cdots, p_m\}$ denote a clustering result to be measured, $Q = \{q_1, q_2, \cdots, q_n\}$ is the classification result with ground truth labels, where m and n are the number of clusters of P and respectively. Both P and Q have O objects. CA is defined as formula (3.3).

$$CA = \frac{1}{N} \sum_{i=1}^{m} max_{q_j \in R} |p_i \cap q_j| \qquad (3.3)$$

(2) NMI

NMI is an increasingly popular criterion for evaluating the cluster quality in terms of the mutual information between the clustering results and the pre-defined labeling of the dataset. Given two clustering members $P = \{p_1, p_2, \cdots, p_m\}$ and $Q = \{q_1, q_2, \cdots, q_n\}$, let $NMI(P, Q)$ denote the NMI value between P and Q, which can be calculated by formula (3.4).

$$NMI = -2 \frac{\sum_{i=1}^{m} \sum_{j=1}^{n} \frac{n_{ij}}{N} \log\left(\frac{n_{ij}N}{n_i n_j}\right)}{\sum_{i=1}^{m} \frac{n_i}{N} \log\left(\frac{n_i}{N}\right) + \sum_{j=1}^{m} \frac{n_j}{N} \log\left(\frac{n_j}{N}\right)} \qquad (3.4)$$

Where N is the number of objects in the dataset; m and n are the number of clusters in the clustering member P, the number of clusters in the member Q respectively. n_i and n_j are the number of objects in cluster p_i and q_i respectively. n_{ij} represents the number of objects that belong to cluster p_i and q_i simultaneously. Intuitively, the greater the NMI value, the more information is shared between P and Q, and the better the quality of the partition.

4.3   Experimental Results

The experiment is performed in R. The geological documents are converted into structure vector representation after the process of text preprocessing. Then, the method as described in Sect. 3 is employed to generate 50 base clustering members for each dataset. The value of the cluster number k is set based on the method described in

Sect. 3. The value of the error ε in random projection method is set to 0.6. Then, a co-association matrix is constructed to record the number of times each object belongs to a cluster. Finally, the ensemble result is determined according to the voting strategy. The process is iterated ten times, and the average results are produced.

The experiment is performed on three datasets. For each dataset, three different methods, including k-means, hierarchical clustering algorithm and the proposed algorithm, are applied to generate the clustering members. These three algorithms are denoted as KE, HE and RKHE respectively. The final ensemble results are evaluated by using CA and NMI measurements. The comparison results are presented in Tables 2 and 3, respectively.

Table 2. Comparison results of three clustering algorithm by using CA measurement.

Dataset	KE	HE	RKHE
Iris	0.7162	0.7422	0.8045
Breast Cancer	0.6923	0.6898	0.7358
Real-world Geological Text	0.6492	0.6623	0.7232

Table 3. Comparison results of three clustering algorithm by using NMI measurement

Dataset	KE	HE	RKHE
Iris	0.7734	0.7515	0.8197
Breast Cancer	0.1559	0.1626	0.1871
Real-world Geological Text	0.5267	0.5448	0.5653

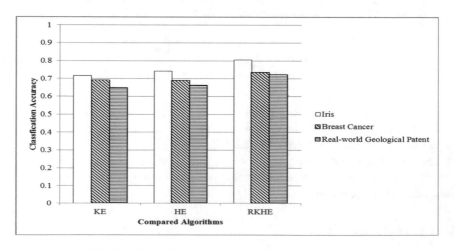

Fig. 1. Clustering ensemble accuracy measured by CA.

We can see from Tables 2 and 3, the proposed algorithm obtains the highest classification accuracy evaluated by CA and NMI on three datasets compared with other methods. In Fig. 1, we also present the final ensemble results with respect to different algorithms. For all dataset, the best clustering result evaluated by CA is always obtained by employing the proposed clustering generation algorithm, because there is a balance between diversity and quality to a certain extent.

5 Conclusion

Clustering analysis of geological text contributes to improve the utilization efficiency of geological text and reveal useful information. The conventional single clustering algorithm often fail to achieve satisfactory results in analyzing geological text. In order to improve the performance of geological text classification, the clustering ensemble algorithm is applied to the process, in which two elements, the diversity among ensemble members and the quality of the ensemble members are experimented to guide the clustering generation process. All clustering members are integrated by the co-occurrence matrix based on voting strategy. The results of the proposed method on two standard datasets taken from the UCI repository and one real-world geological dataset are compared with other generation methods. Experimental results demonstrate the advantage of the proposed method. However, all generated clustering members participate in the ensemble process, which may effect on the clustering results because of some inferior clustering members. So, our future work intends to study how to select parts of superior clustering members to perform the ensemble process.

Acknowledgement. This study is supported in part by National Natural Science Foundation of China (No. 41802247, 41862012), Open Fund of Jiangxi Engineering Laboratory on Radioactive Geoscience and Big data Technology (No. JELRGBDT201708, No. JELRGBDT201705), Key Research Development Foundation of Jiangxi Province Technology Department (No. 20161BBE50063).

References

1. Alizadeh, H., Yousefnezhad, M., Bidgoli, B.M.: Wisdom of crowds cluster ensemble. Intell. Data Anal. **19**, 485–503 (2015)
2. Huang, D., Lai, J.H., Wang, C.D.: Combining multiple clustering via crowed agreement estimation and multi-granularity link analysis. Neurocomputing **170**, 240–250 (2015)
3. Strehl, A., Ghosh, J.: Cluster ensemble-a knowledge reuse framework for combining multiple partitions. In: Proceedings of the 18th National Conference on Artificial Intelligence, pp. 93–98 (2002)
4. Jain, A.K., Topchy, A., Law, M.H., Buhmann, J.M.: Landscape of clustering algorithms. In: Proceedings of the 17th International Conference on Pattern Recognition, pp. 260–263 (2004)
5. Minaei-Bidgoli, B., Topchy, A., Punch, W.F.: Ensembles of partitions via data resampling. In: International Conference on Information Technology: Coding and Computing, p. 188 (2004)

6. Dudoit, S., Fridlyand, J.: Bagging to improve the accuracy of a clustering procedure. Bioinformatics **19**(9), 1090–1099 (2003)
7. Topchy, A., Jain, A.K., Punch, W.: Combining multiple weak clusterings. In: Proceedings of the Third IEEE International Conference on Data Mining, pp. 331–338 (2003)
8. Fern, X.Z., Brodley, C.E.: Random projection for high dimensional data clustering: a cluster ensemble approach. In: Twentieth International Conference on Machine Learning, pp. 186–193 (2003)
9. Hadjitodorov, S.T., Kuncheva, L.I., Todorova, L.P.: Moderate diversity for better cluster ensembles. Inf. Fusion **7**(3), 264–275 (2006)
10. de Amorim, R.C., Hennig, C.: Recovering the number of clusters in data sets with noise features using feature rescaling factors. Inf. Sci. **324**, 126–145 (2015)
11. Meng, J., Hao, H., Luan, Y.S.: Classifier ensemble selection based on affinity propagation clustering. J. Biomed. Inform. **60**, 234–242 (2016)
12. Yu, Z.W., Li, L., Gao, Y.J., You, J., Liu, J.M., Wong, H.S., Han, G.Q.: Hybrid clustering solution selection strategy. Pattern Recognit. **47**(10), 3362–3375 (2014)
13. Yousefnezhad, M., Reihanian, A., Zhang, D.Q., Minaei-Bidgoli, B.: A new selection strategy for selective cluster ensemble based on Diversity and Independency. Eng. Appl. Artif. Intell. **56**, 260–272 (2016)

Research on Cold-Start Problem in User Based Collaborative Filtering Algorithm

Lu Liu$^{(\boxtimes)}$ and Zhiqian Wang

Beijing University of Posts and Telecommunications, Beijing 100876, China
2012213229@bupt.edu.cn

Abstract. In order to solve the cold-start problem existing in traditional user based collaborative filtering algorithm, we propose a novel user clustering based algorithm, which firstly prefills user-item rating matrix, and then considers user characteristics as well as ratings when computing user similarities, and applies optimized k-means algorithm to cluster users. MovieLens is used as the test dataset. It is proved that the algorithm proposed in this paper can solve the cold-start problem and improve the accuracy of recommendation to some extent.

Keywords: Cold-start · Collaborative filtering · *K*-means user clustering · Similarity calculation optimization

1 Introduction

The recommender system arises in the context of information overload [1]. As one of the most successful algorithms, collaborative filtering algorithm has been widely applied to make recommendations. With the rapid growth of the number of items and users, data sparsity becomes increasingly apparent. The cold-start problem is just an extreme manifestation of data sparsity. The problem is that when a new user or a new item enters the recommender system, there are no ratings existing in the system. Hence, it is impossible to recommend items. The problem has seriously affected the quality of recommendation, so it is really urgent to solve it.

Scholars have done deeply research on the cold-start problem in the recent years. We summarize three common ways. The first is matrix dimensionality reduction and decomposition. Liao [2] proposed to transform user-item ratings into user-item attribute ratings. Liu [3] introduced a matrix decomposition model integrating social information. The disadvantage of the two is the loss of partial useful information. The second one is using machine learning. Xu [4] put forward a rating comparison strategy which breaks the barrier between active and cold-start users. And fine-grained iteration is used to calibrate the potential distribution of cold-start users. The weakness is that the quality of recommendation depends largely on the quality and quantity of the training set. The last one is using generalized clustering model. Cao [5] presented to divide user communities by discrete particle swarm optimization algorithm. Hu [6] raised a method to apply optimized *k*-means algorithm to cluster items. Although the method can effectively help find nearest neighbors, the accuracy of partition cannot be guaranteed under the condition of extreme data sparsity.

© Springer Nature Switzerland AG 2020
Q. Liu et al. (Eds.): CENet 2018, AISC 905, pp. 307–315, 2020.
https://doi.org/10.1007/978-3-030-14680-1_34

The above approaches can alleviate the cold-start problem to some extent. However, if the rating matrix is extremely sparse, the quality of recommendation will decrease. In response to the problem, we propose a novel user clustering based algorithm. The algorithm modifies similarity calculation combined with both user characteristics and ratings. And it applies k-means algorithm with meliorated initial center to cluster users which improves the efficiency of nearest neighbor finding. When a new user comes to the recommender system, it can effectively solve the cold-start problem and improve the accuracy of recommendation.

2   Relevant Concepts

2.1   Traditional Collaborative Filtering Algorithm

The first step of traditional collaborative filtering algorithm is to use a $m \times n$ rating matrix to represent user's rating to the item. As is shown in Table 1, $\{u_1, u_2, \ldots, u_m\}$ is the collection of m users, $\{i_1, i_2, \ldots, i_n\}$ is the set of n items. r_{ij} indicates user i's rating to item j.

Table 1. User-item rating matrix.

	i_1	i_2	...	i_n
u_1	r_{11}	r_{12}	...	r_{1n}
u_2	r_{21}	r_{22}	...	r_{2n}
...	...	...	...	...
u_m	r_{m1}	r_{m2}	...	r_{mn}

The second step is to find nearest neighbors. There are three commonly used methods of similarity calculation, cosine, adjust cosine and pearson respectively. Pearson formula [7] is as follows.

$$sim(u, v) = \frac{\sum\limits_{i \in S_{uv}} \left(R_{ui} - \overline{R_u}\right)\left(R_{vi} - \overline{R_v}\right)}{\sqrt{\sum\limits_{i \in S_{uv}} \left(R_{ui} - \overline{R_u}\right)^2} \sqrt{\sum\limits_{i \in S_{uv}} \left(R_{vi} - \overline{R_v}\right)^2}} \tag{2.1}$$

S_{uv} is an item collection in which each item is rated by both user u and v. The meaning of R_{ui} and R_{vi} is user u and v's rating to item i respectively. When a new user comes to the recommender system, there is no or just a few user ratings to each item. Hence, there is no or a few common ratings. The calculation of similarity is not accurate.

The last step is to predict user's rating to items according to the following formula [8] and make recommendations.

$$P(u, i) = \overline{R_u} + \frac{\sum\limits_{v \in NNU(u)} sim(u, v)\left(R_{vi} - \overline{R_v}\right)}{\sum\limits_{v \in NNU(u)} sim(u, v)} \tag{2.2}$$

In the above formula, $NNU(u)$ is the collection of nearest neighbors of the target user, and $sim(u, v)$ is the similarity between user u and v.

2.2 Item Attribute

Each item can be described with attributes. For instance, a movie has the attributes of language, release time, theme etc. Attribute is the characteristic of item itself, which is more stable compared with user-item ratings. The paper is to use a $n \times t$ matrix to represent the attribute of the item. a_{ij} is used to indicate attribute j of item i. The value of a_{ij} is 0 or 1. If item i has attribute j, the value of a_{ij} is 1. On the contrary, the value of a_{ij} equals to 0.

2.3 User Characteristic

User characteristic can reflect the similarity between users. It is easy to find that if two users have similar characteristics, their interests and preferences are also similar. In general, a user has the characteristics of gender, age, occupation etc. We quantify user characteristics with figures in this paper. Gender is expressed with 0 and 1. And age is divided into the following five stages, under 20, 21–30, 31–40, 41–50 and over 50. Occupation is denoted with numbers as well.

3 Design and Description of the Algorithm

3.1 Algorithm Design

The flowchart of the algorithm is shown in Fig. 1. The algorithm introduces item attributes and clusters items on the basis of attribute similarities firstly. Next, it prefills original user-item rating matrix according to the principle of "users tend to give similar ratings to similar items". And then, k-means algorithm with meliorated initial center is used to do user clustering on the filled matrix. After that, it determines the cluster of the target user, and finds nearest neighbors by calculating similarities between the target and other users in the cluster with optimized similarity formula which considers both user characteristics and ratings. The last step is to predict ratings of the target to items based on the ratings of nearest neighbors, and recommend Top-N items with highest ratings.

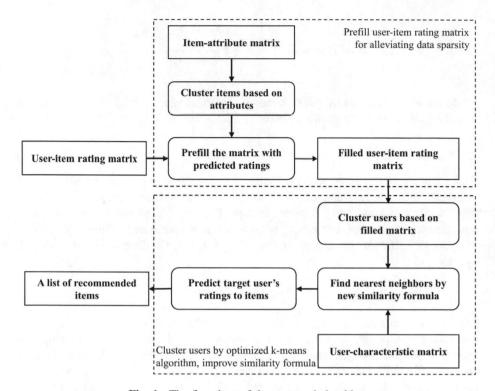

Fig. 1. The flowchart of the proposed algorithm.

3.2   Algorithm Description

K-means Algorithm with Meliorated Initial Center

K-means is an algorithm based on partition in clustering analysis. It is widely used on account of reliability, simplicity and fast convergence. Firstly, the algorithm is to randomly select *k* points as the initial cluster centers. Next is to allocate each remaining point to the nearest cluster. After that, recalculate the mean of each cluster and select them as the new cluster centers. Repeat above two steps until the deviation converges. The drawback of traditional k-means algorithm is that it is sensitive to the initial selected clusters and is unable to determine the value of *k*. To solve the first problem, we propose a new k-means algorithm with initial cluster center optimization combined with *DBSCAN* [9].

Definition 1: Point density. For any point x in the dataset, calculate similarities between x and other points, and use the number of points that exceed the threshold as point density $D(x)$.

$$D(x) = |\{p|sim(x,p) \geq \varepsilon, p \in U\}| \tag{3.1}$$

Definition 2: Core point. For any point x in the dataset, define x whose point density exceeds the threshold as core point. M is the collection of core points.

$$M = \{x|D(x) \geq MinPts, x \in U\} \tag{3.2}$$

The process of the algorithm is as follows.

Step 1: Calculate point density of each point and get the collection of core points;
Step 2: Select the point with the highest point density as the first initial cluster center c_1;
Step 3: Select the point with the lowest similarity to c_1 as the second initial cluster center c_2;
Step 4: Select c_3 which satisfies $min(max(sim(x, c_1), sim(x, c_2)))$;
Step 5: Select c_k which satisfies $min(max(sim(x, c_1), sim(x, c_2), \ldots, sim(x, c_{k-1})))$.

The above algorithm selects a set of points with lowest similarities from the dataset as the initial cluster centers instead of random selection, which can effectively avoid instability of clustering results. Additionally, the selection is in core points but not the whole dataset, so as to greatly narrow the range of selection.

Similarity Calculation Combined with User Characteristics
Traditional similarity calculation is based on ratings. Since there is no or just a few user ratings to items when a new user comes to the system, the result is not ideal. To address the above problem, we propose a new approach to similarity calculation combined with user characteristics. Suppose that $\{c_1, c_2, \ldots, c_q\}$ is the collection of user characteristics. s_{ui} is the value of ith characteristic of user u. The following formula is used to calculate the similarity of user characteristics.

$$sim_c(u, v) = \frac{1}{1 + d(u, v)} = \frac{1}{1 + \sqrt{\sum_{k=1}^{q} (s_{uk} - s_{vk})^2}} \tag{3.3}$$

The similarity calculation of user characteristics (as shown in 3.3) and ratings (as shown in 2.1) has been put forward, next is to select a suitable weight a to combine the two. The final similarity calculation formula is as follows.

$$sim(u, v) = a \times sim_p(u, v) + (1 - a) \times sim_c(u, v) \tag{3.4}$$

Moreover, the calculation of a is shown below.

$$a = \frac{|S_{uv}|}{|S_u| + |S_v|} \tag{3.5}$$

$|S_{uv}|$ is the number of items rated by both user u and v. $|S_u|$ and $|S_v|$ indicate the number of items rated by user u and v respectively. When a new user enters the system, the value of a is 0 since there is no commonly rated items. With the growth of the number of ratings, the value of a is increasing which means that the similarity calculation of ratings plays an increasingly important role.

Complexity

The complexity of k-means is $O(NKT)$. N is the number of users, K and T is the number of clusters and iterations respectively. Nearest neighbor finding is narrowed to a cluster rather than the whole dataset. Hence, the complexity of this part is $O(NK)$. Since K is a constant, the time complexity of the proposed algorithm is $O(N(T+1))$.

4   Experiments

MovieLens is one of the widely used datasets in the research on recommender system. 100 k dataset, which contains 100000 ratings to 1682 movies from 943 users, is used in the experiment. The data sparsity of the dataset is 94.96%. The dataset is divided into the training set and test set with the ratio of 4:1.

4.1   Evaluation Metrics

Evaluation metrics of the recommender system vary from accuracy, coverage, diversity etc. Accuracy is used in this paper to measure the quality of the algorithm. There are two commonly used ways to evaluate the accuracy of the recommendation algorithm, *MAE* and *RMSE* respectively. *MAE* is introduced firstly. Suppose that the collection of predicted ratings is $\{p_1, p_2, \ldots, p_n\}$, and the actual rating set is $\{q_1, q_2, \ldots, q_n\}$. The calculation of *MAE* [10] is as follows.

$$MAE = \frac{\sum\limits_{i=1}^{n} |p_i - q_i|}{n} \tag{4.1}$$

Next is *RMSE*. It can also be used for accuracy measurement. The formula is given below.

$$RMSE = \sqrt{\frac{\sum\limits_{i=1}^{n} (p_i - q_i)^2}{n}} \tag{4.2}$$

Both of them can be used to calculate the deviation. Moreover, the smaller the deviation is, the closer the predicted and actual ratings are. In this paper, *MAE* is used to evaluate the accuracy.

4.2   Experimental Results

The value of e, *MinPts* and k need to be determined before the experiment. The value of e can neither be too small nor too large. The paper calculates different values of *MAE* when the value of e varies from 0.48 to 0.52. And the result is in Fig. 2. Thus, we make the value of e to be 0.52 and *MinPts* to be 90.

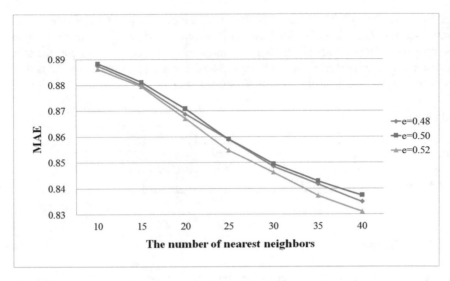

Fig. 2. Accuracy comparison under different values of *e*.

The next step is to determine the value of *k*. The paper calculates different values of *MAE* when the value of *k* varies from 8 to 10. It is clear to find from Fig. 3 that when *k* equals to 10, *MAE* takes smallest values under different number of nearest neighbors.

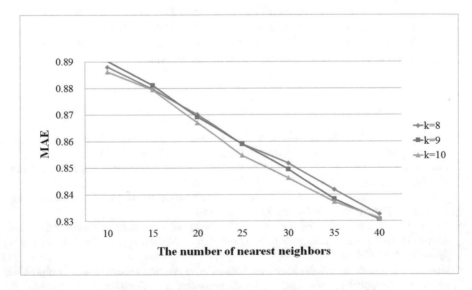

Fig. 3. Accuracy comparison under different values of *k*.

Under the condition that *e* equals to 0.52, *MinPts* is 90 and *k* is 10, the paper calculates *MAE* of the original and improved algorithm when the number of nearest neighbors is 10 to 40. The result is in Table 2. Compared with two algorithms, we can draw a conclusion that *MAE* obtained under different number of nearest neighbors of the proposed algorithm are all smaller than those of the original one. And as the number of nearest neighbors increases, *MAE* presents a decreasing trend which reflects that the recommendation result of the proposed one is better.

Algorithm	MAE						
	10	15	20	25	30	35	40
UCF	0.8910	0.8913	0.8891	0.8916	0.8891	0.8891	0.8892
Proposed algorithm	0.8863	0.8795	0.8671	0.8548	0.8463	0.8372	0.8311

5 Conclusion

When a new user comes to the recommender system, there are no user ratings to items. To address the cold-start problem, we come up with a new collaborative filtering algorithm based on user clustering in this paper. The algorithm firstly prefills user-item rating matrix based on the similarity of item attributes to alleviate data sparsity. And then, it synthetically considers user characteristics and ratings when computing user similarities, and applies k-means algorithm with meliorated initial center to cluster users. The experimental results show that the proposed algorithm can solve the cold-start problem and improve the accuracy of recommendation to some extent. In practice, time factors may have an influence on user interests. As time goes by, user preferences will change. So the next step is to consider the effect of time factors on user similarity calculation and further improve the accuracy of the algorithm.

References

1. Wang, Q., Xie, Y., Yu, S.: Collaborative filtering algorithm combined with the user clustering and item types. Appl. Comput. Syst. **25**(12), 132–137 (2016)
2. Liao, F., Lin, S., Guo, K.: A cold-starting personalized recommendation algorithm. J. Chin. Comput. Syst. **36**(8), 1723–1727 (2015). (in Chinese)
3. Liu, H., Jing, L., Yu, J.: Survey of matrix factorization based social recommendation methods. J. Softw. **29**(2), 1–24 (2018). (in Chinese)
4. Xu, J., Yao, Y., Tong, H., Tao, X., Lu, J.: Ice-breaking: mitigating cold-start recommendation problem by rating comparison. In: International Conference on Artificial Intelligence, pp. 3981–3987 (2015)
5. Cao, C., Ni, Q., Zhai, Y.: An improved collaborative filtering recommendation algorithm based on community detection in social networks. In: Conference on Genetic & Evolutionary Computation, pp. 1–8 (2015)
6. Hu, X., Lu, H., Chen, X., Zhou, G.: *K*-means item clustering recommendation algorithm based on initial clustering center optimization. J. Air Force Early Warn. Acad. **28**(3), 203–207 (2014). (in Chinese)

7. Ren, C.: Improved algorithm of alleviating item cold starting. Comput. Eng. Softw. **37**(8), 11–15 (2016). (in Chinese)
8. Yu, H., Li, J.: Algorithm to solve the cold-start problem in new item recommendations. J. Softw. **26**(6), 1395–1408 (2015). (in Chinese)
9. Ji, Z., Ge, W.: Research and optimization of k-means clustering algorithm in network recommendation systems. Inf. Commun. **175**, 34–36 (2017). (in Chinese)
10. Bouras, C., Tsogkas, V.: Clustering to deal with the new user problem. In: IEEE 15th International Conference in Computational Science and Engineering (2012)

Behavior Similarity Awared Abnormal Service Identification Mechanism

Yingyan Jiang[1], Yezhao Chen[1], Feifei Hu[2], Guoyi Zhang[2], and Peng Lin[3(✉)]

[1] Power Grid Dispatching Control Center of Guangdong Power Grid Co., Ltd., Guangzhou, Guangdong Province, China
[2] Power Grid Dispatching Control Center of China Southern Power Grid, Guangzhou, China
[3] Beijing Vectinfo Technologies Co., Ltd., Beijing, China
linpeng@vectinfo.com

Abstract. In order to maintain network security, it is very important to identify services with abnormal behavior and take targeted measures to prevent abnormal behaviors. We propose abnormal service identification mechanism based on behavior similarity. This method proposes a formula for service behavior similarity calculation of flow ports for services with correlation. And then k-similarity clustering algorithm is proposed to find abnormal service behaviors. Meanwhile, we analyse outliers to improve the accuracy of clustering results. At last, the experimental results show that k-similarity clustering algorithm can differentiate abnormal services accurately.

Keywords: Behavior similarity · Abnormal service · K-similarity clustering algorithm

1 Introduction

End-hosts are the important part of communication networks. If abnormal behaviors occur on an end-host, it might pose a great risk to other end-hosts. As a result, area in which abnormalities occur in the network continues to expand. In order to maintain network security and improve network reliability, it is necessary to effectively identify services with abnormal behaviors.

At present, many research methods in the field of service behavior identification have been proposed, including methods based on DPI/DFI [1, 2], data mining [3–5], and on port numbers. Methods based on port numbers are the simplest and fastest. However, it is pointed out in [6–8] that with the continuous increase of network applications, the accuracy of methods based on port number in a single flow significantly decreases. Hence, existing methods based on port numbers are impossible to meet requirements of service behavior identification. Although those methods based on port number cannot accurately identify service behaviors, port numbers used by services with the same behavior type (especially services with the same abnormal behavior type) have a certain correlation at the same time-space [9, 10]. Literature [11] proposes a method of measuring service behavior similarity. They use the difference in

© Springer Nature Switzerland AG 2020
Q. Liu et al. (Eds.): CENet 2018, AISC 905, pp. 316–324, 2020.
https://doi.org/10.1007/978-3-030-14680-1_35

lists of port numbers among different services as one of the measurement factors of behavior similarity. However, this method doesn't further divide source port numbers and destination port numbers. Some correlation of source port numbers and destination port numbers will be ignored. Literature [12] proposes the share of the number of the same ports in total number of ports between two services. They use the share to calculate behavior similarity. But they only consider source port or destination port, and don't take the difference of flow volume into consideration. Therefore, the results of behavior similarity need to be improved.

In this paper, to improve the accuracy of identifying abnormal services, we present behavior similarity awared abnormal service identification mechanism. The contributions of this paper are summarized as follows:

(a) We propose the measurement factors of service behavior similarity and behavior similarity calculation formula. And then based on MATLAB, we use grayscale matrix to verify the feasibility of behavior similarity formula.
(b) We propose k-similarity clustering algorithm based on service behavior similarity, taking outliers into consideration. Meanwhile, we improve selection process of initial centroids and clustering process of sample points. Experimental results show that the k-similarity algorithm is better than k-means algorithm in terms of identification rate and misjudgment rate.

The remainder of this paper is structured as follows: In Sect. 2, we propose measurement factors of service behavior similarity and model of service behaviors. Section 3 presents calculation formula of service behavior similarity. Section 4 presents k-similarity clustering algorithm. Section 5, we verify performance of the method proposed by this paper through simulation experiments, and Sect. 6 concludes the paper.

2 Measurement Factors of Service Behavior Similarity

In this paper, we study service behavior similarity based on flows from a sink node of a LAN. To calculate behavior similarity, we build a behavior model for all the services. We divide flows into different aggregated flows according to network prefixes and then analyze behavior similarity of services in same aggregated flows.

3 Behavior Similarity Formula

3.1 Discretization of Difference of Flow Volume

Because the difference of flow volume is continuous, it needs to be discretized. The flow volume set of all services in set U is $C = \{C_1, C_2, \ldots, C_N\}$. The difference of flow volume between service U_i and service U_j is expressed as $\Delta C_{i,j} = |C_i - C_j|$. Assuming that the gradient of flow volume difference is β, the discretization approach of $\Delta C_{i,j}$ is as follows:

$$
\alpha_{i,j} = \begin{cases}
\lambda_1, & \Delta C_{i,j} < \beta \\
\lambda_2, & \beta \leqslant \Delta C_{i,j} < 2\beta \\
\lambda_3, & 2\beta \leqslant \Delta C_{i,j} < 3\beta \\
\lambda_4, & 3\beta \leqslant \Delta C_{i,j} < 4\beta \\
\lambda_5, & \Delta C_{i,j} > 4\beta
\end{cases} \tag{1}
$$

We divide difference of flow volume into five levels, $\lambda_1 \sim \lambda_5$. The value of $\lambda_1 \sim \lambda_5$ is set as 1–5. The value of β can be adjusted according to actual flow volume of active services in the network.

3.2   Difference in Proportion of Behaviors with Correlation

Taking service U_i and U_j as an example, behavior models of service U_i and U_j are expressed as $\mathcal{U}_i = \{C_i, \mathcal{F}_i\}$ and $\mathcal{U}_j = \{C_j, \mathcal{F}_j\}$, respectively. Each element in set $\mathcal{F}_i$ and $\mathcal{F}_j$ are expressed as $\mathcal{E}_{i,k} = \{port_{i,c,k}, port_{i,s,k}, \omega_{i,k}\}$ and $\mathcal{E}_{j,k} = \{port_{j,c,k}, port_{j,s,k}, \omega_{j,k}\}$, respectively. To calculate the difference in proportion of behaviors with correlation, we need to divide network behaviors with correlation into a same set.

We set relationship pattern of network behaviors of service U_i as $R_i = (P_i, F_i)$. P_i is the set of all service-end port numbers and server-end port numbers used by service U_i, $P_i = \{port_{i,c,1}, port_{i,s,1}, \ldots, port_{i,c,k}, port_{i,s,k}, \ldots, port_{i,c,n}, port_{i,s,n}\}$. F_i is the dependency set of port numbers, $F_i = \{port_{i,c,1} \rightarrow port_{i,s,1}, port_{i,s,1} \rightarrow port_{i,c,1}, \ldots, port_{i,c,k} \rightarrow port_{i,s,k}, port_{i,s,k} \rightarrow port_{i,c,k}, \ldots\}$. Based on F_i, we construct closures. Every closure $\eta_{i,l}$ is the set of port numbers used by network behaviors with correlation. Port number division set of service U_i is expressed as follows:

$$
\zeta(U_i) = \{\eta_{i,1}, \eta_{i,2}, \ldots, \eta_{i,l}, \ldots\} \tag{2}
$$

According to the correlations of different services' behaviors and port number partition set of service U_i, we divide port numbers of service U_j. And behavior labels which have no correlation with all network behaviors of service U_i will be discarded. Port number partition set of service U_j is expressed as follows:

$$
\zeta(U_j) = \{\eta_{j,1}, \eta_{j,2}, \ldots, \eta_{j,l}, \ldots\} \tag{3}
$$

The building process of $\eta_{j,l}$ is as follows:

I. $\forall port_{i,c,k}, port_{i,s,k} \in \eta_{i,l}, \exists (port_{j,c,p} = port_{i,c,k} \wedge port_{j,s,p} = port_{i,c,k})$
$\vee (port_{j,c,p} \neq port_{i,c,k} \wedge port_{j,s,p} = port_{i,c,k}) \vee (port_{j,c,p} = port_{i,c,k} \wedge port_{j,s,p} \neq port_{i,c,k})$
$\Rightarrow \eta_{j,l} = \eta_{j,l} \cup port_{j,c,p} \cup port_{j,s,p}$

II. $\forall port_{i,c,k}, port_{i,s,k} \in \eta_{i,l}, \nexists(port_{j,c,p} = port_{i,c,k} \wedge port_{j,s,p} = port_{i,c,k})$
$\wedge(port_{j,c,p} \neq port_{i,c,k} \wedge port_{j,s,p} = port_{i,c,k}) \wedge (port_{j,c,p} = port_{i,c,k}$
$\wedge port_{j,s,p} \neq port_{i,c,k}) \Rightarrow \eta_{j,l} = \emptyset$

The flow volume involved by each element of port number division set needs to be calculated. Taking port number set $\eta_{i,l}$ of service U_i as an example, the formula is as follows:

$$\mathcal{N}(\eta_{i,l}) = \sum_{\substack{port_{i,c,k} \in \eta_{i,l} \wedge \\ port_{i,s,k} \in \eta_{i,l}}} \omega_{i,k} \tag{4}$$

The proportion of the flow volume involved in set $\eta_{i,l}$ is as follows:

$$\mathcal{P}(\eta_{i,l}) = \mathcal{N}(\eta_{i,l})/C_i \tag{5}$$

The difference in proportion of behaviors with correlation between service U_i and service U_j is as follows:

$$\mathcal{Q}(\eta_{i,l}, \eta_{j,l}) = \begin{cases} \mathcal{P}(\eta_{i,l})/\mathcal{P}(\eta_{j,l}), \mathcal{P}(\eta_{i,l}) < \mathcal{P}(\eta_{j,l}) \\ \mathcal{P}(\eta_{j,l})/\mathcal{P}(\eta_{i,l}), \mathcal{P}(\eta_{i,l}) \geq \mathcal{P}(\eta_{j,l}) \end{cases} \tag{6}$$

3.3   Behavior Similarity Formula

Definition: Behavior Similarity Formula. Element number of port number division set is $|\zeta(U_i)|$. Behavior similarity between service U_i and service U_j is calculated as follows:

$$S_{i,j} = \frac{1}{\alpha_{i,j}} \cdot \frac{1}{|\zeta(U_i)|} \cdot \sum_{\substack{\eta_{i,l} \in \zeta(U_i), \\ \eta_{j,l} \in \zeta(U_j)}} \mathcal{Q}(\eta_{i,l}, \eta_{j,l}) \tag{7}$$

We definite that $S_{i,j}$ is equal to $S_{j,i}$ and $S_{i,i}$ is equal to 1. Behavior similarity matrix of active N services in an aggregated flow is as follows:

$$s = \begin{bmatrix} S_{1,1} & S_{1,2} & \cdots & S_{1,i} & S_{1,j} & \cdots & S_{1,N} \\ S_{2,1} & S_{2,2} & \cdots & S_{2,i} & S_{2,j} & \cdots & S_{2,N} \\ \vdots & \vdots & \ddots & \vdots & \vdots & \ddots & \vdots \\ S_{i,1} & S_{i,2} & \cdots & S_{i,i} & S_{i,j} & \cdots & S_{i,N} \\ S_{j,1} & S_{j,2} & \cdots & S_{j,i} & S_{j,j} & \cdots & S_{j,N} \\ \vdots & \vdots & \ddots & \vdots & \vdots & \ddots & \vdots \\ S_{N,1} & S_{N,2} & \cdots & S_{i,N} & S_{N,j} & \cdots & S_{N,N} \end{bmatrix} \tag{8}$$

We use grayscale matrix to represent behavior similarity matrix S. For example, the cell for service 1 and service 4 is close to white, indicating that behavior similarity between the two services is relatively large.

4 k-Similarity Clustering Algorithm

Based on the similarity of service behaviors, we can cluster the services. K-means clustering algorithm regards each column as an entity and the value of similarity in a column as an attribute. It clusters services by using distance formula.

4.1   Selection Process of Initial Centroids

Assuming that service set is U, set of initial centroids is Z and the number of elements in Z is k. Selection process of initial centroids is as follows: (1) At the beginning, set Z is empty. We randomly select a non-outlier U_a as the first initial centroid and add it to set Z. (2) We perform remaining k-1 selections of initial centroid. For each selection, we calculate the average of behavior similarity between every non-outlier and all initial centroids in set Z. If the service for the minimum average $S_{\min}$ is U_m, we select U_m as an initial centroid and add it to set Z. The calculation formula of $S_{\min}$ is as follows:

$$S_{\min} = \min_{U_k \in U} \frac{1}{|Z|} \sum_{U_b \in Z} S_{b,k} \tag{9}$$

4.2   Clustering Process of Sample Points

Clustering process of sample points is key to ensuring the accuracy of k-similarity clustering algorithm. K-similarity algorithm divide all services into k clusters and outlier set. We assume that set of cluster is L, $L = \{L_1, L_2, \ldots, L_g, \ldots, L_k\}$, and set of outlier is O. At the beginning, set L is equal to set Z. Taking service U_j as an example, clustering process of services is as follows: (a) Calculate the average of behavior similarity between service U_j and each service in a same cluster. There are k clusters and we will get k averages. (b) If the cluster for the maximum average $S_{\max}$ is L_h, service U_j is added to cluster L_h. Calculation formula of $S_{\max}$ is as follows:

$$S_{\max} = \max_{1 \leq g \leq k} \frac{1}{|L_g|} \sum_{U_b \in L_g} S_{b,j} \tag{10}$$

5   Experiment Analysis

We collect flows of the network sink node from July 18, 2017 to July 25, 2017. These flows are as experimental background flows. In this case, flows can be divided into 559 aggregated flows. To make the analyzed aggregated flows contain attack flows fully, we set timescale T as 10 min.

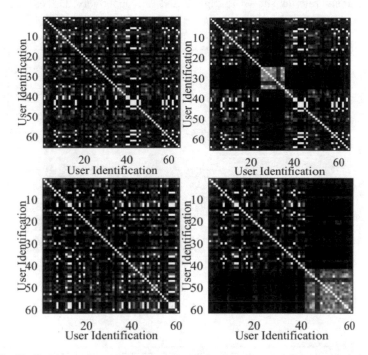

Fig. 1. Similarity matrix with port scanning behaviors & network scanning behaviors.

5.1   Feasibility Analysis of Behavior Similarity Formula

In this experiment, we randomly select five aggregated flows with prefix 10.103.67.*, 10.210.71.*, 10.110.77.*, 10.203.35* and 10.108.85.*, respectively. In each aggregated flows selected, we randomly select continuous 10 min of flows for experimental analysis. Types of abnormal flows generated by simulation include port scan, network scan and etc.

In Fig. 1 the figures on the left show the grayscale matrix of service behavior similarity without abnormal flows, those on the right show the grayscale matrix of service behavior similarity with abnormal flows. In Fig. 1 (2 pics on the left), there are 65 active services at the moment, and services labeled as 25–35 suffer port scanning. In Fig. 1 (2 pics on the right), there are 61 active services, and services labeled as 42–61 suffer from network scanning.

5.2   Feasibility Analysis of *K*-Similarity Algorithm

This experiment simulates 15 abnormal time points, including port scanning, network scanning, DDoS attacks, worm virus and Flash Crowd. We inject abnormal flows into different aggregated flows and calculate service behavior similarity matrix in the same aggregated flow. Then we use *k*-similarity algorithm and *k*-means algorithm to cluster the services respectively. We set *k* as 2–7 to observe the change of identification rate and misjudgment rate, as shown in Figs. 2 and 3, respectively. Dotted line and solid

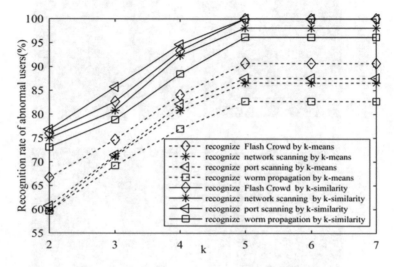

Fig. 2. Identification rate of abnormal services.

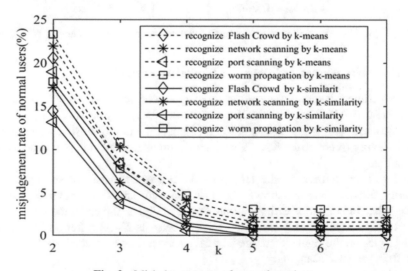

Fig. 3. Misjudgment rate of normal services.

line represent the results obtained by using k-means algorithm and k-similarity algorithm, respectively. Figure 2 shows that identification rate of k-similarity algorithm is overall higher than that of k-means algorithm. Figure 3 shows that misjudgment rate of k-similarity algorithm is overall lower than that of k-means algorithm. Moreover, when k-similarity algorithm clusters services with port scanning and Flash Crowd behaviors, identification rate can achieve 100% and misjudgment rate can drop to 0. Therefore, k-similarity algorithm proposed in this paper is superior to k-means algorithm in identifying abnormal services.

6   Conclusion

We propose behavior similarity awared abnormal service identification mechanism. And then experimental results show that behavior similarity formula and k-similarity algorithm proposed are feasible in identifying abnormal services. In next study, we will set up targeted solutions for each abnormal behavior. After identifying services with abnormal behaviors, we can automatically feedback targeted solution to network administrators or related services.

Acknowledgment. This work was financially supported by Research and Application on Intelligent Operation Management Technology in Voice Exchange Network (036000KK52160009) hosted by Power Grid Dispatching Control Center of Guangdong Power Grid Co., Ltd., China Southern Power Grid.

References

1. Guo, Y.T., Gao, Y., Wang, Y.: DPI & DFI: a malicious behavior detection method combining deep packet inspection and deep flow inspection. Procedia Eng. **174**, 1309–1314 (2017)
2. Parvat, T.J., Chandra, P.: Performance improvement of deep packet inspection for Intrusion Detection. In: Wireless Computing and Networking, pp. 224–228. IEEE (2015)
3. Zhou, Y., Wang, Y., Ma, X.: A service behavior anomaly detection approach based on sequence mining over data streams. In: International Conference on Parallel and Distributed Computing, Applications and Technologies. IEEE (2017)
4. Shi, Q., Xu, L., Shi, Z., Chen, Y., Shao, Y.: Analysis and research of the campus network service's behavior based on k-means clustering algorithm. In: 2013 Fourth International Conference on Digital Manufacturing & Automation, pp. 196–201 (2013)
5. Parwez, M.S., Rawat, D.B., Garuba, M.: Big data analytics for service-activity analysis and service-anomaly detection in mobile wireless network. IEEE Trans. Ind. Inform. **13**(4), 2058–2065 (2017)
6. Cao, J., Chen, A., Widjaja, I., et al.: Online identification of applications using statistical behavior analysis. In: Global Telecommunications Conference, pp. 1–6. IEEE (2008)
7. Bernaille, L., Teixeira, R., Akodkenou, I., Soule, A., Salamatian, K.: Traffic classification on the fly. ACM SIGCOMM Comput. Commun. Rev. **36**(2), 23–26 (2006)
8. Moore, A.W., Papagiannaki, K.: Toward the accurate identification of network applications. In: International Conference on Passive and Active Network Measurement, pp. 41–54. Springer (2005)

9. Nychis, G., Sekar, V., Andersen, D.G., Kim, H., Zhang, H.: An empirical evaluation of entropy-based traffic anomaly detection. In: Internet Measurement Conference, pp. 151–156 (2008)
10. Zhou, Y.J.: Behavior analysis based traffic anomaly detection and correlation analysis for communication networks. University of Electronic Science and Technology of China (2013)
11. Wei, S., Mirkovic, J., Kissel, E.: Profiling and clustering internet hosts. In: International Conference on Data Mining, Las Vegas, Nevada, USA, pp. 269–275. DBLP (2006)
12. Xu, K., Wang, F., Gu, L.: Behavior analysis of internet traffic via bipartite graphs and one-mode projections. IEEE/ACM Trans. Netw. **22**(3), 931–942 (2014)
13. Gordeev, M.: Intrusion detection techniques and approaches. Comput. Commun. **25**(15), 1356–1365 (2008)
14. Rajaraman, A., Ullman, J.D.: Bigdata: large scale data mining and distributed processing. China Sci. Technol. Inf. (22), 26 (2012)

An Energy-Balancing Based Clustering Method in Sensor Networks for IoT Sensing and Monitoring

Hong Zhu[1], Yangling Chen[1(✉)], Mingchi Shao[1], Nan Xiang[1], and Lin Peng[2]

[1] State Grid Jiangsu Electric Power CO., Ltd., Nanjing Power Supply Company, Nanjing 210019, China
448405204@qq.com
[2] Beijing Vectinfo Technologies CO., Ltd., Beijing, China

Abstract. The Internet of Things (IoT) sensors have been widely applied into human social activities for a long time. To better use the limited energy resource, an energy-balancing based cluster-head alternatively selecting and cluster-node evenly clustering strategy is proposed in this paper. First, the optimal number of cluster nodes is obtained through simulation. Then, the weights for each cluster-head node are set according to the proposed strategy to balance the remaining energy of IoT sensor in the entire network. The simulation results show that, compared with the traditional clustering algorithm, the proposed clustering algorithm can balance energy consumption and extend life-time of the whole sensor network.

Keywords: Internet of Things sensors · Energy-balancing · Clustering strategy

1 Introduction

The Internet of Things (IoT) is a network which combines various information sensing devices with the Internet. The purpose of IoT is to have all items connected to the network in order for identification, management and control. The sensors of IoT have been used in many fields such as industrial production, smart home, marine exploration, environmental protection, medical diagnosis, and so on. This paper put forwards a sensor network clustering algorithm which is used to sensor and monitor the transmission line in the IoT.

Transmission lines are distributed in various complex terrains and are difficult to check manually. The equipments of transmission lines are easily damaged. The establishment of a sensor network oriented to transmission lines sensing and monitoring has important theoretical and practical significance. The establishment of the network can improve the efficiency of maintaining and managing transmission lines [1]. In order to achieve optimal allocation of network resources and extend network life-time, it is important to propose a wireless sensor network path planning algorithm based on energy balancing.

© Springer Nature Switzerland AG 2020
Q. Liu et al. (Eds.): CENet 2018, AISC 905, pp. 325–334, 2020.
https://doi.org/10.1007/978-3-030-14680-1_36

Most traditional network clustering protocols are based on dynamic clustering algorithms. Heinzelman et al. proposed the LEACH: Low Energy Adaptive Clustering Hierarchy algorithm [2]. The algorithm is the most classical distributed clustering algorithm. The basic idea of LEACH is selecting clusters randomly in a circular manner in order to balance the energy load in the entire network. Thereby the network life is extended. Based on LEACH, Parul et al. proposed the K-LEACH: Kmedoids-LEACH protocol [3]. The protocol proposed a clustering method based on K-medoids algorithm. Beiranvand et al. proposed the I-LEACH: Improved-LEACH algorithm [4]. The algorithm proposed a cluster-head selection mechanism that considered distance, number of nodes and remaining energy. Upadhyay et al. proposed the T-LEACH: Threshold-based LEACH protocol [5]. The protocol proposed a cluster-head replacement scheme which considered remaining energy based on threshold. This paper proposes an algorithm named Alternative Cluster-head and Evenly Clustering algorithm based on energy balancing (ACEC). ACEC only performs clustering one time in the network life-time. After a certain period of time, cluster-head nodes are replaced according to the rules which the paper proposes.

2   Energy Model

The energy required for the sending node to send k bit data is [6]:

$$E_{send} = \begin{cases} k(E_0 + \varepsilon_{fs}d_{ij}^2), & d < d_0 \\ k(E_0 + \varepsilon_{amp}d_{ij}^4), & d \geq d_0 \end{cases} \tag{2.1}$$

Where E_0 is a fixed value depending on the physical properties of node itself, d is the Euclidean distance between the node and its next hop, d_0 is the threshold of d.

The energy required for the receiving node to send k bit data is [7]:

$$E_{receive} = kE_0 \tag{2.2}$$

3   Algorithm Description

3.1   Algorithm Analysis

(1) The optimal number of cluster nodes

This paper proposes a fixed clustering approach. The network only changes cluster-head nodes after a period of time. This paper will estimate the number of cluster-head nodes by the formula (3.1):

$$k_{ch} = \frac{\sqrt{N}}{2\pi} \sqrt{\frac{\varepsilon_{fs}}{\varepsilon_{amp}}} \frac{L \cdot W}{\overline{d}_{toBs}^2} \tag{3.1}$$

Where N is the number of nodes, L is the length of the area, W is the width of the area, and $\overline{d}_{toBs}$ is the average distance between the distance node and the base station.

The next section will perform simulation analysis based on the established energy model to determine the optimal number of cluster nodes with certain parameters.

(2) Selecting cluster-head nodes

When selecting a cluster-head node, two factors are considered. One is the remaining energy of sensor node. The other is the communication cost between the node and other nodes in the cluster. Since the communication energy consumption is expressed to the square of distance between two nodes, the communication cost is quantified as the sum of distances between the node and all other nodes in the cluster.

The probability that a node becomes a cluster-head node:

$$p(i) = \alpha \frac{1}{s_d(i)} + \beta \log(\frac{E_{left}}{E_{init}}) \tag{3.2}$$

s_d is the sum of distances between the node and other nodes in the cluster. It is expressed as the formula (3.3):

$$s_d(i) = \sum_{j=1}^{n} d_{ij}, \quad i \neq j \tag{3.3}$$

E_{left}/E_{init} is the ratio of remaining energy and initial energy. α and β are the weight factors of communication cost and remaining energy respectively.

(3) Updating time of cluster-head nodes

Because cluster-head nodes consume much more energy than cluster nodes, the network load should be balanced as much as possible to maximize the network life-time. The algorithm make each node become cluster-head node in a time before the network energy is depleted. It has been known that there are $k+1$ nodes in each cluster. Ideally, each node should be cluster node k times and cluster-head node once.

The energy of each node should satisfy:

$$k \cdot \frac{E_{total} - E_{total-ch}}{k} + E_{total-ch} = E_{init} \tag{3.4}$$

That is:

$$E_{total} = E_{init} = E_s + \frac{S - T_s}{T_f} E_f \tag{3.5}$$

Therefore, the best interval for reselecting cluster-head nodes is:

$$S = T_s + \frac{T_f(E_{init} - E_s)}{E_f} \tag{3.6}$$

T_s and E_s are the time delay and energy consumption respectively in the clustering phase, T_f and E_f are the time delay and energy consumption of each frame respectively in the stable phase, E_{init} is the initial energy of each sensor node, E_{total} is the energy consumption of cluster nodes $E_{total-ch}$ is the energy consumption of the cluster-head.

3.2   Algorithmic Process

ACEC's algorithm flow is as Fig. 1:

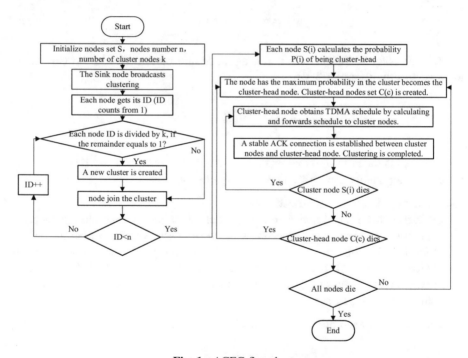

Fig. 1. ACEC flowchart.

As shown in flow chart, the implementation of ACEC is as follows:

Step 1: The sink node broadcasts message in order to initialize all nodes in the network;
Step 2: All nodes join clusters according to their IDs and the variable k;
Step 3: All nodes calculate their probability of becoming cluster-head nodes. In a cluster, the node which has the highest probability is elected as a cluster-head node;

Step 4: Cluster-head nodes broadcast messages that they have become cluster-head nodes. The k cluster nodes in a cluster contact with their cluster-head node. The cluster-head node communicates with the cluster node to share TDMA scheduling;

Step 5: Cluster-head nodes and their own cluster nodes establish stable transmissions of ACK in the TDMA time slot. If a cluster node dies, its cluster-head node discovers and uploads the position of dead node. Then, the cluster-head node updates the TDMA schedule. If a cluster-head node dies, the first cluster node which discovers the case broadcasts this message immediately in the cluster. A new cluster-head node in the cluster is selected. The new cluster-head node uploads the information which is about the location of the previously dead cluster-head node, then goes to the fourth step;

Step 6: After the interval calculated in Sect. 3.1(3), all clusters in the network reselect cluster-head nodes. Before the network dies, cluster-head nodes will be reselected at periodic intervals.

4 Simulation Analysis

4.1 Simulation Parameters Settings

In simulation, we set locations of target nodes, communication distances between nodes, initial energy, transmission energy consumption, and number of nodes, etc.

Chart 1: Simulation parameters settings

Parameters	Values	Parameters	Values
Number of nodes	90	Energy consumption of idle status/(J/bit)	$10\,E_0$
Size of district nodes distribute A/m^2	2×3000	Energy consumption of convergence/(J/bit)	$5\,E_0$
Maximum transmission radius R/m	100	Threshold of the distance between nodes d_0 /m	20
Initial energy of nodes /J	1000	Producing speed of data packets /(bit/s)	1
The location of sink node /m	(3000,1)	Control information /bit	32
Energy consumption of data transmission E_0 /(J/bit)	2×10^{-8}	Data information /bit	4000

4.2 Simulation Results

(1) Comparing energy consumption of ACEC in the case where the variable is the number of cluster nodes

It is necessary to determine the number of cluster nodes that can optimize energy performance when clustering. Therefore, under the conditions that total number of the network nodes is the same and the initial energy of each node is the same, the situation

Rounds

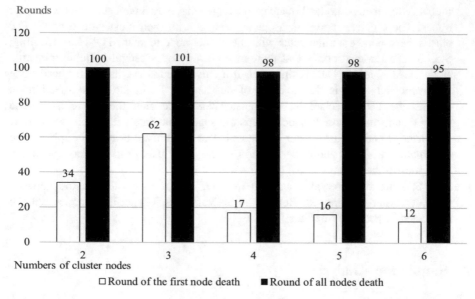

Fig. 2. Energy consumption with cluster nodes of different numbers (mean of 50 experiments).

of nodes death is compared by changing the number of cluster nodes. The rounds of the first node death and all nodes death are recorded. The results are shown in Fig. 2.

It can be seen from this figure that the time of the first node death and all nodes death is the latest when the number of cluster nodes is 3. Therefore, the network has the best energy performance when the number of cluster nodes is 3.

(2) Comparing energy consumption between ACEC and I-LEACH

The following two figures are simulations of I-LEACH and ACEC respectively. The sink node is located at the right-most end. Cluster-head nodes are indicated by "*", cluster nodes are indicated by "o", and death nodes are represented by solid circles. The cluster node and the cluster-head node are connected by a straight line. Two algorithms are executed to record the case that is the death of nodes in the first and 50th rounds. The results are shown in Figs. 3 and 4.

It can be seen that the number of dead nodes using ACEC is smaller than the number of dead nodes using I-LEACH in each round. In addition, death nodes are more evenly distributed in ACEC. With the simulation running, the connectivity of network has not been impacted badly in ACEC.

In order to compare the energy performance and load balancing of two clustering algorithms, during the simulation, the numbers of death nodes are recorded in each round. In order to ensure the reliability of simulation results, all data is derived from the average of 50 experiments results. These results can be summarized into Fig. 5.

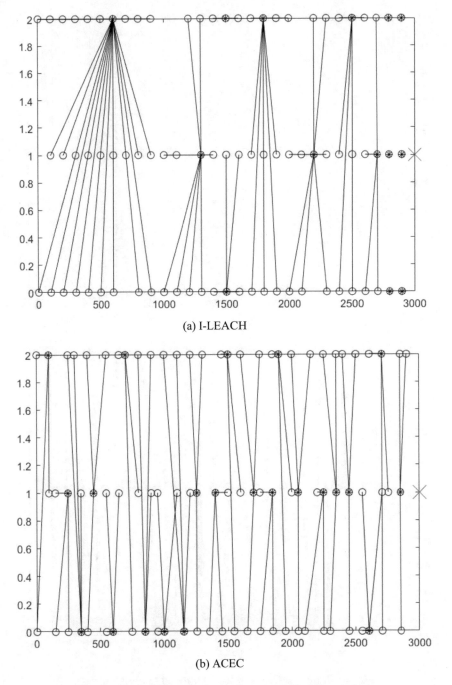

(a) I-LEACH

(b) ACEC

Fig. 3. Clustering results in the first round (3 cluster nodes).

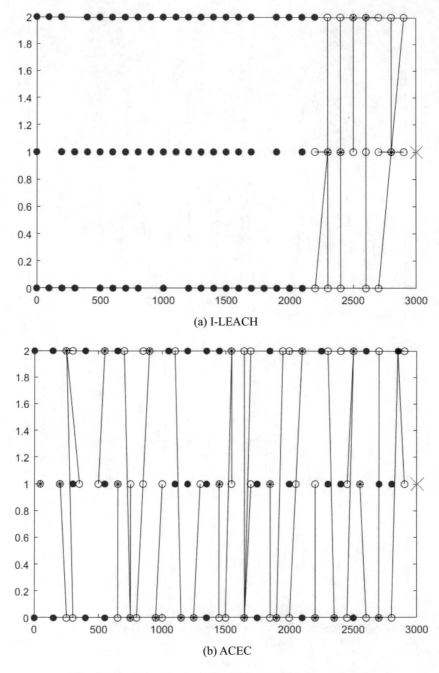

(a) I-LEACH

(b) ACEC

Fig. 4. Clustering results in the 50th round (3 cluster nodes).

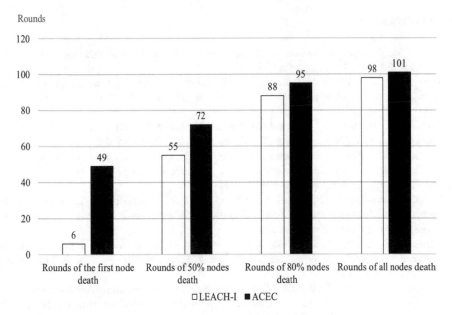

Fig. 5. Rounds of nodes death using different clustering algorithms (mean of 50 experiments).

It can be seen from this figure that compared with two algorithms. The algorithm proposed in this paper consumes less energy than I-LEACH algorithm in each round. So ACEC is more energy-efficient and its load balancing is better. Thus, compared with I-LEACH, ACEC effectively extends the network life-time, improves load balancing and the network connectivity during the network life-time.

5 Conclusion

In order to extend the life cycle of sensor networks, the Alternative Cluster-head and Equal Clustering algorithm based on energy balancing (ACEC) is presented in this paper. The number of cluster nodes is determined through presetting parameters to achieve the best network energy performance. Then the rotation order of cluster-head is decided according to the cluster-head selection function. Finally, the optimal interval time for cluster-head switching is obtained through calculation. The simulation experiments show that, compared with the I-LEACH algorithm, the proposed ACEC algorithm balances the energy consumption and reduces the communication energy consumption, as well as extends the network life time of the entire sensor network.

Acknowledgement. This work was supported by the Science and Technology Project of State Grid Jiangsu Electric Power Co., Ltd. (J2017072) and the National Science and Technology Major Project (2017ZX03001013).

References

1. Yang, X.: In-depth discussion on transmission line condition monitoring system. Sci. Technol. Inf. (30), 20 (2008). (in Chinese)
2. Heinzelman, W.B., Chandrakasan, A.P., Balakrishnan, H.: An application specific protocol architecture for wireless microsensor networks. IEEE Trans. Wirel. Commun. **1**, 660–670 (2002)
3. Bakaraniya, P., Sheetal, M.: K-LEACH: an improved LEACH protocol for, lifetime improvement in WSN. Int. J. Eng. Trends Technol. **4**(5), 1521–1526 (2013)
4. Beiranvand, Z., Patooghy, A., Fazeli, M.: I-LEACH: an efficient routing algorithm to improve performance & to reduce energy consumption in Wireless Sensor Networks. In: Information and Knowledge Technology, pp. 13–18. IEEE (2013)
5. Upadhyay, A., Kumar, R., Tiwari, S.K.: Modified LEACH protocol for sensor network. In: International Conference on Futuristic Trends on Computational Analysis and Knowledge Management, pp. 301–304. IEEE (2015)
6. Sarma, H.K.D., Rai, P., Deka, B.: Energy efficient communication protocol for wireless sensor networks with mobile node. In: Recent Advances and Innovations in Engineering, pp. 1–6. IEEE (2014)
7. Paul, A.K., Sato, T.: Effective data gathering and energy efficient communication protocol in wireless sensor network. In: International Symposium on Wireless Personal Multimedia Communications, pp. 1–5. IEEE (2011)

MOPSO-Based Research on Manufacturing Process Optimization in Process Industry

XueSong Jiang[1(✉)], DongWang Li[1], XiuMei Wei[1], and Jian Wang[2]

[1] Qilu University of Technology (Shandong Academy of Sciences),
Jinan 250000, China
jxs@qlu.edu.cn
[2] Shandong College of Information Technology, Weifang 261061, China

Abstract. To deal with the conflict between multiple targets involved in the manufacturing process of the process industry, multi-objective particle swarm optimization (MOPSO) is used to solve the optimization problem among multiple objectives. Basing on the manufacturing process analysis of the process industry and with the background of the cement manufacturing process of a process industry, two objective functions, which are the total processing cost and the integrated error of the mineral content in the cement compared to the standard, are established, and the concrete realization process of algorithm is given. The results of the example analysis show that when using the results by means of MOPSO algorithm to guide the production, it not only can improve the product performance index, but also can reduce the cost required as much as possible with the same performance indicators. Therefore, it is feasible to use the MOPSO algorithm to optimize the multi-objectives involved in the manufacturing process.

Keywords: Process industry · Particle swarm optimization ·
Manufacturing process optimization

1 Introduction

The optimization of the manufacturing process in the process industry is a multi-objective optimization problem which is dynamic, large-scale and non-disruptive. In the traditional process industry, the manufacturing process is usually controlled by the traditional process industry through the experience of the operators, but this may cause big errors of calculation results and cause waste. As for the two objective functions established in this paper, the costs should be reduced as much as possible when ensuring that the products meet the specification.

Reference [1] puts forward an estimation model based on neural network to solve the problem of multi-objective optimization of cement raw meal ingredients. References [2] and [3] adopt the reference furnace method and regression analysis respectively to predict the amount of additive needed in the production process. Although these three methods have higher accuracy, it is impossible to guarantee that the required cost will be as low as possible when meeting the product specification standards. The traditional methods will face more difficulties in dealing with such multi-objective problems.

© Springer Nature Switzerland AG 2020
Q. Liu et al. (Eds.): CENet 2018, AISC 905, pp. 335–343, 2020.
https://doi.org/10.1007/978-3-030-14680-1_37

This paper takes the production process of cement as the background and applies the MOPSO algorithm to optimize the multi-objectives of the manufacturing process.

The multi-objective particle swarm optimization (MOPSO) [4] was proposed by Coello et al. in 2004. Its objective is to apply the PSO algorithm which can only be used on single targets to multiple targets. The reference [5] applies the basic particle swarm algorithm to solving the AGV dynamic scheduling problem and improves the FMS logistics scheduling efficiency. The reference [6] combines the MOPSO algorithm with the problem of reservoir's flood control to better solve the problem reservoir flood control. The reference [7] uses MOPSO algorithm to solve the problem of optimal scheduling of hydropower stations. On the basis of the researches of these scholars, this paper applies the MOPSO algorithm to the manufacturing process optimization in the process industry and verifies the feasibility of the algorithm with specific examples.

2   Multi-Objective Particle Swarm Optimization Algorithm

Compared with the ordinary particle swarm optimization algorithm, MOPSO algorithm was improved in the option of individual extremum and group extremum. For the option of individual extremum, we can adopt a random option method when we cannot strictly decide which the best one is. In the option of global extremum, the Archive set [8] was used to store the non-inferior solution, and then the global extremum was chosen according to the degree of congestion. In order to guarantee the ability to explore the unknown areas, the one with lower particle density value was usually chosen. In that case, the grid method was used when dividing the density of particles. Its concrete implementation steps are as follows:

(1) Initializing the position of the particle and the entire external archive set
   Firstly, we need to assign the initial value to each parameter, generate an initial group G1, and copy the non-inferior solution in G1 to the external archive set which then generates community A1.
   Suppose that the current number of iterations is t, and actions (2)–(4) are repeated when t is less than the total number of iterations.
(2) Deriving new population by iteration
   Suppose that the number of particles in the current iteration is q, and actions (a)–(c) are repeated when q is less than the population size.
   (a) Collecting particle density information in the archive set
      The available population space is equally divided into different rooms by grid, and the particle density information is calculated according to the number of particles in each room. The more the number of particles in a room is, the greater the density of the particles is, and vice versa.
   (b) Choosing the global optimal extreme value for at in the population.
      The quality of the particles determines the convergence performance of the MOPSO algorithm and the diversity of the non-inferior solution sets based on the option of the density information of the concentrated particles in the archive set. Specifically, the less the density of particles in each divided room is, the greater the probability of being chosen is, and vice versa. The particles

in the archive are used to evaluate the search potential ability in the population. The greater the number is, the stronger the search ability is, and vice versa.

(c) Updating the position and velocity of particles in the population
      The particles in the population search for the optimal solution under the guidance of gBest and pBest.

(3) Updating the external archive set
      After a new evolved generation of group Gt + 1 is available, the non-inferior solutions in Gt + 1 are saved to an external archive set.

(4) Truncation of External Archive Sets

When the number of particles in the archive set is more than the specified size, redundant individuals will be deleted to guarantee the stability of the external archive set scale. For each room n with the number of particles more than 1, it will calculate the number GN of the particles which need to be deleted according to the following Eq. (1), and then GN particles will be randomly deleted in the grid n.

$$GN = Int\left(\frac{|A_{t+1}| - \overline{N}}{|A_{T+1}|} \times Grid[n, 2] + 0.5\right) \qquad (1)$$

Among them, Grid[n] represents the number of particles in the grid n.

3 Manufacturing Process Description of the Process Industry

Taking the cement production process as an example, this section described a general analysis of its manufacturing process. A person who even only knows a little about the cement production process could mention "two grinding and one burning" in the production of cement. They are raw material preparation (one grinding), clinker burning (one burning), and cement grinding (second grinding). In a cement plant, there are several stages in the production of cement: raw material preparation, raw material grinding, clinker firing, cement grinding, storage and shipping. The grinding of cement is the last process of cement production, and it is also the process that consumes the most electricity. Prior to this step, clinker, gypsum and additives will be mixed to ensure that the minerals and finished products in the clinker meet the demands of the performance indicators. In this paper, we will establish a model with the multi-Agent [9] technology for the cement manufacturing process system of a specific process industry [10, 11]. At the same time, the mathematical model is established and optimized to verify the effectiveness of the algorithm, aiming at the above mixing stage.

4 Optimization Based on MOPSO Algorithm

4.1 The Establishment of Mathematical Model

The mixing stage of the cement mainly includes two production targets: First, mix the gypsum and additives before the clinker enters the clinker mill, so that the minerals in the clinker (for example: tricalcium silicate 3CaO · SiO$_2$, can be abbreviated as C3S; dicalcium silicate 2CaO · SiO$_2$, can be abbreviated as C2S; and tricalcium aluminate 3CaO · Al$_2$O$_3$.) can reach the specified range and minimize the error with the specified content, which ensures that the properties of the cement after oxidation, such as compression resistance, flexural resistance, hydration, and durability, meet the specifications. Second, there is a minimum cost ratio for the input of gypsum and additives. The total cost of the added additive is taken as the second optimization goal.

For the goal to be optimized, we represent its objective function as follows:

$$h_1(x) = \sum_{k=1}^{M_1} (N_k - n_k(x)) \tag{2}$$

The purpose of Eq. (2) is to minimize the error $h_1(x)$ as much as possible to enhance cement performance. M_1 indicates the amount of additive added. N_k is the amount of additive required to meet the demand of standard content. $n_k(x)$ is the amount of additive required to achieve the minimum range of required cement performance. The content of all the minerals, the durability of the current cement, and the durability of the concrete can be determined by the detector in the specific manufacturing process.

For goal 2 to be optimized we will express its objective function as follows:

$$h_2(x) = \sum_{t=1}^{M_2} g_t p_t \tag{3}$$

Equation (3) is the total cost required. We need to gain the minimum cost through optimization. M_2 is the amount of additive required. g_t is the amount of additive added in t. p_t is the unit price of each additive.

The above two objective functions can be sorted out to obtain the multi-objective optimization issue of the amount of additives added. Its form is shown in the following formula:

$$\min H(x) = (f_1(x), f_2(x))^T \tag{4}$$

s.t x α min < x α < x α max

x α min, x α max are the lower bound value and upper bound value of the additive input respectively.

4.2 The Specific Implementation of the Optimization Algorithm

Task1: set the population size Z of the particle and the termination condition of the algorithm. The randomly given population contains the particle position x_i, the velocity v_i and the iteration number t.

Task2: add the non-inferior solution to the external archive set and use Eqs. (2) and (3) to find the fitness value of each particle in the population.

Task3: when the number of iterations is less than the given iteration number t, repeat step 4–7. Otherwise, the algorithm comes to end.

Task4: choose the global extremum according to the fitness value of the particles in the archive and update the position and velocity of each particle according to the Eqs. (5) and (6) and find the fitness value of the contained particles once again.

$$v_i(t+1) = w * v_i + c_1 r_1 (h_i(t) - x_i(t)) + c_2 r_2 (k_i(t) - x_i(t)) \tag{5}$$

$$x_i(t+1) = v_i(t+1) + x_i(t) \tag{6}$$

Where w is the inertia weight. We use the dynamic weight to reset rules in this test, which specifies that the maximum weight is 0.9 and the minimum weight is 0.4. In the test process, the value between 0.4 and 0.9 is chosen dynamically for calculation. $c_1 c_2$ is a non-negative constant which we usually set a value of 2 during the test. $r_1 r_2$ is a random number between 0 and 1.

Task5: repeat the process of Task2 and remove the relatively ineligible solutions based on the dominance relationship, then form the external file for the next iteration.

Task6: suppose that the best extremum of the currently available particle in its history is worse than the current position, then use the current latest particle as the optimal extremum. Otherwise, keep the current state unchanged.

Task7: if the position of a random particle in the population is better than the optimal extreme value of the entire population previously chosen, take the same solution as Xigb.

Task8: after the iteration, the particles in the external archive set are the valid solution set for the entire algorithm.

4.3   Analysis of Optimization Results

After setting up the objective function to be optimized, the MOPSO algorithm was used to optimize it. In the simulation experiment, we verified the production data of a cement with the concrete model (A1) in a cement factory.

During the manufacturing process of cement type A1, the required mineral and oxide content are shown in Table 1:

Table 1. Contents and ranges of minerals and oxides in cement.

Contained minerals	Specified content	Allowable range
C3S	50%	45%–60%
C2S	25%	20%–33%
C3A	10%	7%–15%
CaO	65%	62%–67%

The unit price of various additives is shown in Table 2:

Table 2. Price of various additives.

Mineral species	Unit price (yuan/ton)
Early strength agent	25000
Antifreeze	5000
Dry solution	1800
Retarder	33150

When optimizing the data, we choose 200 iterations, and the scale of the group and external archive set was 100. First, we select a set of data and after optimizing the objective function, the front results of the Pareto shown in Fig. 1 are obtained.

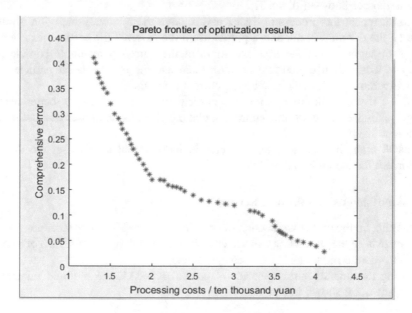

Fig. 1. Pareto solution set of optimization results.

In Fig. 1, the abscissa is the cost of the additives needed in the manufacturing process, and the ordinate is the combined error of the various mineral contents and standards. Each of the points represents a group of optimization solutions. It can be seen from the figure that when the overall error decreases from 0.42 to 0.03, the cost in the manufacturing process increases from 113,000 yuan to 42,000 yuan, which indicates that the inverse relationship between the error when the mineral content is reduced as much as possible and the manufacturing cost is obvious. In the actual manufacturing process, the decision maker can select the optimization solution from the pareto solution according to the specific application requirements in life. For example, under the condition of higher requirements on the performance index of cement, the decision-making scheme can be selected at the position where the integrated error is the smallest.

In addition, multiple sets of data are taken and the MOPSO algorithm is used to optimize so that the results obtained are shown in Figs. 2 and 3.

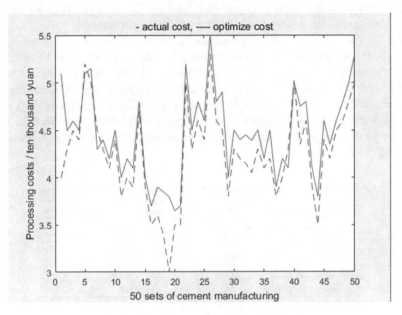

Fig. 2. Actual costs and optimized costs.

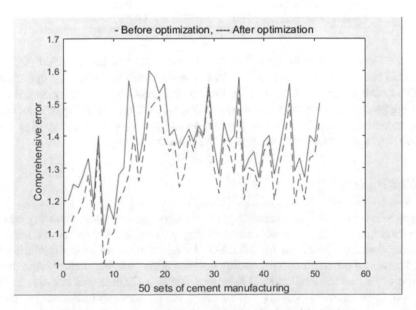

Fig. 3. Error comparison that before and after optimization.

Figure 2 shows the comparison of the costs of the 50 groups before and after the MOPSO algorithm optimization. The conclusion can be drawn clearly that the cost after optimization is lower than the cost before optimization, so the optimized non-inferior solution set can be used as reference in actual production. Figure 3 is a comparison of the combined errors before and after optimization. From the above two figures, it can be seen that the comprehensive error of both the production cost and the performance index has been greatly improved after optimization.

The item-by-item analysis of the errors in various mineral contents and specified values shows that the MOPSO algorithm has achieved good results in the optimization during the manufacturing process. By calculating the average error of the standard content of the four kinds of minerals in the cement products such as tricalcium silicate (C3S), dicalcium silicate (C2S), tricalcium aluminate (C3A), calcium oxide (CaO) and the qualification rate of the product, the effectiveness of the algorithm is verified. According to the range of various content requirements in Table 1, the statistical results are shown in Table 3 below.

Table 3. Comparison of performance aims before and after optimization.

Optimization aims	Before optimization	After optimization
Cost/ten thousand yuan	5.30	5.00
C3S average error	1.32	1.01
Qualification rate/%	91	96
C2S average error	1.43	1.16
Qualification rate/%	74	90
C3A average error	1.57	1.33
Qualification rate/%	67	91
CaO average error	1.176	1.044
Qualification rate/%	89	93

From Table 3, it can be seen that the average error and the qualified rate of the optimized cement mineral content have been greatly improved before the optimization.

The above results show that using the MOPSO algorithm to optimize the manufacturing process of the process industry can improve the quality of the product and the costs, meet the needs of the actual production and can play a certain guiding role in optimizing the manufacturing process of the process industry.

5   Conclusion

The paper based on the manufacturing process analysis of the process industry and with the background of the cement manufacturing process of a process industry, two objective functions based on the MOPSO algorithm to optimize the minimum integrated error and the minimum processing cost of various mineral content. And through specific experimental simulation. It is proved that it is feasible and effective to apply multi-objective particle swarm optimization to process optimization of process industry.

Acknowledgments. This work was supported by Key Research and Development Plan Project of Shandong Province, China (No. 2017GGX201001).

References

1. Pang, Q., Wan, M., Wu, X.G., Wang, J.Y.: Multi-objective optimization design method based neural network for raw meal proportioning of cement. J. Southeast Univ. **39**, 76–81 (2009)
2. Gong, W., Jiang, Z.H., Zheng, W., Chen, N.Z.: Component controlling model in BOF steelmaking process. J. Northeast. Univ. **23**(12), 1155–1157 (2002). (in Chinese)
3. Lv, X.W., Bai, C.G., Qiu, G.B., Ouyang, Q., Huang, Y.M.: Research on sintering burdening optimization based on genetic algorithm. Iron Steel **46**(4), 12–15 (2007). (in Chinese)
4. Cocllo, C.C.A., Pulido, U.T., Lcchunga, M.S.: Handling multiple objectives with particle swarm optimization. IEEE Trans. Evol. Comput. **8**(3), 256–279 (2004)
5. Bian, P.Y., Li, D., Bao, B.J., Lu, Y.: Application of particle swarm optimization in production logistics scheduling. Comput. Eng. Appl. **46**(17), 220–223 (2010). (in Chinese)
6. Xing, X.H., Lu, J.G., Xie, J.C.: Reservoir flood control operation based on improved multi-objective particle swarm optimization algorithm. Comput. Eng. Appl. **48**(30), 33–39 (2012). (in Chinese)
7. Zhang, J., Cheng, C.T., Liao, S.L., Zhang, S.Q.: Application of improved particle swarm optimization in the optimal scheduling of hydropower stations. J. Hydraul. Eng. **40**(4), 435–441 (2004). (in Chinese)
8. Mostaghim, S., Teich, J.: Strategies for finding good local guides in multi-objective particle swarm optimization (MOPSO). In: IEEE 2003 Swarm Intelligence Symposium, pp. 26–33 (2003)
9. Huang, N., Liu, B.: An overview of multi-agent technology. Microprocessors **31**(2), 1–4 (2010). (in Chinese)
10. Zhao, X.Z., Song, B., Yu, C.M.: Multi-agent complex system modeling method based on BDI. Inf. Technol. **10**, 121–123 (2015). (in Chinese)
11. Ni, J.J.: Theory and Application of Multi-agent Modeling and Control for Complex Systems. Publishing House of Electronics Industry, Beijing (2011). (in Chinese)

A Model for Job-Shop Scheduling Based on NN-PSO Technique

Shuaishuai Yao, Xuesong Jiang$^{(\boxtimes)}$, and Xiumei Wei

Qilu University of Technology (Shandong Academy of Sciences),
Jinan 250353, China
jxs@qlu.edu.cn

Abstract. Job-shop scheduling problem (JSP) has been one of the NP-Hard problems. Now, the most advanced algorithm is only limited to solve small scale problems effectively. For large-scale and super large scale job-shop scheduling problem, there is still no effective way to find its optimal solution quickly. In recent years, it is the main trend to solve the problem of job-shop scheduling with the combination of neural network and generic algorithms (GA). However, there will be complex operations such as crossover and mutation, and slow convergence or even stop convergence when approaching the optimal solution with GA. As an alternative, in this paper, we proposed a model for job-shop scheduling based on NN-PSO technique-neural network trained by the particle swarm optimization algorithm, which is simple, less parameter and easy to implement, greatly accelerates the convergence speed. Experiments show that job-shop scheduling based on NN-PSO technique is superior to most traditional scheduling rules.

Keywords: Job-shop scheduling · NN-PSO

1 Introduction

With the proposition of Industry 4.0 in Germany, intelligent manufacturing has become a research hotspot in academia and industry. Job-shop scheduling is a key problem in intelligent manufacturing. It is of practical significance to study such problems. Since the 50 s of the last century, the scheduling problem has begun to attract people's attention and be introduced into the academic research because of its own universality and practicality [1]. The job-shop scheduling problem has already been proved to be a NP-Hard class problem, academic workers for solving NP-Hard problems this classic has paid more than half a century of efforts, and now the most advanced algorithms only can effectively solve the problem of small scale, for large scale job-shop scheduling problem, so far still no effective method can quickly find the optimal solution quickly. In addition, it is also of great significance to study the algorithm of job shop scheduling problem to answer the question whether "P is equal to NP". So, it has great theoretical significance in studying the job shop scheduling problems in depth.

The research method of traditional scheduling is mainly based on operational research, including optimization, heuristic and simulation. With the rapid development of computer technology, intelligent scheduling algorithms have been developed,

Q. Liu et al. (Eds.): CENet 2018, AISC 905, pp. 344–350, 2020.
https://doi.org/10.1007/978-3-030-14680-1_38

including intelligent search algorithm, expert system, Multi-Agent method, neural network and so on. This paper focuses on artificial neural network to solve the problem of job shop scheduling.

In 1988, the application of neural networks to the process of shop floor scheduling problems for the first time has aroused widespread concern [2, 3]. When using neural network method to solve job-shop scheduling problem, a single neural network algorithm solves job-shop scheduling problem, the solution obtained is generally of the best solution and the running time is longer. At present, it is a hot topic to solve the job-shop scheduling problem with the combination of neural network algorithm and other heuristic algorithms.

In recent years, it is the main trend to solve the problem of job-shop scheduling with the combination of neural network and generic algorithms (GA). However, there will be complex operations such as crossover and mutation, and slow convergence or even stop convergence when approaching the optimal solution with GA. Research on the PSO has been more popular, many improved algorithms have been proposed to train the neural network, and some results have been achieved. PSO could effectively optimize neural network in terms of parameter training, structure optimization, learning rule adjustment, rule extraction, initialization of weights and attribute selection. A large number of literature shows that the optimized neural network model can be used to solve most of the problems.

2 Methodology

Because the traditional neural network adopts the gradient descent learning method, it inevitably has some defects and deficiencies:

First, because the learning rate is fixed, network convergence is slow and training time is long;

Second, it is sensitive to the initial value and it is easy to fall into a local optimum.

Third, there is no theoretical guidance on the choice of the number of layers and neurons in the network's hidden layers.

In response to these problems, in this paper, the particle swarm optimization algorithm is applied to neural network (NN-PSO) to overcome the defect that the algorithm is easy to fall into the local minimum. The number of layers and neurons of the hidden layer is determined by experience. Simulation experiments show that the particle swarm optimization algorithm greatly improves the neural network optimization speed, precision, generalization ability and learning ability, and can avoid the local minimum value problem.

The main technical routes of the research are as follows:

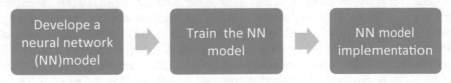

2.1   NN Modeling

Three layers are divided of the whole neural network, input layer, hidden layer and output layer. There are several neurons in each layer. The input layer is mainly used to receive the input signal of the neural network. The number of neurons of input layer depends on the length of the input signal vector. The hidden layer is a signal processing layer of neural network, the hidden layer can be formed by single neurons can also be formed by a plurality of neurons, depending on the complexity of the problem, the number of neurons of hidden layer is usually based on experience value, there is no fixed limit, generally obtained through repeated experiments in the process of building the neural network in.

The output layer is mainly used to output the desired signal of the neural network, and the number of the output layer neurons depends on the length of the output signal vector.

Taking into account the characteristics of the job shop scheduling problem, refer to Weckman [4], the following attributes of the process are extracted as neural network input: Process number, processing time, remaining time, machine load. Neural network structure as shown in Fig. 1 below.

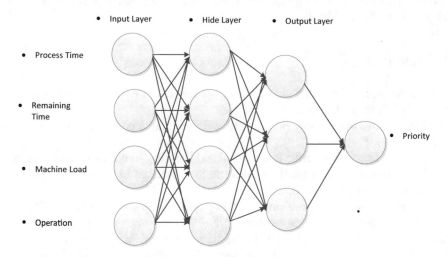

Fig. 1. Multi-layered perceptron.

2.2   Training of NN-Particle Swarm Optimization Algorithm

Particle swarm optimization algorithm (PSO), same as the SA algorithm (Simulated annealing), it starts from the random method to find the optimal method through iteration. It also assesses the quality of the method by calculating the fitness. Different to GA, PSO has its own information sharing mechanism which only the particle with optimal fitness value passes information to others. The renewed process of whole search follows the optimal method at presrent. Therefore, the entire particles can converge to the optimal method faster than GA. At the same time, it does not have the operations such as "crossover" and "mutation" of GA. In general, PSO has received extensive attention from academia due to the advantages of well-implementation, high precision, and quick convergence, and its superiority was verified in addressing realistic problems.

In the PSO algorithm, a particle represent a latent method. The aim of the particle is to get the optimal method. The particle goes to the method by locating itself closer to the flying conductor G^{best}, which is the particle closest to the optimal method in a particular iteration. Particle's position is calculated according to Eqs. (1), (2):

$$V_{id}^{t+1} = W \times V_{id}^{t} + C1 \times Rand \times \left(P_{id}^{best} - P_{id}^{t}\right) + C2 \times Rand \times \left(p_{gd}^{best} - P_{id}^{t}\right) \qquad (1)$$

$$P_{id}^{t+1} = P_{id}^{t} + V_{id}^{t+1} \qquad (2)$$

where V_{id}^{t+1} represent id[th] particle speed of current iteration; W is the initial weight; V_{id}^{t} represent id[th] particle speed of previous iteration; Rand represent random function in [0, 1]; C1, C2 are learning factors, and the general value is 2; P_{id}^{best} represent the id[th] particle's best local position and p_{gd}^{best} is the best global position among all particles; and P_{id}^{t} is the position of id[th] particle in the previous iteration.

Algorithm of PSO is designed as follows:

Start
for eatch particle i
Initialize particle's position and velocity
end for
If swarm size <n
Do for each particle
Calculate the fitness value
If current value > P^{best}
Set P^{best}=current value
Choose best P^{best} as G^{best}
Calculate particle velocity according to equation (1)
Calculate particle position according to equation (2)
End

---

3 NN-PSO Technique

In recent years, the NN-PSO technique-neural network trained by the particle swarm optimization algorithm has been used in various aspects [5–10]. For example, M. Sridhar and Tuan Linh Dang have achieved success on the Ionospheric scintillation forecasting and Hardware/Software Co-design based on this model respectively.

In the NN-PSO, the neural network parameters need to be coded. Regarding the coding strategy, the position of each particle in the group is considered as a vector representing all parameters (weights and deviations) of the neural network. The name of the vector in our program is Q, which is represented as Q = {qi1, qi2 ... qin} where qi is the position of the ith particle in the swarm. The dimension of the Q vector is the number of parameters of the neural network. Each parameter in the NN corresponds to one element in the PSO. When the elements in Q are changed during training, the corresponding neural network parameters of the element are also updated.

The PSO training is executed after the encoding. The target of the PSO training is to find a position of the particles so that the actual outputs from the NN match the desired outputs of the NN. For all particles, the actual output data are compared to the desired output data of the NN by using an objective function which is minimized by the PSO algorithm. The objective for JSP issues is to minimize the maximums of the makespan of every job and to measure the effectiveness of particles when their positions at the end of an iteration are updated. The objective function can be seen in Eq. (3) as follows.

$$F_p = \frac{1}{M} \sum_{i=1}^{M} (D(j) - R(j))^2 \tag{3}$$

Where F_p is the objective function of particle p, M is the number of training samples in our program, D(j) and A(j) are the jth component of particle p in the desired output and the real output of the NN, respectively.

The minimum learning error is calculated by Eq. (3). The movements of the particles mean the weights updating and the biases updating of the NN. The new position of each particle is influenced by P^{best} and G^{best} as presented in Eqs. (1) and (2). The training process based on PSO-NN technique is as follows.

1. Initialize random values for all particles of PSO.
2. Send the data to the input nodes of the NN and the results are received from the output nodes of the NN.
3. Calculate the fitness values from the outputs based on (3).
4. Evaluate the new P_{id}^{best}, and p_{gd}^{best}.
5. Update new velocity V_{id}^{t+1} according to (1).
6. Update the new position P_{id}^{t+1} of particle p according to (2).
7. Return to Step 2 until all particles are processed.
8. Return to Step 2 until the stopping criteria are satisfied.

4   Experimental Results and Discussion

A benchmark problem ft06 was taken as instance. According to the experience and experimental test conditions, the related parameters of PSO are set as follows: The number of particles N is 40, W is 0.9, C1 and C2 are 2. And compared to other rules in addressing ft06 instance reported from the results of [4], the results are shown in Table 1. The deviation in Table 1 is the error between the minimum makespan of the current schedule and the minimum makespan (55) of the known optimal solution.

Table 1. Makespans and deviations of schedules for ft06.

Scheduler	Makespan	Deviation (%)
NN	59	7.27
NN-PSO	58	6.90
AOI	67	21.81
SPT	83	50.90
MWKR	67	21.81
SRMPT	84	52.72
SPT-TWORK	71	29.09

Table 2. Comparison of makespan of different schedulers.

Data set	Size	Best known solution	SPT	NN	NN-PSO
la24	200	935	1874	1564	1515
abz7	300	665	1030	940	925

Later, in order to test the performance of large scale problem, the 200 size of la24 [11] and 300 size of abz7 [12] data sets were tested. The test results are shown in Table 2.

5   Conclusions

In this paper, we proposed a model for job-shop scheduling based on NN-PSO technique to solve the problem of job-shop scheduling. It can be seen from Tables 1 and 2 that job-shop scheduling based on NN-PSO technique is superior to most traditional scheduling rules.

Acknowledgments. This work was supported by Key Research and Development Plan Project of Shandong Province, China (No. 2017GGX201001).

References

1. Rodammer, F.A., White, K.P.: A recent survey of production scheduling. IEEE Trans. SMC **18**(6), 841–851 (1988)
2. Simon, Y.P., Takefuji, Y.: Stochastic neural networks for solving job-shop scheduling. I. Problem representation. In: IEEE International Conference on Neural Networks, vol. 2, pp. 275–282. IEEE (1988)

3. Simon, Y.P., Takefuji, Y.: Stochastic neural networks for solving job-shop scheduling. II. Architecture and simulations. In: IEEE International Conference on Neural Networks, vol. 2, pp. 283–290. IEEE (1988)
4. Weckman, G.R., Ganduri, C.V., Koonce, D.A.: A neural network job- shop scheduler. J. Intell. Manuf. **19**(2), 191–201 (2008)
5. Shen, X., Chen, C., He, T., Yang, J.: Postgraduate entrant and employment forecasting using modified BP neural network with PSO. In: International Conference on Computer Science & Education, pp. 1699–1703. IEEE (2009)
6. Sallehudin, R., Shamuddin, S.M.H.: Grey relational with BP_PSO for time series foreasting. In: IEEE International Conference on Systems, Man and Cybernetics, pp. 4895–4900. IEEE (2009)
7. Hajizadeh, E., Mahootchi, M., Esfahanipour, A., Kh, M.M.: A new NN-PSO hybrid model for forecasting Euro/Dollar exchange rate volatility. Neural Comput. Appl. 1–9 (2015)
8. Chatterjee, S., Sarkar, S., Hore, S., Dey, N., Ashour, A.S.: Particle swarm optimization trained neural network for structural failure prediction of multistoried RC buildings. Neural Comput. Appl. **28**(8), 2005–2016 (2017)
9. Sridhar, M., Ratnam, D.V., Raju, K.P., Praharsha, D.S.: Ionospheric scintillation forecasting model based on NN-PSO technique. Astrophys. Space Sci. **362**(9), 166 (2017)
10. Dang, T.L., Hoshino, Y.: Hardware/software co-design for a neural network trained by particle swarm optimization algorithm. Neural Process. Lett. 1–25 (2018)
11. Adams, J., Balas, E., Zawack, D.: The shifting bottleneck procedure for job shop scheduling. Manage. Sci. **34**(3), 391–401 (1988)
12. Lawrence, S.: Resource constrained project scheduling: an experimental investigation of heuristic scheduling techniques (supplement). Graduate School of Industrial Administration, Carnegie-Mellon University, Pittsburgh, PA (1984)

An Improved ORB Image Matching Algorithm Based on Compressed Sensing

Yijie Wang and Songlin Ge[(⊠)]

School of Information Engineering, East China Jiaotong University,
Nanchang 330013, China
masterx@126.com

Abstract. Aiming at the problems such as large amount of computation, high complexity and slow speed in feature extraction of the existing algorithms, this paper presents an improved ORB image matching algorithm based on compressed sensing. Firstly, compressed sensing is used to compress the target image and the matched image, and obtain sparse matrices of wavelet coefficient respectively. Secondly, the ORB algorithm is used to extract the feature points of the image. Finally, the KNN algorithm is used as a matching strategy to perform image matching. Experimental results show that the algorithm realizes fast image matching and guarantees the matching accuracy.

Keywords: Image matching · Compressed sensing · ORB algorithm · KNN

1 Introduction

Image matching is a process of identifying similarities between two or more images by matching algorithm. Generally, the image to be detected is relatively large, and the target image is slightly smaller. Due to different viewing angles, different shooting equipment will cause the same object to be imaged at different times. The noise, rotation and image preprocessing expand between the matched image and the target image, and increase the difficulty of matching. Therefore, how to accurately detect the exact position of the target image has become a research focus in image processing, and has been widely used in machine vision.

Recently years, many researchers have proposed different image matching algorithms. The commonly used matching algorithms are SIFT, SURF and ORB algorithms. Among them, SIFT algorithm is a scale invariant feature transform algorithm proposed by David in 2004, who extracts feature points in multi-scale space [1]. SURF algorithm is an improved algorithm by SIFT, which replaces scale space decomposition in SIFT by integral images and box filters, so it greatly reduces computation [2]. However, these two algorithms are still unable to meet the requirements in the area of real-time application. The ORB algorithm proposed by Rublee in ICCV 2011 is greatly improved in speed on the basis of keeping scale and rotation constant [3]. As a kind of local invariant feature matching algorithm, the ORB algorithm combines FAST keypoint detector with BRIEF descriptor. The use of binary local feature descriptor real-time is greatly superior to floating-point local feature descriptors such as SIFT and SURF, so it has been widely used nowadays.

© Springer Nature Switzerland AG 2020
Q. Liu et al. (Eds.): CENet 2018, AISC 905, pp. 351–359, 2020.
https://doi.org/10.1007/978-3-030-14680-1_39

Compressed sensing (CS) is a new information acquisition theory put forward by Donoho [4]. This theory breaks through the limitations of Nyquist sampling theorem based on sparse representation of signal. Many scholars have applied compressed sensing into the field of image processing.

Karami et al. applied different types of transformations on images to compare the performance of three different image matching techniques, i.e., SIFT, SURF, and ORB [5]. Ma et al. proposed an improved novel object tracking algorithm, the computation is effectively reduced through compressed sensing [6]. Liu et al. proposed a remote sensing image fusion algorithm combined with distributed compressed sensing to consider the features of the source image information correlation [7]. Aiming at the dynamic scene moving target detection, Li et al. proposed a matching algorithm based on the features of the ORB and effectively improve the detection speed [8]. Liu et al. proposed a k-nearest neighbor improved algorithm, which effectively improved the matching accuracy. However, the detection effect of the algorithm in complex scenes was not ideal [9]. Liu et al. proposed an improved image matching algorithm based on the ORB feature, but the accuracy is not ideal if the image feature points are not obvious [10].

In view of the above problems, compressed sensing is introduced in this paper to describe the spatial transform signal, and the image data is compressed, which can effectively improve the acquisition speed. Then, the ORB algorithm is used to extract the feature points. Finally, image matching is carried out with KNN matching strategy so as to achieve accurate image matching, and it effectively improves the precision and speed of image matching [11].

2 Theory of Compressed Sensing

Compressed sensing develops the discrete samples of the signal by random sampling and reconstructs the signal perfectly through the nonlinear reconstruction algorithm. Compressed sensing combines the L1 norm minimized sparse constraint with the random matrix to obtain the best result of a sparse signal reconstruction performance [12]. In this paper, compressed sensing is introduced to describe the spatial transform signal, and the data is compressed on the premise that the signal is reconstructed sufficiently.

Firstly, suppose X represents 1-dimension compressible signal of R^N. The transformation coefficient vectors described by the signal $X \in R^N$ by the orthogonal basis $\Psi = \{\psi_i\}_{i=1}^N$ can be represented as follow:

$$\Theta = \Psi^T X \tag{1}$$

Where Ψ is an $N \times N$ transform matrix, Θ is the equivalent sparse representation of Ψ.

Secondly, design an observation matrix Φ with a size of $M \times N$ ($M \leq N$) dimensions that is irrelevant to the transform base. When Θ is observed, observation vector $Y = \Phi\Theta$ is obtained. The process can be described that X conducts adaptive observation by matrix A^{CS} ($A^{CS} = \Phi\Psi^T$), and it can be expressed as follow:

$$Y = A^{CS}X \tag{2}$$

Where A^{CS} is a sensing factor, Y is an observation set of $M \times 1$ dimensions.

Finally, perform wavelet transform on the preprocessed images and obtain the sparse coefficient matrix. The observation value $M \times N$ ($M \leq N$) is obtained by the suitable sparse random observation matrix P, and it is value is far smaller than the original signal N. The sparse random observation matrix P is defined as follow:

$$P^{ij} = \sqrt{s} \times \begin{cases} 1 & \text{Probability is } \frac{1}{2S} \\ 0 & \text{Probability is } \frac{1-1}{s} \\ -1 & \text{Probability is } \frac{1}{2s} \end{cases} \tag{3}$$

In formula (3), the value of s ranges between 2 and 4, the image feature points can be obtained by compressed sensing fast and accurately.

3 ORB Algorithm

3.1 ORB Feature Detection

The ORB algorithm uses pyramid to produce multiscale-features [13, 14]. In order to compensate for the limitation that FAST doesn't compute the orientation, the gray scale centroid method is used to computes the intensity weighted centroid [15]. For feature point P, we define that the neighborhood pixel of P is m_{pq}, and the centroid of the image is C. The calculation is shown in formula (4):

$$m_{pq} = \sum_{x,y} x^p y^q I(x, y) \tag{4}$$

Where $I(x, y)$ is the gray value of point (x, y), the centroid of the image is as follow:

$$C = \left(\frac{m_{10}}{m_{00}}, \frac{m_{01}}{m_{00}} \right) \tag{5}$$

The angle θ between the feature point and the centroid is as follow:

$$\theta = \arctan(m_{01}, m_{10}) \tag{6}$$

3.2   ORB Feature Description

The BRIEF descriptor is used to describe the feature points. The calculation of BRIEF descriptor is simple and fast. Take an image of size $S \times S$ neighborhood P. The binary test is defined as:

$$\tau(p; x, y) = \begin{cases} 1, & p(x) < p(y) \\ 0, & otherwise \end{cases} \tag{7}$$

Where $p(x)$ is the gray value of P in the neighborhood of $x = (u, v)^T$ after smoothing. The BRIEF descriptor is an n-dimensional binary string descriptor, as shown in formula (8):

$$f_n(p) = \sum_{1 \leq i \leq n} 2^{i-1} \tau(p; x_i, y_i) \tag{8}$$

Select n pairs feature at the feature points and get the $2 \times n$ matrix S, as shown in formula (9):

$$S = \begin{bmatrix} x_1, x_2, \cdots, x_n \\ y_1, y_2, \cdots, y_n \end{bmatrix} \tag{9}$$

The $2 \times n$ matrix is rotated by the rotation matrix R_θ of the feature direction θ, as shown in formula (10):

$$R_\theta = \begin{bmatrix} \cos\theta & \sin\theta \\ -\sin\theta & \cos\theta \end{bmatrix} \tag{10}$$

Use the matrix R_θ to rotate the S matrix, a new description matrix S_θ is obtained, as shown in formula (11):

$$S_\theta = R_\theta S = \begin{bmatrix} \cos\theta & \sin\theta \\ -\sin\theta & \cos\theta \end{bmatrix} \begin{bmatrix} x_1, x_2, \cdots, x_n \\ y_1, y_2, \cdots, y_n \end{bmatrix} \tag{11}$$

Combined with formual (9), get the transformed BRIEF descriptor, namely Steered BRIEF.

$$g_n(p, \theta) = f_n(p)|(x_i, y_i) \in S_\theta \tag{12}$$

3.3   Image Matching Strategy

The descriptor of ORB is a 256-dimensional vector, and just seek the nearest distance will make mismatch. In this paper, K-Nearest Neighbors (KNN) is used as the matching strategy. The matching degree can be represented by the euclidean distance d. As shown in formula (13):

$$d = \sqrt{(x_1 - x_2)^2 + (y_1 - y_2)^2} \tag{13}$$

4 Improved ORB Algorithm Based on Compressed Sensing

Firstly, the local features of two images are obtained quickly with compressed sensing. Then the ORB algorithm is used to extract the feature points, and the KNN algorithm is adopted as the matching strategy. The improved algorithm flow is shown in Fig. 1.

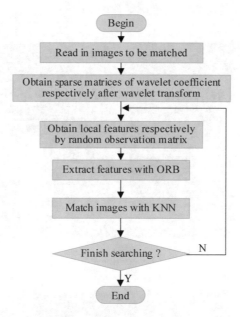

Fig. 1. The improved image matching algorithm flow.

5 Simulation

To compare the proposed algorithm's superiority, two different groups images are selected as experimental objects in this paper. Figure 2 is the first set of two images to be matched. Figure 3 is the second set of images to be matched.

Figures 4 and 5 respectively show two images using improved ORB feature extraction algorithm and KNN algorithm matching strategy for feature point matching.

In order to prove the faster matching speed, the proposed algorithm is compared with the SIFT and SURF algorithms. As shown in Fig. 6.

Fig. 2. Group a.

Fig. 3. Group b.

Fig. 4. Result of group a.

Fig. 5. Result of group b.

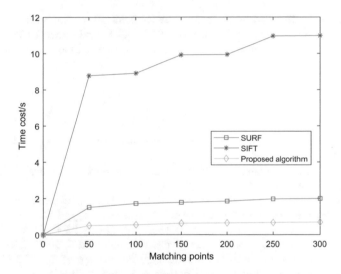

Fig. 6. Time cost of the different algorithms.

Table 1. The comparison of several matching algorithms.

Algorithm	Feature points	Matches	Accuracy (%)
SURF	198/97	148/73	74.9
SIFT	184/89	171/83	93.1
Proposed algorithm	138/68	136/66	97.8

This paper has also compared the proposed algorithm with the classical algorithms in the number of feature points, matches and the accuracy. As shown in Table 1.

The number of feature points and matches in this paper is less than the classical algorithms, but the matching accuracy is better than the other two classical matching algorithms.

Combined with the analysis of the speed of the algorithm in Fig. 6, it fully shows that the proposed algorithm has higher precision and faster speed.

6 Conclusion

In this paper, an improved ORB matching algorithm based on compressed sensing is proposed. It greatly reduces the irrelevant feature points and makes the feature points matching more accurate. The experimental results show that the proposed algorithm achieves higher matching precision and can improve the efficiency of image feature extraction effectively.

Acknowledgement. The work was supported by the National Natural Science Foundation of China (No. 61762037), Science and Technology Project of Jiangxi Provincial Transport Bureau (No. 2016D0037) and Innovation Fund Designated for Graduate Students of Jiangxi Province (No. YC2017-S253).

References

1. Lowe, D.G.: Distinctive image features from scale-invariant keypoints. Int. J. Comput. Vision **60**(2), 91–110 (2004)
2. Bay, H., Ess, A., Tuytelaars, T., Bradski, G.: Speeded-up robust features (SURF). Comput. Vis. Image Underst. **110**(3), 346–359 (2008)
3. Rublee, E., Rabaud, V., Konolige, K., Bradski, G.: ORB: an efficient alternative to SIFT or SURF. In: IEEE International Conference on Computer Vision, pp. 2564–2571. IEEE (2011)
4. Donoho, D.L.: Compressed sensing. IEEE Trans. Inf. Theory **52**(4), 1289–1306 (2006)
5. Karami, E., Prasad, S., Shehata, M.: Image matching using SIFT, SURF, BRIEF and ORB: performance comparison for distorted images. In: Proceedings of the 2015 Newfoundland Electrical and Computer Engineering Conference, St. Johns, Canada (2015)
6. Ma, D., Yu, Z., Yu, J., Pang, W.: A novel object tracking algorithm based on compressed sensing and entropy of information. Math. Probl. Eng. **2015**, 18 (2015). Article ID 628101. https://doi.org/10.1155/2015/628101
7. Liu, J., Li, X.C., Zhu, K.J., et al.: Distributed compressed sensing based remote sensing image fusion algorithm. J. Electron. Inf. Technol. **39**(10), 2374–2381 (2017)
8. Li, X.L., Xie, C.M., Jia, Y.X., et al.: Fast object detection algorithm based on ORB feature. J. Electron. Meas. Instrum. **5**, 455–460 (2013)
9. Liu, W., Zhao, W.J., Li, D.J., et al.: A feature point matching algorithm based on ORB detection. Laser Infrared **45**(11), 1380–1384 (2015)
10. Liu, T., Zhang, J.: Improved image stitching algorithm based on ORB features by UAV remote sensing. Comput. Eng. Appl. **54**(2), 193–197 (2018)
11. Larose, D.T.: k-nearest neighbor algorithm. In: Larose, D.T. (ed.) Discovering Knowledge in Data: An Introduction to Data Mining, pp. 90–106. Wiley, Hoboken (2005)

12. Xie, X., Xu, Y., Liu, Q., et al.: A study on fast SIFT image mosaic algorithm based on compressed sensing and wavelet transform. J. Ambient Intell. Humaniz. Comput. **6**(6), 835–843 (2015)
13. Rosten, E., Drummond, T.: Machine learning for high-speed corner detection. In: European Conference on Computer Vision, pp. 430–443. Springer, Heidelberg (2006)
14. Calonder, M., Lepetit, V., Strecha, C., Fua, P.: Brief: binary robust independent elementary features. In: European Conference on Computer Vision, pp. 778–792. Springer, Heidelberg (2010)
15. Rosin, P.L.: Measuring corner properties. Comput. Vis. Image Underst. **73**(2), 291–307 (1999)

Crime Prediction Using Data Mining and Machine Learning

Shaobing Wu[1], Changmei Wang[2(✉)], Haoshun Cao[1], and Xueming Jia[1]

[1] Institute of Information Security, Yunnan Police College, Kunming 650223, China
[2] Solar Energy Institute, Yunnan Normal University, Kunming 650092, China
823804919@qq.com

Abstract. In order to predict the crime in YD county, data mining and machine learning are used in this paper. The aim of the study is to show the pattern and rate of crime in YD county based on the data collected and to show the relationships that exist among the various crime types and crime Variable. Analyzing this data set can provide insight on crime activities within YD county. By introducing formula and methods of Bayesian network, random tree and neural network in machine learning and big data, to analyze the crime rules from the collected data. According to the statistics released by the YD county From 2012-09-01 to 2015-07-21, The crime of smuggling, selling, transporting and manufacturing drugs, Theft, Intentional injury, Illegal business crime, Illegal possession of drugs, Rape, Crime of fraud, Gang fighting, manslaughter, Robbery made the top ten list of crime types with high number of crimes. The crime rate of drugs was the highest, reaching 46.86%, farmers are the majority, accounting for 97.07%, people under the age of 35 are the subject of crime. Males accounted for 90.17% of crimes committed, while females accounted for 9.83%. For ethnic groups, the top five were han, yi, wa, dai and lang, accounting for 68.43%, 23.43%, 1.88%, 1.67% and 1.25% respectively. By adopting random forest, Bayesian networks, and neural network methods, we obtained the decision rules for criminal variables. By comparison, the classification effect of Random Trees is better than that of Neural Networks and Bayesian Networks. Through the data collection of the three algorithms, the validity and accuracy of the random tree algorithm in predicting crime data are observed. The performance of the Bayesian network algorithm is relatively poor, probably due to the existence of certain random factors in various crimes and related features (the correlation between the three algorithms is low).

Keywords: Crime prediction · Data mining · Machine learning

1 Introduction

For almost everyone, machine learning (ML) is still a very mysterious field that sounds complicated and difficult to explain to a person without any skills [1]. However, this is very important today and will continue in the next few years.

© Springer Nature Switzerland AG 2020
Q. Liu et al. (Eds.): CENet 2018, AISC 905, pp. 360–375, 2020.
https://doi.org/10.1007/978-3-030-14680-1_40

ML is a fairly multidisciplinary field that deals primarily with programming and mathematics (mainly involving probability and density functions). In addition, because it is new and quite complex, it requires good research skills.

For crime detection issues, the game organizer provided a huge database of crime training in San Francisco. The database is tagged, that is, it contains the correct category for each entry (e.g. theft, assault, bribery, etc.), so it is a supervised learning problem. With this in mind, the algorithms used to solve this problem are: Random tree, neural network, Bayesian network.

The US Federal Bureau of Investigation (FBI) defines violent crime as a crime involving violence or threats. The United States Federal Bureau of Investigation (FBI) Unified Crime Report (UCR) program defines each type of criminal behavior as: (i) Murder-intentional (non-faulty) murder. UCR does not include deaths caused by accidents, suicides, negligence, proper homicides, and attempts to murder or assault murder (all of which are classified as serious attacks), in this crime classification [2]. (ii) Forced rape-rape is a sexual assault that violates the will of women. While attempting or attacking rape by threat or force is considered a crime under this category, statutory rape (without the use of force) and other sexual offences are excluded from [3]. (iii) Robbery-threatening or violent by force or force and/or placing the victim in fear, gaining or attempting to obtain anything of value from the care, custody or control of one or more persons. Crimes that aggravate the crime of personal assault and theft are crimes of robbery. Unfortunately, these types of crimes seem to have become commonplace in society. Law enforcement officials have turned to data mining and machine learning to help fight crime prevention and enforcement.

Miquel Vaquero Barnadas [13] proposed machine learning applied to crime prediction. In this paper, he plans to use different algorithms (such as K-Nearest neighbour, Parzen windows and Neural Networks) to solve the real data classification problem (the San Francisco crime classification).

Gaetano Bruno Ronsivalle [14] presented Neural and Bayesian Networks to Fight Crime: The NBNC Meta-Model of Risk Analysis. In his paper, he used this tool with the specific goal of providing an effective model for Italian bank security managers to "describe" variables and define "robbing" phenomena; "interpret" calculations (i) "exogenous", (ii) "Endogenous and (iii) methods of global risk index for each branch; through simulation modules to "predict" composite risk and changes in different branch security systems.

Jeffrey T. Ward, James V. Ray, Kathleen A. Fox [15] developed a MIMIC model for Exploring differences in self-control across sex, race, age, education, and language, and draw a conclusion that apart from race, testing group differences in self-control with an observed scale score is largely unbiased. Testing group differences in elements using observed subscores is frequently biased and generally unsupported.

In this research, we developed the Random Trees, Neural Networks, and Bayesian Networks algorithms using the same finite set of features, on the communities and crime un normalized dataset to conduct a comparative study between the violent crime patterns from this particular dataset and actual crime statistical data for the state of YD County. The crime statistics used from collected data. Some of the statistical data that was provided by YD County people's procuratorate such as the population of YD

County, population distribution by age, number of violent crimes committed, and the rate of those crimes are also features that have been incorporated into the test data to conduct analysis.

The rest of the paper is organized as follows: Sect. 2 gives an overview of data mining and machine learning. Section 3 provides information about the Crime Classification in YD County. Section 4 presents the results from each of the algorithms and Sect. 5 concludes with the findings and discussion of the paper results.

2   Data Mining and Machine Learning Algorithms

2.1   Data Mining

Data mining is part of the interdisciplinary field of knowledge discovery in databases [9]. Data mining consist of collecting raw data and, (through the processes of inference and analysis); creating information that can be used to make accurate predictions and applied to real world situations. It is the application of techniques that are used to conduct productive analytics. The five tasks that these types of software packages are designed for are as follows: (i) Association-Identifying correlations among data and establishing relationships between data that exist together in a given record [9, 10]. (ii) Classification Discovering and sorting data into groups based on similarities of data [6]. Classification is one of the most common applications of data mining. The goal is to build a model to predict future outcomes through classification of database records into a number of predefined classes based on a certain criteria. Some common tools used for classification analysis include neural networks, decisions trees, and if-then-else rules [10]. (iii) Clustering-Finding and visually presenting groups of facts previously unknown or left unnoticed [6]. Heterogeneous data is segmented into a number of homogenous clusters. Common tools used for clustering include neural networks and survival analysis [10]. (iv) Forecasting-Discovering patterns and data that may lead to reasonable predictions [9].

2.2   Machine Learning

Arthur Samuel is a pioneer in the field of machine learning and artificial intelligence. He defines machine learning as a field of study that allows computers to learn without explicit programming [11]. In essence, machine learning is a way for computer systems to learn through examples. There are many machine learning algorithms available to users that can be implemented on data sets. The algorithm has a better understanding of the data set because it has more examples to implement. In the field of data mining, there are five machine learning algorithms for analysis: (i) Classification analysis algorithms-these algorithms use attributes in the data set to predict the value of one or more variables taking discrete values. (ii) Regression analysis algorithms-These algorithms use the properties of the data set to predict the value (e.g. profit and loss) of one or more variables taking continuous values. (3) Segmentation analysis algorithm - divide data into groups or groups with similar attributes.

2.3 Algorithms Selected for Analysis

Random Trees-Aldous [12, 13] discussed scaling limits of various classes of discrete trees conditioned to be large. In the case of a Galton-Watson tree with a finite variance critical offspring distribution and conditioned to have a large number of vertices, he proved that the scaling limit is a continuous random tree called the Brownian CRT. Their main result (Theorem 2.1) stated that the rescaled height function associated with a forest of independent (critical, finite variance) Galton-Watson trees converged in distribution towards reflected Brownian motion on the positive half-line.

In order to derive the Theorem 2.1, they first state a very simple "moderate deviations" Lemma 1.1 for sums of independent random variables.

Lemma 1.1: Let Y_1, Y_2, ... be a sequence of i.i.d. real random variables. We assume that there exists a number $\lambda > 0$ such that $E\left[\exp(\lambda|Y_1|)\right] < \infty$, and that $E[Y_1] = 0$. Then, for every $\alpha > 0$, we can choose N sufficiently large so that for every $n \geq N$ and $l \in \{1, 2, 3, ..., n\}$

$$P\left[|Y_1 + \cdots Y_l| > n^{\alpha + \frac{1}{2}}\right] \ll e^{-n^{\alpha/2}} \tag{2.1}$$

According to Lemma 1.1, they get the Theorem 2.1 as following:

Theorem 2.1: Let θ_1, θ_2, ... be a sequence of independent μ-Galton-Watson trees, and let $(H_n; n \geq 0)$ be the associated height process. Then

$$\frac{1}{\sqrt{p}} H_{[pt]}, t \gg 0 \underset{p \to \infty}{\longrightarrow} \left(\frac{2}{\sigma} \gamma_t, t \gg 0\right) \tag{2.2}$$

Where γ is a reflected Brownian motion. The convergence holds in the sense of weak convergence on the Skorokhod space D(R+; R+).

In their papers, they introduce the exit measure from a domain D, which is in a sense uniformly spread over the set of exit points of the Brownian snake paths from D. they then derive the key integral equation (Theorem 2.2) for the Laplace functional of the exit measure. In the particular case when the underlying spatial motion ε is d-dimensional Brownian motion, this quickly leads to the connection between the Brownian snake and the semilinear PDE(Partial differential equation) $\Delta u = u^2$.

Theorem 2.2: Let g be a nonnegative bounded measurable function on E. For every $x \in E$, set

$$u(x) = N_x\left(1 - exp - Z^D, g\right), x \in D \tag{2.3}$$

The function u solves the integral equation [20]

$$u(x) + 2\prod_x \left(\int_0^T u\varepsilon_s^2 ds\right) = \prod_x \left(1_{\{\tau < \infty\}} g(\varepsilon_\tau)\right) \tag{2.4}$$

Random continuous trees can be used to model the genealogy of self-similar fragmentations [14]. The Brownian snake has turned out to be a powerful tool in the study of super-Brownian motion: See the monograph [15] and the references therein.

Since Random Trees was proposed, the algorithm has become a popular and widely used tool for nonparametric regression applications.

Bayesian Networks-A Bayesian network (BN) approximates the joint probability distribution for a multivariate system based on expert knowledge and sampled observations that are assimilated through training [16, 17]. A tractable scoring metric, known as K2, is obtained from $P(F, T)$ using the assumptions in [6], which include fixed ordering of variables in X:

$$g = \log\left(\prod_{j=1,\ldots,q_i} \frac{(r_i - 1)!}{(N_{ij} + r_i - 1)!} \prod_{k=1,\ldots,r_i} N_{ijk}!\right) \quad (2.5)$$

where r_i is the number of possible instantiations of X_i, and q_i is the number of unique instantiations of pi. N_{ijk} is the number of cases in T. The BN (K, H) represents a factorization of the joint probability over a discrete sample space,

$$p(X) = p(X_1, \ldots, X_n) = \prod_{i=1,\ldots,n} p(X_i|\pi_i) \quad (2.6)$$

for which all probabilities on the right-hand side are given by the CPTs. Therefore, when a variable X_i is unknown or hidden, Bayes' rule of inference can be used to calculate the posterior probability distribution of Xi given evidence of the set of l variables, that are conditionally dependent on X_i,

$$P(X_i|\overline{\mu_i}) = \frac{P(\overline{\mu_i}|X_i)P(X_i)}{P(\overline{\mu_i})} \quad (2.7)$$

Bayesian networks are particularly well suited for crime analysis, as they learn from data and use the experience of criminologists to select nodes and node sequencing. The confidence level provided for the criminal files informs the detective about the possible accuracy of each prediction. In addition, BN's graphical structure represents the most important relationship between criminal behavior and crime scene behavior, which may help develop new scientific assumptions about criminal behavior.

Neural Networks-Artificial Neural Networks (ANN) have been developed as generalizations of mathematical models of biological nervous systems. The basic processing elements of neural network are called artificial neurons, or simply neurons or nodes. The neuron pulse is then calculated as the weighted sum of the input signal of the transfer function transformation. The artificial neurons' learning ability can be realized by adjusting the weight according to the selected learning algorithm [12].

Architectures: An ANN consists of a set of processing elements, also known as neurons or nodes, which are interconnected. It can be described as a directed graph in which each node performs a transfer function of the form

$$y_i = f\left(\sum\nolimits_{i=1,\ldots,n} w_{ij} x_j - \theta_i\right)_i \qquad (2.8)$$

where y_i is the output of the node i, x_j is the th input to the node, and w_{ij} is the connection weight between nodes i and j. θ_i is the threshold (or bias) of the node. Usually, f_i is nonlinear, such as a heaviside, sigmoid, or Gaussian function.

In (11), each term in the summation only involves one input x_j. High-order ANN's are those that contain high-order nodes, i.e. nodes in which more than one input are involved in some of the terms of the summation. For example, a second-order node can be described as

$$y_i = f_i\left(\sum\nolimits_{j,k=1,\ldots,n} w_{ijk} x_j x_k - \theta_i\right) \qquad (2.9)$$

where all the symbols have similar definitions to those in (11).

3 Crime Classification in YD County

As it has been said previously, this project is based on a Project of national social science foundation about the causes and countermeasures of ethnic minority crimes in the county of YD. In this chapter, the principle and formula of Bayesian network, random tree and neural network are given briefly.

3.1 Description of the Problem

In this problem, a training dataset with nearly 35 months of crime reports from all across YD county was provided. This dataset contains all crimes classified in categories, which are the different crime typologies. The main goal of the challenge is to predict these categories of the crimes that occurred.

For the algorithm evaluation, another unlabelled dataset is provided (the test one). It is used to evaluate the algorithm accuracy with new unclassified data.

How Is the Problem Going to be Solved?
In this contest, we will use a different algorithm to get a good result. Each one will be explained, tested and tested, and finally we will see which of them is best for this case.

Cross-validation will be used to validate the model, so the database must be divided into subsets of tests, training, and verification. This division must be layered to ensure that the proportion of the original components is maintained in each segment (the number of crimes per category is the same).

All development and testing is done on a server provided by the university department. In this way, the death penalty can last a whole day without worrying about them, and execution is faster.

Results Submission Format and Evaluation
The data submitted to the contest evaluation must be in a specific format that meets the requirements. To properly evaluate the data, the resulting data set must contain sample

ids that contain a list of all categories and the probability that each sample belongs to each category. Remind the training dataset to label the crime types of all samples (10 different).

Then, instead of predicting which category a given sample belongs to, the output will always be the probability vector.

3.2 Dataset Analysis

Data

The data in this article involves the reported cases of the crime of smuggling, selling, transporting and manufacturing drugs, Theft, Intentional injury, Illegal business crime, Illegal possession of drugs, Rape, Crime of fraud, Gang fighting, manslaughter, Robbery in YD county between the years 2012 and 2015. The summary of the data is as provided in Table 1.

Table 1. Summary statistics of the data set on crime activities.

Crime types	Numbers of crime
The crime of smuggling, selling, transporting and manufacturing drugs	224
Theft	86
Intentional injury	85
Illegal business crime	23
Illegal possession of drugs	18
Rape	18
Crime of fraud	13
Gang fighting	11

The data provides insight into criminal activity, and its research can help reduce crime (protecting communities) and decision-making. Part of the analysis provided can be used to explain the relationship between certain criminal activities.

The data can further be analyzed using other statistical methods like Random Trees, Bayesian, Quasi-neural networks and so on.

From Table 1 and Fig. 3, The crime of smuggling, selling, transporting and manufacturing drugs is the most important crime types, the following is theft, intentional injury and so on.

Data Analysis

The provided dataset has different "features", each one being of a different relevance. In this chapter we will proceed to analyze this database and extract the useful information out of it. There are 478 samples of crime analysis. These data were collected from 2012-09-01 to 2015-07-21 in YD.

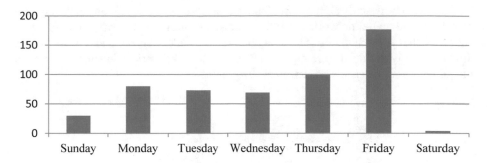

Fig. 1. Number of crimes by day of the week (YD County).

Another interesting analysis is to count the number of crimes that occur every day of the week so that we know if this is relevant information. Figure 1 shows that the day when the offenders choose the most is Friday, and the rest of the week is distributed differently.

We chose the top eight of the crime categories as the basis for this analysis and discussion.

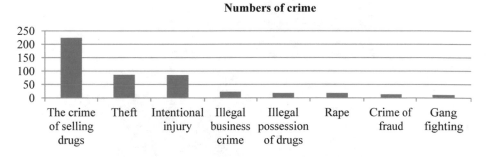

Fig. 2. Numbers of crime types.

From the Fig. 2, we can see that the crime of smuggling, selling, transporting and manufacturing drugs 224; Theft 86; Intentional injury 85; Illegal business crime 23, which are the main parts of crime types.

4   Results

The data set selected in this study is a denormalized data set for community and crime. Including the social and economic data from 2012-09-01 to 2015-07-21, the law enforcement data of the YD County People's Procuratorate, and the crime data from 2012-09-01 to 2015-07-21. It also includes 478 cases or criminal cases and 12 total attributes reported from the entire YD County, often referred to as features.

This section describes all the implementation results of the random tree, Bayesian algorithm, and neural network algorithm. The algorithm is run to predict the following characteristics of each data set: he smuggles crimes, sells, transports and manufactures drugs, illegally holds drugs, rapes, steals, murders, robberies, intentional injuries, illegal business crimes, fraud, criminal gang fights.

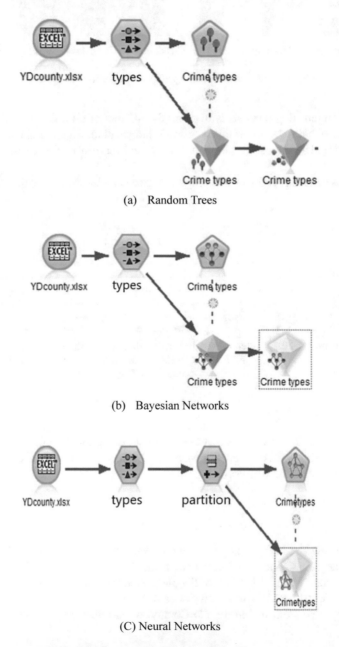

Fig. 3. The modeling of machine learning algorithm.

4.1 Modeling Based on Machine Learning Algorithm

In this section, we build three models based on Random Trees, Bayesian, Quasi-neural networks algorithms. Figure 3 shows the modeling of machine learning algorithms.

4.2 Relation Between Sex (or Gender), Ethnic, Age, Education and Crime

The relationship between sex and crime as following: first, the number relationship is as Table 2 and Fig. 4 shows the relationship between sex and crime. From Fig. 4 and Table 2, number of the males more than females.

Table 2. Convictions according to gender for YD.

Year	Males	Females
2012	6	0
2013	169	20
2014	221	16
2015	89	11

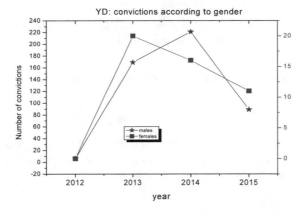

Fig. 4. Relationship between sex and crime.

In the 478 samples, the relationship between age and Crime numbers for drugs See Fig. 5. From Fig. 5, it shows that people for the age from 16 to 35, accounting for about half of the total sample, are the main criminal group.

The relationship between education and Crime numbers for drugs See Fig. 6. From Fig. 6, it shows that people for illiteracy or semi-illiteracy, Primary school and Junior high school, are the main criminal group.

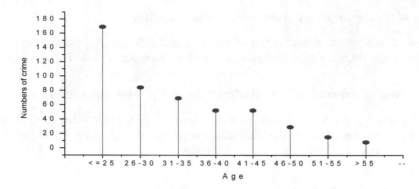

Fig. 5. The relationship between age and Crime numbers for crime types.

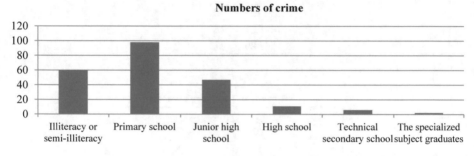

Fig. 6. The relationship between education and crime.

4.3   The Conditional Probability Table for Bayesian Networks

The Conditional probability based on for The crime of smuggling, selling, transporting and manufacturing drugs and Gang fighting for professional, education, gender, ethnic and age are followed as Tables 3, 4, 5, 6, 7 and 8.

Table 3. Conditional probability for crime types.

The crime of smuggling, selling, transporting and manufacturing drugs	Gang fighting
0.95	0.05

Table 4. Conditional probability for gender.

Crime types	Male	Female
The crime of smuggling, selling, transporting and manufacturing drugs	0.86	0.14
Gang fighting	1.00	0.00

Table 5. Conditional probability for professional.

Crime types	Gender	Individual worker	Migrant worker	Farmers	Foreigners	Student	Teacher
The crime of smuggling, selling, transporting and manufacturing drugs	Male	0.01	0.00	0.97	0.02	0.00	0.00
The crime of smuggling, selling, transporting and manufacturing drugs	Female	0.00	0.00	0.97	0.03	0.00	0.00
Gang fighting	Male	0.00	0.00	1.00	0.00	0.00	0.00

Table 6. Conditional probability for education.

Crime types	Gender	High school	Illiteracy or semi-illiteracy	Junior high school	Primary school	Technical secondary school	Specialized subject graduates
The crime of smuggling, selling, transporting and manufacturing drugs	Male	0.05	0.22	0.23	0.46	0.03	0.01
The crime of smuggling, selling, transporting and manufacturing drugs	Female	0.06	0.53	0.09	0.28	0.03	0.00
Gang fighting	Male	0.00	0.00	0.82	0.18	0.00	0.00

Table 7. Conditional probability for ethnic.

Crime types	Gender	bai	Blang	Buyi	dai	Deang	Han	Hui	Lisu	man	miao	Other	wa	yi
The crime of smuggling, selling, transporting and manufacturing drugs	Male	0.00	0.01	0.01	0.02	0.00	0.55	0.02	0.00	0.01	0.02	0.01	0.00	0.39
The crime of smuggling, selling, transporting and manufacturing drugs	Female	0.00	0.00	0.00	0.00	0.00	0.47	0.00	0.00	0.00	0.00	0.03	0.00	0.50
Gang fighting	Male	0.00	0.00	0.00	0.00	0.00	1.00	0.00	0.00	0.00	0.00	0.00	0.00	0.00

Table 8. Conditional probability for age.

Crime types	Gender	<=24.6	24.6–35.2	35.2–45.8	45.8–56.4	>56.4
The crime of smuggling, selling, transporting and manufacturing drugs	Male	0.22	0.38	0.28	0.10	0.02
The crime of smuggling, selling, transporting and manufacturing drugs	Female	0.12	0.44	0.32	0.12	0.00
Gang fighting	Male	0.82	0.18	0.00	0.00	0.00

4.4 The Comparison of Result for Random Trees, Neural Networks and Bayesian Networks

The Comparison of Data Analysis Results Predictive Variable Importance
The comparison of data analysis results predictive variable importance based on different algorithms is as following:
Predictive variable importance of crime types for Random Trees, Bayesian Network and Neural Network See Fig. 7. From Fig. 7, it shows that the age is important variable for Random Trees and Neural Networks, and the education is important variable for Bayesian Networks.

The Comparison of Model Accuracy
The data analysis results based on Random Trees and Bayesian Networks. The comparison of model accuracy based on Random Trees Classification and Bayesian Networks Classification are as follows:
From Fig. 8, it shows that Model accuracy for Random Trees is much higher than Bayesian Networks.

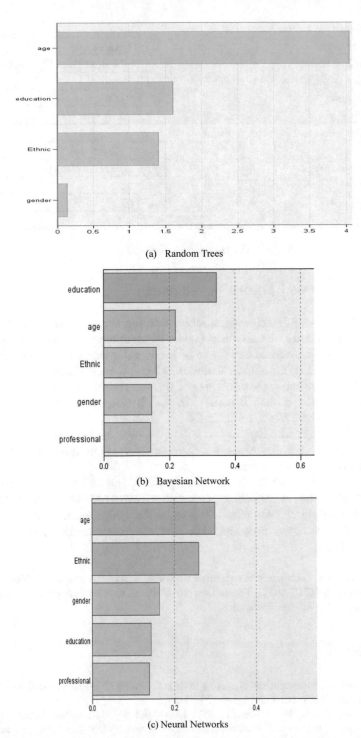

(a) Random Trees

(b) Bayesian Network

(c) Neural Networks

Fig. 7. Predictive variable importance of crime types for Random Trees, Bayesian Network and Neural Networks.

Model information for Random Trees

The target field	Crime types
Model building method	Random Trees Classification
The number of predictive variables entered	5
Model accuracy	0.974
Misclassification rate	0.026

Model information for Bayesian Networks

The target field	Crime types
Model building method	Bayesian Networks Classification
The number of predictive variables entered	5
Model accuracy	0.537
Misclassification rate	0.463

Fig. 8. The comparison of model accuracy.

5   Conclusions and Future Development

In the field of artificial intelligence, machine learning is a very powerful field. If the model is done correctly, the accuracy that some algorithms can achieve can be surprising. Of course, the current and future of intelligent systems are subject to ML and big data analysis. From the above discussion, we can draw the following conclusions:

In the data we selected, the crime rate of was the highest, reaching 46.86%, which was the main crime type in YD county. In fact, it was theft and intentional injury, reaching 17.99% and 17.78% respectively.

For ethnic groups, the top five were han, yi, wa, dai and lang, accounting for 68.43%, 23.43%, 1.88%, 1.67% and 1.25% respectively. From this, the han nationality is the main criminal in the nation.

Through the data collection of the three algorithms, the validity and accuracy of the random tree algorithm in predicting crime data are observed. The performance of the Bayesian network algorithm is relatively poor, probably due to the existence of certain random factors in various crimes and related features (the correlation between the three algorithms is low).

Acknowledgements. This study is supported by scientific research projects of National Social Science Foundation (13CFX038). The authors would like to express their gratitude to the Office of the national social science foundation.

References

1. McClendon, L., Meghanathan, N.: Using machine learning algorithms to analyze crime data. Mach. Learn. Appl. Int. J. **2**(1), 1–2 (2015)
2. Murder. http://www.fbi.gov/ucr/cius2009/offenses/violent_crime/murder_homicide.html
3. Forcible Rape. http://www.fbi.gov/ucr/cius2009/offenses/violent_crime/forcible_rape.html
4. Robbery. http://www.fbi.gov/ucr/cius2009/offenses/violent_crime/robbery.htm

5. Assault. http://www.fbi.gov/ucr/cius2009/offenses/violent_crime/aggravated_assault.html
6. Vaquero Barnadas, M.: Machine learning applied to crime prediction. http://upcommons. upc.edu/bitstream/handle/2117/96580/machine%20learning%20applied%20TO%20crime% 20precticion.pdf
7. Ronsivalle, G.B.: Neural and bayesian networks to fight crime: the NBNC meta-model of risk analysis. In: Artificial Neural Networks-Application, pp. 29–42 (2011)
8. Ward, J.T., Ray, J.V., Fox, K.A.: Developed a computer model exploring differences in self-control across sex, race, age, education, and language: considering a bifactor MIMIC model. J. Crim. Justice **56**, 29–42 (2018)
9. Nirkhi, S.M., Dharaskar, R.V., Thakre, V.M.: Data mining: a prospective approach for digital forensics. Int. J. Data Min. Knowl. Manag. Process **2**(6), 41–48 (2012)
10. Ngai, E.W.T., Xiu, L., Chau, D.C.K.: Application of data mining techniques in customer relationship management: a literature review and classification. Expert Syst. Appl. 2592–2602 (2008)
11. McCarthy, J.: Arthur samuel: pioneer in machine learning. AI Mag. **11**(3), 10–11 (1990)
12. Aldous, D.: The continuum random tree I. Ann. Probab. **19**(1), 1–28 (1991)
13. Aldous, D.: The continuum random tree III. Ann. Probab. **21**(1), 248–289 (1993)
14. Haas, B., Miermont, G.: The genealogy of self-similar fragmentations with negative index as a continuum random tree. Electron. J. Probab. **9**, 57–97 (2004)
15. Le Gall, J.F.: Spatial Branching Processes, Random Snakes and Partial Differential Equations. Birkhauser, Boston (1999)
16. Heckerman, D.: A Bayesian approach to learning causal networks. Technical report MSR-TR-95-04, pp. 1–23 (1995)
17. Jensen, F.V.: Bayesian Networks and Decision Graphs. Springer, New York (2001)

Improve Link Prediction Accuracy with Node Attribute Similarities

Yinuo Zhang[✉], Subin Shen, and Zhenyu Wu

School of IOT, Nanjing University of Posts and Telecommunications,
Nanjing 210003, China
1016071603@njupt.edu.cn

Abstract. Link prediction is one of the significant research problems in social networks analysis. Most previous works in this area neglect attribute similarity of the node pair which can easily obtain from real world dataset. Traditional supervised learning methods study the link prediction problem as a binary classification problem, where features are extracted from topology of the network. In this paper, we propose a similarity index called Attribute Proximity. The set of features are similarity index we proposed and four others well-known neighbourhood based features. We then apply a supervised learning based temporal link prediction framework on DBLP dataset and examine whether attribute similarity feature can improve the performance of the link prediction. In our experiments, the AUC performance is better when attribute similarity feature is considered.

Keywords: Link prediction · Social networks · Node attributes

1 Introduction

With the advance of information technology, researches of complex networks have attracted much attention. As one branch of researches in complex networks, many things in real life can be modelled and analyzed using social networks, such as co-authorship network and friend relationship network. In this paper, we focus on one of the significant problems in social networks, this is the link prediction problem.

Link prediction has been widely applied in many domains. (1) In social domains, link prediction is used to analyze whether two authors in a co-authorship network will cooperate in the future [1] or recommend friends to users in friend relationship networks. (2) In biomedical domains, it will cost too much to verify whether two proteins will interact [2]. And therefore, link prediction is the smart method to cut down the cost. (3) In e-commerce domains, recommending commodity is regarded as link prediction in user-item bipartite networks [3].

Traditional link prediction method first takes two glances successively at the network in order to get two snapshot of the network, that are called training period and test period, and then assign a topological $score(x, y)$ to each non-connected node pair $<x, y>$ in training period and finally produces a list in descending order of $score(x, y)$ [1]. However, this method ignores two important characteristics that social networks naturally possess. One characteristic is that social networks changes over time

© Springer Nature Switzerland AG 2020
Q. Liu et al. (Eds.): CENet 2018, AISC 905, pp. 376–384, 2020.
https://doi.org/10.1007/978-3-030-14680-1_41

so one snapshot cannot reflect a period of evolution of network. Another characteristic is that the nodes of social networks are people. The extracted information ought to contain not only the relationship, but also their personal attributes. To solve these two problems, we propose a similarity index called attribute proximity and apply it in a supervised learning based temporal link prediction framework.

In real life, people who have more common interests will be more likely to interact with each other. When an author in co-authorship networks publishes an article, the title or the keywords may reflect his interests or focuses. If two authors in different research area share more common focuses, they will be more likely to cooperate to publish an article. Inspired by this idea, we want to define a similarity feature between two nodes according to their attribute proximity and combine with topological features which will be introduce in Sect. 3 to make predictions. Compared to other features, the feature we proposed can be easily computed, just count the same attributes between two nodes (Fig. 1).

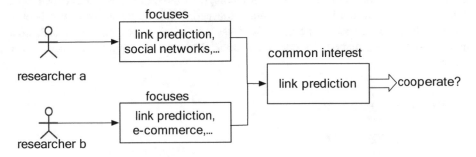

Fig. 1. The motivation of our method.

Figure 1 shows how we come up with our method. In co-authorship network, two researchers who have more common focuses are more likely to cooperate. In daily life, two people who have more common interests are more likely to be friends.

The rest of this paper is arranged as follows. Section 2 introduces related works in supervised link prediction. Section 3 illustrates the feature we proposed and the method we taken. Section 4 shows our experiments on DBLP dataset and analyze the experimental result. We make a conclusion in the last section.

The main contributions of this work are briefly summarized as follows:

(1) We propose a similarity index called Attribute Proximity which is often ignored in supervised link prediction methods.
(2) We apply the forecast models in supervised link prediction which consider the dynamics of the networks. These models make networks closer to social networks in real world than those static ones.
(3) We use Receiver Operating Characteristic (ROC) curve to evaluate the performance of our supervised link prediction.

2 Related Work

2.1 Supervised Link Prediction

Hasan et al. [4] proposed using supervised learning method in link prediction. They study the link prediction problem as a binary-classification problem. In their paper, they divide node pairs into two classes, the linked ones are in positive class, the unlinked are in negative class. They take the use of proximity features, aggregated features and topological features between two nodes in training set to train a classifier and predict whether two nodes will interact in the future. However, this method ignores the fact that social networks change over time.

2.2 Temporal Link Prediction

Soare et al. [5] study that whether temporal information can bring any performance improvement to link prediction task. They compute topological metrics between node pairs in training set and build a temporal sequence. Then, they use forecast model to process the sequence and generate a final score between node pairs. And last, both unsupervised and supervised method can be applied to make predictions using these scores.

3 Proposed Approach

3.1 Notation Representations

Before starting to propose our method, we will introduce some notations that we later use in our paper. $G(V, E)$ represents a graph, where V represent the set of nodes and E represent the set of edges. Lowercase letters $u, v \in V$ represent nodes in the graph. $(u, v) \in E$ represents an edge or a node pair in the graph. When temporal information are considered, G_t represents graph observed at time t. $(G_1, \ldots G_T)$ represents a graph sequence in which graph is observed periodically.

3.2 The Approach

Let G_T be the graph representation of a network at time t. Let $[G_1, G_2, \ldots, G_T]$ be the frames formed by periodically observation of the network. Therefore, we get T frames of the network. Once the network structure is split, we follow the steps bellow:

(1) Choose some similarity metrics to build a feature set (e.g. attribute proximity, preference attachment, common neighbors, …)
(2) For each two-hop node pair in training period:
   (a) Create a time series by computing each feature in the feature set of the node pair on each frame.
   (b) The features must be null for a frame in which any of the nodes in the pair does not exist.

(c) Once the time series was built, its forecast is computed by using a forecasting model.

(3) Choose a forecasting model (e.g. moving average, average, random walk…)

(4) For each time series, set the feature set of node pair to be the one-step ahead prediction given by the chosen forecasting model when applied to the series.

(5) Assign a label to each node pairs according to the formula bellow:

$$label(u,v) = \begin{cases} 1, (u,v) \in E \\ 0, (u,v) \notin E \end{cases} \qquad (3.1)$$

(6) Choose a classifier (e.g. SVM, KNN, …) and start training.

(7) Predict whether two nodes in the node pair will interact in the test period and evaluate the results.

3.3 Feature Set

The features extracted from dataset are essentially important to the performance of classification and they should reflect the proximity between a node pair to some extent. This section proposes a simple introduction of features we use in our method. Due to most features are extracted from the topology of network, we propose a similarity index called attribute proximity which measures how two nodes are close to each other in sight of attribute. We define a function $\Gamma(\cdot)$ to denote the set of neighbors and a function $|\cdot|$ to denote the size of a set.

(1) Preference Attachment [6] (PA):

$$PA(u,v) = |\Gamma(u)| \cdot |\Gamma(v)| \qquad (3.2)$$

(2) Common Neighbors [7] (CN):

$$CN(u,v) = |\Gamma(u) \cap \Gamma(v)| \qquad (3.3)$$

(3) Jaccard Coefficient [8] (JC):

$$JC(u,v) = \frac{|\Gamma(u) \cap \Gamma(v)|}{|\Gamma(u) \cup \Gamma(v)|} \qquad (3.4)$$

(4) Adamic-Adar [9] (AA):

$$AA(u,v) = \sum_{w \in \Gamma(u) \cap \Gamma(v)} \frac{1}{log|\Gamma(w)|} \qquad (3.5)$$

(5) Attribute Proximity (AP)

Attribute Proximity is the metric we proposed which measures how close two nodes are in sight of their attributes. It comes from the reality that two people are more likely to interact if they share more common interests. It can be easily computed as we just consider the number of the attributes two nodes both possess as the metric. For a node pair (u, v), if we use $attr_u$ and $attr_v$ to denote node u and node v's attribute sets, this metric can be defined as follows:

$$AP(u, v) = |attr_u \cap attr_v| \tag{3.6}$$

Since keywords are not available in DBLP dataset, we can extract other information from the dataset. For an author in this dataset, the attribute can be the word in his article title or the conference his article is published. The research area and focuses of the author can be easily observed from this information. If two authors have more focuses in common, they will be more likely to cooperate to publish an article in the future.

3.4   Forecast Model

All these features proposed above can be directly applied in supervised link prediction if not consider temporal information. In this section, we introduce some forecast models which take temporal information into consideration.

We here introduce some notations. We use $Z_t(t = 1, \ldots, T)$ to denote a temporal sequence with T observations and $\hat{Z}_t$ to denote the forecast at time t.

(1) Moving Average [10] (MA)

$$\hat{Z}_t = \frac{Z_{t-1} + Z_{t-2} + \ldots + Z_{t-n}}{n} \tag{3.7}$$

(2) Average (Av)

$$\hat{Z}_t = \frac{Z_{t-1} + Z_{t-2} + \ldots + Z_{t-T}}{T} \tag{3.8}$$

(3) Random Walk (RW)

$$\hat{Z}_t = Z_{t-1} \tag{3.9}$$

Except for these models, there are other forecast models such as Linear Regression, Simple Exponential Smoothing, Linear Exponential Smoothing. Due to the expensive time cost of these forecast models, we only consider the first three simple forecast models.

4   Experiments

4.1   Dataset

DBLP is an integrated database system for computer-based English papers that focuses on the author's research results in the computer field. Until 2009, DBLP has integrated 2.1 million papers. It provides search services for scientific literature in computer field, such as title, author, publication date and so on. Unlike the popular situation, DBLP does not use a database but uses XML to store data.

We extract several conference papers with publication dates ranging from 2001 to 2016 in DBLP database. For each paper, the authors are listed in a specific order as $a_1, a_2, \ldots a_k$, and undirected author pairs (a_i, a_j) are generated as collaboration edges $(1 \leq i < j \leq k)$. Due to the fact that a network with n nodes contains $O(n^2)$ possible links [11], we only consider two nodes which are two-hop apart.

4.2 Experiments and Analysis

ROC (Receiver Operating Characteristic) curve and AUC (Area Under Curve) are often used to evaluate the performance of a binary-classifier [12]. In this paper, we use AUC to evaluate the performance of the link prediction. We use two classifiers (SVM and KNN) and three forecast models (Moving Average, Average and Random Walk). We make a comparison between the method which excludes Attribute Proximity and the method which includes Attribute Proximity. The figure of ROC curve with different classifiers and forecast models is shown as follows (Figs. 3, 4, 5, 6 and 7):

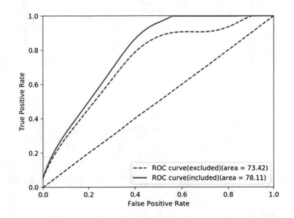

Fig. 2. The ROC curve using SVM and moving average

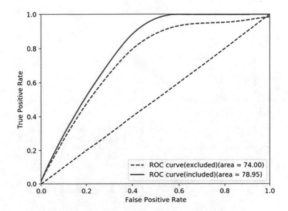

Fig. 3. The ROC curve using SVM and average

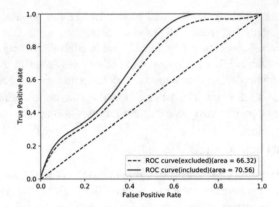

Fig. 4. The ROC curve using SVM and random walk

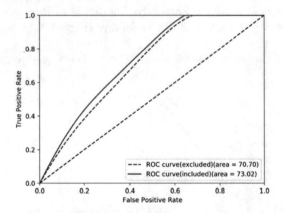

Fig. 5. The ROC curve using KNN and moving average

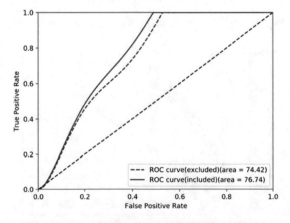

Fig. 6. The ROC curve using KNN and average

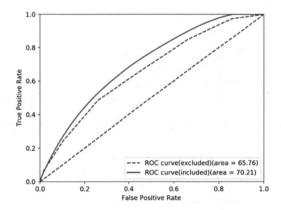

Fig. 7. The ROC curve using KNN and random walk

These figures above show that the performance of our method is better than the method which does not take attribute proximity into consideration. We summarize the AUC values using different classifiers and forecast models in Table 1 below:

Table 1. The AUC value using different classifiers and forecast models.

Models classifiers	Forecast		
	Moving average	Average	Random walk
SVM	0.7811(0.7342)	0.7895(0.7400)	0.7056(0.6632)
KNN	0.7302(0.7070)	0.7674(0.7442)	0.7021(0.6576)

From the table above, we can see that the performance of classifier SVM is better than KNN regardless of the forecast model it used. The performance of forecast model Average is better than other forecast models regardless of the classifier it used.

5   Conclusions

In this paper, we study the link prediction as a binary-classification problem. We first propose a similarity feature called Attribute Proximity. Except for this similarity index, we also introduce four well-known neighborhood based features. We extract four topological features along with the feature we proposed and then apply a forecast model to build the final feature set. Then we use the feature set to train different classifiers. Finally, we make comparisons in experiments using different feature set (exclude or include the feature we proposed), different classifiers and different forecast models. The results show that our method provides higher accuracy than those which don't take node attribute into consideration.

Acknowledgement. This work is supported by National Natural Science Foundation of China (No. 61502246), NUPTSF (No. NY215019).

References

1. Liben-Nowell, D., Kleinberg, J.: The Link-Prediction Problem for Social Networks, vol. 58, pp. 1019–1031. Wiley, New York (2007)
2. Yu, H.Y., Braun, P., Yildirim, M.A., Lemmens, I., Venkatesan, K.: High-quality binary protein interaction map of the yeast interactome network. Science **322**(5898), 104–110 (2008)
3. Schafer, J.B., Konstan, J.A., Riedl, J.: E-commerce recommendation applications. Data Min. Knowl. Disc. **5**(1–2), 115–153 (2001)
4. Hasan, M.A.: Link prediction using supervised learning. In: Proceedings of SDM Workshop on Link Analysis Counterterrorism & Security, vol. 30(9), pp. 798–805 (2006)
5. Soares, P.R.D.S., Prudêncio, R.B.C.: Time series based link prediction. In: International Joint Conference on Neural Networks, vol. 20, pp. 1–7. IEEE (2012)
6. Barabási, A.L., Jeong, H., Néda, Z., Ravasz, E., Schubert, A., Vicsek, T.: Evolution of the social network of scientific collaborations. Phys. A **311**(3), 590–614 (2001)
7. Newman, M.E.J.: Clustering and preferential attachment in growing networks. Phys. Rev. E **64**(2), 025102 (2001)
8. Salton, G., Mcgill, M.J.: Introduction to Modern Information Retrieval, vol. 41, pp. 305–306. McGraw-Hill (1983)
9. Adamic, L.A., Adar, E.: Friends and neighbors on the Web. Soc. Netw. **25**(3), 211–230 (2003)
10. Makridakis, S.G., Wheelwright, S.C.: Forecasting Methods for Management, vol. 15, pp. 345–349. Wiley (1985)
11. Duan, L., Aggarwal, C., Ma, S., Hu, R., Huai, J.: Scaling up link prediction with ensembles. In: ACM International Conference on Web Search and Data Mining, pp. 367–376. ACM (2016)
12. Fawcett, T.: An introduction to ROC analysis. Pattern Recogn. Lett. **27**(8), 861–874 (2006)

Web Services Tagging Method Based on Weighted Textual Matrix Factorization

Guoqiang Li$^{(\boxtimes)}$, Yifeng Cui$^{(\boxtimes)}$, Haifeng Wang, Shunbo Hu,
and Li Liu

School of Information Science and Engineering, Linyi University,
Linyi City 276005, China
lgq_2005@126.com, cuiyifeng@lyu.edu.cn

Abstract. Web services are important technique basis of the Service-oriented Architecture. Services discovery is the prior to use the services which are published on the internet accurately. Tagging technique is widely used to assist the searching of service currently. To solve the time-consuming and error-prone problem in manual tagging, we propose a novel automatic approach to tag web services in this paper. There are two steps consisting of WSDL (Web Services Description Language) documents extracting and tag recommendation using the weighted textual matrix factorization. Experiments on real dataset are conducted and the results prove that our approach is effective.

Keywords: Service-oriented Architecture · Services tagging ·
Weighted textual matrix factorization · Services discovery

1 Introduction

Service-oriented computing (SOC) is a new computing model, which aims to construct a rapid, low-cost reliable business application, where web services as components are composed to meet the real requirement. So, a problem needed to be solved is how to select right services from a huge services candidate set.

To implement the services selection or discovery, one method is to make use of the semantic information from the services description, e.g. WSDL document which only provides limit information. So, in recent years, tag information are introduced as an important semantic information source, which becomes a hot research topic accompanied with the emergence of web 2.0. Considering the time-consuming and error-prone problem to tag the services manually, automatic tagging should be studied.

In this paper, we firstly propose to extract a set of terms from a WSDL description and tag description at the same time. We use the TF-IDF model to construct the occurring matrix of terms, which are only used step by step in other works, e.g. [1]. Then we use WTMF (weighted textual matrix factorization) to determine the similar services to enrich the candidate tags for a service. At last, we conduct experiments to validate our approach.

© Springer Nature Switzerland AG 2020
Q. Liu et al. (Eds.): CENet 2018, AISC 905, pp. 385–390, 2020.
https://doi.org/10.1007/978-3-030-14680-1_42

The rest of this paper is organized as follows. Section 2 gives the related work about the tagging method for web services. Section 3 details our architecture to tag enriching. Section 4 validates our proposed algorithm. Section 5 concludes our work and presents the future work.

2   Related Work

A novel multi-label active learning approach is proposed for web service tag recommendation. This approach is able to identify a small number of most informative web services tagged by domain experts. Active learning process is improved by learning and leveraging the correlations among tags [1]. To realize the complete and refinement of tags, matrix completion approach is proposed which not only models the low-rank property of service-tag matrix, but also simultaneously integrates the content correlation consistency and the tag correlation consistency to ensure the correct correspondence between web services and tags [2]. To facilitate the process of API management, a graph-based recommendation approach is proposed to automatically assign tags to unlabeled APIs by exploiting both graph structure information and semantic similarity. It first leverages the multi-type relations to construct a heterogeneous network. A random walk with restart model is applied to alleviate the cold start problem where no API has ever been tagged [3]. Two tagging strategies are proposed: tag enriching and tag extraction, the first step is to cluster web services using WSDL documents, and the second step is that tags are enriched for a service with the tags of in the same cluster. Tags are extracted from WSDL documents and related descriptions [4]. To address tag relevancy issue, three parameters are recommended to each tag: score, popularity and occurrence. WordNet dictionary is used to take into account the synonyms of the tags in the service search [5]. Voting and sum strategies are proposed to compute the candidate tags to realize recommendation [6]. Text mining and machine learning are adapted to extract tags from WSDL descriptions, and then these tags are enriched by extracting relevant synonyms using Word net [7]. Carrot search clustering and k-means are used to group similar services to generate tags [8].

3   Architecture

3.1   Preprocessing

As we know, WSDL documents, which are constructed using an xml style, describe the services' functionality. Several important elements include: type, message and port. Tag information occurs also in many service webs, e.g. Seekda.

In this paper, we extract these information from the web services documents using information retrieval method same as the work [7]. So we can get the whole words occurred in the documents, and its number is defined N.

3.2 TF-IDF Model

In our scenario, all the WSDL documents whose number is M can be seen as the corpus. In information retrieval, TF-IDF short for term frequency–inverse document frequency is a numerical statistic that is intended to reflect how important a word is to a document in a collection or corpus. Two terms are included in the TF-IDF weight: TF and IDF. TF stands for term frequency, presenting the number of times a word appears in a document. IDF stands for the inverse document frequency, and it diminishes the weight of terms that occur very frequently in the document set and increases the weight of terms that occur rarely. After the TF-IDF values are computed, we can get an N*M matrix X. The row entries of the matrix are the M words. The columns are the service ids. Each X_{ij} of X is the TF-IDF value, namely that TF-IDF value of word w_i in service s_j. For D documents, a term t, a documents $d \in D$, $f_{t,d}$ is defined as raw count of a term in a document, we using the following equations to compute these two weights:

$$\mathbf{tf}(t,d) = log\left(1 + f_{t,d}\right) \tag{1}$$

$$\mathbf{idf}(t,D) = log\left(1 + \frac{M}{m_t}\right) \tag{2}$$

where $m_t = |d \in D : m \in d|$, So the value of TF-IDF is tf(t, d) * idf(t, D).

3.3 WTMF Model

WTMF is designed by Weiwei Guo. The key idea of this model is the semantic space of both the observed and missing words in the sentence makes up the complete sematic profile of a sentence and missing words can help to prune the hypotheses which related to the missing words [9]. In WTMF, the matrix $X(N*M)$ will be factorized into two matrices such that $X \approx P^T Q$, where P is a K*M matrix, and Q is a K*N matrix. In our paper, we use the same objective function equation as [9] to estimate the latent vectors of words P sentences Q as follows:

$$\sum_{i=1,\dots,n} \sum_{i=1,\dots,m} w_{ij}\left(P_i Q_j - X_{i,j}\right)^2 + lambda\|P\|^2 + lambda\|Q\|^2 \tag{3}$$

Where w_m is the weight for missing words cells, and

$$W_{ij} = \begin{cases} 1, & if\ X_{ij} \neq 0 \\ w_m, & if\ X_{ij} = 0 \end{cases},$$

we also set the w_m = 0.01 and lambda = 20 same as [9].

Similarly, we use the same iterative formula to update the matrix P and Q same as [9].

$$P_i = \left(QW^iQ^T + lambad \times I\right)^{-1}QW^iX_i^T, Q_i = \left(PW^iP^T + lambad \times I\right)^{-1}PW^iX_j.$$

We use the same method to finish the experiment as [9]. After the running of WTMF, the similarity of two services s_j and s_k can be computed by the inner product of Q_j and Q_k.

4   Experiments

To evaluate the performance of web service tagging approaches, we use the data about 15968 real web services from web service search engine Seeka same as [6].

4.1   Metrics

In the evaluation, we used precision and recall as metrics. For each service $s \in T$, where T is the training corpus. We introduce the metrics from [7] as follows:

$$precision(s) = \frac{|I|}{|A|}, recall(s) = \frac{|I|}{|E|}$$

$$precision(T) = \frac{\sum precision(s)}{|T|}, recall(T) = \frac{\sum recall(s)}{|T|},$$

where T is the training corpus, A is the set of tags produced by the proposed method, different from [7], we determine A by getting some services with the similarities making use of the WTMF. M is the tags given by Seekda users and W is the set of words appearing in the WSDL, $I = A \cap M, E = M \cap W$.

4.2   Comparison and Analysis

Currently, we compare with the tag enriching method based K-means. In our experiments, we conduct the K-means with the value of K equaling to 4. Eighty percent of data are used trained and twenty percent are to be tested. To find the similar services, we set the similarity threshold value 0.75. From the following figures, we can see that the precision and recall achieved by WTMF are better than K-means. Especially, for the precision, the result of WTMF is about 92%, and there is about 3% improvement for the k-means. After analysis, the reason is that WTMF find the better similar services than the K-means which introduces some irrelevant tags by clustering some irrelevant services (Figs. 1 and 2).

The running time of our algorithm is concentrated in the matrix decomposition process, and the running time of K-MEANS is mainly concentrated in the clustering process. Because the data set used in this paper is smaller, the running time of the two algorithms is approximately equal. The more running efficiency of WTMF may be referred to the work [9].

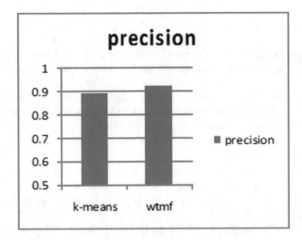

Fig. 1. Precision result.

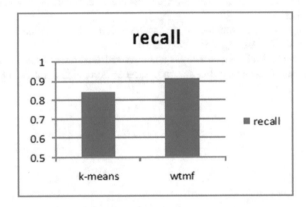

Fig. 2. Recall result.

5   Conclusion

In this paper, we use WTMF to enrich the tags of web services. As an important precondition, we preprocess the WSDL documents to gain all information including tag, which are model in a weighted matrix at the same time. Then we conduct experiments to verify its validity. In the future, we will compare with more methods and improve our method.

Acknowledgement. This work was supported by Ph.D. research foundation of Linyi University LYDX2016BS077, LYDX2015BS022), Project of Shandong Province Higher Educational Science and technology program (No. J17KA049, J13LN84), Shandong Provincial Natural Science Foundation (No. ZR2015FL032, ZR2016FM40, ZR2017MF050), and partly supported by National Natural Science Foundation of China (No: 61771230).

References

1. Shi, W., Liu, X., Yu, Q.: Correlation-aware multi-label active learning for web service tag recommendation. In: Proceeding of International Conference on Web Services, pp. 229–236 (2017)
2. Chen, L., Yang, G., Chen, Z., Xiao, F.: Tag completion and re finement for web service via low-rank matrix completion. In: Proceeding of PAKDD, pp. 269–278 (2014)
3. Liang, T., Chen, L., Wu, J., Bouguettaya, A.: Exploiting heterogeneous information for tag recommendation in API management. In: Proceeding of International Conference on Web Services, pp. 436–443 (2016)
4. Fang, L., Wang, L., Li, M., Zhao, J., Zou, Y., Shao, L.: Towards automatic tagging for web services. In: Proceeding of International Conference on Web Services, pp. 528–535 (2012)
5. Sellami, S., Becha, H.: Web services discovery based on semantic tag. In: Proceeding of OTM Conference, pp. 465–472 (2015)
6. Chen, L., Hu, L., Zheng, Z., Wu, J., Yin, J.: WTCluster: utilizing tags for web services clustering, vol. 7048, pp. 204–218. Springer Berlin Heidelberg (2011)
7. Azmeh, Z., Falleri, J., Huchard, M., Tibermacine, C.: Automatic web service tagging using machine learning and wordnet synsets. In: Lecture Notes in Business Information Processing, pp. 46–59 (2011)
8. Lin, M., Cheung, D.W.: Automatic tagging web services using machine learning techniques. In: Proceeding of WI-IAT, pp. 258–265 (2014)
9. Guo, W., Diab, M., Wei, W.: A simple unsupervised latent semantics based approach for sentence similarity. In: Joint Conference on Lexical & Computational Semantics, pp. 586–590 (2012)

Soft Frequency Reuse of Joint LTE and eMTC Based on Improved Genetic Algorithm

Jingya Ma[✉], Jiao Wang, and Dongbo Tian

Nari Group Corporation, Nanjing, China
majingya6613@163.com

Abstract. Enhanced machine-type (eMTC) communications for cellular Internet of Things (IoT) are expected to make up the shortcomings of traditional LTE technologies. In this paper, LTE and eMTC are jointly used in the cell, and the frequency is efficiently used through soft frequency reuse (SFR). After that, a 0-1 knapsack problem with the purpose of maximize access ratio was constructed. We construct an improved genetic algorithm to solve the problem. The simulation results show that the joint LTE and eMTC methods increase the access ratio by 8%, and the improved genetic algorithm also has a mostly 5% improvement compared to the traditional dynamic programming method.

Keywords: Soft frequency reuse · eMTC · Genetic algorithm · 0-1 knapsack problem

1 Introduction

The Internet of Things (IoT) is regarded as a technology and economic wave in the global information industry after the Internet. With the increasing of IoT applications and equipment, base stations (BSs) need to have higher access capabilities and greater coverage capabilities. The enhanced machine-type communication (eMTC) is one of the most promising technologies that fits the requirements of IoT. The eMTC can be directly upgraded based on the existing LTE network and can be used together with the existing LTE BSs. Therefore, eMTC can implement low-cost rapid deployment and can be used to supplement the deficiencies of LTE technology.

Soft frequency reuse (SFR) is a spectrum reuse method proposed to increase spectrum efficiency and reduce inter-cell interference (ICI) in LTE/LTE-A networks. In the SFR, the subcarriers are divided into a cell center group and a cell edge group. The ICI is controlled by setting different power levels for cell center and cell edge subcarrier groups used by users in the cell center and cell edge regions, respectively [1]. At present, there are many research results for the SFR solution in the LTE network, and various parameters of the SFR solution are analyzed. In [2, 3], the authors optimized the division of the center and the edge of the cell. In [4], the authors optimized the transmit power over the system bandwidth. In [5, 6], joint power allocation algorithms and self-organizing bandwidth allocation schemes were proposed. Unfortunately, none of the existing literature considers soft frequency reuse of both LTE and eMTC protocols.

© Springer Nature Switzerland AG 2020
Q. Liu et al. (Eds.): CENet 2018, AISC 905, pp. 391–398, 2020.
https://doi.org/10.1007/978-3-030-14680-1_43

This article considers using both LTE and eMTC protocols simultaneously in a cell. LTE provides high-speed, reliable connections for IoT equipment in the cell center regions, and eMTC provides stronger access and coverage capabilities for IoT equipment in the cell edge regions. LTE and eMTC use SFR technology to increase the utilization of spectrum and increase the capacity of the cell. The spectrum within the cell is divided into three parts. The cell center uses two parts, and the cell edge uses a part. The center cell is interfered by all areas of the neighboring cell, and the cell edge is only interfered by the neighboring center cell. Without loss of generality, this paper can characterize the problems of the entire network by studying the problems in a cell.

Due to system bandwidth limitations, the system wants to satisfy as many IoT equipment accesses as possible. Therefore, in this paper, the access ratio of active IoT equipment in the cell is adopted as the optimization target. We model this problem as a classic 0/1 knapsack problem, which is an NP-hard problem. At present, there are many research results on this issue, such as the design of traditional methods, heuristic methods, and intelligent calculation methods. The dynamic programming method analyzes the optimal substructure properties of the optimal solution of the problem and establishes the optimal recurrence relation. Intelligent calculation methods such as genetic algorithm, artificial fish algorithm, simulated annealing algorithm, particle swarm optimization algorithm, immune algorithm and so on.

The rest of this article is organized as follows. The Sect. 2 introduces the cellular network scenario of joint LTE and eMTC using SFR, and models the problem as a 0/1 knapsack problem. The Sect. 3 proposes the solution algorithm of this paper. In Sect. 4, the performance of the SFR is analyzed using the solution algorithm proposed in this paper to verify its feasibility. The fifth section summarizes the full text.

2   System Model

Ordinary LTE protocol has been difficult to satisfy large-scale IoT equipment to access. This paper considers using the eMTC technology to supplement the disadvantages of weaker coverage and weak access capabilities in the LTE protocol.

This article considers using the LTE protocol in the cell center region and the eMTC protocol at the cell edge region. Considering the sparseness of IoT equipment in the edge region, LTE uses a wider subcarriers and lower transmit power relative to the eMTC. The cell edge region uses the allocated subcarriers and a higher transmit power. The way of dividing the system bandwidth of LTE and eMTC and the way of deployment are shown in Fig. 1.

$$R_j = B_j^{edge} \cdot \log\left(1 + \frac{g_j^{edge} \cdot p^{edge}}{ICI_j^{edge} + N_0 \cdot B_j^{edge}}\right). \tag{2.1}$$

Where B_i^{center} indicates the subcarriers used by the i-th center-IoT, and B_j^{edge} indicates the subcarriers used by the j-th edge-IoT. In addition, p^{center} and p^{edge} represent the transmit power of the cell center regions and cell edge regions, respectively. The g_i^{center} indicates the channel gain of the BS to the i-th center-IoT. And, ICI_i^{center} and

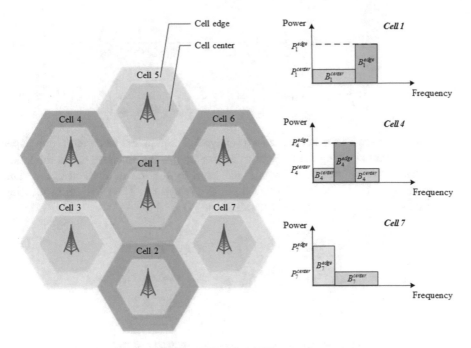

Fig. 1. LTE and eMTC scenario.

ICI_j^{edge} represent the inter-cell interference received at the i-th center-IoT and the j-th edge-IoT, respectively.

$$ICI_i^{center} = \sum_{k \in C} g_{k,i}^{edge} \cdot p^{edge} + \sum_{k \in C} g_{k,i}^{center} \cdot p^{center} \qquad (2.2)$$

$$ICI_j^{edge} = \sum_{k \in C} g_{k,j}^{center} \cdot p^{center} \qquad (2.3)$$

Where, C denotes a set of neighboring cells. The $g_{k,i}^{edge}$ denotes the channel gain that the edge region of k-th BS to the i-th center-IoT, and $g_{k,j}^{center}$ denotes the channel gain that the edge region of k-th BS to the j-th center-IoT.

Because $ICI_i^{center} \gg N_0 \cdot B_i^{center}$ and $ICI_j^{edge} \gg N_0 \cdot B_j^{edge}$. In order to solve this problem, this paper simplifies the calculation of noise power as follows:

$$N_0 \cdot B_i^{center} \approx N_0 \cdot B^{center} / \tilde{M} \qquad (2.4)$$

$$N_0 \cdot B_j^{edge} \approx N_0 \cdot B^{edge} / \tilde{N} \qquad (2.5)$$

Where $\tilde{M}$ and $\tilde{N}$ are the average number of active equipment calculated based on the general IoT equipment's rate requirements. According to the above formula, we can obtain the required subcarrier length for each active IoT device.

$$B_i^{center} = R_i \cdot \log^{-1}\left(1 + \frac{g_i^{center} \cdot p^{center}}{ICI_i^{center} + N_0 \cdot B^{center}/\tilde{M}}\right) \quad (2.6)$$

$$B_j^{edge} = R_j \cdot \log^{-1}\left(1 + \frac{g_j^{edge} \cdot p^{edge}}{ICI_j^{edge} + N_0 \cdot B^{edge}/\tilde{N}}\right) \quad (2.7)$$

In order to increase the access ratio of active IoT equipment in the cell, this paper sets the goal of maximizing the number of access equipment. Because of the limitation of the bandwidth, the sum of the subcarriers used by the IoT equipment at a certain moment cannot exceed the bandwidth length of the system. Therefore, the following objective functions and constraints are obtained.

$$max_{X,Y} \sum_{i=1}^{M} \omega_i^{center} \cdot x_i + \sum_{j=1}^{N} \omega_j^{edge} \cdot y_j \quad (2.8)$$

$$\text{s.t.} \quad \sum_{i=1}^{M} x_i \cdot B_i^{center} \leq B^{center}$$
$$\sum_{j=1}^{N} y_j \cdot B_j^{edge} \leq B^{edge} \quad (2.9)$$

Where ω_i^{center} and ω_j^{edge} represent the priorities of different IoT equipment, and the larger the number, the higher the priority. For general problems, we think that different terminals have the same priority. In this case, ω_i^{center} and ω_j^{edge} are the vectors whose elements are one. $x_i, y_j \in (0, 1)$ denotes whether the i-th center-IoT is accessed and the j-th edge-IoT is accessed, 1 represents the access, and 0 represents the non-access.

3   Improved Genetic Algorithm

3.1   Dynamic Programming

This article describes the solution of the 0/1 knapsack problem as a vector $(x_1, \ldots, x_n)$, so the problem can be seen as a decision of an n-variable 0/1 vector. The decision for any one of the classifications x_t is $x_t = 1$ or $x_t = 0$. After x_{t-1} is decided, the sequence $(x_1, \ldots, x_{t-1})$ has been determined. When deciding on x_t, the problem is in one of the following two states: (i) The backpack capacity is insufficient to load the object t, then $x_t = 0$, and the total value of the backpack does not increase. (ii) The backpack capacity is sufficient to hold the object t, then $x_t = 0$, and the total value of the backpack increases by v_t. In both cases, the most valuable person to put in a backpack should be the value after x_t decision. The pseudo code of the dynamic specification method is given below.

3.2   Genetic Algorithm

In order to solve the 0/1 knapsack problem, this paper improves the traditional genetic algorithm. In order to satisfy the constraints of the problem, this paper modified the cross-operation and mutation operation in the genetic algorithm. We use a multi-point crossover method to randomly set multiple crossover points in a maternal pair's gene string. The mutation operation in the algorithm uses single-point mutation.

In a word, the pseudo-code of the improved genetic algorithm is as follows.

Algorithm 1. Improved Genetic Algorithm

INPUT: Two sequences $v = (v_1, ..., v_n)$ and $\omega = (\omega_1, ..., \omega_n)$ the number of items n and the maximum weight W.

OUTPUT: The optimal of the knapsack and the sequence ω_{final}.

1 Dynamic programming was used to generate the initial population, and population size is Q.

2 Determine the iteration number k, the crossover probability p_c, and the mutation probability p_m. $it \leftarrow 1$.

3 **while** $it < MAX\_ITERATION$

4   Use Algorithm 2 for crossover.

5   Use Algorithm 3 for mutation.

6    $it \leftarrow it + 1$.

7 **end while**

---

4 Simulations

For ease of analysis, the number of active IoT equipment in the simulation scenario is 500. The number of center-IoT is 300, and the number of edge-IoT is 200. The other simulation parameters are shown in Table 1.

Table 1. Simulation parameters.

Parameter	Value	Parameter	Value
Cell center area	0 ~ 250 m	Transmit power of cell center	30 dBm
Cell edge area	250 ~ 300 m	Transmit power of cell edge	35 dBm
System bandwidth	5 MHz	Noise power	−174 dBm
Bandwidth used by cell center	3.5 MHz	LTE path loss model	$143 + 35 \cdot 1\ g(d[km])$
Bandwidth used by cell edge	1.5 MHz	eMTC path loss model	$136 + 34 \cdot 1\ g(d[km])$

As shown in Fig. 2, as the occupancy ratio of the bandwidth of the center cell increases, the access ratio of the center cell increases almost linearly. The access ratio of the cell edge regions decreases slowly when the bandwidth ratio is less than 0.5, and then decreases rapidly. The maximum of total access ratio is 0.34, when bandwidth ratio is 0.5.

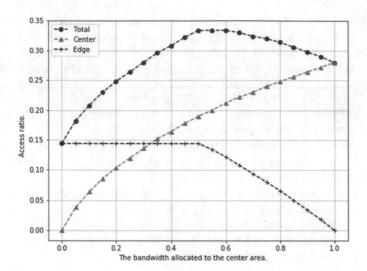

Fig. 2. Center region's bandwidth occupation ratio *vs.* access rate.

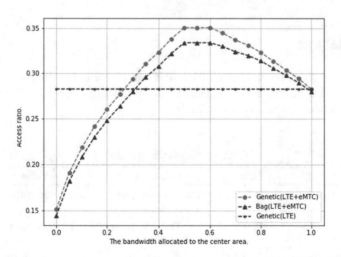

Fig. 3. Center region's bandwidth occupation ratio *vs.* access ratio of different algorithms.

Figure 3 compares the performance of the dynamic programming algorithm and the improved genetic algorithm. According to the simulation calculations, our algorithm improves 3% to 5% compared to the dynamic programming algorithm. Figure 4 also compares the case of using only the LTE protocol in the cell. It can be seen that the method of combining LTE and eMTC has improved 7% to 8% compared to the method using LTE.

As can be seen from Fig. 4, as the transmit power of the BS to the center-IoT increases, the access ratio of the cell center regions start to rise rapidly. When the access ratio reaches 31%, the access ratio starts to converge. In addition, the transmit

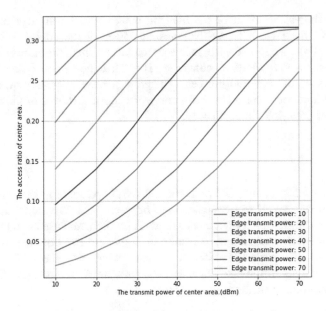

Fig. 4. Transmit power of center regions vs. access ratio of center regions.

power of the BS to the cell edge region also has a certain influence on the access ratio of the cell center region. As the transmit power for the cell edge region rises, the interference to the cell center region increases, and the center-IoT needs higher transmit power to achieve a similar access ratio.

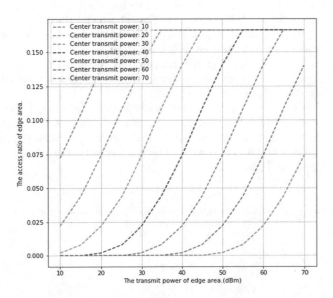

Fig. 5. Transmit power of cell edge regions vs access ratio of cell edge regions.

Figure 5 shows the effect of the transmit power of the cell edge regions on the access ratio of the edge-IoT. As the transmit power of the edge-IoT increases, the access ratio of the edge-IoT also increases rapidly. When the access ratio reaches 17.5%, the access ratio starts to converge. The cell edge region is interfered by the center regions of the neighboring cell. Therefore, as the transmit power of the cell center region increases, the access ratio curve of edge-IoT moves to the right, and the access ratio of edge-IoT drops sharply.

5 Conclusion

This article considers using LTE protocol and eMTC protocol in the cell at the same time, and then using SFR to frequency reuse between cells. We model the problem of maximizing access ratio as a classic 0/1 knapsack problem and use an improved genetic algorithm to solve it. The simulation results show that our solution increases the access ratio by 7% to 8% compared with the scheme using only the LTE protocol. The improved genetic algorithm also has a 3–5% improvement compared to the traditional dynamic programming method.

Acknowledgement. This study is supported by 2017 state grid science and technology project "Research and Application for Adaption of Power Business based on LTE Wireless Private Network".

References

1. Qian, M., Hardjawana, W., Li, Y., Vucetic, B.: Adaptive soft frequency reuse scheme for wireless cellular networks. IEEE Trans. Veh. Technol. **64**(1), 118–131 (2015)
2. Assaad, M.: Optimal fractional frequency reuse (FFR) in multicellular OFDMA system. In: Vehicular Technology Conference, pp. 1–5. IEEE (2008)
3. Hassan, N.U.L., Assaad, M.: Optimal fractional frequency reuse (FFR) and resource allocation in multiuser OFDMA system. In: International Conference on Information and Communication Technologies, pp. 88–92. IEEE (2009)
4. Bohge, M., Gross, J., Wolisz, A.: Optimal power masking in soft frequency reuse based OFDMA networks. In: Wireless Conference, pp. 162–166. IEEE, European (2009)
5. Rengarajan, B., Stolyar, A.L., Viswanathan, H.: Self-organizing dynamic fractional frequency reuse on the uplink of OFDMA systems. In: 44th Annual Conference on Information Sciences and Systems, pp. 1–6. IEEE (2010)

Mobile Internet Application Classification Based on Machine Learning

Yangqun Li[(✉)]

College of Internet of Things, Nanjing University of Posts
and Telecommunications, Nanjing 210003, Jiangsu, China
yqli@njupt.edu.cn

Abstract. Service classification technology helps service suppliers understand how to use network services that offer personalized services to users. In this study, machine learning technology is used to classify such services, called mobile Internet applications (APPs). Firstly traffic of different application is collected and is used to build model by using C4.5 and SVM algorithm. Secondly, the model is used to classify the application type from the mobile internet traffic. Then traffic of applications is merged as major class by application type such as web browsing, e-commerce and the above two algorithms are applied to classify the major class that different applications belong to. Finally the Precision and Recall ration of two algorithms are compared and analyzed. Mobile stream video/audio is better recognized than Mobile app using http by using machine learning method. In order to improve the mobile application classification result, the only machine learning algorithm is not enough.

Keywords: Service classification · Machine learning · C4.5 · SVM ·
Mobile app

1 Introduction

As smartphones and mobile communication technology has developed, mobile Internet technology has rapidly increased in popularity. End users use the mobile Internet and its available APPs via mobile terminals, causing large mobile Internet management challenges for service suppliers. Due to the diversity of APPs and, consequently, their differences in network traffic usage, it is difficult for network operators to obtain detailed information concerning mobile Internet usage; thus, they are unable to provide users with personalized services while simultaneously maintaining fine-grained network management. Under this circumstance, being able to identify the types of APPs being used on the mobile Internet is critical. Based on this information, operators could better provide personalized and customized services for these APPs. At present, a variety of methods exist to identify various Internet traffic types. These methods can be divided into methods based on network protocol characteristics and machine learning methods based on statistical traffic characteristics.

The methods based on network protocol characteristics identify applications using information such as their port number, IP address, or characteristics of their network payload fields. However, there is no guarantee that these ports are exclusive to the

© Springer Nature Switzerland AG 2020
Q. Liu et al. (Eds.): CENet 2018, AISC 905, pp. 399–408, 2020.
https://doi.org/10.1007/978-3-030-14680-1_44

abovementioned APPs. In fact, these ports are commonly used by many different APPs. At the same time, APPs such as WeChat may support several service types, including voice, video, games, messages, and so on. These different service types cannot be identified accurately because within a given APP, all the service types may use the same IP address and port number.

2 Related Work

2.1 DPI Technology

Based on the definition in the ITU-T standard, the purpose of DPI technology [1, ITU-T Y.2770] is to analyze the header information in the network protocol layers (L2/L3/L4), the payload content, and the message attributes based on the network protocol hierarchy to perform application category classification. DPI technology requires an analysis of the features of the network data payloads produced by the applications. The key of this method lies in the selection of features and the pattern matching algorithms. Currently, DPI technology is already used in the classification of many types of APPs, for example, the classification of peer-to-peer (P2P) traffic, including Skype, BitTorrent, Network TV and Thunder downloads [2–4], and the Voice over Internet Protocol (VoIP) [5]. The statistical features of network traffic were analyzed manually to distinguish between the HTTP and P2P protocols based on the features in [6].

2.2 Traffic Classification Technology Based on Machine Learning

Machine learning is a process that simulates the learning process of humans, who reason based on incomplete information to discover new rules or reach conclusions [7]. Many scholars have applied machine learning approaches to classify Internet application traffic, and some advancements have been made. Using the C4.5 decision tree algorithm, Li et al. [8] classified network traffic by collecting 12 features from the network flows. The scheme considers not only accuracy but also latency and system throughput, and it is able to classify traffic for several major types of Internet applications, e.g. HTTP, FTP, P2P, multimedia and games. Moreover, their approach is semi-automatic and includes three main stages: Feature selection, classification model construction, and traffic classification. In [9, 10], an unsupervised Bayesian network, a naïve Bayesian network and a C4.5 decision tree were employed to classify network traffic, and the effects of different feature on the results were demonstrated. Currently, traffic classification based on machine learning is not intended to inspect network packets, but to compare the statistical features of the network traffic itself. Based on the imbalanced aspect of network flows, Zhang [11] used a decision tree, an SVM and Bayesian algorithms to classify network traffic and were able to classify traffic types from some common Internet APPs. They also adopted the Bagging algorithm, an integrated machine learning method for classification.

Machine learning approaches can classify network traffic based on statistical traffic features instead of depending on network packet payload features. Therefore, they provide the basis for automatic network traffic classification without human intervention because they can dynamically adjust to accommodate changes to or newly deployed network APPs.

3 Machine Learning Algorithms

3.1 Traffic Classification Based on the Decision Tree Algorithm

The decision tree algorithm is a supervised learning algorithm that constructs a tree through sample data training. The tree consists of nodes and branches, where each node represents a test on some data attribute, and the output branch of the node represents the test conditions. The data attribute and test condition is compared; when they satisfy the requirement, the next node on this branch will continue to be compared. This process is repeated until a leaf node is reached. Then, the leaf node represents the data classification result. The C4.5 algorithm builds a decision tree based on the division of the gain ratio of the information entropy of the data attributes. Suppose that there are n categories in traffic training sample set S: $C_i(i = 1 ..., n)$. In training sample set S, the expected information (namely, information entropy) needed for classification is shown below:

$$H(S) = - \sum_{i=1}^{n} p_i log_2 p_i \qquad (1)$$

where p_i is the probability that the traffic sample in sample training set S falls into category C_i.

Consider that attribute T of the traffic sample has been chosen to split the tree and assume that the attribute T has m possible choices that divide the attribute into m subsets S_k ($k = 1, 2 ... m$). The information entropy required to divide this attribute by this classification set is shown in Eq. (2):

$$H_T(S) = - \sum_{k=1}^{m} \frac{|S_k|}{|S|} \times H(D_k) \qquad (2)$$

Given the above two equations, the information gain from this division can be derived and as follows:

$$Gain(T) = H(S) - H_T(S) \qquad (3)$$

The above information gain can be standardized as

$$SplitH_T(S) = - \sum_{k=1}^{m} \frac{|S_k|}{|S|} \times log_2 \frac{|S_k|}{|S|} \qquad (4)$$

According to the Eq. (2), the formula for calculating the information gain is

$$GainRation = \frac{Gain(T)}{SplitH(T)} \qquad (5)$$

Therefore, different attributes can be selected based on the size of the information gain to construct a decision tree.

3.2   Traffic Classification Based on Support Vector Machine Algorithm

The support vector machine (SVM) algorithm can be effectively applied to categorize high-dimensional data. The SVM adopts the maximal-margin hyperplane approach to search the data set and categorize the data.

The SVM algorithm categorizes data as follows:

$$f(x) = sign\left(\sum_{i=1}^{n} w_i y_i K(x_i, x) + b\right) \tag{6}$$

where w_i, b is the hyperplane parameter; the value of y_i, which represents the sample category, is −1 or 1; x_i is the sample; x is the test data; f(x) represents the SVM-identified result of data x in the sample set $x_i (i = 1, 2 \ldots n)$, where the value is 1 or −1; and $K(x_i, x)$ is a kernel function that can be selected based on different functional requirements.

4   Experiment Scheme and Results

4.1   Data Acquisition and Tool Selection

To construct the classification model for different APPs, the traffic related to a single APP should be acquired. For this test, we captured the network traffic from various applications, including web browsing, online gaming, news and entertainment, etc. During data acquisition, only one application was running on the mobile terminal. The results of the sample data acquisition are shown in Table 1. The data in Table 1 are merged by application type. For example, microblogs, news and shopping are categorized into Web Browsing, as shown in Table 2. The Weka 3.6.12 package applied here contains several machine learning algorithms, providing tools like data preprocessing, classification and visualization.

Table 1. Traffic acquisition for each application.

Source	Application type	Traffic (KB)	Number of flows	Proportion to total number of flows
Ifeng news	HTTP	36162	400	16.89903
Sina Weibo	HTTP	25568	270	11.40684
Sohu video	Video	211706	369	15.58935
QQGame	Game	2739	321	13.56147
Suning YiGou	HTTP	7150	400	16.89903
People	HTTP	91347	238	10.05492
Kugo radio	Audio	8497	369	15.58935

Table 2. Traffic data after merging.

Application type	Percentage
Web	54%
Video	16%
Audio	16%
Game	14%

4.2   Feature Selection

Currently there are several hundred features that can be applied for flows, several dozen of which are commonly applied, as shown in Table 3. The traffic sample data is captured into .pcap files. Then, the Netmate tool is used to obtain the statistical attributes from the pcap files listed in Table 3.

Table 3. Statistical features of flow.

Feature num	Feature name
1	Minimum forward packet length
2	Mean average forward packet length
3	Forward packet length variance
4	Minimum length of a backward packet
5	Mean backward packet length
6	The maximum backward packet length
7	Backward packet length variance
8	Forward packets minimum interval
9	Mean forward packets interval
10	Forward packets maximum interval
11	Forward packets interval variance
12	Backward packets minimum interval
13	Backward packets mean interval
14	Backward packets maximum interval
15	Backward packets interval variance
16	Flow duration
17	Forward packets total number
18	Forward packets bytes

4.3   Evaluation Model

In classification based on machine learning method, recall ratio and precision are generally used as metrics to judge the result. These two metrics are calculated as follows:

$$Recall = \frac{TP}{TP + FN} \tag{7}$$

$$Precision = \frac{TP}{TP + FP} \tag{8}$$

where TP (true positive) is the number of correctly classified samples in the traffic sample set, and FN (false negative) is the number of samples incorrectly classified as another type. FP (False Positive) refers to the number of flows that are of some other type but are incorrectly classified as the correct type.

4.4 Classification Model

(1) Decision-Tree-Based Classification Model

The traffic from 7 individual APPs was tested respectively based on the model of C4.5 algorithm, and the results are listed in Table 4.

Table 4 shows the following:

(a) The recall ratios of Kugou Audio and QQGame are satisfactory, which means that these two APPs are well classified;
(b) The recall ratio of Sina Weibo is the worst;
(c) The classification precision of QQGame is the lowest;
(d) The shopping website Suning has the highest classification precision, which means that the possibility of other APPs being incorrectly classified as this class is low;
(e) QQGame represents the highest proportion of the traffic incorrectly classified as Sina Weibo, and vice versa. This result indicates that the features of these two APPs are relatively similar and difficult to distinguish.

The possible reasons for this phenomenon are listed below.

(a) QQGame is encrypted; therefore, its traffic features are not accessible. Consequently, a considerable portion of the traffic of these APPs is incorrectly classified as QQGame traffic. Table 4 shows that the precision of QQGame is quite low, which means that some of the traffic from other APPs is incorrectly classified as QQGame traffic;
(b) A considerable portion of Sina Weibo traffic is incorrectly classified as QQGame traffic. This may have occurred because the transmitted data in each Sina Weibo flow is small and similar to that of QQGame. This similarity stems from the traffic optimization of mobile Internet applications, which is intended to reduce traffic volume. In fact, most of the QQGame traffic comes from game commands consume little traffic;
(c) Among the streaming media APPs, the recall ratio of Kugou Audio was higher than that of Sohu Video. The most likely reason is that the audio stream is purer and is not mixed with HTTP traffic. However, when acquiring the Sohu Video traffic, both video and webpage information are shown at the same time; therefore, HTTP traffic is partially mixed with the video traffic in this application.

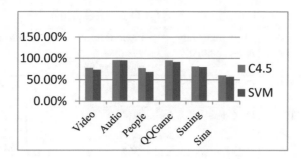

Fig. 1. Comparison of the recall ratios of the C4.5 and SVM algorithms.

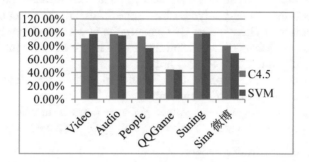

Fig. 2. Comparison of the precision of the C4.5 and SVM algorithms

Table 4. Result of C4.5-based recognition.

App	Number of test	Most frequently mis-classified	Proportion of incorrect classification	Precision	Recall
Sohu video	440	QQGame	10.9%	90.9%	77.5%
Kugou Audio	370	QQGame	4.1%	97.5%	95.7%
People	145	QQGame	20.7%	94.1%	77.2%
QQGame	283	Microblog	3.5%	44.6%	95.4%
Suning	538	QQGame	9.3%	98.2%	81%
Sina Weibo	441	QQGame	37.9%	80.6%	60.3%

In the following, web browsing, multimedia stream and games are regarded as three different types and classified accordingly. The number of samples and the classification results are shown in Table 5.

Table 5. Result of C4.5-based recognition.

App	Test num	Most frequently mis-classified	Proportion of incorrect classification	Precision	Recall
Multimedia stream	810	QQGame	8.8%	94.4%	85.8%
Web Browsing	1124	QQGame	22%	93.8%	74.4%
QQGame	283	web	3.9%	45.6%	95.1%

(a) The recall ratio of multimedia streams is good, but lower compared with the independent test of the two types of multimedia stream. The proportion of traffic that is incorrectly classified as QQGame traffic is comparatively high. The possible reason is that many web browsing traffic is mixed into video traffic and is incorrectly classified as QQGame traffic.

(b) The precision of QQGame is low. Substantial part of web browsing traffic is incorrectly classified as QQGame traffic.

(c) It is difficult to distinguish encrypted QQGame flows from web browsing traffic, and it is necessary to resort to other technology to improve the precision.

(2) SVM-based categorization model

Test parameter configuration: The algorithm is executed on LibSVM software platform; Kernel function is radial basis function (RBF); Parameters coat and gamma are optimized by GridSearch and cross validation method. First, classification model is built based on sample data; then, this model is applied to the test data. Table 6 is the classification result of SVM algorithm.

The recall ratios of Kugou Audio and QQGame are higher, which is similar to the result of C4.5 algorithm. The recall ratio of Sina Weibo is still the lowest. The precision of QQGame ranks the last. After comparing the results of SVM algorithm and C4.5 algorithm, as shown in Figs. 1 and 2, it can be seen that the precision and recall ratio of the two algorithms are quite similar.

Table 6. Result of svm-based recognition.

Application	Number of test	Most frequently mis-classified	Proportion of incorrect classification	Precision	Recall
Sohu video	440	QQGame	15.2%	97.6%	73.2%
Kugou Audio	370	QQGame	3.5%	95.4%	95.4%
People	145	QQGame	21.4%	76.7%	68.3%
QQGame	283	Microblog	3.2%	44%	91.5%
Suning	538	Microblog/QQGame	10.03%	98.6%	79.7%
Sina Weibo	441	QQGame	38.3%	69.1%	56.9%

Likewise, the traffic from the above APPs is classified by SVM algorithm after being merged into one main category. The results are shown in Table 7. From the results, the recall ratio for QQGame is lower when using the SVM algorithm compared to the results of the C4.5 decision tree algorithm. However, in the other two APPs, the recall ratio is higher compared with that of the C4.5 algorithm. The results of SVM and C4.5 algorithm on the main categories are compared and shown in Figs. 3 and 4.

Table 7. Result of SVM-based recognition.

Application	Number of test	Most frequently mis-classified	Proportion of incorrect classification	Precision	Recall
Multimedia Stream	810	web	13.3%	95.2%	83.8%
Web Browsing	1124	QQGame	1.6%	85.3%	97.3%
Game	283	web	28.3%	81.5%	64%

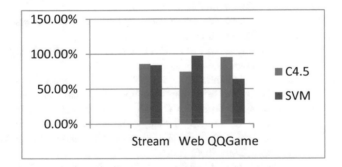

Fig. 3. Comparison of recall of C4.5 and SVM in classifying major categories.

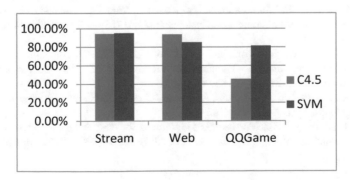

Fig. 4. Comparison of the precision of the C4.5 and SVM algorithm in classifying major categories.

5   Conclusions

In this article, machine learning methods are applied to mobile application network traffic classification. Because the APPs used for web browsing all employ the HTTP protocol, it is difficult to distinguish between them. However, the HTTP web-traffic protocol is obviously different from that of media streaming and gaming. The C4.5 and SVM algorithm are both used to classify mixed traffic. This approach is effective for some APPs, but not as effective for others. In our classification process, traffic from QQGame is difficult to distinguish because of its encryption. Thus, determining how to apply machine learning to the classification of encrypted traffic in a mobile Internet environment still requires further research. In addition, classifying different types of service traffic from the same application also requires further research.

References

1. ITU-T Y.2770 (Electronic Version). https://www.itu.int/rec/dologin_pub.asp?lang=e&id=T-REC-Y.2770-201211-I!!PDF-C&type=items
2. Jiang, W.L.: Research and design of P2P traffic monitoring system based on DPI. BeiJiing University of Posts and Telecommunications (2013). (in Chinese)
3. Zhang, H.: Design of a P2P traffic detection based on DPI. BeiJiing University of Posts and Telecommunications (2013). (in Chinese)
4. Zhu, Y.N.: Identification and control of P2P business based on DPI. Shanghai Jiao Tong University (2008). (in Chinese)
5. Han, Y.M.: Design and implementation of a VoIP traffic detection system based on DPI. BeiJiing University of Posts and Telecommunications (2010). (in Chinese)
6. Hurley, J., Garcia-Palacios, E., Sezer, S.: Classification of P2P and HTTP using specific protocol characteristics. In: Oliver, M., Sallent, S. (eds.) Proceedings of the 15th Open European Summer School and IFIP TC6.6 Workshop on the Internet of the Future (EUNICE 2009), pp. 31–40, Springer, Heidelberg (2009)
7. Michalski, R.S., Carbonell, J.G., Mitchell, T.M., et al.: Machine Learning: An Artificial Intelligence Approach. Springer, Berlin (1983)
8. Li, W., Moore, A.W.: A machine learning approach for efficient traffic classification. In: Proceedings of the 15th International Symposium on Modeling, Analysis, and Simulation of Computer and Telecommunication Systems (MASCOTS 2007), Washington, DC, USA, pp. 310–317 (2007)
9. Zander, S., Nguyen, T., Armitage, G.: Automated traffic classification and application identification using machine learning. In: Proceedings of IEEE Conference on Local Computer Networks, Sidney, Australia (2005)
10. Singh, K., Agrawal, S., Sohi, B.S.: A near real-time IP traffic classification using machine learning. Intell. Syst. Appl. 3, 83–93 (2013)
11. Zhang, H.L., Lu, G.: Machine learning algorithms for classifying the imbalanced protocol flows: Evaluation and comparison. J. Softw. 23(6), 1500–1516 (2012). (In Chinese)

A Local Feature Curve Extraction Algorithm for Dongba Hieroglyphs

Yuting Yang[1] and Houliang Kang[2(✉)]

[1] Culture and Tourism College, Yunnan Open University, Yunnan, China
[2] College of Humanities and Art, Yunnan College of Business Management,
Yunnan, China
`kangful979110@163.com`

Abstract. Dongba hieroglyph is a kind of very primitive picture hieroglyphs; they express meaning by using pictures. This makes the same type words have some similarities in morphology and structure, and they are distinguished only by local differences. In this paper, we analyse the basic structural of the single graphemes in Dongba hieroglyphs. By analysing the writing methods and habits, we give a local feature curve extraction algorithm for Dongba Hieroglyphs based on CDPM (Connected Domain Priority Marking Algorithm) and DCE (Discrete Curve Evolution Algorithm). The algorithm can extract local feature curves which can contain one or more glyphic features, and it can also combine redundant curves based on template curves to facilitate the comparison between the local curves and template. Extracting the local feature curves of Dongba hieroglyph is helpful to analysis the local similarities in different hieroglyphs, classify and retrieve hieroglyph. Moreover, it also has very important significance in studying the creation method of Dongba hieroglyph.

Keywords: Local curve extraction · Local feature curve ·
Curve extraction algorithm · Dongba hieroglyphs feature curve

1 Dongba Hieroglyph and Their Features

Dongba hieroglyph is a kind of very primitive picture hieroglyphs [1]; they use pictures to express meaning. So, many Dongba characters of the same type have some similarities in morphology and structure, and they are distinguished only by local differences [2]. In 2003, Dongba scriptures written by Dongba hieroglyphics are listed into Memory of the World Heritage List by UNESCO [3, 4].

Through analyzing the hieroglyphic structures, we find that grapheme is the smallest unit to compose Dongba hieroglyphs. It has obvious pictographic feature [5–7], and it is distinguished by local difference only [8], as shown in Table 1.

In order to obtain a curve segment with one or more local features of glyphs, firstly, we use the Connected-Domain Priority Marking algorithm to obtain the feature curves of Dongba hieroglyphs. Then, we divide the feature curve into local curve segments under some rules, and merge them which do not contain character features. Finally,

© Springer Nature Switzerland AG 2020
Q. Liu et al. (Eds.): CENet 2018, AISC 905, pp. 409–416, 2020.
https://doi.org/10.1007/978-3-030-14680-1_45

Table 1. The classification of Dongba hieroglyphs.

Single Graphemes	Hieroglyph						
	Meaning	Bird	Sparrow	Peace Bird	Goose	Yak	Deer
	Hieroglyph						
	Meaning	People	Sit	Up	Bear load	Grandson	God
Compound Graphemes	Hieroglyph						
	Meaning	Weka	Macaw	Ring	Fowling	Aix galericula	Incubate
	Hieroglyph						
	Meaning	Weed	Stir	Kick	Drop	Sister	Tea

under the requirement of the template character, we combined the local curve segments into the specified number. This method can help to analysis the relationships and different between feature curves of different hieroglyphs more closed.

2 Feature Extraction and Simplification of Dongba Hieroglyphs

Dongba Master writes Dongba hieroglyph by using bamboo pen in line drawing method. Since the bamboo pen is a hard pen, the line width of the hieroglyph is basically the same. So, we binarize and refine the words, and use the Connected Domain Priority Marking (CDPM) algorithm [9] to extract the feature curves of hieroglyphs, firstly. Then we use Discrete Curve Evolution (DCE) [10, 11] to remove the potentially redundant and noise points in the curve. The feature curve used to represent the feature of Dongba hieroglyphs has been completely simplified, as shown in Fig. 1. Among them, Fig. 1(1) shows the original character, Fig. 1(2) shows the character refinement, Fig. 1(3) shows the character feature curve extracted by using CDPM algorithm, and Fig. 1(4) shows the simplified feature curve by using DCE algorithm.

3 Extraction of Local Feature Curve of Dongba Hieroglyphs

3.1 Segmentation of Feature Curves

In the shape, the convex arcs determine parts of object boundary that enclose visual parts of the object. Therefore, we divide the feature curve by using the concave vertices as a demarcation point. Dividing the feature curve can be roughly divided into two steps: First, we judge and extract the concave vertices in the feature curve [13, 14]. Then, divide the feature curve by using concave vertex as the dividing point.

In the process of segmentation, we should use the neighboring concave vertices as the starting and end point of the local curve to maintain the continuity of the original

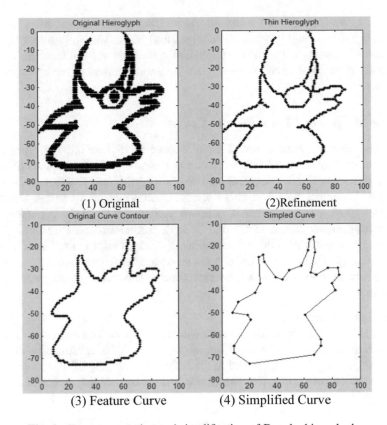

Fig. 1. Feature extraction and simplification of Dongba hieroglyphs.

curve. We use the Dongba hieroglyph shown in Fig. 1 as an example. The feature curve includes 13 concave vertices. And the feature curve is divided into 13 independent segments with the concave vertices, as shown in Figs. 2(2) and (3).

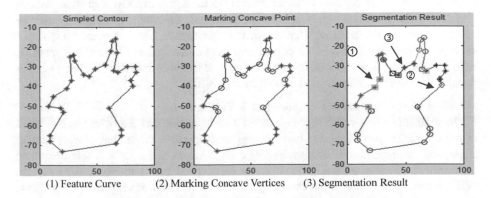

Fig. 2. The segmentation of Dongba hieroglyph.

In Fig. 2, we find that too many concave vertices lead to more local curve segments. This will not only increase the computation of the similarity measure, and the invalid curve segments will also interfere with the analysis of local features. Therefore, we need to merge local curve to ensure that each independent curve contains at least one feature.

3.2   The Mergence of Local Feature Curves

The local curves that can be merged in feature curve are divided into two types, one is a curve segment composed of two consecutive concave vertices, as shown in ① of Fig. 2(3). It is actually a line segment, and does not contain any features. So, we should combine it. The other type is a curve segment composed of 3 vertices which are 1 concave vertex+1 convex vertex+1 concave vertex. Some of these curves contain character features, as shown in ② of Fig. 2(3). It contains the character's jaw and should not be merged. And the curve segment ③ which adjacent to curve ② has the same structure of ②, but it does not contain any character features, so it can be merged. Therefore, when the local curve segment satisfies the condition, we merge it; otherwise, we keep it.

Merging Line Segment
We merge line segment by using Discrete Curve Evolution. The main idea is: In every evolution step, a pair of consecutive line segments s1, s2 is substituted with a single line segment joining the endpoints of s1 ∪ s2. The key property of this evolution is the order of the substitution [19]. The substitution is done according to a relevance measure K given by:

$$K(s_1, s_2) = \frac{\beta(s_1, s_2) l(s_1) l(s_2)}{l(s_1) + l(s_2)}$$

Where $\beta(s_1, s_2)$ is the turn angle at the common vertex of segments s1, s2 and l is the normalized length. The higher the value of $K(s_1, s_2)$, the larger is the contribution to the shape of the curve of arc s1 ∪ s2. And, we achieve curve simplification by continuously replace the segments that contribute the least. So, in DCE algorithm, we can find that if the common vertex is convex, the larger the weight, the more potential character feature information is included; if it is concave, the larger the weight, the stronger independence of the two curved segments connected to the vertex. Therefore, when we merge a line segment, it should be merged to the endpoint which has a smaller weight $K(s_1, s_2)$. The result of is shown in Fig. 3(2).

Merging Curve Segments Composed of 3 Vertices
If a local feature curve which is composed of 3 vertices contains character features (such as mouth, ear or horn, etc.), the two line segments in curve would have larger turning angle, and the angle corresponds to the only convex vertex in the curve. Therefore, when the turning angle is larger than 90°, we retain the original curve; otherwise, we merge it. When we merge it, because the curve segment contains two line segments, we separate the two line segments and merge them with the respective adjacent curve segments. As shown in Fig. 3(3).

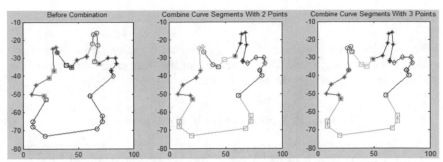

(1) Curve Dividing (2) Merging Line Segment (3) Merging Curve with 3 points

Fig. 3. Merging local curve segment

By combining twice, the number of local curve segments is decrease from 13 to 7 and each curve includes at least one or more local features.

3.3   Combination of Feature Curves of the Same Type

For each type of Dongba hieroglyphs, characters of the same type tend to have similar or identical features. When the test character contains a large number of local feature curves, merging them to the specified number of template can help them to compare both globally and locally. In the process of merging, we judge the angle between the center points of the two feature curve segments as shown in Fig. 4(2). The smaller the angle, the closer the two curve segments are, and the more they should be merged. Therefore, based on the number of feature curves of this type, we merge the 7 feature curves to 5, and the result is shown in Fig. 4(3).

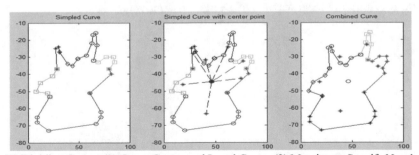

(1) Dividing Curves (2) Curve Center and Local Center (3) Merging to Specify Number

Fig. 4. Merging the local curves to specify number of template.

4 Experiment

In order to fully test the validity of the local feature curve extraction algorithm, we select 10 types of Dongba hieroglyphs and count the number of local feature curves included in each category, as shown in Table 2.

Table 2. 10 Types of Dongba hieroglyphic templates.

Type	Contoured Dongba Hieroglyph	Number of Feature Curve	Type	Contoured Dongba Hieroglyph	Number of Feature Curve
Hill		1	Mountain		2
House		1	Bird		3
Water		2	Fish and insect		4
Hand		2	Flower		4
Dongba Master		2	Livestock		5

Take the Dongba character in Fig. 1 as a test character. Since its type is unknown, we use the number of local characteristic curves of the template character as the standard, and combine the curves in the test character into the specified number. The combined effect is shown in Fig. 5. Among them, when the feature curve of the test character is merged into five, it is the closest to the form of the template character. It shows that extracting the local feature curve of characters is very effective for analyzing the external form and local features of Dongba Hieroglyph.

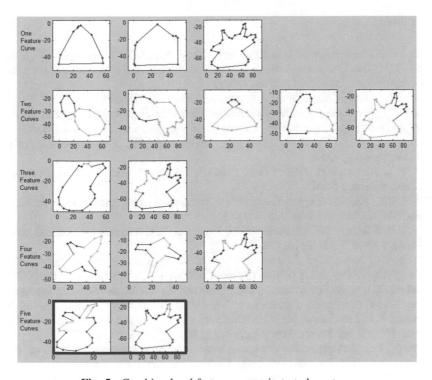

Fig. 5. Combine local feature curves in test characters.

5 Conclusion

Dongba hieroglyph is a kind of very primitive picture hieroglyphs; it expresses meaning by using pictures. By analyzing the writing methods and habits, we give a local feature curve extraction algorithm for Dongba Hieroglyphs based on CDPM and DCE. The algorithm can extract local feature curves containing one or more glyphic features, and it can also combine redundant curve according to the number of template curves to facilitate the comparison between the sample and template. Extracting the local feature curves of Dongba hieroglyph is helpful to analysis the local similarities in different hieroglyphs, and can also help to classify and retrieve hieroglyph with local similar features. Moreover, it has very important significance in studying the creation method of Dongba hieroglyph.

Acknowledgement. This research was partially supported by the Scientific Research Fund of Yunnan Education Department (2018JS748).

References

1. He, L.M.: The research of Dongba culture heritage. Soc. Sci. Yunnan **1**, 83–87 (2004)
2. He, J.G.: The development trend of Dongba culture studies of the Naxi nationality. J. Yunnan Natl. Univ. **24**(1), 81–84 (2007)
3. Ge, A.G.: Dongba culture review. J. Natl. Artist. Res. **2**, 71–80 (1999)
4. He, Z.W.: Discussing the characteristics of Naxi hieroglyphic including the different of original pictograph, hieroglyphic and ideograph. Soc. Sci. Yunnan **3**, 71–82 (1982)
5. Fang, G.Y., He, Z.W.: The Dictionary of Naxi Hieroglyphic. Yunnan People's Publishing House, Kunming (2005)
6. Li, L.C.: Pronunciation Dictionary of Naxi Hieroglyphic. Yunnan Nationalities Publishing House, Kunming (2001)
7. Rock, J.F., Khi, N.: English Encclopedic Dictionary (Part I). Roma Istituto Italiano Peril Medio ed Estreme Prientale, Roma (1963)
8. He, L.F.: Hieroglyphic Structure of Naxi Dongba and Its International Standard. Shanghai Normal University, Shanghai (2016)
9. Yang, Y.T., Kang, H.L.: A novel algorithm of contour tracking and partition for Dongba hieroglyph. In: Processing of 13th Chinese Conference IGTA 2018, pp. 453–463 (2018)
10. Latecki, L.J., LakaÈmper, R.: Polygon evolution by vertex deletion. In: Processing in International Conference of Scale-Space Theories in Computer Vision, pp. 398–409 (1999)
11. Latecki, L.J., LakaÈmper, R.: Shape similarity measure based on correspondence of visual parts. IEEE Trans. Pattern Anal. Mach. Intell. **22**(10), 1–6 (2000)
12. Latechi, L.J., Lakamper, R.: Convexity rule for shape decomposition based on discrete contour evolution. Comput. Vis. Image Underst. **73**(3), 441–454 (1999)
13. Chen, B.F., Qian, Z.F.: An algorithm for identifying convexity-concavity of a simple polygon. J. Comput. Aided Des. Comput. Graphics **2**(14), 214–217 (2002)
14. Song, X.M., Cheng, C.X.: An analysis and investigation of algorithms for identifying convexity concavity of a simple polygon. Remote Sens. Land Resour. **9**(13), 25–31 (2011)

Adaptive Learning for Correlation Filter Object Tracking

Dongcheng Chen$^{(\boxtimes)}$ and Jingying Hu

Jiangsu Automation Research Institute, Lianyungang 222000, China
chendongcheng8710@163.com

Abstract. To solve the real-time quality and adaptive quality of traditional correlation algorithm. An adaptive correlation filter tracking algorithm is proposed. First, training the filter using machine learning, make the algorithm be adaptive to the object changing. Then, weighting the image patch with a cosine window, the object region has larger weighting value than the edge region, which ensures the continuity of the Cyclic matrix. At last calculate the response matrix using the convolution of the input image patch and the filter matrix in Fourier domain. Experiment on various videos shows that for 26 pixel × 24 pixel object, filter the image in a 130 × 120 pixel patch, the processing speed could be 210 fps. The proposed tracking algorithm can track object with good timing quality and robustly.

Keywords: Correlation filter · Weighted · Fourier domain · Cyclic matrix

1 Introduction

Object tracking is widely used in industry and life, and is always a study topic and hard point of machine learning. Object tracking calculate object position using only current video frame usually, and the most popular object tracking algorithm was object correlation matching. In recent years, with the fast development of computer technology and tracking algorithm, many researchers develop tracking algorithm using machine learning, which not only use current frame data but also using previous frame data when calculating position. As tracking going on, the tracking algorithm is enhanced, which make the computer has the ability of learning. The tracking algorithm is adaptive to the position changing, shape changing and shading and tracks the object well. When tracking, the well tracking information of current frame can be used in the following frames tracking, and current tracking result can influent the following tracking [1–3].

Duda first rise model correlation matching algorithm in 1973 [4]. It is studied by many researchers all over the world and widely used in engineering although the shortcomings of the algorithm. In order to overcome the shortcomings of the correlation algorithm, many researchers develop a serial of correlation filter algorithm [5]. Hester used correlation filter algorithm in several, which was used in missile guidance navigation and control, and can recognize object position [6]. Kumar rise a method of using SDF, using filter model and the input training image, together with least variance SDF, which can restraining the noise in the input image [7]. Mahalanobis proposed a MAGE filter, which can enhance the correlation peak and increase the recognition

© Springer Nature Switzerland AG 2020
Q. Liu et al. (Eds.): CENet 2018, AISC 905, pp. 417–424, 2020.
https://doi.org/10.1007/978-3-030-14680-1_46

precision [8]. Bolme proposed a synthetizing accurate classifier method, using lot of image training the classifier, and confirming the recognition object filter. The algorithm was used to recognize and locate eye [9]. Bolme also propose using the state of the object in every frame to train the filter, the filter was used in object real-time tracking [10]. Henriques proposed an object tracking algorithm based on kernel correlation, together with the Cyclic matrix and linear classification theory in pattern recognition, finally moving object tracking real-time tracking was met [11].

An object tracking algorithm based on online correlation filter learning is proposed. Train the filter in Fourier domain, using the Fourier transforming and inverse Fourier transforming to calculate response of the input image. The experiment result shows, the proposed algorithm has good real-time property. For a 261 × 24 pixel object tracking in 130 × 120 pixel searching domain, the tracking rate could reach 210 fps, and the algorithm can track the object steadily. The tracking algorithm can be adaptive to the object gesture changing and similar object disturbing.

2   Spatial Domain Correlation Tracking

Traditional correlation object tracking processes the video in spatial domain. Sliding window pixel by pixel is used to find the object. The processing load is very large, it is hard to track object real-time for large object tracking in large searching domain. The tracking step of traditional correlation tracking is selecting the object in the first video frame, search the object in the specific domain when the new frame coming. The object model is used to match with every possible object patch using correlation processing. When tracking a larger object in larger searching domain, the processing load will be larger. For large object tracking in large domain, it is hard to track object real-time. Even tracking in Fourier domain, large object cannot be tracked real-time. Traditional sliding window tracking calculating the correlation coefficient is shown in (1).

$$R(i,j) = \frac{\sum_{m=1}^{M} \sum_{n=1}^{M} S^{i,j}(m,n) \times T(m,n)}{\sum_{m=1}^{M} \sum_{n=1}^{M} [S^{i,j}(m,n)]^2} \tag{1}$$

T represent the template patch, S_{ij} is a patch that has the same size as template T. according to formula (1), the processing load of traditional correlation tracking is $O(N^4)$.

3   Correlation Filter Tracking

3.1   Cyclic Matrix

Cyclic matrix is a special Toeplitz matrix [12], its every row vector comes from the upper row and cyclic right shifting. Cyclic matrix is widely used in discrete Fourier transforming. The cyclic matrix property can make the image filtering faster. As u is a one dimension $n \times 1$ vector. Its cyclic matrix $C(u)$ can be defined as formula (2).

$$C(u) = \left\{ \begin{array}{c} u_0 u_1 \cdots u_{n-1} \\ u_{n-1} u_0 \cdots u_{n-2} \\ \cdots\cdots\cdots\cdots \\ u_1 u_2 \cdots u_0 \end{array} \right\} \tag{2}$$

Cyclic matrix can be conveniently used calculating the vector convolution. As v is a $n \times 1$ vector, according to the property of convolution, spatial correlation can be calculated by a function's Fourier transforming multiplying the conjugate of another function's Fourier transforming. Thus, the convolution of vector v and vector u can be represented by the cyclic of vector u $C(u)$ multiply vector v, just as formula (3) shown.

$$C(u)v = F^{-1}(F^x(u) \circ F(v)) \tag{3}$$

In formula (3), the symbol ∘ represents the corresponding element multiplying of two matrixes. F and F^{-1} represent the Fourier transforming and inverse Fourier transforming of the vector. * represents the conjugate of Fourier transforming. Because the sum, product and inversing of cyclic matrix are all cyclic matrix, the applying of cyclic matrix can be spread to other image processing. Henriques has proved the good properties of cyclic matrix can also be used in 2-D vector. The cyclic of 2-D vector is the cyclic shifting of row vector or column vector. It is similar to the sliding window in object tracking, which can search every possible object in the searching domain. Using the property of convolution, together with the fast Fourier transforming (FFT) and inverse fast Fourier transforming (IFFT), the calculating load can be reduced a lot, thus the object tracking using correlation algorithm can be done real-time.

Each vector of cyclic matrix $C(u)$ can be represented by Piu. Where P is a permutation matrix, Piu means to cyclic shifting vector u for i times. For input image x, the cyclic shifting of image x is x^i, $x^i = P^i x$. It is equal to sliding the middle part of the image in the image. The sliding can search every point of the image domain. Because of the discontinuous of the image edge, there may be some segmentation lines after cyclic shifting. The discontinuous edge shifting can enlarge the bandwidth of the Fourier domain of the image. So cosine window or Hamming window can be used to restrain the searching domain edge. Thus the cyclic shifting image is continuous, and it equals to weighting to the searching domain. The area near the middle of the image has larger weighting. As Fig. 1 shows.

3.2 Object Tracking Based on Correlation Filter

In correlation object tracking, if the object keeps unchanging, and there's no similar object in the searching domain, the tracking result will be very well. But if the object shape changing, shading or light changing when tracking, the tracker can distinguish the object from background, which will lead to tracking failure. Traditional object tracking processed the image in time domain, the processing load is quite large, when the searching domain is large enough, the tracker can not track the object real-time. While in Fourier domain, time domain convolution transforms to point to point multiplying. With the using of Fourier transforming, the real-time quality can be met.

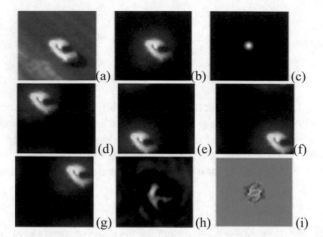

Fig. 1. (a) Region with target (b) Weighted image (c) Ideal output (d)–(g) Cyclic shifted image (h) Response output (i) Fourier domain of the filter.

Here an adaptive correlation filtering tracking algorithm is proposed. Achieve correlation using convolution. Using Fourier transforming to calculate the correlation result, the processing rate is accelerated a lot. Training the filter model using the tracking image, then use the trained filter model to search object in the new coming image. Multiply the input image with the filter model, the Fourier transforming of responding is get. Then inverse Fourier transform the getting result, the result matrix is the desired responding matrix. The position where the largest number located is the object location [9].

Figure 1 is the information used in the proposed algorithm. The input training image is f_i, the expected correlation coefficient matrix is g_i. The responding function is 2 dimension function at point (x_i, y_i), as Fig. 1(c) shows. The formula is shown as (4).

$$g_i(x,y) = \exp\left(\frac{-(x-x_i)^2 + (y-y_i)^2}{\sigma^2}\right) \tag{4}$$

Where σ is variance, which control the tolerance the system can adapt.

According to the convolution theory, spatial convolution can convert to multiplying in Fourier domain, as formula (5) shows.

$$g(x,y) = (f * h)(x,y) = F^{-1}(F(w,v) \circ H(w,v)) \tag{5}$$

In formula (5), f is the input image, h is the filter block and g is the output responding in the spatial domain. F, H, G is the 2-D Fourier transforming of f, h and g. The proposed algorithm uses FFT and IFFT to improve the calculating speed. For an $N \times N$ object, the calculating complexity of FFT or IFFT is $O(N^2 \log N^2)$, which is much smaller than the traditional correlation algorithm complexity $O(N^4)$ when object is large enough. In order to get the filter, according to the convolution theory, we have formula (6).

$$G(w, v) = F(w, v) \circ H^x(w, v) \tag{6}$$

Then the filter H can be represented using formula (7).

$$H^x(w, v) = \frac{G(w, v)}{F(w, v)} \tag{7}$$

In order to get the best filter block H, which can make the responding the most expected. We have formula (8).

$$min_{H^x}|F_i \circ H^x - G_i|^2 \tag{8}$$

To get the best filter block H, do partial derivation on H^*, as formula (9) shows.

$$\frac{\partial}{\partial H^x}|F_i \circ H^x - G_i|^2 = 0 \tag{9}$$

Then we have best filter H.

$$H^x = \frac{\sum G_i \circ F_i^x}{\sum F_i \circ F_i^x} \tag{10}$$

3.3   On Line Learning of Filter H

The best filter block H as formula (10) shows. As tracking goes on, the object and background is always changing, the object in current frame is different from the object selected in the first frame. If the object keeps unchanging, when the object change bit by bit, the tracker may lose the object. According to the changing object and background, an on-line learning algorithm of the filter is proposed. Because of the parameter changing of the filter when tracking, and the frame before current frame can influent the tracking parameters, all the information current and before will be used, which will make the track be adaptive to the object changing.

$H_i^x = \frac{A_i}{B_i}$, where Ai and Bi are as formula (11), (12) show.

$$A_i = \eta G_i \circ F_i^x + (1 - \eta)A_{i-1} \tag{11}$$

$$A_i = \eta F_i \circ F_i^x + (1 - \eta)B_{i-1} \tag{12}$$

The η in the formula is learning factor, which control the influence the current frame makes. If the learning factor is large, the current frame can make more influence on the filter block. The learning factor is often a small factor, thus the previous frames can make more inflation on the tracker. Even though the current frame is shaded, the tracker will not change a lot. When the object appears, the tracker can get it immediately. The learning step of the filter is as Table 1 shows.

Table 1. Step of proposed algorithm.

Input: f: Sample image, g: expected responding, η: Learning factor
For i = 1…frame
F_i=fft(f)
G_1=fft(g)
if i=1
$A_1=G_1 \circ F_1^*$, $B_1= F_1 \circ F_1^*$
else
$A_i=G_{i-1} \circ F_i^* \eta +(1-\eta)A_{i-1}$
$B_i= G_{i-1} \circ F_i^* \eta +(1-\eta)B_{i-1}$
$H_i=A_i/B_i$
$G_i=F_i \circ H^*$
g_i=ifft(G_i)
end if
output: The coordinate where the largest value located in g_i
End for

4   Experiment Result and Analysis

To evaluate our approach, we perform comprehensive experiments on three benchmark databases. In all the experiments, parameters remain unchanged. In all experiment, in order to improve the robustness of the tracking, when the tracking object moving step is large, the tracker should choose a large searching domain. During training and tracking, the object in the first frame is selected by hand, and the training set come from the first frame. As the tracking goes on, the training in every frame use current frame and previous frames. The influence previous frames perform are controlled by learning factor η.

Experiment 1: Fig. 2 is the tracking result of Running Car. There are 1821 frames in the video. The tracking rectangle is 26×24 pixel, the searching domain is 5 times larger than the object rectangle. The learning factor $\eta = 0.0125$, the expected responding variance of Gaussian function is $\sigma = 0.1$. The processing rate can reach 210 fps. In the tracking result figure, we can see, when using traditional correlation tracking, when the object does not change a lot, the tracker can track the object well. But if the object changes in shape or being shaded, the tracker would lose the object. In the figure, the object is located at the car head by hand, the first 150 frames is tracked robustly. From frame 155 on, the car starts turning, the tracking rectangle start to move to the back of the car. The reason is the small learning factor, the tracker can't update in time. As tracking goes on, the tracking rectangle track the back of the car steady. At frame 280–500, the object turning again, as the learning filter block, the tracking rectangle does not shift. The camera shooting point changed during tracking, and after frame 550, the object surpasses 3 similar cars, the surpassed car in frame 1359 is almost the same as the tracking object, the tracking rectangle does not shift because of the weighting window.

Fig. 2. Tracking result of Running Car (frame 100, 186, 238, 280, 518, 1283, 1359, 1477, 1678, 1818)

Experiment 2: Fig. 3 is the tracking result of RedTeam. There are 1918 frames in the video. The proposed algorithm can track the red car steady. The tracking rectangle is 21 × 31 pixel, the searching domain is 4 times larger than the object rectangle. The learning factor is $\eta = 0.0125$, the expected responding variance of Gaussian function is $\sigma = 0.0625$. The processing rate can reach 162 fps. The object scale change several times during tracking, and the background is very complex. From frame 1 to frame 600, the object is always enlarging. As the learning ability of the algorithm, the object scale changes, the tracker can adapt the object changing. After frame 600, the object begin to go small in scale, the tracker can also tracks well. At frame 1520, the object is shaded, the algorithm can also track the object.

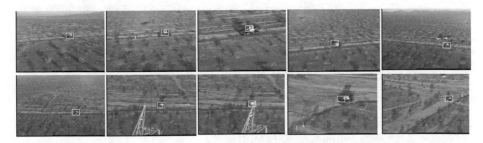

Fig. 3. Tracking result of RedTeam (frame 1, 293, 542, 700, 890, 1285, 1514, 1525, 1717, 1895).

Tracking object using traditional correlation algorithm, when doing experiment on experiment 1, the object lost when object turning, and never get object again. In experiment 2, the traditional tracker can only track object for 18 frames. The proposed algorithm can process video much faster than traditional algorithm.

5   Conclusion

As traditional correlation tracking algorithm can't track object real-time and cannot be adaptive to object changing, an adaptive learning for correlation filter object tracking algorithm is proposed. Weight the tracking window when training the filter, which can make the object weight more when tracking. Using the cyclic matrix theory to calculate the filter, it will reduce the processing load. The experiment result shows, the proposed algorithm can track object robustly when object change in position, scale and shading. When object rectangle is 26×24 pixel, the searching region is 130×120 pixel, the tracking rate can reach 210 fps.

References

1. Zhang, T., Xu, C., Yang, M.H.: Multi-task correlation particle filter for robust object tracking. In: IEEE Conference on Computer Vision and Pattern Recognition, pp. 4819–4827. IEEE Computer Society (2017)
2. Yang, Y., Bilodeau, G.A.: Multiple object tracking with kernelized correlation filters in urban mixed traffic. In: IEEE Computer Society, pp. 209–216 (2017)
3. Koubaa, A., Qureshi, B.: DroneTrack: cloud-based real-time object tracking using unmanned aerial vehicles over the internet. IEEE Access **6**(99), 13810–13824 (2018)
4. Duda, R., Hart, P.: Pattern Classification and Scene Analysis, pp. 276–284. Wiley, New York (1973)
5. Sharma, S., Ansari, J.A., Murthy, J.K., Krishna, M.: Beyond pixels: leveraging geometry and shape cues for online multi-object tracking. In: IEEE International Conference on Robotics and Automation. IEEE (2018)
6. Hester, C.F., Casasent, D.: Multivariant technique for multiclass pattern recognintion. Appl. Opt. **19**(11), 1758–1761 (1980)
7. Vijaya Kumar, B.V.K.: Minmum-variance synthetic discriminant functions. Opt. Eng. **31**, 915 (1992)
8. Mahalanobis, A., Vijaya Kumar, B.V.K., Casasent, D.: Minimum average correlation energy filters. Appl. Opt. **26**(17), 3633 (1987)
9. Bolme, D.S., Draper, B.A., Beveridge, J.R.: Average of synthetic exact filters. In: IEEE Conference on Computer Vision and Pattern Recognition (2009)
10. Bolme, D.S., Beveridge, J.R., Draper, B.A., Lui, Y.M.: Visual object tracking using adaptive correlation filters. In: IEEE Conference on Computer Vision and Pattern Recognition (2010)
11. Henriques, J.F., Caseiro, R., Martins, P., et al.: Exploiting the circulant structure of tracking-by-detection with kernels. In: Europe Conferance on Computer Vision (2014)
12. Hon, S., Wathen, A.: Circulant preconditioners for analytic functions of Toeplitz matrices. Numer. Algorithms (suppl. C), 1–20 (2018)

Face Recognition Algorithm Based on Nonlinear Principal Component Feature Extraction Neural Network

Jinlong Zhu[✉], Fanhua Yu, Guangjie Liu, Yu Zhang, Dong Zhao,
Qingtian Geng, Qiucheng Sun, and Mingyu Sun

Department of Computer Science and Technology,
Changchun Normal University, Changchun, China
zhujinlong19840913@126.com

Abstract. In order to solve the face recognition problem with different factors such as light, human expression and angle of shooting, a nonlinear principal component feature extraction neural network model is used in this paper. Researchers have proposed many algorithms for identifying applications, but the limitations of single algorithms cannot meet the requirements of practical application. By introducing the nonlinear principal component analysis (PCA), the proposed algorithm deletes a large amount of nonlinear data on the premise that a small amount of information can be guaranteed after linear dimensionality reduction. And the computational complexity of the neural network algorithm (BPNN) is large, so we use the nonlinear principal component analysis method to ensure the amount of information of the original data, and combined with the neural network algorithm to improve the efficiency of the algorithm. Empirical results show that PCA method can effectively reduce redundant data and ensure data integrity. Data after dimensionality reduction can improve the computational efficiency of neural network algorithm. The BPNN based PCA model can effectively solve the time consumption and recognition accuracy of the face recognition.

Keywords: Face recognition · Neural network ·
Nonlinear principal component

1 Introduction

Face recognition combines computer vision and pattern recognition, and is widely used in robotics and other disciplines [1]. The technology of computer face recognition has been developed in recent 20 years. In 90 s, it became a hot spot of research. Between 1990 and 1998, the related literature of EI can be retrieved as many as thousands of [7], because face database used in face recognition experiment is usually not large, and the most common face Library includes only about 100 images. Such as the MIT library, the Yale library, the CMU library, and the FERET library.

In recent years, the application of face recognition has been widely applied in all walks of life. Therefore, face recognition technology has developed rapidly, and its development can be roughly divided into three stages.

© Springer Nature Switzerland AG 2020
Q. Liu et al. (Eds.): CENet 2018, AISC 905, pp. 425–433, 2020.
https://doi.org/10.1007/978-3-030-14680-1_47

The first stage is to study the facial features [4] required for face recognition. Bertillon puts forward a simple sentence to establish a corresponding relationship with a face image in the database, combined with fingerprint analysis. This method can provide a recognition result with high accuracy. Parke generates a high quality facial grayscale model through computer.

The second stage is the man-machine interactive recognition system. The method proposed by Kaya and Kobayashi belongs to the method of statistical identification. They use Euclidean distance to characterize the facial features, such as the distance between the lips and the nose, the height of the lips, and so on. Goldstein proposed a geometric feature parameter identification method. They use geometric feature parameters to represent face frontal images. This method uses 21 dimensional feature vectors to represent facial features. T. Kanad designs a semi automatic backtracking recognition system with certain knowledge guidance. It creatively uses integral projection method to calculate a group of facial features from a single image, and then uses the pattern classification technique to match the standard face.

The third stage is the automatic identification stage. Researchers have produced various automatic identification systems. According to the different face representation methods, the face recognition technology can be divided into three categories: recognition based on geometric features, recognition based on Algebraic Features and recognition based on connection mechanism.

Researchers often use neural network and principal component analysis to identify methods. The face recognition algorithm based on neural network avoids the complex feature extraction, and can obtain the implicit expression of the rules and rules of face recognition which is difficult to obtain by other methods. This method is slow in training and may fall into local optimum. The principal component analysis (PCA) constructs the main eigenvector space based on a set of face training samples. These feature vectors are composed of the eigenvectors corresponding to the main eigenvalues of the generation matrix of the image. The PCA method alone has a large amount of computation, so PCA is often used in conjunction with other identification methods. Our method uses PCA to reduce the dimension of neural network training, so as to improve the slow training speed.

2   Face Recognition Model

2.1   Face Feature Extraction Method Based on PCA

PCA is mathematically defined as an orthogonal linear transformation that transforms the data to a new coordinate system such that the greatest variance by some projection of the data comes to lie on the first coordinate (called the first principal component), the second greatest variance on the second coordinate, and so on [3].

We have N face class samples, and each sample has M feature (variable), and its feature matrix can be obtained.

$$X = \begin{bmatrix} x_{11} & \cdots & x_{1m} \\ \vdots & \ddots & \vdots \\ x_{n1} & \cdots & x_{nm} \end{bmatrix} = (X_1, X_2, \ldots, X_M) \tag{1}$$

For the M vectors of matrix X, the characteristic linear combination (i.e. the integrated eigenvector) is composed of:

$$\begin{cases} F_1 = a_{11}X_1 + a_{21}X_2 + \cdots A_{m1}X_m \\ F_2 = a_{12}X_1 + a_{22}X_2 + \cdots A_{m2}X_m \\ \quad \cdots \cdots \\ F_n = a_{1n}X_1 + a_{2n}X_2 + \cdots A_{mn}X_m \end{cases} \tag{2}$$

The formula can be abbreviated as

$$F_i = a_{1i}X_1 + a_{2i}X_2 + \cdots A_{mi}X_m \quad i \in [1, n] \tag{3}$$

When we extract principal components, we estimate the covariance matrix by estimating the value s, assuming that the original characteristic matrix is:

$$X = \begin{bmatrix} x_{11} & x_{12} & \cdots & x_{1m} \\ x_{21} & x_{22} & \cdots & x_{2m} \\ \vdots & \vdots & \ddots & \vdots \\ x_{n1} & x_{n2} & \cdots & x_{nm} \end{bmatrix} \tag{4}$$

The formula of variance is obtained as follows:

$$s_{ij} = \frac{1}{n} \sum_{l=1}^{n} (x_{li} - \bar{x}_i)(x_{lj} - \bar{x}_j) \tag{5}$$

The formula of correlation coefficient is obtained as follows:

$$R = \frac{s_{ij}}{\sqrt{s_{ii}}\sqrt{s_{jj}}} \tag{6}$$

2.2 Neural Network Algorithm Based on Nonlinear Principal Component Analysis

The overall flow of the algorithm is to analyze the data of the face image samples with nonlinear principal components. The algorithm transfers the extracted principal component features to the neural network (BPNN) for training learning sample data. In this way, we retain the main feature information of the variables of the original face image, and reduce the number of variables, thus improving the efficiency of the neural network operation.

First, the gray value of the face image file is assigned to the principal component analysis method. According to formula (7), the value of training output node is calculated.

$$\text{Net}_{output_i} = \sum \text{Net}_{input_j} * \text{Net}_{weight_{ji}} \tag{7}$$

We calculate the error value of each point according to the input node and output node generated by the weight of the previous step. The formula (8) is shown below (Fig. 1).

$$error = Net_{input_i} - Net_{output_j} * Net_{weight_{ij}} \tag{8}$$

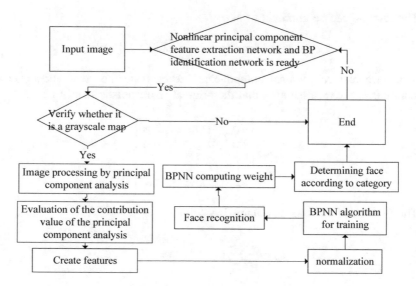

Fig. 1. Algorithm flow chart.

The error value is input into the adjustment weight function as input parameter, and we get the line number aw according to formula (9).

$$aw = Net_{outputn} / \sum_{i=1}^{n} (Net_{output_i})^2 \tag{9}$$

All the output nodes are traversed, and they are input to the nonlinear function as input parameters, and the middle line value s is obtained. Then the weights of each node of the input layer are adjusted according to the formula (10).

$$Net_{weight} = u * s * Net_{error_{ij}} \tag{10}$$

After 100 iterations, the weights of the final nodes are obtained, and normalization of the features is normalized. The normalization formula is followed by formula (11).

$$normalized\ vector = fv/Net_{norm} \tag{11}$$

Among them,

$$Net_{norm} = \sqrt{ss/n} \tag{12}$$

$$ss = \sum fv^2 \tag{13}$$

The computation process of the nonlinear principal component analysis is finished, and the normalized vector is passed to the BPNN algorithm as input parameter.

Based on the formula (14), we calculate the nonlinear principal component analysis method and calculate the output error for all the samples. The BP network correctly identifies the number of images and the average error limit of the recognition results.

$$err_p = \sum_p 0.5 * (Net_{output_i} - Net_{T\,\arg et_i})^2 \tag{14}$$

In the formula, P represents the number of samples, and then the average error is calculated.

$$err = \frac{1}{p} \sum_p err_p \tag{15}$$

The BPNN algorithm is achieved by constantly reducing the err and modifying the weights to make the error value close to the given value of the sample. The adjustment function and the weight correction function are called as follows.

$$new\_dw = 0.3 * Net_{output\_delta} * Net_{hidden\_units} + 0.3 * Net_{ho\_deltaweights} \tag{16}$$

The whole iteration process of the BPNN algorithm is shown in the following figure (Fig. 2).

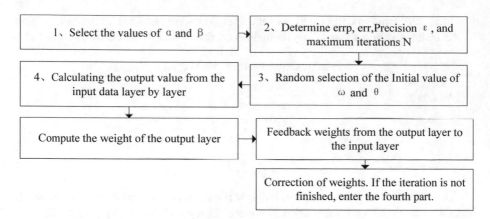

Fig. 2. Neural network flow.

3 Result

In actual experiments, we designed two strategies to automatically modify parameters to achieve the best recognition results. Namely:

1. If the error rate exceeds the threshold, we set the threshold to 5%, then we will clear the weight and increase the training rate by *0.7.
2. If the error rate is less than 5%, the current weight is maintained (Table 1).

Table 1. Face recognition results table.

	Correct	Error	Total	Accuracy [%]
PCA [2]	417	623	1040	40.10%
LDA + PCA [5]	720	320	1040	69.23%
EBGM [6]	922	118	1040	88.65%
ANN [8]	721	319	1040	69.33%
Bayesian_ML [9]	885	155	1040	85.10%
Bayesian_MAP [10]	884	156	1040	85.00%
PCA + BPNN	992	48	1040	95.38%

We used 40 face categories and 400 face images to train the algorithm to test the performance of the algorithm. The recognition rate is shown in the following chart, in which the red cross represents the PCA principal component analysis recognition method, the green difference representative LDA + PCA recognition method, the blue star representing the EBGM elastic map matching algorithm, the purple box representing the Bayesian_ML Bayes maximum likelihood algorithm, the light blue solid block frame generation neural network algorithm, and the yellow circle representing the original. The nonlinear PCA + BPNN algorithm proposed in this paper is based on

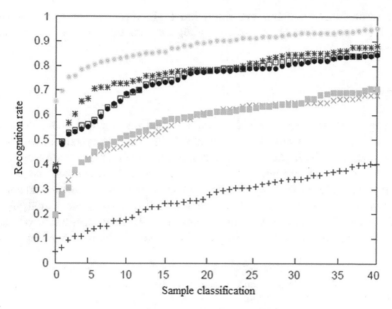

Fig. 3. Recognition rate diagram.

the BPNN algorithm based on the feature extraction of nonlinear principal component analysis, and the black solid circle represents the Bayesian_MAP Bayesian matching algorithm (Fig. 3).

(1) When the number of training samples is gradually increased from 1 to 11, the correct recognition rate of the 7 recognition algorithms is also gradually increasing. This shows that the recognition rate of face recognition algorithm increases with the increase of training samples.

(2) The comparison between the light blue solid box and the green difference, the correct curve of the recognition rate is similar, we can see that the recognition phase of the neural network is compared with LDA + PCA, only one algorithm is much more advanced.

(3) In addition, when the number of samples ranges from 17 to 26, the recognition accuracy of Bayesian_MAP, Bayesian_ML and EBGM elastic graph matching algorithms almost coincides. This shows that the effectiveness of the three algorithms in this view is similar to face recognition, but the recognition rate of the EBGM elastography matching algorithm is always on the top of Bayesian_MAP and Bayesian_ML.

(4) Finally, we can see from the graph that NPCA + BPNN algorithm shows strong momentum, and its recognition rate is much higher than other algorithms. Because PCA simplifies data, the complexity of neural network algorithm is reduced, and time efficiency and accuracy are improved.

(5) The deviation of recognition is mainly caused by wearing ornaments. The illumination is faint when the illumination is not ambiguous.

The results show that our algorithm has a significant advantage in recognition accuracy and recall rate, and F-score is higher than the value of neural network algorithm.

4 Conclusion

In this paper, we propose a BPNN algorithm based on principal component analysis (PCA), which is an improved scheme of face recognition system based on neural network. Based on nonlinear principal component analysis and neural network, the algorithm is more efficient and practical. We compare the proposed face recognition algorithms with the other 6 algorithms, and show that the NP neural network face recognition algorithm based on the nonlinear principal component analysis feature extraction algorithm has a certain improvement in the face recognition rate and stability. The application of custom samples also achieves a high recognition rate.

Acknowledgment. This work is supported by National Natural Science Foundation of China (Grant No. 61604019), Science and technology development project of Jinlin Province (20160520098JH, 20180201086SF), Education Department of Jilin Province (JJKH20181181KJ, JJKH20181165KJ), Jilin provincial development and Reform Commission (2017C031-2).

References

1. Zhou, J., Lu, C.Y., Zhang, C.S.: A survey of automatic face recognition. Acta Electronica Sinica **28**, 102–106 (2000)
2. Mukhedkar, M.M., Powalkar, S.B.: Fast face recognition based on Wavelet Transform on PCA. In: International Conference on Energy Systems and Applications, pp. 761–764. IEEE (2016)
3. Zhou, Z.H., Geng, X.: Projection functions for eye detection. Pattern Recogn. **37**, 1049–1056 (2004)
4. Zhang, Y., Zhao, D., Sun, J., Zou, G., Li, W.: Adaptive convolutional neural network and its application in face recognition. Neural Process. Lett. **43**(2), 389–399 (2016)
5. Borade, S.N., Deshmukh, R.R., Ramu, S.: Face recognition using fusion of PCA and LDA: Borda count approach. In: Control and Automation, pp. 1164–1167 (2016)
6. Chen, X., Zhang, C., Dong, F., Zhou, Z.: Parallelization of elastic bunch graph matching (EBGM) algorithm for fast face recognition. In: IEEE China Summit & International Conference on Signal and Information Processing, pp. 201–205. IEEE (2013)
7. Lecun, Y., Bottou, I., Benuio, Y., Haffner, P.: Gradient-based learning applied to document recognition. In: Proceedings of the IEEE, vol. 86, no. 11, pp. 2278–2324. IEEE (1998)
8. Mageshkumar, C., Thiyagarajan, R., Natarajan, S.P., Arulselvi, S., Sainarayanan, G.: Gabor features and LDA based face recognition with ANN classifier. In: International Conference on Emerging Trends in Electrical and Computer Technology, pp. 831–836. IEEE (2011)

9. Wallis, T.S.A., Taylor, C.P., Wallis, J., Jackson, M.L., Bex, P.J.: Characterization of field loss based on microperimetry is predictive of face recognition difficulties. Invest. Ophthalmol. Vis. Sci. **55**(1), 142 (2014)
10. See, J., Eswaran, C., Fauzi, M.F.A.: Dual-feature bayesian MAP classification: exploiting temporal information for video-based face recognition. In: International Conference on Neural Information Processing, pp. 549–556. Springer-Verlag, Heidelberg (2012)

A Weld Seam Dataset and Automatic Detection of Welding Defects Using Convolutional Neural Network

Wenming Guo[1], Lin Huang[1(✉)], and Lihong Liang[2]

[1] Beijing University of Posts and Telecommunications, Beijing 100876, China
woshiellen2@163.com
[2] China Special Equipment Inspection and Research Institute,
Beijing 100029, China

Abstract. In this paper, we propose a dataset which contains 13006 digitalized x-ray images of welds. And do some preparation work on the original images which can be used to put into the convolutional neural network. Firstly, because of the feature of the input is the original image, but defects of welds are very small in the whole diagram, so that cut up the welds area. Secondly, do some picture preprocessing which includes data enhancement, image regularization, mean subtraction, normalization, etc. Then a model which is built to train and then test the weld images cropped from x-ray images is constructed based on convolutional neutral network. Different from the results ever achieved, this model directly using the feature between pixels and pixels of images without extra extraction of the image feature. Finally, tell the procedure when it comes to train dataset and test dataset, compare the different result of the different image preprocessing, we propose several experiments and results. The results demonstrate that what kind of preprocessing method is better and to do the classification of the picture.

Keywords: Weld image · Dataset · Deep learning · Weld seam detection · Convolutional neural network

1 Introduction

X-ray image classification is becoming popular. There are various kinds of methods are used on human x-ray images to detect focus of infection of many kinds of deseases. The welds detection only relies on the manual feature extraction or semi-automatic feature extraction with the scan on fixed position. Previous studies on welds detection using the feature they extracted from the image as the input of the neutral network model [1, 2]. It depends on precise instruments and human resources [3].

Convolution neutral networks (CNNs) [4] use relatively little pre-processing compared to other image classification algorithms. Although fully connected neural networks can be used to learn features of the picture and then can be used to classify which label the picture belongs to, but when it comes to reality, the model works not well on images. CNN is equivalent to sparse the inputs and parameters [5], focus on the feature on the picture but not every pixel. When it comes to the fully connected layer,

© Springer Nature Switzerland AG 2020
Q. Liu et al. (Eds.): CENet 2018, AISC 905, pp. 434–443, 2020.
https://doi.org/10.1007/978-3-030-14680-1_48

the model combines all the features that the last convolution layer outputs, without leaving any feature behind. Using CNN can release the requirement of outside devices and humans, it is a more automatic way for getting input features.

Besides, the open dataset of welds images is very small. GDXray [6] is an open dataset which contains x-ray images that in nature or luggage and welds. The welds image we can find in GDXray is one of the small groups in the whole dataset, only 88 pictures are welds. With the compare of GDXray, our dataset gives more capability to get the different size and different kinds of weld defect.

2 Convolutional Neural Network

2.1 Convolution Layer

The input of model is the pixel values of whole image, it is far more than the features that people extract previously. If we put such a large feature box into the model, there will be a large number of parameters to be trained, so that the quantity of samples is highly demanded. Compared to basic deep learning method, convolution neutral network gives a model that exists weight sharing. The weights and bias of the same layer are identical, so that we can reduce a large quantity of parameters [7].

The features of a complete image depend on the position and pixel values, CNN keeps the position feature, so that the feature of adjacent pixels are kept.

The deeper the model, the expressive of model will be better. Because CNN is sparse on parameter, so that we can use limited sample train a deeper model to get a higher Recognition rate.

2.2 Max Pooling Layer

The purpose of this layer is to reduce parameter of next layer, at the same time, leave out useless parameters, find out the valuable one output to the next layer.

2.3 Fully Connect Layer

After the convolution part and pooling part have collected enough features of a picture, we use fully connect layer in order not to lose any details of the parameters [8].

2.4 Model

$$INPUT \rightarrow [[CONV] * N \rightarrow POOL?] * M \rightarrow [FC] * K$$

As is shown in the Fig. 1, INPUT is the sample image, CONV is convolution part and POOL is max pooling, the (N, M, K) of this sample is (1, 2, 2).

In the convolution layer, an indicates the input matrix, when it comes to the first convolutional layer, z is a piece of pixels in the image, w is the weights of filter, b is the

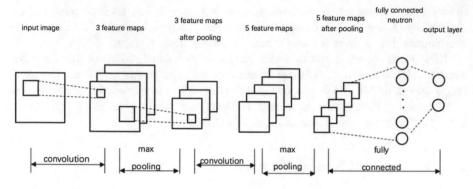

Fig. 1. A simple structure of CNN.

bias of the filter, a is the output of the first convolution layer. The calculation formula is as follows:

$$a_j^l = \sigma\left(\sum_k w_{jk}^l z_k^l + b_j^l\right) \tag{2.1}$$

Also, a_j^l is in only one feature map of the 3 feature maps, and the shape of the filter is 3 s (equals to the amount of feature maps).

Pooling layer pick out the maximum feature of n*n filter size.

Fully connected layer get all features together and give a logit score that whether these features should belong to one class or not.

The model of CNN is easy to learn, but how do train dataset and train labels do on this model, the answer is back propagation.

The process of the back propagation is from output to the input, calculate the cost between the logit score and true label, then calculate how much does every parameter adjust itself. We begin with the cost function [9]:

$$C = \frac{1}{2n}\sum_x \|y(x) - a^L(x)\|^2 \tag{2.2}$$

x is the input data, y(x) is the true label of the data, L is the output layer, aL(x) is the output of the L layer. Then, definite the error of the j neutron of L layer is:

$$\delta_j^l = \frac{\partial C}{\partial z_j^l} \tag{2.3}$$

we can conclude the way to calculate δ_j^l by basic mathematical methods:

$$\delta^L = \nabla_a C \odot \sigma'(z^L) \tag{2.4}$$

The relationship of C and a can be obtained from Eq. (2.2). z^L is the output of a layer, we can conclude from Eq. (2.1) about the relationship between z and a.

Next step is use δ^L to calculate the dissimilarity between prediction and true value of the former layer:

$$\delta^{l-1} = (w^l)^T \delta^l \odot \sigma'(z^{l-1}) \tag{2.5}$$

After that, we use δ^{l-1} calculate the gradient that a weight parameter changes:

$$\frac{\partial C}{\partial w_{jk}^l} = a_k^{l-1} \delta_j^l \tag{2.6}$$

See Eq. (2.6) on the above, we can use the output of last layer and the dissimilarity of the current layer to update the current weight. In like manner, we can calculate the bias:

$$\frac{\partial C}{\partial b_j^l} = \delta_j^l \tag{2.7}$$

Every time after calculating loss value, the parameter of the whole network will be updated one time.

3 Experiment

3.1 The X-Ray Image Set

The digitalized welds x-ray images are 16bit tiff pictures which have 65536 levels of pixel value, so that we cannot check on the personal computer. So we transfer these images to 8 bit pictures which easy to analyze.

First, find the max and min pixel values of a picture, then stretch the pixel value from 0 to 255, extend minimum value to 0, and maximum value to 255, execute this process on all images. Then transfer them to png format.

3.2 Pre-processing of the Images

Firstly, we choose useful data that are clear enough to recognize weld area and any defect on the weld. Check the imbalances of train samples. Our real situation is that, in training dataset, picture with circle defect far more than picture without circle defect. So we remove a picture labels circle defect every 3 circle defect pictures, and get a dataset that are balance enough to train and test. The final target of this experiment is to classify whether an image exists circle defect. Then, because of the image we want to put into the CNN model is very big, we cannot compress the picture to a smaller one, so that if we want to get a better classification, we need to deeper neutral network, then use huge number of data to train the model, we need to accomplish the enhancement of picture. Mirror the image upside down, mirror the left and right turn and so on. After that, we have data four times bigger than before.

Secondly, because the low signal-to-noise ratio of the weld seam, we cut off the weld area from the big image [10]. The mean idea is to do the longitudinal projection of whole image, then, do some ordinary fitting of projection value [11], calculate derivative solution of fitting function, and find which solution is the left side and right side of the weld.

The way to do projection is:

First, find a threshold value, we use two methods and compare the result of every method. First is use the mean pixel and an offset value of an exact image, and second is appoint a threshold (we have tried many times and find the most accurate one).

The weld image is shown on Fig. 2. We consider the image as a matrix. The mean pixel value is:

$$mean = \frac{1}{m * n} \sum_{j=0}^{n-1} \sum_{i=0}^{m-1} image[i][j] \qquad (3.1)$$

m represents width and n represents height of image.

As the experiment goes, we choose the best offset of mean method and the best threshold value which is an unchangeable integer (Table 1).

Table 1. The accuracy of excision using different value.

Value	Accuracy
Mean + offset(offset = −16)	82.8%
Threshold value = 40	76.6%

Fig. 2. Weld x-ray image.

Second, ergodic the whole image, calculate how many pixels larger than threshold value of every column. The weld image is like Fig. 2.

After calculated, we get an array, whose length is the size of x. Put the array to a chart, like Fig. 3 shows.

Consider x direction, the region between two red points are the region we want to cut off.

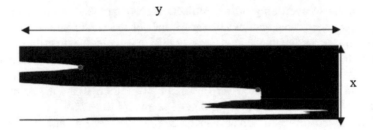

Fig. 3. Array chart image.

Thirdly, find the X coordinate value of the two red points mentioned above. Do least squares polynomial fit, which means fit the array to a function:

$$p(x) = p[0] * x^{deg} + \ldots + p[deg] \tag{3.2}$$

$$y = p(x) \tag{3.3}$$

Point(x, y) can be calculated from Eq. (3.2) deg is degree of the fitting polynomial, the larger the degree, the better it fitted. Fitting function image works like Fig. 4.

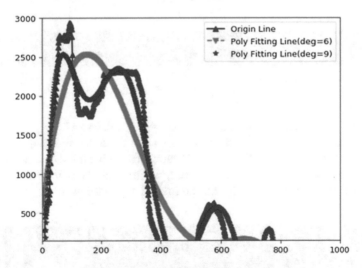

Fig. 4. Fitting function image.

As Fig. 4 shows, the blue line based on array values before fitting. The green line is the fitting function whose degree is 6, obviously this fitting cannot satisfy our demand. The red line is a fitting function whose degree is 9, From observing Fig. 4 we find out it approximately fits the blue line, this means, we can calculate left and right side of weld seam from the fitting function. But when degree increasing, cutting effect does not work well, because overfitting of this projection array can lead to change of function within the weld seam area, causes cutting off half of the weld seam.

To find the coordinate value of res points, calculate the first-order derivative function of the fitting function, then, calculate the zero-solution. There are many zero-solutions, just chose the left and right one of the lowest fitting function value. Look at the red line in Fig. 4, point A(255, 2307) and point B(583,610) are the points we want to find, A and B are the left and right one of point C(470, −750).

We have tested several different degrees cutting effect when cutting off weld seam region:

Table 2. The efficient of different degree.

Degree	Efficiency
9	79.3%
12	82.8%
18	77.5%

From Table 2 we find out that cutting work does better when fitting degree is 12.

Finally, we use the value of two points as left and right side to cut off original image, after cutting the picture looks like Fig. 5.

Fig. 5. Image after cutting.

From Fig. 5 we can see the result of auto-cutting works well.

Next, to get a better classification, we tried not to stretch the shape of an origin picture, then, we do some filling to the picture that with different ratio of length to width. With the observation of whole dataset, we find that the ratio of length to width roughly close to 8:1, so that we fill all the images to a ratio as 8:1.

Fig. 6. Image1 before filling.

Image 1 and image 2 are the weld seams cutting off from the origin images, the ratio of length to width is larger than 8 in Fig. 6, so we fill image1 like Fig. 7. The ratio of length to width is smaller than 8 in Fig. 8, so fill image2 like Fig. 9. Filling way in the Figs. 7 and 9 are one of the three ways. Others are with image mirrors and fill edges

Fig. 7. Image1 after filling.

Fig. 8. Image2 before filling.

Fig. 9. Image2 after filling.

with edge pixels, the way in Fig. 9 using value 0 to fill. We use these filled pictures as input, different filling way lead to different results, classification degree like Table 3 shows.

Table 3. The efficient of different degree.

Filling way	Classification accuracy
No filling	58.3%
Filling with 0	59.5%
Filling with image mirror	60.2%
Filling with edge pixels	59.8%

The pro-processing is the most important part of the whole experiment, until then, we have done with all the procedure, next part will be the training part.

3.3 Training and Testing

We use three ways putting images into the convolutional neutral network. First one is to resize the image to 448*448, and put the image with the shape of (448,448,1) into CNN model. Second one is to segment the image to 8cuts, then transfer 8cuts to a shape of (224,224,8). The last one is to put the whole image to the CNN model, the shape is (1792,224,1). After adjusting the CNN model again and again, until now, the best results is as follows (Table 4):

Table 4. The accuracy of test sets.

Filling way	Classification accuracy
(448,448,1)	56.7%
(224,224,8)	59.5%
(1792,224,1)	60.2%

4 Conclusions

From this paper, we find out that, due to the industrial welding seam X-ray imaging is a 16-bit TIFF figure, and the sample image is too large so that the image has large noise and weak signal.

In order to make the sample data suitable to convolutional neutral network, we have optimized the data samples as much as possible, and the signal-to-noise ratio of the input images are improved, and we get an entire procession from an original weld seam image to an image which can be put into CNN model, and do the compare between different processing method, then, put the processed sample data to the CNN model, after training and testing, we have got an initial achievement.

Acknowledgement. This research was supported by the National key foundation for exploring scientific instrument of China (Project No. 2013YQ240803).

References

1. Gatys, L.A., Ecker, A. S., Bethge, M.: A neural algorithm of artistic style (2015). https://arxiv.org/abs/1508.06576
2. Gatys, L., Ecker, A.S., Bethge, M.: Texture synthesis using convolutional neural networks. In: NIPS 2015 Proceedings of the 28th International Conference on Neural Information Processing Systems, vol. 70, no. 1, pp. 262–270 (2015)
3. Valavanis, I., Kosmopoulos, D.: Multiclass defect detection and classification in weld radiographic images using geometric and texture features. Expert Syst. Appl. **37**, 7606–7614 (2010)
4. Raschman, E., Záluský, R., Ďuračková, D.: New digital architecture of CNN for pattern recognition. J. Electr. Eng. **61**(4), 222–228 (2010)
5. Yu, W., Sun, X.S., Yang, K.Y., Rui, Y., Yao, H.X.: Hierarchical semantic image matching using CNN feature pyramid. Comput. Vis. Image Underst. **169**, 40–51 (2018)
6. Mery, D., Riffo, V., Zscherpel, U., Mondragón, G., Lillo, I., Zuccar, I., Lobel, H., Carrasco, M.: GDXray: the database of X-ray images for nondestructive testing. J. Nondestr. Eval. **34**(4), 1–12 (2015)
7. Hassan, J., Awan, A.M., Jalil, A.: Welding defect detection and classification using geometric features. In: 10th International Conference on Frontiers of Information Technology, pp. 139–144 (2012)

8. Rathod, V.R., Anand, R.S.: A comparative study of different segmentation techniques for detection of flaws in NDE weld images. J. Nondestr. Eval. **31**, 1–16 (2012)
9. Bai, X., Huang, F.Y., Guo, X.W., Yao, C., Shi, B.G.: Training method and apparatus for convolutional neutral network model. Patent: WO2016155564 (2016)
10. Zapata, J., Vilar, R., Ruiz, R.: An adaptive-network-based fuzzy inference system for classification of welding defects. NDT and E Int. **43**, 191–199 (2010)
11. Zhang, Y., Zhang, E., Chen, W.: Deep neural network for halftone image classification based on sparse auto-encoder. Eng. Appl. Artif. Intell. **50**, 245–255 (2016)

Prediction Model of Wavelet Neural Network for Hybrid Storage System

Haifeng Wang, Ming Zhang, and Yunpeng Cao[✉]

School of Information Science and Engineering, Linyi University,
Linyi 276000, China
caoyunpeng@lyu.edu.cn

Abstract. The Hybrid storage system needs to distinguish the data state to manage data migration. The frequently data may be placed solid-state hard disk to improve the accessing performance. Here a novel prediction model of the frequently accessing data that is called hot access data is proposed. This model extracts the workload features and is built based on the wavelet neural network to identify the data state. The prediction model is trained by the sampling data from historical workloads and can be applied in the hybrid storage system. The experimental results show that the proposed model has better accuracy and faster learning speed than BP neural network model. Additionally, it has better independent on training data and generalization ability to adapt to various storage workloads.

Keywords: Hybrid storage system · Wavelet Neural Network ·
Prediction model · Data migration

1 Introduction

The hybrid storage system is consisted of multiple storage devices [1], such as solid-state drive (SSD), hard disk driver (HDD), magnetic disk driver, et al. In this hybrid storage system, SSD is a flash-based electronic storage device, which has better read performance than HDD. So the frequently read data called hot data may be placed in SSD to improve the accessing performance [2]. And the data with frequently writing tendency may be assigned to HDD to improve the performance of write operations in hybrid storage system. Then the key issue is to identify the data state. This is the problem of data state classification [3]. The data of storage workloads should be divided into hot and cold data. Currently, the researchers focus on the statistical methods. Kgil [4] and Koltsidas [5] proposed a classification model based on the statistics information of I/O data page to adjust the data migration strategy. However this approach has a shortcoming that is the time accumulation effect. This defect is from the statistical method and may cause the classification model to be insensitive to the variation of random accessing workloads. But there exist a lager number of I/O random access workloads in the hybrid storage system. To improve the accuracy of classification, Yang Puyuan proposed a method of increasing intermediate state to reduce occasional read tendency page classification error and the caused false migration [6].

To deal with the cumulative effect of statistical method and reduce the classification errors, the prediction model to identify the hot data in the hybrid storage system is built based on the wavelet neural network. This model extracted the features of workloads and was trained by the historical accessing data from the hybrid storage system. This prediction model will guide the optimization of data migration strategy.

2 The Prediction Model

2.1 The Features of Input Vector

In this section, we considered the accessing features from the data block perspective to describe the variation of workloads. The features of workloads are the vectors $<T$, ς, S, $L>$. T is the recent time span of a data block. As for one data block, the set of time for each access is $<t_1, t_2, \ldots, t_n>$ from the creation. And the time span for each access operation is as follows: $t_2 - t_1$, $t_3 - t_2$, $t_n - t_{n-1}$. For simplicity, we use the latest accessing interval as the time span of data block. For example, When the current time is t_i, the time span of data block is $T = t_i - t_{i-1}$. ς is the number of processes which access data block. It reflects the degree of concurrent access to data block to a certain extent. S is the size of attached file, which is the file including the data blocks. L represents the life of data block. It is the time interval between current time and the creation time of data block. Generally speaking, the longer the life of data block is, the lower probability of being accessed is.

2.2 The Hot Degree of Model

The output of prediction model is the data state, which represents the frequently degree of accessing. It is called hot degree, which is a continuous value in [0, 1]. In Eq. (1), the hot degree h is defined.

$$h = w\delta + (1 - w)\theta \; , \; h \in [0, 1] \tag{1}$$

δ is the accessing frequency in each unit period. θ is the access density in each unit period. The ratio of the number of unit times to the length of the statistical period. It reflects the distribution of data block access. In Eq. (1), the larger the value is, the greater h of data block is. Note that the weight $w = 0.7$, which describes the importance of these two indexes δ and θ.

2.3 The Structure of Prediction Model

The prediction model is to identify a nonlinear function. The input of this function is feature vector and the output is the hot degree of data block $h \in [0, 1]$. The wavelet neural network (WNN) is used to construct the prediction model [7]. WNN is suitable for fitting the nonlinear relation between the workload features and the hot degree of data block. The wavelet neural network excellent nonlinear mapping and generalization ability due to that it integrates the advantages of wavelet analysis and artificial neural network [8]. The prediction model is to combine wavelet function with back propagation BP neural network (BP). And the sigmoid function in BP neural network is replaced with nonlinear wavelet function [8]. The structure of wavelet neural network is shown in Fig. 1.

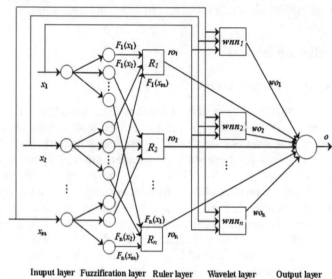

Inuput layer   Fuzzification layer   Ruler layer   Wavelet layer   Output layer

Fig. 1. Architecture of wavelet neural networks.

The wavelet neural network includes input layer, fuzzy layer, inference layer, wavelet layer and output layer. The number of neuron nodes in each layer is n, $n \times M$, M, M and 1 respectively. ① Input layer. Each node of this layer is directly connected to each input component x_j of the feature vector $<T, \varsigma, S, L>$, and passes the input value $X = [x_1, x_2, x_3, \ldots, x_n]$ to the next layer. ② Fuzzy layer. Through the fuzzy rules, the input vector is transformed into fuzzy values. ③ Rule layer. Rule layer is also known as inference layer. Each neuron corresponds to a fuzzy rule. ④ Wavelet layer. The wavelet network layer output is mainly used for output compensation. ⑤ Output layer. This layer is the final output layer of wavelet neural network. It can generate results of prediction model.

2.4   The Training of Model

In this section, the detailed training of model is discussed. The prediction model of wavelet neural network can be expressed by Eq. (2).

$$Q(X) = \sum_{j=1}^{N} W_{ij}^{(J)} \Psi_j(z) + \bar{\theta}, \tag{2}$$

where $X = [x_1, x_2, x_3, \ldots, x_M]^T$ is input vector. $\Psi_j(z)$ is the neuron wavelet activation function of wavelet layer. W_{ij} is the weight of the connection from the j-th node of wavelet layer to output layer. $\overline{\theta}$ is the average value of output. $Q(X)$ is the output value of wavelet neural network [9].

The purpose of training is to use sample data to determine the important parameters in WNN. Those parameters include Z_{ij}, W_j, a_j, b_j, $\overline{\theta}$. Firstly, the parameters to be trained are constructed into parameter vector with string structure by orders. Each vector is a chromosome for genetic operation. Each chromosome is encoded with a real number [10]. The initial value of the parameter is determined by the following steps: ① Determination of stretching and translation parameters. According to the nature of wavelet function, the window center position and width of the wavelet function are fixed values. Given that the initial center of the wavelet window is x_{0j} and the window width is Δx_{0j}, then the scaling factor a_j is given by Eq. (3):

$$a_j = \frac{\sum_{j=1}^{M} x_{jmax} - \sum_{j=1}^{M} x_{jmin}}{\Delta x_{0j}} \tag{3}$$

The translation factor b_j is determined by Eq. (4):

$$b_j = 0.5 \times \left(\sum_{j=1}^{M} x_{jmax} + \sum_{j=1}^{M} x_{jmin} \right) - a_j \times x_{0j} \tag{4}$$

M in Eqs. (3) and (4) are the number of input vectors. x_{jmax} and x_{jmin} are the maximum and minimum values of the j-th neuron in the input layer, respectively. ② Determination of network weights. The initial value of the weight from input layer to wavelet layer Z_{ij} and the weight from wavelet layer to output layer W_j is to select a uniformly distributed random number in $[-1, 1]$ and to ensure that the various values are not zero. ③ Determination of the parameter $\overline{\theta}$. It is obtained according to the partial sample data. This parameter should be constantly updated and corrected during calculation.

The training algorithm is designed based on evolutionary algorithm [11]. Before the start of training, the genetic population size was set to $L = 200$. The maximum number of iterations was $J = 300$. The network convergence accuracy was $\varepsilon = 0.003$. The probability of selection was $p_s = 0.65$. The crossover probability was $p_c = 0.8$. The probability of variation is $p_m = 0.03$. The specific training algorithm is as follows.

Algorithm: The training algorithm for Prediction model
① Set the initial value of the iteration variable, $J = 0$, and then base on the determination method of parameter initial value to create L initial parent classes $T_1^{(0)}$, $T_2^{(0)}$, ..., $T_L^{(0)}$;
② Calculate fitness function, as shown in Eq. (5). When the smaller the value of the adaptive function is, it means the better the network training effect.

$$E(T_l^{(J)}) = 1/2 \sum_{k=1}^{K} [Q(X_k, T_l^{(J)}) - Q'_k]^2, \tag{5}$$

where $Q(X_k, T_l^{(J)})$ is the wavelet neural network output value calculated by Eq. (2), Q'_k is the expected output value of prediction model, K is the number of elements in training sample set.

③ Cross and mutate the j-th generation of chromosomes, and select N individuals to enter the next generation of evolution.

④ Determine convergence. When the condition $(E_{max} - E_{min})/E_{avg} \leq \varepsilon$ or $J > J_{max}$ is satisfied, the training algorithm ends; otherwise update the variable J, $J = J + 1$ and then return to Step (3). E_{max}, E_{min} and E_{avg} represent the maximum, minimum and average value of the calculated fitness function, respectively.

⑤ Select the best combination of parameters that has reached best fitness accuracy in the previous step and then perform real-time prediction.

3 Experiment and Analysis

The data access trace of hybrid storage system in production environment is used to train the prediction model and perform the comparison of performance. The experimental environment is Linux operating system and file system is ext2, whose data block is 4 KB. The software Blktrace is used to collect I/O data of disk data blocks in Linux system. It can monitor the I/O events of a particular disk block device, capture and record events such as read operation, write operation, synchronous operation. Blkparse is used to analyze blktrace log data.

3.1 Performance Analysis

In this section, we mainly validate the advantages of the prediction model constructed by wavelet neural network. Here the BP neural network is used as the benchmark model to compare. In BP model, the excitation function of neuron is S function. The weights and coefficients in BP model are adjusted adaptively. The minimum mean square error energy function is used to optimize network parameters. The error is defined in Eq. (6).

$$E = 1/2 \sum_{l=1}^{L} (y_l - \widehat{y}_l)^2, \tag{6}$$

where l is the number of samples, y_l and $\widehat{y}_l$ is the actual and the predicted value, respectively. Then the gradient descent method is used to calculate the instantaneous gradient vector; the learning rate parameter in the reverse propagation phase $\eta = 0.05$. In order to facilitate distinction, the proposed prediction model is denoted as WNN and the BP model is denoted as BP. Firstly, the training speed of the two models is compared. 300 sample data are randomly selected to train the models.

The error of two models is shown in Fig. 2. The error accuracy of WNN is lower than the pre-set value after 57 iterations. The BP model is not lower than the pre-set value until after 217 iterations.

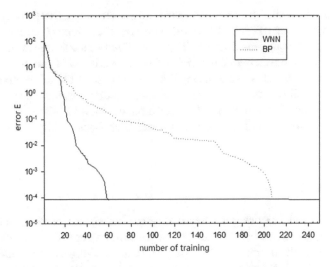

Fig. 2. Training comparison of wavelet neural network and BP neural network.

In addition to adjusting network weights, the wavelet neural network model can adjust the scaling factor and translation factor of wavelet function. So it has better adaptability and elasticity than BP neural network.

3.2 Comparison of Prediction Accuracy

After the train phase, these two prediction models are used to predict the hot degree of data block online. The main purpose is to compare the prediction accuracy of the two models. The accuracy of wavelet neural network and BP neural network is compared by selecting 10 actual data which are shown in Table 1. Table 1 lists the predicted values and the variance of the predicted values of WNN and BP.

Table 1. Comparison of predicted results.

Ord	Actual	WNN		BP	
		Predicted	Variance	Predicted	Variance
1	0.592	0.583	0.019	0.610	0.044
2	0.081	0.082	0.003	0.010	0.001
3	0.453	0.446	0.015	0.467	0.034
4	0.403	0.409	0.013	0.415	0.03
5	0.102	0.103	0.003	0.098	0.007
6	0.073	0.071	0.002	0.075	0.005
7	0.415	0.421	0.014	0.402	0.029
8	0.524	0.515	0.017	0.540	0.039
9	0.173	0.175	0.006	0.167	0.012
10	0.393	0.386	0.013	0.380	0.027

The results show that the prediction error of wavelet neural network is 1.6%, while the prediction error of BP neural network is 3.1%. This indicates that wavelet neural network has better nonlinear fitting ability. Additionally, the variance of the predicted value of wavelet neural network is 3.3%, while the variance of the predicted value of BP neural network is 7.1%. This means that the proposed prediction model has better fault tolerance ability due to the fact that BP neural network uses conventional S function as excitation function. The excitation function is relatively smooth. So it cannot quickly respond to those sample data which changes rapidly from wave peak to wave trough.

4 Conclusion

The prediction model is designed to identify the hot degree of data block in hybrid storage system. The kernel of this model is wavelet neural network. The fuzzy theory is applied to deal with the input vector of prediction model. And this may enable the prediction model to process the uncertain data. Additionally, the wavelet neural network has better nonlinear fitting ability for small training data set. The experimental results show that the wavelet neural network not only has better prediction accuracy than BP neural network, but also has faster speed of training and learning than BP neural network. However, the prediction model is not be applied to data migration in hybrid storage system. In the future, we integrate this prediction model with data block migration and verify the prediction accuracy in practice.

Acknowledgement. This project is supported by Shandong Provincial Natural Science Foundation, China (No. ZR2017MF050), Project of Shandong Province Higher Educational Science and technology program (No. J17KA049), Shandong Province Key Research and Development Program of China (Nos. 2018GGX101005, 2017CXGC0701, 2016GGX109001) and Shandong Province Independent Innovation and Achievement Transformation, China (No. 2014ZZCX0 2702).

References

1. Yamato, Y.J.: Use case study of HDD-SSD hybrid storage, distributed storage and HDD storage on OpenStack. IEEE Trans. Electr. Electron. Eng. **11**(5), 674–675 (2016)
2. Huang, D.M., Du, Y.L., He, Q.: Research on ocean large data migration algorithm in hybrid cloud storage. J. Comput. Res. Dev. **51**(1), 199–205 (2014). (In Chinese)
3. Luo, J.J., Fan, L.Y., Li, Z.H., Tsu, C.: A new big data storage architecture with intrinsic search engines. Neurocomputing **181**(C), 147–152 (2016)
4. Kgil, T., Roberts, D., Mudge, T.: Improving NAND flash based disk caches. In: Proceedings of the 35th International Symposium on Computer Architecture, pp, 327–338, Beijing, China (2008)
5. Koltsidas, I., Viglas, S. D.: Designing a flash-aware two-level cache. In: Proceedings of the 15th International Conference on Advances in Databases and Information Systems, pp, 153–169, Berlin, Heidelberg (2011)

6. Yang, P.Y., Jin, P.Q., Yue, L.H.: A time-sensitive SSD and HDD efficient hybrid storage model. Chin. J. Comput. **35**(11), 2294–2305 (2012). (In Chinese)
7. Guo, Y.C., Wang, L.H.: Hybrid wavelet neural network blind equalization algorithm controlled by fuzzy neural network. ACTA Electronica Sin. **39**(4), 975–980 (2011). (In Chinese)
8. Lin, F.J., Hung, Y.C., Ruan, K.C.: An intelligent second-order sliding-mode control for an electric power steering system using a wavelet fuzzy neural network. IEEE Trans. Fuzzy Syst. **22**(6), 1598–1611 (2014)
9. Do, H.T., Zhang, X., Nguyen, N.V., Li, S.S., Chu, T.T.: Passive-Islanding detection method using the wavelet packet transform in grid-connected photovoltaic system. IEEE Trans. Power Electron. **31**(10), 6955–6967 (2016)
10. Alessia, A., Sankar, K.: Rough sets, kernel set, and spatiotemporal outlier detection. IEEE Trans. Knowl. Data Eng. **26**(1), 194–207 (2014)
11. Wang, X.Y., Liu, Q., Fu, Q.M., Zhang, L.: Double elite co-evolutionary genetic algorithm. J. Softw. **23**(4), 765–775 (2014). (In Chinese)

Particle Swarm Optimization Algorithm Based on Graph Knowledge Transfer for Geometric Constraint Solving

Mingyu Sun, Qingliang Li[✉], Jinlong Zhu, and Yu Zhang

College of Computer Science and Technology, Changchun Normal University,
Changchun 130032, China
lql_321@126.com

Abstract. In order to more effectively solve complicated geometric constraint problems by applying swarm intelligence technologies, a particle swarm optimization (PSO) algorithm based on graph knowledge transfer for geometric constraint solving (GCS) is proposed. By fusing the graph knowledge transfer mechanism into the PSO algorithm to select parameters deciding the algorithm performance, avoiding getting stuck in a local extreme value and then making the algorithm stagnating when solving a practical complicated geometric constraint problem. Empirical results show that using the graph knowledge transfer mechanism to select the parameters of PSO can obtain high-quality parameters of GCS. It improves the efficiency and reliability of GCS and possess better convergence property.

Keywords: Geometric constraint solving · Particle swarm optimization · Graph knowledge transfer

1 Introduction

Geometric constraint solving (GCS) roots in the introduction of artificial intelligence in computer aided design (CAD) field. Whether for parametric design system or for variational design system, GCS is undoubtedly the core in these design systems, it is considered as the automation of geometric drawing, that is a user assembles geometric primitives and geometric constraints to construct a sketch of modeling, the sketch is automatically transformed into a modeling satisfying constraint by GCS. The process of transforming the sketch into a modeling satisfying constraint is a process of solving a geometric constraint problem.

A geometric constraint problem is formalized as $GCP = (P, C)$, in which $P = (p_1, p_2, \ldots, p_n)$ is a set of geometric primitives such as points and lines in 2D or points, lines and planes in 3D, $C = (c_1, c_2, \ldots, c_m)$ a set of geometric constraints such as distance constraints between two points and angle constraints between two lines. The constraint equations can be constructed to represent a geometric constraint problem. Thus, solving the geometric constraint problem is namely solving the constraint equations, which is

© Springer Nature Switzerland AG 2020
Q. Liu et al. (Eds.): CENet 2018, AISC 905, pp. 452–462, 2020.
https://doi.org/10.1007/978-3-030-14680-1_50

usually nonlinear. Let $X = (x_1, x_2, ..., x_n)$ be the unknowns of the equations, so a geometric constraint problem is expressed as follows:

$$F(X) = 0 \tag{1}$$

In practice, a geometric constraint problem is complicated, causing the geometric constraint equations are also, there is still no an efficient and accurate method of solving complicated equations so far, some existing solving methods are various. Converting the equations into an optimization problem and then solving the equations by using swarm intelligence algorithms, such as particle swarm optimization (PSO) algorithm, ant colony (AC) algorithm and artificial bee colony (ABC) algorithm and so on, to solve the optimization problem is a feasible method of GCS. We make a transformation of geometric constraint equations, as follows:

$$\dot{F}(X) = \sum\nolimits_{i=1,...,m} |f_i(X|) \tag{2}$$

If there is a vector X satisfying $F'(X) = 0$, then X can also satisfy Eq. (1). Therefore, GCS is converted into the solving of optimization problem, which is a solving process of minimizing objective function in the case of satisfying constraints of independent variable, the equivalent mathematical model may be expressed as follows:

$$Min\dot{F}(X) = 0 \tag{3}$$

The mathematical model must satisfy:

$$\dot{F}(X) < \varepsilon \tag{4}$$

ε is the accuracy threshold satisfying GCS demand, used to judge whether the optimization procedure should be terminated.

Cao Chunhong proposed a crossbreeding PSO algorithm [1], a PSO based on quantum behavior [2] and a niche improved PSO algorithm [3], to solve geometric constraint problems, respectively, Yuan Hua proposed a tuba PSO algorithm [4] to solve geometric constraint problems. Many scholars also proposed various improved PSO algorithms to solving optimization problems, typically, such as, Coello's PSO algorithm based on hypergrid and mutation [5], Chi Yuhong's PSO algorithm based on search space zoomed factor and attractor [6], Zhou Xinyu's PSO algorithm based on elite opposition [7]. Tan's chaotic PSO algorithm [8], Shen Yuanxia's parallel-cooperative bare-bone PSO algorithm [9], Zhao Ji's cooperative quantum behavior PSO algorithm with adaptive mutation based on entire search history [10], Zhou Lingyun's PSO algorithm based on neighborhood centroid opposition [11], Lv Boquan's fuzzy PSO algorithm based on filled function and transformation function [12], and so on. The performance of a PSO algorithm mainly depends on its parameters, for different types of optimization problems, different parameter values ought to be selected to produce swifter effects. However, these improved PSO algorithms did not try to begin with a change of parameter value so that the optimal results are unsatisfactory when facing new types of optimization problems. In this paper, we propose a PSO algorithm based on graph knowledge transfer to solving geometric constraint problems, the training samples of parameter set can be learned by graph knowledge transfer to generate better parameter set of PSO algorithm. Whether the parameter set of problem that to be optimized is good depends on the training samples. Graph knowledge transfer can make a parameter set obtained maintain good stability in quality, although it is not ensured that the parameter set obtained by learning is optimal. The experiment shows that applying the algorithm of this paper to solve geometric constraint problems improves the efficiency and reliability of solving and this algorithm possess better convergence property.

2   Basic Particle Swarm Optimization Algorithm

Particle Swarm Optimization (PSO) algorithm was proposed by Kennedy and Eberhart firstly in 1995, which is an evolutionary computation technology inspired by the results of the artificial life and simulates, simulating the migrating and clustering behaviors that birds forage. Like genetic algorithm, PSO algorithm has been widely applied to solve many complicated optimization problems since it is structurally simple and easy to realize [13].

2.1   Basic Principle

According to the pre-set principle, a group of particles having search capacity are scattered in the D-dimensional solution space of the problem that to be optimized, they cooperatively search the solution space to find an optimal solution meeting accuracy. There is no an absolutely central control individual in all particles and simultaneously, the more diversity particles have, the more capable of dealing with complicated problems this algorithm is.

Each of particles possess a good memory feature such that it has a certain cognitive capacity. As a flying individual having simple intelligence, it represents a potential solution of optimization problem, having no mass and volume. During searching, it records the optimal location that it has arrived itself each time, called a personal best location Pb, the corresponding fitness value is called a personal extreme value Pe. Besides, for obtaining the flying experience of particles in neighborhood, the optimal location that the whole particle swarm has arrived is also recorded, called a global best location Gb, the corresponding fitness value is called a global extreme value Ge. A particle is capable of learning from others, thus it can adjust its own flying velocity according to the flying experience of itself and neighborhood particles to speed up the local convergence of this algorithm.

2.2   Mathematical Model of PSO

According to the basic principle of PSO algorithm, constructing the corresponding mathematical model is the key of solving practical problems. Using the PSO algorithm to find an optimal solution is essentially an iterative process of self-organization learning. It is assumed that there a population $S = \{X_1, X_2, ..., X_n\}$ consisting of n particles representing the potential solution of the problem in the D-dimensional solution space, so the mathematical model that the iteration of each of particles follows is formulated as:

$$V_i^{k+1} = wV_i^k + c_1 r_1 \left(Pe_i^k - X_i^k\right) + c_2 r_2 \left(Ge_i^k - X_i^k\right), \begin{cases} V_i = V_{max}, \text{if } V_i > V_{max} \\ V_i = V_{min}, \text{if } V_i < V_{min} \end{cases} \quad (5)$$

$$X_i^{k+1} = X_i^k + V_i^{k+1}, \begin{cases} X_i = V_{max}, \text{if } X_i > X_{max} \\ X_i = V_{min}, \text{if } X_i < X_{min} \end{cases} \quad (6)$$

Therein, $V_i^k = \left(V_i^{k1}, V_i^{k2}, ..., V_i^{kD}\right)$ is the velocity of the k-th times-iteration of the i-th particle, $X_i^k = \left(x_i^{k1}, x_i^{k2}, ..., x_i^{kD}\right)$ the location of the i-th particle in the D-dimensional solution space after k times-iteration, $Pb_i^k = \left(pb_i^{k1}, pb_i^{k2}, ..., pb_i^{kD}\right)$ and $Gb^k = (gb^{k1}, gb^{k2}, ..., gb^{kD})$ the history best location and the global best location of the i-th particle, respectively, w the inertia weight that used to maintain flying state, generally w belongs to [0.9, 1.2] [14], c_1 and c_2 the nonnegative learning factors, r_1 and r_2 two random numbers that read on the interval [0,1]. The velocity and location of each particle is restricted in an allowed scope, in other words, $v^j \in \left[v_{Min}^j, v_{Max}^j\right], x^j \in \left[x_{Min}^j, x_{Max}^j\right]$, ($j = 1, 2, ..., D$), ensuring these particles can find an optimal solution meeting accuracy in an established feasible region.

The formula (5) is composed of three parts, the first part wV_i^k describes the ability of a particle to maintain its original flying state, is the inertia indicator that it flies, playing an important role in balancing the global and local search, the second part $c_1 r_1 \left(Pe_i^k - X_i^k\right)$ is the self-cognitive ability of this particle, indicates its subsequent searches depend on its own flying experience obtained, avoiding to stuck in a local extreme value, the third part is the social cognitive ability, directly reflects the cooperation and knowledge sharing between particles. The Fig. 1 shows the optimizing pattern of basic PSO algorithm.

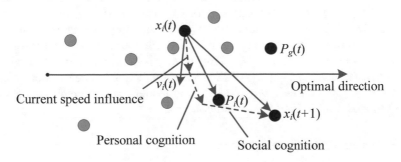

Fig. 1. The optimizing pattern of basic PSO algorithm pattern.

3   PSO Algorithm Based on Graph Knowledge Transfer

3.1   Parameter Analysis of PSO Algorithm

The parameter set of PSO algorithm mainly consists of the inertia weight w, the learning factors c_1 and c_2 in the velocity and the population size NP. w determines the proportion of the last velocity obtained by iterating that need to be retained, c_1 and c_2 determine the self-cognitive ability and the social cognitive ability, respectively. If c_1 is too large, then all of the particles tend to search on the local scale, which is adverse to search globally. And if c_2 is also, then all of them tend to prematurely get stuck in a local extreme value such that the accuracy of solution is reduced. In further, if NP is too large, then the convergence speed is low, otherwise, all of the particles remain at a point such that the algorithm stagnates. Therefore, for enhance the performance of PSO algorithm to optimize different problems, the parameter set $\{w, c_1, c_2, NP\}$ must be reasonably selected.

3.2   Adjacent Transfer Matrix

Given the source task set $T = \{t_{i,i=1,\dots,n}\}$ that solve an optimization problem, in which for each specific source task t_i, the PSO algorithm model m_i is adopted to solve the optimization problem. The model system m_i contains parameter set $P_i = \{w_i, c_{1i}, c_{2i}, NP_i\} \in R^\theta$ that needed when the PSO algorithm runs. The parameter set P_i can migrate among different tasks. Firstly, the source task set T is mapped into a high dimensional space R^θ, which is the transfer space of T that used to describe the transfer performance among different tasks, θ is equal to the dimension of P_i, that is $\theta = 4$, each P_i represents a different task. Secondly, the parameter set P_i is used to represent the θ-dimensional space coordinate of source task t_i. Thirdly, the transfer weight w_{ij} is used to connect t_i with t_j. The transfer weight of t_i and t_j is represented as $w_{ij} = (Tlength_{jj} - Tlength_{ij})/Tlength_{jj}$, in which $Tlength_{ij}$ and $Tlength_{jj}$ are the average running time of the model m_i and m_j under the control of the parameter set P_i and P_j, respectively. w_{ij} describes quantitatively the performance that source task t_i transfers t_j, the transfer between t_i and t_j is asymmetric, that is it is generally true $w_{ij} \neq w_{ji}$. And it is required that $Tlength_{ij}$, $Tlength_{jj} \in (0, +\infty)$, so w_{ij} *belongs to* $(-\infty, 1)$. The transfer performs worse if w_{ij} decreases. The transfer is positive for $w_{ij} > 0$, showing that the current transfer parameter set is better than the original, it is invariant after transferring for $w_{ij} = 0$ and the transfer is negative for $w_{ij} < 0$, showing that it is worse than the original. An adjacent transfer matrix $A_n = \{a_{i,j}\}$ of T is created by merging these transfer weights, $a_{i,j} = 0$ if $i = j$, $a_{i,j} = Min(w_{ij}, w_{j,i})$ if $i \neq j$. The model transfer graph $G = (P, A_n)$ is adopted to model R^θ [15], P is the task point set.

3.3   Extended Model Transfer Graph

An extended model transfer graph $G_e = (P_e, A_{n+1})$ can be constructed to optimize the target task t_{n+1}, whose parameter set is $P_{n+1} = \{w_{n+1}, c_{1,n+1}, c_{2,n+1}, NP_{n+1}\} \in R^\theta$. It is assumed that P_{n+1} is the parameter set of t_{n+1}, $w_e = (w_{e1}, w_{e2}, \dots, w_{en})$ the transfer weight sequence that each source task t_i in T transfer t_{n+1}, then the extended model transfer graph of G is $G_e = (P_e, A_{n+1})$, in which $P_e = P \cup P_{n+1}$, $A_{n+1} = ((A_n, w_e^T),$ $(w_e, 0))^T$. After obtaining the extended model transfer graph, knowledge transfer can be realized by using it to construct the transfer function, which refers to a mapping f_T: Te $\rightarrow R^\theta$ of a task set $T_e = \{t_{i,i=1,\dots,n+1}\}$ of containing the task t_{n+1} into the parameter set P_e, this mapping distributes each parameter set P_i to the task t_i.

Each known parameter set $P_i \in P$ in the source task T may be taken as the learning samples of a transfer function f_T that to be obtained. The function may transfer knowledge of P from T to t_{n+1} such that t_{n+1} obtains an optimized parameter set P_{n+1} that used in the PSO algorithm model m_{t+1} by learning knowledge contained in P.

3.4 Parameter Optimization

The model transfer graph $G = (P, A_n)$ of the source task set $T = \{t_{i,i=1,\dots,n}\}$ is a θ-dimensional undirected connected graph, containing n vertices and n^2 transfer weights, $d_i = \sum_{j=1,\dots,n} a_{ij}$ is the degree of each vertex P_i, $D = diag(d_1, d_2, \dots, d_n)$ is the diagonal matrix of the degree. G can be uniquely mapped into the Laplace transform matrix $L = D - A_n$. It is assumed that Q is the eigenvector matrix of the Laplace transform matrix L, $\Delta = diag\{lambda_1, \dots, lambda_n\}$ the diagonal matrix of Q, satisfying $\{0 = lambda_1 \ll \dots \ll lambda_n\}$, and $f_m = (P_1, P_2, \dots, P_n)^T$ the transfer function matrix, so $L = Q \Delta Q^T$ is obtained by the eigen decomposition of L. Similarly, the Laplace operator L_e, eigenvalue diagonal matrix Δ_e and eigenvector matrix Q_e of the extended model graph G_e are obtained. There is a $(n + 1) \times \theta$-dimensional weight matrix W such that $f_m = Q_eW$, the least square method is used to obtain the practical weight matrix $W_{*,i}$ as follows:

$$W_{*,i} = \underbrace{argmin}_{w} \left\| f_{*j-Qw} \right\|_2^2 + \left\| \sqrt{lambdaw} \right\|^2 \tag{7}$$

$Q = Q_{e1,2,\dots,n*}$ is the first n rows of Q_e, $f = (P_1, P_2,\dots, P_n)^T$, $\left\| \sqrt{lambdaw} \right\|^2$ is the penalty term, with a good refrain effect on high-frequency component in f to avoid over-fitting, of a second derivative, it may be obtained by the following formula:

$$\langle \nabla f_m, \nabla f_m \rangle = \langle f_m, L_e f_m \rangle = (Q_ew)^T (L_eQ_ew) = \left\| \sqrt{\Delta_e}w \right\| \tag{8}$$

Then, the following weight matrix W is further obtained:

$$W = \left(Q_e^T Q_e + \Delta \right)^{-1} Q_e^T f \tag{9}$$

Then $f_m = Q_eW$ is used to obtain the transfer function f_m. The optimal parameter set P_{n+1} of the target task t_{n+1} is the transposition of f_m's (n + 1)-throw.

3.5   Algorithm Description

We use the graph knowledge transfer to learn training samples of PSO algorithm parameter set, generating better parameter set of problem that to be optimized, if the training samples is good, then the parameter set obtained by learning is also. The PSO based on graph knowledge transfer may maintain a better stability of parameter set although it is not always optimal.

Algorithm: PSO based on graph knowledge transfer

Begin
1   Obtain the parameter set of the algorithm;
1.1 Extract the necessary source task set T and task point set P from the training samples;
1.2 Compute each transfer weight w_{ij} of transfering parameter set P_i to P_j according to $w_{ij}=$ $(Tlength_{jj} - Tlength_{ij}) / Tlength_{jj}$;
1.3 Construct the adjacent transfer matrix A_n;
1.4 Randomly generate the parameter set P_{n+1} of the target task t_{n+1};
1.5 Compute each transfer weight $w_{i,n+1}$ of transfering parameter set P_i to P_{n+1};
1.6 Construct the transfer weight sequence w_e^T;
1.7 Construct the extended model transfer graph G_e according to $A_{n+1}=((A_n, w_e^T), (w_e, 0))^T$;
1.8 Compute the weight matrix W according to the formula (7), (8) and (9);
1.9 Transpose the element of the $n+1$th line of W to obtain the parameter set P_{n+1} of the target task t_{n+1};
2   Optimize
2.1 Compute the fitness value of each particle. Set the fitness value of the i-th particle as its current personal extreme value Pe_i, the best fitness value of all particles as the global extreme value G_e;
2.2 Compute the velocity and location of each particle according to the formula (5) and (6);
2.3 Compare the personal extreme value Pe_i of each particle with its current fitness value and set the better one as its new personal extreme value Pe_i, then compare the global extreme value Ge with the personal extreme value Pe_i of each particle and set the best one as the new global extreme value Ge;
2.4 Judge if the given terminal condition is satisfied, then stop searching and putout the solution, else go to 2.2.
End

It is assumed that there are $m - 1$ resource tasks and 1 target task. For obtaining the parameter set of the target task, the time complexity has the highest power 3 that obtained by solving the inverse matrix in the formula (9) of Step 1.8, the power of time complexity of every other step is smaller than 3, thus the time complexity of obtaining the parameter set of the target task is $O(m^3)$. Also assumed that the number of particles is n and the number of iterations of each particle is m_i ($1 \leq i \leq n$). For optimizing, the iteration time complexity of each particle is $O(m_i)$, so the time complexity of optimizing is $O\left(\sum_{i=1,\ldots,n} m_i\right)$. It is obtained that the time complexity of the algorithm is $O\left(m^3 + \sum_{i=1,\ldots,n} m_i\right)$. Because m is relatively small, the time that the algorithm takes mainly focus on optimizing.

4 Application Instance of Geometric Constraint Solving

With a parametric two-dimensional dimension-driven sketch, for example, by using system dimensioning commands to mark a sketch, you can always modify parameter values of the sketch. As is shown in Fig. 2, the modeling consists of three circles C_1, C_2 and C_3, nine lines L_1, L_2,..., L_9 and fifteen points P_1, P_2,..., P_{15}, the freedom degree of each geometric primitive is 2, containing 51 geometric constraints, the constraint degree of each geometric constraint is 1, such that it is a rigid body, that is, it is undeformable.

We change the angle constraint between L_3 and L_8 from $\pi/4$ to $\pi/3$, the distance constraint between P_1 and P_2 from 140 to 120 and the radius constraint of C_3 from 36 to 20. Except three geometric constraints that used to fix the location of the modeling, each of other 48 geometric constraints is converted into a nonlinear equation, thus the geometric constraint equations that to be solved consist of 48 nonlinear equations.

Figure 2 is a sketch in engineering design. After some sizes are modified, a new sketch may be obtained automatically by the PSO based on graph knowledge transfer. Once a series of geometric constraints are defined by a design, these geometric constraints will be satisfied by selecting proper state after the parameter set is modified by this algorithm.

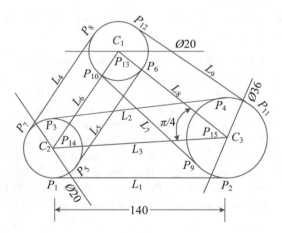

Fig. 2. A design instance of geometric modeling.

We compare the performance of the PSO algorithm based on graph knowledge transfer (abbr. PSO-GKT) with those of the tuba PSO(abbr. TPSO) algorithm, the crossbreeding PSO (abbr. CPSO) algorithm and the basic PSO algorithm through solving the above geometric constraint problem 20 times, the parameter sets of the CPSO, TPSO and PSO algorithms are set as $\{w = 0.95, c_1 = 2.00, c_2 = 2.00, NP = 70\}$, while the parameter set of the PSO-GKT algorithm is P_7: $\{w = 1.03, c_1 = 2.23, c_2 = 2.03, NP = 89\}$, which is the parameter set of the target task T_7 that obtained by

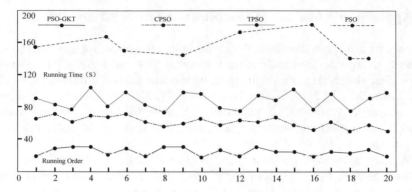

Fig. 3. Performance comparison curve of PSO-GKT, CPSO, TPSO and PSO algorithms.

using the graph knowledge transfer to learn from the following parameter sets of source task $T_j(1 \ll j \ll 6)$:

P_1: $\{w = 1.10, \quad c_1 = 2.12, \quad c_2 = 2.15, \quad NP = 90\}$, P_2: $\{w = 0.95, \quad c_1 = 2.10,$
$c_2 = 2.15, NP = 75\}$
P_3: $\{w = 0.97, \quad c_1 = 2.10, \quad c_2 = 2.03, \quad NP = 82\}$, P_4: $\{w = 1.04, \quad c_1 = 2.14,$
$c_2 = 2.13, NP = 73\}$
P_5: $\{w = 0.90, \quad c_1 = 2.15, \quad c_2 = 2.00, \quad NP = 86\}$, P_6: $\{w = 1.12, \quad c_1 = 2.09,$
$c_2 = 2.17, NP = 94\}$

Therein, each parameter set is obtained by learning from all those in front of it and with it the PSO algorithm has a better performance for solving a geometric constraint problem similar with that of Fig. 2. Initially, P_7 is manually set as $\{w = 1.00,$ $c_1 = 2.05, c_2 = 2.10, NP = 90\}$, under control of P_i, the average running time of task T_j is shown in Table 1, which is used to obtain transfer weight. And as it can be seen from Fig. 3, the PSO-GKT algorithm is best, having least running time, it is stable since the fluctuation of running time is small. The TPSO and PSO algorithms are relatively unstable since the fluctuation is bigger than that of the PSO-GKT and CPSO algorithms, and the stagnation times of the PSO algorithm are many. Using the PSO-GKT algorithm relative the others can get better performance and better convergence.

Table 1. The average running time of the task Tj under control of parameter set Pi: (S)

	T_1	T_2	T_3	T_4	T_5	T_6	T_7
P_1	34.18	40.24	46.30	39.01	50.24	70.20	62.21
P_2	90.24	32.03	81.02	50.26	71.33	60.20	55.40
P_3	87.36	50.13	33.91	40.29	62.14	44.10	51.27
P_4	70.39	61.29	77.12	30.20	57.33	67.36	39.13
P_5	90.27	40.60	48.23	70.36	29.34	52.73	80.37
P_6	75.34	59.20	60.44	91.17	39.16	30.02	64.13
P_7	54.20	70.32	46.07	38.49	60.01	60.46	80.13

Table 2 shows the average running time, the contents in each parenthesis are the minimum and maximum running time, respectively. If one algorithm of them stagnates at one time, then the running of this time will not be counted in average running time.

Table 2. Comparison of PSO-GKT, PSO, CPSO and TPSO algorithm.

Algorithm	Average solving time (S)	The number of stagnation times
PSO-GKT	31.24 (20.30, 35.87)	0
CPSO	72.51 (44.74, 74.33)	0
TPSO	90.16 (76.02, 108.24)	0
PSO	164.41 (143.74, 187.30)	12

5 Conclusions

As an evolutionary computation technology of swarm intelligence, PSO algorithm has fast convergence rate and high convergence accuracy. Its performance depends on parameter values selected, an improper selection of parameter is easy to get it stuck in a local extreme value and slow its convergence rate later. In order to apply PSO algorithm to efficiently and reliably solve complicated geometric constraint problems, we propose the PSO algorithm based on graph knowledge transfer, the parameter set that used to solve a geometric constraint problem is obtained by using the graph knowledge transfer to learn from existing parameter sets which are good for solving geometric constraint problem, such that a good parameter set can be obtained. The experiment shows that the algorithm of this paper can improve the efficiency and reliability of GCS, possessing better convergence property.

Acknowledgement. This work is supported by National Natural Science Foundation of China (Grant No. 61604019), Science and Technology Development Project of Jilin Province (20160520098JH, 20180201086SF), Education Department of Jilin Province, (JJKH2018118KJ, JJKH20181165KJ), Talent Introduction Scientific Research Project of Changchun Normal University, China (RC2016009).

References

1. Cao, C.H., Zhang, Y.J., Li, W.H.: The application of crossbreeding particle swarm optimizer in the engineering geometric constraint solving. Chin. J. Sci. Instrum. **25**, 397–400 (2004)
2. Cao, C.H., Tang, C., Zhao, D.Z., Zhang, B.: Application of the quantum particle swarm optimization approach in the geometric constraint problems. J. Northeast. Univ. (Nat. Sci.) **32**(9), 1229–1232 (2011)
3. Cao, C.H., Wang, L.M., Zhao, D.Z., Zhang, B.: Geometric constraint solving based on niche improved particle swarm optimization **33**(9), 2125–2129 (2012)
4. Yuan, H., Li, W.H., Li, N.: Tabu-PSO algorithm for solving geometric constraint problem. Microelectron. Comput. **27**, 26–29 (2010)
5. Coello, C.A.C., Pulido, G.T., Lechuga, M.S.: Handling multiple objectives with particle swarm optimization. IEEE Trans. Evol. Comput. **8**(3), 256–279 (2004)

6. Chi, Y.H., Sun, F.C., Wang, W.J., Yu, C.M.: An improved particle swarm optimization algorithm with search space zoomed factor and attractor. Chin. J. Comput. **34**(1), 115–130 (2011)
7. Zhou, X.Y., Wu, Z.J., Wang, H., Li, K.S., Zhang, H.Y.: Elite opposition-based particle swarm optimization. Acta Electronica Sin. **41**(8), 1647–1652 (2013)
8. Tan, Y., Tan, G., Deng, S.: Hybrid particle swarm optimization with chaotic search for solving integer and mixed integer programming problems. J. Central South Univ. **21**(7), 2731–2742 (2014)
9. Shen, Y.X., Zeng, C.H., Wang, X.F., Wang, X.Y.: A parallel-cooperative bare-bone particle swarm optimization algorithm. Acta Electronica Sin. **44**(7), 1643–1648 (2016)
10. Zhao, J., Fu, Y., Mei, J.: An improved cooperative QPSO algorithm with adaptive mutation based on entire search history. Acta Electronica Sin. **44**(12), 2901–2907 (2016)
11. Zhou, L.Y., Ding, L.X., Peng, H., Qiang, X.L.: Neighborhood centroid opposition-based particle swarm optimization. Acta Electronica Sin. **45**(1), 2815–2824 (2017)
12. Lv, B.Q., Zhang, J.J., Li, Z.P., Liu, T.Z.: Fuzzy particlel swarm optimization based on filled function and transformation function. Acta Automatica Sin. **44**(1), 74–86 (2018)
13. Lee, S., Park, H., Jeon, M.: Binary particle swarm optimization with bit change mutation. IEICE Trans. Fundam. Electron. Commun. Comput. Sci. **E90-A**(10), 2253–2256 (2007)
14. Shi, Y. H., Russell, E.: Modified particle swarm optimizer. In: Proceedings of the 1998 IEEE International Conference on Evolutionary Computation. IEEE, Piscataway (1998)
15. Erie, E., Marie, D. J., Terran, L.: Modeling transfer relationships between learning tasks for improved inductive transfer. In: Lectures Notes in Artificial Intelligence, vol. 5211, pp. 317–332 (2008)

Image-Based Human Protein Subcellular Location Prediction Using Local Tetra Patterns Descriptor

Fan Yang[✉], Yang Liu, and Han Wei

School of Communications and Electronics,
Jiangxi Science and Technology Normal University, Nanchang 330003, China
kooyang@aliyun.com

Abstract. Protein subcellular location has a huge positive influence on understanding protein function. In the past decades, many image-based automated approaches have been published for predicting protein subcellular location. However, in the reported literatures, there is a common deficiency for diverse prediction models in capturing local information of interest region of image. It motivates us to propose a novel approach by employing local feature descriptor named the Local Tetra Patterns (LTrP). In this paper, local features together with global features were fed to support vector machine to train chain classifiers, which can deal with multi-label datasets by using problem transformation pattern. To verify the validity of our approach, three different experiments were conducted based on the same benchmark dataset. The results show that the performance of the classification with LTrP descriptor not only captured more local information in interest region of images but also contributed to the improvement of prediction precision since the local descriptor is encoded along horizontal and vertical directions by LTrP. By applying the new approach, a more accurate classifier of protein subcellular location can be modeled, which is crucial to screen cancer biomarkers and research pathology mechanisms.

Keywords: Human protein · Subcellular location prediction ·
Local tetra patterns · Multi-label learning

1 Introduction

Protein subcellular location prediction is a principally important field in molecular cell biology, proteomics and system biology [1]. In recent years, it has become a hot spot of research in bioinformatics field. The main reason is that knowing protein subcellular localization will be propitious to comprehend protein functions [2], and is essential for the screening of cancer markers, the early detection, diagnosis and prognosis of tumors.

In the past decade, protein subcellular localization has gained great international attention. Traditional approaches such as electronic microscopy and fluorescence microscopy are expensive, time consuming and laborious. To improve and refine the method of protein subcellular location, many efforts have been done on amino acid sequence-based [3].

© Springer Nature Switzerland AG 2020
Q. Liu et al. (Eds.): CENet 2018, AISC 905, pp. 463–473, 2020.
https://doi.org/10.1007/978-3-030-14680-1_51

Moreover, pattern recognition and machine learning methods has been widely used in this field based on biological images. For instance, Zhao group built an novel machine learning model to recognize the subcellular organelles of proteins with more than one organelle, and their experiment result demonstrated that the organization accuracy of subcellular organelles of proteins can increase over 80% [4].

Peng group proposed a method by training labeled images for each class instead of co-localization with organelle-specific fluorescent markers to measure protein ratio in different cellular compartments in [5], which demonstrated that the statistic information of image can be use for quantifying drug-dependent protein translocations.

These image-based methods can be easier automated analysis by using effective feature descriptor which can capture the intrinsic feature of interest region of target image [6]. Subcellular Location Feature (SLF) sets have developed by Murphy group and a back propagation neural network (BPNN) them trained to predict subcellular localization [7], and the results of corresponding experiments had shown that automated approaches using numerical features can achieve better result of protein localization classification than human observers in some cases. Ojala group proposed to employ the Local Binary Pattern (LBP) texture feature to train a classification model [8]. Hu and Murphy have proposed Zernike moment-based features which is a relatively inexpensive texture features [9]. Hamilton group have proposed threshold adjacency statistics (TASs) [10] and trained a Support Vector Machine (SVM) to prove the performance of these features [11]. In recent years, Zhenhua Guo group have proposed the completed LBP descriptor (CLBP) [12], and their results of experiment shows its effectiveness.

Although most of the existing feature descriptors are effective in building prediction models to a certain extent [13], there are still margin in improving the prediction accuracy. Hence, a new approach by employing a feature descriptor named local tetra patterns (LTrP) [14] was proposed in this paper. In details, LTrP can improve deficiencies in existing feature descriptors and effectively capture information of bio-images to enhance the performance of classifier. In this way, a more accurate classifier can be modeled to predict protein subcellular location which is of great significance to the study of protein structure.

2 Materials and Methods

We built the benchmark dataset by collecting IHC images from the Human Protein Atlas (HPA) database, which devoted to provide the distribution information of cell and tissue of all human protein. The whole HPA database is divided into two categories according to the dying method, i.e., Immunohistochemistry (IHC) and Immunocytochemistry (IF). Moreover, IHC images in HPA contain normal and cancer images.

There are high quality, medium quality, low quality and very low quality which represent the IHC image staining intensity in the HPA database respectively [15] and it denotes the staining extent of protein in IHC image. Another crucial index is the subcellular annotation of each protein with a reliability score, which including enhance score, supportive score, approve score and uncertain score based on the consistency of image subcellular annotation with other reported literatures. To improve the general applicability of the proposed methods, the protein image sets consisted of IHC images with high-level protein expression and "enhance" reliability annotation from HPA were employed in this study. The details has been described in the following sections.

2.1 Benchmark Datasets Preparation

In this study, we sort out a collection of datasets from IHC database with the highest Protein expression level and reliability score in Fig. 1. The seven main cellular organelles are contained in these images Cytoplasm, Endoplasmic Reticulum, Golgi Apparatus, Centrosome, Mitochondria, Nucleus and Vesicles. Among the whole datasets we used in this study, of which over 30% are multi-label images. The datasets contain 12 multi-class proteins and 58 signal-class proteins, and the total number of images is 3240.

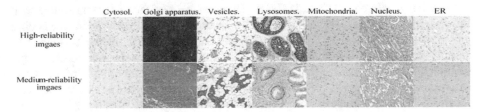

Fig. 1. Comparison of protein images with different reliability levels in corresponding subcellular locations.

2.2 Image Pre-preprocessing and Extraction of Global Feature

IHC images from HPA database were saved in RGB mode and a few images were stained badly. For the rest of images, we used spatial transformation to make it normalization [16]; changing the original IHC images from RGB space to HSV space to gain the hue value of image, which determines whether the hue values of image exceeded a certain threshold. Finally, we converted the images back to RGB space [17].

Moreover, due to the IHC images consisted of two major channels: DNA and protein, the linear spectral separation was applied to gain the protein channel and the information of DNA distribution respectively. We utilized Daubechies wavelet decomposition by vanishing moments from 1 to 10 levels when extracting Haralick features in protein channel, and by applying inertia, isotropy and so on to describe images intuitively. Haralick features employed to capture image texture information from intuitive aspects of image [18]. Additionally, four DNA distribution features utilized in this paper [16], which combined with Haralick feature. Finally, the total number of features for each image are 840 on each Daubechies wavelet decomposition level.

2.3   Difference Between LBP and CLBP

In recently years, some local descriptors were proposed and have achieved better effects in describing local texture features of interest region of target images, such as Local Binary Patterns (LBP) and completed LBP (CLBP) descriptor. Comparing with global features, local features can detect edges spots and flat areas of images, which can effectively represent the image since reducing redundancy of information to a large extent. In most of existing works, local texture information of target image is extracted by the LBP or the CLBP descriptor, which has achieved preferable result in training a prediction model [19, 20]. However, these two methods only relies on two directions, namely positive and negative direction, which may miss some crucial information of images. Thus, it motivated us to consider the gradient information of more directions. Due to LTrPs descriptor is local tetra patterns and presents the local image texture information by four directions, thus it employed as local feature descriptor in this work.

Comparience between LBP and CLBP. Different from LBPs, CLBP constituted with three descriptors, namely, sign, magnitude and center pixel level descriptor, The coding methods of sign component like LBP, namely comparing between the gray level of central pixel and its neighboring pixels. The magnitude components considers the relationship between the local neighboring pixels of given central pixel and the average gray level value for the whole image. The magnitude components coded the relationship between the given central pixel and the average gray level value for the whole image. The different details of formula between these two descriptors were given in Table 1.

Table 1. The comparison of LBP and CLBP descriptor.

Pattern descriptor	Formula and parameters[a,b]	Details
Local binary pattern (LBP)	$LBP_{n,r} = \sum_{i=1,\dots,n} 2^{(i-1)} f_1(g_i - g_c)$ (2.1) $f_1(x) = \begin{cases} 1, x \geq 0 \\ 0, else \end{cases}$ (2.2)	Computing the gray level of central pixel c and its neighboring pixel generates the LBP descriptor
Complete Local binary Pattern (CLBP)	$CLBP_{M_{n,r}} = \sum_{i=0,\dots,n-1} t(m_n, g_c) 2^n$ $t(x,p) = \begin{cases} 1, x \ll p \\ 0, x > p \end{cases}$ (2.3) $CLBP_{n,r} = \sum_{i=1,\dots,n-1} 2^{(i-1)} f_1(g_i - g_c)$ $CLBP_{n,r} = \sum_{i=1,\dots,n-1} 2^{(i-1)} f_1(g_i - g_c)$ (2.4) $CLBP_{n,r} = t(g_c, c_A)$ (2.5)	Unlike LBP, CLBP has three descriptors including CLBP-Center, CLBP-Sign and CLBP-Magnitude. The one of observations is that CLBP supplements the LBP which only uses CLBP_S

[a.] In LBP, the gray level of the center pixel and its neighbors are represented by g_c and g_i. respectively. i is the number of neighood pixel, and r is the radius of the neighborhood.

[b.] In CLBP, where m_n is the mean value of whole image, and n as well as r are same as what represent in LBP. t in Expression (2.5) is defined as (2.3). Here, threshold C_A represents average gray level of image.

Local Tetra Patterns. The descriptor of local tetra patterns (LTrP) was first proposed by Murala group, [14] comparing with other local descriptors, it has been proven that achieved considerable results in content-based image retrieval and employed for medical image retrieval and classification [21, 22].

The LTrP image descriptor coded the difference between each central pixel and its neighboring pixels in horizontal and vertical directions. Then, we make the $(n-1)$th-order derivatives in these two directions in the central pixel for a specific local region. Hence, these derivatives of two directions of LTrP image descriptor can be formulated as:

$$I_{hori}^{n-1}(g_c) = I_{hori}^{n-2}(g_h) - I_{hori}^{n-2}(g_c) \tag{2.6}$$

$$I_{vect}^{n-1}(g_c) = I_{vect}^{n-2}(g_h) - I_{vect}^{n-2}(g_c) \tag{2.7}$$

Then, the $(n-1)^{th}$-order direction of the center pixel can be calculated as Expression (2.8).

$$I_{Dir}^{n-1}(g_c) = \begin{cases} 1, I_{hori}^{n-1}(g_c) \gg 0, I_{vect}^{n-1}(g_c) \gg 0 \\ 2, I_{hori}^{n-1}(g_c) < 0, I_{vect}^{n-1}(g_c) \gg 0 \\ 3, I_{hori}^{n-1}(g_c) < 0, I_{vect}^{n-1}(g_c) < 0 \\ 4, I_{hori}^{n-1}(g_c) \gg 0, I_{vect}^{n-1}(g_c) < 0 \end{cases} \tag{2.8}$$

Afterwards, the nth-order LTrP descriptor for each central pixel can be calculated as follows:

$$LTtP^n(g_c) = \{s_1(I_{Dir}^{n-1}(g_c), I_{Dir}^{n-1}(g_1)), \ldots, s_1(I_{Dir}^{n-1}(g_c), I_{Dir}^{n-1}(g_8))\} \tag{2.9}$$

$$I_{Dir}^{n-1}(g_p) = \begin{cases} 0, I_{Dir}^{n-1}(g_c) = I_{Dir}^{n-1}(g_p) \\ I_{Dir}^{n-1}(g_p), else \end{cases} \tag{2.10}$$

Where P represents the number of neighborhood pixels of the central pixel, and the *Dir.* is the reference direction, which has four potential quadrant values for each central pixel. Based on the Expression (2.8) and the Expression (2.10), the result of $s_1(\cdot, \cdot)$ is four different quadrant values by comparing the differences of direction between the central pixel and its neighborhood pixel as being shown in the Fig. 2, which obtained by comparing central and its neighborhood pixel. Thus, from Expression (2.9) and Expression (2.10), the $LTtP^n(g_c)$ is coded as "0" when the direction of central pixel is the same as the direction of neighborhood pixels, and if not, the $LTtP^n(g_c)$ is coded as the directions of neighborhood pixels. Finally, we got an 8-bit tetra patterns for each center pixel.

For instance, taking $LTtP^n(g_c) = 1$ as an example, the possible values of tetra pattern are "1, 2, 3, 4", seeing Fig. 2. As the direction of central pixel is "1", the potential values of tetra patterns are "0, 2, 3, 4" and the pattern "0" means $LTtP^n(g_c) = LTtP^n(g_p)$, which is defined in Expression (2.10). Then, three binary tetra patterns in four directions of center pixels can be generated, so we finally get 12 binary

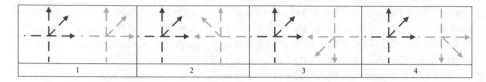

Fig. 2. Representation the possible tetra pattern based on center-pixel direction "1".

patterns. Furthermore, to extract more effective image texture features, the magnitude of LBP is employed as the 13th binary patterns.

2.4   Chain Multi-label Classification

To select essential subset of features, stepwise discriminant analysis (SDA) approach that the most discriminating features were judged by Wilk's statistic with iteratively method [16] was used in this study. Numerous methods aimed to the multi-label problem can be summarised into two sorts: Algorithm Improvement Approach and Problem Transformation Pattern. The chains classification (CC) is a widely known framework for multi-label classification which is the representative method. It transforms the multi-label problem into the single-label problem and takes account into the correlation among labels. There is a great challenge in predicting subcellular localization directly by classificator in the multi-label images, but if we can attain more prior acknowledge, this problem can be solved efficiently. Therefore, for each classifier, chains classification increases the feature space by adding the label (0 or 1) of previous classifiers and these binary classifiers referring to different labels are formed a chain. The order of chains classification was generated randomly. Because of the different label order can cause slightly different results. In this article, the SVM classifier with an RBF kernel implemented by applying LIBSVM toolbox, which can be download (https://www.csie.ntu.edu.tw/~cjlin/libsvm/), and parameter $\sigma = 0.2$ was selected by the parameter tuning from 0.1 to 1.2 at a step size of 0.1 by using grid search on the training data. A seven dimensionals score vector $[S_1, \ldots, S_{17}]$ can be gained for each testing images, in which each vector element corresponds to a special subcellular location. Each element of score vector $[S_1, \ldots, S_{17}]$ represents the confidence of belonging to the corresponding class. When the binary classifiers correctly predict the label of test samples, the score is positive, otherwise it will be negative. To decide the label set of testing images, top criterion and threshold criterion were employed [23]. In this paper, the computational complexity of algorithm is comprised of four factors i.e.: m (number of training examples), d (features dimensiona), q (number of potential classes) and s (number of support vectors). Then, the algorithm has computational complexity $O(q \times (s^3 + s \times m + s \times d \times m))$ for training and $O(q \times (s^3 + s + s \times d))$ for testing.

3   Experimental Results

In this paper, to invalidate the robustness of model, we used 5-fold cross validating strategy and compared the proposed method in this article based on the SVM classifier.

3.1   Comparison SDA Selection Results for Different Local Features

Table 2 shows the different pattern features proportion by SDA feature selection. Various local features incorporated with global features, and after SDA feature selection, the number of local and global features increases and decreases respectively. It demonstrates that local features are more discriminating than global features. For instance, the biggest proportion was occupied by the LTrP features after SDA feature selection, which is around $\sim 70\%$ in 5-fold. It illustrates that LTrP features can capture the more image texture information than other local descriptors.

Table 2. Comparing the SDA results originated from incorporating with different local descriptors and global features based on db6.

Fold	SLFs_LBP feature fed into SDA	Ratio of LBP feature	The subset accuracy of model	SLFs_LTrP feature fed into SDA	Ratio of LTrP feature	The subset accuracy of model
Fold1	72 (G36-L36)	50.00%	56.91%	86 (G32-T54)	62.79%	61.79%
Fold2	67 (G33-L34)	50.75%	59.13%	84 (G21-T63)	75.00%	61.83%
Fold3	75 (G37-L38)	50.67%	59.96%	86 (G22-T64)	74.42%	62.03%
Fold4	71 (G32-L39)	54.93%	59.18%	100 (G28-T72)	72.00%	62.63%
Fold5	75 (G34-L41)	54.67%	60.25%	88 (G30-T58)	65.91%	63.15%

In Table 2, Fold1-Fold5 is each fold of 5-fold cross-validation. G denotes Global features (Haralick and DNA), L denotes LBP feature and T denotes LTrP feature. All percentages stand for the proportion of the local descriptor in the whole feature set.

3.2   Classification Results Based on Feature Level Fusing

In this paper, five multi-label classification evaluation metrics, which were generously used in other works [24], are applied to evaluate the performance with 5-fold cross-validation and the five evaluation metrics are subset accuracy, accuracy, recall, precision, and average label accuracy respectively.

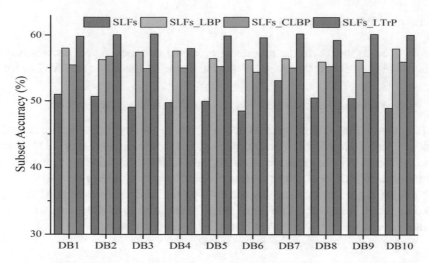

Fig. 3. The subset accuracy obtained by combining different features which contain SLFs, SLFs_LBP, SLFs_CLBP, SLFs_CLBP, SLFs_LTrP on db1–db10.

From Fig. 3, the comparison of subset accuracy among the three local pattern features combining with global features is given. The performances of SLFs_LTrP (symbol "_" denotes combination) are better than SLFs, SLFs_LBP and SLFs_CLBP on 10 db. Since the subset accuracy is the most significant evaluation index for judging performance of classification and illustrates test images by predicted label result entirely consisted with the true label set in this work. The increments of subset accuracy by integrating with the LTrP feature indicated that the local treta patterns can efficiently capture more the texture information of image. By contrast, the LBP just extracts a part of texture information and CLBP may contain irrelevant information leading to a lower result.

Prediction results of entire dataset (70 proteins) are shown in Fig. 4 based on the average of five-fold in combination with global features and local pattern features on db6. For order to demonstrate the significance of LTrP descriptor, the whole of evaluation metrics is given for comparison.

According to the results of Fig. 4, we can draw a conclusion that the LTrP feature extracts more distinguishable information which is helpful for training a desirable classification model and achieves better effects than LBP and CLBP descriptors. The results (five evaluation indexes) shown that all of the five evaluation indexes are improved. For instance, the subset accuracy of SLFs_LTrP is improved validly among three different combinations of global and local pattern features, over 3%.

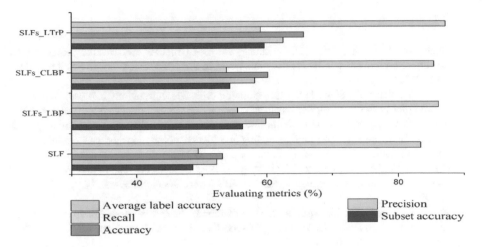

Fig. 4. The results of different evaluation metrics for SLFs, SLFs_LBP, SLFs_CLBP, SLFs_CLBP and SLFs_LTrP on db6.

4   Conclusion

In this study, a novel framework by employing a feature descriptor named local tetra patterns(LTrP) is used for predicting human protein subcellular localization based on the IHC image. The results of our experiment shown that LTrP descriptor outperforms LBP and CLBP descriptors, because the coding pattern of four directions of LTrP features can help capture more information of images to train a high-performance predictor. Besides, by making a comparison of the global features and the combination with the global and local features, we found that the more desirable performance of predictor using local and global features together can achieved than just utilizing the former only.

To develop a more effective classificator, some efforts will be made in further. A crucial label correlation will be taken into consideration when doing feature selection, as the feature selection approach of SDA which we used in our experiment is just regarded to select features independently. Moreover, just high level IHC images were used in this study, but there are a large number of medium and low level images in the HPA dataset, which can also be utilized to train a prediction model. Thus, we will aim to construct a semi-supervise experiment framework by utilizing these lower-quality images to enlarge training datasets to achieve better predictive performance.

Acknowledgements. This study was supported by the National Natural Science Foundation of China (61603161), and the Science Foundation of Artificial Intelligence and Bioinformatics Cognitive Research Base Fund of Jiangxi Science and Technology Normal University of China (2017ZDPYJD005) and the Key Science Foundation of Educational Commission of Jiangxi Province of China (GJJ160768).

References

1. Shao, W., Ding, Y., Shen, H.B., Zhang, D.: Deep model-based feature extraction for predicting protein subcellular localizations from bio-images. Front. Comput. Sci. 11(2), 243–252 (2017)
2. Xu, Y.Y., Yang, F., Shen, H.B.: Incorporating organelle correlations into semi-supervised learning for protein subcellular localization prediction. Bioinformatics 32(14), 2184 (2016)
3. Zhou, H., Yang, Y., Shen, H.B.: Hum-mPLoc 3.0: prediction enhancement of human protein subcellular localization through modeling the hidden correlations of gene ontology and functional domain features. Bioinformatics 33(6), 843 (2016)
4. Zhao, T., Velliste, M., Boland, M.V., Murphy, R.F.: Object type recognition for automated analysis of protein subcellular location. IEEE Trans. Image Process. 14(9), 1351–1359 (2005)
5. Peng, T., Vale, R.D.: Determining the distribution of probes between different subcellular locations through automated unmixing of subcellular patterns. Proc. Natl. Acad. Sci. U.S.A. 107(7), 2944–2949 (2010)
6. Hamilton, N.A., Wang, J.T., Kerr, M.C., Teasdale, R.D.: Statistical and visual differentiation of subcellular imaging. BMC Bioinf. 10(1), 94 (2009)
7. Boland, M.V., Murphy, R.F.: A neural network classifier capable of recognizing the patterns of all major subcellular structures in fluorescence microscope images of HeLa cells. Bioinformatics 17(12), 1213–1223 (2001)
8. Ojala, T., Pietikainen, M., Maenpaa, T.: Multiresolution gray-scale and rotation invariant texture classification with local binary patterns. IEEE Trans. Pattern Anal. Mach. Intell. 24 (7), 971–987 (2002)
9. Hu, Y., Murphy, R.F.: Automated interpretation of subcellular patterns from immunofluorescence microscopy. J. Immunol. Methods 290(1), 93–105 (2004)
10. Hamilton, N.A., Pantelic, R.S., Hanson, K., Teasdale, R.D.: Fast automated cell phenotype image classification. BMC Bioinform. 8(1), 1–8 (2007)
11. Chang, C.C., Lin, C.J.: Libsvm. (2), 1–27 (2011)
12. Guo, Z., Zhang, L., Zhang, D.: A completed modeling of local binary pattern operator for texture classification. IEEE Trans. Image Process. 19(6), 1657–1663 (2010)
13. Nanni, L., Brahnam, S., Lumini, A.: Novel features for automated cell phenotype image classification. Adv. Exp. Med. Biol. 680, 207–213 (2010)
14. Murala, S., Maheshwari, R.P., Balasubramanian, R.: Local tetra patterns: a new feature descriptor for content-based image retrieval. IEEE Trans. Image Process. 21(5), 2874–2886 (2012)
15. Xu, Y.Y., Yang, F., Zhang, Y., Shen, H.B.: Bioimaging-based detection of mislocalized proteins in human cancers by semi-supervised learning. Bioinformatics 31(7), 1111 (2015)
16. Newberg, J., Murphy, R.: A framework for the automated analysis of subcellular patterns in human protein atlas images. J. Proteome Res. 7(6), 2300–2308 (2008)
17. Chebira, A., Barbotin, Y., Jackson, C., Merryman, T., Srinivasa, G., Murphy, R.F., Kovačević, J.: A multiresolution approach to automated classification of protein subcellular location images. BMC Bioinf. 8(1), 1–10 (2007)
18. Xu, Y.Y., Yang, F., Zhang, Y., Shen, H.B.: An image-based multi-label human protein subcellular localization predictor (iLocator) reveals protein mislocalizations in cancer tissues. Bioinformatics 29(16), 2032–2040 (2013)
19. Zhao, G., Pietikainen, M.: Dynamic texture recognition using local binary patterns with an application to facial expressions. IEEE Trans. Pattern Anal. Mach. Intell. 29(6), 915–928 (2007)

20. Ahmed, F., Hossain, E., Bari, A.H., Shihavuddin, A.S.M.: Compound local binary pattern (CLBP) for robust facial expression recognition. In: IEEE International Symposium on Computational Intelligence and Informatics, pp. 391–395 (2011)
21. Oberoi, S., Bakshi, V.: A framework for medical image retrieval using local tetra patterns. Int. J. Eng. Technol. **5**(1), 27 (2013)
22. Thangadurai, K.: An improved local tetra pattern for content based image retrieval. J. Global Res. Comput. Sci. **4**(4), 37–42 (2013)
23. Boutell, M.R., Luo, J., Shen, X., Brown, C.M.: Learning multi-label scene classification. Pattern Recogn. **37**(9), 1757–1771 (2004)
24. Yang, F., Xu, Y.Y., Shen, H.B.: Many local pattern texture features: which is better for image-based multilabel human protein subcellular localization classification? Sci. World J. **2014**(12), 429049 (2014)

Information Analysis and Application

This chapter aims to examine innovation in the fields of bioinformatics, semantic information, software design and information retrieval. This chapter contains 24 papers on emerging topics on information analysis and application.

Social Book Recommendation Algorithm Based on Improved Collaborative Filtering

Hanjuan Huang and Qiling Zhao[✉]

Wuyi University, Wuyishan 354300, China
215995504@qq.com

Abstract. Recommendation is one of the new personalized services in social book search system that assists the user to find the appropriate books. This paper proposes an improved collaborative filtering recommendation algorithm based on the user interest objects, which try to settle the cold start problems of traditional collaborative filtering. Firstly, set up a book retrieval system. Then, use the traditional collaborative filtering algorithm to calculate the recommended collection of books (Bc). Next, get the collection of user interest objects, calculate the similarity, and get the recommended collection of books (Bi). Finally, reorder the books of the set Bc and the set Bi according to certain weights, and get the first ten books as the books that are finally recommended to users. Experiments on real-world datasets results show that the proposed algorithm improves the performance of the collaborative filtering algorithms.

Keywords: Social book · Recommendation · Collaborative filtering · Cold start

1 Introduction

With the arrival of information age, people are more prefer to obtain books from the social networks and share their reading reviews, rates, and tags on the internet. At the same time, users also like to establish personal library, which contain users' interest and user reading log. At present, professional metadata is often used to build index library in most of the books retrieval system. Therefore, if users want to retrieve books from these systems, they need to accept the relevant professional training. But in reality, most users cannot provide corresponding professional search terms for the retrieval system by lack of recognition of the profession. Furthermore, it requires user previous reading log in traditional collaborative filtering recommendation algorithm, however, the majority of users in the process of searching for books is in essence a process of learning, looking for the books that they never read.

In conclusion, we propose an improved collaborative filtering recommendation algorithm based on the user interest object, which settle the cold start problems of traditional collaborative filtering to some extent. The rest of this paper is organized as follows: In Sect. 2, we introduce several used methods in social book search. In Sect. 3, we describe the specific improved collaborative filtering recommendation algorithm. Section 4 introduces the datasets and presents experiments on the overall. Section 5 is conclusion.

Q. Liu et al. (Eds.): CENet 2018, AISC 905, pp. 477–484, 2020.
https://doi.org/10.1007/978-3-030-14680-1_52

2   Related Work

In view of the lack of professional training for most users, paper [1–4] created an index library containing user-generated metadatas. The user-generated metadatas are generated by users, which is unprofessional and arbitrary. Paper [1–4] investigated and proved the relative value of user-generated content for social book search and recommendation. Therefore, we borrowed the method that used to establish an index library from paper [1–4], setting up a book retrieval system, in which every book record is an XML document, which contains professional metadata and social metadata.

Collaborative filtering methods predict what users will like based on their similarity with other users by collecting and analyzing a large amount of information on users' preferences, activities, and behaviors. In the early 2000s, collaborative filtering became widely adopted by internet companies and online retail stores like Amazon [5]. Viewed From One Perspective, collaborative filtering recommenders fall into two main categories: One is neighborhood based [6, 7] that calculates the similarity between system users or items and leverages that to estimate how much a user like an item and the other is matrix factorization [8, 9] that uses linear algebra techniques to analyze the user/item ratings data and then forecast how much a user would like an unseen item.

Traditional collaborative filtering recommendation algorithm can help users to obtain the corresponding books to some extent. But it should be noted that the book retrieval system is different from other retrieval systems. The traditional books retrieval system requires the user to have professional quality. They seek retrieval process is a learning process and search the book what they have never read, therefore, traditional collaborative filtering recommendation algorithm has the problem of cold start [10]. We found that the user query intent has a similar relationship with the user interest object query log by analyzing the user's profile and query intention in test set. Therefore, we propose an improved collaborative filtering recommendation algorithm based on the user interest objects.

3   Social Book Recommendation Algorithm Based on Improved Collaborative Filtering

3.1   Related Definitions and Techniques

Collaborative Filtering. In real life, people always like to ask their friends or trusted people when facing unfamiliar problems or things, and make choices according to their opinions.

Collaborative filtering to simulate the process of book retrieval, it analysis of users and other users of ratings on a certain project to calculate the most similar neighbors, according to the most similar to the neighbour's interest in direction to predict the user's interest, to help users a way to make decisions.

Similarity Calculation Method. The methods of similarity calculation include Euclidean distance, Pearson correlation coefficient, cosine similarity and so on. This chapter adopts Euclidean calculation similarity method to calculate the nearest neighbor to the target user. The definition is as follows:

Euclidean Distance (Euclidean Distance): supposing that the n-dimensional space is a and b, the Euclidean Distance between these two points as follows:

$$d(a,b) = \sqrt{\left(\sum (a_i - b_i)^2\right)} \tag{1}$$

Euclidean Distance (Euclidean Distance): supposing that the n-dimensional space is a and b, the Euclidean Distance between these two points as follows:

$$sim(a,b) = \frac{1}{1 + d(a,b)} \tag{2}$$

Social Book Tag Frequency. In this paper, we confirm the type of book which the user interest in according to the social book tag frequency that the user marked in all the history books. It indicates that the user is more interested in a book type if the tag of the book type has highest frequency. Frequency formula as follows:

$$M(t) = \frac{frequency(t)}{\sum\limits_{\sigma \in V} frequency(\sigma)} \tag{3}$$

Where the frequency(t) represents the frequency of book tag t appears on the document F, that is the collection of all the social tags of history books for a user's interested object. σ is the social book tag for any book in document F, V represents the social tagging of all books in document F.M(t) appears the number of book tag t appears on the document F.

3.2   Social Book Recommendation Algorithm Based on Improved Collaborative Filtering

The Traditional User-Based Collaborative Filtering Algorithm Is Used to Recommend Books. In general, users with certain behavioral similarities also have similarities when choosing a target item. Therefore, we can use the similar users of target users to evaluate a project to generate project recommendations to target users. There are three main steps in the user-based collaborative filtering method: Data presentation; Discover nearest neighbors; Generate recommendations.

Algorithm:

(1) Input object user e.g. AnnieMod
(2) Obtain AnnieMod's user profile and others' user profile.
(3) Object the book ratings, and establish the user-rating matrix.
(4) Select the users who have read the same book with object user, and re-form new user-rating matrix.

(5) Calculate the similarity by formula (2), and select the first K users as the nearest neighbors.
(6) Use the scoring formula (4) of the former K user's nearest neighbor to predict the rating of the book for the object user, the scoring formula as follows:

$$P_{a,u} = \overline{R_a} + \frac{\sum\limits_{b \in N} sim(a,b) \times \left(R_{b,u} - \overline{R_b}\right)}{\sum\limits_{b \in N} sim(a,b)} \qquad (4)$$

In the formula, $R_{a,u}$ represents user a graded the book u, $\overline{R_a}$ $\overline{R_b}$ represent the average of scores of books that user a and user b has scored. $sim(a,b)$ represents the similarity between user a and user b. $P_{a,u}$ represents the score that the user a will grade.

Choose the top 10 books with the highest forecast, and form a boo recommendation collection Bc.

User Interest Objects Collaborative Filtering Algorithm Is Used to Recommend Books. The object user annotated some users as his interest objects recently, indicating that he has been interested in the types of books what the user interest objects are looking at. Therefore, we first find out the types of books that are concerned in the historical book social annotation of these user interest objects, and then recommend books to the object user according to these types of books.

Algorithm:

(1) Collect the collection of user interest objects from the object user's personal profile and integrate them into the collection O;
(2) Analyze the personal profile of each user interest object Oi in the collection O, and obtain the historical book social tags of each user interest object, and save them to the corresponding document F_i;
(3) Using the formula (3) to calculate the frequency $M(t_j)$ of each book tag t_j in document F_i;
(4) Repeat Steps (2) and (3) until the collection O is finished;
(5) Arrange all frequency $M(t)$ values from high to low, select the highest N $M(t_j)$, and obtain the book Bt_j corresponding to the annotation word t_j;
(6) Obtain all the scores of these objects on book B_{tj}, and take the average scores, and rank them according to the average score from high to low, and selected 10 books to make up the recommendation book collection B_I.

The Improved Collaborative Filtering Recommendation Algorithm. Select the first $\alpha\%$ of the results of the recommended book collection B_I and the first $\beta\%$ of the book recommendation set B_C as the final book recommendation result. When the book appears in the set B_I and appears in the set B_C, it ranks high in the results.

4 Experiment and Analysis

4.1 Dataset

There are three data sets in this study: One is Amazon/LibraryThing (A/LT) corpus, one is Training set, and the last one is User profile.

The Amazon/LibraryThing (A/LT) corpus comes from the INEX Social book search task. The collection consists of 2.8 million book records from Amazon, extended with social metadata from LT, and we select 100000 book records to set up the book retrieval library. Each book is identified by an ISBN. Each book record is an XML file with fields like ISBN, title, author, publisher, dimensions, number of pages and publication date. The social metadata from Amazon and LT is stored in the tag, rating, and review fields.

The User profile dataset consists of 308 user profiles from Library Thing. Each user profile also is an XML file with fields like user name, about library, friends, groups, interesting_library, library. The library fields contains fields like work id, title, author, year, ratting, tags.

In the experiment, 100,000 book metadata were used. The Training set include 220 requests. The user's ratings in the three dataset range from 0 to 10.

4.2 Experimental Evaluation

MAE (Mean Absolute Error) [11]. MAE is adopted as the evaluation method. The MAE is calculated by summing these absolute errors of the corresponding rating-prediction pairs and then computing the average. The MAE is defined as shows:

$$MAE = \frac{\sum_{i=1}^{N} |p_i - q_i|}{N} \quad (5)$$

Where N is the number of total books in test set, p_i represents the user's predicted ratings on book i. q_i represents the user's actual rating on the book i. A smaller value for MAE indicates higher accuracy of recommendation algorithm.

4.3 Experimental Results and Discussion

This section presents the results of a performance comparison of the different algorithms. The method is as follows: The traditional user-based collaborative filtering algorithm is used to recommend books and our method. In our method, we select the first $\alpha\%$ of the results of the recommended book collection B_I and the first β of the book recommendation set B_C as the final book recommendation result. We would find the best weight of α and β. Set α to range from 0.1 to 0.9. We Choose 15 as the result display. The results are shown below.

In Figs. 1, 2 and 3, the MAE value of our method as a whole is smaller than that of traditional collaborative filtering algorithm.

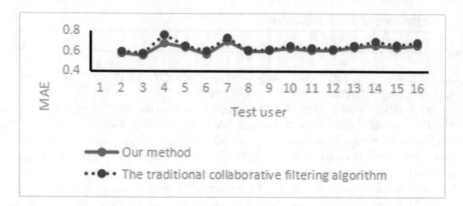

Fig. 1. α = 0.9, β = 0.1.

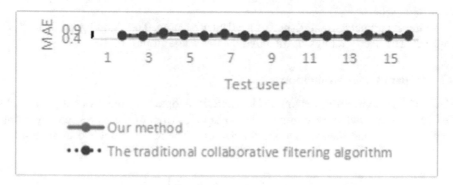

Fig. 2. α = 0.6, β = 0.4.

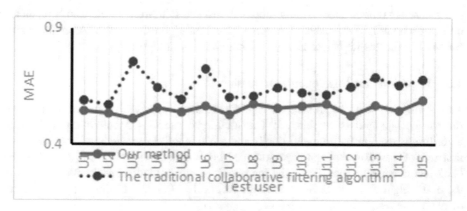

Fig. 3. α = 0.7, β = 0.3.

Form Fig. 4, where $\alpha = 0.1$, $\beta = 0.9$, our method's MAE is greater than that of traditional collaborative filtering recommendation algorithm, the reason is that the user interest objects in user profile are not so many.

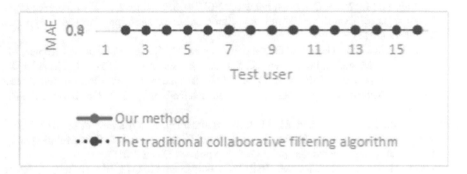

Fig. 4. $\alpha = 0.1$, $\beta = 0.9$.

As can be seen from the above results, the MAE value of our method as a whole is smaller than that of traditional collaborative filtering algorithm, it means our method is more effective. In particular, U4 and U7, MAE value gap is particularly large. We delve into the fact that the types of books currently being searched by users U4 differ greatly from their reading history, and the reand records in user U7's user profile are almost empty. These show that our method can solve the problem of cold start intraditional collaborative filtering algorithm to some extent. The value of P@10 and nDCG@10 show that our method improves the satisfaction of user queries and precision.

5   Conclusions

In this paper, we proposes an improved collaborative filtering recommendation algorithm based on the user interest objects, the experiments on real-world datasets results show that the proposed algorithm improves the performance of the collaborative filtering algorithms. Meanwhile, our method can solve the problem of cold start intraditional collaborative filtering algorithm to some extent. Our method improves the satisfaction of user queries and precision.

Acknowledgement. This work was partly supported by Research on Stability Analysis of a Type of PWM Feedback Time Delay System No. XL201510, Matlab PWM system simulation and application JAT160507 and Research on Information Hiding Technology based on DeepLearning JT180546.

References

1. Bellot, P., Doucet, A., Geva, S., Gurajada, S., Kamps, J., Kazai, G., Koolen, M., Mishra, A., Moriceau, V., Mothe, J., Preminger, M., SanJuan, E., Schenkel, R., Tannier, X., Theobald, M., Trappett, M., Wang, Q.Y.: Overview of INEX 2013. In: International Conference of the Cross-Language Evaluation Forum for European Languages, pp. 269–281. Springer, Heidelberg (2013)
2. Bellot, P., Bogers, T., Geva, S., Hall, M., Huurdeman, H., Kamps, J., Kazai, G., Koolen, M., Moriceau, V., Mothe, J., Preminger, M., SanJuan, E., Schenkel, R., Skov, M., Tannier, X., Walsh, D.: Overview of INEX 2014. In: Kanoulas, E., et al. (eds.) Information Access Evaluation. Multilinguality, Multimodality, and Interaction, pp. 212–228. Springer, Cham (2014)
3. Koolen, M., Bogers, T., Gäde, M., Hall, M., Huurdeman, H.: Overview of the CLEF 2015 social book search lab. In: Mothe, J., et al. (eds.) Experimental IR Meets Multilinguality, Multimodality, and Interaction, pp. 545–564. Springer, Cham (2015)
4. Koolen, M., Bogers, T., Gäde, M., Hall, M., Huurdeman, H.: Overview of the CLEF 2016 social book search lab. In: Fuhr, N., et al. (eds.) Experimental IR Meets Multilinguality, Multimodality, and Interaction, pp. 351–370. Springer, Cham (2016)
5. Linden, G.: Amazon.com recommendations. IEEE Internet Comput. **4**(1), 76–80 (2003)
6. Resnick, P., Iacovou, N., Suchak, M., Bergstrom, P.: GroupLens: an open architecture for collaborative filtering of netnews. In: ACM Conference on Computer Supported Cooperative Work, pp. 175–186. ACM (1994)
7. Wei, S., Ye, N., Zhang, S., Bergstrom, P.: Item-based collaborative filtering recommendation algorithm combining item category with interestingness measure. In: International Conference on Computer Science and Service System, pp. 2038–2041. IEEE Computer Society (2012)
8. Breese, J.S., Heckerman, D., Kadie, C.: Empirical analysis of predictive algorithms for collaborative filtering. In: Fourteenth Conference on Uncertainty in Artificial Intelligence, pp. 43–52. Morgan Kaufmann Publishers Inc. (1998)
9. Shoaran, M., Thomo, A., Weber, J.: Social Web Search. Springer, New York (2014)
10. Wei, J., He, J., Chen, K., Zhou, Y., Tang, Z.: Collaborative filtering and deep learning based recommendation system for cold start items. Expert Syst. Appl. **69**, 29–39 (2017)
11. Willmott, C.J., Matsuura, K.: Advantages of the mean absolute error (MAE) over the root mean square error (RMSE) in assessing average model performance. Climate Res. **30**(1), 79 (2005)

Design and Realization of 3D Virtual Campus Roaming System Based on Unity3D

Jia Chen[1], Peishan Hu[1], and Zengxi Huang[2(✉)]

[1] School of Educational Information Technology,
Central China Normal University, No. 152 Luoyu Road, Wuhan, Hubei, China
[2] School of Computer and Software Engineering, Xihua University,
No. 999 Jinzhou Road, Jinniu District, Chengdu, China
Huangz@mail.xhu.edu.cn

Abstract. In order to let the fresh students who have not entered university quickly understand the campus environment, we design and develop a virtual Central China Normal University (CCNU) campus roaming system. Firstly, collecting a large number photos of the campus, such as buildings and plants. Secondly, building three-dimensional (3D) models based on the photographs and maps, in the meantime, using Photoshop to deal with photos, letting the building looks as real as possible. Thirdly, putting the model into the Unity platform, and enabling it to achieve the interactive function. Finally, using computer to export the program, then the virtual operation roaming system has been finished. Compared with Baidu's 3D map, our roaming system shows the real scene better, and also integrates multimedia introductory resources such as video and audio of each building on campus, which greatly improves the user experience.

Keywords: Virtual roaming · Unity3D · Virtual reality · 3D map · Digital campus

1 Introduction

Today, with the rapid development of internet technology, virtual reality technology has become more and more important. The interactive audio-visual immersive experience brought by virtual reality further promote the popularity of the virtual reality field. At the same time, virtual reality has brought unlimited fun to the audience. The most widely used virtual roaming product is Baidu's 3D map, but it does not perfectly show the real scene and its interact function is not what you like.

We design and realize a 3D virtual campus roaming system [1]. This work is designed for students who are going to enter university and freshmen who are not familiar with campus. With this online system, users can realize 3D virtual roaming of Central China Normal University (CCNU), reappear scenes of the campus, which is convenient for users to be familiar with routes and all the architectures in the campus soon. Additionally, there are some sections concerning vital information of CCNU, such as *Tips for Freshmen, Contact Information of School Systems, Brief of CCNU* and *School Calendar* and so on. These sections not only help users save time without

© Springer Nature Switzerland AG 2020
Q. Liu et al. (Eds.): CENet 2018, AISC 905, pp. 485–492, 2020.
https://doi.org/10.1007/978-3-030-14680-1_53

asking teachers or looking up online for information and screening out unreliable message, but also ensure students to adapt to college life as soon as possible. Meanwhile, the background music (BGM) is the original song of CCNU "Autumn Moon of Southern Lake". The BGM reveals exclusive atmosphere of CCNU, which makes users have a deeper understanding of CCNU and then blend in the new school becoming a qualified student rapidly. Compared with Baidu's 3D map, our roaming system shows the real scene better, and also integrates multimedia introductory resources such as video and audio of each building on campus, which greatly improves the user experience. Last but not least, this system has played an active role in enhancing the reputation, constructing digital campus, promoting school culture power, ensuring campus safety and establishing intellectual campus etc.

2   Design of 3D Virtual Campus Roaming System

With the popularity of virtual reality technology, a variety of derivative products which based on virtual reality are also widely loved by people. CCNU is a century-old institution, it has been known for its beautiful campus environment and rigorous studying atmosphere. How to let new students or foreign guests visit the beautiful campus is a problem. Virtual reality technology can achieve this goal. By using the modeling software and Unity platform, we can build a digital school environment, it can give users the freedom to "roaming" in the campus, feeling the academic culture, and enjoying the scenic spots in the virtual world.

The campus roaming system is designed to meet the following conditions:

1. The whole scene should be as close as possible to the real campus of CCNU, and it needs to be 3D. It is important to try to ensure the fidelity within the technical scope, and the position of buildings and plants should also be consistent with the real campus.
2. The user is allowed to move up and down, and left and right in the scene, so that the user can observe the virtual campus scene from different views.
3. The scene is ensured to be rendered smooth, and the material, texture, light settings in the entire scene should be as natural as possible.
4. Allowing users to get a real roaming experience, and allowing them to move, rotate and do other operations, the virtual roaming system should has real-time interactive function.

3   Realization of 3D Virtual Campus Roaming System

3.1   Capture Textures for 3D Models

In the process of developing a virtual campus roaming system, the most important component is the buildings and landscapes in the virtual scene [2]. In order to build virtual scenes as realistic as possible and let it rich in detail, a large number of landmarks on the campus's iconic buildings and landscapes are required as shown in Fig. 1.

Fig. 1. Capture textures for 3D models.

3.2 Import 3D Models

Unity3D only approves 3D models of *fbx* format [3], thus, when we were designing, we adopt Autodesk Maya, the world's top 3D animation software, to export our 3D models into *fbx* format. Adding file folder named Object to Asset folder in Unity3D, the software will automatically identify models, depart Material and Texture, then separately import them in the Object folder. By now, when we import our models, we can just click the menu "Assets-import new asset" to choose the particular files which are able to display all the texture maps. After designing models, we need integrate all the models and scenes according to our planar graph. In addition, in order to add background music, our program is supported by multi-media files. Besides, we set up Directional Light and Sky Box to make scenes more bright and realistic as shown in Fig. 2.

3.3 Create Campus Terrain

There is an excellent Terrain Editor in Unity3D [4] that basically can establish complex terrain in real life. Clicking the menu "GameObject-Create Other-Terrain", a terrain will be set up. However, it is just a simple panel. We need to build a hypsogram according to the collecting data, which can be done by Photoshop or Maya, to make it more accurate and more realistic. This map must be an 8 bit or 16 bit grey-scale map in RAW format, where brighter parts mean higher terrain, darker parts mean lower terrain. After building up the map, we import it into the attribution of Terrain as shown in Fig. 3. If you need to make some little adjustment, you can use Lifting tool, Drawing tool, Smooth Height tool in the Examination Panel to do some fine adjusting.

3.4 Realization of Role-Based Control

As shown in Fig. 4, when users are roaming in campus, they can observe the whole virtual scene in Character Controller's perspective. To make our virtual roaming

Fig. 2. Import 3D models.

system more realistic, Character Controller is supposed to be equipped with Translation function, Camera Tracking function, and Camera Viewport Rotating function [4]. This software adopts the Third Person Controller, which means that there is always a character that can move in all directions leading in front of our roaming character.

Fig. 3. Create campus terrain.

Fig. 4. Realization of role-based control.

Here, we set up three Script Components, AICharacterControl.cs, ThirdPersonCharacter.cs, ThirdPersonUserControl.cs, to realize Mouth Control Perspective function, Listening Direction Button function and Character Control Coefficient setting function.

3.5 Collision Detection and Little Navigation Map

When the character is roaming, the collision is usually happen between himself and other architectures, ground or trees. Therefore, we add Collision Controller to realize collision detection [5]. The Unity platform has the function settings of adding collision for objects by default. Here we mainly use Box Collider, which is mainly used for collision between cuboid and cube. There are also other colliders, such as sphere, capsule, mesh, and wheel that are used less frequently. The collision detection method used in this work is mainly the AABB bounding box method [6], which uses set box information to calculate whether a collision will occur.

When the character is roaming, a navigation map is needed to show the position in the whole scene. We use a little dot to mark the position in the map. If the character moves, the dot changes. Adding KGFMapSystem little Map plug-in to the project, setting the character as itsTarget, making it a target in the center of little map every minute, and then, little navigation Map is completed successfully.

4　Results of 3D Virtual Campus Roaming System

The 3D virtual campus roaming system which has been designed by us has the following functions:

1. As shown in Fig. 5, the whole scene of the 3D virtual campus roaming system is as close as possible to the real campus of CCNU. The virtual scene has a enough fidelity that the position of buildings and plants can be consistent with the real campus.
2. As shown in Fig. 6, the user is allowed to move up and down, and left and right in the scene during 3D virtual roaming, so that the user can observe the virtual campus scene from different views. And the scene has day and night patterns.
3. As shown in Fig. 7, the scene is rendered smooth, and the material, texture, light settings in the entire scene are designed as natural as possible.
4. It has real-time interactive function, allowing users to get a real roaming experience, and also allowing users to move, rotate and do other operations.

Fig. 5. 3D campus of CCNU.

Fig. 6. 3D virtual roaming.

Fig. 7. Smooth rendering.

5   Conclusion

The 3D virtual campus roaming system is designed for students who are going to enter university and freshmen who are not familiar with campus. With this online system, users can realize 3D virtual roaming of Central China Normal University (CCNU), reappear scenes of the campus, which is convenient for users to be familiar with routes and all the architectures in the campus soon. Additionally, there are some sections concerning vital information of CCNU, such as *Tips for Freshmen*, *Contact Information of School Systems*, *Brief of CCNU* and *School Calendar* and so on. These sections not only help users save time without asking teachers or looking up online for information and screening out unreliable message, but also ensure students to adapt to college life as soon as possible. Meanwhile, the background music (BGM) is the original song of CCNU "Autumn Moon of Southern Lake". The BGM reveals exclusive atmosphere of CCNU, which makes users have a deeper understanding of CCNU and then blend in the new school becoming a qualified student rapidly. Last but not least, this system has played an active role in enhancing the reputation, constructing digital campus, promoting school culture power, ensuring campus safety and establishing intellectual campus etc.

Acknowledgement. This work is financially supported by National Natural Science Foundation of China (Nos. 61605054, 61602390), Hubei Provincial Natural Science Foundation (No. 2014CFB659), and self-determined research funds of CCNU from the colleges' basic research and operation of MOE (No. CCNU15A05023). Specific funding for education science research by self-determined research funds of CCNU from the colleges' basic research and operation of MOE (No. CCNU16JYKX039).

References

1. Xie, Q., Ge, L., Wang, Z.: Design of the virtual reality roam system for ancient city. In: International Conference on Systems and Informatics, pp. 119–123. IEEE (2015)
2. Li, W., Yang, R.: Fast texture mapping adjustment via local/global optimization. IEEE Trans. Visual Comput. Graphics **99**, 1–7 (2018)
3. Huang, Y.: Virtual reality technology based on Unity3D yuelu academy 3D roaming design. Hunan University (2016)
4. Nelimarkka, P.: Teaching Unity3D in game programming module. In: International Conference on Serious Games Development and Applications, pp. 61–71. Springer (2011)
5. Tang, M., Tong, R., Wang, Z., et al.: Fast and exact continuous collision detection with Bernstein sign classification. ACM Trans. Graph. **33**(6), 1–8 (2014)
6. Kuang, Y., Jiang, J., Shen, H.: Study on the virtual natural landscape walkthrough by using Unity3D. In: IEEE International Symposium on VR Innovation, pp. 235–238. IEEE (2013)

Chinese Information Retrieval System Based on Vector Space Model

Jianfeng Tang$^{(\boxtimes)}$ and Jie Huang

School of Software Engineering, Tongji University, Shanghai 201804, China
`billtangjf@tongji.edu.cn`

Abstract. This paper designs and implements a Chinese information retrieval system based on vector space model. The system will be able to quickly retrieve 50 thousand news documents to help users find the information they want in a short period of time, so as to help their own production and life. The beginning of this system is data capture, that is, data acquisition module, and then the preprocessing of the document, including Chinese word segmentation, discontinuation of words, document feature words extraction, and document vector representation and so on. The system first carries out the data capture, forms the document, then carries out Chinese word segmentation to the document, and handles the discontinuation of words in the Chinese word segmentation. After that, the document word item is extracted, and then vectors are represented and stored on disk. At the same time, the document word items are inverted indexed and stored on disk. When the user inquires, the user query statement is regarded as a document, the document is preprocessed, then the index is read, the related documents read to the vector similarity are calculated. Finally, the comprehensive results are sorted according to the calculated scores. In the experimental result and test part, several examples are given to analyze the recall rate and precision rate, and one of the practical problems is solved by interpolation method.

Keywords: Chinese information retrieval · Vector space model · Rule execution · Data acquisition

1 Introduction

The Chinese information retrieval system is essentially a full-text retrieval system. It provides the full text with the function of searching and querying documents in plain text format.

This paper implements a Chinese information retrieval system based on vector space model. The system will be able to quickly retrieve 50 thousand news documents to help users find the information they want in a short period of time, so as to help their own production and life.

The beginning of this system is data capture, that is, data acquisition module, and then the preprocessing of the document, including Chinese word segmentation, discontinuation of words, document feature words extraction, and document vector representation and so on.

© Springer Nature Switzerland AG 2020
Q. Liu et al. (Eds.): CENet 2018, AISC 905, pp. 493–503, 2020.
https://doi.org/10.1007/978-3-030-14680-1_54

2   Chinese Information Retrieval System Based on Vector Space Model

2.1   System Architecture Design

Generally speaking, full-text retrieval requires two basic functions, namely indexing and querying, which is the core and most important part of full-text retrieval. In addition, in order to meet different needs, the two basic functions will be optimized and expanded. Figure 1 is a system structure diagram in this paper.

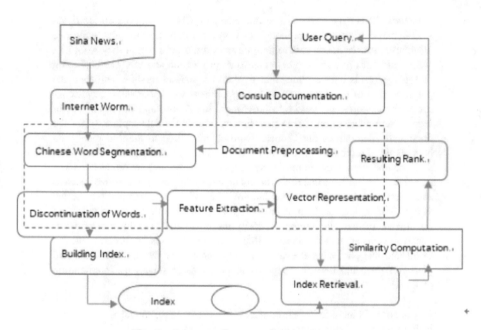

Fig. 1. Structure diagram of retrieval system.

As you can see in Fig. 1, the system first carries out the data capture, forms the document, then carries out Chinese word segmentation to the document, and handles the discontinuation of words in the Chinese word segmentation. After that, the document word item is extracted, and then vectors are represented and stored on disk. At the same time, the document word items are inverted indexed and stored on disk.

When the user inquires, the user query statement is regarded as a document, the document is preprocessed, then the index is read, the related documents read to the vector similarity are calculated. Finally, the comprehensive results are sorted according to the calculated scores.

The structure flow of the system can be seen clearly in the diagram, and Chinese word segmentation, discontinuation of words, feature extraction and vector representation are unified as document preprocessing.

The data acquisition module is customized in the data accumulation stage, the content of Sina News Channel is collected, the news is extracted, and then the pre-processing modules, such as segmentation, feature extraction and document vector, are used to construct index module, storage module, result sorting module and so on.

2.2 Data Acquisition Module

This acquisition module belongs to the vertical focused crawler. It is based on Sina network news gathering procedure, concentrating on gathering news information of Sina News Channel.

According to the seed link in Table 1, we can collect the content of the news and get the corresponding classification of the page, which will automatically be stored on the disk according to the news category when the collection program is run.

Table 1. Acquisition module seed link table.

Seed link	Channel
http://news.sina.com.cn/china/	Domestic News
http://news.sina.com.cn/world/	International News
http://news.sina.com.cn/society/	Social News

The acquisition module implemented in this paper adopts breadth first traversal method in grabbing strategy. The idea of breadth first traversal is directly appending the links contained in the newly downloaded pages to the queue to be grabbed. For example, there are three hyperlinks in a HTML page, which are A, B, and C. First, the first link A is taken and the corresponding HTML page is processed, and the linked D is obtained. Then the policy does not deal with the link D immediately, but chooses the B processing, and then returns to the C processing. In this loop, all the links are processed hierarchically, which ensures that the shallow links will be prioritize, and can also avoid the situation that deep links cannot be dealt with when they encounter deep links.

2.3 Internet Worm System Flow

Because the pages to be captured in this system are only Sina News, and the link rules of Sina News generally put important news in the shallow layer, so it is appropriate to use the breadth priority traversal strategy. First, set the seed URL, and then take out a URL. Since the seed URL contains no news text content, it contains only links to each news text, so parse it first and get page links.

Add links to the grabbing queues and queues to be grabbed, if it is not empty, a URL gets a news page for disk storage, at the same time gets the URL in the page, and determines whether the grabbing queue has been contained. If it does not contain, it is added to the queue to be grabbed and the grabbing queues.

This cycle is completed until the queue is empty, and all the grabbing tasks under a seed URL are completed.

The Internet worm system flow is as follows (Fig. 2).

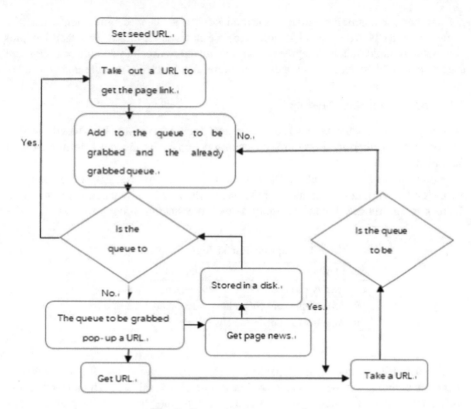

Fig. 2. Internet worm system flow.

2.4   Text Extraction Method

Text extraction is a specific screening of Web information, and eliminates the meaningless web page content according to the needs.

The text extraction flow is as follows:

This paper adopts the method of text extraction based on HTML code density, and extracts the webpage text from Baidu news in search of key names. Because the content of the news page is relatively centralized, so it can be used as the basis for exporting text blocks with large text density (Fig. 3).

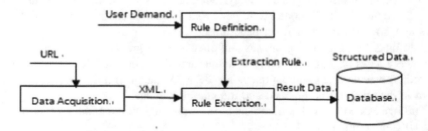

Fig. 3. Text extraction flow.

2.5   Chinese Word Segmentation Module

The segmentation system in this paper adopts the technology of word segmentation combined with forward maximum matching and reverse maximum matching, and eliminates ambiguities appropriately.

In order to reduce the probability of error, the length of the string is reduced first, and the string is decomposed into a substring of smaller length. For example, the string is decomposed into a new short string according to the sentence notation in the string (such as a comma, a period) (Fig. 4).

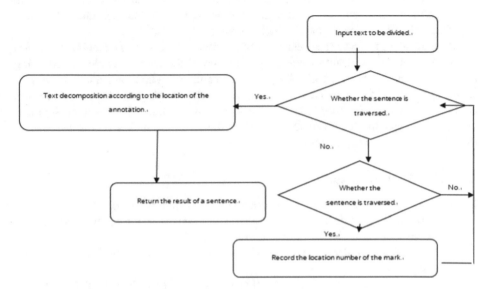

Fig. 4. Word segmentation flow.

2.6   Dictionary Construction

The hash table in the data structure has a great advantage in the search speed, but because the dimension of the dictionary is very large, it will occupy a large amount of memory only using the hash structure, so it is not feasible to use the hash structure only. After research, we use the hash table and the binary tree structure to store the dictionary.

Through statistics, there are 5180 commonly used Chinese characters in Chinese, and most of the text in the document is contained in this scope. So we can store these 5180 words in hash. Make the first word and then store the second words sequentially. We can use the array form, and then store the word in the form of the binary tree after the second word. The left is the brother, the right is the child.

In a dictionary, for a word, the same number of entries in the first two words is few, and the three words are fewer. Therefore, after using this method to organize the structure of the dictionary, the second layer of the method, that is, the first layer of the tree structure, can be used to improve the matching speed by the way of binary search.

From the third level, the number of sibling nodes of each sub node is very small, and the sequential search can meet the requirements.

2.7   Bi-directional Matching Strategy Presented by This Paper

The flow of the forward maximum matching algorithm is shown in the following figure.

When Bi-directional matching is used, ambiguity will inevitably arise. The principle of handling differences in the word segmentation system is as follows.

If the result of the word segmentation of the forward maximum matching algorithm is the same as the result of word segmentation of the reverse maximum matching algorithm, then take one of them, which shows no ambiguity.

If the result of the word segmentation of the forward maximum matching algorithm is not the same as the result of word segmentation of the reverse maximum matching algorithm, then return the result of a word segmentation with fewer numbers of substring. This principle is based on such a consideration: the amount of information contained in a word is proportional to the length of a word. After the word segmentation, the longer the length of the substring, the less the number of substrings, and therefore the more information is contained (Fig. 5).

2.8   Feature Item Extraction

In order to reduce the feature representation dimension of document vectors and improve the ability of feature representation documents, appropriate methods are needed to select feature words. The solution is try to get rid of the characteristics of weak contribution to document content, and retain the characteristics of strong contribution.

This system has constructed a discontinuation words list to filter out the commonly used function words and punctuation marks in Chinese characters from the feature set.

According to the word frequency statistics, the words in the literature are arranged according to word frequency from high to low. According to the law of Zipf, we have:

Word frequency * sequence number ≈ constant

Zipf law is applied to estimate the importance of words in the literature according to word frequency characteristics.

Given a literature set consisting of N documents$\{D_i, 1 \le i \le n\}$, we calculate the frequency of each word tk in each document Di, that is, the frequency in the literature freqik, which is denoted as tf_{ik} for convenience. We calculate the total frequency totalfreq$_k$ of each word tk in the literature set.

$$totalfreq_k = \sum_{i=1}^{n} tf_{ik}$$

Do descending sort according to totalfreq$_k$. Select a suitable threshold UpperCutoff and delete all words that totalfreq$_k$ values are above the threshold, that is, delete common words. Similarly, select a smaller threshold LowerCutoff, and delete all words

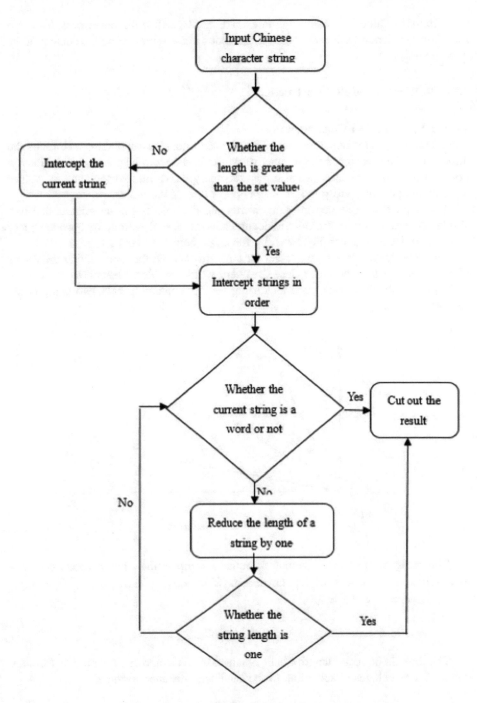

Fig. 5. Forward maximum matching algorithm flow.

that totalfreq$_k$ values are below the threshold, that is, delete the rare words. Take the remaining moderate frequency words as the index items representing the content of the document.

2.9 Retrieval Module Construction

Vector Space Model Construction

Each document in the document set is transformed into a corresponding vector, and the dimension of the vector is the number of words in the word segmentation dictionary. For user input query words, we also map the query words into vectors, and the values of each component correspond to the weight of one of the words.

The approximate degree of query words and documents can be transformed into similarity between query vectors and document vectors, so the similarity between query words and documents can be shown by the angle between vectors.

The smaller the angle is, the larger the similarity is, and the more similar the vector is, which shows that the more close the query words and documents are.

Figure 6 illustrates this basic idea using vectors containing only two components (A and B).

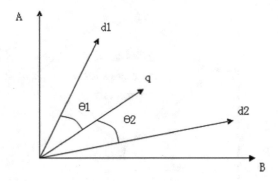

Fig. 6. Document similarity.

Linear algebra provides a simple formula to compute the angle between two vectors. Given two vectors of $|v|$ dimension, $\vec{x} = <x_1, x_2, \ldots \ldots x_{|v|}>$, $\vec{y} = <y_1, y_2, \ldots \ldots, y_{|v|}>$, we have:

$$\vec{x} \cdot \vec{y} = |\vec{x}| \cdot |\vec{y}| \cos(\theta) \tag{1}$$

The formula of vector length can be obtained from the above point product formula, which can be calculated according to the Euclidean distance formula.

$$|\vec{x}| = \sqrt{\sum\nolimits_{i=1}^{|v|} x_i^2} \tag{2}$$

To replace the above formula, we can get the following formula.

$$\cos(\theta) = \frac{\vec{x}}{|\vec{x}|} \cdot \frac{\vec{y}}{|\vec{y}|} = \frac{\sum_{i=1}^{|v|} x_i y_i}{\left(\sqrt{\sum_{i=1}^{|v|} x_i^2}\right)\left(\sqrt{\sum_{i=1}^{|v|} y_i^2}\right)} \tag{3}$$

Document Similarity Calculation

We use the cosine value directly as the similarity index. When the cosine value of the two vectors is 1, the angle between them is 0°, which indicates that the two vectors are collinear and similar. If the cosine value is 0, its angle is 90°, indicating that its similarity is not enough. Given a document vector d and a query vector q, the cosine similarity sim (d, q) formula is as follows.

$$sim(d, q) = \frac{d}{|d|} \bullet \frac{q}{|q|} \tag{4}$$

The dot product of document vector and query vector can be normalized to unit length. Assuming that every component of the vector is nonnegative, the cosine similarity value range is 0–1, and the higher the similarity is, the greater the value is.

Even in a medium-sized document set, the vector space model yields millions of dimensions. High dimensionality may reduce efficiency, but in many cases, query vectors are sparse, with only a few nonzero components. There may be thousands of words in the document, and every non-zero vector in the document vector corresponds to each word. However, the length of the document vector is not related to the query, so the length of the document vector can be precalculated and stored in the word frequency index or the position index with other document description information.

In general, the query words entered by the user are shorter. So when dealing with user query words, it is also treated as an ordinary document. Therefore, the similarity between users' retrieval words and documents can be transformed into similarity between documents.

3 Experimental Result

There are two commonly used indicators for evaluating the effectiveness of retrieval system, namely, precision rate and recall rate. Precision rate is the ratio of the number of related documents to the total number of documents. Recall rate refers to the ratio of the number of related documents to the total number of relevant documents.

The following examples are given to illustrate the method.

Suppose that there is a related document set

R = {d1, d6, d11, d19, d29, d56, d63, d84, d91, d112}, |R| = 10, which is corresponding to the query topic I. The sort result of the I's result document set A is: {d63, d49, d29, d95, d73, d4, d6, d31, d72, d131, d45, d1, d79, d55, d22}.

If only d63 is relevant, and the retrieval result is only the first, the recall rate is 10% and the precision rate is 100%. If d63 and d29 are related, the recall rate is 20% and the precision rate is 66.7%. At that time, two documents in the three documents are related. In turn, when d6 and d1 appear, the corresponding recall/precision rates are 30%, 42.9% and 40%, 33.3% respectively, as shown in Table 2.

Table 2. Recall rate/precision rate

Recall rate	Precision rate
0.0%	100%
10%	100%
20%	66.7%
30%	42.9%
40%	33.3%
50%	0
60%	0
70%	0
80%	0
90%	0
100%	0

Since the result set of I does not contain all the relevant documents, the corresponding precision rate is 0 when the recall rate is greater than 40%.

If the number of related document sets R corresponding to the query topic I is not 10, that is, the recall rate of a query is not necessarily a standard value, for example, when |R| = 5, the recall rate of the first related document is 1/5. The general way to solve this problem is to use interpolation method. If RJ is a parameter of the jth standard recall rate, the interpolation precision rate corresponding to the jth standard recall rate is defined as:

$$P(r_j) = \max_{r_j \leq r \leq r_{j+1}} P(r) \tag{5}$$

That is, the interpolation accuracy rate corresponding to the jth standard recall rate is the maximum value corresponding to any recall rate between jth and (j + 1)th standard recall rates.

For example: Suppose

R = {d6, d29, d63}, A = {d1, d63, d5, d7, d20, d29, d21, d23, d56, d121, d6}, then the recall rates are: 33.3%, 66.7%, 100%, and the corresponding precision rates are: 50%, 33.3%, 27.3%. The recall/precision rate obtained by interpolation method is shown in Table 3.

Table 3. Standard recall rate/interpolation precision rate.

Standard recall rate	Interpolation precision rate
0.0%	50%
10%	50%
20%	50%
30%	50%
40%	33.3%
50%	33.3%
60%	33.3%
70%	27.3%
80%	27.3%
90%	27.3%
100%	27.3%

4 Conclusion

This paper implements a Chinese information retrieval system based on vector space model. The system will be able to quickly retrieve 50 thousand news documents to help users find the information they want in a short period of time, so as to help their own production and life.

The beginning of this system is data capture, that is, data acquisition module, and then the preprocessing of the document, including Chinese word segmentation, discontinuation of words, document feature words extraction, and document vector representation and so on.

In the experimental result and test part, several examples are given to analyze the recall rate and precision rate, and one of the practical problems is solved by interpolation method.

References

1. Christopher, D.M., Prabhakar, R., Hinrich, S.: Introduction to Information Retrieval, pp. 14–92. Cambridge University Press, Cambridge (2009)
2. Gao, J.F., Li, M., Huang, C.N., Wu, A.: Chinese word segmentation and named entity recognition: a pragmatic approach. Comput. Linguist. **4**, 19–20 (2005)
3. Zheng, Z.H., Wu, X.U.Y., Srihari, R.: Feature selection for text categorization on imbalanced data. ACM SIGKDD Explor. Newsl. **1**, 17–22 (2004)
4. Jing, L.P., Huang, H.K., Shi, H.B.: Improved feature selection approach TFIDF in text mining. In: Proceedings of 1st Information Conference on Machine Learning and Cybernetic, vol. 6, pp. 22–23 (2002)

Adjustment and Correction Demarcation Points in Dongba Hieroglyphic Feature Curves Segmentation

Yuting Yang[1(✉)] and Houliang Kang[2]

[1] Culture and Tourism College, Yunnan Open University,
Kunming, Yunnan, China
tudou-yeah@163.com
[2] College of Humanities and Art, Yunnan College of Business Management,
Kunming, Yunnan, China

Abstract. Dongba hieroglyph is a kind of very primitive picture hieroglyphs. Its picture features allow us to analyse the overall and local features of Dongba characters in combination with the existing feature extraction, simplification and segmentation algorithms for shape in computer vision. Moreover, the analysis of the local features of Dongba hieroglyphs play an important role in the study of Dongba hieroglyph's writing, the evolution process of hieroglyphs and the comparison between similar hieroglyphs. Therefore, in this paper, we first use the Chain-Code Based Connected Domain Priority Marking Algorithm (CDPM) and the Discrete Curve Evolution Algorithm (DCE) to obtain the simplified feature curve of Dongba hieroglyph. Then, we focus on the selection and adjustment of demarcation points and local curve segmentation. And, the experiment proves that our method can further correct the potential errors in the curve segmentation, which is helpful to improve the correct rate of hieroglyph's local feature extraction.

Keywords: Dongba hieroglyphic feature curve · Curve segmentation · Demarcation points adjustment · Demarcation points correction

1 Introduction

Dongba hieroglyph is a kind of very primitive picture hieroglyphs [1, 2]; they use pictures to express meaning. So, many Dongba hieroglyphs of the same type have some similar attributes in morphology and structure, and they are distinguished only by local differences [3, 4]. Now, there are only a few of algorithms to solve the problem of recognizing ancient words which include hieroglyphs, pictographs [5–7] and petroglyphs [8, 9]. And the existing algorithms are not entirely suitable for processing Dongba hieroglyphs. But, the picture features of Dongba hieroglyph allow us to analyze the overall and local features of Dongba hieroglyphs in combination with the existing feature extraction, simplification and segmentation algorithms [10] for shape in computer vision. Among them, the complete feature curve is good for us to analyze the external and overall features of hieroglyphs, while the local feature curve is beneficial

© Springer Nature Switzerland AG 2020
Q. Liu et al. (Eds.): CENet 2018, AISC 905, pp. 504–512, 2020.
https://doi.org/10.1007/978-3-030-14680-1_55

for us to analyze the local composition of hieroglyphs and the commonalities and differences between different hieroglyphs in the same type.

In order to realize the extraction of local feature curves, the selection and judgment of demarcation points are the basis of curve segmentation and local feature curve extraction, and also the key to obtaining local feature curves with good independence, clear meaning and completeness. At the same time, the accurate selection of the demarcation point plays an important role in the automatic analysis, comparison and recognition of the local feature curves. It is also the basic work of Dongba character recognition research.

2 Extraction and Simplification of Dongba Feature Curve

In order to realize the local feature curve segmentation, we should first extract and simplify the feature curves of Dongba hieroglyph. Among them, the Chain-Code Based Connected-Domain Priority Marking Algorithm (CDPM) [11] provides us a feature curve extraction algorithm for Dongba hieroglyph. The Discrete Curve Evolution (DCE) [12] provides an intuitive, fast and easy to implement method for feature curve simplification, and also provides us with the theoretical basis for local feature curve segmentation [13, 14]. Therefore, first we use the CDPM algorithm to extract the feature curve of Dongba hieroglyph. Its main idea is:

(1) Let G denote a single hieroglyph, and P_i is a feature point of G. The steps of extracting feature curve are:
(2) Let P_1 denote the start point, and the direction of the initial vector is $\overrightarrow{OP_1}$. We traversal a feature curve in counter-clockwise.
(3) Let P_1 denote a known point. If the direction vector $\overrightarrow{OP_1}$ satisfies v.x $\neq 0$ and v.y $\neq 0$, then the 8 neighborhood priority weights of point $\overrightarrow{OP_1}$ satisfy in clockwise order locally, and the neighboring point with the largest priority value is the next feature point P_2.
(4) Let P_2 denote a known point, and $\overrightarrow{P_1P_2}$ is a new vector. Repeat Step (2) until we return to the starting point of the curve.

Taking the glyph in Fig. 1(1) as an example, we use CDPM algorithm to extract the feature curve as shown in Fig. 1(2).

Then, we use DCE algorithm to simplify the feature curve. It can remove a large number of redundant points in the feature curve from both global and local aspects. The main idea of DCE is: In every evolution step, a pair of consecutive line segments s1, s2 is substituted with a single line segment joining the endpoints of s1 ∪ s2. The key property of this evolution is the order of the substitution [12]. The substitution is done according to a relevance measure K given by:

$$K(s_1, s_2) = \frac{\beta(s_1, s_2)l(s_1)l(s_2)}{l(s_1) + l(s_2)}$$

Where $\beta(s_1, s_2)$ is the turn angle at the common vertex of segments s1, s2 and l is the length function normalized with respect to the total length of a polygonal curve C. The higher the value of $K(s_1, s_2)$, the larger is the contribution to the shape of the curve of arc s1 ∪ s2. And we achieve curve simplification by continuously replace the segments that contribute the least. The result of simplifying the feature curve in Fig. 1 (2) is shown in Fig. 1(3).

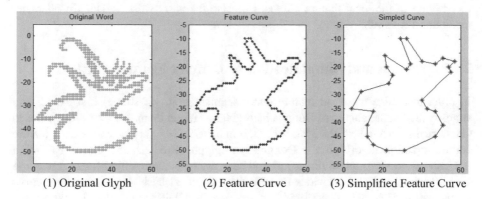

(1) Original Glyph   (2) Feature Curve   (3) Simplified Feature Curve

Fig. 1. Extraction and simplification of the feature curve of Dongba hieroglyph.

3   Selection and Adjustment of Demarcation Points

Combined with the DCE algorithm, local features are generally included in the convex curve segment, so we use the concave vertex as the demarcation point. The way we judge the concavity of the vertices in the curve is as follows:

For a feature vertex V_i, if its turning angle $\beta_i \in (0, \pi)$, the vertex is convex; if $\beta_i \in (\pi, 2\pi)$, the vertex is concave [15].

Combined with the definition, first we use the cross-multiplication of the adjacent edge vectors to determine the direction of the curve, and then we determine the turning angle based on the angle of the adjacent edge vectors. Finally, we determine the concavity and convexity of the vertex based on turning angle β_i. Therefore, for a given curve, its feature vertices sequence are $P_i(i = 2, 3, \ldots, n-1)$, we select 3 adjacent vertices P_{i-1}, P_i and P_{i+1} where [16]:

$$P_{i-1} = (x_{i-1}, y_{i-1}), P_i = (x_i, y_i), P_{i+1} = (x_{i+1}, y_{i+1})$$

Let

$$A = \overrightarrow{P_{i-1}P_i} = (x_i - x_{i-1}, y_i - y_{i-1}), B = \overrightarrow{P_iP_{i+1}} = (x_{i+1} - x_i, y_{i+1} - y_i)$$

And the direction of the curve is:

$$N = A \times B = \overrightarrow{P_{i-1}P_i} \times \overrightarrow{P_iP_{i+1}} = [(x_i - x_{i-1})(y_{i+1} - y_i) - (x_{i+1} - x_i)(y_i - y_{i-1})]k$$
$$= n_i k$$

Then, the vector angle $\alpha_i (\alpha_i \in [0, \pi])$ which is corresponding to P_i can be calculated by the vertices set of the vector, so that the sorted direction of the curve is clockwise (reverse direction), and the vector angle α_i is:

$$\alpha_i = \begin{cases} \cos^{-1}\left(\dfrac{A \cdot B}{|A||B|}\right) & n_i \geq 0 \\ \pi - \cos^{-1}\left(\dfrac{A \cdot B}{|A||B|}\right) & n_i < 0 \end{cases} \quad \text{and } \alpha_i \in [0, \pi]$$

And, in the reverse curve, the turning angle β_i corresponding to the feature vertex $P_i(i = 1, 2, 3, \ldots, n)$ is:

$$\beta_i = \begin{cases} \infty & i = 1 \text{ or } i = n \\ \alpha_i + \pi & n_i > 0 \text{ and } i = 2, 3, \ldots, n - 1 \\ \alpha_i & n_i < 0 \text{ and } i = 2, 3, \ldots, n - 1 \end{cases}$$

Therefore, based on the value of turning angle, we can judge the concavity and convexity of the vertex $P_i(i = 1, 2, 3, \ldots, n)$ is:

$$P_i = \begin{cases} \text{convex vertex} & 0 \leq \beta_i < \pi \\ \text{concave vertex} & \pi \leq \beta_i < 2\pi \end{cases}$$

Taking the glyph shown in Fig. 1 as an example, we first mark the concave vertices and convex vertices in the feature curve. Then, we divide feature curve with concave vertices, and the processing results are shown in Fig. 2(1) and (2). Obviously, the partial curve segment in Fig. 2(2) is too small, which results in the generation of some invalid curve segments that do not contain glyphic features. Therefore, we remove the invalid curve segments by combining the curve segments consisting of two vertices and three vertices. The merge rules are:

(1) **Combine the curve segments with two points:** It can be known from DCE algorithm is that for convex vertices, the larger the K value, the more significant the glyphic features it contains, and the more it should be retained; but for concave vertices, the larger the K value, the more independent of the two curved segments connected to it and the more suitable it is be a demarcation point [13]. Since the end points of the curve segments must be concave, we merge the curve segments to the smaller end of the K and the other end as the demarcation point. The combined effect is shown in Fig. 2(3).

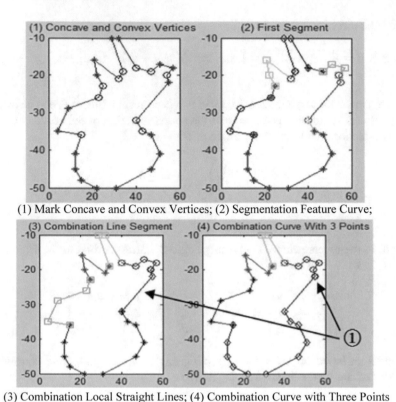

(1) Mark Concave and Convex Vertices; (2) Segmentation Feature Curve;

(3) Combination Local Straight Lines; (4) Combination Curve with Three Points

Fig. 2. Marking concave and convex vertices and segmentation feature curve.

(2) **Combine the curve segments with three points:** The curve contains two straight line segments. So, before the combination, we divide the curve segment from the middle vertex into two straight lines, and then merge the two straight lines with their adjacent curved segments respectively. The combined effect is shown in Fig. 2(4). However, this combination may cause the demarcation point to move, and cause the point to change from a concave vertex to a convex vertex, as shown in comparison ① of Figs. 2(3) and (4). Therefore, we need to further judge and correct the demarcation point.

4   Adjustment and Correction Demarcation Point

In order to ensure the continuity of the curve, the starting point of the local feature curve coincides with the end point of the adjacent previous curve. And its end point coincides with the starting point of the adjacent next curve. Then, if one or both endpoints in the curve to be tested are convex vertices, it would have the same glyphic characteristics as the adjacent one or two curve segments. In order to ensure the local independence of each curve and avoid the repeated judgment of the endpoint, we specify that the starting point of each curve is the demarcation point between it and the previous curve segment. And the demarcation point of the next curve adjacent to it is actually the termination point. It means that each local curve has only one demarcation point. Therefore, in the process of adjustment, each curve only judges or corrects one demarcation points that it has. So, even if there are common points in adjacent curves, they will not cause confusion when adjusting the demarcation points. As shown in Fig. 3(1), in the clockwise direction, the local curve segments constituting the characters are as follows:

line1: P1 (demarcation point) → P5
line2: P5 (demarcation point) → P9
line3: P9 (demarcation point) → P12
line4: P12 (demarcation point) → P17
line5: P17 (demarcation point) → P28(P1), The starting point P1 coincides with the ending point P28 in feature curve.

After we merged the local curves, not all the demarcation points need to be re-adjusted. Only when the demarcation points are convex vertices, we adjust them. Therefore, the steps for correcting demarcation points in each curve segment are as follows:

Step 1.   Traverse the local curve to determine the concavity and convexity of the demarcation point; If the demarcation point is convex, we will find adjacent concave vertices;

Step 2.   If both vertices adjacent to demarcation point are concave, we choose the larger K point as the new demarcation point;

Step 3.   If only one adjacent vertex is concave, we use it as the new demarcation point, directly;

Step 4.   If the adjacent vertices are convex, we will continue to find the adjacent points on their left (or right) side, and repeat Step 2 to Step 3;

Step 5.   Repeat Step 1 to Step 4 until all the demarcation point are judged and corrected.

For example, we adjust the demarcation point P17 of line 5 in Fig. 3(1), and the effect is shown in Fig. 3(2). At this time, the original convex demarcation points have been adjusted, so that the "mouth" feature of the glyph is fully presented.

Let n denote the total number of feature points. The number of times is uses to find demarcation points and adjust them to concave. The computational complexity of finding demarcation points is $O(n)$, and adjusting them to concave is $O(n)$, also. Therefore, the total computational complexity of the whole algorithm is $2 * O(n) \approx O(n)$, and it is linear.

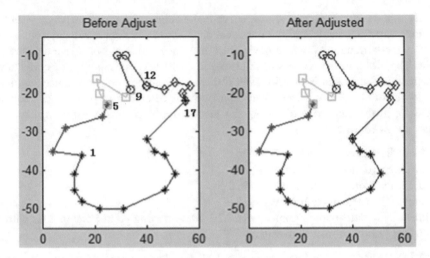

Fig. 3. Adjustment and correction demarcation point.

5   Experiments

In order to verify the validity of our algorithm, we randomly select 100 glyphs from 1593 Dongba hieroglyphs, and use CDPM and DCE algorithms to get their simply feature curves. Then, we use our algorithm to judge and correct the demarcation points. Among them, there are 87 glyphic feature curves containing convex demarcation points. We corrected 161 demarcation points in 87 different glyphs by using our selection and correction algorithm, which accounted for 35.94% of the total demarcation points.

Figure 4 shows the total number of demarcation points included in different glyphs and the number of demarcation points which has been corrected in each glyph. The experimental results show that our algorithm can make the local feature curve get more accurate segmentation by judging and correcting the demarcation point in local curve.

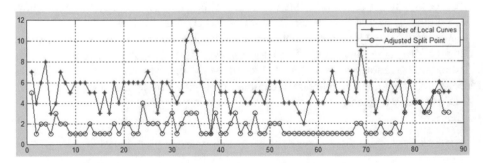

Fig. 4. The total number and corrected number of demarcation points.

6   Conclusions

The segmentation and correction of local feature curve is the basic work of classification, retrieval and recognition of Dongba hieroglyphics. It plays a very important role in studying creation methods, researching the process of hieroglyph evolution and comparing the similar words of Dongba hieroglyphs. Therefore, this paper focuses on the demarcation point selection and correction, local feature curve segmentation in glyphic feature curves. And, we use experiments to prove that this method can further correct potential errors in curve segmentation and improve the accuracy of glyphic local feature extraction.

Acknowledgements. This research was partially supported by the Scientific Research Fund of Yunnan Education Department (2018JS748).

References

1. He, L.M.: The research of Dongba culture heritage. Social Sciences in Yunnan (1), pp. 83–87 (2004)
2. Rock, J.F., Khi, N.: English encclopedic dictionary (Part I). Roma Istituto Italiano Peril Medio ed Estreme Prientale, Roma (1963)
3. Fang, G.Y., He, Z.W.: The Dictionary of Naxi Hieroglyphic. Yunnan People's Publishing House, Kunming (2005)
4. Li, L.C.: Pronunciation Dictionary of Naxi Hieroglyphic. Yunnan Nationalities Publishing House, Kunming (2001)
5. Guo, H., Zhao, J.Y.: Segmentation method for NaXi pictograph character recognition. J. Convergence Inf. Technol. **5**(6), 87–98 (2010)
6. Guo, H., Zhao, J.Y.: Research on feature extraction for character recognition of Naxi pictograph. J. Comput. **6**(5), 947–954 (2011)
7. Guo, H., Yin, J.H., Zhao, J.Y.: Feature dimension reduction of naxi pictograph recognition based on LDA. Int. J. Comput. Sci. **9**(1), 90–96 (2012)
8. Zhu, Q., Wang, X., Keogh, E., Lee, S.H.: Augmenting the generalized hough transform to enable the mining of petroglyphs. In: Proceeding of ACM SIGKDD International Conference on Knowledge Discovery & Data Mining, pp. 1057–1066 (2009)
9. Zhu, Q., Wang, X., Keogh, E., Lee, S.H.: An efficient and effective similarity measure to enable data mining of petroglyphs. Data Min. Knowl. Disc. **23**(1), 91–127 (2011)
10. Zhou, Y., Liu, J.T., Bai, X.: Research and perspective on shape matching. Acta Automatica Sin. **38**(6), 889–910 (2012)
11. Yang, Y.T., Kang, H.L.: A novel algorithm of contour tracking and partition for Dongba hieroglyph. In: Processing of 13th Chinese Conference IGTA 2018, pp. 453–463 (2018)
12. Latecki, L.J., LakaÈmper, R.: Polygon evolution by vertex deletion. In: Processing in International Conference of Scale-Space Theories in Computer Vision, pp. 398–409 (1999)
13. Latecki, L.J., LakaÈmper, R.: Convexity rule for shape decomposition based on discrete contour evolution. Comput. Vis. Image Underst. **73**(3), 441–454 (1999)

14. Latecki, L.J., LakaÈmper, R.: Shape similarity measure based on correspondence of visual parts. IEEE Trans. Pattern Anal. Mach. Intell. **22**(10), 1–6 (2000)
15. Chen, B.F., Qian, Z.F.: An algorithm for identifying convexity-concavity of a simple polygon. J. Comput. Aided Des. Comput. Graph. **2**(14), 214–217 (2002)
16. Song, X.M., Cheng, C.X.: An Analysis and Investigation of Algorithms for Identifying Convexity-Concavity of a Simple Polygon. Remote Sens. Land Resour. **9**(13), 25–30 (2011)

The Hot Spots Components from Drawings of Subway Using Template Matching

Jinlong Zhu[✉], Guangjie Liu, Fanhua Yu, Yu Zhang, Dong Zhao,
Qingtian Geng, Qiucheng Sun, and Mingyu Sun

Department of Computer Science and Technology,
Changchun Normal University, Changchun, China
zhujinlong19840913@126.com

Abstract. In order to solve the labelling component hotspots of drawings problem with database for power plant, an Optical character recognition (OCR) hot spot recognition model based on template matching (OCRTM) is used in this paper. In our opinion a very important percent of text used to be "OCRed" is coming from labeling components of drawings and realizing the links and matching function according to the method of operating hotspot, thus also encapsulates method of editing hotspot in this component which include the add operation, modify operation and delete operation of hotspot. Empirical results show that the relationship structure can be established from the top floor drawings to the lowest level drawings, and the relationship structure can facilitate users to get drawings information through hot spots. The proposed architecture based on template matching. Performance can effectively solve the recognition hot spots and the relationship between drawing and hot spots problem.

Keywords: Hot spots · Template matching · Drawings of subway

1 Introduction

OCR generally consists of two steps: 1. Detection- Identification region containing the text; 2. Classification- Identification text in the region. First, detection models, the most popular object detection model [3]. In the last two years, the hottest object detection model has faster-rcnn [2] and Yolo [1]. The two models are based on CNN giving proposed regions and classifying object region. Yolo is faster than faster-rcnn, but there are some losses on accuracy. The more famous classification model [4–8] is the Multidigit Number Recognition from Street View Imagery using Deep Convolutional, which was proposed by Ian Goodfellow in 13 years. The method is also based on the deep CNN identification method, but its deficiency is to select the maximum length of the predictable sequence in advance, more suitable for the door number or the license plate number (a small number of characters, and each character can be regarded as independent). Another common method is RNN/LSTM/GRU+CTC, which was first proposed by Alex Graves in 06 years for speech recognition. The advantage of this approach is to identify any length of text, and the nature of the model determines its ability to learn the connection between text and text (temporal relations/dependencies).

© Springer Nature Switzerland AG 2020
Q. Liu et al. (Eds.): CENet 2018, AISC 905, pp. 513–519, 2020.
https://doi.org/10.1007/978-3-030-14680-1_56

The disadvantage is that the sequential natural determines its computational efficiency is not CNN high, and there is a potential gradients exploding/vanishing problem. Due to the above problems, we propose a hot spot recognition method based on OCR template matching to improve the recognition efficiency of the system. The template matching method is used to compare the characters with the prefabricated character templates one by one. If the result is less than a threshold, the result is the character on the template.

2 The Hot Spots Components Recognition Model

2.1 Layout Algorithm

We used the bottom-up layout method to divide the layout complexity of the drawings into signs, texts, graphs and tables. First, we mark all connected domains in the drawing and store the location and area of the connected domain. According to the area distribution of all the basic connected domains in the image, we analyze the layout of the drawing and classify the connected domain to symbol, text, chart and table. For text connected domain, the gradual merging method is used to generate rows from words and generate regions by lines, and finally get the largest independent text area. For graph and table type connected domain, we analyze whether there are regular row structures in connected domain, which distinguishes between graphs and tables.

2.2 Text Extraction Algorithm

We construct a text recognition area for the structure of text connected domain. Text recognition area uses statistical method to extract text. It includes two steps: line segmentation and word segmentation. The line segmentation gets all the lines of the text area. Then, all text lines are segmented to get all independent text in the text area.

Line Segmentation Algorithm
We take into account the gap between rows and rows in the text area to segment the text. The Horizontal projection gets the number of black pixels per pixel row. If the row of I pixels is a row gap, the horizontal projection of the pixel row is 0. The model traverses the horizontal projection results of all pixel rows to complete the line segmentation of the text area.

Word Segmentation Algorithm
1. Analysis the height and width of the text;
2. Getting the right edge of the text based on the left margin of the text and the width of the estimated text;
3. Getting the smallest rectangle in the text according to the location of the text;
4. Extraction word for each text line to find all word and storage in the word area.

2.3   Word Recognition Algorithm

The process of text recognition includes text normalization, text refinement, text feature extraction and template matching to complete character recognition.

Text Size Normalization

The text is recognized to transform the text of different sizes into the same size text. When text is normalized, there are two cases of text enlargement and text reduction. Set to identify the size of the text/normalized text size = percent, if percent > 1, text reduction, else text magnification. The text coordinates are transferred to the (x', y') of the normalized text at (x, y) literal pixels. Then the conditions for x, y, x', y' are to be satisfied:

$$\begin{cases} x/x' = \text{percent} \\ y/y' = \text{percent} \end{cases}$$

That is, $x' = x/\text{percent}$, $y' = y/\text{percent}$. We needed to do $x' = x/\text{percent}$, $y' = y/\text{percent}$ to round up and round down numbers.

Text Thinning

The theoretical basis of the thinning algorithm used in this paper is as follows:

For each image in a 3 * 3 region, the tag names are P1, P2, P3,… P9. As shown in Figs. 1 and 2, 1 is used for black and 0 for white. Set P1 to be a black spot. If P1 is a black spot and satisfies the following conditions, the method sets P1 to be white.

P3	P2	P9
P4	P1	P8
P5	P6	P7

Fig. 1. Template structure.

1. $2 \leq NZ(P1) \leq 6$
2. $Z0(P1) = 1$;
3. $P2 * P4 * P8 = 0$ or $Z0(P2) \neq 1$
4. $P2 * P4 * P6 = 0$ or $Z0(P4) \neq 1$

The algorithm repeats the above steps to each point in the text image until all points are not deleted, that is, image thinning is completed. NZ (P) represents the number of black pixels in the 8 neighborhood of the point, while Z0 (P) is calculated as follows:

1. $n = 0$
2. if $P(-1, 0) = 0$ and $P(-1, -1) = 1$, n++
3. if $P(-1, -1) = 0$ and $P(0, -1) = 1$, n++
4. if $P(0, -1) = 0$ and $P(1, -1) = 1$, n++
5. if $P(1, -1) = 0$ and $P(1, 0) = 1$, n++

6. if P(1,0) = 0 and P(1, 1) = 1, n++
7. if P(1, 1) = 0 and P(0, 1) = 1, n++
8. if P(0, 1) = 0 and P(−1,1) = 1, n++
9. if P(−1, 1) = 0 and P(−1, 0) = 1, n++
10. Z0(P) = n

P(-1,-1)	P(-1,0)	P(-1,1)
P(0,-1)	P	P(0,1)
P(1,-1)	P(1,0)	P(1,1)

Fig. 2. The location of template structure.

Text Feature Extraction

We choose text density, mesh and peripheral features to extract text features. The density of text refers to the percentage of black pixels in the text image, which belongs to the overall characteristics of the text. Because of the fact that text density is not strong enough to distinguish characters, we use text density feature to coarse classification.

The text grid features divide the text image into n * n parts, and N usually takes 8. The number of pixels in each grid occupies the percentage of the whole text pixels, and the image is refined and classified according to the grid characteristics of the text.

The peripheral features of the text can be found out of the outer frame of the text, and then the text is divided into n * n parts. The N usually takes 8, scanning from the four borders to the opposite side, calculating the ratio of the area of the non text part and the whole area of the text, which is initially touched with the written stroke, and the ratio of the 4N dimension is a peripheral feature of the text. Then the ratio of the 4N dimension of the non text area and the total text area of the second times to the word line is the second peripheral feature, and the feature vectors of 8N dimensions are formed, and the peripheral features of the text are formed.

Template Matching Recognition

Word density matching: (1) Extraction of text density. (2) Calculate the absolute value of text density and five kinds of density difference. (3) Sort out the results of Step (2) and find out the smallest density difference. The density matching result of text is the density.

Text grid and peripheral features match: (1) Extract grid features and peripheral features of text. (2) Calculate the absolute value of the grid character and the peripheral features of the text respectively, and the difference between the characters and the peripheral features of all the characters in the template library. (3) Sort out the results of Step 2, and find out the smallest 10 of the absolute values of grid characteristics and peripheral differences respectively. (4) Merge grid differences and peripheral features differences, choose the smallest word template is the result of word recognition.

3 Result

We divide the blueprint into 20 types: circuit diagram and subway structure drawing. The circuit diagram is divided into 7 types, and the subway structure diagram is divided into 13 types. In what follows, accuracy is dened as the total number of correctly classied words divided by the total number of words classied. Accuracy and recall are two measure values in information retrieval and statistical classification, and are used to evaluate the quality of results. The accuracy is to retrieve the ratio of the number of relevant documents to the total number of documents retrieved, to measure the precision of the retrieval system. The recall rate refers to the ratio of the number of relevant documents to the retrieved documents and the number of all relevant documents in the document library, and the recall rate of the retrieval system.

The recognition result is shown in Fig. 3, The hot spot information of the green area is the connection point string; the red area is the line numbered string and the component numbered symbol respectively; the proposed method can find the circuit diagram through the numbering.

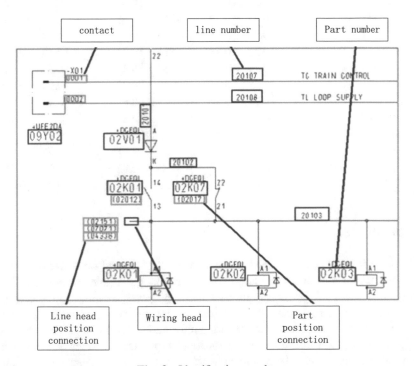

Fig. 3. Identification results.

The results show that our algorithm has a significant advantage in recognition accuracy and recall rate, and F-score is higher than the value of neural network algorithm in Table 1. Table 1 shows that the recognition rate of the proposed method is higher than neural network. Template matching method is characterized by higher accuracy than other methods, but the replacement of other font drawings will reduce the recognition rate. Recall show that OCRTM can identify more hot spots. In ensuring the accuracy of the premise, the proposed method has superior performance.

Table 1. Identification result table.

Type	OCRTM				Neural network			
	Time [s]	Accuracy [%]	Recall	F-score	Time [s]	Accuracy [%]	Recall	F-score
1	46152.16	0.97	0.96	0.96	48737	0.89	0.94	0.92
2	39741.77	0.94	0.97	0.95	43352	0.87	0.94	0.90
3	37357.09	0.99	0.98	0.99	40718	0.84	0.97	0.90
4	36162.91	1.00	0.96	0.98	39978	0.91	0.94	0.92
5	39062.57	0.92	0.98	0.95	47910	0.89	0.95	0.92
6	33150.13	0.96	0.97	0.97	44970	0.85	0.96	0.90
7	40248.04	0.92	0.97	0.94	43248	0.89	0.95	0.92
8	38982.97	0.99	0.97	0.98	49774	0.88	0.96	0.92
9	37085.07	0.95	0.96	0.95	39573	0.85	0.93	0.89
10	40344.97	0.92	0.97	0.94	44511	0.86	0.96	0.90
11	37143.60	0.95	0.96	0.95	42099	0.90	0.94	0.92
12	43024.37	0.99	0.96	0.98	49611	0.88	0.95	0.92
13	35090.17	0.97	0.96	0.97	41667	0.89	0.93	0.91
14	38948.86	0.95	0.95	0.95	42255	0.85	0.93	0.89
15	36622.99	0.97	0.98	0.97	44513	0.85	0.95	0.90
16	40034.46	0.98	0.96	0.97	47150	0.86	0.95	0.90
17	46651.37	0.95	0.95	0.95	52418	0.87	0.93	0.90
18	37582.14	0.97	0.96	0.97	41483	0.87	0.95	0.90
19	41890.49	0.98	0.97	0.97	48163	0.86	0.95	0.91
20	35602.55	0.92	0.96	0.94	44132	0.86	0.95	0.90

4   Conclusion

In this paper it has been discussed an OCR technology based on template matching. The main task of the drawing classifier is to group the characters that need to be "OCRed" in classes of similar hotspot forms. The proposed method mainly divides the information in the circuit diagram into four groups of words, symbols, graphs and tables, and determines whether the identified information is a hot spot according to the hot spot data. The algorithm saves the location, type and character of the hot information to the database, and the relationship structure can facilitate users to get drawings information through hot spots. This algorithm detects the hot results of the subway drawings compared with the neural network algorithm, and finds that the accuracy and time consumption have more advantages.

Acknowledgment. This work is supported by National Natural Science Foundation of China (Grant No. 61604019), Science and technology development project of Jinlin Province (2016052 0098JH, 20180201086SF), Education Department of Jilin Province (JJKH20181181KJ, JJKH20181165KJ), Jilin provincial development and Reform Commission (2017C031-2).

References

1. Goodfellow, I.J., Bulatov, Y., Ibarz, J., Arnoud, S., Shet, V.: Multi-digit number recognition from street view imagery using deep convolutional neural networks. Comput. Sci. **2013**, 1–12 (2013)
2. Redmon, J., Divvala, S., Girshick, R., Farhadi, A.: You only look once: unified, real-time object detection. In: Computer Vision & Pattern Recognition, pp. 779–788 (2015)
3. Kissos, I., Dershowitz, N.: OCR error correction using character correction and feature-based word classification. In: Document Analysis Systems, pp. 198–203 (2016)
4. Reul, C., Dittrich, M., Gruner, M.: Case study of a highly automated layout analysis and OCR of an incunabulum: 'Der Heiligen Leben' (1488). Computer Vision and Pattern Recognition, pp. 155–160 (2017)
5. Fink, F., Schulz, K. U., Springmann, U.: Profiling of OCR'ed historical texts revisited. In: International Conference on Digital Access to Textual Cultural Heritage, pp. 61–66. ACM (2017)
6. Thompson, P., Mcnaught, J., Ananiadou, S.: Customised OCR correction for historical medical text. Digit. Herit. **2016**, 35–42 (2016)
7. Joshi, Y., Gharate, P., Ahire, C., Alai, N., Sonavane, S.: Smart parking management system using RFID and OCR. In: International Conference on Energy Systems and Applications, pp. 729–734. IEEE (2016)
8. Mei, J., Islam, A., Wu, Y., Moh'D, A., Milios, E.E.: Statistical learning for OCR text correction. Comput. Vis. Pattern Recogn. **2**(3), 2–14 (2016)

QoS-PN Ontology Model for Service Selection

MaoYing Wu and Qin Lu$^{(\boxtimes)}$

Qilu University of Technology (Shandong Academy of Sciences), Jinan, China
`wu1992mao@163.com`

Abstract. In order to solve the existing QoS (Quality of Service) model lack of descriptions on the dynamic changes of attributes as well as user personalization QoS-PN, a new ontology model for QoS, is presented to provide users with a sharable and unified service quality description framework. By introducing User Personalize class and Status class. The QoS-PN model can rationally describe user's personalized requirements and dynamic changes of attributes, which solve the problem of service reliability and service personalization requirement. QoS-PN is described as three-layer ontology model: QoS configuration layer, QoS attributes layer, and QoS metric layer. QoS-PN can describe detailed information related to QoS attributes, metric methods; QoS attribute values, and units for the QoS, which improve the accuracy of service matching. An example of travel route inquiry is presented to show the feasibility of the QoS-PN model. Empirical results show that this model supports rich semantic description and has strong extensibility. The stability and accuracy of selected service are effectively improved when QoS-PN model is used to describing QoS information. The QoS-PN enhances descriptive ability for QoS information and the description of dynamic Attribute and personalization has been expanded. The problem of QoS stability and user personalization in service selection is solved.

Keywords: Web service · Semantic web · QoS · QoS ontology model

1 Introduction

The QoS ontology model is the basis of semantic services selecting. The QoS-PN model aim to allow service providers and requesters to publish QoS information and describe QoS requests, consequently, the QoS semantic information can be shared. The QoS ontology model should provide rich semantic description capabilities and have good scalability. Currently, we only consider the result of service selection, neglecting whether the service provided by service provider is stable [1]. In order to solve the problem, the QoS ontology model provides the description of dynamic change of attributes to calculate whether the service is stable and reliable. In order to help the requesters to find more approving services, the QoS ontology model considers. Many scholars and research institutions have proposed their own QoS ontology models. DAML-QoS [2] provides a clear description of QoS. But the DAML-QoS has weak extensibility. OSQS [3] is an expanded version of OWL-S. The advantage of OSQS ontology is that the authors attempt to integrate quality standards to provide an accurate definition of quality attributes, and through expand the SPARQL query language, it increase the accuracy and efficiency of service discovery. ML-QoS [4] ontology can

© Springer Nature Switzerland AG 2020
Q. Liu et al. (Eds.): CENet 2018, AISC 905, pp. 520–529, 2020.
https://doi.org/10.1007/978-3-030-14680-1_57

support the measurement and matching of heterogeneous QoS parameters. It also provides a description of the trend, unit of measurement, combination parameters, and metric types. But it does not support the dynamics description of the attribute. In [5], it is a higher-level ontology that extends OWL-S. But the support in terms of personalization, QoS value types an so on is relatively weak.

Compared to the existing model, the QoS-PN model provides a richer semantic description for QoS, but also supports the description of dynamic attribute. This paper propose User Personalize class. Through this class, the description of personalized is more flexible. The next section is a detailed explanation of this model.

2   Design of QoS-PN

2.1   The Advantage of QoS-PN

The QoS ontology model designed in this paper has the following several characteristics: (1) The model provides capabilities of semantic description that can describe semantic information and numerical information of QoS attributes, and it has good scalability. (2) The description of the attribute is more structural and logical by classifying QoS attributes. (3) The previous ontology model merely provides a single description method for the weight. We provide a description method of combined with the subjective empower and objective empower. (4) We provide a description of the dynamic attribute. It can describe dynamic change of attribute and provide a basis for studying service stability.

This paper uses OWL as a modeling language and uses the ontology editor protege to construct the QoS ontology model QoS-PN. We divide ontology models into three layers. That is QoS configuration layers, QoS attribute layers, and QoS metric layers.

2.2   QoS Configuration Layer

The QoS configuration layer describes some information of QoS configuration. As shown in Fig. 1, we define a main class QoSProfile and five subclasses QoSAttribute, QoSProvide, QoSConsumer, QoSInput and QoSOutput in the QoS configuration layer.

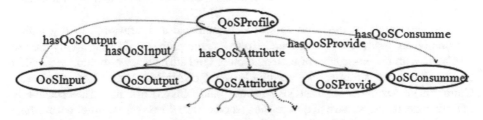

Fig. 1. Ontology structure of QoS configuration layer.

The QoSInput and QoSOutput, which are used to describe QoS input information and QoS output information. The QoSConsummer and QoSProvide classes describe relational information about service requesters and service providers, such as some identity information, company information, and so on. The QoSAttribute class describes the detailed information related to the QoS attributes. It is also the main part of the QoS-PN. The description of the QoSAttribute class is described in detail in the following section.

2.3   QoS Attribute Layer

The QoS attribute layer is mainly composed of a QoSAttribute as superclass and six subclasses:Tendency, Aggregated, Relationship, Metric, Status, User personalize. As shown in Fig. 2.

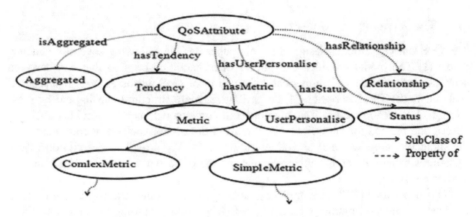

Fig. 2. Ontology structure of the QoS attribute layer.

The QoSAttribute class is a description of the QoS attribute. In [6], five QoS attributes such as price, availability, response time, reliability, and reputation are proposed according to need. In [7], Through summarizing other QoS attributes, it defines all possible QoS requirements of Web services, including performance, reliability, scalability, availability and security, etc. It also points out some compound attributes. For example, performance is composed of delay, throughput, fault tolerance and response time.

The description of the attribute should expandable and structural. In [8], author provided the classification of attribute. However, the problem of expansibility are not resolved. With the emergence of more and more QoS attributes, we should have ability to describe more attribute. So we divide the QoSAttribute into three subclasses: CommonQoSAtrribute, SpecialQosAttribute and GivenAttribute. The scalability mainly reflects the service requsters and service providers can self-define attribute through GivenAttribute. The CommonQoSAtrribute class describes common attributes. The SpecialQosAttribute class is used to describe the QoS properties in the professional field. Requesters and providers have different requirement for attribute, we divide the attributes into service requester-oriented attributes and service-oriented provider attributes.

We provide userPersonalise class to describe the user personalized requirements. The previous ontology model merely provides the description for the weight. However, sometimes the concept of QoS attributes is ambiguous for requester. It is entirely empowered by the requester, which is too subjective, it may result in unreasonable weight distribution. The subclass of QoSdefaultis RestricAttributeis used to bind partial attributes that the requester prefer. When the service provider provides weight, it must consider the binding attributes and thus new default weights are given, which can make the user personalized description more flexible.

The original model can only describe static attribute. We provide Status class to describe dynamic attribute. If Status is true, then the value of attribute is static, if is False, then it is dynamic. For the numerical description for the dynamic attribute, we use the method similar to log file in the database to describe the change of the attribute value. Transaction updates in the log file are identified by $<T_i, A_j, V_1, V_2>$. Ti represents the transaction T, A_j represents the operation object, and V_1 represents the value before update. V_2 represents the updated value. We provide the following two classes - BeforConditional and AfterConditional, like V_1 and V_2 in the log file, to describe the numerical changes. We can also calculate the service stability dynamic numerical values that have saved by us.

At the same time, this layer also provides a description of attribute trend and a description of the relationship between attributes.

2.4   QoS Metric Layer

The metric layer provides a measuring method of the QoS attribute and description of QoS attribute value. As shown in Fig. 3. Metric has two subclasses of Complex Metric and Simple Metric. The Complex Metric class implements complexity of metrics. The Simple Metric class Provides four measurement methods: fuzzy, precise, interval, and set.

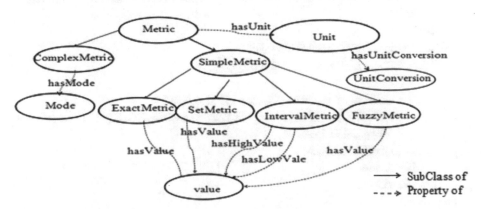

Fig. 3. Ontology structure of QoS metric layer.

The Metric class also describes the units of attributes through the Unit attribute, the Unit Conversion can transform the different units.

The Value class is used to describe the type of the QoS attribute value, there are three types of Value: numerical type, language type and data type. The numeric type describes the QoS attribute value is integer and the real number. The time data type describes the QoS attribute value is time. The language type describes the value is language, and we divide the language type into five levels of description: high, high, low, low, and medium. We also add a function of conversion between different values. For example, the language data is converted to numeric data.

3   Experiment and Implementation

3.1   Example Application

This section uses concrete examples to verify the semantic description capabilities of the QoS-PN model.

If a requester wants to inquire related services for the travel route, the request information is described using QoS-PN model firstly. Figure 4 shows the description of part of request information. The description of requestor's personalization requirements are shown in Fig. 5. The other attribute description of the service requester is similar to Figs. 4 and 5.

```
<price rdf:ID="Price_case">
<hasMetric>
  <Metric rdf:ID="Metric_case">
    <hasUnit>
        <Unit rdf:ID="Unit_case"
          < unit rdf:datatype="http://www.w3.org/2001/XMLSchema#string ">yuan</unit>
        </Unit>
    </hasUnit>
    <ExactMetric rdf:ID="ExactMetric_case">
      <rdfs:subClassOf rdf:resource="#Metric_case"/>
        <hasValue>
            <NumericalValue rdf:ID="NumericalValue_case">
            <numericalValue rdf:datatype="http:// www.w3.org/2001/XMLSchema#int">
            50</numericalValue>
            </NumericalValue>
        </hasValue>
    </ExactMetric>
</hasMetric>
```

Fig. 4. The description of price by QoS-PN

```
<hasUserPersonalise>
    <UserPersonalise rdf:ID="Personalise_case">
        <hasQoSdefault>
            <QoSdefault rdf:ID="default_case" >
                <qoSdefault   rdf:datatype="http://
                    www.w3.org/2001/XMLSchema#boolean">true</weightvalue>
                <RestricAttributerdfs:subClassOf rdf:ID="Restric_case" >
                    <restric rdf:datatype="http://www.w3.
                        org/2001/XMLSchema#string">price</restric>
                </RestricAttribute>
            </QoSdefault>
        </hasQoSdefault>
    </UserPersonalise>
</hasUserPersonalise>
```

Fig. 5. The description of user personalization.

We use the XML Parser to parse information described by QoS-PN, then we obtain requesting parameters (See Table 1).

Table 1. The requester request parameters parameters of service.

Attributes	Service requester
Performance	High
Reliability	high
Delay	100–200
Pice(bind)	50

The requesting parameters and the remaining candidate service are matched by the Jene inference engine and the traditional semantic service matching algorithm. Five service provider is retrieved.

We get price's historical data through Status class (See Table 2). Although we observe the historical records, the average value of service 2, 5 is small. However, the price fluctuates greatly, and even exceeds other service two times, which is obviously not expected for users. So based on history data of QoS, we should use the Cloud Model [8] to determine whether the service is stable. We remove unstable services to improve service stability.

When taking into account service stability, we obtain three candidate services (See Table 3).

Table 2. The historical records of price.

Historical record	Provider 1	Provider 2	Provider 3	Provider 4	Provider 5
A1	50	50	52	51	55
A2	52	40	51	51	45
A3	49	55	52	50	53
A4	50	54	53	52	52
A5	50	45	52	50	42

Table 3. The service provider provide parameters of service.

Attributes	Service provider1	Service provider2	Service provider3
Performance	High	High	Low
Reliability	low	High	High
Delay	90–100	90–100	100–200
Price	50	52	50

The weights of this model are not as completely as the traditional model is given by the requester. In the experiment, the service requester adopts the automatic weighting method and binds the price attribute. Obviously, the price is the preference attribute of the requester, and the service provider must consider Based on the preferences of the requester, new weights are given (See Table 4).

Table 4. Weights of the attributes.

Attributes	Neglect user personalization	Take into account user personalization
Performance	0.32	0.29
Reliability	0.25	0.23
Delay	0.19	0.18
Price	0.24	0.30

If the search does not take into account of user personalization. The result: service provider 3 > service provider 1 > service provider 5.
When taking them into account. The result:
service provider 1 > service provider 3 > service provider 5.
The services that best meet consumer needs are selected through weighted calculations. The services that best meets the needs of consumers are selected through weighted calculations. The price of Service 1 and Service 5 are lower than Service 2, although in terms of stability, Service 1 is inferior to Service 3, but the price is prioritized, obviously Service 1 is better than Service 3. And in terms of Service 3 and Service 5, although Service 5 is less expensive than Service 3, Service 3 has higher performance and stability than Service 5, and Service 3 is more suitable. Obviously the result of the experiment is more in line with the needs of consumers.

3.2   Case Study

We use the data set is OWL-t4c. The Onology editor is Protege. The development tool is Eclipse. We compare the traditional model with QoS-PN for the number of candidate services. The result is shown in the Fig. 6.

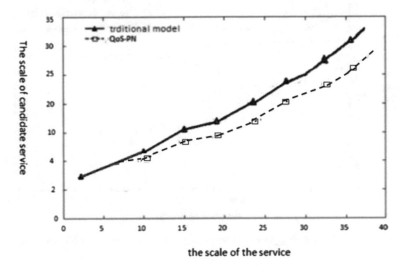

Fig. 6. Comparison between QoS-PN and other models on the number of candidate services.

The Fig. 6 shows that with the expansion of service scale the numble of the candidate service is decrease.

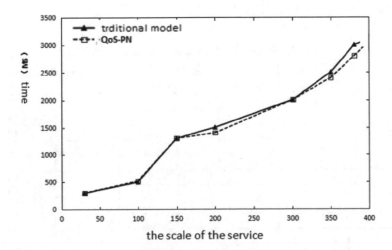

Fig. 7. Comparison between QoS-PN and other model on the time of service selection.

We compare the traditional model with QoS- PN for time and accuracy of service selection. The results are shown in the Figs. 7 and 8.

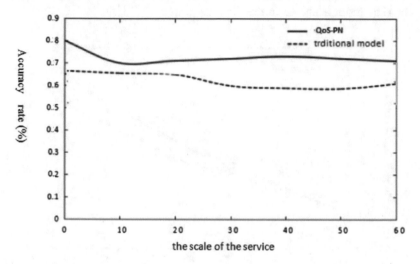

Fig. 8. Comparison between QoS-PN and other model on accuracy of service selection.

The Fig. 8 shows that accuracy of service selection has been significantly improved.

We compare the QoS-PN ontology with the existing QoS ontology from scalability, value type, unit conversion, trend, value conversion, dynamic and personalization. The ontology was evaluated in three levels: no support, weak support, and support (represented by 1, 2, and 3 respectively. The result show QoS-PN provides richer description than others. (see Table 5).

Table 5. The result of comparing with other models.

QoS ontology	Scalability	Value type	Unit conversion	Trend	Value conversion	Personalize	Dynamic	Metric type
QoS-MO	1	1	1	3	1	1	1	3
OSQS	2	3	3	3	1	2	2	3
ML-QoS	2	3	2	3	1	2	1	3
onQoS	2	2	1	3	1	1	2	3
QoSOnt	2	2	1	3	1	1	2	2
MoQ	2	3	1	1	1	1	1	3
DAML-QoS	1	1	1	1	1	1	1	3
QoS-PN	3	3	3	3	3	3	3	3

4   Conclusion

QoS-PN model provides a unified description framework for service providers and service requesters. We can use it to describe QoS information. The model is divided into three layers: QoS configuration layer, QoS attribute layer, and QoS metric layer. The model has greater support metric, value type, and scalability. The functions of numerical conversion, dynamics and personalization have been expanded and innovated. The result show that we use QoS-PN to describe QoS information, which can improve the accuracy and Shorten the time for service selection.

References

1. Du, W.J.: Research on Web service selection method based on uncertainty QoS. Chongqing University, Chongqing (2014). (In Chinese)
2. Chen, Z., Chia, L.T., Lee, B.S.: DAML-QoS ontology for Web services. In: IEEE International Conference on Web Services, pp. 472–479 (2014)
3. Jean, S., Losavioy, F., Matteoy, A., Levyz, N.: An extension of OWL-S with quality standards. In: Fourth International Conference on Research Chall, pp. 483–494 (2010)
4. Gu, J.N.: Research on a three-layer QoS model of QoS in web service selection. Inf. Comput. **14**, 40–41 (2015). (In Chinese)
5. Farzi, P., Akbari, R.: Improving semantic web service discovery method based on QoS ontology. In: IEEE Swarm Intelligence and Evolutionary Computation, pp. 343–350 (2017)
6. Zhang, L.F., Dong, C.L., et al.: A service selection method of supporting mixed QoS attributes. Comput. Appl. Softw. **33**(09), 15–19 (2016). (In Chinese)
7. Wang, X., Vitvar, T., Kerrigan, M., et al.: A QoS-aware selection model for semantic web services. In: Lecture Notes in Computer Science, vol. 4294, pp. 390–401 (2006)
8. Wang, X.M.: Research on improved cloud service selection algorithm based on QoS history data. Chongqing University, Chongqing (2016). (In Chinese)
9. Ling, Y.H.: Research on web service selection method based on uncertainty of QoS User preference. Chongaing University, Chongqing (2016). (In Chinese)

A Research About a Conflict-Capture Method in Software Evolution

Tilei Gao[1,2], Tong Li[3], Rui Zhu[1], Rong Jiang[2], and Ming Yang[2(✉)]

[1] School of Software, Yunnan University, Kunming 650000, China
[2] School of Information, Yunnan University of Finance and Economics,
Kunming 650000, China
83300934@qq.com
[3] Key Laboratory in Software Engineering of Yunnan Province,
Kunming 650000, China

Abstract. Software evolution is an important research direction and research focus in the field of software engineering. During the process of software dynamic evolution, the existence of conflicts between evolutionary components and original components is one of the key factors determining whether the dynamic evolution process is corrected or not. To solve this problem, resource management method is introduced and the concept of resource in software evolution process is redefined. According to the different patterns of resources used, sequence, selection and conflict relationship models among components are defined. And at last, a conflict-capture algorithm, which is used to capturing the components involved in the conflict relationships, based on recourse management. The results of this paper lay a theoretical foundation for maintaining consistency before and after evolution and ensuring the correctness of dynamic evolution.

Keywords: Software evolution · Resource management · Conflict capture

1 Introduction

Evaluability is one of the most important attributes of software [1]. The phenomenon of software evolution has attracted much attention since its discovery in the last century [2]. With the deepening and development of software engineering discipline, the importance and universality of software evolution are becoming stronger and stronger [3]. Software evolution refers to the process of software changing and achieving the desired form. At present, dynamic evolution is not only a hot topic in academic circles, but is a research difficulty as well. As the development of software engineering, the ability of software encapsulation is increasing, but the emergence of new and complex computing model such as cloud computing and service computing brings new challenges and difficulties to the research of software dynamic evolution. One of the challenges and difficulties is how to ensure the correct implementation of dynamic evolution and the maintenance of consistency before and after evolution.

© Springer Nature Switzerland AG 2020
Q. Liu et al. (Eds.): CENet 2018, AISC 905, pp. 530–537, 2020.
https://doi.org/10.1007/978-3-030-14680-1_58

To solve the problems above, existing researches include:

In the field of software evolution research, literature [3] and [4] have studied the evolution process from the aspects of process description and implementation respectively, and the method of building and evolving the whole evolution framework is realized. However, there is no in-depth analysis of the underlying causes of incorrect evolution, which means the lack of application resources. In the field of conflicts detection and conflicts control, in the related literature of the operating system [5] and [6], it is reported that resources are the root cause of the conflicts and PV primitives and monitor are used to control the conflict processes. The deadlock problems are solved and the programs can be executed normally. However, there is no in-depth analysis in the reason causing conflicts, and PV primitives and monitors have not been widely applied in other research fields. Researches in conflict capture and conflict detection is widely distributed in all directions of computer science, including: security policy based on directed graph [7], collaborative design of business process [8], the design and implementation of database [9] and so on. Conflict detection and management have been achieved through different research methods, but few of them have been analyzed and studied from the root causes (resources) of conflicts. Also, these research results of conflict detection and capturing are rarely applied in software engineering field. In anticipation of resource management, literature [10] has introduced the importance of resources, kinds of resource management patterns from resource acquisition to release. But they are only used in business process carding, and rarely mentioned how to use in software.

Learning from predecessors' research results about resource management patterns, conflict detection and management process method, from the point of view of software engineering and software evolution, this paper analyzes the root cause of conflicts in the process of evolution. According to the use of inter-process resources, three relationships are defined with Petri net as the main formalization tool: order, concurrency and conflict. Then, algorithms for resource capture between inter-program and evolutional processes are proposed and according to the shared resources, conflict relationships between the evolution processes and the original processes can be captured.

2 Related Work

2.1 Resources Management Pattern

In software systems, resource management refers to the process of controlling the availability of resources by resource users. Resource users can be entities that acquire, access, or release resources, and resource management covers the following three aspects: (1) Ensure that resources are available when needed, and specific implementation modes include: Lookup pattern, lazy acquisition pattern, eager acquisition pattern and partial acquisition; (2) Ensure that the lifecycle of resources is determined, and the specific implementation modes include: caching pattern, pooling pattern, coordinator pattern and resource lifecycle manager pattern; (3) Ensure that resources are released in time, so as not to affect the response speed of the system using resources and specific implementation modes include: Leasing pattern and evictor pattern [10]. In general, resource management is the three operations of resource acquisition, resource lifecycle management and resource release. These three main aspects have a far-reaching impact on the overall performance, scale, scalability and service life of the running software. In any field, from traditional embedded small applications [11], large enterprise systems [12], grid computing [13] to the current cloud computing, Internet of things, large data fields, resource management is an important method to ensure the normal operation of all kinds of software in form, granularity, performance, security and so on.

2.2 Bernstein's Condition

Bernstein's condition is an set of principles used to define irrelevant processes' concurrency based on resource requirements, and shared resources causing conflicts can be captured by using the principles reversely. Suppose R_i (W_i) is the set of variables read (written) by a section of code i. Bernstein's Condition states that sections i and j may be executed in an arbitrary order, or concurrently if there are no dependences among the statements in the sections, i.e. if $R_i \cap W_j = \Phi$, $W_i \cap R_j = \Phi$ and $W_i \cap W_j = \Phi$. Dependence is the relationship of a calculation B to a calculation A if changes to A, or to the ordering of A and B, could affect B. If A and B are calculations in a program, for example, then B is dependent on A if B uses values calculated by A. There are four types of dependence: true dependence, where B uses values calculated by A; anti dependence, where A uses values overwritten by B; output dependence, where both A and B write to the same variables and control dependence, where B's execution is controlled by values set in A [14, 15].

3   Definitions

Resources are all inputs and prerequisites needed by the processes to accomplish their own functions, and are not part of any process using them.

Definition 1 resource [16]. A resource is a two tuple, s = <sort, num> , which consists of two parts: type and quantity. According to whether recyclable, resources are divided into reusable resource and non-reusable resources.

Definition 2 Reusable resources. Reusable resource are usually obtained from resource providers and released after use. Such resources can be reacquired and used after release. For example, the memory in computer is such kind of resource. In computer, reusable resources are the most important resource categories, because the resources owned by the resource providers are usually limited.

Definition 3 Non-reusable resources. Non reusable resources will be consumed. Therefore, after obtaining such resources, either it will not be released or implicitly released. For example, processing time in a computer grid [17] is such kind of resource. Processing time is lost after being acquired and used, and cannot be returned.

Basic relationships among evolution processes include sequential relations, concurrent relations and conflict relations. Petri Net [18] is used to define these three relationships and the definitions are as below:

Definition 4 sequential relation. Let M be a mark of Petri Net (PN). If there exist t1 and t2 makes M[t1>M' and ¬M[t2>, M'[t2>. That is, under M, t1 is enabled and t2 is not enabled, but triggered t1 will enable t2, which means t1 is the condition to trigger t2. The relationship between t1 and t2 is called sequential relation, as shown in Fig. 1.

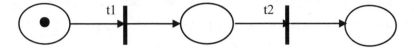

Fig. 1. Sequencial relation.

Theorem 1. Let s be a resource, if si $\subseteq$ output (t1), and si $\subseteq$ input (t2), the relationship between process t1 and process t2 is the sequential relation.

Theorem 1 means that: If one process's inputs belong to another one's outputs, the relationship between the two processes is sequential relation.

Definition 5 Concurrent relation. Let M be a mark of PN. If there exist t1 and t2 makes M[t1> and M[t2> , M[t1 > M1 $\Rightarrow$ M1[t2> , M[t2> M2 $\Rightarrow$ M2[t1> . The relationship between t1 and t2 is called concurrent relation, that is under mark M, t1 and t2 are enabled and each one's triggering has nothing to do with the other one's triggering, as shown in Fig. 2.

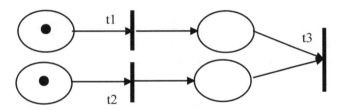

Fig. 2. Concurrent relation.

Theorem 2. Let S be a set of resources and input(t1) $\subseteq$ S and input(t2) $\subseteq$ S. $\forall$ x $\in$ S, if x.num $\geq$ count(input(t1)) + count(input(t2)), the relationship between process t1 and process t2 is the concurrent relation.

Theorem 2 means that: if two or more processes use some or more resources at the same time and the resources needed are sufficient, they belong to concurrent relation.

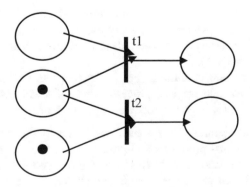

Fig. 3. Conflict relation.

Definition 6 Conflict relation. Let M be a mark of PN. If there exist t1 and t2 makes M[t1> and M[t2> , M[t1 > M1 $\Rightarrow$ ¬M1[t2 > , M[t2 > M2 $\Rightarrow$ ¬M2[t1> . The relationship between t1 and t2 is called sequential relation, that is under mark M, t1 and t2 are enabled but each one's triggering will cause the other one's failing, as shown in Fig. 3.

Theorem 3. Let S be a set of resources and input(t1) $\subseteq$ S and input(t2) $\subseteq$ S. $\exists$ x $\in$ S, if:

x.num $\geq$ Max(count(input(t1)), count(input(t2))),
and also x.num $\leq$ count(input(t1)) + count(input(t2)).

The relationship between process t1 and process t2 is the conflict relation.

Theorem 3 means that: If two or more processes use some or more resources at the same time but the resources needed are deficient for all, some of the processes cannot get enough resources and suspended. Then, the processes suspended belong to conflict relation. Actually, the number of resources (x.num) is enough for each of processes to run, but not enough for all the processes to run at the same time.

4   The Conflict-Capture Method

4.1   The Conflict-Capture Method

Algorithm 1 shared resource set acquisition algorithm:
Inputs: set of processes P={p1, p2, ... , pn}.
Outputs: shared resource set S={s1, s2, ... , sn}.
Steps:
(1)   S=∅;
(2)   Investigate any two processes pi, pk, if input(pi)∩input(pk)≠∅, then si=input(pi) ∩ input(pk);
(3)   If input(pi) ∩ output(pk) ≠ ∅, then si = si ∪ (input(pi) ∩ output(pk));
(4)   If output(pi) ∩ input(pk) ≠ ∅, then si = si ∪ (output(pi) ∩ input(pk));
(1)   Repeat steps (2)-(4), until the set of P is empty;
(2)   S={s1, s2, ... , sn};
(3)   Return S.

Complexity analysis of this algorithm: the complexity mainly depends on the scale of processes mentioned. In Algorithm 1, every two processes should be computed and execute Steps (2)–(4). So, the program includes a double circulation and the complexity belongs to the scale of $O(n^2)$.

In the case of software evolution, there exist two kinds of processes which are processes to be evolved and original processes and the situations between evolution processes and original processes are made up of three kinds:

(1)   Always adequate resources. In this case, there is no conflict between the evolution processes and original processes and processes can execute normally;
(2)   Always inadequate resources. In this case, there is conflict relations before evolution, and evolution processes often aggravate the deficiency.

(3) Original adequate resources but inadequate resources when evolving. Before evolving, there are enough resources, but as the evolution processes considered, the amount of resources increases with the new processes and the number of resources cannot meets all processes' needs, so, conflict relations occur.

Our research focuses on the situation (3) and part of the situation (2). In case (2), if the exasperated situation can also guarantee processes' running sequentially, we will deal with this kind of conflicts the same as situation (3).

Algorithm2 shared resources of evolution acquisition algorithm:
Inputs: executing process set $P = \{p1, p2, \ldots, pn\}$, evolution process ep.
Outputs: relevant process set shared the same resources with evolution process $P' = \{ep, pm, \ldots, pr\}$.
Steps:
$ES = \emptyset$;
(1) $P' = \{ep\}$;
(2) Investigate any process pi, if $input(pi) \cap input(ep) \neq \emptyset$, then $esi = input(pi) \cap input(ep)$; $P' = P' \cup \{ pi \}$;
(3) Repeat step (3), until P is empty;
(4) $ES = \{es1, es2, \ldots, esk\}$;
(5) $P' = \{ep, pm, \ldots, pr\}$;
(6) Return P'.

Complexity analysis of this algorithm: the complexity also mainly depends on the scale of processes mentioned. Algorithm 2 is based on Algorithm 1. Suppose that each evolution process only include on process to be evolved, and in Algorithm 2, in step (2), only one process should be investigated and compared with other original processes. So, the main Step (2) only needs to be executed n times (n is the number of processes) and the complexity belongs to the scale of $O(n)$.

(a) Simulation

In the operating system, the dining philosophers is a classic problem in conflict research. The philosopher's dining problem is reformed here, and Algorithm 1 is introduced to capture resources in traditional philosophers' dining problem and Algorithm 2 is used to capture the conflict processes after a new philosopher joined. As sequentially adding philosopher or the philosopher added with chopsticks does not cause conflicts, we only simulated and analysed the situation which produces conflicts and the whole procedure is as shown in Fig. 4.

Supposed that there are 5 philosophers at first which belong to the set $P = \{p0, p1, p2, p3, p4\}$, $pi \in P$ means one of the philosophers and now the resource set C is empty, $C = \emptyset$. By calling Algorithm 1, investigate pi and pj, and shared resource set C' will be built $C' = \{c0, c1, c2, c3, c4\}$.

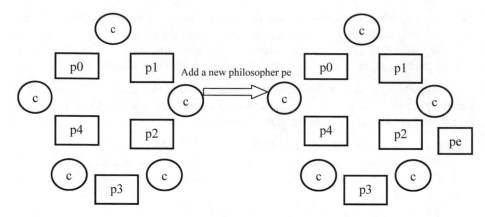

Fig. 4. The evolution procedure.

As a new philosopher pe comes, the original structure should be evolved. Algorithm 3 is used and the inputs becomes pe $\cup$ P. Through comparing the resources used by pe and all other philosophers with the method proposed in Algorithm 3 steps (3), we will get the resources c1 and c2 and the conflict philosophers with pe, which are p1, p2 and p3.

From the simulation, the feasibility and correctness of Algorithms 1 and 3 are verified. Capturing conflicts in evolution is an important part of software evolution, which plays an important role in the correlation analysis of software evolution process and the consistency maintenance before and after evolution. The method come up with in this paper, which is used to capture conflict relationships among software evolution processes from the perspective of shared resources, is an effective.

5   Conclusion

With the emergence and rapid development of new technologies such as cloud computing, Internet of things, big data, etc. the principles, techniques and methods of software engineering and software dynamic evolution are progressing. At the same time, new challenges have been put forward in the development of these fields. In the new and complex environment, ensuring correctness is the top priority of dynamic evolution. To ensure the correctness of the evolution process, we must find out the conflict relationship caused by the evolution and the related components involved. This is the object of this paper, and a conflict relationship method or algorithm is proposed. In this paper, resources used in computer and programs are redefined and classified. On this basis, a relationships model of processes caused by resource is put forward and conflict relationship is defined. Finally, a method of capturing conflict components or processes based on shared resources is presented. Identify the components involved in evolutionary conflict, builds a theoretical foundation for the following research about optimizing the evolution process and ensuring consistency before and after evolution.

Acknowledgement. This work was supported by National Natural Science Foundation of China (Nos. 61379032, 61763048, 61263022, 61303234, 61662085), National Social Science Foundation of China (No. 12XTQ012), Science and Technology Foundation of Yunnan Province (No. 2017FB095), Yunnan Province Applied Basic Research Project (No. 2016FD060), Science Research Project of Yunnan Education (Nos. 2017ZZX001, 2017ZZX227), Key Project of Scientific Research of Yunnan Education (2015Z018), Provincial Scientific and Technological Innovation Team Project of Yunnan University (2017HC012), the 18th Yunnan Young and Middle-aged Academic and Technical Leaders Reserve Personnel Training Program (No. 2015HB038).

References

1. Yang, F.Q.: Thinking on the development of software engineering technology. J. Softw. **16**(1), 1–7 (2005). (in Chinese)
2. Cook, S., Harrison, R., Lehman, M.M., Wernick, P.: Evolution in software systems: foundations of the SPE classification scheme: research articles. J. Softw. Maint. Evol. Res. Pract. **18**(1), 1–35 (2012)
3. Li, T.: An Approach to Modelling Software Evolution Processes. Springer, Berlin (2008)
4. Li, C.Y.: Research on Architecture-Based Software Dynamic Evolution. Zhejiang University (2005). (in Chinese)
5. Tang, X.D., Liang, H.B., Zhe, F.P.: Computer Operation System. Xidian University Press, Xi'an (2007). (in Chinese)
6. Ma, D.Y., Yu, S.Q., Hu, H.M.: Two Petri net models for solving philosophers' dining problems. J. Shanghai Teach. Univ. (Nat. Sci.) **31**(3), 37–40 (2002)
7. Yao, J., Mao, B., Xie, L.: A dag based security policy conflicts detection method. J. Comput. Res. Dev. **42**(7), 1108–1114 (2005)
8. Zhao, H.S., Tian, L., Tong, B.S.: Constraint-based conflict detection and negotiation in collaborative design. Comput. Integr. Manuf. Syst. **8**(11), 896–901 (2002)
9. Deng, Z.M., Wang, S., Meng, X.F.: Conflict detection and resolution strategy in replicated mobile database systems. Chin. J. Comput. **25**(3), 297–305 (2002)
10. Kircher, M., Jain, P.: Patterns for Resource Management. Post & Telecom Press, Beijing (2013)
11. Weir, C., Weir, C.: Small Memory Software: Patterns for Systems with Limited Memory. Addison-Wesley Longman Publishing Co., Inc., Boston (2001)
12. Fowler, M., Rice, D., Foemmel, M.: Patterns of Enterprise Application Architecture. Pearson Schweiz Ag, Zug (2002)
13. Baker, M., Buyya, R., Laforenza, D.: Grids and grid technologies for wide-area distributed computing. Softw. Pract. Exp. **32**(15), 1437–1466 (2002)
14. Bernstein, A.J.: Analysis of programs for parallel processing. IEEE Trans. Electron. Comput. **15**(5), 757–763 (1966)
15. Hawick, K.A.: High performance computing and communications glossary (1994). http://wotug.org/parallel/acronyms/hpccgloss
16. Gao, T.L., Li, T., Yang, M., Jiang, R.: An approach to conflict resolution in software evolution process. In: ICTCS 2018, Colombo, Sri Lanka, pp. 105–111 (2018)
17. Grid Computing Info Centre: (2004). http://www.gridcomputing.com
18. Gu, T.L.: Formal Methods of Software Development. Higher Education Press, Beijing (2005). (in Chinese)

A Technique to Predict Software Change Propagation

Leilei Yin$^{(\boxtimes)}$, Yanchen Liu, Ziran Shen, and Yijin Li

The 28th Research Institute of China Electronics Technology Group Corporation,
Nanjing 210007, China
yllwyp@163.com

Abstract. Software change can occur frequently and it is inevitable in the lifecycle of software development. It can cause tremendous impact on software engineering including cycle, cost, workload, etc. This work proposes an advanced technique to assess the software changeability on the class level. From the perspective of member variable and function, mutual relations between classes are introduced to record the change likelihood and impact. Change propagation model between classes is developed to transform the change impact and the total impact is calculated to analyse the class changeability. Based on the analysis results, class changes with high risk should be avoided and those with low change risk should be given priority in the software change process. Finally, a simplified software system is applied for the initial evaluation of the method.

Keywords: Class change · Mutual relation · Change propagation

1 Introduction

Rapid development of software technologies enables the software to handle various problems. However, software change can occur frequently and it is inevitable in the lifecycle of software development [1], which has tremendous impact on software engineering including cycle, cost, workload, etc. Since the software is composed of numerous subsystems and modules through mutual relations among them, change originating from a subsystem or module will propagate within the software. It may further aggravate the change influence and complicate its impact analysis.

Software change influence has aroused extensive attention of many scholars. Leung and White [2] proposed the concept of "firewall" to assist the tester in focusing on possible errors which may have been introduced by a correction or a design change. Rothermel and Harrold [3] applied class dependence graph and program dependence graph to analyze change influence for object-oriented software. And change influence tracing method based on the program dependence graph was developed in reference [4] to track the change elements and different change propagation paths. Member function dependence graph was applied to describe the relations between classes by Jang et al. [5]. It can be seen that dependence graph is widely used in the software change management. However, the dependence graph can be complex when the software system is large. Besides, it is difficult to describe the relation between software elements in detail.

© Springer Nature Switzerland AG 2020
Q. Liu et al. (Eds.): CENet 2018, AISC 905, pp. 538–543, 2020.
https://doi.org/10.1007/978-3-030-14680-1_59

Matrix has been widely used to record relations between items. It is a good tool mapping information flow and recording its impact in product [6]. Wang et al. [7] proposed three types of matrix (i.e. trace-matrix of change-source, change-object, change component and reachability) to record the propagation of function-requirement change. A calculating approach based on the SA (software architecture) relation matrix and reachability matrix was developed to analyze the ripple influence of software architecture evolution [8]. By present, research on software change has been paid more and more attention. However, optimal change scheme of software has rarely been studied, which is the key to implementation of software change.

In view of the above, this paper applies the matrix-based approach to record the software dependence on the class level. The class dependence is detailed through the member variable, member function, and expressed in the form of matrix named by CCM (class-class matrix). Class change is regarded as the initial change and propagated to other classes through the class dependence. An initial change class can correspond to several change propagation paths. This paper develops a method for searching the optimal change propagation path, which includes acquiring the change propagation paths on the class level, calculating the change risk, and further selecting the optimal change scheme.

2   Outline of the Method

Software change can originate from infrastructure change, functional requirement change, new algorithm, technical environment and so on. It is difficult to carry out the software changes as most of them are caused by the mutual relation of classes.

In this paper, dependence between classes is demonstrated through member variable and function. For example, class C_1 uses the member variables or functions of C_2 to achieve a requirement. Thus it is feasible to learn about the dependence between classes through the member variable and function. A model to analyze the change risk is outlined as shown in Fig. 1. First, the dependence between classes in terms of member variable and function is specified. For example, class C_5 can be affected by member variable or function c_{21}, c_{22} and c_{23} of class C_2, and class C_7 can be affected by member variable or function c_{31}, c_{32} and c_{33} of class C_3. After that, the change propagation paths on the class level can be acquired with a developed algorithm automatically. Then impacts of member variable or function on classes are valued for the calculation of the change risk of different change propagation paths.

Taking Fig. 2 as an example, an initial change class corresponds to several change propagation paths. Based on the risk, the optimal change propagation path with the least change risk can be acquired. Besides, the combined change risk for the initial change class can be analyzed to obtain the software change scheme with low duration, cost, etc.

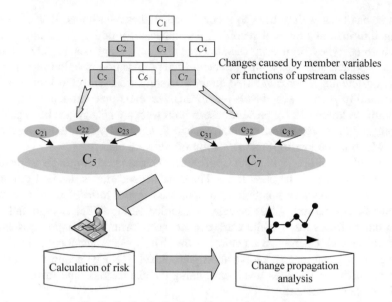

C_1, C_2, C_3, C_4, C_5, C_6, C_7 are classes of software and
c_{ij} are decomposed member variables or functions from the components C_2, C_3.

Fig. 1. Outline of the change risk analysis.

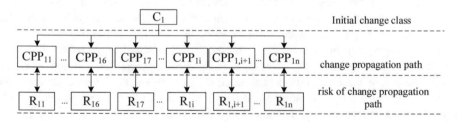

Fig. 2. Process of the change risk analysis.

3   Change Propagation Analysis

Initial change of software is aimed at satisfying the change requirement, which is usually favored. Mutual relations between classes are the primary causes for the change propagation, which can make the system reach a new stable status. However, change propagation path with huge impact should be avoided. The traceability achievement for change propagation on the class level can be depicted and interpreted by answering the following questions:

- What are the classes that can be changed during the software lifecycle?
- What member variable and function does each class contain?
- What are the class-class relations constructed through member variable and function?
- What are the change possibility and impact of affected classes?

3.1   Mutual Relations Between Classes

In this article, relations between classes are acquired through member variable and function, which are arranged in the class-class matrix. As shown in Fig. 3, relations among the C_1, C_2, C_3 and C_4 are taken as examples. Components in each column are member variables or functions from the column class, which may affect the row class. Take the relation between classes C_3 and C_4 as an example, C_4 can be affected by C_3 through c_{34} and c_{35}. Accordingly, C_3 can be affected by C_4 through c_{41} and c_{42}.

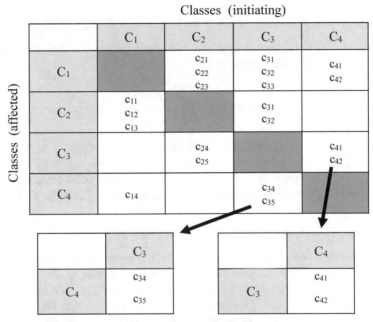

Member variables c_{34}, c_{35} of C_3 affecting C_4 Member functions c_{41}, c_{42} of C_4 affecting C_3

Fig. 3.   Class-class matrix.

In order to make quantitative analysis of change impact, the relations between classes are quantified. The impact of member variable and function from a class on another class can be described as shown in Eq. (1).

$$(C_i, C_j) = [(l_{1j}, i_{1j}), \ldots, (l_{ij}, i_{ij}), \ldots, (l_{nj}, i_{nj})] \tag{1}$$

where (C_i, C_j) is the relation between classes C_i and C_j. l_{ij} and i_{ij} are the likelihood and impact of member variable or function c_{ij} causing C_j to change respectively. l_{ij} and i_{ij} can be valued as 0.25, 0.50 or 0.75, which are estimated according to low, medium or high possibility and impact.

3.2   Change Propagation Model

The impact between classes is assumed as a linear transformation and is diminishing along the change propagation path. Besides, four assumptions for the linear algebraic analysis are made as follows:

1. Class change is defined as the activity of changing the member variables or functions.
2. The change impact is transformed through member variables or functions.
3. The change impact decreases progressively with the increase of change propagation step.
4. The influence coefficient do not vary with time.
5. Change propagation is ended when the influence coefficient is no more than 1%.

According to the above assumptions, combined impact (i.e. influence coefficient) between any two classes is demonstrated as shown in Eq. (2).

$$CI(C_i, C_j) = 1 - \prod_{k=1}^{n(C_i,C_j)} \left(1 - \frac{1}{N(C_i)} l_{kj} \times i_{kj}\right) \qquad (2)$$

where $CI(C_i, C_j)$ is the combined impact from C_i to C_j. $n(C_i, C_j)$ is the number of member variables or functions from C_i that cause C_j to change. $N(C_j)$ is the amount of member variables or functions from class C_i that may affect other classes. l_{ij} and i_{ij} are the likelihood and impact of member variable or function c_{ij} causing C_j to change.

3.3   Change Propagation Analysis and Results

A simplified software system is applied for the initial evaluation of the method. Four main classes and mutual relations are concluded as shown in Fig. 4. Class C_1 is regarded as the initiating change class. The change propagation paths and the combined impact are shown in Fig. 5. It can be concluded from Fig. 5 that the average change risk is 1.1084. Besides, the average change risk from C_2, C_3 and C_4 can be acquired similar to that of C_1. They are 1.2807, 1.2342 and 1.0135 respectively. Thus C_4 should be given priority to implement the changes. And the change propagation paths can aid decision making in the software change process.

Classes (initiating)

Classes (affected)		C₁	C₂	C₃	C₄
	C₁		0.50 0.75 0.25 0.50 0.50 0.25	0.25 0.50 0.75 0.75 0.25 0.25	0.50 0.75 0.25 0.50
	C₂	0.25 0.50 0.50 0.50 0.75 0.25		0.25 0.50 0.50 0.50	
	C₃		0.25 0.50 0.50 0.50		0.50 0.50 0.75 0.25
	C₄	0.50 0.50		0.25 0.50 0.50 0.25	

Fig. 4. Valued class-class matrix.

Path1	CI	...	...	Path5	CI	...	...	Path10	CI
C_3	1			C_3	1			C_3	1
↓				↓				↓	
C_1	0.1028			C_2	0.0519			C_4	0.0603
↓				↓				↓	
C_2	0.0138	...	...	C_1	0.0714	...	...	C_3	0.0064
↓				↓					
C_3	0.0073			C_4	0.0045				
Total impact	1.1239	...	...	Total impact	1.1278	...	...	Total impact	1.0667

Fig. 5. Change impact originating from class C_3.

4 Conclusion

A theoretical method is proposed to analyze the software change propagation. Class-class matrix is applied to demonstrate the class relations in terms of member variable and function. Change propagation path originating from a class can be acquired and its change risk can be analyzed with the proposed mathematical model. Through comparing the change risk of classes, class with low change risk should be given priority in the software change process. On the contrary, class with high change risk should be avoided. By referring to this method, a simplified software system is applied to demonstrate the method initially.

References

1. Ramzan, S., Ikram, N.: Making decision in requirement change management. In: International Conference on Information and Communication Technologies, pp. 309–312. IEEE (2005)
2. Leung, H.K.N., White, L.: A study of integration testing and software regression at the integration level. In: Proceedings of Conference on Software Maintenance, pp. 290–301. IEEE (1990)
3. Rothermel, G., Harrold, M.J.: Selecting regression tests for object-oriented software. In: Proceedings of International Conference on Software Maintenance, pp. 14–25. IEEE (1994)
4. Badri, L., Badri, M., St-Yves, D.: Supporting predictive change impact analysis: a control call graph based technique. In: Asia-Pacific Software Engineering Conference, pp. 167–175. IEEE Computer Society (2005)
5. Jang, Y.K., Chae, H.S., Yong, R.K., Bae, D.H.: Change impact analysis for a class hierarchy. In: Proceedings of Software Engineering Conference, Asia Pacific, pp. 304–311 (1998)
6. Tang, D., Xu, R., Tang, J., He, R.: Design structure matrix-based engineering change management for product development. Int. J. Internet Manuf. Serv. 1(3), 231–245 (2008)
7. Wang, Y.H., Wang, L.F., Zhang, S.K., Wang, Q.F.: Tracing approach of software requirement change. Acta Electron. Sin. 34(8), 1428–1432 (2006)
8. Wang, Y.H., Zhang, S.K., Liu, Y., Wang, Y.: Ripple-effect analysis of software architecture evolution based on reachability matrix. J. Softw. 15(8), 1107–1115 (2004)

An Improved Cyclostationary Feature Detection Algorithm

Tianfeng Yan, Fuxin Xu$^{(\boxtimes)}$, Nan Wei, and Zhifei Yang

School of Electronic and Information Engineering, Lanzhou Jiaotong University,
Lanzhou 730070, China
17794268568@163.com

Abstract. For the question of the limits for practical application in the actual problems about complicated calculation and time-consuming of cyclostationary feature detection in the spectrum detection process, this paper proposes a new improvement algorithm based on the spectrum correlation characteristics and test statistics of the cyclostationary detection, and deduced its detection probability and false alarm probability. The improved algorithm starts from the spectral correlation characteristics and proves that the cyclic spectrum is conjugate symmetry about the relevant axis, which reduces the computational complexity. Starting from the test statistics, after discarding the more complex correlation factors, the computation complexity and detection performance of the improved test statistics are analyzed. It is concluded that the complexity is obviously reduced and the performance loss is minimal after discarding the correlation factors. The simulation results show that the improved algorithm detection performance is slightly reduced, but the computational complexity is greatly reduced, which further satisfies the requirement of fast and accurate spectrum detection and has strong practicality.

Keywords: Spectrum correlation · Test statistics · Computational complexity · Cyclostationary feature detection

1 Introduction

The rapid development of cognitive radio, making accurate and fast detection of the spectrum become an important technology. The compilation of relevant literature, the spectrum detection technology comes down to three categories: local detection, cooperative spectrum sensing and external spectrum sensing. Comparison, testing and coordination increased external detection control channel overhead, the overhead traffic, hardware cost, etc. and thus the scope of use is limited. Document [1] mentioned binding energy detection and use up cyclostationary feature detection algorithm to achieve accurate and efficient detection. Reference [2] proposed an algorithm combining adaptive dual-threshold detection and cyclostationary feature detection, which improved the detection performance. Literature [3] proposed a cyclostationary feature detection algorithm based on feedback superposition principle. Effectively improve the system's detection performance. However, the above-mentioned improved methods based on local detection generally have problems such as large hardware overhead.

© Springer Nature Switzerland AG 2020
Q. Liu et al. (Eds.): CENet 2018, AISC 905, pp. 544–555, 2020.
https://doi.org/10.1007/978-3-030-14680-1_60

Literature [4–6] proposes to use cooperative detection and cooperative detection to reduce the computational complexity. However, in reality, a lot of times is not easy to satisfy users to collaborate; and the number of times users greater control channel also will lead to greatly increase the cost of broadband, thus affecting the spectral efficiency. In order to avoid the above problems, this paper starts from reality and studies how to reduce the computational complexity without increasing hardware requirements and without reducing performance.

2 Cyclostationary Feature Detection Algorithm

2.1 Principles of the Algorithm

The detailed steps of cyclostationary feature detection [7] are shown in Fig. 1. The main discriminating method is: to detect whether the received signal $r(t)$ has a cyclostationary feature, so as to discriminate whether there is a main user [8]. The spectral correlation function is denoted by $S_X^\alpha(f)$, the cyclic autocorrelation function is denoted by $R_X^\alpha(f)$, which are a pair of Fourier transform pairs, $S_X^\alpha(f)$ and $R_X^\alpha(f)$ are respectively expressed as follows:

$$S_X^\alpha(f) = \int\limits_{-\infty}^{+\infty} R_X^\alpha(\tau)e^{-j2\pi f\tau}d\tau \tag{1}$$

$$R_X^\alpha(\tau) = \lim_{T\to\infty}\frac{1}{T}\int_{-\frac{T}{2}}^{\frac{T}{2}} R_X(t+\frac{\tau}{2}, t-\frac{\tau}{2})e^{-j2\pi\alpha t}dt \tag{2}$$

Among them, α is the cycle frequency. If the spectral correlation function only shows a peak at $\alpha = 0$, it indicates that the primary user does not exist; if a peak appears at the cyclic frequency, it indicates that the primary user exists.

Fig. 1. Cyclostationary feature detection execution flow.

Short time Fourier transform may be defined as a signal, the signal within a time window of the Fourier transform, so another equivalent way to define the spectral correlation function is defined as:

$$X_W(t,f) = \int\limits_{t-\frac{w}{2}}^{t+\frac{w}{2}} X(u)e^{-j2\pi fu}du \tag{3}$$

In the above equation, the time window is $[t - W/2, t + W/2]$. The nature of short-time Fourier transform is an estimate of the signal spectrum in the limit of time window. Through the short-time Fourier transform, the spectral correlation function obtained is:

$$S_x^\alpha(f) = \lim_{W,T\to\infty} \frac{1}{T} \int_{-\frac{T}{2}}^{\frac{T}{2}} \frac{1}{W} X_W(t,f+\frac{\alpha}{2})X_W^*(t,f-\frac{\alpha}{2})dt \tag{4}$$

2.2 Design of Test Statistics

System Model
Spectrum detection can be seen as a binary hypothesis discriminant problem:

$$\begin{cases} H_0 : r(t) = n(t) \\ H_1 : r(t) = hs(t) + n(t) \end{cases} \tag{5}$$

Among them, h0 and h1 represent the idle state and active state of the primary user, respectively; At time t, the transmit and receive signals of the primary and secondary users are denoted by s and r, respectively; n represents the secondary user receiving white noise [9]; h represents the channel gain between primary and secondary users.

The Existing Test Statistic Design
The cyclostationarity feature detector based on maximum likelihood estimation in [10] is represented as follows:

$$Y_{ML} = \int_{-\frac{T}{2}}^{+\frac{T}{2}} \int_{-\frac{T}{2}}^{+\frac{T}{2}} R_S(u,v)r(u)r*(v)dudv \tag{6}$$

Where T is the observation duration and $R_S(u,v) = E[s(u)s*(v)]$ is the autocorrelation function. If the signal has a cyclostationary feature, perform the Fourier transform of Eq. (6):

$$Y_{ML} = \sum_{k=1}^{N_\alpha} \int_{-\infty}^{+\infty} S_s^{\alpha_k}(f) * S_r^{\alpha_k}(f)df, \alpha_k = k/T_c \tag{7}$$

Where N_α is the number of cycle frequencies; It is α_k cycle frequency of k; $S_s^{\alpha_k}(f)$ denotes the spectral correlation function of the main user signal $s(t)$ at α_k; $S_r^{\alpha_k}(f)$ denotes the sub-user signal $r(t)$ spectral correlation function at α_k. In a cognitive radio, a rectangular window function $S_s^{\alpha_k}(f)$ with a bandwidth of Δf can be used, then Eq. (7) can be expressed as:

$$Y_{MC} = \sum_{k=1}^{N_\alpha} \int_{f-\frac{\Delta f}{2}}^{f+\frac{\Delta f}{2}} S_r^{\alpha_k}(v)dv \tag{8}$$

Thus the sum of the power of the cyclostationary signal spectrum is:

$$Y_{MC} = \sum_{k=1}^{N_\alpha} R_r^{\alpha_k}(\tau = 0) \tag{9}$$

Where $R_r^{\alpha_k}(\tau = 0)$ is the power spectrum of the cyclostationary signal at the k th cycle frequency.

Since $Y_{MC} = \text{Re}(Y_{MC}) + j\,\text{Im}(Y_{MC})$ is a complex random variable, the test statistic can be expressed as:

$$T_{MC} = |Y_{MC}|^2 = \sum_{k=1}^{N_\alpha} \left| R_r^{\alpha_k}(\tau = 0) \right|^2 + \sum_{k=1}^{N_\alpha} \sum_{n=1,n\neq k}^{N_\alpha} R_r^{\alpha_k}(\tau = 0) R_r^{\alpha_n*}(\tau = 0) \tag{10}$$

Comparing the test statistic T_{MC} with the threshold λ, it can be concluded that the primary user is active or idle.

3 Simplified Cyclic Spectrum Calculated

3.1 Simplified Spectral Correlation Function

According to the definition of cycle spectrum (4), the frequency f is reversed:

$$S_x^\alpha(-f) = \lim_{W,T\to\infty} \frac{1}{T} \int_{-\frac{T}{2}}^{\frac{T}{2}} \frac{1}{W} X_W\left(t, -f + \frac{\alpha}{2}\right) \times X_W^*\left(t, -f - \frac{\alpha}{2}\right) dt \tag{11}$$

For most of the terms of the signal, which is the Fourier transform on the spectrum of the conjugation or even function of frequency, and therefore, $X_W(t,f) = X_W^*(t,-f)$ is substituted into (11) yields:

$$S_x^\alpha(-f) = \lim_{W,T\to\infty} \frac{1}{T} \int_{-\frac{T}{2}}^{\frac{T}{2}} \frac{1}{W} X_W\left(t, -f + \frac{\alpha}{2}\right) X_W^*\left(t, -f - \frac{\alpha}{2}\right) dt$$
$$= \lim_{W,T\to\infty} \frac{1}{T} \int_{-\frac{T}{2}}^{\frac{T}{2}} \frac{1}{W} X_W\left(t, f - \frac{\alpha}{2}\right) X_W^*\left(t, f + \frac{\alpha}{2}\right) dt = S_x^\alpha(f)^* \tag{12}$$

From Eq. (12), it can be seen that the cyclic spectrum is symmetric about the α axis. Because the conjugate symmetry means that the amplitudes are equal, only the cyclic spectrum at $\alpha \geq 0$ is calculated, which reduces the computational complexity to half. In the same way, it can be proved:

$$S_x^{-\alpha}(f) = \lim_{W,T\to\infty} \frac{1}{T} \int_{-\frac{T}{2}}^{\frac{T}{2}} \frac{1}{W} X_W(t,f-\frac{\alpha}{2}) X_W^*(t,f+\frac{\alpha}{2}) dt = S_x^\alpha(f)^* \tag{13}$$

So the cyclic spectrum is also symmetric about the f axis. Because only the amplitude of the cyclic spectrum is concerned, only the cyclic spectrum in the $(f > 0, \alpha > 0)$ quadrant is calculated, which reduces the computational complexity to a quarter 8.

3.2   Simplification of Test Statistics

Observation of formula (10) can be found in the conventional test statistic T_{MC} detection algorithm is divided into two portions: $\sum_{k=1}^{N_\alpha} \sum_{n=1,n\neq k}^{N_\alpha} R_r^{\alpha_k}(\tau=0)R_r^{\alpha_n*}$ $(\tau = 0)$ and $\sum_{k=1}^{N_\alpha} |R_r^{\alpha_k}(\tau=0)|^2$.

We boldly assume that if $\sum_{k=1}^{N_\alpha} \sum_{n=1,n\neq k}^{N_\alpha} R_r^{\alpha_k}(\tau=0)R_r^{\alpha_n*}(\tau=0)$ is discarded, it will only cause a small loss of detection performance. Therefore, we will focus on the simulation analysis of the detection performance after removing this item.

Complexity Analysis
Let the test statistic of the improved algorithm be T_{sim}:

$$T_{sim} = \sum_{k=1}^{N_\alpha} |R_r^{\alpha_k}(\tau=0)|^2 \tag{14}$$

When T_{sim} is calculated, N_α complex multiplications and $N_\alpha-1$ complex additions are performed. Therefore, the total number of calculations for T_{sim} is $2N_\alpha-1$. The computational complexity is $\Theta(N_\alpha)$.

When T_{MC} is calculated, N_α^2 complex multiplications and N_α^2-1 complex additions are performed. Therefore, the total number of calculations for T_{MC} is $2N_\alpha^2-1$. The computational complexity is $\Theta(N_\alpha^2)$.

Among the actual testing, the number of cycles of frequency, thus improving the computational complexity is greatly reduced.

Performance Analysis
The final rule for improving the algorithm is:

$$T_{sim} = \sum_{k=1}^{N_\alpha} |R_r^{\alpha_k}(\tau=0)|^2 \mathop{\gtrless}_{>}^{<} \lambda \tag{15}$$

λ is the detection threshold, then:

$$P_d = \int_\lambda^{+\infty} P(T_{sim}|H_1)dT_{sim} \tag{16}$$

$$P_f = \int_{\lambda}^{+\infty} P(T_{sim}|H_0)dT_{sim} \tag{17}$$

Next, we derive the closed expressions of Eqs. (16) and (17):
First, in order to get the probability density function of the two conditions, let

$$Y_1 = R_r^{\alpha_k}(\tau = 0) \tag{18}$$

Then, $R_r^{\alpha_k}(\tau = 0)$ is expressed as:

$$R_r^{\alpha_k}(\tau) = \lim_{T \to \infty} \int_{t_0-\frac{T}{2}}^{t_0+\frac{T}{2}} r(t+\frac{\tau}{2})r*(t-\frac{\tau}{2})e^{-j2\pi\alpha_k t}dt \tag{19}$$

Then the discrete time Y_1 is:

$$Y_1 = \frac{1}{N_S}\sum_{w=1}^{N_S} |r[w]|^2 e^{-j2\pi\alpha_k w} \tag{20}$$

Therefore, the binary hypothesis discrimination under the Gaussian white noise channel is expressed as:

$$\begin{cases} H_0 : Y_1 = \dfrac{1}{N_S}\displaystyle\sum_{w=1}^{N_S} |n[w]|^2 e^{-j2\pi\alpha_k w} \\[4mm] H_1 : Y_1 = \dfrac{1}{N_S}\displaystyle\sum_{w=1}^{N_S} |s[w]+n[w]|^2 e^{-j2\pi\alpha_k w} \end{cases} \tag{21}$$

Therefore, Eq. (21) can be regarded as the detector detecting only the k th cycle frequency, then:

$$T_1 = |Y_1|^2 = |R_r^{\alpha_k}(\tau = 0)|^2 \underset{\underset{H_0}{>}}{\overset{\overset{H_1}{<}}{}} \lambda \tag{22}$$

Then, the false alarm probability and detection probability detected at the k th cycle frequency are:

$$P_{f,\alpha_k} = \int_{\lambda}^{+\infty} P(T_1|H_0)dT_1 \tag{23}$$

$$P_{d,\alpha_k} = \int_{\lambda}^{+\infty} P(T_1|H_1)dT_1 \tag{24}$$

According to the Central Limit Theorem [11], $N_S, N_\alpha \gg 1$, Y_1 is considered to be approximately Gaussian. The complex Gaussian random variable $(T_1|H_0)$ obeys the centroid chi-square distribution with a degree of freedom of 2, the probability density function is [12]:

$$P(T_1|H_0) = \frac{1}{2\sigma_1^2} e^{\frac{T_1}{2\sigma_1^2}}, \sigma_1^2 = \frac{\sigma_0^4}{N_S} \tag{25}$$

Then, the false alarm probability detected for the k th loop frequency is:

$$P_{f,\alpha_k} = e^{\frac{\lambda}{2\sigma_1^2}} \tag{26}$$

Similarly, it can be obtained that the complex random variable $(T_1|H_1)$ obeys the non-center chi-square distribution with a degree of freedom of 2, and the conditional probability density function is [13]:

$$P(T_1|H_1) = \frac{1}{2\sigma_2^2} e^{-\frac{T_1+u_1}{2\sigma_2^2}} I_0\left(\frac{\sqrt{T_1 u_1}}{\sigma_2^2}\right) \tag{27}$$

Among them, $u_1 = \sqrt{\{E[\mathrm{Re}(Y_1|H_1)]\}^2 + \{E[\mathrm{Im}(Y_1|H_1)]\}^2} = |P_{\alpha_K}|$, $\sigma_2^2 = 2\sigma_0^2 \frac{P}{N_S} + \frac{\sigma_0^4}{N_S}$, then the detection probability of the k th cyclic frequency is:

$$P_{d,\alpha_k} = Q_1\left(\frac{u_1}{\sigma_2}, \frac{\sqrt{\lambda}}{\sigma_2}\right) \tag{28}$$

To simplify the expression, let $u_2 = \sum_{k=1}^{N_\alpha} u_1 = \sum_{k=1}^{N_\alpha} |P_{\alpha_k}|$. Derive a closed expression for detection probability and false alarm probability:

$$P_d = \int_\lambda^{+\infty} P(T_{sim}|H_1) dT_{sim} = Q_{N_\alpha}\left(\frac{\sqrt{u_2}}{\sigma_2}, \frac{\sqrt{\lambda}}{\sigma_2}\right) \tag{29}$$

$$P_f = \int_\lambda^{+\infty} P(T_{sim}|H_0) dT_{sim} = \frac{\Gamma\left(N_\alpha, \frac{\lambda}{2\sigma_1^2}\right)}{\Gamma(N_\alpha)} \tag{30}$$

Let the signal-to-noise ratio at the receiver be γ, then $\gamma = \frac{u_2}{\sigma_0^2}$. The closed expression of the detection probability is:

$$P_d = Q_{N_\alpha}\left(\frac{\sigma_0 \sqrt{\gamma}}{\sigma_2}, \frac{\sqrt{\lambda}}{\sigma_2}\right) \tag{31}$$

4   The Simulation Results Analysis

In order to verify the accuracy of the improved algorithm, we use a sampling rate of 10 times the carrier frequency for BPSK signal simulation. The performance of the improved algorithm is analyzed using the -SNR curve and receiver operating characteristic (ROC).

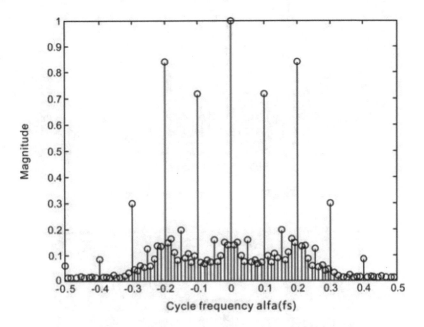

Fig. 2. Cyclic spectrum projection of BPSK signal.

Figure 2 is a cyclic spectral projection of a BPSK signal, and Fig. 3 is a projection of a BPSK cyclic spectrum obtained by calculating only a quarter of the range. Further observed that the simplified algorithm still peaks at integer multiples of the carrier frequency, illustrates a simplified algorithm does not lose information.

Figure 4 shows the operating characteristics of the receiver before and after the improvement of the SNR conditions. The simulation conditions are: The SNR at the receiver is 5 dB, 10 dB, and 15 dB, respectively, and the number of loop frequencies is $N_\alpha = 100$. With a certain number of loop frequencies, the SNR at the receiver is set to 5 dB, 10 dB, and 15 dB, respectively. The existing algorithm is represented by "old", the improved algorithm is represented by "new", the theoretical value is represented by a solid line, and the simulation value is represented by a dashed line. From the analysis of Fig. 4, on the one hand, the theoretical value is in agreement with the simulation value, so the expression of the detection probability and the false alarm probability of the improved algorithm is accurate. On the other hand, comparing the performance of the two algorithms, it can be seen that the detection performance of the simplified algorithm is slightly reduced, but the loss of detection performance is very small, and the specific performance loss analysis is shown in Fig. 5.

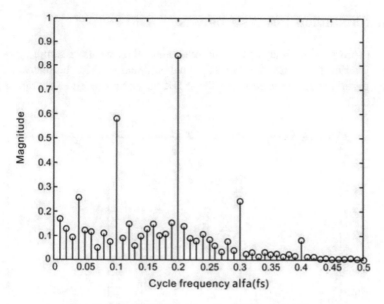

Fig. 3. Simplified algorithm for BPSK projection.

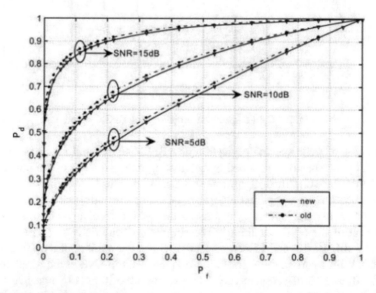

Fig. 4. Detection performance of existing algorithms and improved algorithms under different SNR ($N_\alpha = 100$).

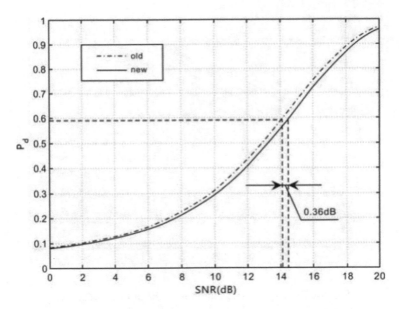

Fig. 5. Quantitative analysis of differences in detection performance between existing and improved algorithms ($P_f = 0.01$ $N_\alpha = 100$).

Quantitative analysis, based on the same detection probability P_d, the difference between the required SNR before and after the improvement algorithm is ΔSNR.

Analyzing Fig. 5, different detection probabilities P_d, the difference in SNR required for the algorithm is different. $P_d = 0.1$, $\Delta SNR \approx 0$, $P_d = 0.59$, $\Delta SNR = 0.36$ dB. Then: $0 \leq \Delta SNR \leq 0.36$ dB By comparison, the detection performance of the improved algorithm attenuates at most s. In terms of computational complexity, the simulation parameter $N_\alpha = 100$ is taken as an example, the complexity of the improved algorithm is $\Theta(100)$ and the complexity of the original algorithm is $\Theta(10000)$. Therefore, the efficiency and accuracy of the improved algorithm have been affirmed.

The summation of the power spectrum at N_α cycle frequencies is the essence of the cyclic feature detection. Therefore, the larger N_α is, the more accurate the statistic is, and the more accurate the judgment result is. Thus by setting different values to verify whether the improved algorithm has this property. Figure 6 shows the ROC curve of the improved algorithm at $N_\alpha = 20, 40, 60, 80, 100$. It can be seen that N_α increases and the performance increases gradually. This is due to the nature of the cyclic feature detection, since its test statistic is the sum of spectral power at N_α cycle frequency, Therefore, the greater the number of N_α, the more accurate the test statistic obtained and the more accurate the judgment result. Consistent with the characteristics of the existing algorithms, the accuracy of the improved algorithm is more proved.

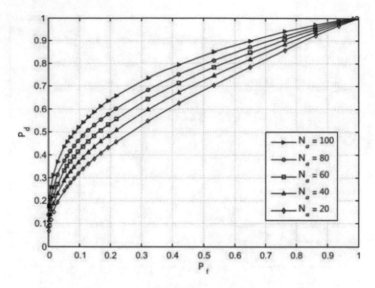

Fig. 6. The improved algorithm of different cycle frequency.

5   Conclusion

The improved cyclic spectrum detection algorithm proposed in this paper, under the condition of not adding additional hardware overhead, innovative from the spectral correlation function characteristics and test statistics to verify the efficiency and accuracy of the improved algorithm. Through the calculation of complexity, the efficiency of the improved algorithm is proved from the data. In the simulation environment, the characteristic curves of the algorithm before and after improvement are compared, and the accuracy of the improved algorithm is proved. The simulation results show that the improved algorithm proposed in this paper maintains high detection performance with low computational complexity, and effectively solves the problem that existing algorithms have high computational complexity and cannot be widely used. Effectively broaden the use of the cyclostationary feature detection.

Acknowledgement. This study is supported by Scientific research plan projects of Gansu Natural Science Foundation (1508RJZA071); Lanzhou Jiaotong University Youth Fund (2015008).

References

1. Liu, X.L., Zhu, Q.: Joint frequency spectrum detection method based on energy-cycling stationarity feature. J. Nanjing Univ. Posts Telecommun. **30**(3), 34–38 (2010)
2. Yuan, H.Y., Hu, Y.: Spectrum sensing algorithm combining cyclostationarity and adaptive dual threshold detection. J. Comput. Aided Des. Comput. Graph. **25**(4), 573–577 (2013)

3. Ma, B., Fang, Y., Xie, X.Z.: An improved cyclostationary feature detection algorithm for master user at random arrival. J. Electron. Inf. Technol. **37**(7), 1531–1537 (2015)
4. Derakhshani, M., Le-Ngoc, T., Nasiri-Kenari, M.: Efficient cooperative cyclostationary spectrum sensing in cognitive radios at low snr regimes. IEEE Trans. Wirel. Commun. **10**(11), 3754–3764 (2011)
5. Chen, X.Y., Yang, R.J., Li, X.B., Xie, C.: Collaborative detection of cyclo-stationary spectrum based on maximum ratio merging. Mod. Def. Technol. **41**(4), 113–117 (2013)
6. Yang, M., Li, Y., Liu, X., Tang, W.: Cyclostationary feature detection based spectral sensing algorithm under complicated electromagnetic environment in cognitive radio networks. China Commun. **12**(9), 35–44 (2015). (English)
7. Lu, G.Y., Ye, Y.H., Sun, Y., Mi, Y.: Spectrum-aware algorithm based on goodness-of-fit test for overcoming noise uncertainty. Telecommun. Technol. **56**(1), 26–31 (2016)
8. Gao, Y.L., Chen, Y.P., Guan, X., Zhang, Z.Z., Sha, X.J.: Spectrum sensing algorithm based on cyclic spectrum symmetry. In: National Cognitive Wireless Network Conference (2011)
9. Zhang, J., Zhang, L., Huang, H., Zhang, X.J.: Improved cyclostationary feature detection based on correlation between the signal and noise. In: International Symposium on Communications and Information Technologies, pp. 611–614. IEEE (2016)
10. Trees, H.L.V.: Detection, estimation, and modulation theory. Part III. A Papoulis Probability Random Variables & Stochastic Processes, vol. 8(10), pp. 293–303 (2001)
11. Qi, P.H., Si, J.B., Li, Z., Gao, R.: A novel spectrum-aware spectrum sensing algorithm for anti-noise uncertainty spectrum. J. Xidian Univ. **40**(6), 19–24 (2013)
12. Zhu, Y.: Research on Key Technologies for Efficient Spectrum Detection in Cognitive Wireless Networks. Beijing University of Posts and Telecommunications (2014)
13. Wang, H., Yuan, X.B., Zhang, H.L.: Statistical analysis based on energy detection method. Aerosp. Electron. Warf. **30**(6), 41–44 (2014)

Chinese News Keyword Extraction Algorithm Based on TextRank and Word-Sentence Collaboration

Qing Guo[(✉)] and Ao Xiong

Institute of Network Technology,
Beijing University of Posts and Telecommunications,
No. 10, Xitucheng Road, Haidian District, Beijing 100876, China
guoqingbupt@163.com

Abstract. TextRank always chooses frequent words as keywords of a text. However, some infrequent words may also be keywords. To solve the problem, a keyword extraction algorithm based on TextRank is proposed. The algorithm takes the importance of sentences into consideration and extracts keywords through word-sentence collaboration. Two text networks are built. One network's nodes are words where the diffusion of two words is defined to calculate the correlation between words. Another's nodes are sentences where BM25 algorithm is used to calculate the correlation between sentences. Then a sentence-word matrix is constructed to extract the keywords of a text. Experiments are conducted on the Chinese news corpus. Results show the proposed algorithm outperforms TextRank in Precision, Recall and F1-measure.

Keywords: Keyword extraction · TextRank · BM25 · Word-sentence collaboration

1 Introduction

The keywords are a brief summary of an article's topic, which is of great significance to readers' mastery of the article's main idea. In addition, the keywords have important practical value in literature retrieval, text classification, recommendation systems, etc. Although many research articles may come with several keywords when published, many other texts such as news or blogs articles don't. Therefore, the automatic extraction of keywords has become a hot topic in the field of natural language process and many algorithms have been proposed in the past decades, which can be divided into supervised methods and unsupervised methods. In the unsupervised methods, a graph-based ranking algorithm called TextRank has been studied by many researchers [1]. It constructs a text network based on the co-occurrences of words within a text window and then ranks the words according to their scores. The top-ranked words are chosen as keywords. TextRank always give high-frequent words high scores. However, those with low frequency may also be keywords.

To improve the situation, this paper proposes an algorithm based on TextRank. We construct two text networks: word network and sentence network. In the word network,

© Springer Nature Switzerland AG 2020
Q. Liu et al. (Eds.): CENet 2018, AISC 905, pp. 556–563, 2020.
https://doi.org/10.1007/978-3-030-14680-1_61

we introduce the diffusion of two words which considers how the words spread in the text. It would adjust the correlation between two words, which would raise scores of the words with low frequency but spreading widely in different sentences. In the sentence network, we use BM25 as the correlation between two sentences and get the sentence scores. Then we build a sentence-word matrix to calculate the final scores of words, where words in high-score sentences would get higher scores. This paper focuses on the Chinese news keyword extraction. Experiments are conducted on news corpus. Results show the proposed algorithm performs better than TextRank.

The rest is organized as follows. Section 2 briefly browses some related work on keyword extraction. Section 3 describes the details of our algorithm. Section 4 presents and discusses the experiment results. Finally, Sect. 5 will make the conclusion of our work.

2   Related Works

The keyword extraction algorithms can be divided into two categories: the supervised and the unsupervised. With respect to supervised algorithms research, Witten et al. designed the KEA system and proposed a naive Bayesian machine learning method for keyword extraction [2]. The algorithm uses only two features of the candidate words: TF-IDF (Term Frequency-Inverse Document Frequency) and the first occurrence in the text. After training a large number of annotated corpus, a keyword extraction model is obtained. Turney et al. developed the GenEx system, which uses C4.5 decision tree model and a genetic algorithm to extract keywords [3]. In the field of unsupervised algorithms, Sparck first proposed the IDF (Inverse Document Frequency) concept [4]. Salton et al. discussed the application of TF-IDF in the field of information retrieval [5]. Since TF-IDF considers the words as mutually independent in an article, in 2004, Mihalcea et al. proposed the TextRank algorithm [1]. The algorithm was inspired by Google's PageRank algorithm for web page ordering [6]. It builds a text network, namely a grpah where words are nodes, and calculate the scores of the words iteratively.

Although TextRank avoids the weakness of TF-IDF, it still tends to choose keywords that occurs frequently in a text as keywords. However, some words that may not occur frequently might also be selected as keywords. Many scholars have proposed improvements to the TextRank algorithm from different perspectives. For example, Li et al. used the open database provided by Wikipedia to represent each term in Wikipedia as a vector, where each element in the vector is the TF-IDF of its corresponding term, using the cosine similarity of two vectors to represent the edge's weight [7]. Additionally, the first N words, whether the word appears in the first sentence in the paragraph and other statistical features can also be taken into consideration for keyword extraction [8].

Based on TextRank, this paper introduces the concept of diffusion between two words, constructs a new edge weight calculation formula, and reranks the words by word-sentence collaboration. The experiment uses Chinese news as the corpus. Comparisons will be made between the proposed algorithm and the original TextRank.

3   The Algorithm Description

3.1   Text Preprocess

First, segment the text into words and sentences. Then the stop words are removed and only verbs, nouns and English words are reserved as candidate words. When the text is segmented into sentences, the punctuation marks such as the period, the exclamation, the ellipsis and the semicolon are considered as the end of a sentence.

3.2   Diffusion of Candidate Words

The following statistics are made on the text: (1) For each candidate word W_i, count the number of sentences containing W_i as N_i; (2) For each pair of candidate words (W_i, W_j), count the number of sentences that contain both W_i and W_j as N_{ij}. Thus $N_{ij} = N_{ji}$.

The definition of the diffusion of two candidate words is as follows.

$$d_{ij} = \frac{N_i + N_j - 2N_{ij} + 0.5}{N + 0.5} \tag{3.1}$$

where N represents the total number of sentences of the text.

The diffusion of two words indicates how dispersed the two words are in the article. It will be used to define the correlation of two words in the following part.

3.3   Intermediate Scores of Candidate Words

In a text, let l $(l \geq 2)$ be the length of the co-occurence window, the distance between every two adjacent words be 1, and c_{ij} be the times where the distance between word W_i and word W_j is less than l. Then we define the correlation of two candidate words as

$$w_{ij} = c_{ij} \cdot d_{ij} \tag{3.2}$$

from (3.2) we know that w_{ij} is a balance of the co-occurence time and the diffusion between two candidate words.

There are two ways to calculate the distance between candidate words. One is to count the non-candidate words (such as stop words, non-verbs, non-nouns and non-English words, etc.) between candidate words into the distance. The other is to calculate the distance after removing non-candidate words. Considering that the non-candidate words can provide distance information to determine the correlation between two words, the former is used [9].

Then a text network will be constructed. All the candidate words are the nodes in the network, and the edge's weight between two nodes can be got by (3.2). Every node has an randomly initial score which is close to zero. And then calculate the score of each node iteratively according to the following formula.

$$S(W_i) = (1 - d) + d \cdot \sum_{W_j \in In(W_i)} \frac{w_{ji}}{\sum_{W_k \in Out(W_j)} w_{jk}} S(W_j) \tag{3.3}$$

where $S(W_i)$ denotes the score of word W_i; d is a damping factor between $[0, 1]$, which ensures the algorithm can converge; $In(W_i)$ represents a set of nodes that point to W_i and $Out(W_i)$ represents the set of nodes pointed to by W_i. Since the text network we constructed above is an undirected graph, $In(W_i)$ and $Out(W_i)$ means the same set.

The iteration stops when all the scores converge or it comes to the maximum number of iterations. Then we normalize the scores as follows.

$$S'(W_i) = \frac{S(W_i) - \min}{\max - \min} \tag{3.4}$$

where min and max are the lowest and the highest score of all nodes.

Thus we obtain the intermediate scores of the candidate words.

3.4 Correlation Between Sentences Using BM25

We also consider the importance of sentences has influence to the keywords of a text. To assess the importance of a sentence we have first to calculate the correlation between sentences.

BM25 is a ranking function for information retrieval [10]. For a query string Q, which contains words $q_1, q_2, \ldots, q_n$, the BM25 score of a document D is

$$Score(Q, D) = \sum_{i=1}^{n} IDF(q_i) \cdot R(q_i, D) \tag{3.4}$$

where

$$IDF(q_i) = \log_2 \frac{N - n(q_i) + 0.5}{n(q_i) + 0.5} \tag{3.5}$$

$$R(q_i, D) = \frac{f(q_i, D) \cdot (k_1 + 1)}{f(q_i, D) + k_1 \cdot \left(1 - b + b \cdot \frac{|D|}{avgdl}\right)} \tag{3.6}$$

In (3.5), N represents the total number of documents in the corpus; $n(q_i)$ is the number of documents containing words q_i. In (3.6), $f(q_i, D)$ is the times that q_i occurs in document D; $|D|$ is the length of document D, which is the total number of words in it; $avgdl$ is the average length of documents in the corpus; k_1 and b are free parameters and in this paper we choose $k_1 = 0.5$ and $b = 0.75$.

In a text, if the i-th sentence is used as the query string, the j-th sentence as document D, we can get $Score(S_i, S_j)$, which will be used as the correlation between two sentences.

3.5   Final Scores of Candidate Words Using Word-Sentence Collaboration

Another text network will be constructed, where all the sentences in the text are nodes, and the BM25 scores between the sentences are the edge's weight between two nodes. The following formula which is quite similar to (3.3) is used to iteratively calculate the scores of each node.

$$S(S_i) = (1 - d) + d \cdot \sum_{j=1}^{N} \frac{Score(S_j, S_i)}{\sum_{k=1}^{N} Score(S_j, S_k)} S(S_j) \tag{3.7}$$

where N is the total number of sentences in the text; S_i is the i th sentence in the text; $S(S_i)$ is the score of sentence S_i.

The iteration stops when it reaches the maximum number of iteration or when it converges. Same as (3.4), we normalize the sentence scores and make it into a row vector $\mathbf{SV} = [S'(S_1), S'(S_2), \ldots, S'(S_n)]^{\mathrm{T}}$.

Suppose we have m candidate words in the text which are $W_1, W_2, \ldots, W_m$, for any sentence S_i, it can be expressed as a row vector according to the number of candidate word occurrence in the sentence. So the vector can be expressed as $[tf_{i1} \cdot S'(W_1), tf_{i2} \cdot S'(W_2), \ldots, tf_{im} \cdot S'(W_m)]^{\mathrm{T}}$, where tf_{ij} represents the number of occurrences of word W_j in sentence S_i.

In a text containing n sentences and m candidate words, a sentence-word matrix can be built as

$$\mathbf{SWM} = \begin{pmatrix} tf_{11} \cdot S'(W_1) & \cdots & tf_{1m} \cdot S'(W_m) \\ \vdots & \ddots & \vdots \\ tf_{n1} \cdot S'(W_1) & \cdots & tf_{nm} \cdot S'(W_m) \end{pmatrix} \tag{3.8}$$

The final score vector of the candidate words is got in by the following calculation.

$$\begin{aligned} \mathbf{WV} &= \mathbf{SV} \cdot \mathbf{SWM} \\ &= [WS(W_1), \ldots, WS(W_m)] \end{aligned} \tag{3.9}$$

Then we sort the candidate words descendently according to their scores, and pick the first k words as the keywords of the text.

4   Experiments and Results

4.1   Corpus and Evaluation

We choose news articles released in September 2017 from **South Daily** as the corpus. The editor would tag most news articles with keywords before publishing, and we take these keywords as reference keywords. 500 articles with no less than 5 reference keywords are randomly chosen as the test corpus.

We use Precision (*P*), Recall (*R*) and F1-measure (*F1*) to evaluate the quality of results [11]. Here are the definitions.

$$P = \frac{|A \cap B|}{|A|}, R = \frac{|A \cap B|}{|B|}, F1 = \frac{2PR}{P+R} \qquad (4.1)$$

where *A* is the set of keywords extracted by the algorithm, *B* is the set of the reference keywords and $|A|$ is the number of elements in *A*.

4.2 Results

We use TR short for TextRank and WSCTR short for our proposed algorithm, Word-Sentence Collaboration TextRank.

After experiments, we found in different co-occurrence window lengths (*l*) and damping factors (*d*), WSCTR outperforms in Precision, Recall and F1-measure compared to TR. When $l = 10$ and $d = 0.7$, the two algorithms both perform best. So in this condition, we compare the results when the number of extracted keywords varies from 1 to 15.

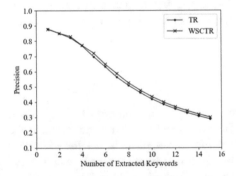

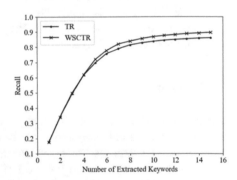

Fig. 1. Precision curves when the number extracted keywords varies from 1 to 15.

Fig. 2. Recall curves when the number of of extracted keywords varies from 1 to 15.

As can be seen from Figs. 1, 2 and 3, when the number of extracted keywords (here in after referred shortly as "the number") is small, the Precision, Recall and F1 curves of the two algorithms almost overlap. With the number increasing, WSCTR's Precision, Recall and F1 are higher than TR. This is because when the number is small, the candidate words with the highest frequency are likely to become the center of the text network. At the same time, the most frequent candidate words are often used as reference keywords. So the candidate words with the highest scores in the two algorithms tend to be the same. When the number becomes large, WSCTR can improve the scores of the candidate words which are of low frequency but in the important sentences in the text, and thus the result is significantly improved than TR's.

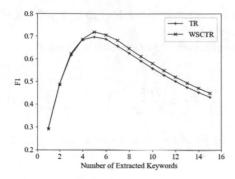

Fig. 3. F1 curves when the number of extracted keywords varies from 1 to 15.

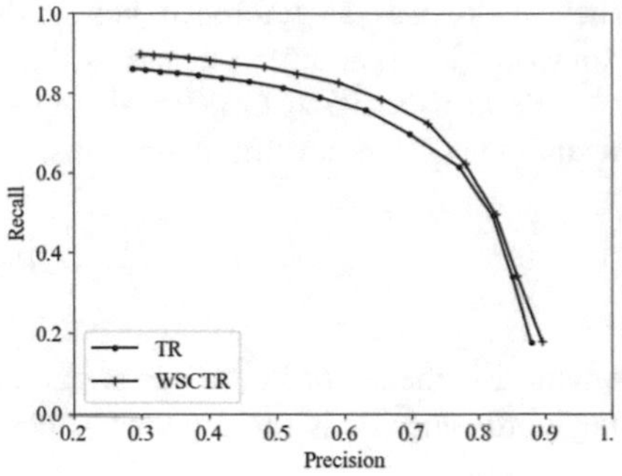

Fig. 4. Precision-Recall curves.

We can also see in Fig. 3 that F1 reaches the peak value when the number is 5 and then eases down after 5. This is because most of the news articles have 5 reference keywords. When the algorithms extract 5 keywords, the accuracy rate and recall rate can reach relatively high values, causing F1 to reach the peak value.

In Fig. 4, we can see that Precision and Recall are in a contradictory relationship. The TR's curve is inside the WSCTR's curve, which means that at the same Precision (or Recall), the WSCTR's Recall (or Precision) is higher than the TR's. So the result shows WSCTR outperforms TR.

5   Conclusion

The paper optimizes the TextRank algorithm by bringing the concept of diffusion and the word-sentence collaboration algorithm. How the words spread in the text and the importance of sentences are taken considered into the algorithm. So the algorithm actually extracts keywords in the level of the whole text instead of only within a co-occurrence window. Results show that the proposed algorithm outperforms TextRank in Precison, Recall and F1-measure.

References

1. Mihalcea, R., Tarau, P.: TextRank: bringing order into text. In: Proceedings of the 2004 Conference on Empirical Methods in Natural Language Processing, pp. 404–441. ACL, Stroudsburg (2004)
2. Frank, E., Paynter, G.W., Witten, I.H., Gutwin, C., Nevill-Manning, C.G.: Domain-specific keyphrase extraction. In: 16th International Joint Conference on Artificial Intelligence (IJCAI 99), pp. 668–673. Morgan Kaufmann Publishers Inc., San Francisco (1999)
3. Turney, P.D.: Learning to extract keyphrases from text. Inf. Retr. 2(4), 303–336 (2000)
4. Jones, K.S.: A statistical interpretation of term specificity and its application in retrieval. J. Doc. 28(1), 11–21 (1972)

5. Wu, H., Salton, G.: A comparison of search term weighting: term relevance vs. inverse document frequency. In: Proceedings of the 4th Annual International ACM SIGIR Conference on Information Storage and Retrieval, pp. 30–39. ACM Press, New York (1981)
6. Wu, X., Kumar, V., Quinlan, J.R., Ghosh, J., Yang, Q., Motoda, H., McLachlan, G.J., et al.: Top 10 algorithms in data mining. Knowl. Inf. Syst. **14**(1), 1–37 (2008)
7. Li, W., Zhao, J.: TextRank algorithm by exploiting Wikipedia for short text keywords extraction. In: 2016 3rd International Conference on Information Science and Control Engineering (ICISCE), pp. 683–686. IEEE, Piscataway (2016)
8. Siddiqi, S., Sharan, A.: Keyword and keyphrase extraction techniques: a literature review. Int. J. Comput. Appl. **109**(2), 18–23 (2015)
9. Liu, Z.Y.: Research on Keyword Extraction Using Document Topical Structure. Tsinghua University, Beijing (2011). (in Chinese)
10. Géry, M., Largeron, C.: BM25t: A BM25 extension for focused information retrieval. Knowl. Inf. Syst. **32**(1), 1–25 (2012)
11. Powers, D.M.W.: Evaluation: from precision, recall and F-measure to ROC, informedness, markedness and correlation. J. Mach. Learn. Technol. **2**(1), 37–63 (2011)

A New Balanced Greedy Snake Algorithm

Le Cheng[1,2(✉)], Liping Zheng[1], Haibo Wang[1], Yanhong Song[1], and Jihong Gao[1]

[1] Department of Computer Science and Engineering,
Huaian Vocational College of Information Technology, Huaian 223003, China
cl211282@163.com
[2] College of Computer and Information,
Hohai University, Nanjing 210098, China

Abstract. The existing greedy snake algorithm (GSA) suffers from some problems, such as three forces are unbalance and the extracting contour on concave region is unsatisfactory. This paper presents an algorithm, called balanced greedy snake algorithm (BGSA), for solving objective contour extraction problem. BGSA is compose of continuity force, curvature force and image force, which is similar to the origin GSA. Whereas, BGSA improved the computing rule of GSA to balance the influence of above three forces. Especially, BGSA can process the image with concave region well. The results of experiment show that BGSA is efficient and outperform the existing GSA.

Keywords: Greedy snake algorithm · Concave region · Objective contour extraction · Balanced

1 Introduction

The active contour, or snake, is a successful algorithm to extract object boundary, which has been used in many applications such as object tracking [1], medical imaging [2–4], image segmentation [5], etc. Snake model was first introduced by Kass et al. [6] in 1988. The traditional snake regards boundary detection as an optimization problem and all the candidate boundary points, called snaxel, are evaluated by an energy function that is compose of internal contour forces, image forces and external forces.

Whereas, Kass's method suffers from some problems, such as numerical instability and bunched snaxels. In order to make up the deficiencies, a greedy snake algorithm (GSA) was proposed in [7]. The GSA is a considerable improvement for snake and adopts a different energy function that is compose of continuity force, curvature force and image force. By using the new energy function, GSA can get the smoother contour and the snaxels are more evenly spaced. Later on, the fast greedy snake algorithm (FGSA) and skippy greedy snake algorithm (SGSA) are proposed in [8] and [9], respectively. The two variations of GSA increase the speed of convergence by improve the search patterns or neighborhood pattern. The FGSA employs two alternate pixel search in 3×3 neighborhood pattern during the iteration process, and the SGSA adopts two step sizes in two kinds of neighborhood patterns to achieve better convergence. Whereas, these variations of GSA suffer from the same problem that is

© Springer Nature Switzerland AG 2020
Q. Liu et al. (Eds.): CENet 2018, AISC 905, pp. 564–570, 2020.
https://doi.org/10.1007/978-3-030-14680-1_62

out-off-balance on the forces. Practical experience shows that the value of image force is much bigger than the value of continuity force, curvature force. That means the movements of snaxels is almost completely controlled by image forces.

Based on the GSA, a new balanced greedy snake algorithm (BGSA) is proposed in this paper. The primary focus of our research is at follow aspect: (1) Improve the energy function of traditional GAS and implement the better tradeoff of three forces; (2) BGSA can extract the object boundary of image with concave region.

2 Balanced Greedy Snake Algorithm

2.1 Description on Balanced Greedy Snake Algorithm

The original version of GSA is compose of continuity force, curvature force and image force, which is similar to that of BGSA. The description of GSA can be found in literature [7]. The energy function (EF) of BGSA is described as follow:

$$EF = \min\left\{\sum_{i=1}^{N} E_{i,j}\right\}$$

(1)

where :

$$E_{i,j} = \alpha_i E\_cont_{i,j} + \beta_i E\_curv_{i,j} + \gamma_i E\_image_{i,j}$$

Here, $E_{i,j}$ is the energy value of the jth neighboring pixel of the ith snaxel. N defers to the scale of snaxels. α_i, β_i and γ_i are called as weighting factors that control the important of above three forces. The neighbor modeling of GSA is defined as Fig. 1.

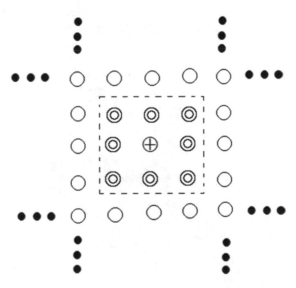

Fig. 1. Neighbor modeling of GSA ($\leftarrow$ snaxel; ◎ $\leftarrow$ neighboring pixel; ○ $\leftarrow$ pixel)

Essentially, $E_{i,j}$ represents the movement of the ith snaxel. That is, the ith snaxel move to the jth neighboring pixel position. The snaxel movement is to minimize the Eq. (1). E_cont, E_curv and E_image refer to the continuity force, curvature force and image force, respectively. Continuity force is defined as follow:

$$E\_cont_{i,j} = \frac{\left|\bar{d} - d_{i+1,j}\right|}{\left\{\left|\bar{d} - d_{i+1,j}\right|\right\}} \tag{2}$$

Here, $E\_cont_{i,j}$ is the continuity force of the jth neighboring pixel of the ith snaxel. $d_{i+1,j}$ is the average distance among the N snaxels. $d_{i+1,j}$ refers to the distance between the (i + 1)th snaxel and the jth pixel. When i = N, the di + 1 denote d1. $\max\left\{\left|\bar{d} - d_{i+1,j}\right|\right\}$ is the maximum of $\left|\bar{d} - d_{i+1,j}\right|$, i = 1, $\cdots$, N and j = 1, $\cdots$, 8. The is computed as followed:

$$\bar{d} = \frac{\sum\limits_{i=1}^{N}\left|d_{i+1} - d_i\right|}{N} \tag{3}$$

Continuity force is to make the snaxels to be evenly spaced.

$p_i(i = 1, \cdots, N)$ refers to the position of ith snaxel and $p_j(i = 1, \cdots, 8)$ refers the position of jth neighbor of pi. Thus, Curvature force is defined as follow:

$$E\_curv_{ij} = \frac{\left|\overrightarrow{p_jp_{i+1}} + \overrightarrow{p_jp_{i-1}}\right|}{\max\left\{\left|\overrightarrow{p_jp_{i+1}} + \overrightarrow{p_jp_{i-1}}\right|\right\}} \tag{4}$$

$\overrightarrow{p_jp_{i+1}}$ is the vector. The processing of $\left|\overrightarrow{p_jp_{i+1}} + \overrightarrow{p_jp_{i-1}}\right|$ is described as Fig. 2: Curvature force is to make the extracted contour to be smoother.

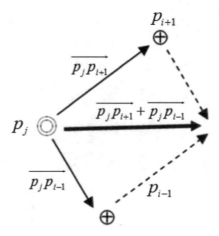

Fig. 2. Neighbor modeling of GSA ($\leftarrow$ snaxel; $\odot$ $\leftarrow$ neighboring pixel; $\bigcirc$ $\leftarrow$ pixel)

Image force is defined as follow:

$$E\_image_{i,j} = \frac{\min\{I_{i,j}\}}{|I_{i,j}|} \tag{5}$$

Where, Ii,j represents the gradient of the jth neighboring pixel of the ith snaxel.

2.2 Flow Chart on Description of BGSA

In this section, we give the whole description on BGAS. According follow flow chart, this algorithm can be implemented easily (Fig. 3).

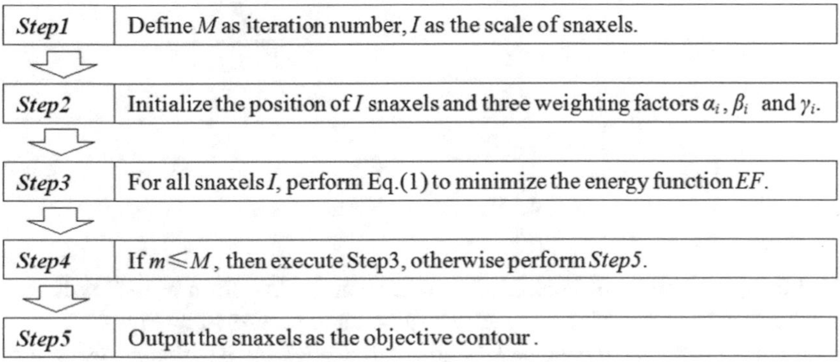

Step1	Define M as iteration number, I as the scale of snaxels.
Step2	Initialize the position of I snaxels and three weighting factors α_i, β_i and γ_i.
Step3	For all snaxels I, perform Eq.(1) to minimize the energy function EF.
Step4	If $m \leqslant M$, then execute Step3, otherwise perform $Step5$.
Step5	Output the snaxels as the objective contour .

Fig. 3. Flow chart of BGSA

3 Process on Concave Region

When handling on the image with concave region, the origin GSA cannot get a good contour. To overcome this problem, BGSA perform a post-processing for the contour. First, the center point C is computed according to Eqs. (6) and (7).

$$C_x = \frac{\sum\limits_{i=1}^{N} p_{i,x}}{N} \tag{6}$$

$$C_y = \frac{\sum\limits_{i=1}^{N} p_{i,y}}{N} \tag{7}$$

Here, pi,x and pi,y are the x-coordinate and y-coordinate of the ith snaxel, respectively. C is the center point of contour.

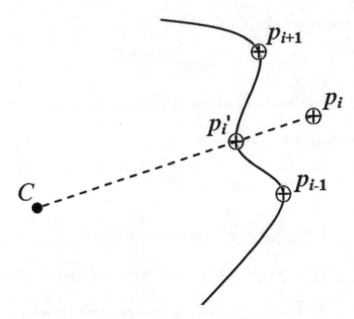

Fig. 4. Post-processing of BGSA

And then, the point with maximum gradient on the line from pi to C will be found and as the final position of snaxel. Above process is described as Fig. 4.

Figure 4 shows that the snaxel pi move the position p I′ that is the maximum gradient position on the line from pi to C.

4   Simulation Experiment

This section gives a group of experiments for evaluating BGSA. The test environment is Intel(R) Core(TM) i5-4300U CPU and 4 GB memory. The simulation software is JAVA. BGSA and FGSA are compared in this section. The parameters is set as: $\propto = 2.0$, $\beta = 2.10$ and $\gamma = 1.0$. The scale of snaxels is $N = 500$ and the iteration number is 150. The testing results are shown is Figs. 4, 5 and 6.

The symbol "⊕" marks a part of snaxels on the snakes. Figure 4(b) is an image with one concave region, Fig. 5(b) is an image with complex concave regions and Fig. 6(b) is to extract for head boundaries. Among three tests, BGSA outperforms FGSA. Especially, the BGSA can converge to the concave region well.

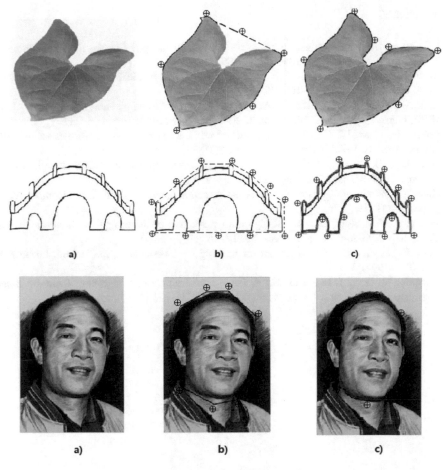

Fig. 5. Results of simulation experiment

5 Conclusion

This paper proposes a new greedy snake algorithm for objective contour extraction. BGSA have new energy function and can deal with concave region. A group experiments are design to test the performance of BGSA. The test results show that BGSA can be applied to extracting the contours that have complex concaves.

Acknowledgement. This work is supported by the National Natural Science Foundation of China (Grant No. 71301078), the Natural Science Foundation of Education Bureau of Jiangsu Province (Grant No. 16KJB520049) and the Natural Science Foundation of Huaian City (Grant No. HAB201709).

References

1. Lie, W.N.: Automatic target segmentation by locally adaptive image thresholding. IEEE Trans. Image Process. **4**(7), 1036–1041 (1995)
2. McInernery, T., Terzopolous, D.: Deformable models in medical image analysis: a survey. Med. Image Anal. **1**(2), 91–108 (1996)
3. Jain, A.K., Smith, S.P., Backer, E.: Segmentation of muscle cell pictures: a preliminary study. IEEE Trans. Pattern Anal. Mach. Intell. **2**(3), 232–242 (1980)
4. Fok, Y.L., Chan, J.C.K., Chin, R.T.: Automated analysis of nerve cell images using active contour models. IEEE Trans. Med. Image **15**(3), 353–368 (1996)
5. Ginneken, B., Frangi, A.F., Staal, J.J.: Active shape model segmentation with optimal features. IEEE Trans. Med. Image **21**(8), 924–933 (2002)
6. Kass, M., Witkin, A., Terzopoulos, D.: Snakes: active contour models. Int. J. Comput. Vis. **1**(4), 321–331 (1987)
7. Williams, D.J., Shah, M.: A fast algorithm for active contours and curvature estimation. Comput. Vis. Graph. Image Process. **55**, 14–26 (1992)
8. Lam, K.M., Yan, H.: Fast algorithm for locating head boundaries. J. Electron. Image **3**(4), 352–359 (1994)
9. Mustafa, S., Lam, K.M., Yan, H.: A faster converging snake algorithm to locate object boundaries. IEEE Trans. Image Process. **15**(5), 1182–1191 (2013)

Predicting Drug-Target on Heterogeneous Network with Co-rank

Yu Huang[1], Lida Zhu[2], Han Tan[3], Fang Tian[1], and Fang Zheng[1,2(✉)]

[1] College of Informatics, Huazhong Agricultural University,
Wuhan 430070, China
zhengfang@mail.hzau.edu.cn
[2] Hubei Key Laboratory of Agricultural Bioinformatics, College of Informatics,
Huazhong Agricultural University, Wuhan 430070, China
[3] Department of Mechanical Engineering,
Wuhan Vocational College of Software and Engineering, Wuhan 430070, China

Abstract. Heterogeneous network can bring more information compare with homogeneous network, so it has been extensively employed in many research field. In this research, we use Co-rank frame to predict the heterogeneous network of drug targets. First, we construct separate networks according the drug and target data information, and then merge these networks with the bipartite graph network. We have designed an Intra-network RWR and an Inter-network RWR to combine the heterogeneous network of drug and target. We compared our algorithm to the RWR and the ROC and Recall curves of the algorithm are all superior to the RWR.

Keywords: Heterogeneous network · Co-rank · Drug-target

1 Introduction

In the past decades, drug target prediction has attracted a lot of scholars' attention, and many classic algorithms and models have also been generated. For example, machine learning is also used for drug target prediction, based on the biological hypothesis of similar target genes associated with similar drugs, the known DTI is used as a label, and the chemical structure of the drug and the sequence of the target are used as a feature to train and predict. Bleakley [1] apply support vector machine framework to predict drug target interaction (DTI) based on bipartite local model (BLM). [2, 20] is able to expand the framework by combining the BLM (called BLMNII), which can learn from the neighbor's DTI features and predict the interaction of new drugs or target candidates.

However, the above prediction method is mainly based on homogeneous network information, these approaches ignore a rich variety of information on physical and functional relationships between biological macromolecules. Compared to the network which has only one type of node, heterogeneous networks can bring more information. Heterogeneous network models have been widely in many fields such as the web services [3–5], social networks [6, 7] and biological networks [8, 9]. A network-based approach is to fuse heterogeneous information through a network diffusion process, and

© Springer Nature Switzerland AG 2020
Q. Liu et al. (Eds.): CENet 2018, AISC 905, pp. 571–581, 2020.
https://doi.org/10.1007/978-3-030-14680-1_63

predict the scores of DTIs [21]. A meta-path based approach has also been proposed to extract the semantic features of DTIs from heterogeneous networks [22]. The Co-rank can utilize all kinds of objects in heterogeneous networks for ranking. For example, it has been used to predict authors and their publications and the documents and authors in a bibliographic network [10–12].

In this research, we use Co-rank frame to predict the heterogeneous network of drug targets. First, we construct separate networks according the drug and target data information, and then merge these networks with the bipartite graph network. The matrix of transition probability of the merged network is calculated according the transition probabilities from drug and target networks. Random walk with start algorithm runs on the separate networks and the merged network. We test our method on a benchmark data set with 606 drugs, our Co-rank algorithm is better than RWR (Random Walk with Restart) of ROC and Recall value.

2 Methods

2.1 The Drug-Target Bipartite Graph Heterogeneous Network Model

A bipartite graph heterogeneous network containing bipartite associations between them [13]. We take the Drug-Target as an example to describe bipartite heterogeneous network model as Fig. 1.

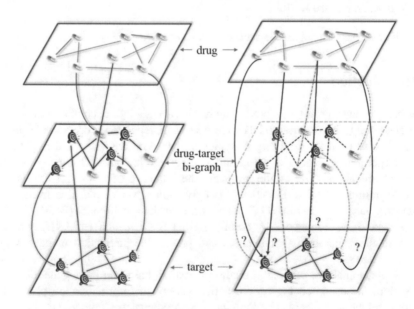

Fig. 1. Heterogeneous network model of drug-target.

This heterogeneous network was composed of a drug-drug similarity network, a target network, and a bipartite graph containing drug-target associations. The walk on the above three networks, four state transition probability matrices (P_D, P_T, $P_{DT(n \times m)}$ and $P_{TD(m \times n)}$) are establish for jump between the two networks.

Suppose the $A_{D(n \times n)}$, $A_{T(m \times m)}$, A_B are the adjacency matrix of D (drug-drug similarity network), T (target network) and B (bipartite graph network containing drug-target associations), then P_D, P_T, $P_{DT(n \times m)}$ and $P_{TD(m \times n)}$ can be defined as following formula (2.1), (2.2), (2.3), (2.4).

$$P_{D(i,j)} = P(d_j | d_i) \begin{cases} A_{D(i,j)} / \sum_{k=1}^{n} A_{D(i,k)} & \text{if } e(i,j) \in E \\ 0, & \text{otherwise} \end{cases} \tag{2.1}$$

$$P_{T(i,j)} = P(t_j | t_i) \begin{cases} A_{T(i,j)} / \sum_{k=1}^{m} A_{T(i,k)} & \text{if } e(i,j) \in E \\ 0, & \text{otherwise} \end{cases} \tag{2.2}$$

$$P_{DT(i,j)} = P(t_j | d_i) \begin{cases} A_{B(i,j)} / \sum_{j} A_{B(j,i)} & \text{if } \sum_{j} A_{B(j,i)} \neq 0 \\ 0, & \text{otherwise} \end{cases} \tag{2.3}$$

$$P_{TD(i,j)} = P(d_j | t_i) \begin{cases} A_{B(i,j)} / \sum_{j} A_{B(i,j)} & \text{if } \sum_{j} A_{B(i,j)} \neq 0 \\ 0, & \text{otherwise} \end{cases} \tag{2.4}$$

2.2 Co-rank on the Drug-Target Bipartite Graph Heterogeneous Network Based on RWR

Ranking is an important data mining task in network analysis [14], Random walk with restart (RWR) as a ranking algorithm, which has been used in biology in the past [15, 16].

RWR method has been proposed in many homogeneous networks, but ranking in heterogeneous information networks is also an important and meaningful task. We extend the RWR algorithm to operate on bipartite graph heterogeneous network of drug-target to predict a set of candidate targets. Co-rank framework [10] based on the bipartite graph and RWR. We take drug-target network as an example to describe the Co-rank framework.

The Co-rank consists of two types of random walk. If the random seed walk within G_D or G_T network, it is called an intra-subnet-work walk. For example, there is a link between node d_i and d_j, which are exactly from G_D. The random walker can move from d_i to its link neighbor d_j node with probability $\alpha(\alpha \in (0, 1))$ or move back to its home node with the restart probability $1 - \alpha$. The other is a random walk between networks based on the bipartite graph G_B network. If a random walker on the G_D and moves to a bridging node, it can jump to the other network G_T with a probability $\lambda(\lambda \in (0, 1))$ and takes $2k + 1$ inter-network steps or move back to the other nodes in its home subnet with the probability $1 - \lambda$ take m intra-network steps walk on G_D.

The transition probability matrix P for bipartite graph heterogeneous network of drug-target is as following formula (2.5)

$$P = \begin{bmatrix} (\alpha)P_T^m & (1-\alpha)P_{TD}(P_{DT}P_{TD})^k \\ (1-\alpha)P_{DT}(P_{TD}P_{DT})^k & (\alpha)P_D^n \end{bmatrix} \quad (2.5)$$

A initial vector y^0 is defined as the $y^0 = \begin{bmatrix} d_0 \\ t_0 \end{bmatrix}$, then the Co-rank can be rewritten as formula (2.6).

$$y^{t+1} = (1-\lambda)Py^t + \lambda y^0 \quad (2.6)$$

3   Experiments and Results

In this paper, we used drug and target dataset to test our approach. And we compared our algorithm with RWR (Random Walk Restart) to evaluate accuracy and calculation performance.

3.1   Datasets

Drug dataset gives the structure similarity between the 606 drugs. Drug-target bipartite graph dataset contains drug and its corresponding target. Drug dataset and drug-target bipartite graph dataset were obtained from Drug Bank, DGIDB and TTD databases, which were descripted in document [17]. We used PPI (Protein-Protein Interaction) as target-target interaction dataset, PPI dataset was obtained from the Human Protein Reference Database [18] which included 39240 interactive information between 9673 proteins. So n of adjacency matrix $A_{D(n \times n)}$ is 606, m of adjacency matrix $A_{T(m \times m)}$ is 9673, Table 1 shows the sizes of these networks.

Table 1. Sizes of the networks.

Network	Number of nodes	Number of edges
Drug	606	367236
Target	9673	39240
Drug-target	5759 – 2813	363102

3.2   Test of Parameters on the Co-rank

In this paper, we used drug and target dataset to test our approach. And we compared our algorithm with RWR (Random Walk Restart) to evaluate accuracy and calculation performance. In this research, we use Leave-One-Out Cross-Validation (LOOCV) method in document [19] to evaluate the accuracy of the algorithm. Co-rank framework parameters will be estimated according to the experimental results, and a better combination of parameters will be selected. Finally, the co-rank algorithm and RWR method are compared, and case analysis is carried out to show the effectiveness of the algorithm.

In this method, there are five network parameters, m, n, k, λ and α. Among them, m, n is the walk steps in the respectively network of drug and target. λ is the jumping probability of Inter-net. α is called restart probability of Intra-net. A random walker can jump to the other network with a probability α or move back to the other nodes in its home sub-network with the probability $1 - \alpha$. k is the jump steps between the networks.

We tested the five parameters respectively, showing the influence of the parameters from two aspects, one is the average rank of all the predicted results, another one is the number of the top 1% and top 5% target in all 2960 drug-target relationships.

1. m, n (walk steps in the intranet of drug and target network)

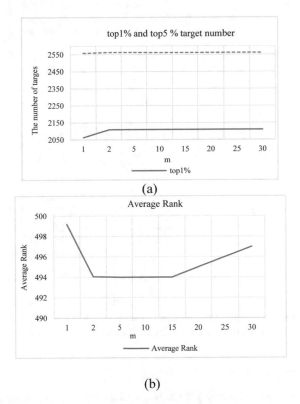

(a)

(b)

Fig. 2. (a) Test of average rank with variation of m ($n = 20$, $k = 1$, $\lambda = 0.5$, $\alpha = 0.5$) (b) Test of top 1% and top 5% target number with variation of m ($n = 20$, $k = 1$, $\lambda = 0.5$, $\alpha = 0.5$)

We test of parameter m, and the results are shown in Fig. 2. The value of m has little effect on the average rank, and the difference between the maximum rank and the minimum rank is not significant. According to the test results, when $m = 5$, top 1% and top 5% reached the maximum and average rank on the highest, so we chose $m = 5$ to test other parameters. We also tested the effect of parameter n on network stability, and the test results are shown in Fig. 3. The n has a great impact on the average rank and top 1% and top 5% number. From Figs. 3(a) and (b), we can see that when the value of n is 6, the network performance is the best.

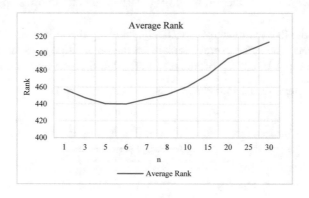

(a)

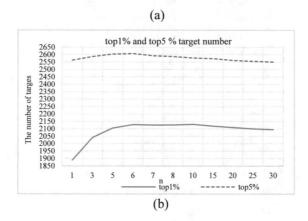

(b)

Fig. 3. (a) Test of average rank with variation of n ($m = 5$, $k = 1$, $\lambda = 0.5$, $\alpha = 0.5$) (b) Test of top 1% and top 5% target number with variation of n ($m = 5$, $k = 1$, $\lambda = 0.5$, $\alpha = 0.5$)

2. λ (the jumping probability of Inter-net)

In this test, m takes 5, n takes 6, k takes 1, and α takes 0.5. From the Fig. 4, it can be seen that the value of λ has a great influence on the experimental results. From the Fig. 4, it is found that the average rank is significantly improved with the increase of λ, and the number of top 1% and top 5% targets is also greatly improved. When λ fluctuates from 0.5 to 0.9, the curse is stable, so we set λ with 0.5.

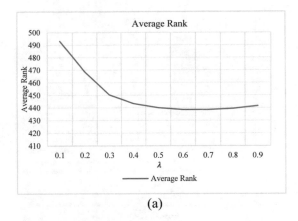

(a)

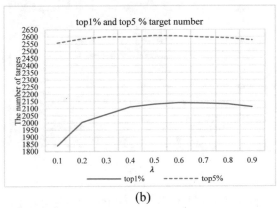

(b)

Fig. 4. (a) Test of average rank with variation of λ ($m = 5$, n $= 6$, $k = 1$, $\alpha = 0.5$) (b) Test of top 1% and top 5% target number with variation of λ ($m = 5$, $n = 6$, $k = 1$, $\alpha = 0.5$).

3. α (restart probability of Intra-net)

From the Fig. 5, it can be seen that the value of α also has a great influence on the experimental results. When $\alpha = 1$, this is the classical RWR algorithm, which does not take into account the information of heterogeneous networks. When the value of α is between 0.1 and 0.3, the algorithm is relatively stable, so α is selected as 0.3.

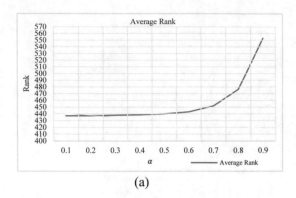

(a)

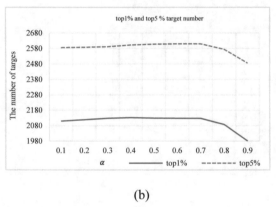

(b)

Fig. 5. (a) Test of average rank with variation of α ($m = 5$, $n = 6$, $k = 1$, $\alpha = 0.5$) (b) Test of top1% and top5% target number with variation of α ($m = 5$, $n = 6$, $k = 1$, $\lambda = 0.5$).

3.3   Results

Through the experiments above, we can see that the selection of parameters has a great influence on the effect of the algorithm. Considering the complexity of the algorithm, we set the parameter values as followings: $m = 5$, $n = 6$, $k = 1$, $\lambda = 0.5$, $\alpha = 0.6$. We can get positive and negative data by setting the threshold of LOOCV, which produced the following two kinds of results we wanted to use:

(1) The true positive data which predicted by the algorithm are also positive. We call this result is True Positive (TP).
(2) The real negative data which predicted by the algorithm are also positive. We call this data False Positive (FP).

The number of true negative data represented by N, and M is the number of positive data. Therefore, FPR = FP/N and TPR = TP/M, and we draw the ROC and Recall curves based on TPR and FPR. By observing the ROC diagram shown in Fig. 6, we find that with the increase of threshold, the prediction accuracy of Co-rank algorithm

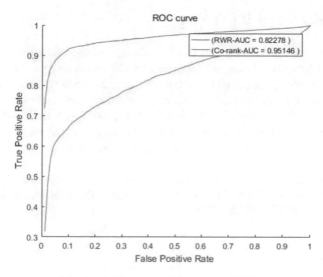

Fig. 6. ROC curve of Co-rank and RWR

has been greatly improved compared with the RWR algorithm. The result showed that the Co-rank algorithm did better at mining the biological rules hidden in the network relationship.

From Fig. 7 of the recall curve, we find that the result of the Co-rank algorithm is better than RWR algorithm. The Co-rank algorithm obtain 0.95146 accuracy, and the RWR algorithm is only 0.82278. And the recall of the Co-rank algorithm is also better than RWR. The results show that the Co-rank algorithm was more effective in predicting.

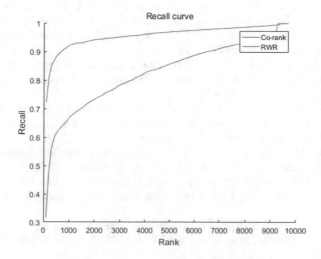

Fig. 7. Recall curve of Co-rank and RWR

4   Conclusions and Discussion

In this paper, we have proposed a Co-rank algorithm based on bipartite graph heterogeneous network for drug-target prediction. We have designed an Intra-network RWR and an Inter-network RWR to combine the heterogeneous network of drug and target. We provide a way to integrate heterogeneous network node information into a coherent network. And we use CUDA parallel method to speed up the algorithm which makes the Co-rank algorithm could use in massive datasets. The results show that our algorithm was effective in predicting drug-target interactions.

Acknowledgement. This study is supported by Natural Science Foundation of Hubei Province of China (Program No. 2015CFB524), the Fundamental Research Funds for the Central Universities (Program No. 2015BQ023).

References

1. Bleakley, K., Yamanishi, Y.: Supervised prediction of drug–target interactions using bipartite local models. Bioinformatics **25**, 2397–2403 (2009)
2. Mei, J.P., Kwoh, C.K., Yang, P., Li, X.L., Zheng, J.: Drug-target interaction prediction by learning from local information and neighbors. Bioinformatics **29**, 238–245 (2013)
3. Schattkowsky, T., Loeser, C.: Peer-to-peer technology for interconnecting web services in heterogeneous networks. In: International Conference on Advanced Information Networking and Applications, p. 611. IEEE Computer Society (2004)
4. Skjervold, E., Hafsøe, T., Johnsen, F.T., Lund, K: Delay and disruption tolerant Web services for heterogeneous networks. In: Military Communications Conference, Milcom, pp. 1–8 (2009)
5. Zhu, D., Zhang, Y., Chen, J., Cheng, B.: Enhancing ESB based execution platform to support flexible communication web services over heterogeneous networks. In: IEEE International Conference on Communications, pp. 1–6. IEEE Xplore (2010)
6. Dong, Y., Tang, J., Wu, S., Tian, J.L., Chawla, N.V., Rao, J.H., Cao, H.H.: Link prediction and recommendation across heterogeneous social networks. In: International Conference on Data Mining, pp. 181–190. IEEE (2013)
7. Huang, J., Nie, F., Huang, H., Tu, Y.C.: Trust prediction via aggregating heterogeneous social networks. In: ACM International Conference on Information and Knowledge Management, pp. 1774–1778. ACM (2012)
8. Blin, G., Fertin, G., Mohamed-Babou, H., Rusu, I., Sikora, F., Vialette, S.: Algorithmic aspects of heterogeneous biological networks comparison. In: International Conference on Combinatorial Optimization and Applications, pp. 272–286 (2011)
9. Li, J., Zhao, P.X.: Mining functional modules in heterogeneous biological networks using multiplex pagerank approach. Front. Plant Sci. **7**, 903 (2016)
10. Zhou, D., Orshanskiy, S.A., Zha, H., Giles, C.L.: Co-ranking authors and documents in a heterogeneous network. In: IEEE International Conference on Data Mining, pp. 739–744. IEEE (2007)
11. Soulier, L., Jabeur, L.B., Tamine, L., Bahsoun, W.: On ranking relevant entities in heterogeneous networks using a language-based model. J. Assoc. Inf. Sci. Technol. **64**(3), 500–515 (2013)

12. Ng, K.P., Li, X., Ye, Y.: MultiRank: co-ranking for objects and relations in multi-relational data. In: ACM SIGKDD International Conference on Knowledge Discovery and Data Mining, Diego, CA, USA, pp. 1217–1225 (2011)
13. Sarma, A.D., Molla, A.R., Pandurangan, G., Upfal, E.: Fast distributed pagerank computation. In: Distributed Computing and Networking, pp. 113–121. Springer, Heidelberg (2013)
14. Shi, C., Li, Y., Zhang, J., Sun, Y., Yu, P.S.: A survey of heterogeneous information network analysis. IEEE Trans. Knowl. Data Eng. **29**(1), 17–37 (2017)
15. Blatti, C., Sinha, S.: Characterizing gene sets using discriminative random walks with restart on heterogeneous biological networks. Bioinformatics **32**(14), 2167–2175 (2016)
16. Li, J.R., Chen, L., Wang, S.P., Zhang, Y., Kong, X., Huang, T., Cai, Y.D.: A computational method using the random walk with restart algorithm for identifying novel epigenetic factors. Mol. Genet. Genomics MGG **293**(1), 1–9 (2017)
17. Quan, Y., Liu, M.Y., Liu, Y.M., Zhu, L.D., Wu, Y.S., Luo, Z.H., Zhang, X.Z., Xu, S.Z., Yang, Q.Y., Zhang, H.Y.: Facilitating anti-cancer combinatorial drug discovery by targeting epistatic disease genes. Molecules **23**(4), 736 (2018)
18. Human protein reference database. http://www.hprd.org/
19. Liu, X.Y.: Heterogeneous Network Model Based Method for Disease Gene Prediction. XiDian University, Xi'an (2013)
20. Buza, K., Peška, L.: Drug-target interaction prediction with Bipartite Local Models and hubness-aware regression. Neurocomputing **260**, 284–293 (2017)
21. Wang, W., Yang, S., Zhang, X., Li, J.: Drug repositioning by integrating target information through a heterogeneous network model. Bioinformatics **30**(20), 2923–2930 (2016)
22. Fu, G., Ding, Y., Seal, A., Chen, B., Sun, Y.Z., Bolton, E.: Predicting drug target interactions using meta-path-based semantic network analysis. BMC Bioinformatics **17**(1), 160 (2016)

A Collaborative Filtering Algorithm for Hierarchical Filtering of Neighbors Based on Users' Interest

Shuwei Lei[✉], Jiang Wu, and Jin Liu

College of Information Science and Technology,
Northwest University, Xi'an 710127, China
lsw2012117060@163.com

Abstract. In order to alleviate the sparsity of rating data and single rating similarity, from the perspective of optimizing similarity, this paper proposes a collaborative filtering algorithm for filtering neighbors hierarchically based on user interest (HFNCF). Firstly, this paper adopts the user's rating number and rating size of each category to calculate user interest, and then improved similarity calculation, and join the interest coincidence degree factor to find similar users with the same interest of the target user, then find similar users with the target user having the largest interest cross-domain, and obtain the neighbor candidate set. Secondly, this paper considers the relative time changes between user ratings and adding time factors to improve the similarity of ratings to select the final similar neighbors. Finally, the comprehensive similarity is obtained by combining the two similarities, so that the target user's ratings are predicted based on the rating of the final similar neighbors, and generate recommendations by the rating. Experimental proof, the algorithm proposed in this research achieves a lower MAE on MovieLens dataset compared to the traditional algorithm. Therefore, the HFNCF algorithm effectively selects neighbors, and improves the recommendation quality and accuracy.

Keywords: HFNCF · User interest · User interest coincidence degree · Time factor

1 Introduction

Personalized recommendation is an effective way to relieve "information overload" [1]. Collaborative filtering (CF) algorithm is the best known recommendation algorithm in personalized recommendation [2]. However, due to large-scale increase of data, the sparsity of rating data is becoming more and more serious. Simply using the user's single rating information does not essentially solve the above problem.

To solve sparsity of rating data and single rating similarity, many studies incorporate item category, calculate the user's interest similarity by the user's subjective rating reflecting the user's interest in different categories of items, and based on this improvement. Some methods converted the item rating matrix into an item attribute rating matrix based on user preferences, and then calculate similarity on the new rating

© Springer Nature Switzerland AG 2020
Q. Liu et al. (Eds.): CENet 2018, AISC 905, pp. 582–591, 2020.
https://doi.org/10.1007/978-3-030-14680-1_64

matrix [3]. Others considered that the more users rate the category of item, the more interested users are in this category of item [4, 5].

In fact, user interest changes irregularly over time. In order to alleviate the problem of user interest changes, one approach adopted the linear function or forgetting curve to add the interest weight to the new similarity method and update the user's interest [6, 7]. The other approach assigned each fractional weight function to increase the user's interest in recent, long-term, and periodical periods [8].

This paper proposes a collaborative filtering optimization algorithm for filtering neighbors hierarchically based on user interest. Firstly, obtaining user interest from the user's rating number of each category item and the rating size, and get the interest of coincidence between users, improve the similarity calculation of interest, then find similar users as candidate set. Secondly, regardless of the change of individual user's interest over time, starting with the relative interest changes between users, integrate time factor to improve the traditional person similarity. The algorithm proposed in this paper optimizes the similarity calculation, and improves the quality of the recommendation algorithm.

2   Related Work

CF recommendation algorithms memory-based can be divided into user-based (UBCF) and item-based (IBCF) [9]. The research of this paper is based on the former. Explain the main mathematical symbols in UBCF. Where m and n represent the total number of users and items. U and I represent the set of total users and items. $R(m, n)$ is the user-item rating matrix. $R_{u,i}$ represents the rating of $user_u$ on $item_i$. The average of all $R_{u,i}$ is $\bar{R}_u$. R_{uv} represents the $user_u$ and $user_v$ common rating item set.

At present, mining user interest to optimize similarity calculation is an effective means to alleviate sparseness. According to the analysis of the literature, the existing research still has the following two problems in mining user interest. Firstly, when analyzing user interest through the category information of the item, it is not practical to consider a single user's rating size or the user's rating of each category of item. Secondly, user interest changes over time, most of the current algorithm principles for time improvement are closer to the current time and more consistent with user interest. But measure user interest using time context information is very difficult, such as year, quarter, month, day, and even every morning, midday, evening have a certain influence on user interest. Therefore, it is difficult for individual user's interest to change with time, is necessary to take into account changes in the relative interests between users.

3   Improved Collaborative Filtering Algorithm

3.1   Calculate Interest Similarity with Integrate Interest of Coincidence

The common rating item category includes both rating items that are common among users and their respective ratings, which reflect the largest crossover area of user interest. The degree of user attention to a certain category of items can be reflected in two aspects. First, the number of ratings for such category item, and the second is the rating size of it.

Therefore, this paper optimizes the similarity computation based on item category, and selects the candidate nearest to the target user. The principle of calculation is shown in Fig. 1.

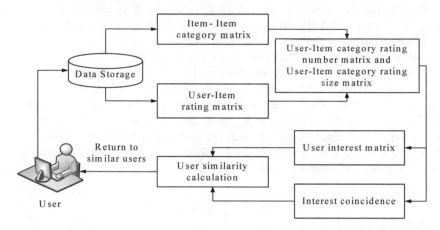

Fig. 1. Interest similarity calculation.

Definition1: User-Item Category Rating Number Matrix. It describes the number of user's ratings for each category of item. For $\forall item_j \in I$, C represents items category set, $item_j$'s category vector is $C_j = \{category_1^j, category_2^j, \ldots, category_t^j\}, t = 1, 2, \ldots,$ obtain the category vector for each item as formula (3.1), and then compose item-category matrix.

$$category_k^j = \begin{cases} 1, & item_j \in category_k \\ 0, & item_j \notin category_k \end{cases}, k = 1, 2, \ldots, t \tag{3.1}$$

Calculate user-item category rating number matrix $N(m, t)$ by user-item rating matrix and item-category matrix, the meaning of the matrix element is the number of ratings each user has for each category of item. Where m and t represent the total number of users and item categories, the element $N_{i,j}$ represents the rating number of $user_i$ on $category_j$.

Then, obtain matrix $H(m, t)$ by dividing the row vector of the $N(m, t)$ by the total number of ratings for each user. $H_{m,t}$ represents the rating number of $user_m$ on $category_t$ proportion of the total number of $user_m$, as in (3.2), where N_m refers to the total number of $user_m$ rating on all items. The greater the number, the higher the interest.

$$H_{m,t} = \frac{N_{m,t}}{N_m} \tag{3.2}$$

Definition 2: User-Item Category Rating Size Matrix. It describes the user's rating size for each category of item. Calculate user-item category rating size matrix $S(m, t)$

by user-item rating matrix and item-category matrix, the meaning of the matrix element is the average rating each user has for each category of item, it is calculated from the user's total ratings for each category of item divided by the number of ratings. The higher the average rating, the more interested users are for such items. The element $S_{i,j}$ refers average rating of $user_i$ rating on $category_j$.

Definition 3: User Interest Matrix. The user's interest in a certain category of the item is not only reflected in the number of ratings for such items, but also reflected in the rating size of such items. Therefore, considering two reflections on the interests of users, get the user's interest matrix $Ins(m, t)$ in the item category, is shown in Table 1.

Table 1. User interest matrix.

	$category_1$	$\cdots$	$category_j$	$\cdots$	$category_t$
$user_1$	$Ins_{1,1}$	$\cdots$	$Ins_{1,j}$	$\cdots$	$Ins_{1,t}$
$\cdots$	$\cdots$	$\cdots$	$\cdots$	$\cdots$	$\cdots$
$user_i$	$Ins_{i,1}$	$\cdots$	$Ins_{i,j}$	$\cdots$	$Ins_{i,t}$
$\cdots$	$\cdots$	$\cdots$	$\cdots$	$\cdots$	$\cdots$
$user_m$	$Ins_{m,1}$	$\cdots$	$Ins_{m,j}$	$\cdots$	$Ins_{m,t}$

The row vector of the matrix represents the user's vector of interest for each category of item. The meaning of the matrix element $Ins_{m,t}$ is multiplied by the degree of interest on the number of ratings and rating size, as in (3.3).

$$Ins_{m,t} = H_{m,t} \times S_{m,t} \tag{3.3}$$

Definition 4: User Interest Coincidence Degree. When calculating the similarity between users, the user's interest coincidence degree (Icd) is introduced to find similar users with the largest cross-domain. The principle is that the more common rating category items are, the less non-common rating category items are, and the more similar the user is.

The Jaccard similarity coefficient is used to define the degree of user interest coincidence degree [10]. Calculate the user Icd using formula (3.4), where $C_u \cap C_v$ represents the common rating item category between $user_u$ and $user_v$, and $C_u \cup C_v$ refers all rating item categories between $user_u$ and $user_v$.

$$Icd = \frac{|C_u \cap C_v|}{|C_u \cup C_v|} \tag{3.4}$$

Definition 5: Improved User Interest Similarity. Each user's interest degree in each category of item is considered as a vector of the t-dimensional item category space, this vector is the row vector in the user interest matrix, t is the total number of item categories, and the similarity of interest between the two users can be seen as the cosine of the angle between the two user interest vectors. The larger the cosine, the higher the similarity, and then integrate the user's *Icd* to optimize it, and calculate the improved user interest similarity using formula (3.5).

$$sinIns(u,v) = \frac{Icd \cdot \sum_{k=1}^{t} Ins_{u,k} Ins_{v,k}}{\sqrt{\sum_{k=1}^{t} Ins_{u,k}^2} \sqrt{\sum_{k=1}^{t} Ins_{v,k}^2}} \tag{3.5}$$

3.2   Person Similarity Calculation of Integrate Time Factor

For the problem of user interest changes, relative user interest changes do not need to consider complex time context information, the calculation weight given to the user's similarity is based only on the time difference of the user's rating, improving the user's similarity calculation and indirectly alleviating the problem of changes in the user's interest over time.

Definition 6: Time Factor. This paper considers changes in users' relative interests, and adds time factors (*tf*) to similarity calculation of UBCF algorithms. The principle is that if $user_u$ acts on the $item_i$ time $t_{u,i}$ is farther away from the $user_v$ acts on the $item_i$ time $t_{v,i}$, the smaller the time factor, the smaller the similarity weight, as in (3.6), and rating time is in days.

$$tf = \frac{1}{1 + |t_{u,i} - t_{v,i}|} \tag{3.6}$$

Definition 7: Improved Rating Similarity. When calculating the person similarity according to the rating information, integrate time factor reflects the change of user interest over time, and calculate the improved rating similarity using formula (3.7).

$$simRating(u,v) = \frac{\sum_{i \in R_{uv}} (R_{u,i} - \bar{R}_u) \times (R_{v,i} - \bar{R}_v) \times \frac{1}{1 + |t_{u,i} - t_{v,i}|}}{\sqrt{\sum_{i \in R_{uv}} (R_{u,i} - \bar{R}_u)^2} \times \sqrt{\sum_{i \in R_{uv}} (R_{v,i} - \bar{R}_v)^2}} \tag{3.7}$$

3.3   Rating Prediction Based on Synthesize Similarity

According to the formula (3.5) and (3.7) are calculated respectively with improved interest and rating similarity, and hierarchical select the target users' nearest neighbors, then linear weighted this two similarity. Thus, the comprehensive similarity of users as in formula (3.8).

$$compsim(u,v) = \beta * simIns(u,v) + (1 - \beta) * simRating(u,v) \tag{3.8}$$

The value of the parameter β is between 0 and 1, which is to adjust the weight between the improved interest similarity and the rating similarity. According to the similarity calculation formula obtained a new prediction rating formula, as shown in (3.9).

$$P(u, i) = \bar{R}_u + \frac{\sum_{v \in U_N} compsim(u, v) \times (R_{v,i} - \bar{R}_v)}{\sum_{v \in U_N} compsim(u, v)} \qquad (3.9)$$

Among them, the set of U_N is hierarchical select the final neighbor of the target user by two similarity calculations.

3.4   Hierarchical Filtering Neighbor Collaborative Filtering Algorithm

The degree of user's interest in each category of item reflects the user's interest partition and depth. The user interest vector is constructed by the user's degree of preference for each category, and the user having the greatest interest intersection area is divided into the target user's candidate nearest neighbor set. The user's rating for item reflects the intensity of the user's interest in the item, this intensity changes over time, and corrects the similarity using the time factor, so that the similar candidate user is double-selected. Therefore, this paper proposes a CF algorithm to filter neighbors hierarchically based on mining user interests.

The algorithm has four phases. (1) Data description; (2) Integrate interest coincidence to calculation interest similarity; (3) Integrate time factor to calculate rating similarity; (4) Rating prediction, generate recommendations. The detailed description is shown in Algorithm 1.

Algorithm 1.   Hierarchical filtering neighbor collaborative filtering algorithm

---

Input: Rating matrix $R(m.n)$, item-category matrix, rating time information, the target user u, the number of neighbors K, the equilibrium factor β, recommended list size N.
Output: The set of top-N recommendations of the target user u .

Step 1: For $\forall u \in U, \forall category \in C$, counting the number of ratings for each *category* by u to establish $N(m, t)$, use formula (3.2) to calculate $H_{m,t}$, and establish $H(m, t)$
Step 2: For $\forall u \in U, \forall category \in C$, calculate the average rating for each category of item by u to establish $S(m, t)$
Step 3: For $\forall u \in U, \forall category \in C$, calculate $Ins_{m,t}$ using formula (3.3), establish $Ins(m, t)$
Step 4: For $\forall u \in U, \forall v \in V, u \neq v$, calculate Icd using formula (3.4)
Step 5: For $\forall u \in U, \forall v \in V, u \neq v$, integrate the user's Icd calculate improved interest similarity $simIns(u, v)$ using formula (3.5)
Step 6: In descending order of $simIns(u, v)$, select the candidate neighbors set U_C of the former $2K$ as the target user u
Step 7: For $\forall u \in U, \forall v \in V, u \neq v, \forall i \in R_{uv}$, calculate tf using formula (3.6), and calculate $simRating(u, v)$ using formula (3.7)
Step 8: Sorts users in candidate set U_C in descending order of $simRating(u, v)$, select the eventually similar neighbors U_N of the former K
Step 9: Select the value of β, calculate $compsim(u, v)$ using formula (3.8)
Step 10: Calculate $P(u, i)$ using formula (3.9), select the value of N, recommend the top-N highly rated item set to target user u

---

Assuming that there are m users and n items, and need to find K similar neighbors. In terms of time complexity, the execution time overhead of the algorithm is mainly in formula (3.5), (3.7) and (3.9). We analyze the time complexity of the algorithm through the following detailed steps.

When Step 1 and Step 2 calculate $N(m, t)$ and $S(m, t)$ by the user rating matrix and item-category matrix, the time complexity is $O(l) < O(m \times n)$. Step 5 and Step 7 respectively calculate the improved interest similarity and rating similarity between the m users, and the time complexity is $O(m^2)$. STEP 6 and STEP 8 sort the m users according to the similarity size respectively, and find the nearest neighbors, and the time complexity is $O(m\log m)$. Step 10 predicts the target user's ratings through K nearest neighbors to generate a recommended result set which time complexity is $O(K \times m)$. Since K is a constant much smaller than m, the time complexity is $O(m)$. In summary, the total time complexity of the algorithm is $O(m^2)$.

4   Experimental Results and Analysis

4.1   Data Set and Evaluation Criterion

The experiment selected the MovieLens 100k dataset for testing [11]. The dataset includes 100,000 ratings for 1,682 movies rated by 943 users. The 1–5 rating reflects the user's preference for movies. Each user rated at least 20 movies, and all movies are divided into 19 categories. The experiment used Mean Absolute Error (MAE) to evaluate the accuracy of the recommended results. MAE make summation over the difference between the predicted rating $P(u,i)$ and the actual rating $R(u,i)$. The less the MAE, the higher the accuracy of the recommendation. Then the calculation of MAE is shown in formula (4.1).

$$MAE = \frac{\sum_{i=1}^{N} |P(u, i) - R(u, i)|}{N} \tag{4.1}$$

4.2   Experimental Results

In the experiment, the data set is divided into two parts at random, 80% as training set and 20% as test set. The Cross-validation was used to test the recommended results of the algorithm.

According to formula (3.9), the parameters affecting the prediction rating $P(u, i)$ are the number of similar neighbors K in the similar nearest neighbor set U_N, and the parameter β in the formula (3.8) of the comprehensive similarity $compsim(u, v)$.

Experiment One: The purpose of this experiment is to verify that the MAE value is affected by the change of the parameter β in formula (3.8), so as to determine the optimal β. The value of the weighting coefficient β is between 0 and 1, which is to adjust the weight between the improved interest similarity and the rating similarity. According to formula (3.8), it can be seen when $\beta = 1$, the interest similarity is improved independently based on the UBCF algorithm (INSBCF). When $\beta = 0$, the algorithm integrates the time factor to improve the rating similarity (TFBCF). Both INSBCF and TFBCF are used for contrast experiments.

In this experiment, in order to avoid the contingency of the number of similar neighbors K on the optimal β value, four sets of K values as 10,20,30,40 were set respectively, and observed the change of MAE with β at each K value. In the range of $0 < \beta < 1$, and β is increased by 0.1 in each time, during which the affect of the value of β on the MAE is observed, and the results are shown in the Fig. 2. It can be observed from the Fig. 2 that when the value of β is around 0.4, the test data show the lowest MAE. Therefore, β in this paper takes a value of 0.4.

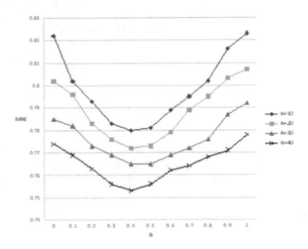

Fig. 2. The effect of β in MAE.

Experiment Two: In order to verify the effectiveness of HFNCF recommendation algorithm proposed in this paper. Comparing with other related algorithms using the same criterion and same parameters. These algorithms include the traditional UBCF and IBCF algorithms, INSBCF and TFBCF mentioned in Experiment one, non hierarchical hybrid filtering algorithm (NHHCF), as well as existing algorithm UICICF that mentioned forward [5]. Referring to the K value setting of UBCF, the number of the similar neighbors K changes from 5 to 50 by 5 each time, as well as the value of β is set as 0.4 by experiment one, and observing changes in MAE as K changes. The experimental results are shown in Fig. 3.

It can be seen that the algorithm MAE in this paper is smaller than other related algorithms. The HFNCF in this paper is better than the traditional UBCF and IBCF, as well as exceeds to the improved single similarity calculation method INSBCF and TFBCF. Which is due to the HFNCF algorithm not only considers the rating, but also considers users' interest on each category of the item. Besides, by comparing HFNCF and UICICF, it shows that, while K is greater than 20, both have a lower MAE. When

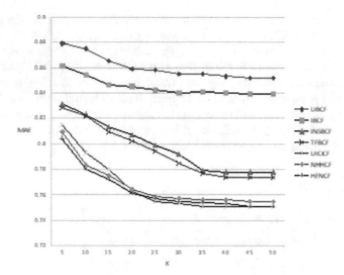

Fig. 3. Contrastive experiment.

K is lesser than 20, the MAE of the former is less than the latter. Finally, comparing with NHHCF, the MAE of HFNCF is lower, which shows that HFNCF algorithm can be better select similar neighbors. In contrast, the algorithm improves the accuracy of recommendation.

5   Conclusion

In order to alleviate the sparsity of rating data and single rating similarity, this paper proposes a collaborative filtering based on analysis user interest. Adopt the user's rating number of each category and the rating size to reflect the interest of the user, and consider the degree of coincidence between users when calculating user similarity. In this way, we can find similar users with the user having the largest interest cross-domain. The algorithm also takes account of the effect of time on the similarity of rating, and considers the relative time difference of the score rating when calculating similarity, so as to adapt to the change of users' interests. This paper combines interest similarity and rating similarity to make double choices for similar users, after analysis and verification, the algorithm can better mine user interest, accurately determine the nearest neighbors, and improve the accuracy of recommendation. However, the cold start problem has not been considered and needs further improvement.

References

1. Adomavicius, G., Sankaranarayanan, R., Sen, S., Tuzhilin, A.: Incorporating contextual information in recommender systems using a multidimensional approach. ACM Trans. Inf. Syst. **23**(1), 103–145 (2005)
2. Adomavicius, G., Tuzhilin, A.: Toward the next generation of recommender systems: a survey of the state-of-the-art and possible extensions. IEEE Trans. Knowl. Data Eng. **17**(6), 734–749 (2005)
3. Yao, P.P., Zou, D.S., Niu, B.J.: Collaborative filtering recommendation algorithm based on user preferences and project properties. Comput. Syst. Appl. (2015)
4. Liu, J., Wang, J.: An approach for recommending services based on user profile and interest similarity. In: IEEE International Conference on Software Engineering and Service Science, pp. 983–986. IEEE (2014)
5. Ye, F., Zhang, H.: A collaborative filtering recommendation based on users' interest and correlation of items. In: International Conference on Audio, Language and Image Processing, pp. 515–520. IEEE (2017)
6. Koychev, I.: Gradual forgetting for adaptation to concept drift. In: Proceedings of ECAI 2000 Workshop Current Issues in Spatio-Temporal Reasoning, pp. 101–106 (2004)
7. Wu, W., Wang, J., Liu, R., Gu, Z., Liu, Y.: A user interest recommendation based on collaborative filtering. In: International Conference on Artificial Intelligence and Industrial Engineering (2016)
8. Gasmi, I., Seridi-Bouchelaghem, H., Hocine, L., Abdelkarim, B.: Collaborative filtering recommendation based on dynamic changes of user interest. Intell. Decis. Technol. **9**(3), 271–281 (2015)
9. Khurana, P., Parveen, S.: Approaches of recommender system: a survey. Int. J. Emerg. Trends Technol. Comput. Sci. **34**(3), 134–138 (2016)
10. Niwattanakul, S., Singthongchai, J., Naenudorn, E., Wanapu, S.: Using of Jaccard coefficient for keywords similarity. Lecture Notes in Engineering & Computer Science, vol. 2202, no. 1 (2013)
11. Kuzelewska, U.: Clustering algorithms in hybrid recommender system on movielens data. Stud. Logic Gramm. Rhetor. **37**(1), 125–139 (2014)

An Improved Algorithm for Dense Object Detection Based on YOLO

Jiyang Ruan and Zhili Wang[✉]

Beijing University of Posts and Telecommunications, Beijing 100876, China
zlwang@bupt.edu.cn

Abstract. The YOLO v3 (you only look once) algorithm based on CNN (convolutional neural network) is currently the state-of-the-art algorithm that achieves the best performance in real-time object detection. However, this algorithm still has the problem of large detection errors in dense object scenes. This paper analyses the reason for the large error, and proposes an improved algorithm by optimizing confidence adjustment strategy for overlapping boxes and using dynamic overlap threshold setting. Experiments show that the improved algorithm has better performance in dense scenes while has little difference in other scenarios compared to the original algorithm.

Keywords: Object detection · Confidence adjustment · Dynamic threshold

1 Introduction

Object detection is an important application of computer vision. The main task is to locate the objects that appear in an image and classify them. Compared to the detection of specific targets (such as face detection, pedestrian detection, and vehicle detection), there are many types of objects that need to be detected for general object detection. The distance between categories is large and the difficulty is increasing greatly, so that traditional sliding window and classifier's detection process is difficult to accomplish this task. In recent years, deep learning has continuously made breakthroughs in image recognition and has attracted the attention of researchers around the world.

The existing object detection algorithm based on convolutional neural network has a complex network structure, and it needs to overcome the shortcomings of slow training speed and large model parameters for practical applications. At the same time, these current state-of-the-art algorithms still have some errors in the detection of dense object scenes. Researchers are required to conduct in-depth research on these problem areas and design a more adaptable object detection network.

2 Related Work

In recent years, deep learning has made a breakthrough in the field of object detection by virtue of the advantages of convolution neural network in extracting high-level features of images and based on the achievements of image classification task. In 2014, Ross Girshick et al. proposed the R-CNN [1] network at the CVPR conference.

Q. Liu et al. (Eds.): CENet 2018, AISC 905, pp. 592–599, 2020.
https://doi.org/10.1007/978-3-030-14680-1_65

On the VOC 2012 dataset, the mean Average Precision (mAP) of object detection was increased from 30% to 53.3%. In 2015, Ross Girshick and Shaoqing Ren successively proposed Fast R-CNN [2] and Faster R-CNN [3], which improved the detection speed while improving the accuracy. The processing speed can reach 5 FPS (frames per second). In 2016, the YOLO algorithm proposed by Joseph Redmon et al. achieved a true video detection rate of 45 FPS. Although YOLO has improved detection speed and sacrificed accuracy, it has come up with a new idea that can integrate classification and positioning. In the same year, on the basis of the YOLO, Wei Liu and Joseph Redmon successively released SSD and YOLO v2 [4] to improve the detection accuracy and detection speed to a satisfactory level. YOLO v2 was slightly better at a speed of 67 FPS and mAP can reach 76.8%. The YOLO v3 put forward in 2018 has a new improvement in both speed and accuracy, but there is no major change in the overall thinking.

3 YOLO Network Structure

The YOLO network structure is a convolutional neural network structure. It consists of an input layer, a convolutional layer, a pooling layer, a fully connected layer, and an output layer (Fig. 1).

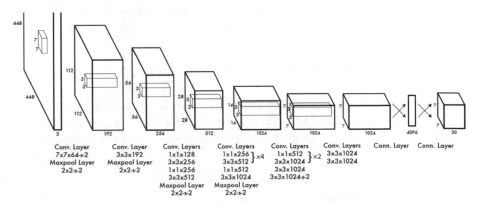

Fig. 1. The structure of the YOLO network [5].

The output layer is the last layer of the convolutional neural network. The role is to classify the input one-dimensional feature vectors, which is equivalent to a classifier. The number of output feature maps of the output layer is the number of classification of the target.

YOLO divides the input image into S × S grids. If the center of an object falls within a grid, the grid is responsible for detecting the object. Each network predicts B target frames, and each target frame corresponds to 5 prediction parameters, the coordinates (x, y) of the center point of the target frame, the width (w, h), and the confidence score s_i:

$$s_i = \Pr(O) * IOU \tag{3.1}$$

$\Pr(O)$ is the possibility of an object existing in the target frame of the current model. O represents the target object, and the IOU (Intersection-Over-Union) represents the accuracy of the target frame predicted by the current model. If there is no object in the target frame, $\Pr(O)$ is 0; if there is an object in the target frame, $\Pr(O)$ is 1, and then the IOU is calculated according to the predicted target frame p and the real target frame t:

$$IOU_p^t = \frac{box_p \cap box_t}{box_p \cup box_t} \tag{3.2}$$

box_t represents the real target border (ground truth), and box_p represents the predicted target frame. It can be seen that the IOU indicates the ratio of the intersection of the predicted target frame and the real target frame to their union (combination ratio). At the same time, it can predict the conditional probability $\Pr(C_i|O)$ that the object belongs to a certain class in the presence of an object. Assume that there is a total of K objects, then the conditional probability that each mesh predicts the i-th object C_i is $\Pr(C_i|O)$, i = 1, 2, ... , K; After calculating $\Pr(C_i|O)$, a certain target frame class-related confidence can be calculated during the detection:

$$\Pr(C_i|O) * \Pr(O) * IOU_p^t = \Pr(C_i) * IOU_p^t \tag{3.3}$$

In particular, the YOLO algorithm divides the input image into 7 × 7 grids (S = 7), each grid predicts 2 target frames (B = 2), and there are 20 types of objects to be detected (K = 20). Then YOLO finally predicts a S × S × (B × 5 + K) = 7 × 7 × 30 vector.

4 Improved Approach

Non-maximum suppression algorithm (NMS) is an important part of object detection process. Its main function is to select the appropriate detection box and suppress some redundant boxes. It first generates the detection box based on the object detection score. The detection box with the highest score, M, is selected, and other detection boxes with obvious overlap with the selected detection box are suppressed. This process is applied recursively to the rest of the detection boxes.

YOLO algorithm also uses the non-maximum suppression algorithm. This paper improves the traditional non-maximum suppression algorithm in terms of confidence adjustment and overlap threshold setting in the overlapping boxes, thereby improving the performance of YOLO algorithm in dense scenes.

4.1 Confidence Adjustment

The traditional non-maximum suppression algorithm first produces a series of detection box B and the corresponding score S in the detected picture. When the maximum score detection box M is selected, it is removed from the set B and put into the final detection result set D. At the same time, any detection box in set B that overlaps with the detection box M greater than the overlapping threshold Nt will also be removed. The biggest problem of this strategy is that it forces the scores of neighboring detection boxes to zero. In this case, if a real object appears in the overlapping area, it will fail to be detected and the average precision (AP) of the algorithm is reduced. This problem is particularly acute in the detection of dense scenarios.

The traditional NMS method can be expressed by the following fractional reset function:

$$s_i = \begin{cases} s_i, & iou(M, b_i) < N_t \\ 0, & iou(M, b_i) \geq N_t \end{cases} \tag{4.1}$$

The paper has improved the adjustment rules for the confidence of overlapping boxes. When the degree of overlap between the adjacent detection boxes and M exceeds the overlapping threshold Nt, the detection score of the detection box is attenuated to some extent according to the degree of overlap (IOU). In this case, the detection box close to M is greatly attenuated, and the detection box far from M is not greatly affected. This method effectively suppresses the redundant frame and reduces the missed detection rate. In this paper, the score of the adjacent detection box is only reduced based on a function related to the degree of overlap with M, instead of being eliminated completely. Although the score is reduced, the adjacent detection box is still in the object detection sequence.

The designed fractional reset function is shown in expression (4.2). When the degree of overlap between the adjacent detection box and M exceeds the overlap threshold Nt, the detection score of the detection box is linearly attenuated.

$$s_i = \begin{cases} s_i, & iou(M, b_i) < N_t \\ s_i(1 - iou(M, b_i)), & iou(M, b_i) \geq N_t \end{cases} \tag{4.2}$$

There may be a case where the function of the linear decay function is interrupted at the threshold point. This paper refers to another set of reset functions, the Gaussian attenuation function, designed in the Soft-NMS [6]:

$$s_i = s_i e^{-\frac{iou(M,b_i)^2}{N_t}}, \forall b_i \notin \mathcal{D} \tag{4.3}$$

In actual experiments, it was found that due to the design of dynamic thresholds, and this kind of threshold interruption rarely occurs and has little effect on the detection results, which hardly affects the detection accuracy.

4.2   Dynamic Threshold Design

In the traditional non-maximum suppression algorithm, the threshold of the overlapping boxes is a fixed value (usually 0.2–0.6), that is, for different types of pictures, the same overlap threshold is used to suppress the redundant frame. This will cause the following problems:

(1) The choice of the threshold will have a great influence on the detection result.
(2) If the threshold value is too large, some redundant boxes may not be efficiently removed.
(3) If the threshold value is relatively small, the actual object boxes may be deleted by mistake for some dense images, resulting in a larger detection error.

The threshold value is a hyper-parameter in the original algorithm, and it needs to be manually set in advance. This will lead to problems such as low adaptability for different types of pictures and low detection accuracy especially in dense scenes.

This paper designed that the thresholds in this algorithm are not manually set, but are set dynamically by the neural network according to the characteristics of the input picture. For more sparse scenes, the improved algorithm sets a relatively large threshold, and relatively small thresholds are set for denser scenes. In this way, the defect of the fixed threshold can be avoided, thereby improving the detection accuracy of the YOLO algorithm.

Inspired by the application of neural network in the crowd density analysis, such as Switch-CNN [7], the data about the sparseness of the objects in the picture can be obtained in the neural network, and the threshold value can be set according to the sparsity degree data (Fig. 2).

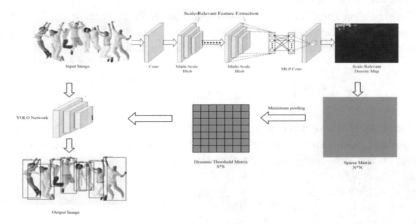

Fig. 2. The flow chart of the improved algorithm by using dynamic threshold.

This paper designed a dynamic threshold matrix with a size of S × S (S × S is the number of grids in which YOLO divides the input picture). In one iteration of the non-maximum suppression process, assuming that the detection box with the highest score is M, and the threshold of center grid of box M is μ, and other detection boxes that overlap with M are used to adjust the confidence of the overlapping boxes by the threshold μ.

The method adopted in Spatial Pyramid Pooling [8] is used to convert the object sparsity matrix (N, N) output by the neural network into the dynamic threshold matrix (S, S) required here. A maximum pooling layer is added that dynamically calculates the window size and step size of the pooling window based on the size of the input. Assuming that the size of the object sparseness matrix is 100×100, then a 7×7 pooling result is to be obtained. The window size can be made to $\lceil 100 \div 7 \rceil = 15$ and the step size is $\lfloor 100 \div 7 \rfloor = 14$.

The reason for using maximum pooling rather than average pooling here is that the threshold for this area needs to be determined based on the maximum density within a region.

5   Experiments

The experimental hardware server configuration is: Intel(R) Xeon(R) CPU E5-2660 v4 processor, NVIDIA Titan XP graphics card, 128 GB RAM. The software environment is Ubuntu 16.04, python 3.6, tensorflow 1.1.0, cuda 6.0 and opencv-python 3.3.0.

The PASCAL VOC and MS COCO are used in this paper, which are standard datasets for classification recognition and detection of visual objects.

5.1   Training

This paper has increased the training rate and effectiveness by adding methods such as partial batch normalization and full convolution. Figure 3 shows the comparison

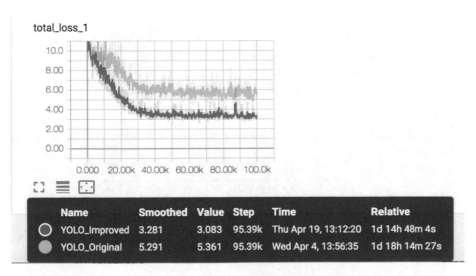

Fig. 3. Loss function gradient descent.

Fig. 4. Comparison of some test results. The left picture is the improved algorithm, and the right one is the original algorithm. It can be seen intuitively that the left improves the detection effect.

between the original YOLO v2 and the improved algorithm. The improved algorithm has a better convergent value during the training period.

5.2   Results

This paper compared the improved algorithm with the original algorithm (YOLO v2) in dense scenes and found that the effect is consistent with the expectations. At the same time, the performance of the algorithm in the rest of the scene is consistent with the original algorithm, so this algorithm is more generic for different scenes when dealing with object detection tasks. Figure 4 is a comparison of the two algorithms in the detection results (Tables 1 and 2).

Table 1.   PASCAL VOC2012 test detection results.

Method	Dataset	mAP(general)	mAP(dense)
YOLO v2	VOC2012	78.6%	65.6%
Improved-YOLO	VOC2012	79.1%	74.6%

Table 2.   MS COCO test-dev results. Because of the design of dynamic threshold, Improved-YOLO only has the AP.

Method	Dataset	AP	AP_{50}	AP_{75}
YOLO v2	COCO test-dev2017	21.7%	44.2%	19.7%
Improved-YOLO	COCO test-dev2017	33.4%	NA	NA

6 Conclusion

The improved algorithm in this paper implements two factors for dense object detection:

(1) Adjustment of scores for overlapping boxes: Adjustments are based on the degree of overlap. The effect is that: Some boxes with actual objects will not be ignored by mistake.
(2) Dynamic thresholds: Thresholds are adjusted according to the object density in the picture. The role is to adjust the extent of processing the overlapping boxes dynamically.

Experiments have verified that the improved algorithm effectively improves the performance of the algorithm for dense scenes, and is more versatile.

References

1. Girshick, R., Donahue, J., Darrell, T., Malik, J.: Rich feature hierarchies for accurate object detection and semantic segmentation. In: IEEE Conference on Computer Vision and Pattern Recognition, pp. 580–587. IEEE Computer Society (2014)
2. Girshick, R.: Fast R-CNN. In: IEEE International Conference on Computer Vision, pp. 1440–1448. IEEE Computer Society (2015)
3. Ren, S., He, K., Girshick, R., Sun, J.: Faster R-CNN: towards real-time object detection with region proposal networks. IEEE Trans. Pattern Anal. Mach. Intell. 39(6), 1137–1149 (2017)
4. Redmon, J., Farhadi, A.: YOLO9000: better, faster, stronger. In: IEEE Conference on Computer Vision and Pattern Recognition, pp. 6517–6525. IEEE Computer Society (2017)
5. Redmon, J., Divvala, S., Girshick, R., Farhadi, A.: You only look once: unified, real-time object detection. In: Computer Vision and Pattern Recognition, pp. 779–788. IEEE (2016)
6. Bodla, N., Singh, B., Chellappa, R., Davis, L.S.: Soft-NMS-Improving object detection with one line of code. In: IEEE International Conference on Computer Vision, pp. 5562–5570. IEEE Computer Society (2017)
7. Sam, D.B., Surya, S., Babu, R.V.: Switching convolutional neural network for crowd counting. In: Computer Vision and Pattern Recognition, pp. 5744–5752. IEEE (2017)
8. He, K., Zhang, X., Ren, S., Sun, J.: Spatial pyramid pooling in deep convolutional networks for visual recognition. IEEE Trans. Pattern Anal. Mach. Intell. 37(9), 1904–1916 (2014)

User Identity Linkage Across Social Networks

Qifei Liu[1]([⊠]), Yanhui Du[1,2], and Tianliang Lu[1,2]

[1] Information Technology and Network Security Institute,
People's Public Security University of China, Beijing 100038, China
cppsu_lqf@163.com
[2] Collaborative Innovation Center of Security and Law for Cyberspace,
People's Public Security University of China, Beijing 100038, China

Abstract. In order to distinguish the accounts that belong to the same person, we propose a method to link user identity across social networks based on user profile and relation. According to similarity calculation algorithms and network embedding, a feature extraction method in multi dimension was designed based on username, location, personal description, avatar and relation. Then a hierarchical cascaded machine learning model (HCML) is proposed to integrate the classifiers in different dimension. The experiment validates that the method in this paper outperforms feature extraction in single dimension, traditional machine learning algorithm and weighting algorithm. The method can be applied to integrate user information across social networks.

Keywords: User identity · Across social networks · User profile · User relation

1 Introduction

A variety of social network platforms have changed people's life. As is shown in the research, 84% of Internet users have more than one social network account [1]. Internet users generally have multiple accounts in different social network platforms. If we can link user identity across different social networks, we will gain more accurate and comprehensive profile of users. It will benefit so many applications in Internet service.

There are some existing works to address user identity linkage. Most methods are based on user profile, user generated content and user relations. There are many researchers paying attention to username which can be used to link user identity [2, 3]. Comprehensive similarity value of user profile can be calculated based on some weight determination such as subjective weight determination or information entropy [4, 5]. A researcher proposes a method based on global view features to align users with crowdsourcing [6]. A method named HYDRA models heterogeneous behavior by long-term behavior distribution analysis and multi-resolution temporal information matching [7].

In previous studies, the general feature extraction is choosing only one specific method. If the one-sided feature is used in the subsequent analysis, it will inevitably result in poor results, especially the problem of low recall. In addition, the machine

© Springer Nature Switzerland AG 2020
Q. Liu et al. (Eds.): CENet 2018, AISC 905, pp. 600–607, 2020.
https://doi.org/10.1007/978-3-030-14680-1_66

learning algorithm is too inflexible in the previous research. In this paper, our contributions can be summarized as follows:

- A feature extraction method is designed for similarity calculation in multi dimension, which is based on username, location, personal description, avatar and relation.
- Taking advantage of the information contained in user profile and relation, a hierarchical cascaded machine learning model (HCML) based on features of different dimension is designed.
- We verify the methods designed in this paper based on real user data in Sina and Douban.

The rest of this paper is organized as follows. Section 2 describes the method of feature extraction. In Sect. 3, user identity linkage model is described more in detail. Section 4 is to show the dataset, experiment design and experimental results. Finally, some conclusions are drawn in Sect. 5.

2   Feature Extraction Method

In this paper, user profile commonly owned by users is selected as the object of feature extraction, including username, location, personal description, avatar, relation.

The feature extraction for user profile is based on similarity calculation algorithms of different types of data. The data includes words, short text, location and image. The features of different data are extracted by similarity calculation algorithm with different meanings. The relation of user is transformed to low-dimensional vectors by network embedding. The processing procedure is shown in Fig. 1. The combination of those features more comprehensively represents the similarity of users from two different social networks.

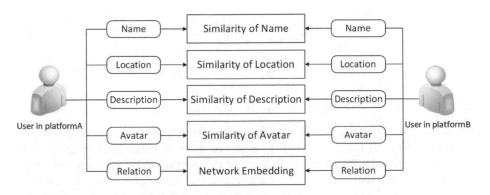

Fig. 1. Method of feature extraction.

2.1   Feature Extraction of Username

Username is the most common basic profile of social network users. It is valuable to link user identity based on username. When users register their accounts on different social network platforms, they tend to make minor adjustments on the basis of the same username, such as replacement, insertion, deletion, transposition, abbreviations and adding some special symbols. So in this paper, we use the following five username similarity calculation methods to generate username feature vector N, N = (n_1, n_2, n_3, n_4, n_5). The five similarity calculation methods are based on Jaro-Winkler Distance, the longest common subsequence (LCS), Levenshtein Distance, Jaccard similarity, Simhash Hamming distance.

2.2   Feature Extraction of Location

Intuitively, location is not a determinant for user identity linkage. However, as a supplementary factor, the location feature is effective. We use the following four location similarity calculation methods to generate location feature vector L, L = (l_1, l_2, l_3, l_4). Because the user location information string is similar to the username, we select Jaro-Winkler Distance, LCS, Levenshtein Distance as the first three methods to calculate location similarity. The fourth method is real distance similarity. We use Baidu API to translate the location information string into longitude and latitude. (lat_i, lat_i) and (lat_j, lat_j) is the longitude and latitude of user U_i^A and U_j^B. The real distance between two users is as follow:

$$d(U_i^A, U_j^B) = 2R \times \arcsin \sqrt{\sin^2(\frac{lat_i - lat_j}{2}) + \cos(lat_i)\cos(lat_j)\sin^2(\frac{lon_i - lon_j}{2})}$$

$$(2.1)$$

The similarity based on real distance is as follow:

$$Sim_{\text{distance}(U_i^A, U_j^B)} = \frac{d(U_i^A, U_j^B)}{\pi R}$$

$$(2.2)$$

where R is the radius of the earth. πR is the farthest distance on the earth.

2.3   Feature Extraction of Personal Description

Users commonly have personal descriptions displayed on the home page in social network platforms. They may publish similar or even same personal descriptions on different social network platforms. We use the following three personal description similarity calculation methods to generate personal description feature vector D, D = (d_1, d_2, d_3). The three similarity calculation methods are based on Word2vec, TF-IDF, Word Mover's Distance [8]. And cosine similarity is used to calculate the similarity of personal description.

2.4   Feature Extraction of Avatar

Avatar is also a kind of common profile. Whether the avatar is the similar is a very important feature for user identity linkage. As a result of various demands for avatar in different platforms, the avatar may be an image that has been stretched, compressed, blurred and cut. In order to detect whether the avatar is similar, we need to quantify the similarity of images. We use the following five methods to calculate avatar feature vector A, $A = (a_1, a_2, a_3, a_4, a_5)$. The five similarity calculation methods are based on pHash, aHash, dHash, SIFT and histogram statistics.

2.5   Feature Extraction of Relation

The topological structure of a same person in different social networks tends to be redundant and similar. In order to utilize relation to link user identity, embedding large information networks into low-dimensional vector spaces is necessary. We embed user relation into vector by LINE [9] which is a method of large-scale information network embedding. As is shown in Fig. 2.

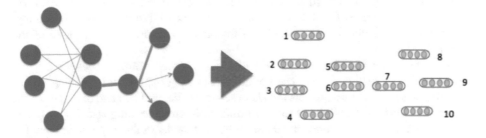

Fig. 2. Large-scale information network embedding.

After network embedding, every user can be represented as a vector of K dimension. K can be determined by researchers. The vector will be utilized in the classifier.

3   User Identity Linkage Model

The user identity linkage across social networks model is as shown in Fig. 3. The main part is a hierarchical cascaded machine learning model, which is designed to select the best classifier in features of different dimension and integrate classification results of different dimension.

The effect of classifier is highly dependent on the characteristics of training sets. Different data is suitable for different classifiers. In order to make best use of the features in five different dimension, a hierarchical cascaded machine learning model is designed in this paper.

In term of user relation, the vectors of different social network accounts are concatenated and they are used to train multi-layer perceptron classifier.

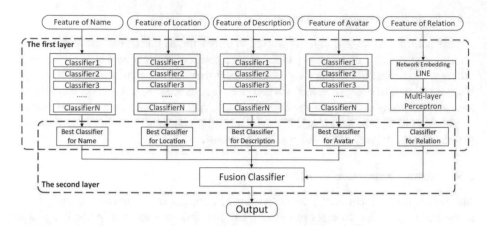

Fig. 3. Framework of user identity linkage model.

In the first layer of this model, we set up a variety of basic classifiers for training and testing in different dimension of user profile and select the best classifier for each dimension. Besides, the classifier based on relation is also trained. In the second layer of this model, the outcome from the first layer is integrated through ensemble learning. The probability of classification results of first layer is regarded as the input of the fusion classifier.

We refer to the Stacking method of traditional ensemble learning between the first and second layer of this model. The traditional Stacking method uses training sets to train classifiers, while training sets are also used to predict and become the input of the next model. This inevitably leads to over fitting problem on training sets. So we use K fold cross validation to generate the input of the next model in this paper to address the problem. K fold cross validation divides training sets into K parts. Each part of the data is used as the test sets and the other k − 1 parts of data as training sets. The above training sets and testing sets are used to train and predict for K times. The combination of K parts of predicting results becomes training sets for the second layer.

4   Experiment and Results

4.1   Dataset

It is difficult to acquire the dataset for user identity linkage experiment. Some researchers propose a method to collect data using the link that links to other personal page on other social networks [10]. Referring to that method we collect 14457 pairs of accounts from Sina Weibo and Douban which belong to the same person and we randomly collect 14457 pairs of accounts that belong to different people. Meanwhile, their user profile is crawled including username, location, personal description and avatar. In addition, a large user relation data including these 28914 pairs of accounts is collected. And 14457 pairs of accounts belonging to the same person are positive samples, whose labels are 1. 14457 pairs of accounts belonging to different people are negative samples, whose labels are −1.

4.2   Experimental Results

The Best Classifier Selection

In the first layer of HCML, we apply 9 kinds of machine learning algorithms to select the best classifier in different dimension. The algorithms include decision trees, logistic regression, support vector machines, K-nearest neighbors, naive Bayes, random forests, extreme random trees, Gradient Boost, Adaboost.

These machine learning algorithms are trained in different dimension. The effect of different machine learning algorithms is obviously different and none of the classifiers works the best in all four dimension. The best classifier is selected based on F1 value and accuracy. In our datasets, the best classifier for username is logistic regression. The best classifier for location is Gradient Boosts. The best classifier for personal description is random forests. The best classifier for avatar is K-nearest neighbors. If richer training dataset is obtained, the selection of the best classifier may be changed.

Ensemble Learning of the Best Classifier

To verify the effectiveness of HCML, we compare each classifier in first layer with the ensemble learning classifier. At the same time, to verify whether feature of each dimension makes contribution to the final effect of the ensemble learning classifier, an experiment of ensemble learning classifier is set up in the absence of each dimension. We use 5 fold cross validation to train the ensemble learning classifier. Table 1 shows the performance of these classifiers.

Table 1. Performance of different classifiers.

Classifier	Precision	Recall	F1	Accuracy
C_{name}	0.8988	0.6125	0.7286	0.7916
C_{loc}	0.8284	0.6570	0.7328	0.7812
C_{des}	0.5249	0.4559	0.4874	0.5649
C_{avatar}	0.4773	0.5197	0.4976	0.5207
$C_{relation}$	0.5930	0.6589	0.6242	0.7362
$C_{name+loc+des+avatar}$	0.8313	0.8707	0.8505	0.8661
$C_{name+loc+avatar+relation}$	**0.9210**	0.6487	0.7612	0.8130
$C_{name+des+avatar+relation}$	0.8690	0.7976	0.8318	0.8522
$C_{loc+des+avatar+relation}$	0.7988	0.7230	0.7590	0.7906
HCML	0.8426	**0.8753**	**0.8586**	**0.8747**

The method (HCML) proposed in this paper achieves the best effect in recall, F1 value and accuracy, but the precision is not the highest. Because when user identity is linked based on a certain dimension or a certain number of dimension, it is always right to determine a pair of accounts belonging to the same person. But it's difficult to discover all pairs of accounts belonging to the same person. So the precision of the HCML proposed in this paper is not the highest, but the recall is much higher than the other classifiers. The F1 value and accuracy are also the highest in the comparison experiment. It is proved that HCML is effective.

Compared with the classifier of each dimension C_{name}, C_{loc}, C_{des}, C_{avatar}, $C_{relation}$, the effect of HCML is obviously improved. It indicates that it is meaningful to integrate features of different dimension through ensemble learning.

There are four classifiers in the absence of each dimension, named $C_{name+loc+des+avatar}$, $C_{name+loc+avatar+relation}$, $C_{name+des+avatar+relation}$, $C_{loc+des+avatar+relation}$. From the comprehensive evaluation criteria F1 value and accuracy, their effects are not as good as the ensemble learning classifier HCML. It indicates that each dimension is helpful to HCML.

Method Comparision

We compare HCML with other four methods. In the method named Alias-Disamb, the researchers utilize avatar, location and signature. The feature extraction methods include downsampling method of digital image, Google Map API and Jensen-Shannon divergence [2]. In the method named Vosecky, the similarity in different dimension is calculated. The weight is determined by subjective weighting method [4]. The weight of different features is learned from large training data by probabilistic modeling in method named HYDRA [7]. The weight of different features is determined by information entropy in method named IE-MSNUIA [5]. Table 2 shows the performance of different methods.

Table 2. Performance of different methods.

Method	Precision	Recall	F1	Accuracy
Alias-Disamb	0.7857	0.7980	0.7915	0.8072
Vosecky	0.8012	0.6342	0.7074	0.7602
HYDRA	0.8281	0.6515	0.7292	0.7791
IE-MSNUIA	**0.8967**	0.5880	0.7103	0.7802
HCML	0.8426	**0.8753**	**0.8586**	**0.8747**

The method (HCML) proposed in this paper achieves the best effect in recall, F1 value and accuracy, but the precision of IE-MSNUIA is higher than our method. The reason is that IE-MSNUIA is a threshold-type judgment method based on comprehensive similarity, when the threshold is high enough, it can be guaranteed that the identified accounts actually belong to the same person. However, this method can not identify all accounts belonging to the same person, so it will result in high precision but low recall. Alias-Disamb represents the methods with feature extraction in single dimension and traditional machine learning algorithm and it performs not so good in experiment. Therefore, it can be concluded that the method proposed in this paper achieves the best effect on actual data.

5 Conclusion

In this paper, we have proposed a method to link user identity across social networks. According to similarity calculation algorithms and network embedding, a feature extraction method in multi dimension was designed. Through the hierarchical cascaded machine learning model (HCML), we can make full use of the information contained in user information of each dimension. The model is trained and tested using dataset collected from Sina Weibo and Douban. Experiment validates the effectiveness and rationality of our method. Our method can be applied to integrate user information across social networks.

Acknowledgement. This work was supported by the National Key R&D Program of China (2017YFB0802804), the National Natural Science Foundation of China (61602489).

References

1. Bartunov, S., Korshunov, A., Park, S., Ryu, W., Lee, H.: Joint link-attribute user identity resolution in online social networks categories and subject descriptors. In: Workshop on Social Network Mining and Analysis, pp. 104–109. ACM (2012)
2. Liu, J., Zhang, F., Song, X.Y., Song, Y., Lin, C.Y., Hon, H.W.: What is in a name? An unsupervised approach to link users across communities. In: ACM International Conference on Web Search and Data Mining, pp. 495–504. ACM (2013)
3. Zafarani, R., Lei, T., Huan, L.: User Identification across social media. ACM Trans. Knowl. Discov. Data **10**(2), 1602–1630 (2015)
4. Vosecky, J., Hong, D., Shen, V.Y.: User identification across multiple social networks. In: First International Conference on IEEE, pp. 360–365 (2009)
5. Wu, Z., Yu, H.T., Liu, S.R., Zhu, Y.H.: User identification across multiple social networks based on information entropy. J. Comput. Appl. **37**(8), 2374–2380 (2017). (in Chinese)
6. Wang, Q., Shen, D.R., Feng, S., Kou, Y., Nie, T.Z., Yu, G.: Identifying users across social networks based on global view features with crowdsourcing. J. Softw. **29**(3), 811–823 (2018). (in Chinese)
7. Liu, S., Wang, S., Zhu, F., Zhang, J., Krishnan, R.: HYDRA: large-scale social identity linkage via heterogeneous behavior modelling. In: ACM SIGMOD International Conference on Management of Data, pp. 51–62. ACM (2014)
8. Matt, J., Yu, S., Nicholas, I., Kilian, Q.: From word embeddings to document distances. In: International Conference on Machine Learning, pp. 957–966 (2015)
9. Tang, J., Qu, M., Wang, M., Zhang, M., Yan, J., Mei, Q.Z.: LINE: large-scale information network embedding. In: International World Wide Web Conferences Steering Committee, pp. 1067–1077. ACM (2015)
10. Maira, V., Carsten, E.: A cross-platform collection of social network profiles. In: ACM SIGIR Conference on Research and Development in Information Retrieval, pp. 665–668. ACM (2016)

Research and Improvement of TF-IDF Algorithm Based on Information Theory

Long Cheng, Yang Yang[✉], Kang Zhao, and Zhipeng Gao

State Key Laboratory of Networking and Switching Technology,
Beijing 100876, China
yyang@bupt.edu.cn

Abstract. With the development of information technology and the increasing richness of network information, people can more and more easily search for and obtain the required information from the network. However, how to quickly obtain the required information in the massive network information is very important. Therefore, information retrieval technology emerges, One of the important supporting technologies is keyword extraction technology. Currently, the most widely used keyword extraction technique is the TF-IDFs algorithm (Term Frequency-Inverse Document Frequency). The basic principle of the TF-IDF algorithm is to calculate the number of occurrences of words and the frequency of words. It ranks and selects the top few words as keywords. The TF-IDF algorithm has features such as simplicity and high reliability, but there are also deficiencies. This paper analyzes its shortcomings for an improved TFIDF algorithm, and optimizes it from the information theory point of view. It uses the information entropy and relative entropy in information theory as the calculation factor, adds to the above improved TFIDF algorithm, optimizes its performance, and passes Simulation experiments verify its performance.

Keywords: Word frequence · Information theory · Information entropy · Relative entropy

1 Introduction

The TFIDF algorithm is relatively simple, with high accuracy and recall, and it has been a concern of many related researchers. The TFIDF algorithm to be improved in this paper has the following defects: (1) Simply link the word frequency to the weight, if the number of candidate keywords in a document appears less in the document, the greater the weight of the word, the word is considered to have a Good discrimination, but some unfamiliar words that are not related to the article will be given too much weight. At the same time, in many cases high-frequency words should be given a higher weight, rather than simply reduce their weight; (2) Does not consider the impact of the different distribution of words in different documents on its weight, when there are two words of the same frequency In the case of a word, if one of the words is concentrated in several documents and the other is distributed in different documents, then the weights of the words in the centralized distribution should be higher than those in the scattered words.

© Springer Nature Switzerland AG 2020
Q. Liu et al. (Eds.): CENet 2018, AISC 905, pp. 608–616, 2020.
https://doi.org/10.1007/978-3-030-14680-1_67

In view of the above defects, the following two improvements have been made: (1) The amount of information of the inquired keyword is added as a calculation factor to the calculation of the weight of the keyword, and the insufficiency caused by calculating the weight of the keyword based on word frequency alone is avoided; (2) The relative entropy of the inquired keyword is added as a calculation factor to the calculation of the weight of the keyword, so as to avoid the deviation of the weight calculation due to the different keyword distribution. In the end, this paper proposes a TFIDF-BOIT keyword extraction algorithm based on information theory.

This article next introduces the related research status of TFIDF; the IFIDF-BOIT algorithm proposed in this paper; the experiment and analysis of the improved algorithm; the conclusion of this paper.

2 Related Work

Unsupervised keyword extraction includes the Topic Model (thematic model), the Text Rank algorithm based on the graph model (text ranking algorithm), and the TF-IDF algorithm based on statistics. According to the overall structure of the document, the topic model obtains the subject of the document by clustering the words and words within the document. This method has a large disadvantage for documents with less space. Generally, it needs to be supplemented with the semantic implicit topic model. The concept of Text Rank based on the graph model comes from Page Rank algorithm. Judging whether there is association between two web pages only needs to judge the links of two web pages, and judges the importance of a web page according to the quality of linked web pages. The advantage is that there is no need to train the sample, but its disadvantage is that it needs to build a network structure and optimize the network topology structure through an iterative algorithm. The calculation is large and the efficiency is low. The TF-IDF algorithm was first proposed by Karen Sparck Jones, a computer scientist at the University of Cambridge in the article titled "Statistical Interpretation of Keyword Specialties and Its Application in Document Retrieval," TF-IDF. The algorithm is mainly applied to information retrieval and text mining. The TF-IDF algorithm filters keywords by counting word frequencies and multiplying the weight of each word. It is simple and practical, and it is more practical in practical applications [1]. However, the TF-IDF algorithm is only applicable to single text, and faced with multiple categories and multiple texts, the feature classification ability is insufficient. For example, in the entire sample database, two words have the same word frequency, but one of the words is concentrated in a certain number of documents, while the other word is scattered in the entire sample database. Obviously, the former has more classification ability than the latter. Should give higher weight.

Jones KS first proposed the idea of IDF [2] in 1972. In 1973, Salton combined the ideas of KS JONES and proposed the TFIDF (Frequency & Inverse Document Frequency Term) algorithm in [3]. Since then, he has repeatedly proved the effectiveness of the algorithm in information retrieval [4]. In 1988 he classified feature words and weights into literature searches and discussed the experimental results [5]. Then he derived the TFIDF algorithm. Have the following conclusions: If the frequency of words or phrases in the TF section is high, and another article is rarely seen, the word or phrase is considered to have good discrimination and the ability to classify; words appear in the document The wider the range, the lower the lower attribute (IDF) it distinguishes between document content. In 1999, Roberto Basils proposed an improved TF IWF IWF algorithm [6]. Depending on the number of different types of documents, there may be an order of magnitude difference between Bong Chih How and Narayanan K, and a category term descriptor (CTD) was proposed to improve TFIDF [7] in 2004. It solves the bad effects on the TFIDF algorithm from different categories of file numbers.

3   Improved Algorithms TFIDF-BOIT

The basic principle of the TF-IDF algorithm is to calculate its ranking based on the number of occurrences of words and word frequency weights, and to select the top few words as keywords.

In the conventional TFIDF, TF is the number of feature items that appear in the document, and IDF is the inverse document frequency. The formula is:

$$\text{TFIDF} = \text{TF} * \text{IDF} \tag{3.1}$$

Taking into account the length of the file content will affect the calculation of the key weight, we use the normalization process and get the following formula:

$$\text{TFIDF} = \frac{TF * \log\left(\frac{N}{n_i} + 0.01\right)}{\sqrt{\sum_{i=1}^{N} TF^2 * \log\left(\frac{N}{n_i} + 0.01\right)^2}} \tag{3.2}$$

Where TF is the number of features that appear in the document, N is the total number of documents, and n_i is the number of documents that are related to the keyword.

The traditional TFIDF algorithm does not consider the allocation information between feature words: If m is the number of documents that contain the word t in a certain category, k is the total number of documents in other categories. Obviously, the number of documents containing t is n = m + k, and m increases as n increases. According to the IDF formula, a very low IDF value can be obtained. This shows that it's classification ability is not strong. However, in reality, the larger m is, which means that the t keyword often appears in the word document. It shows that t is a good distinguisher of the text and should be given a higher weight and be selected as a text feature word. In response to the defects of the traditional TFIDF algorithm, Aizhang Guo et al. proposed an improved TFIDF algorithm [8]. The improved TFIDF is as follows:

$$\text{TFIDF} = tf * \log n * \log\left(\frac{N}{n+k} + 1\right) * \left[1 - \frac{\sqrt{\frac{\sum_{j=1}^{n}\left(tf - \widehat{tf}\right)^2}{k-1}}}{\widehat{tf}}\right] \tag{3.3}$$

Where TF denotes the number of feature words that appear in document j, and (tf) denotes the average number of feature words that appear in each document.

3.1 Weight Model Based on Information Entropy

The relevance of a keyword to its own document cannot be measured simply by the number of occurrences of a word. Measuring the weight of a keyword in a document depends heavily on the size of this keyword in the amount of information in this document [9]. For example, financial documents are always related to economic terms. Sports documents are always related to athletes and sports. Related and so on.

The problem of filtering out candidate keywords in the sample web page document can be equivalent to the problem of matching the query with the web page document being queried through a user query [10]. Make the following assumptions:

The sample space is N (the total number of words in the entire data set). A user query is defined as T, and each T contains multiple keywords. Each query word's corresponding weight should take into account the amount of information the query term has in this document. The probability of defining a query term t in a document is $\mu(t)$, the probability of appearance in all documents is $\varphi(t)$, and the frequency of occurrence of a query term in a document is tf(t). The amount of information for a query word is defined as iv(t), which has the following formula based on the concept of entropy:

$$iv(t) = -\varphi(t) * \log \varphi(t) \tag{3.4}$$

$\varphi(t)$ is the probability that the keyword t appears in sample space N, so the above formula can be written as:

$$iv(t) = -\frac{tf(t)}{N} * \log \frac{tf(t)}{N} \tag{3.5}$$

Considering the amount of information in the keyword, the TFIDF with initial improvement is as follows:

$$\text{TFIDF-BOIT}(t) = iv(t) * tf(t) * \log n * \log\left(\frac{N}{n_t + k} + 1\right) * \left[1 - \frac{\sqrt{\frac{\sum_{t=1}^{n}\left(tf(t) - \widehat{tf}\right)^2}{k-1}}}{\widehat{tf}}\right] \tag{3.6}$$

3.2 Differential Distribution Weight Model

Here, we use the idea of relative entropy [11] to solve the problem of the distribution of keywords. The relative entropy itself is used as a comparison of the similarities of two different distributions. The greater the difference in similarity, the greater the value of entropy, and vice versa. We use a word distribution in the entire document set as a standard distribution. If a word is concentrated in several documents, its difference from the standard distribution is very large, and the corresponding entropy value is large; if the word is widely distributed, Then the difference between it and the standard distribution is very small, and the corresponding entropy value is very small. In this way, the higher the word distribution, the higher the weight.

Set the weight function for differential distribution as:

$$
\begin{aligned}
f(x, y) &= K(F(x), F(y)) \\
&= \sum F(x) \log \frac{F(x)}{F(y)} \\
&= \sum F(x)(\log F(x) - \log F(y))
\end{aligned}
\tag{3.7}
$$

f(x, y) denotes the difference weight function, x denotes the number of occurrences of the keyword, and F(x) denotes the distribution of keywords in the sample document set. y indicates the number of tim es the standard keyword appears, and F(y) indicates the distribution of standard keywords in the sample document set.

Combining the difference distribution weight model with the previous section TFIDF-BOIT(t) results in the final keyword extraction model:

$$
\text{TFIDF-BOIT} = f(tf_i, tf_{avr}) * \text{TFIDF-BOIT}(t)
$$

$$
= \left(\sum_{tf_i} F(tf_i) \log \frac{F(tf_i)}{F(tf_{avr})} \right) * iv_i * tf_i * \log n_i * \log \left(\frac{N}{n_i + k} + 1 \right) * \left[1 - \frac{\sqrt{\frac{\sum_{i=1}^{n}(tf_i - tf_{avr})^2}{k-1}}}{tf_{avr}} \right]
\tag{3.8}
$$

TFIDF-BOIT is an information-based keyword extraction algorithm. Where i denotes a candidate keyword, tf_i denotes the frequency of occurrence of i, tf_{avr} denotes the average frequency, and $F(tf_i)$ denotes the probability distribution of the frequency of i in the entire sample set, $F(tf_{avr})$ represents the probability distribution of the average word frequency in the sample set. N represents the total number of documents in the sample set, n_i represents the number of documents containing the candidate keyword i, and k is the total number of documents in other categories.

4 Experimental Results and Analysis

The types of documents used in the experiment include finance, military, entertainment and sports. These four categories are based on experimental corpus. Each category consists of 1,500 news items, with a total of 6000 documents as the total corpus, 1000 documents from each category, a total of 4,000 documents as a training set; 500 documents left in each category, a total of 2000 documents as a test set, tested the effect of the improved TFIDF algorithm. Table 1 shows the distribution of experimental data in the training set and test set.

Table 1. Distribution of experimental data.

	Finance and economics	Military	Entertainment	Sports
Training set	1000	1000	1000	1000
Test set	500	500	500	500

We usually use the recall rate and accuracy to evaluate the text classification system. For a particular category, the recall rate R is defined as the proportion of all correctly retrieved item (TP) items (TP+FN) that should be retrieved. The accuracy rate P is defined as the ratio of all correctly retrieved item (TP) to all actually retrieved (TP+FP).

The recall rate and accuracy rate reflect two different aspects of the classification quality, but all must be considered comprehensively. The effects of both of the F1 test values are considered together. The higher the F1 test value, the better the classification results. The mathematical formula for the F1 test value is as follows:

$$F_1 = \frac{2 * R * P}{R + P} \tag{4.1}$$

The experimental results of each algorithm in different categories of documents are as follows (Tables 2, 3, 4, 5 and Fig. 1):

Table 2. The experimental results of the algorithm in financial texts.

	Recall rate (R)	Accuracy (P)	F1
TFIDF	85.24%	82.37%	83.78%
TFIDF-BOIT	92.32%	90.41%	91.36%

Table 3. Algorithmic results in military texts.

	Recall rate (R)	Accuracy (P)	F1
TFIDF	83.56%	84.36%	83.96%
TFIDF-BOIT	90.58%	91.49%	91.03%

Table 4. Algorithmic results in entertainment texts.

	Recall rate (R)	Accuracy (P)	F1
TFIDF	87.05%	88.78%	87.91%
TFIDF-BOIT	92.58%	93.47%	93.02%

Table 5. Algorithmic results in sports texts.

	Recall rate (R)	Accuracy (P)	F1
TFIDF	85.07%	87.25%	86.15%
TFIDF-BOIT	91.65%	90.25%	90.94%

From the experimental data of Figs. 2, 3, 4, 5, 6, and 7, it can be seen that for the improved TFIDF algorithm of Aizhang Guo [8], the recall rate R and accuracy of the TFIDF-BOIT algorithm taking into account the amount of keyword information are taken into account. The rate P and F1 values have been significantly improved compared to the original algorithm. The improved algorithm TTFID-BOIT based on the TDIDF-BOIT considers the distribution of keyword differences. Compared with the recall rate R and the precision rate P and F1 values, only the keyword information is considered. The amount of TFIDF-BOIT has increased significantly. The experimental results show that the improved TFIDF-BOIT algorithm has significantly improved the initial TFIDF performance under the conditions of information entropy and relative entropy.

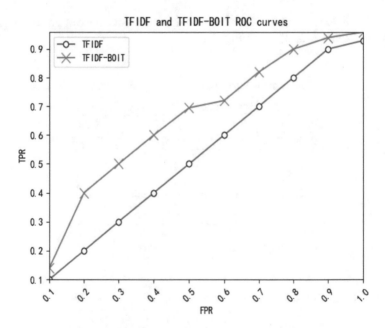

Fig. 1. TFIDF and TFIDF-BOIT ROC curves.

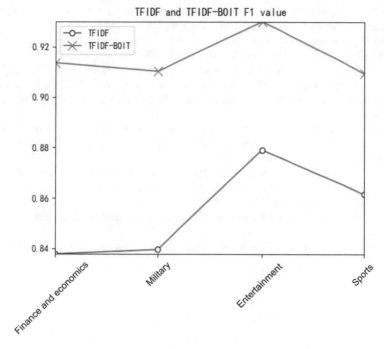

Fig. 2. TFIDF and TFIDF-BOIT F1 value.

5 Conclusion

This article takes the text keyword as the research content, summarizes the existing calculation methods, describes the advantages and disadvantages of the classic algorithm, introduces the research and improvement formulas of the relevant algorithms.

Experiments on the TFIDF algorithm and the improved algorithm using text data show that the performance of the improved TFIDF-BOIT algorithm is better than the previous algorithm.

References

1. Saltong, G., Mcgill, M.J.: Introduction to Modern Information Retrieval. McGraw-Hill, New York (1983)
2. Jones, K.S.: A statistical interpretation of term specificity and its application in retrieval. J. Doc. **28**(1), 11–21 (1972)
3. Saltong, G., Yu, C.T.: On the construction of effective vocabularies for information retrieval. In: Proceedings of the 1973 Meeting on Programming Languages and Information Retrieval, p. 11. ACM, New York (1973)
4. Saltong, G., Fox, E.A., Wu, H.: Extended Boolean information retrieval. Commun. ACM **26** (11), 1022–1036 (1983)

5. Saltong, G., Buckley, C.: Term-weighting approaches in automatic text retrieval. In: Information Processing and Management, pp. 513–523 (1988)
6. Basili, R., Pazienzam, M.: A test classifier based on linguistic processing. In: Proceedings of IJCAIp 1999, Machine Learning for Information Filtering (1999)
7. How, B.C., Narayanan, K.: An empirical study of feature selection for text categorization based on term weight age. In: Proceedings of the 2004 IEEE W/IC/ACM International Conference on Web Intelligence, pp. 599–602. IEEE Computer Society, Washington, DC (2004)
8. Guo, A., Yang, T.: Research and improvement of feature words weight based on TFIDF algorithm. In: Information Technology, Networking, Electronic and Automation Control Conference, pp. 415–419. IEEE (2016)
9. Zuo, R.: Information theory, information view, and software testing. In: Seventh International Conference on Information Technology: New Generations, pp. 998–1003. IEEE Computer Society (2010)
10. Salton, G., Fox, E.A., Wu, H.: Extended Boolean information retrieval. Cornell University (1982)
11. Lin, F.L., Ning, B.: Relative entropy and torsion coupling. Phys. Rev. D **94**(12), 126007 (2016)

A Method for Measuring Popularity of Popular Events in Social Networks Using Swarm Intelligence

Jiaying Chen[(✉)] and Zhenyu Wu

Institution of Internet of Things, Nanjing University of Posts
and Telecommunications, Nanjing, Jiangsu Province, China
876179824@qq.com

Abstract. Social network is a platform for users to post and forward contents that they are paying attention to. So through the measurement of the popularity of the event, it is possible to excavate social focus and predict the development trend of the event. The user is used as the main body to measure the popularity, so as to construct the interaction graph, and the indicator of the graph has a degree distribution sequence, a clustering coefficient and a degree centrality. Among them, the user distribution of the degree distribution sequence of the interaction graph shows the distribution of "power law distribution", and the power index introduced by the degree distribution sequence can effectively reflect the distribution of the user's participation degree in the popular event; the clustering coefficient reflects the user's agglomeration in the popular event; the degree centrality reflects the dominant position of the user participating in the popular event; the number of users reflects the size of the network formed by the popular event. The comprehensive indicator obtained after nondimensionalization and analytic hierarchy process of these four indicators can comprehensively and accurately measure the popularity of popular events. The comprehensive indicator shows that the popular event of the "Ching Ming Festival" is more consistent with the actual situation during the day. Further more, the measure of popularity is more sensitive and can reflect minor changes in the popular spot.

Keywords: Interaction graph · Event popularity · Analytic hierarchy process

1 Introduction

Social networks are the mapping of the real world, where popular events reflect the current public concerns. Through mining popular events, we can discover current interests and topics of the general public and make predictions about the development trend of the events. The measurement of the popularity has an important role in the mining and forecasting of popular events.

Currently, existing studies measure the popularity of popular events by the frequency of keyword occurrences. However, the evaluation of this method has certain incompleteness, because the number of keywords does not reflect the situation of forwarding, that is, there will be the unpopular event which has a high popularity. Therefore, in order to comprehensively evaluate the popularity, in addition to using the

© Springer Nature Switzerland AG 2020
Q. Liu et al. (Eds.): CENet 2018, AISC 905, pp. 617–626, 2020.
https://doi.org/10.1007/978-3-030-14680-1_68

content, it is necessary to pay attention to the interactive relationship between users. In other words, the interaction is more frequent in a popular event. Therefore, it is a good way to calculate a comprehensive indicator by using indicators in the interaction graph to evaluate the popularity of an event.

The structure of this paper is as follows: The second part is the related research; The third part introduces the evaluation indicator of the popular event and the method to obtain it; The fourth part introduces the analysis results on the Weibo experimental data set; the fifth part gives the conclusion of this paper.

2   Related Work

The evaluation of the popularity of a popular event can be evaluated from various aspects, and the following methods for evaluating the popularity are proposed in the existing researches.

From the content of popular events, the number of keywords (or tweets) or proportion of popular events is taken as the evaluation criterion: the popularity rating of popular words in the popular event uses the frequency of occurrence of the popular word as the evaluation criterion. The popularity evaluation of popular words in popular events uses the frequency of the popular words as the evaluation criteria, and the frequency distribution of popular words is also considered [1]. According to the premise that the Internet public sentiment is proportional to the amount of public information, the degree of association between the popularity vector and the maximum popularity vector is defined as the popularity of Internet public events [2]. According to the relevance degree of keywords, the event is extracted, and the popularity is determined by the ratio of the number of microblogs involved in the event to the total number of microblogs [3]. The number of tweets related to the event or the percentage of the total number of those tweets is used as the criterion for the popularity evaluation [4]. The popularity is the number of tweets supporting this event in a time interval [5]. This kind of evaluation only based on its content has a certain limitation and cannot fully evaluate the popularity.

From the perspective of the user: Using the number of participants in an event as a criterion for evaluating popularity [6]; The heat of an event depends on the number of users' friends [7]. These methods only considering the number of users also have certain limitations.

Evaluating popular events comprehensively from multiple angles: the popularity is evaluated using the comprehensive indicator obtained by the AHP using the number of like, comments and forwarding numbers [8]. According to the influence of the Weibo posts included, the score of the popular event is calculated as the popularity, and the influence of each Weibo post is obtained by comprehensively calculating the fan number, the number of forwarding and the number of reviews [9]. The comprehensive indicator is obtained by constructing a indicator system by detecting popularity. The system is decomposed in a one-dimensional dimension and continues to be subdivided in the two-dimensional and so on, so as to obtain first-level, second-level, and third-level indicators, then uses Analytic hierarchy process, Delphi method, BP neural network, etc. to obtain the weight of each indicator [10–14]. This kind of methods considering multiple indicators makes the evaluation more comprehensive, but it is based on a static method to consider the dynamic situation which is not so sensitive to the change in popularity.

3   The Method to Measure the Popularity of Social Events

3.1   Definition of Popular Events

The emergence of events on social networks is because users publish the related content on the social network and other users then forward. When there is a fierce discussion on an event, there must be a large number of users involved and a complicated interactive relationship.

3.2   Interaction Graph

For a popular event, the main target of an event is the use rather than contents, so using the users as nodes and the relationships of users as edges construct the interaction graph. If there is an forwarding between two users, there will be an edge, just as the Fig. 1 shows. The interaction graph is a mapping of interactions between users.

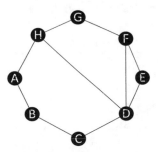

Fig. 1. Interaction graph.

3.3   The Sequence of Indicators

(1)   The sequence of Degree Distribution
Degree Distribution is a description of the distribution of each node's degree in the graph which reflects the overall connection of a network.

The relationship between messages published by users is forwarding and forwarded, and the number of edges connected to a node indicates the number of times the user represented by the node to be forwarding or forwarded. In a time slice, the degree of all the nodes constitutes the sequence of Degree Distribution, i.e. $D = \{d1, d2, \ldots, dm\}$.

(2) The sequence of Clustering Coefficient
    Clustering Coefficient is a description of the degree of integration of cluster. In each time slice, different forwarding conditions will lead to different user clustering conditions. The clustering coefficient is used to represent the degree of clustering, namely CL.

(3) The sequence of Degree Centrality
    Degree Centrality is a description of the importance of a node in the graph. The importance of each user in the network will be different. For example, if a public person publishes a message, and then the message will be forwarded in a large amount, so the user will take a more central position. In a time slice, the degree centrality of all nodes constitutes the sequence of Degree Centrality, $DC = \{dc_1, dc_2, \ldots, dc_m\}$.

3.4   The Method of Measuring Popularity

(1) The conversion of the sequence of Degree Distribution
    In a time slice, each node has its own degree, which means that judging the degree of participation of users' needs to consider all nodes. So it is necessary to covert the degree distribution sequence to evaluate the overall interaction. The degree distribution is shown in the Fig. 2, the abscissa is the node degree, and the ordinate is the number of nodes in such degree.

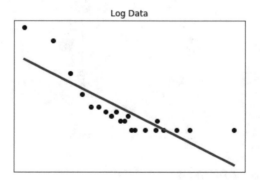

Fig. 2. The sequence of Degree Distribution in a time splice.

The sequence of degree distribution is a power-law distribution, and a one-dimensional linear regression of this distribution can give the slope k, that is power exponent whose value reflects the user's participation: the more users with high participation, the greater the value.

(2) The conversion of the sequence of Degree Centrality

The reason why the degree centrality sequence needs to be converted is similar to that of the degree distribution. The average of the degree centrality represents the average importance of all users, so as to represent the overall situation of the users' centrality, such as:

$$\overline{DC} = \frac{\sum\limits_{i=1}^{m} dc_i}{m} \tag{3.1}$$

(3) The calculation of the comprehensive indicator about the popularity of an popular event

There are four indicators for measuring the popularity: Power exponent (K), clustering coefficient (CL), degree centrality (DC), and number of users (N). For a time point T, four indexes K_t, CL_t, $\overline{DC_t}$, and N_t need to be integrated to obtain a comprehensive indicator.

The steps for obtaining the comprehensive indicators are as follows:

(1) Nondimensionalizing

Due to the different units of measurement and magnitudes, those four indicators are not comparable. Therefore, it is necessary to carry out nondimensionalization, such as

$$x_i' = \frac{x_i}{\max\limits_{1 \le i \le n} (x_i)} \tag{3.2}$$

Where x_i represents the ith value in the indicator sequence.

(2) Establishing hierarchical structure

For the popularity of a popular event, the clustering of users can best represent it, so the clustering coefficient is the first; secondly, the user's participation can better represent it, so the exponent is the second; thirdly, the dominant position of users can express it, so the degree centrality is the third; finally, the number of users reflects the size of the network, so the number of users is the fourth. The order of the four indicators is as follows:

$$CL > K > \overline{DC} > N \tag{3.3}$$

(3) Constructing a comparison matrix

Let $x_1 = CL$, $x_2 = K$, $x_3 = \overline{DC}$, $x_4 = N$. The comparison matrix A is as follows

$$A = \begin{bmatrix} 1 & 3 & 5 & 7 \\ 1/3 & 1 & 3 & 5 \\ 1/5 & 1/3 & 1 & 3 \\ 1/7 & 1/5 & 1/3 & 1 \end{bmatrix} \tag{3.4}$$

Where in the comparison matrix A, a_{ij} represents the result of the importance of the element x_i and the element x_j, the larger the value, the greater the importance of x_i than x_j.

(4) Testing the consistency of the comparison matrix

The average random consistency indicator CI of the comparison matrix A, such as:

$$CI = \frac{\lambda_{max}(A) - n}{n - 1} \tag{3.5}$$

Where n denotes the number of elements, and $\lambda_{max}(A)$ denotes the maximum value of the characteristic value of the comparison matrix A.

And test index CR (where the value of RI is a constant), such as:

$$CR = \frac{CI}{RI} \tag{3.6}$$

According to the formula and the formula, it can be obtained that $CR = 0.0443$. Because $CR < 0.1$, the comparison matrix passes the consistency test.

(5) Obtaining weights of four indicators

The eigenvector w is the maximum eigenvalues of the comparison matrix A, such as:

$$w = \begin{bmatrix} 0.5650 \\ 0.2622 \\ 0.1175 \\ 0.0553 \end{bmatrix} \tag{3.7}$$

(6) Calculating the comprehensive indicator

The eigenvector w is the weight vector of the four indicators, and using it to calculate the comprehensive indicator P of the popularity of the popular event in a certain time, such as:

$$P = 0.5650 * CL + 0.2622 * K + 0.1175 * \overline{DC} + 0.0553 * N \tag{3.8}$$

4   Experiments

4.1   Data Set

The data set used in this experiment was 6416475 microblog data on the day of April 5th, 2015, and there were 228,129 pieces of data on the popular event of "Ching Ming Festival".

4.2   The Result of the Popularity of the Popular Event

The general method for measuring the popularity of a popular event is to measure the popularity using the number of nodes, the number of edges, or the number of nodes and edges. This paper compares these three methods with the method using comprehensive indicators by drawing pictures to show the popularity change of the popular event about "Ching Ming Festival" on April 5, 2015. And the abscissa represents the time t, $t \in (0, 24]$, and the unit is hour; the ordinate represents the value of popularity.

(1) The evaluation of popularity based on general methods
     In Fig. 3(a), the trend of the change of the popularity is that from 0 o'clock to 4 o'clock, falling; and from 4 o'clock to 24 o'clock, basically in a state of continuous rise. And in Figs. 3(b) and (c), the trend is similar to the trend in Fig. 3(a).
(2) The evaluation of popularity based on the comprehensive indicator
     The data for the day of April 5 is divided into 18 time slices, and in Fig. 4, there is a popularity-smoothing curve of April 5, 2015 about the changes of "Ching Ming Festival".

From Fig. 4, the change of popularity is that: from 0 o'clock to 2 o'clock, dropping; from 2 o'clock to 17 o'clock, in a rising state as a whole, but to a certain extent, floating; from 17 o'clock to 24 o'clock, continuing to decline.

(3) The actual situation
     According to the data given by Baidu Index, the trend of popularity of the popular event about the "Ching Ming Festival" on June 4, 2018 is shown in the Fig. 5.

After comparing the actual changes with the above four changes one by one, the conclusion can be drawn that the trend of changes in the comprehensive indicator most closely matches the actual situation, but in the other three methods there is a large deviation.

The actual trend of popularity is: reducing-gradually increasing (but the process has obvious floats)-reducing. Only the trend based on comprehensive indicator most accords with the actual trend, whereby the trend obtained by the other three evaluation methods which does not have the decrease in the third phase is not quite consistent with the actual trend.

In addition, the comprehensive indicator which is more sensitive can detect a minor change, therefore in the trend based on the comprehensive indicator, the popularity has a certain degree of fluctuation that is more obvious, which is also consistent with the actual situation. But the remaining three trends have less fluctuations, which is inconsistent with the actual situation.

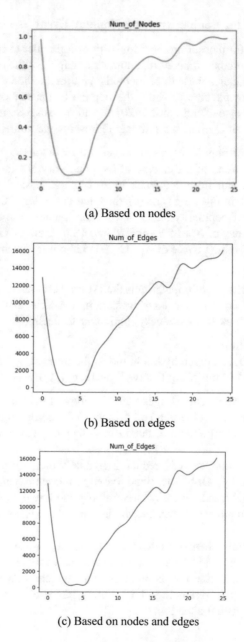

(a) Based on nodes

(b) Based on edges

(c) Based on nodes and edges

Fig. 3. The change of popularity.

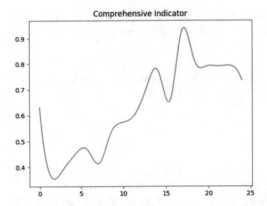

Fig. 4. The change of popularity based on comprehensive indicator.

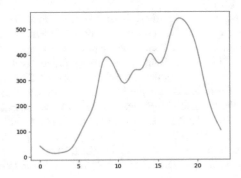

Fig. 5. The change of popularity given by Baidu Index.

5   Conclusions

The main object to promote the development of an event is the user, therefore from the perspective of users, it is reasonable to use the four indicators, which are power exponent, clustering coefficient, degree centrality and the number of users, representing the users' interactions and dominant position as the benchmark for the measurement of popularity.

The following conclusions can be drawn by comparing the experimental results: the trend based on the measurement of popularity calculated by the comprehensive indicator is consistent with the overall actual change trend, and there are similar popularity-floating situations in the detail part; whereby in the other three methods, there are certain shortcomings: the overall trend is not consistent with the actual situation and the floating situations are not similar to the actual. Therefore, the comprehensive indicator consisting of four indicators as a measure of popularity achieves more accurate and comprehensive results than that obtained by the other three methods.

In the measurement of popularity of future popular events, the evaluation methods based on comprehensive indicator mentioned above can be used to measure the popularity.

Acknowledgement. This work is supported by Natural Science Foundation of China (No. 61502246), NUPTSF (No. NY215019).

References

1. Wang, S., Feng, L.X.: Mining chat text based on large data word identification of new hotspots opinion. Electron. Commer. **1**, 60–61 (2018)
2. Lan, Y.X., Liu, B.Y., Zhang, P., Xia, Y.X., Li, Y.Q.: Research on dynamic prediction model of internet public opinion for big data. J. Inf. **36**(06), 105–110+147 (2017)
3. Wang, Z.X.: Based on mining and emotional hot events microblogging analysis. Shanghai Jiaotong University (2013)
4. Gupta, A., Mittal, N., Kohli, N.: Analysis of a sporting event on a social network: true popularity & popularity bond. In: Proceedings of the International Conference on Data Engineering and Communication Technology. Springer, Singapore (2017)
5. Gupta, M., Gao, J., Zhai, C.X., Han, J.: Predicting future popularity trend of events in microblogging platforms. Proc. Assoc. Inf. Sci. Technol. **49**(1), 1–10 (2012)
6. Shen, P., Zhou, Y., Chen, K.: A probability based subnet selection method for hot event detection in Sina Weibo microblogging. In: IEEE/ACM International Conference on Advances in Social Networks Analysis and Mining, pp. 1410–1413 (2013)
7. Jia, R., Liu, T., Zhang, J., Fu, L.Y., Gan, X.Y., Wang, X.B.: Impact of content popularity on information coverage in online social networks. IEEE Trans. Veh. Technol. **99**, 1 (2018)
8. Xu, Y.N.: Based on media spectacle heat network public opinion trends microblogging. Inf. Sci. **2**, 92–97 (2017)
9. Li, H., Zhu, X.: Detection of emerging hot spots of Weibo based on influence. Comput. Appl. Softw. **33**(05), 98–101+165 (2016)
10. Wang, C.N., Chen, W.Q., Xu, H.: Indicator System of microblogging public opinion monitoring and early warning of heat. Comput. Mod. **1**, 126–129 (2013)
11. Lee, L.X., Yang, T.F.: Based on the government network Sina Weibo negative public opinion trend. J. Intell. **10**, 97–100 (2015)
12. Yan, C.X., Zhang, X., Liu, L.: Based on the level of emergency response network public opinion Redu. China Manag. Sci. **22**(3), 82–89 (2014)
13. Wen, L., Yao, Y.Y., Wang, C.K.: Research network public opinion Redu evaluation of food safety incidents: based on BP neural network method. Mod. Manag. Sci. **9**, 30–32 (2016)
14. Sun, X,F., Cheng, S.H., Jin, X.T.: Government quantitative method for evaluating a negative public opinion of heat networks-Sina microblogging, for example. J. Intell. **8**, 137–141 (2015)

Personalized Recommendation Leveraging Social Relationship

Wenchao Shi, Yiguang Lu, Zhipeng Huang, Enbo Du, Yihan Cheng, and Jinpeng Chen[✉]

School of Software Engineering,
Beijing University of Posts and Telecommunications,
Beijing 100876, China
jpchen@bupt.edu.cn

Abstract. The goal of recommender system is recommending products users may be interested in. Recommending content is an important task in many platforms. But many platforms only have implicit feedback information from users. Each single user generates behavior on only a few products lead to sparse datasets from users. And traditional recommending algorithms show poor performance when face the two problems. In this paper, we present a social bayesian personalized ranking based on similar friends (SFSBPR). We alleviate the sparsity problem when recommend by users' implicit feedback. Experiment on real-world sparse datasets show that our method outperforms current state-of-the-art recommendation method.

Keywords: Personalized recommendation · Social relationship · Epinions · SVD

1 Introduction

Personalized recommending system plays an important role in many information systems. At present, almost all large e-commerce systems, such as Amazon, eBay and Taobao have used various forms of recommendation systems in various degrees.

However, there are two major problems in the recommendation of most platforms. The first is that many platforms only have implicit feedback information from users. Explicit feedback requires the user to give a clear rating to the recommended content, which is not user-friendly. The recommendation systems will make predictions for items that the user has not given an evaluation. And most of them will make predictions based on explicit feedback (the users' ratings of items). However, on many platforms, like TopBuzz[1], Weibo[2] and Taobao[3], Implicit feedback collection is much easier. Implicit feedback includes user clicks, browsing, favorites, purchases, etc. It means we can get some positive feedback from almost any information system by recording users' behavior. But negative feedback is mixed with missing data. For example, a user's click on an article may represent a preference, but no click does not mean that the user does not like it at all. We can only guess their intention from users' behavior. Another problem is that a single user generates behavior on only a few

© Springer Nature Switzerland AG 2020
Q. Liu et al. (Eds.): CENet 2018, AISC 905, pp. 627–634, 2020.
https://doi.org/10.1007/978-3-030-14680-1_69

products and the total amount of products is very large, causing the public products selected by users to be very rare and each user's feedback data on the project is also extremely sparse.

Due to the development of social networks, users' social relationships are often easier to obtain. In fact, users' preferences can be obtained from their friends' behaviors in addition to their own behaviors. Friends with similar interests have a stronger influence on each other [9].

In this paper, we try to introduce social relationship between users to alleviate the data sparsity problem in the recommendation system.

The major contributions of this paper are summarized in the following.

1. We focus on the data sparsity problem and one-class recommendation problem [8] in product recommendation. We propose a novel method named Social Bayesian Personalized Ranking based on similar friends (SFSBPR) ranking to alleviate the two problems effectively.
2. We evaluate the proposed method on a sparse datasets, and the empirical results show that our model can improve recommendation performance.

2   Related Work

In this section, we will briefly review related work.

The most widely used method in the recommendation system is KNN collaborative filtering [1]. The general idea of the method is to find people who are similar with user u. Then these similar person's preferences are similar with preference of user u. This is followed by latent factor model. Latent factor model complements the values in the user-item matrix by learning the parameters that are not directly observed. Common matrix factorization has been widely used in the recommendation system [3]. Many algorithms for learning feature matrices have been proposed. Singular value decomposition(SVD) is one of them [2]. The MF models learned by SVD is prone to overfitting. Then the SVD++ is proposed to solve this problem [6]. But when the datasets is very sparse, these methods are not good solutions. The recommendation algorithms based on social network have been proposed [10]. Such as Social Recommendation Using Probabilistic Matrix Factorization (SoRec) [7]. It learns the user latent feature matrix and item latent feature matrix by employing user-item matrix and social network matrix simultaneously. And the method has achieved good result in alleviating the data sparsity problem and cold-start problem. However, this recommendation method focuses on rating estimation problems. It is not good at solving one-class recommendation problem. Then some method solving one-class collaborative filtering was proposed [8, 11]. The main purpose of the recommendation system is to predict the top-K items and to return these items to users. The method for personalized ranking can solve the problem better. This kind of algorithm does not optimize the user's rating on the items, but directly optimize the user's ranking of the item. Such as BPR [4] and SBPR [5], this type of algorithm works better than matrix factorization when applied to a recommendation system that uses implicit feedback as input.

With the increasing number of users and items, the recommendation algorithm based on ranking learning can more effectively reflect the user's preferences. A detailed discussion of the relationship among our approach, the SVD approach, the SoRec approach and the traditional BPR. Compare with the above works, the main advantage of our approach in recommending products is: Our method produces a better result in one class recommendation when the datasets is very sparse.

3   Problem Definition

Definition 1. Let U be the set of users, and let I be the set of items. Implicit feedback matrix $X : U \times I$ can be observed. X is an 0-1 matrix. When X_{ui} is 0, it means that user u has no feedback to i. When X_{ui} is 1, it means that user u has feedback to i.

Definition 2. We have a social network about users $G = (U, E)$, for $u \in U$, $v \in U$, $(u, v) \in E$ indicate user u and v are friends.

Then the problem we face can be described as follows:

We can supplement the matrix by social network G. Then we use this supplementary matrix X to get each user's ranking for all items. And a top-k list of items can be obtained from the ranking. $f(u, G, X) \to rankedList(I_u)$, Given an item set $M \in I$, and a user $u \in U$ our aim is to return a top-k list of item pieces.

4   The Recommendation Algorithm Based on SBPR

4.1   Optimization Criterion

The problem we studied was the one-class recommendation in the case of a scarce datasets. Discover more item-user pairs by introducing social information so that we can train the model on a denser user-item matrix. Our solution is to find user's friends who are similar with the user. Then we supplement the user's preference list by these friends' feedback to items.

This is a notation-list to explain the meanings of mathematical symbols used in this section.

The specific method is as follows: For User $u \in U$, X_u is the feedback list for u. In his friends and their friends of these friends set F_u, User $v \in F_u$, X_v is the feedback list for v.

We define the similarity between u and v as:

$$w_{uv} = Similarity(u, v) = \frac{\sqrt{X_u \times X_v}}{|X_u||X_v|} \tag{4.1}$$

We sort users in F_u according to their similarity. And find the k friends most similar with u. Let SF_u denote the set of the k users most similar with u in F_u.

Then the preference of u to i can be defined as $p_{ui} = \sum w_{uv} X_{vi}$.

We can use p_{ui} as X_{ui}. In this way, we get a new preference list of items for each user so that we get a denser matrix X.

Then we decomposed the matrix X into two low-rank matrices $W : |U| \times n$ and $H : |I| \times n$. Each row w_u in W is a feature vector of user u. And each row h_i in H is the feature vector of item i.

For user u, we divide the total items I into three parts: P_u, FP_u, N_u.

For this we can assume a ranking: Items user has selected $>_u$ items users in SF_u have selected $>_u$ items neither user nor his friends have selected

$$x_{ui} >_u x_{uj} >_u x_{uk}, i \in P_u, j \in SF_u, k \in N_u \qquad (4.2)$$

Then between each user and their similar friends, we specify a S_{uk} coefficient to be used as an evaluation factor for the user's friends to treat an item.

We can find the Bayesian formulation of finding the personalized ranking.

$$p(\theta| >_u) \propto p(>_u |\theta)p(\theta) \qquad (4.3)$$

where θ represents parameter matrix W and parameter matrix H.

The BPR algorithm uses the maximum a posteriori estimation to learn the matrices W and H.

Table 1. List of notations.

Notation	Description
$>_u$	$a >_u b$ represent user u may like a more than b
X_u	The feedback list for u
F_u	User u's friends and their friends of these friends set
SF_u	The set of the k users most similar with u in F_u
W	Users' feature matrix
H	Items' feature matrix
P_u	The set of items which user u has given positive feedback
FP_u	The set of items which users in SF_u have given positive feedback
N_u	The set of items which u and user in SF_u haven't given positive feedback

We assume users' preference is independent of each other. And for a user, the ordering of each pair of items is independent of each other. We find the likelihood function $p(\theta| >_u)$ can be written as a product of probabilities (Some notations are shown in Table 1).

$$\prod_{i \in PSF_u, k \in PSF_u} p(x_{ui} >_u x_{uk})^{\alpha(u,i,k)} \left(1 - p(x_{ui} >_u x_{uk})^{1-\alpha(u,i,k)}\right)$$

$$\prod_{k \in SFN_u, j \in SFN_u} p(x_{ui} >_u x_{uk})^{\beta(u,k,j)} \left(1 - p(x_{ui} >_u x_{uk})^{1-\beta(u,k,j)}\right) \qquad (4.4)$$

Where $PSF_u = P_u \cup SF_u$, $SFN_u = SF_u \cup N_u$.

α and β are indicator functions.

$$\alpha(u,i,k) = \left(\begin{array}{c} 1 \; if \; i \in P_u, k \in SF_u \\ 0 \; otherwise \end{array} \right), \beta(u,k,j) = \left(\begin{array}{c} 1 \; if \; i \in SF_u, k \in N_u \\ 0 \; otherwise \end{array} \right) \qquad (4.5)$$

The above formula can be simplified to:

$$\prod_{i \in P_u, k \in SF_u} p(x_{ui} >_u x_{uk}) \prod_{k \in SF_u, j \in N_u} p(x_{uk} >_u x_{uj}) \qquad (4.6)$$

And we use a sigmoid function $\sigma(x) = 1/(1+e^{-x})$ to approximate the function $p(x)$

Then for all users, we can rewrite the above formula.

$$\sum_{u \in U} \left(\sum_{i \in P_u} \sum_{k \in SF_u} ln(\sigma(x_{ui} - x_{uk})) + \sum_{k \in SF_u} \sum_{j \in N_u} ln\left(\sigma\left(x_{uk} - x_{uj}\right)\right) \right) - \lambda \| \theta \|^2$$
$$(4.7)$$

Our goal is to maximize the above objective function.

4.2 Model Learning

We employ the widely used stochastic gradient descent (SGD) algorithm to maximize the objective function.

We update the parameters $\theta(W$ and $H)$ by walking along the ascending gradient direction.

$$\theta^{t+1} = \theta^t + \frac{\mu \times \partial \vartheta(\theta)}{\partial x} \qquad (4.8)$$

The entire learning process can be summarized as follows:

1. The first step is to initialize W and H as normal distribution matrices.
2. The second part traverses each quad in the training set u, i, k, j, iteratively corresponds to the parameters W and H.
3. The third step is to determine if the iteration converges. If it converges, the algorithm ends otherwise, return to the second step.

5 Experiment

This section compares the experimental results of our recommendation algorithm and matrix factorization algorithm on sparse datasets.

This paper uses the dataset obtained from Epinions. The size of the dataset is shown in the Table 2.

Table 2. Datasets.

Basic meta Users	Items	Reviews	User context Users	Links
40163	139738	664824	49289	487183

In the datasets, 40163 users rated at least one of 139738 different items. The density of the user-item matrix is 0.01186%. We can find the datasets is very sparse.

We will consider the user's rating of 2–5 points as the user's positive feedback, Similarly, we consider the rating of 1 points and no feedback as the user's negative feedback. Then the corresponding social feedback can be obtained.

We adopt Recall@k(R@k), F-Score (F1) and Mean Average Precision (MAP) to evaluate the recommendation quality of our proposed approach in comparison to baseline methods. Among them:

$$F1 = \frac{2 * Recall * Precision}{Recall + precision} \qquad AP = \frac{\sum_{i \in R_u} \frac{\sum_{k:\pi_u(k) \le \pi_u(i)} R_{ui}}{\pi_u(i)} R_{ui}}{\sum_{i \in R_u} R_{ui}} \qquad (5.1)$$

Where R_u represents the recommended list, $\pi_u(i)$ represents the position of item i in R_u. R_{ui} is 1 when i in R_u is appear in test set, otherwise it is 0.

MAP is the mean of AP of all users in test set.

In the experiments, we set the initial learning rate μ as 0.05 and the value of regularization parameters λ as 0.01. Here we let the number of latent factors equals 10.

Figure 1 details the results for Recall@k. We can find that our approach SFSBPR shows significant improvement compared with other algorithms on the sparse datasets. The SVD approach and the SoRec approach show very poor performance on the datasets. The SVD approach and the SoRec approach do not apply to one-class collaborative filtering, especially when the datasets is very sparse. The traditional BPR algorithm and our method are superior to the SVD approach because they directly

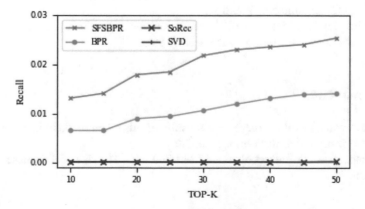

Fig. 1. Recall.

optimize the ordering of different commodities. From Figs. 2 and 3, we can obtain that our proposed method shows better performance than BPR. Our models exceeded the traditional BPR algorithm in terms of various indicators. The possible reason for the results is that our proposed approach introduces the social information to make the datasets denser and produces more item pairs.

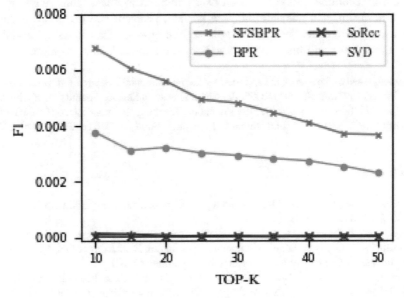

Fig. 2. F1.

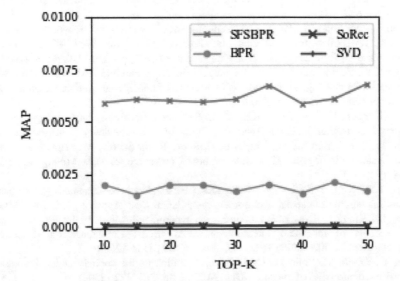

Fig. 3. MAP.

6   Conclusion and Future Work

In this work, we introduce social relationship to alleviate the data sparsity problem and improve the performance on one-class recommendation problem. Experiments on the sparse Conference on Information and Knowledge Management datasets show that our approach effectively improve the accuracy of recommendation. For future work, we are interested in recommendation based on user's implicit feedback. First, we will investigate more features to alleviate the data sparsity problem and use more features to model users' preference. Second, we will explore more method to model users' preference. Finally, we will find more methods to solve the cold start problem.

Acknowledgement. This work is supported by the Fundamental Research Funds for the Central Universities under Grant No. 2017RC55, the National Natural Science Foundation of China under Grant No. 61702043, No. 61502202, the Education Teaching Reform Project of BUPT in 2018 under Grant No. 2018JY-B04, and Research Innovation Fund for College Students of BUPT.

References

1. Deshpande, M., Karypis, G.: Item-based top-n recommendation algorithms. ACM Trans. Inf. Syst. **22**, 143–177 (2004)
2. Sarwar, B., Karypis, G., Konstan, J., Riedl, J.: Incremental singular value decomposition algorithms for highly scalable recommender systems. In: Proceedings of the 5th International Conference in Computers and Information Technology (2002)
3. Rennie, J.D.M., Srebro, N.: Fast maximum margin matrix factorization for collaborative prediction. In: ICML 2005: Proceedings of the 22nd International Conference on Machine Learning, New York, NY, USA, pp. 713–719 (2005)
4. Rendle, S., Freuden-thaler, C., Gantner, Z., Schmidt-Thieme, L.: BPR: Bayesian personalized ranking from implicit feedback. In: Conference on Uncertainty in Artificial Intelligence (2009)
5. Zhao, T., McAuley, J., King, I.: Leveraging social connections to improve personalized ranking for collaborative filtering. In: Proceedings of the 23rd ACM International Conference on Information and Knowledge Management, pp. 261–270. ACM (2014)
6. Koren, Y.: Factorization meets the neighborhood: a multifaceted collaborative filtering model. In: Proceedings of the 14th ACM SIGKDD International Conference on Knowledge Discovery and Data Mining (2008)
7. Ma, H., Yang, H., Lyu, M.R., King, I.: SoRec: social recommendation using probabilistic matrix factorization. In: Conference on Information and Knowledge Management (2008)
8. Pan, R., Zhou, Y., Cao, B., Liu, N.N., Lukose, R.M., Scholz, M., Yang, Q.: One-class collaborative filtering. In: IEEE International Conference on Data Mining, pp. 502–511 (2008)
9. Seo, Y.D., Kim, Y.G., Lee, E., Baik, D.K.: Personalized recommender system based on friendship strength in social network services. Expert Syst. Appl. **69**, 135–148 (2017)
10. Ma, H., King, I., Lyu, M.R.: Learning to recommend with social trust ensemble. In: Proceedings of the 32nd International ACM SIGIR Conference on Research and Development in Information Retrieval, Boston, MA, USA (2009)
11. Hu, Y., Koren, Y., Volinsky, C.: Collaborative filtering for implicit feedback datasets. In: IEEE International Conference on Data Mining, pp. 263–272 (2008)

A Method Based on Improved Cuckoo Search for On-Site Operation and Maintenance Work Order Distribution in Power Communication Network

Mi Lin[✉], Jie Hong, Ming Chen, Bo Li, Huanyu Zhang,
Zhiyuan Long, Mingsi Zhu, Weiming Wu, and Wencheng Deng

Hainan Power Grid Co., Ltd., Haikou 570100, China
linmi@hn.csg.cn

Abstract. The on-site operation and maintenance (O&M) of the power communication network is an important condition for the stability and efficiency of the smart grid. Moreover, the distribution of the work order is the basis for task assignment in the power system, ensuring the efficient unified management and safe stable operation. In order to optimize the process of the work order distribution for on-site O&M of power communication network, the paper proposes a distribution method based on the improved cuckoo search. First of all, according to distribution requirements of the work order, multiple factors are analyzed, the overall strategy and workflow of distribution are designed, and a distribution analysis model for work order is established. Then, a work order distribution algorithm based on improved cuckoo search is proposed, mixing extreme dynamics optimization search, combining with work order distribution characteristics to find the optimal configuration solution under multiple optimization goals. The four goals of less makespan, higher quality, higher personnel utilization, and less wait time for work orders are achieved, which provides theoretical support for improving efficiency, standardizing management and ensuring safety on on-site O&M. Finally, the simulation shows that this method can achieve higher efficiency and quality of on-site O&M.

Keywords: Work orders · Cuckoo search · Multi-objective optimization · On-site operation and maintenance

1 Introduction

Safe and stable operation of the power grid system is necessary to guarantee social stability, industrial production and residents' living. Meanwhile [1], the on-site operation and maintenance (O&M) of the power communication network is an important condition for the stable and efficient operation of the smart grid [2].

Work order distribution is the basis for task assignment, guaranteeing the efficient unified management and safe stable operation of the power system's on-site O&M [3]. In the past, on-site O&M was based on personnel experience, habits, and unquantified techniques, lacking standardized digital work order management, particularly detailed

Q. Liu et al. (Eds.): CENet 2018, AISC 905, pp. 635–645, 2020.
https://doi.org/10.1007/978-3-030-14680-1_70

records and a fixed distribution model, resulting in inefficiency and unequal distribution. With the rapid development of sensor technology, Internet technology and intelligent analysis technology, the Internet of Things (IoT) technology and big data technology have been widely applied and works well in many fields. In order to improve the efficiency of on-site O&M management, and to increase the information exchange and analysis capabilities in the operation process, the IoT technologies and big data analysis technologies are applied to work order distribution of O&M field.

At present, most studies mainly consider the least makespan and the best quality, but few consider optimizing three or more indicators. This dissertation mainly aims to consider the less makespan, the greater quality, the larger personnel utilization rate and the less waiting time of work orders as the four optimization goals, so that the issue of work order distribution optimization could be studied more comprehensively.

The remaining parts of this paper are organized as follows: Sect. 2 presents the related work; Sect. 3 builds a characteristic model for distribution; Sect. 4 proposes a work order distribution algorithm based on improved cuckoo search; Sect. 5 presents simulation results; Sect. 6 concludes.

2   Related Work

Unlike the single-objective optimization problem, the salient features of multi-objective optimization are incommensurability and contradiction between goals. However, the intelligent algorithm could provide multiple solutions for reasonable choice, mitigating the two features. At present, common algorithms for multi-objective optimization include multi-objective particle swarm optimization (MOPSO), non-dominated sorting genetic algorithm (NSGA), firefly algorithm (FA) and multi-objective evolutionary algorithm (MOEA). However, these algorithms have the disadvantages of poor convergence and easy to fall into local optimum, and become inefficient faced with the problem with more than three goals. Besides, MOPSO has high computational complexity, low search accuracy and low versatility. Both NSGA and MOEA [4] have complex solution processes, no ideal diversity of solution sets and bad distribution in high-dimensional problems [5, 6]. FA has a large dependence on excellent individuals, prone to shocks near the peak. This paper uses the improved cuckoo algorithm to solve this problem [7].

3   Characteristic Analysis Model of Work Order Distribution

3.1   Factors of Work Order Distribution

In order to form an efficient work order distribution mechanism, a description of the issues was described as Fig. 1 and the problem model is established. The 4 factors are considered.

(1) The O&M personnel centrally store the respective attributes of the personnel, including capabilities, permissions and location. Capabilities include historical average makespan and historical averages quality for each type of job type.

O&M Personnel Set	O&M Resource Set
Capabilities {job type, average completion time, average quality} Permission {security level, position authority} Status Location	Type Margin

O&M Work Order Set	Job object Set
Object serial number Work type Capability requirements {completion time, quality} Permission requirements Resource requirements {type, amount}	Serial number Position Type

Fig. 1. Factors of work order distribution for on-site O&M of power communication.

(2) The O&M work order includes the object serial number, work type, capability requirements, resource requirements, permission requirements. Capability requirements include the minimum quality and the maximum makespan required.

(3) The Job Object Set collectively store the three attributes of serial number, position, and type. The category attribute facilitates the association of work orders and job instructions.

(4) The O&M resource set includes two attributes: type and margin. Types represent the classification, such as tools, O&M vehicles, and other hardware and software resources. The margin attribute is the available quantity of each type of O&M resource.

3.2 Process of Work Order Distribution

The workflow of the model of work order distribution is shown in Fig. 2. The four sets of factors are respectively performed in three types of operations in sequence. Type matching associates the work order with the corresponding work instruction book. Static selection are based on the constraints imposed by the work order. A person can get more than one work order which will be completed in sequence. The dynamic selection is based on the intelligent algorithm and the optimization goals for distribution.

3.3 Constraints of Work Order Distribution

Due to the limited personnel and resources in the O&M site, the distribution must meet the following constraints:

(1) Permissions

$$L_{P_x} \geq L_{W_x} \tag{3.1}$$

$$M_{P_x} \geq M_{W_x} \tag{3.2}$$

(2) Capability Requirements

$$q_x \leq Q_{W_x} \tag{3.3}$$

$$t_x \geq T_{W_x} \tag{3.4}$$

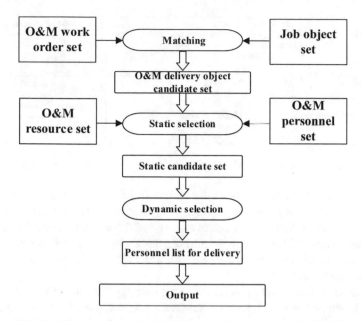

Fig. 2. The workflow of work order distribution for on-site O&M.

(3) Available Resources

$$\sum_{x=1}^{w} Rn_{yW_x} \leq r_y \qquad (3.5)$$

For any x ($0 < x \leq w$, w is the number of work order) and any y ($0 < y \leq a$, a is the number of resource types): L_{W_x} and M_{W_x} respectively represent the value required by the xth work order W_x, while L_{P_x} and M_{P_x} are respectively the own value of the person. q_x and t_x are the requirement for the minimum quality and the maximum time for the work type from W_x, while Q_{W_x} and T_{W_x} represents the historical average quality and the historical average makespan of the O&M person who is assigned to W_x for the same work type. Rn_{yW_x} represents the number of requests for the y-th resource from W_x, and r_y represents the margin of the y-th resource.

3.4   Optimization Goals of Work Order Distribution

(1) The Highest Quality

$$Q = Max \sum_{x=1}^{w} Q_{W_x} \qquad (3.6)$$

Where w represents the total number of work orders to be assigned, and Q_{W_x} represents the historical average quality of work performed by O&M personnel assigned to work order W_x.

(2) The Least Makespan

$$T = Min \sum_{x=1}^{w} T_{W_x} \tag{3.7}$$

Where T_{W_x} represents the historical average makespan of the person assigned to W_x.

(3) Maximum Utilization of Personnel

It equals to the least average free time, as well as the smallest variance in working time.

$$U = Min \sum_{x=1}^{N} \frac{(T_x - \mu_T)^2}{N} \tag{3.8}$$

Where N represents the number of O&M personnel, T_x represents the total makespan of all the work orders for this person, and μ_T represents the average number of staff working hours.

(4) Minimum Waiting Time for Work Orders

It means the shortest lasting time from distribution of work order to arrival of personnel. In the problem, it equals to the smallest sum of time for each person to each target work order.

$$W = Min \sum_{x=1}^{N} TA_x \tag{3.9}$$

$$TA_x = \sum_{y=1}^{n} \left(TA_{x(y-1)} + T_{x(y-1)} + \frac{d_{xy}}{v} \right)^2 \tag{3.10}$$

$$d_{xy} = \begin{cases} \sqrt{[J_{xy} - J_{x(y-1)}]^2 + [K_{xy} - K_{x(y-1)}]^2}, y > 1 \\ \sqrt{[J_{xy} - J_{P_x}]^2 + [K_{xy} - K_{P_x}]^2}, y = 1 \end{cases} \tag{3.11}$$

TA_x means the sum of the waiting time of each work order in the sequence received by the x-th O&M personnel. Assume that the personnel's moving speed is equivalent to v. n means the number of work orders received by the x-th person. For any $y(0 < y \leq n)$, and for the x-th person, $TA_{x(y-1)}$ is the waiting time of the $(y-1)$-th work order in the sequence, $T_{x(y-1)}$ is the makespan of the $(y-1)$-th work order in the sequence, $\frac{d_{xy}}{v}$ is the time that the person moves from the object of the $(y-1)$-th one to that of the y-th one, and d_{xy} is the distance from the former to the latter. J_{P_x} and K_{P_x} are respectively the latitude and longitude of the current position of the x-th person. J_{xy} and K_{xy} are respectively the latitude and longitude of the object position in the y-th work order, $J_{x(y-1)}$ and $K_{x(y-1)}$ are respectively for the $(y-1)$-th one.

4 Work Order Distribution Algorithm Based on Improved Cuckoo Search

O&M work order distribution algorithm based on the improved cuckoo (CS-EO) algorithm introduce the extreme dynamics optimization (EO) algorithm into the Cuckoo algorithm (CSA), utilizing the powerful local optimization of EO to help CSA jump out of the local extreme point, improving accuracy and speed. The CS-EO algorithm performs an EO search without CS every N_e generations, without increasing complexity of the original algorithm. The interval counter N_e could be appropriately selected according to the function's complexity, for example turning small to increase local search if the function is relatively complex or the local optimum is quite large.

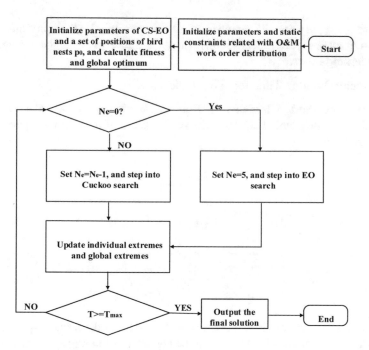

Fig. 3. The workflow of the work order distribution algorithm based on the improved cuckoo search.

The algorithm workflow is as follows and Fig. 3:

(1) Objective function $F(x)$, and $x = (x_1, \cdots, x_d)^T$. Set the parameters of distribution.
(2) Set related parameters of CS-EO algorithm: detection probability P and the maximum number of iterations T_{max}; Initialize the iteration counter $T = 0$ and interval counters $N_e = 5$. Initialize n host nests, with dimensions $d = 4$ and $p_0 = [x_1^0, x_2^0, \cdots, x_n^0]^T$, and then find the optimal position x_b^0 and optimal solution f_{min}, where $b \in \{1, 2, \cdots, n\}$.

(3) Determine if N_e is equal to 0. If yes, go to (5) and set it to the initial value of 5; otherwise, make it $N_e = N_e - 1$ and proceed to the next step.
(4) Optimize each nest according to the cuckoo search then transfer to (7).
(5) Optimize each nest according to the EO search.
(6) Compare the fitness of each nest position with that of the global optimum of the previous generation. If the current one is better than the latter, update the position x_b^t as the current global optimum and the corresponding global extreme f_{min}.
(7) Determine whether T_{max} has been reached. If $T \geq T_{max}$, terminate the iteration and output the optimum x_b^t and optimal value f_{min}; otherwise, let $T = T + 1$ and return (4).

The algorithm core is described as follows:

(1) Initialization of birds' nest
   According to the characteristics of the work order distribution model and the Levy flight, the attributes of the O&M personnel are encoded by real numbers. Each dimension vector x_{ij} where $j \in \{1, 2, \cdots, d\}$ in a set of sequential position vectors $x_i = [x_{i1}, \cdots, x_{id}]$ of host's nests selected by CS is converted into a sequence of dispatched work orders $\pi = (w_1, w_2, \cdots, w_s)$ for each personnel. That is, each nest uniquely corresponds to an O&M work order distribution Π.
(2) Fitness function
   Due to particularity of O&M of power communication, quality is the primary consideration. With the quality measured as a positive number from 1 to 20 and the time calculated in hours, considering the 4 optimization goals, the fitness function is therefore expressed as follows:

$$F(x) = \frac{T + \sqrt{U} + W}{Q} \tag{4.1}$$

(3) Cuckoo search
   Do the following for the current individual $p_{t-1} = [x_1^{t-1}, x_2^{t-1}, \cdots, x_n^{t-1}]^T$:
   (a) Create a new location for each nest according to Levy's Flight, obtain a new set p_t, calculate its fitness, compare it with the fitness of the previous generation of nests p_{t-1}, and replace bad one by a better-fit location to obtain a better set set $g_t = [x_1^t, x_2^t, \cdots, x_n^t]^T$.
   (b) Generate a uniform random number $r \in [0, 1]$ for each nest and compare them with P. If $r > P$, change the position to create a new nest, to get a new set of in this way and calculate their fitness. Compare the fitness with that of each location in g_t, replace the nest location by that with better fitness, and obtain better group $p_t = [x_1^t, x_2^t, \cdots, x_n^t]^T$.
   For the Step (a), the path and position updating formula for Levy's Flight is as follows:

$$x_i^{t+1} = x_i^t + \alpha \oplus Levy(\lambda) \tag{4.2}$$

$$Levy(\lambda) = \frac{u}{|v|} \tag{4.3}$$

$$\delta_u = \left[\frac{\Gamma(1+\beta)sin\frac{\pi\beta}{2}}{\Gamma\left(\frac{1+\beta}{2}\right)2^{(\beta-1)/2}\beta}\right]^{1/\beta} \tag{4.4}$$

$\oplus$ is a point-to-point multiplication. $\alpha = \alpha_0(x_i^t - x_b^t)$. Usually $\alpha_0 = 0.01$. $u \sim N(0, \delta_u^2)$, $v \sim N(0, 1)$, $\lambda(1 < \lambda \le 3)$ and $\lambda = 1 + \beta$. In the CS, usually $\beta = 1.5$. Γ is the Gamma function.

(4) EO search

Do the following for the current individual $p_{t-1} = [x_1^{t-1}, x_2^{t-1}, \cdots, x_n^{t-1}]^T$:

(a) Calculate the fitness λ_i^{t-1} of each component x_i^{t-1}, where $i \in \{1, 2, \cdots, n\}$;

(b) Sort n fitness levels and find the worst component x_{imin}, which has $\lambda_{imin} \le \lambda_i^{t-1}$;

(c) Find a neighbor p_τ of the current individual p_{t-1}, forcing the worst-case x_{imin} to change;

(d) Unconditionally make $p_{t-1} = p_\tau$ and calculate f_{t-1} of the mutant.

5 Simulation

To verify the effectiveness, the algorithm is implemented in simulation and compared with the original Cuckoo Search (CS) algorithm and Particle Swarm Optimization (PSO) algorithm. The parameters of factors required for the mission is set to: 10 of O&M personnel, from 10 to 100 of O&M work orders changed in order, and 50 of the margin of resource. The main parameters in the algorithm are set as follows: discovery probability P = 0.25 and maximum iteration number T_{max} = 200; initial value of iteration counter Ne = 5. The initial size of the host nest is n = 30.

From Fig. 4, 5, 6 and 7, we can see that for the distribution of O&M work orders the CS-EO algorithm performs better, converging faster, without limit of local optimal, easy to obtain a better solution to meet the requirement of higher quality, less makespan, less waiting time and higher utilization rate. As the task volume is small, the three algorithms are not much different. With the increase in tasks, the CS-EO algorithm is superior to the others. The reason may be small solution space so that the gap is unobvious with few tasks but the advantages are reflected with much more tasks for the better local search, retaining good global search capabilities.

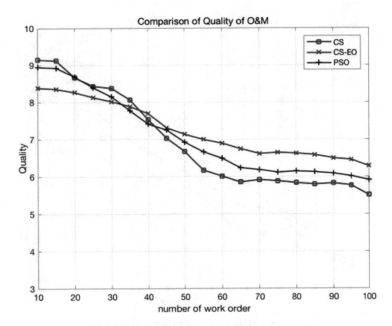

Fig. 4. Comparison of quality of O&M.

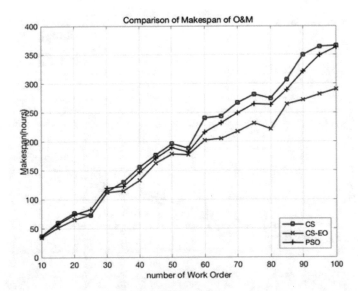

Fig. 5. Comparison of makespan.

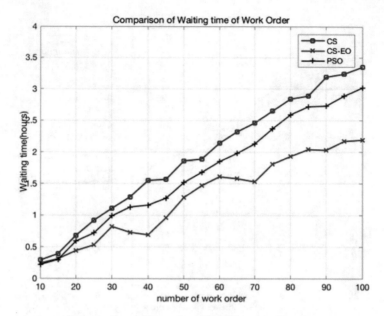

Fig. 6. Comparison of waiting time of work order.

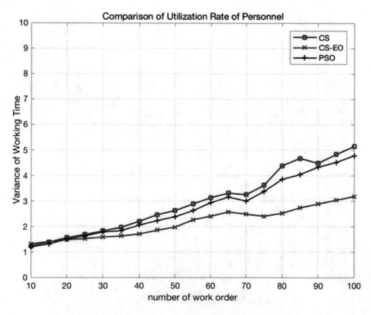

Fig. 7. Comparison of utilization rate of personnel.

6   Conclusion

This paper focuses on the work order distribution of on-site O&M in power communication network. First of all, according to distribution requirements of work orders, multiple factors were analyzed, the overall strategy and flow of distribution were designed, and a distribution analysis model of work order was established. Then a work order distribution algorithm based on improved cuckoo search is proposed. The cuckoo search is improved by mixing extreme dynamics optimization search, and combined with characteristics of work order distribution to find the optimal configuration under the four optimization goals of less makespan, less waiting time, higher utilization and higher completion quality, providing theoretical support for improving on-site O&M efficiency, standardizing management and ensuring safety. Finally, the simulation shows that the method could achieves the four goals for work orders in the same time.

Acknowledgement. This work was supported by the Science and Technology Foundation of the China Southern Power Grid Company Limited: Research and demonstration application of wearable intelligent operation and maintenance technology for power communication field (070000KK52170008).

References

1. Zhao, M., Shen, C., Liu, F., Huang, X.Q.: A game-theoretic approach to analyzing power trading possibilities in multi-microgrids. In: Proceedings of the CSEE, vol. 35, no. 4, pp. 848–857 (2015). (in Chinese)
2. Chen, J.L.: IoT-based on-site operation and maintenance standardized operation management system. Electrotechnical Appl. **1**(S1), 264–268 (2010). (in Chinese)
3. Zhou, Y.: Study on the mobile operation and maintenance system for the electric power communication network. Jiangsu University, Nanjing (2016). (in Chinese)
4. He, Q.: Implementation and application of a modified multi-objective particle swarm optimization algorithm. Beijing University of Chemical Technology, Beijing (2016). (in Chinese)
5. Elarbi, M., Bechikh, S., Gupta, A., Ben Said, L., Ong, Y.S.: A new decomposition-based NSGA-II for many-objective optimization. IEEE Trans. Syst. Man Cybern. Syst. **48**(7), 1–20 (2018)
6. He, Z., Yen, G.G.: Diversity improvement in decomposition-based multi-objective evolutionary algorithm for many-objective optimization problems. In: 2014 IEEE International Conference on Systems, Man, and Cybernetics, San Diego, CA, USA, pp. 2409–2414. IEEE (2014)
7. Ling, M.G., Liu, F.J., Tong, X.J.: A new fuzzy clustering algorithm based on firefly algorithm. Comput. Eng. Appl. **50**(21), 35–73 (2014). (in Chinese)

Publication Topic Selection Algorithm Based on Association Analysis

Qingtao Zeng[1,2,3], Kai Xie[1(✉)], Yeli Li[1], Xinxin Guan[1],
Chufeng Zhou[1], and Shaoping Ma[2]

[1] Beijing Institute of Graphic Communication, Beijing 102600, China
xiekai@bigc.edu.cn
[2] Department of Computer Science and Technology,
Tsinghua University, Beijing 100084, China
[3] Postdoctoral Management Office of Personnel Division,
Tsinghua University, Beijing 100084, China

Abstract. In the process of traditional education publishing, publication topic selection is completed by subjective experience of editorial team, which is difficult for editorial team to take into account complex factors such as needs of readers, knowledge update and market change. A large number of teaching materials are unsalable and publications can hardly meet actual needs of reader and market. Theme and content of traditional textbooks are lagging behind, which are difficult to meet the needs of today's education development. To solve these problems, this paper focuses on Publication topic selection algorithm based on association analysis. First of all, an algorithm for automatically acquiring data and information from web pages is designed. Then, this paper designs similarity degree calculation method, score prediction algorithm and prediction score updating algorithm. Finally, effectiveness of the algorithm is verified by experiments.

Keywords: Publication · Topic selection · Association analysis

1 Introduction

With development of Internet technology, the speed of knowledge renewal and dissemination is becoming faster. Theme and content of traditional textbooks are lagging behind, which are difficult to meet the needs of today's education development. Education and publishing industry is facing unprecedented challenges and opportunities [1, 2]. Difficulties faced by educational publishing are mainly related to the way of selecting subjects for publishing editors. In the process of traditional education publishing, topic is completed by subjective experience of editorial team, which is difficult for editorial team to take into account the complex factors such as needs of readers, knowledge update and market change [3]. In addition, due to professional restrictions,

Q. Liu et al. (Eds.): CENet 2018, AISC 905, pp. 646–652, 2020.
https://doi.org/10.1007/978-3-030-14680-1_71

most editorial teams' ability to use information technology is limited, which seriously hampered the support role of data and information for educational publishing decisions. This will lead to deviation between topic selection of editorial team and readers' needs, knowledge update and market change, which makes current textbooks have phenomenon of blind follower and content stagnation. At the same time, a large number of teaching materials are unsalable and publications can hardly meet actual needs of reader and market [4, 5].

In this context, the key to reverse unfavorable situation is to introduce big data and data mining technology into education and publishing industry, and make full use of objective data to assist editorial team to complete topic planning [2, 6]. Through data and information automatic acquisition technology of readers' demand, knowledge update and market change, which can make publishing institutions accumulate massive publication related data, combine them with large data and data mining technology to break through data and information bottlenecks of educational publishing institutions, realize planning of editorial team and readers' needs, and knowledge update and market change are highly compatible. Finally, giving full promoting effect of publishing in education.

2 Data and Information Collection

This technology mainly involves web page automatic extraction program, starting from one or several initial web pages URL and using initial web page URL to extract new URL from current page and put it into queue until specific conditions are met, which can enable editorial team to download a large amount of data from website. However, workflow of the technology is rather complicated. It is necessary to design a specific web page analysis algorithm to filter out independent URL of topic and put useful URL into queue. After screening, web page data will be extracted and stored in database to establish index, which will provide a scientific and objective database for subsequent analysis and mining.

```
def get_mark(url):
    response = url_open(url).decode('utf-8')
    html = etree.HTML(response, parser=None, )
    # Get a list of total labels
    categories = html.xpath('//a[@class="tag-title-wrapper"]/@name')
    literature_string     =     re.compile('(<a     name="    teaching    material"
    class="tag-title-wrapper">\s (\s|.)*?\s</div>)').findall(response)[0][0]
    # Instantiated object
    html = etree.HTML(literature_string, parser=None, )
    literature_string = html.xpath('//td/a/text()')
```

Algorithm steps are as follows:

1) set up URL connection

```
def url_open(url):
    head = {"User-Agent": generate_user_agent()}
    req = urllib.request.Request(url, headers=head)
    response = urllib.request.urlopen(req).read()
return response
```

2) import classification values and store main categories, such as teaching materials

```
def save_main_mark(row, value):
# create a table
if main_row_start == 1:
wb = Workbook()          # Create a worksheet
    ws1 = wb.active
ws1.title = " Label "
    ws1.sheet_properties.tabColor = "1072BA"
else:
    wb = load_workbook(filename="teaching material.xlsx",)
```

3) storage subclassification

```
def save_mark(x):
# Define global variables to facilitate automatic storage in order every time
global row_start
global main_row_start
tag = 0
wb = load_workbook(filename="teaching material.xlsx",)
ws1 = wb.active
for col in 'ABCD':
ws1.column_dimensions[col]. width = 20
# Store it as a table
for row in range (row_start, row_start+10):
    for col in range (1, 5):
```

3 Recommended Algorithm

Get readers' comments on books from review website, then extract keywords and ratings from website, and build the following table (Table 1).

Table 1. Reader and keyword association table.

Keyword	Reader					
	Reader 1	Reader 2	......	Reader j	......	Reader n
Keyword 1	r_{11}	r_{12}	......	r_{1j}	......	r_{1n}
Keyword 2	r_{21}	r_{22}	......	r_{2j}	......	r_{2n}
......	......	......	......	......	......	......
Keyword i	r_{i1}	r_{i2}	......	r_{ij}	......	r_{in}
......	......	......	......	......	......	......
Keyword n	r_{n1}	r_{n2}		r_{nj}		r_{nn}

Next, we use Pearson similarity algorithm to calculate similarity between readers' preferences.

$$Interest_{ij} = \frac{\sum_{k \leq n} (r_{ki} - \overline{r}_i) \cdot (r_{kj} - \overline{r}_j)}{\sqrt{\sum_{k \leq n} (r_{ki} - \overline{r}_i)^2} \cdot \sqrt{\sum_{k \leq n} (r_{kj} - \overline{r}_j)^2}} \tag{1}$$

In formula (1), r_{ki}, r_{kj} represent the reader K's score for keywords i and j, respectively, $\overline{r}_i, \overline{r}_j$ indicating average value of reader k's for keywords i and J. Subsequently, the keywords with larger similarity (less than a specific threshold value) are classified as one class.

In prediction grading stage, method is based on the similarity of items and neighbor set N(i) of keyword i, which assesses a reader's score on target keyword. The weighted average method (formula (2)) is usually used to evaluate collaborative filtering method based on keyword. Then, according to the size of prediction score, the recommended sequence of N keywords are generated. According to the similarity of keyword and neighbor set N(i) of keyword i, the method is used to evaluate a reader's score on target keyword. The weighted average method (formula (2)) is usually used to evaluate the collaborative filtering method based on keyword, and then, N keywords are generated recommended sequence according to reader's score on keyword from large to small.

$$Score_{u,i} = \overline{r}_i + \frac{\sum_{j \in N(i)} (r_{u,j} - \overline{r}_j) \cdot (Interest_{ij})}{\sum_{j \in N(i)} |Interest_{ij}|} \tag{2}$$

In formula (2), Score$_{u,i}$ represents user u's prediction score for keyword i. $\overline{r_i}$ indicates average score of keyword i. N(i) shows neighborhoods of keyword i. Score$_{u,i}$ expresses user u's score on keyword j. $\overline{r_j}$ indicates average score of keyword j. *Interest$_{ij}$* represents similarity degree of keyword i and j.

$$\text{Score}'_{u,i} = \frac{\overline{r_i} + r_{v,i}}{2} + \frac{\sum\limits_{j\in N(i)} \left(r_{u,j} - \frac{\overline{r_j}+r_{v,j}}{2}\right) \cdot (\textit{Interest}_{ij})}{\sum\limits_{j\in N(i)} |\textit{Interest}_{ij}|} \tag{3}$$

In formula (3), Score$'_{u,i}$ represents user u's prediction score for keyword i, when get reader v's score on the keyword i and j. The flow chart of recommended algorithm is as follows, time complexity is $O(n^2)$ (Fig. 1).

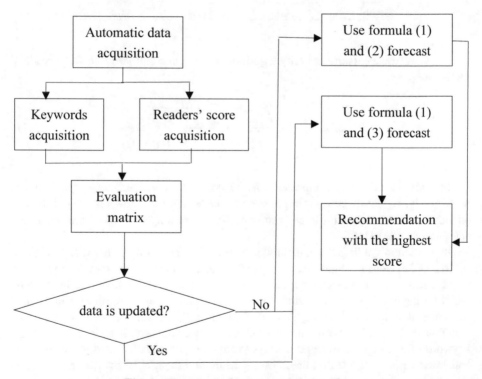

Fig. 1. Flow chart of recommended algorithm.

4 Analysis of Experimental Results

Comparison of Person algorithm and the improved algorithm in this paper between 10K data and 100k data is carried out, which shows that error of the improved algorithm is lower. In Figs. 2 and 3, Y-axis represents error of Person algorithm and the improved algorithm, X-axis represents the size of neighborhood node set of keyword i. In Fig. 3, the performance gain is more remarkable than that in Fig. 2, which is due to the more keywords are used, the fewer errors caused by prejudice of a few readers against keyword i as average value is used in formula (2).

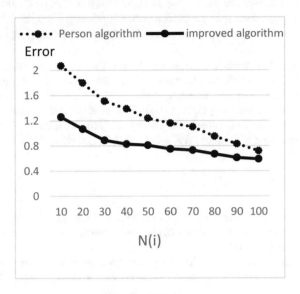

Fig. 2. 10K data.

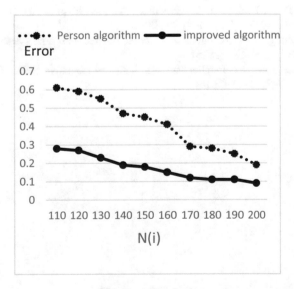

Fig. 3. 100K data.

5 Conclusion

This paper focuses on Publication topic selection algorithm based on association analysis. First of all, an algorithm for automatically acquiring data and information from web pages is designed. Then, this designs similarity degree calculation method, score prediction algorithm and prediction score updating algorithm. Finally, experimental results show that the improved algorithm designed in this paper has lower error.

Acknowledgement. This work was supported by the Curriculum construction project-Linux Program design (22150118005/014), doctoral research funding (04190117003/044), Construction of school teachers- doctoral research funding (27170118003/007), Construction of the publication data asset evaluation platform (04190118002/039) and Construction of computer science and technology in predominant construction (22150118010/006) and Printing electronic technology (Collaborative innovation) (PXM2017_014223_000055).

References

1. Wintrode, J., Sell, G., Jansen, A.: Content-based recommender systems for spoken documents. In: IEEE International Conference on Acoustics, Speech and Signal Processing, pp. 5201–5205. IEEE (2015)
2. Wu, M.L., Chang, C.H., Liu, R.Z.: Integrating content-based filtering with collaborative filtering using co-clustering with augmented matrices. Expert Syst. Appl. **41**(6), 2754–2761 (2014)
3. Lamothe, A.R.: Factors influencing the usage of an electronic book collection: size of the e-book collection, the student population, and the faculty population. Coll. Res. Libr. **1**, 39–59 (2013)
4. Stone, R.W., Baker-Eveleth, L.J.: Student's intentions to purchase electronic textbooks. J. Comput. High. Educ. **4**(25), 27–47 (2013)
5. Bobadilla, J., Ortega, F., Hernando, A., Gutiérrez, A.: Recommender systems survey. Knowl. Based Syst. **46**(1), 109–132 (2013)
6. Karydi, E., Margaritis, K.: Parallel and distributed collaborative filtering: a survey. ACM Comput. Surv. **49**(2), 1–41 (2016)

Gait Recognition on Features Fusion Using Kinect

Tianqi Yang$^{(\boxtimes)}$ and Yifei Hu

Jinan University, Guangzhou 510632, China
tytq@jnu.edu.cn

Abstract. The paper proposes a gait recognition method which is about multi-features fusion using Kinect. The data of 3D skeletal coordinates is obtained by Kinect, and the multi-features are as follows. Firstly, the human skeletal structure is treated as a rod-shaped skeletal model, and it can be simple and convenient to reflect the static characteristics of structure of the human body from the overall. Secondly, the angle of the hip joint is observed during walking so that the dynamic characteristics of the gait information are reflected from the local area. Thirdly, the key gait body postures features are selected from a special gait to reflect the characteristics of walking. Then, the three gait features information is fused, which improves the overall recognition rate. After removing the noise from the bone data, in order to fully reflect the uniqueness of the individual, gait features are extracted from multiple angles, including both static and dynamic features. For finding the center of mass, the distance from the center of mass to the main joint point are calculated to measure the change in the center of mass, and the characteristics of hip joints reflecting the changes of the lower extremity joints while walking. The paper classify each feature separately by using the Dynamic Time Warping (DTW) and K-Nearest Neighbor (KNN) algorithm, which is used at the decision level. The experimental results show that the proposed method achieves a better recognition rate and has a good robustness.

Keywords: Gait recognition · Features fusion · Kinect

1 Introduction

Gait recognition is an emerging technology in biometrics [1], Lee et al. [2] proposed to represent the different parts of the human body using seven ellipses to model and extracted the ellipse parameters as features. Cunado et al. [3] proposed a method in which the human thigh was modeled as a pendulum, and gait features were obtained from the frequency components of the tilt angle signal. Yoo et al. [4] construct a 2D human rod model. They usually require a well-calibrated multi-camera system and are computationally expensive.

The paper proposes a method of multi-features fusion, using Kinect 3D skeletal data collected from three aspects of gait recognition research. Firstly, propose a new human static rod-shaped skeletal model based on the theory of Yoo [4], from the human skeletal structure. Secondly, when human is walking, the hip angle changes are

Q. Liu et al. (Eds.): CENet 2018, AISC 905, pp. 653–661, 2020.
https://doi.org/10.1007/978-3-030-14680-1_72

extracted from the dynamic angle information. Finally, the important information in the key frame of the human gait sequence is extracted as the gait feature from the information of the special gait. Then, use DTW and KNN algorithm to classify these features respectively, which will be fused at the decision level. Experimental results on the widely used UPCV gait database are presented to demonstrate the effectiveness of the proposed method. Meanwhile, it also has better robustness.

2 Features Fusion

2.1 Kinect and Gait Cycle

Kinect is Microsoft's launch of a new somatosensory device in 2010, which is equipped with a RGB camera, a depth sensor composed of an infrared light emitter and a camera sensible to this infrared. Because it can simultaneously capture RGB images, depth images of the scene, human skeletal data stream and voice, researchers use to develop different applications. The sensor and its API simplify the steps of capturing video stream data, image pre-processing and application of computer vision algorithms to create a skeleton model from a subject [5]. The Kinect sensor is a human movement tracker that does not require special markers or direct contact with the subject. The proposed method only uses kinect to capture the skeletal data stream. With 30 fps speed, Kinect provides 20 skeletal joints of the three-dimensional coordinates of the coordinate system.

In the previous work, it is found that people walking under normal circumstances is considered a cyclical movement, and the limbs of the state of motion show a cyclical change, which is the most prominent cyclical changes in the distance between the feet (i.e. the foot pitch). Thus, features extracted from a single gait cycle may represent complete gait information. In this paper, the definition of gait cycle is to select the interval between the pitch changes, the interval between the three consecutive peaks of the data between the peaks as a gait cycle sequence, that is selected as the starting point of the maximum foot spacing, walking in the natural state until restore the original state, as shown in Fig. 1. The abscissa represents the number of frames, and the ordinate represents the step size of the person walking.

2.2 Features

The characteristics of this paper are divided into three parts: The rod-like model of the human body, the angle of the hip joint and the posture of the key sequence of the gait.

Static Attributes

For each frame captured the measurements of several body segments, shown in Fig. 2, are calculated using the Euclidean distance between joints, in a similar fashion to the methodology employed in [2]. The parameters of which are six kind of length of the skeletal structure with more significant differences, which are head and neck length, shoulder width, hip width, upper torso length, arm length and leg length. As the symmetry between human body arm and leg, therefore, it selects one of the both sides as its length. The mean and standard deviation of each body segment and height over all frames of a walk are calculated.

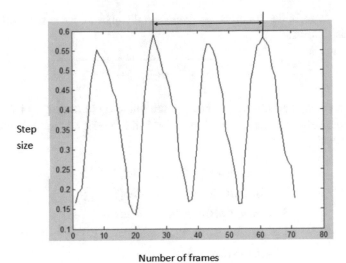

Step size

Number of frames

Fig. 1. Gait cycle.

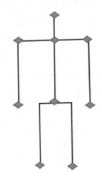

Fig. 2. Stick model.

Fig. 3. Key gait sequence.

The distance formula chosen is the Euclidean distance between two joints, which is calculated as follows:

$$dist(i,j) = \sqrt{(x_i - x_j)^2 + (y_i - y_j)^2 + (z_i - z_j)^2} \tag{2.1}$$

Where (x_i, y_i, z_i) and (x_j, y_j, z_j) represent the i-th and j-th human skeletal coordinate values, respectively. The parameters for the rod-like model of the skeleton are calculated as follows:

$$
\begin{aligned}
a1 &= \{mean(dist(1,2)), std(dist(1,2))\}; \\
a2 &= \{mean(dist(3,4)), std(dist(3,4))\}; \\
a3 &= \{mean(dist(13,14)), std(dist(13,14))\}; \\
a4 &= \{mean(dist(2,12)), std(dist(2,12))\}; \\
dist(4,6) &= dist(4,5) + dist(5,6); \\
a5 &= \{mean(dist(4,6)), std(dist(4,6))\}; \\
dist(14,16) &= dist(14,15) + dist(15,16); \\
a6 &= \{mean(dist(14,16)), std(dist(14,16))\}; \\
SF &= \{a1, a2, a3, a4, a5, a6\};
\end{aligned}
\tag{2.2}
$$

Angle Attributes

For the study of gait recognition, many people have studied the characteristics of the angle. Ahmed et al. [7] proposed a method of gait which utilized non-directly connected bone points between the joint relative distance (JRD) and joint relative angle (JRA). This paper presents a method for angle change of hip from the aspect of dynamic angle information. It gets the angle θ between joints i, j and k in the following way:

$$
\begin{aligned}
A &= \sqrt{(x_i - x_j)^2 + (y_i - y_j)^2 + (z_i - z_j)^2} \\
B &= \sqrt{(x_i - x_k)^2 + (y_i - y_k)^2 + (z_i - z_k)^2} \\
C &= \sqrt{(x_j - x_k)^2 + (y_j - y_k)^2 + (z_j - z_k)^2} \\
\cos \theta &= \frac{A^2 + C^2 - B^2}{2AC}
\end{aligned}
\tag{2.3}
$$

where joints i, j and k represents the hip joint, and knee joint respectively.

Key Gait Attributes

According to the human body model, the whole division is divided into 15 parts, and the dividing method and its parameters are shown in [9]. By calculating the distance from the center of mass to the key joint point, the key frames is indicated. The most important thing is to extract the features from the 3D joint data in the key frame, and use the extracted features to identify the human gait. The relative distance between the coordinate data of the joints is used as the feature for human gait recognition. Skeleton representation and joint index, as shown in Fig. 3, the feature extraction method is as follows:

$$bx1 = mean(\text{abs}(x(3) - x(4)));$$
$$bx2 = mean(\text{abs}(x(6) - x(9)));$$
$$bx3 = mean(\text{abs}(x(13) - x(14)));$$
$$bx4 = mean(\text{abs}(x(16) - x(19)));$$
$$bx5 = mean(\text{abs}(x(6) - x(19)));$$
$$bx6 = mean(\text{abs}(x(9) - x(16))); \quad (2.4)$$
$$by1 = mean(\text{abs}(y(12) - (y(17) + y(20))/2));$$
$$by1 = mean(\text{abs}(y(12) - (y(15) + y(18))/2));$$
$$b1 = mean(dist(12, 16));$$
$$b2 = mean(dist(12, 19));$$
$$KF = \{bx1, bx2, bx3, bx4, bx5, bx6, by1, by2, b1, b2\};$$

3   Fusion Classifier

The paper chooses the additive rule for merging. It uses the KNN algorithm to match the characteristics of the skeletal rod model, the hip joint and the key gait, respectively, and gets the matching scores. The final recognition rate is acquired by the Decision-level fusion. DTW and KNN classification algorithm and weighted matching fusion implementation steps are as follows:

Input: 3D gait skeletal data set.
Output: recognition rate.
Step 1: Enter the skeletal coordinate sequence to get a cycle of data. (Training set and test set).
Step 2: The characteristic values of SF, AF and DF are obtained according to the above formula. The characteristics of the training sample and the test sample are set as R and T.
Step 3: The following operations were performed on the characteristics of the above-described test samples, and the matching scores and weights of each feature are calculated.

(1) Initialization K;
(2) The distance metric was calculated using the DTW algorithm. Find the dimensions of Rand T respectively, denoted by q and p. Suppose the distance matrix of p * q was d, and the Euclidean distance between two points was chosen as the distance metric. The path cumulative cost function D(p, q) represents the minimum accumulated cost distortion value from the starting point to the point (p, q).

According to the following formula, the final cumulative cost distortion value can be obtained as the distance measure of KNN [8].

If $p = 1, q = 1, D(1, 1) = d(1, 1)$
Else if $p > 1, q = 1, D(p, q) = d(p, q) + (p - 1, q)$
Else if $p = 1, q > 1, D(p, q) = d(p, q) + d(p, q - 1)$
Else

$$p > 1, q > 1, D(p, q) = d(p, q) + \min\{d(p - 1, q - 1), d(p - 1, q), d(p, q - 1)\} \tag{3.1}$$

From the above DTW algorithm formula can be calculated between the training sample and the test sample the best match distance.

(3) The distance obtained by (2) is classified according to the nearest neighbor classification algorithm

(4) The recognition rate:

$$R = (\text{Total number of correct matches}/\text{total number of test samples}) * 100\% \tag{3.2}$$

(5) Assume that the weight w_l of Tl is the reciprocal of its error rate, i.e.

$$w_l = R_l/(1 - R_l) \tag{3.3}$$

where l represents the number of features.

Step 4: The matching scores obtained for each eigenvalue of the test sample are normalized using the Min-Max criterion:

$$r^* = \frac{r - \min(R)}{\max(R) - \min(R)} \tag{3.4}$$

Where the matching score set of the classifier R, $r \in R$, r^* is the normalized score.

Step 5: Decision-level fusion. This paper uses the weighted addition for fusion, to get the final recognition rate R, the specific calculation is shown below.

$$R = \sum_{l=1}^{n} (w_l * r^*) \tag{3.5}$$

4   Experiments and Results

In this paper, a simulation experiment is carried out in Matlab environment using the UPVC skeleton gait dataset published by Kastaniotis et al. [6]. The UPCV database contains data for 30 people, each containing five gait sequences, each sample contains four gait sequences, a total of 120 gait sequences. The paper choose the first three sample sequences as the training set and the fourth sample sequence as the test set for each sample.

At the beginning of this paper, three classifiers are used: KNN, BP, SVM. Based on the recognition rate of SF feature, the recognition rates obtained by the three classifiers are shown in Table 1, and the recognition rates of different classifiers under SF feature are given. The experimental results show that the recognition rate of KNN is higher

Table 1. Recognition rate of SF by different classifier.

Classifier	SFRecognition rate
KNN	86.67%
BP	76.66%
SVM	50.00%

than that of SVM and BP algorithm, because the training samples are few. So the final classifier is KNN.

KNN algorithm, the choice of k values to SF and DF, for example, k take 1–10 values, as is shown in Fig. 4 there are different values of SF and DF recognition rate. Some experiments have shown that the KNN recognition rate of SF and DF is the highest when k = 1. Thus, the value of k chosen in this paper is 1.

In order to evaluate the classification results, the correct classification rate CCR (Correct Classification Rate) is used as the evaluation index. Equation (3.2) defines the

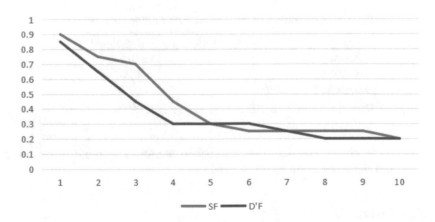

Fig. 4. Selection of K value.

Table 2. DTW and KNN algorithm to get the correct recognition rate (CCR).

The database size	SF	DF	AF	Fusion	Number of errors
10	100%	90%	70%	100%	0
20	90%	85%	50%	90%	2
25	88%	76%	44%	92%	2
30	86.67%	66.67%	43.33%	90%	3

correct classification rate CCR. Table 2 shows the correct classification rate for DTW+KNN recognition using the gait feature selected in this experiment.

It can be seen from Table 2 that when the number of samples is small, the recognition rate of static feature (SF) is equivalent to the recognition rate after fusion; when the number of samples is large, the role of fusion begins to appear.

In order to estimate the performance of using decision-making fusion, the fusion rate and the average of the correct recognition rate (CCR) of each feature are compared

Table 3. Comparison of feature fusion.

The database size	CCR average	CCR MAX	Fusion	Up and down (average)	Up and down (MAX)
10	86.67%	100%	100%	↑	–
20	75%	90%	90%	↑	–
25	69.33%	88%	92%	↑	↑
30	65.55%	86.67%	90%	↑	↑

with the maximum value (MAX). From Table 3, it can be seen that the recognition rate after fusion is much better than the average CCR of each feature.

Table 4 shows the comparison of the recognition rate of this method with other methods. The experiment uses the same number of database samples as the method.

Table 4. Comparison of recognition rates of different methods.

Method	The database size	Recognition rate
Literature [5]	9	85.10%
Our work		100.00%
Literature [6] (MNPD)	30	77.27%
Our work		90.00%

The comparison shows that the method is better than other methods in overall performance, and the recognition result is satisfactory.

5   Conclusion

The paper present a multiple characteristics based approach to gait recognition based on Microsoft Kinect. The features included a rod model of the human skeleton, the angle of the hip joint, and key gait features. These features are good for the dynamic and static characteristics of the human gait. The DTW and KNN algorithm is used to match the three gait features respectively, and the fusion is achieved at the decision level. The comparison of different recognition methods and different databases shows that the method can effectively recognize the gait.

Acknowledgement. This study is supported by Science and Technology Project of Guangdong (2017A010101036).

References

1. Lee, T.K.M., Belkhatir, M., Sanei, S.: A comprehensive review of past and present vision-based techniques for gait recognition. Multimed. Tools Appl. **72**(3), 2833–2869 (2014)
2. Lee, L., Grimson, W.E.L.: Gait analysis for recognition and classification. In: Proceeding of Fifth IEEE International Conference on Automatic Face and Gesture Recognition, pp. 148–155 (2002)
3. Cunado, D., Nixon, M.S., Carter, J.N.: Automatic extraction and description of human gait models for recognition purposes. Comput. Vis. Image Underst. **90**(3), 1–41 (2003)
4. Yoo, J.H., Nixon, M.S., Harris, C.J.: Extracting gait signatures based on anatomical knowledge. In: Proceeding of The British Machine Vision Association and Society for Pattern Recognition, London (2002)
5. Preis, J., Kessel, M., Linnhoff-Popien, C.: Gait recognition with Kinect. In: 1st International Workshop on Kinect in Pervasive Computing (2012)
6. Kastaniotis, D., Theodorakopoulos, I., Theoharatos, C., Economou, G., Fotopoulos, S.: A framework for gait-based recognition using Kinect. Pattern Recognit. Lett. **68**, 327–335 (2015)
7. Ahmed, F., Paul, P.P., Gavrilova, M.L.: DTW-based kernel and rank-level fusion for 3D gait recognition using Kinect. Vis. Comput. Int. J. Comput. Graph. **31**(6–8), 915–924 (2015)
8. Luo, Z.P., Yang, T.Q.: Gait recognition by decomposing optical flow components. Comput. Sci. **43**(9), 295–300 (2016)
9. Chen, X., Yang, T.Q.: Cross-view gait recognition based on human walking trajectory. J. Vis. Commun. Image Represent. **25**, 1842–1855 (2014)

Automatic Retrieval of Water Chlorophyll-A Concentration Based on Push Broom Images

Yingying Gai[1(✉)], Enxiao Liu[1], Yan Zhou[1], and Shengguang Qin[2]

[1] Institute of Oceanographic Instrumentation,
Qilu University of Technology (Shandong Academy of Sciences),
Shandong Provincial Key Laboratory of Marine Monitoring Instrument
Equipment Technology, National Engineering and Technological Research
Center of Marine Monitoring Equipment, Qingdao 266061, China
gyygyy1234@163.com
[2] Qingdao Leicent Limited Liability Company, Qingdao 266100, China

Abstract. In order to fill the domestic blank in automatic retrieval of water Chlorophyll-a concentration using push broom images, an automatic retrieval system of Chlorophyll-a concentration based on push broom images was designed and implemented on ENVI redevelopment platform in this paper. An airborne push broom hyper-spectrometer called Pika L with high spectral and spatial resolutions provides hardware support for retrieving more accurate and real-time Chlorophyll-a concentration. According to the characteristics of Pika L images, the automatic retrieval system mainly includes geometric correction and mosaicking, radiometric calibration, atmospheric correction and Chlorophyll-a concentration retrieval. The results show that the automatic processing of Pika L push broom images based on the ENVI redevelopment is a feasible technical solution and it provides technical support for automatic and real-time retrieval of water Chlorophyll-a concentration.

Keywords: Push broom image · Chlorophyll-a · Automatic retrieval · ENVI · Redevelopment

1 Introduction

Chlorophyll-a concentration is an important parameter for water quality monitoring. Large-scale, accurately and real-time monitoring of water Chlorophyll-a concentration is of great significance for mastering the pollution status timely [1]. Remote sensing has been proved to be an effective high-tech means for obtaining large-scale water quality information in real time. With the continuous improvement of application requirements, people have put forward higher requirements for the timeliness and ease of operation of water quality monitoring. It is hoped that water quality information can be obtained more conveniently and intuitively, and data can be processed more timely and accurately. Therefore, all these expectations give rise to higher requirements for the spectral, spatial and temporal resolutions of sensors and the automation of retrieval process [2].

© Springer Nature Switzerland AG 2020
Q. Liu et al. (Eds.): CENet 2018, AISC 905, pp. 662–670, 2020.
https://doi.org/10.1007/978-3-030-14680-1_73

Pika L is a hyper-spectrometer designed and produced by Resonon Company of USA. It has a high spectral resolution of 2.1 nm in spectral range of 400–1000 nm. The image acquired from Pika L can be geometrically corrected to a spatial resolution of 0.1 m, and due to its small size, it is convenient for being boarded on the UAV (Unmanned Aerial Vehicle). So it has now become a new way of water quality monitoring for remote sensing. However, the push broom data obtained by Pika L need to be pre-processed before retrieval. Currently, there is no domestic water quality monitoring system based on Pika L data, and it is difficult to meet the requirements for automatic and rapid data processing. Chen designed and implemented a water pollution monitoring visual system, but it does not include push broom image mosaicking [3]. SpectrononPro software for Pika L can do mosaicking, but it cannot do water quality monitoring and it is not free for users [4]. In this paper, an automatic Chlorophyll-a concentration retrieval system based on Pika L push broom data is designed and developed, and some core functions are realized in this system. It provides technical support and application demonstration for promotion of Pika L data in water quality monitoring.

2 System Structure Design

According to the demands of water Chlorophyll-a concentration retrieval by remote sensing images, and characteristics of line push broom imaging, the automatic retrieval system mainly includes data input, data processing and data storage. Data processing module consists of four parts: image geometric correction and mosaicking, radiometric calibration, atmospheric correction and Chlorophyll-a concentration retrieval. The functional structure of the system is shown in Fig. 1.

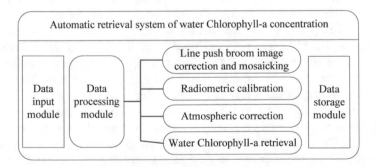

Fig. 1. Functional structure of the system.

Data input module mainly includes direct input and indirect input. Image files and GPS/INS navigation files use an indirect input method, which uses the rich data reading functions provided by IDL (Interface Description Language) to read them, as shown in Fig. 2(a). The original image obtained from Pika L is in BIL format, which can be read by the header file based on band combinations. The GPS/INS navigation file is in LCF format and can be converted to TXT format for reading. The parameters which need to be set in the data processing process are input by direct method, that is, the parameters are set and read through GUI (Graphical User Interface). Then, the input images and navigation data are processed according to the input parameters from GUI, as shown in Fig. 2(b).

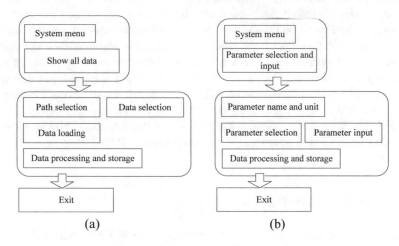

(a)   (b)

Fig. 2. Module design of data input.

The data processing module includes four sub-modules: line push broom image correction and mosaicking, radiometric calibration, atmospheric correction and water Chlorophyll-a retrieval.

The line push broom image correction and mosaicking module mainly completes the geometric correction of the image. During push scanning, the hyperspectral image recorded by CCD (Charge Coupled Device) detector is a line along the direction of flight. Because the flight platform equipped with the spectrometer cannot always maintain an ideal attitude during flight, the position, yaw angle, pitch angle and roll angle of the flight platform are changing constantly and randomly. Therefore, the geometrical distortion of the image is caused by the inconsistency of external orientation elements corresponding to each scanning line. So each line needs to be geometrically corrected, and after geography information is added, they can be mosaicked into an image. The original Pika L image is a hyperspectral image that is arranged in the order of push broom sequence and the bands have been synthesized. ENVI (The Environment for Visualizing Images) redevelopment functions can be used to read the data line by line. Combined with GPS/INS data, each line can be geometric corrected. At last, Geographic information can be used to mosaic them together. Mosaicking images must have overlapped and missed parts. The overlapped parts are well processed during mosaicking, and the missed parts can be filled by interpolation algorithm. The flow design is shown in Fig. 3.

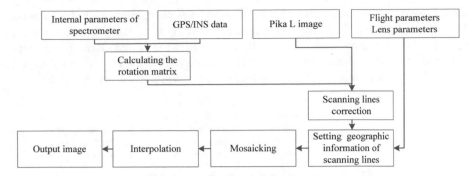

Fig. 3. Flow design of geometric correction and mosaicking module.

The radiometric calibration module completes the radiometric calibration of the image, that is, DN (Digital Number) value is converted into radiance by radiometric calibration. The radiometric calibration parameters are provided with the spectrometer and file format is ICP, which can be interpreted as TXT format. Different bands need to be processed by different calibration parameters.

FLAASH (Fast Line-of-sight Atmospheric Analysis of Spectral Hypercubes) module is used to do atmospheric correction. IDL provides FLAASH batch interface called FLAASH_batch, so that FLAASH engineering parameters, water vapor and aerosol retrieval parameters, image parameters and position of study area can be easily set by this interface. FWHM (Full Width at Half Maximum) of the image is an essential parameter in FLAASH module, while the original header file acquired by Pika L does not contain FWHM parameter. Therefore, before running the FLAASH module, FWHM needs to be set. That is, the spectral resolution of the spectrometer must be written into the header file. The flow design of atmospheric correction module is shown in Fig. 4.

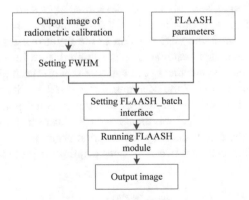

Fig. 4. Flow design of atmospheric correction module.

Chlorophyll-a retrieval is based on the remote sensing reflectance of water after geometric correction, radiometric calibration and atmospheric correction. The retrieval model is established through repeated verification of image retrieval results and laboratory measurement results. The model establishes the relationship between remote sensing reflectance and Chlorophyll-a concentration. By processing reflectance among different bands, the concentration can be reflected on the image. That is, reflectance obtained after atmospheric correction can be calculated into concentration by retrieval model. The flow design of Chlorophyll-a retrieval module is shown in Fig. 5.

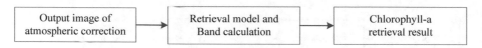

Fig. 5. Flow design of Chlorophyll-a retrieval module.

The data storage module is mainly for saving the processing results according to the user's needs, including the setting of saving path and saved image name and format. ENVI provides visualization interface to hyperspectral images and spectral curves, so the image display can be implemented on ENVI platform without redevelopment on IDL platform.

3 System Key Technologies

3.1 Use of GPS/INS Data

The instability of flight platform attitude causes the image scanning lines to be inter-laced [5]. GPS/INS data can provide geographic information for the correction of scanning lines. Usually, the body coordinate system and the geographic coordinate system, respectively expressed as $ox_by_bz_b$ and $ox_ny_nz_n$, are introduced to describe the attitude of the flight platform, as shown in Fig. 6. The change of flight platform attitude causes angular displacements between two coordinate systems, which are represented by yaw angle, pitch angle, and roll angle. These three angles are expressed as ψ, θ and ϕ, respectively. They are recorded in the GPS/INS navigation file as the flight time changes. At the same time, the GPS position of the flight platform is also recorded in the navigation file once each scanning line is acquired. It is in one-to-one correspondence with the angular displacement data and varies with the flight time.

The distortion correction of the scanning line caused by the attitude of the flight platform is finally a projection mapping of two-dimensional plane points from body coordinate system to geographic coordinate system. According to plane homography [6] and three-dimensional coordinate system rotation principle [7], the plane coordinate conversion relationship between two scan lines acquired from the same imager center can be expressed as:

$$q_b = MC_n^b M^{-1} q_n \tag{3.1}$$

Here, q_b and q_n represent the plane points in body coordinate system and geographic coordinate system respectively, M is internal parameter matrix of the imager,

C_n^b is the rotation matrix from the geographic coordinate system to the body coordinate system and expressed as follows:

$$C_n^b = \begin{bmatrix} \cos\theta\cos\psi & \cos\theta\sin\psi & -\sin\theta \\ \sin\theta\sin\phi\cos\psi - \cos\phi\sin\psi & \sin\theta\sin\phi\sin\psi + \cos\phi\cos\psi & \cos\theta\sin\phi \\ \sin\theta\cos\phi\cos\psi + \sin\phi\sin\psi & \sin\theta\cos\phi\sin\psi - \sin\phi\cos\psi & \cos\theta\cos\phi \end{bmatrix}$$

$$(3.2)$$

Distorted images in the body coordinate system can be corrected into orthographical images in the geographic coordinate system by Eq. (3.1). When all scanning lines are all corrected into orthographical images in the same geographic coordinate system, the mosaicking can be performed according to the GPS information of the imaging center. At this time, the latitude and longitude of the imaging center obtained from GPS can be considered as those of the orthographical point on the ground, that is, the latitude and longitude of the center point of scanning line after correction. First, ground spatial resolution and projection method are properly set. Then, the image parameters (including row number and column number) and geographic information after first two scanning lines are mosaicked are calculated, and the first two scanning lines are mosaicked by ENVI's mosaic batch function called MOSAIC_DOIT. After that, the image parameters and geographic information after mosaicked image obtained from previous step and the third scanning line is mosaicked are calculated, and the third scanning line is mosaicked. Mosaicking is performed in turn until all the scanning lines are mosaicked. At last, a regional image is obtained. The mosaicking process is shown in Fig. 7.

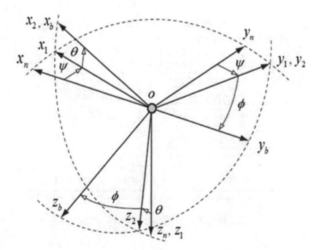

Fig. 6. Angular displacement and coordinate system rotation.

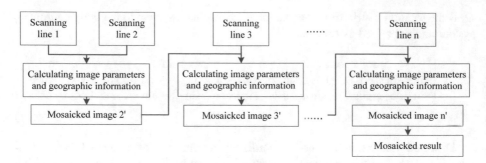

Fig. 7. Mosaicking process of scanning lines.

3.2   FLAASH Interface and Parameters Setting

On ENVI redevelopment platform, atmospheric correction can be realized through the FLAASH interface called FLAASH_batch. FLAASH module needs to set a large number of input parameters, including input and output files, water vapor absorption band, atmospheric model, aerosol model, visibility, image acquisition time, image latitude and longitude, sensor height, ground average elevation, ground spatial resolution and so on. According to the needs of water Chlorophyll-a concentration retrieval by Pika L sensor, not all parameters need to be set or modified before each execution, only input and output files, image acquisition time, image latitude and longitude, atmospheric model, sensor height and spatial resolution need to be set. Input and output files are set according to the design of data input and storage modules using the IDL data operation function. Other parameters are read through GUI, as shown in Fig. 8.

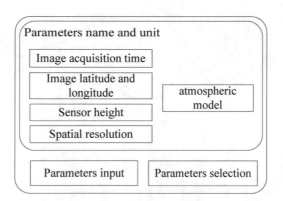

Fig. 8. FLAASH parameters design.

FWHM of image needs to be written into the header file in advance in order to ensure the successful running of FLAASH_batch. FWHM of the image acquired by Pika L sensor is 2.1 nm. The flight height of UAVs equipped with Pika L sensors is approximately 100–150 m. Due to the low flight height, the influence of atmosphere to the image radiance can be almost negligible. Water vapor is a major factor affecting the calculation of the remote sensing reflectance. Therefore, water vapor retrieval is required. Usually, water vapor absorption band is set to be 820 nm. According to the conversion relationship between the unit of radiance after radiometric calibration and the required unit of FLAASH module, a unit conversion factor needs to be set of 1000. The image format after atmospheric correction is the same as that of original input image and corrected image is saved in the specified path and folder.

4 Conclusion

In this paper, an automatic system for retrieval of water Chlorophyll-a concentration based on line push broom images is studied. The visual interface, image processing functions and interfaces provided by ENVI redevelopment platform greatly simplify system implement complexity. Some complex remote sensing image processing processes can be easily implemented by functions, and only key parameters need to be set. At present, the system has been completed and realized, and will be used as supporting data processing software for the marine airborne hyperspectral imager researched and developed by Institute of Oceanographic Instrumentation, Qilu University of Technology (Shandong Academy of Sciences). The marine airborne hyperspectral imager has the same data acquisition principle as Pika L, and data formats are also the same. However, the marine airborne hyperspectral imager is still in the late experimental stage and there is no relevant image to use. Therefore, Pika L data is selected as the subject of this study. In multiple application demonstrations, this system provides to users with standardized seawater Chlorophyll-a concentration retrieval products, reducing the application difficulty of line push broom remote sensing data in marine water quality monitoring, and promoting application of the marine airborne hyperspectral imager in the marine monitoring.

Because the retrieval model is regional, and retrieval parameters in one area cannot be directly applied to another area. Therefore, the retrieval model needs further research and verification. Besides this, we currently only implement several core functions that users need, and the system is still in the software coding stage. There are still a lot of work for function improvement and testing. The system also needs continuous improvement and upgrading, and finally reaches the degree that fully meets the actual needs of users.

Acknowledgement. This work is supported by Special Fund for Marine Public Welfare Scientific Research (201505031), National Key Research and Development Program of China (2016YFC1400300), Shandong Natural Science Foundation for Youths (ZR2017QD009), National Natural Science Foundation of China (61701287), and Key Research and Development Plan of Shandong Province (2017GGX10134).

References

1. Cao, Y.: Remote Sensing Monitoring Technology and Application of Water Quality in Macrophytic Lake. Donghua University, Shanghai (2016). (in Chinese)
2. Zhu, M.H.: Data Processing Method Based on Water Quality Monitoring System. Shanghai University, Shanghai (2008). (in Chinese)
3. Chen, T.: Visual design and implementation of water quality pollution monitoring based on IDL. University of Electronic Science and Technology of China, Chengdu (2010). (in Chinese)
4. SpectrononPro: Hyperspectral imaging system software manual.http://docs.resonon.com/spectronon/pika_manual/html/
5. Ji, Y.: Research on Geometric Correction Method of Pushing and Scanning Pavement Images by Carborne CCD Camera Aided by IMU/DGPS. Capital Normal University, Beijing (2008). (in Chinese)
6. Cheng, Z.G., Zhang, L.: An aerial image mosaic method based on UAV position and attitude information. Acta Geodaetica Cartogr. Sin. **45**(6), 698–705 (2016). (in Chinese)
7. Wang, S.M., Zhang, A.W., Hu, S.X., Sun, W.D., Meng, X.G., Zhao, W.J.: Geometric correction of linear push-broom hyperspectral camera side-scan imaging. Infrared and Laser Engineering **43**(2), 579–585 (2014). (in Chinese)

Hybrid Theme Crawler Based on Links and Semantics

Kang Zhao, Yang Yang$^{(\boxtimes)}$, Zhipeng Gao, and Long Cheng

State Key Laboratory of Networking and Switching Technology,
Beijing University of Posts and Telecommunications, Beijing 100876, China
yyang@bupt.edu.cn

Abstract. Common theme crawler generally analyses the page content or link structure, without solving the problem of computational complexity and easy "myopia", resulting in the page of recall and precision is not high. This paper introduces a mixed theme decision strategy, which fully considers the text content and link structure of the page. By introducing knowledge map database and entity database, the computational complexity is simplified and the judgment accuracy is increased. The experiment shows that the rate of inspection and precision is greatly improved.

Keywords: Crawler · HowNet · Links · Semantics · Similarity

1 Introduction

As an important tool for searching network resources, search engine plays an irreplaceable role in network life. Existing search engines, such as Google and Baidu, start from the seed page when collecting web page information, traversing the entire network as much as possible in the way of graph traversing, aiming to reach all users and cover all the returned results. However, with the rapid development of network technology, the network resources become diversified, heterogeneous, open and the amount of web data far exceeds the coverage of general search engines [1].

Theme web crawler is a crawler with the function of theme correlation determination. The crawling of theme related pages is completed by theme web crawler. Theme web crawler is very different from general web crawler in grasping strategy and analyzing web pages. General Web crawler way to traverse the entire Web network, indiscriminate crawled on all links, regardless of the scraping of the Web page content. Theme web crawler uses a certain search strategy to traverse web resources based on predetermined themes starting with seed links. If the captured page is not related to the theme, it is discarded. The theme crawler improves the quality of grasping resources and the utilization rate of network resources, saving hardware and software resources.

In our study, we propose a hybrid theme crawler. First, we use HowNet and knowledge map named zhishi.me to calculate the text similarity. Then, the improved PageRank algorithm is used to analyze the web page link structure and combine it with text similarity. The hybrid algorithm overcomes the local optimal solution and improves the crawler speed.

© Springer Nature Switzerland AG 2020
Q. Liu et al. (Eds.): CENet 2018, AISC 905, pp. 671–680, 2020.
https://doi.org/10.1007/978-3-030-14680-1_74

The rest of this paper is organized as follows. Section 2 discusses related work. Section 3 describes the traditional similarity calculation method and introduces the proposed approach. Simulation results are presented in Sect. 4 and this paper is concluded in Sect. 5.

2   Related Work

De Bra et al. proposed the fish-search algorithm, which vividly compared the theme network crawler to the Fish in the sea. When the Fish found the food (related website), they could continue to reproduce (to crawl forward) [2]. When the Fish could not find the food, they would gradually die (to stop crawling). M Hersorici proposed shark-search algorithm, which mainly makes up for the shortcomings of the fish-search algorithm.

Chakrabati et al. were the first researchers to put forward the generic structure of theme network crawler, and the proposed generic structure of theme network crawler laid the foundation for later research of theme crawler.

Jon Kleinberg proposed the HITS algorithm, which added two attribute values of Authority and Hub for each web page. The algorithm iterated and calculated two weight values, giving priority to return the web pages with large weight for users. The evaluation algorithm based on link structure only considers the distribution characteristics of web pages in the network, and does not consider the text content of web pages.

Ahmed Patel proposed a thematic crawler strategy based on t-graph (Treasure Graph) algorithm. The algorithm is based on this algorithm, and by building a model of t-graph, it guides the web crawler.

Shokouhi et al. proposed an intelligent crawler called Gcrawler, which uses genetic algorithms to estimate the optimal path and extend the initial keyword during the crawling process.

It can be seen that in the aspect of theme crawler, many researches have been done before, and research on improving the performance of theme crawler is still a hot topic at present.

3   Problem Formulation

3.1   Semantic Similarity Calculation

HowNet Semantically Similarity

The semantic similarity calculation method based on information content (IC) makes use of the information contained in the original sememe node itself for comparison and calculation, which greatly improve the accuracy of calculation [3]. The accuracy of IC calculation is crucial to the result of semantic similarity.

Traditional IC Calculation Method

(1) A sememe node located on a non-leaf node has less information content than a conceptual node located on a leaf node [4].

$$IC(c) = 1 - \frac{log(hypo(c) + 1)}{log(Node_{max})} \tag{3.1}$$

$hypo(c)$ represents the total number of the lower bits of the sememe node c in HowNet, and $Node_{max}$ represents the total number of the sememe nodes in HowNet.

(2) The depth of the concept node and the number of lower words affect the size of IC [5].

$$IC(c) = k \times (1 - \frac{log(hypo(c) + 1)}{log(Node_{max})} + (1 - k) \times \frac{log(Depth(c))}{log(Depth_{max})} \tag{3.2}$$

$Depth(c)$ represents the depth of the sememe in HowNet, $Depth_{max}$ represents the maximum depth of HowNet, and k is the parameter.

(3) In addition to the number of lower words and their own depth, the depth of each lower word will also affect the IC size [6].

$$IC(c) = \frac{log(Depth(c))}{log(Depth_{max})} \times (1 - \frac{log(\sum_{\alpha \in hypo(c)} \frac{1}{Depth(\alpha)} + 1)}{log(Depth_{max})} \tag{3.3}$$

Improved IC Computing Methods

(1) The IC value of the upper word has an effect on the lower word. The more information it contains, the more information it points to.

$$IC(c)_1 = \frac{e^{\lambda * IC(b)} - e^{-\lambda * IC(b)}}{e^{\lambda * IC(b)} + e^{-\lambda * IC(b)}} * \frac{log(Depth(c))}{log(Depth_{max})} * (1 - \frac{log(\sum_{\alpha \in hppo(c)} \frac{1}{Depth(\alpha)} + 1}{log(Depth_{max})}) \tag{3.4}$$

(2) The IC value of polysemy is smaller. Because polysemous words tend to have more uncertainty, they contain less information.

$$IC(c)_2 = \frac{e^{\lambda * IC(b)} - e^{-\lambda * IC(b)}}{e^{\lambda * IC(b)} + e^{-\lambda * IC(b)}} * \frac{log(Depth(c))}{log(Depth_{max})} * (1 - \frac{log(\sum_{\alpha \in hppo(c)} \frac{1}{Depth(\alpha)} + 1}{log(Depth_{max})}) \\ * \frac{2 + N}{1 + N} \tag{3.5}$$

N is the number of synonyms in the sememe.

(3) The IC values of synonyms and synonyms are the same. In the process of calculating IC value, synonyms and synonyms should not differ greatly due to their similar semantics, so the set of synonyms and synonyms should have the same IC value.

$$IC(c)_3 = \frac{\sum_{i=0}^{m} IC(c)_{2i}}{M} \tag{3.6}$$

M is the number of synonyms and synonyms.

Text Similarity

The traditional method divides the webpage text into words, constructs the webpage text vector, compares it with the theme text vector, and obtains the text similarity by calculating the cosine similarity.

However, due to too many text words, the vector dimension is too large and the calculation is complicated. So we need to do vector reduction.

(1) First screening using TF-IDF
   Studies have shown that the feature items that are really useful for classification often account for only 5% to 10% of the total feature items. Using tf-idf, only take the first 20% of words.
(2) The synonym and near synonym are deweighted using HowNet.
   Real texts there are a large number of synonyms and near synonyms, in large-scale corpus, they show the approximate statistical rule, with the same characteristics of the evaluation function is applied to a set of synonyms or near-synonyms, their function value is approximately equal. In feature selection, this set of synonyms or synonyms is either filtered out to cause the loss of semantic information or selected into feature subset. Become a set of "morphological" independent rather than "semantic" independent feature items. This is in conflict with the assumption that the feature term of VSM is orthogonal, which seriously affects the accuracy of classification.
   HowNet uses W_C, W_E, and DEF to distinguish synonyms and near synonym. If only W_C is different, W_E and DEF are the same, they are synonyms. If W_C and W_E are different, but DEF is the same only, it is a synonym. Using this feature, synonyms and near synonyms can be merged into the same feature term.
(3) Repeated synonyms and synonyms should not be treated as one word only, but should be given higher weight.
   For example, if synonyms and near synonyms of a word have appeared three times, their weight is higher than that of only one time.

$$\alpha = \alpha_0 * log(1+N) \tag{3.7}$$

α_0 represents the original IC value of the sememe, and N represents the number of synonyms (including itself).

(4) According to the relationship of concepts in HowNet, words with correlation are classified into one category.

The words in HowNet are labeled, and the adjacent concepts have similar labels. The difference in the size of the labels can be used to determine whether they are adjacent words. If adjacent words are divided into one category, all words are divided into several categories.

For example, vector A is divided into A1, A2, and A3. Vector B is going to be B1, B2, B3. Each vector is only compared with the closest vector in the other group. If A1 is closest to B2, A2 is closest to B1, and A3 is closest to B3. The calculation can be simplified to A1A2, A2B1 and A3B3.

Theme Comparison Text

Theme comparison text is extracted from the existing knowledge map entity library.

The vector comparison process requires reference text, which is retrieved from the local zhishi.me entity library. The searcher retrieves the relevant text from the entity library based on the theme keyword. After theme decision module decision related web pages will be processed to store to build knowledge map, with the establishment of the self-built knowledge map, theme keyword will change accordingly, which affect contrast text that was obtained from the entities in the library.

3.2   Theme Crawler Based on Link Structure and Semantics

At present, there have been a lot of studies on theme reptiles, but there are some shortcomings. The method based on content evaluation does not take into account the link relation between web pages, so some very authoritative web pages are often neglected in the process of crawling [7]. At the same time, such as PageRank theme crawler algorithm based on link structure, although can through the analysis of web page link between the relationship from the perspective of global calculation of importance, but because it does not take into account in the process of crawl links to the theme of the correlation, so frequently, the theme "offset" [8].

To solve the above problems, a theme crawler based on content and links is proposed (Figs. 1, 2).

Correlation calculation formula:

$$\mathrm{PRi(Pi)} = (1-\mathrm{d}) + \mathrm{d}\sum_{i=0}^{n}\frac{PR(Pj)}{C(Pj)} * Sim(\alpha 0, \alpha i) \qquad (3.8)$$

There is negative feedback on the threshold of correlation determination:

If the judgment of more than 100 consecutive pages is irrelevant, the threshold is decreased. If the judgment of more than 40 consecutive pages is relevant, the threshold is increased

$$\beta = \begin{cases} \beta - \textit{offset} & N > 100 \\ \beta + \textit{offet} & M > 40 \end{cases} \qquad (3.9)$$

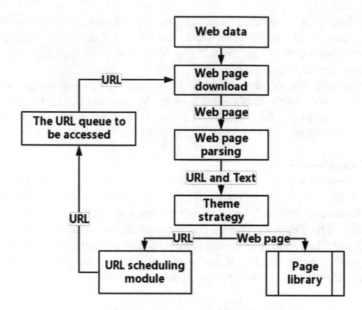

Fig. 1. Crawler workflow.

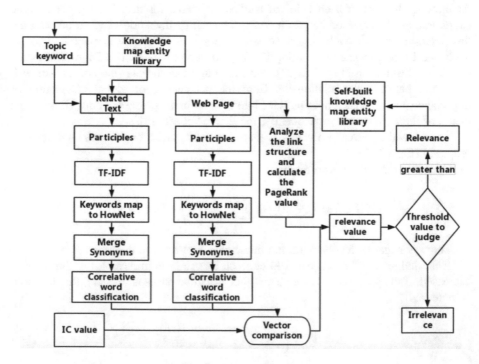

Fig. 2. Theme strategy.

4   Experimental Results and Analysis

The commonly used performance indexes to evaluate theme crawler include the precision rate, the completeness rate, the speed of crawling and average score related to theme. Among them, the accuracy rate refers to the ratio of the number of pages actually crawled to the total number of pages already crawled. Completion rate refers to the ratio of the number of theme-related web pages that have been crawled to the total number of theme-related web pages within the crawling range. Crawler speed is a measure of the number of crawlers crawling web pages per unit of time. Average score related to theme can well measure the overall performance of theme crawlers.

The Chinese classification training set provided by the computer network and distributed system laboratory of Peking University was used as the crawler data set. Compared with the word vector clustering weighted shark-search algorithm and PageRank algorithm based on tags. The experimental results are shown as follows:

Figure 3 shows that the decision algorithm based on link structure and content of the subject of the precision compared with other two kinds of algorithms are promoted, and the increase in the number of download page precision change is not big. This means that the theme judgment algorithm based on link structure and content can overcome the myopia of the original algorithm. Figure 4 shows that while with the passage of time the crawler system of recall the last will be close to 65%, but the decision algorithm based on link structure and content of the subject of recall all period, compared with other two kinds of algorithm are ascending. And this advantage is particularly significant in the short running time of the system. This means that theme determination algorithms based on link structure and content can better distinguish more relevant web pages in limited time. Figure 5 shows that the hybrid strategy is better than the word vector clustering weighted shark-search algorithm at crawling speed and is similar to PageRank algorithm based on tags. This is because the mixed

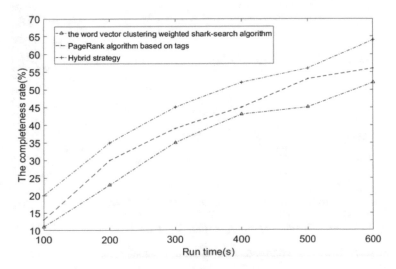

Fig. 3. The precision rate.

theme crawler reduces the amount of computation through the link structure and HowNet's prejudgment. Figure 6 shows that the mixed strategy is more accurate in terms of correlation, and PageRank algorithm based on tags is less effective because it is regardless of the text.

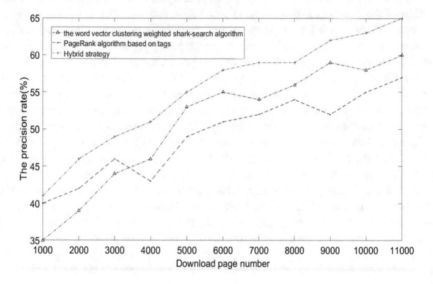

Fig. 4. The completeness rate.

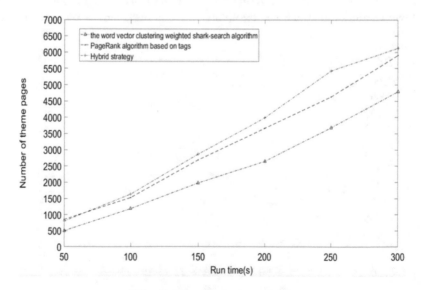

Fig. 5. The speed of crawling.

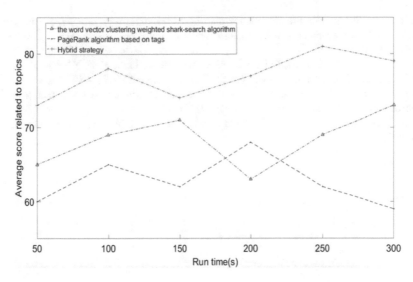

Fig. 6. Average score related to theme.

5   Conclusion

The acquisition of Internet subject information is of great value to both enterprises and governments, so it has been extensively studied in recent years. In this paper, the theme judgment crawler strategy based on link structure and content is adopted, and the role of knowledge map ontology library and entity library in the process of crawling is emphasized. The efficient subject crawler system is realized, and the computational complexity of subject crawler is reduced by vector dimensionality reduction. Through the existing knowledge map entity database and the self-built knowledge map, the theme comparison text can be obtained dynamically, and the theme page of the theme related websites can be obtained efficiently and comprehensively. The experiment shows that the subject reptile studied in this paper has high efficiency and coverage.

References

1. Yang, X.U., Wang, W.Y.: Research on Subject relevance algorithm of theme crawler. Mod. Comput. (2016)
2. Qiu, L., Lou, Y., Chang, M.: Research on theme crawler based on shark-search and PageRank algorithm. In: International Conference on Cloud Computing and Intelligence Systems, pp. 268–271. IEEE (2016)
3. Zhang, Y.F., Zheng, S.H.: Heritrix based theme crawler design. J. Chang. Univ. Technol. (2016)
4. Qiu, L., Lou, Y.S., Chang, M.: An improved shark-search algorithm for theme crawler. Microcomput. Appl. (2017)

5. Kumar, N., Singh, M.: Framework for distributed semantic web crawler. In: International Conference on Computational Intelligence and Communication Networks, pp. 1403–1407. IEEE (2016)
6. Wu, T., Liang, Y., Wu, C., Piao, S.F., Ma, D.Y., Zhao, G.Z., Han, X.S.: A chinese topic crawler focused on customer development. Procedia CIRP **56**, 476–480 (2016)
7. Wu, L., Wang, Y.B.: The research of the topic crawler algorithm based on semantic similarity aggregation. J. Commun. Univ. China (Sci. Technol.) (2018)
8. Yuan, Y.W., Lu, P.J.: Design of a topic crawler for college bidding announcement. Softw. Guid. (2018)

Network Traffic Forecasting Using IFA-LSTM

Xianbin Han and Feng Qi[(✉)]

Institute of Network Technology, Beijing University of Posts
and Telecommunications, No. 10, Xitucheng Road, Haidian District,
Beijing 100876, China
qifeng@bupt.edu.cn

Abstract. A network traffic prediction model is built based on LSTM neural network optimized with the improved firefly algorithm (IFA-LSTM). Aiming at some disadvantages of firefly algorithm including slow convergence, and easy to fall in local optimal values, we introduce a location update strategy based on the diversity of population, to avoid the optimization to fall into local optimal values. A dynamic step length updating measure is proposed to improve accuracy of the optimization, and to avoid the optimal solutions' oscillation problem. Simulation examples show that the prediction accuracy and convergent speed of the IFA-LSTM method are obviously improved,it can be used to predict network traffic.

Keywords: Network traffic · Firefly algorithm · LSTM

1 Introduction

With the development of network technology, people can access network resources more conveniently, this bring new challenges to the management of network resources. Higher standard quality of service and the monitoring requirement for network anomalies are strongly needed for modern network service. All of above require researchers to propose a new prediction models for internet traffic which is more accurate. Well trained model will be applied to many aspects such as network flow control, discovering network anomalies. Besides, network traffic data is growing exponentially, more complex, and gradually forms the big data of network traffic. The traditional network traffic predict model cannot meet the demand of big data, so a new traffic predict model for big data is urgently necessary.

For network traffic forecasting, researchers had done a lot of work and had achieved some results. According to the modelling method, it can be divided into two categories. One is a prediction method based on linear time series modelling, and the other is a nonlinear predicting method using neural network [1]. Linear time series modelling mainly include: Auto-regressive moving average model (ARMA) [2], autoregressive integrated moving average mode (ARIMA) [3]. These methods are simple and easy to implement, and have high short-term prediction accuracy. However, this method assumed that the network traffic is a stationary time series, ignored the non-linear characteristics of network traffic such as randomness, time-varying and chaotic. Therefore, it is difficult to predict the long-term network traffic.

© Springer Nature Switzerland AG 2020
Q. Liu et al. (Eds.): CENet 2018, AISC 905, pp. 681–692, 2020.
https://doi.org/10.1007/978-3-030-14680-1_75

Recent years, some researchers have introduced simple neural network and support vector machine (SVM) [4] into the prediction of network traffic, which have greatly improved prediction accuracy of network traffic. The prediction results and parameters of the neural network model are relatively large. If the parameters are not properly chosen, it is easy to converge too quickly and fall into a local optimal situation. For this reason, scholars have proposed using ant colony Algorithm [5], genetic Algorithm [6] to optimize the neural network parameters, but these algorithms have the disadvantages of high computational cost.

In this paper, aiming at shortcomings of slow convergence and adjustment difficulties of hyper parameters in LSTM (Long Short Term Memory networks) [7], we bring out firefly Algorithm [8] to optimize neural network structure to improve the performance of the models' forecast accuracy. For some disadvantages of firefly algorithm including slow convergence, and easy to fall in local optimal values, we introduce a location update strategy based on the diversity of population to avoid the optimization falling into local optimal values. Also, a dynamic step length updating measure is proposed to improve the accuracy of the optimization, and to avoid the optimal solutions' oscillation problem. With actual data of university network traffic, it is proved that the accuracy of improved predictive model can be effectively improved and can meet actual requirements.

2 Firefly Algorithm and Improvement

2.1 Basic Firefly Algorithm

Firefly algorithm is a metaheuristic algorithm inspired by behaviour of firefly flicker. It is a random search algorithm based on group search. Algorithm uses glowing fireflies to simulate the solution point in solution space. The value of fluorescing in each firefly indicates the quality of solution. Fireflies use flashing behaviour to attract other fireflies, iteratively finds the optimal position. Basic firefly algorithm steps are as follows.

Step 1 Initialize basic parameters of the algorithm: number of fireflies N, maximum attractiveness β_0, light absorption coefficient γ, step factor α, maximum number of iterations MAX or wanted search accuracy ε.

Step 2 Randomly initialize the location of fireflies, calculate each firefly's objective function value as their maximum fluorescence brightness I_0.

Step 3 Calculate relative brightness I and the attractiveness β of all fireflies according to formulas 2.1 and 2.2 (r_{ij} represents the distance between firefly I and j). Relative brightness determines the direction of movement of fireflies.

$$I = I_0 e^{-\gamma r_{ij}} \tag{2.1}$$

$$\beta(r_{ij}) = \beta_0 e^{-\gamma r_{ij}^2} \tag{2.2}$$

Step 4 Recalculate the new position of firefly after being attracted according to the formula 2.3. α is the step length factor, it normally is a constant in [0, 1], and rand stands for random factor in [0, 1].

$$x_t(t+1) = x_t(t) + \beta * (x_j - x_i) + \alpha * \left(rand - \frac{1}{2}\right) \qquad (2.3)$$

Step 5 Recalculate each firefly's brightness based on the new position. If maximum number of iterations is reached or the search accuracy ε is satisfied, go to Step 6, otherwise increase the number of iterations by 1, and go to Step 3.
Step 6 Output the result.

2.2 Improved Firefly Algorithm

Basic firefly algorithm is simple, with many shortcomings. In order to improve the accuracy of solution, this paper alter the position update iteration and step length factor update strategy of algorithm to improve its performance.

Population Diversity Location Update Strategy
In basic firefly algorithm, firefly mainly relies on the nearby fluorescent firefly's attraction to its position to update itself. After several iterations, due to lack of randomness, entire search space may converge to local optimum.

Therefore, we introduce the population diversity characteristic, expression is shown in (2.4).

$$div(t) = \frac{1}{|P|} \sum_{i=1}^{|P|} \sqrt{\sum_{j=1}^{N} (x_{ij} - \bar{x})^2} \qquad (2.4)$$

where $|P|$ represents the dimension of solution space. x_{ij} represents the value of jth component of ith individual, and $\bar{x}_j$ denotes the jth dimension component of group average centre $\bar{x}$. Based on the population diversity, we introduce a new strategy for updating the location of fireflies as follows:

$$x_t(t+1) = x_t(t) + \beta * (x_j - x_i) + \omega * (x_i - x_{best}) + \alpha * \left(rand - \frac{1}{2}\right) \qquad (2.5)$$

ω stands for the weight that varies according to population diversity and number of iterations with some randomness. It is calculated by (2.6) and (2.7).

$$\omega = \begin{cases} -rand * p & div(t) \leq p * div(0) \\ 0 & div(t) > p * div(0) \end{cases} \qquad (2.6)$$

$$P = \frac{Max - t}{Max} \qquad (2.7)$$

where div(t) represents the population diversity index of t-th iteration. Max is maximum iteration number, and t is current iteration number. In early stage of the algorithm, ω tends to be a negative number. Fireflies would have a high probability of moving away from the best individual in an irregular direction, which gives algorithm a wider range for global optimal search. In later stages of the whole iterations, ω would be a small number due to low diversity requirements, so that each firefly tends to fly to best individual. Dynamic adjustment balances the need of global search in beginning and local search in the end, which helps algorithm to avoid the convergence to local optimal.

Adaptive Step Length Factor

In basic firefly algorithm, the step length factor α is a constant. In early stage α can help algorithm for global optimum searching, but when algorithm reach near the max iteration number, due to the constant value α, most fireflies tends to be moving back and forth around the best solution and could not converge to it.

In order to balance the contradiction between early and late stage of converging process of whole calculating process, we introduce a dynamic adjustment to the step length factor as shown in (2.8), where p is calculated by formula (2.7).

$$\alpha = \alpha_0 * p \tag{2.8}$$

3   Network Traffic Forecast Model

LSTM

LSTM is a special kind of RNN (Recurrent Neural Networks) [9] that adds long-term and short-term memory to the network, allowing model to memorize history information for a long time. Compared to standard RNN, LSTM is more suitable for time sequence forecast problem.

The core of LSTM consists of three inputs: last moment's inputs, cell status, and input of the current moment. Cell state is the key element of LSTM networks. Cell state preserves history information and participates in calculating the new states. For each LSTM core, each round of calculation is divided into following steps.

Step 1 Decide what information we are going to forget from the cell state. This decision is made through a sigmoid layer which is called the "forget gate layer".

$$f_t = \sigma(W_f \cdot [h_{t-1}, x_t] + b_f) \tag{3.1}$$

Step 2 Next step is to decide what new information we are going to store in the cell state. This has two parts. First we decide which values we are going to update through a sigmoid layer called the "input gate layer". Next, we use a tan layer to create a vector of new candidate values, $\tilde{C}_t$, that could be added to the state.

$$i_t = \sigma(W_i \cdot [h_{t-1}, x_t] + b_f) \tag{3.2}$$

$$i_t = \tanh(W_c \cdot [h_{t-1}, x_t] + b_c) \tag{3.3}$$

Step 3 Multiply the old state by f_t, forgetting things we decided to forget earlier. Then we add $i_t * \tilde{C}_t$, which is the new candidate values.

$$C_t = f_t * C_{t-1} + i_t * \tilde{C}_t \tag{3.4}$$

Step 4 The final step is to decide what we're going to output. The output will be based on our cell state, but will be a filtered version. All using the formula showed as follows:

$$\sigma_t = \sigma(W_0[h_{t-1}, x_t] + b_0) \tag{3.5}$$

$$h_t = \sigma_t * \tanh(C_t) \tag{3.6}$$

Because LSTM uses history information to calculate the next state, it works well for time sequence forecasting task.

IFA-LSTM

The training process of neural network is to gradually adjust and reduce error between predicted state and real state, and finally achieve needed accuracy. Through whole process, the topological structure of each layer in network, the learning rate and the random proportion of each layer are part of hyper parameters that needs to be adjust, different hyper parameters configurations would influence a lot on its training result of the model.

The model we build consists of three-tier structure. The hidden layer contains two layers of LSTM network. The input information's dimension, the number of nodes in each hidden layer, and the dropout ratio are the hyper parameters we choose in this model. Whole IFA-LSTM model training process is showed as follows (Fig. 1).

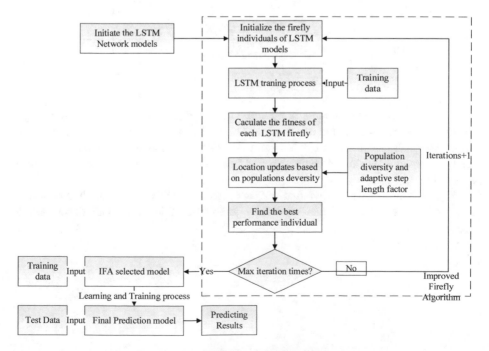

Fig. 1. IFA-LSTM model training process.

We use IFA for hyper parameters configuration optimise. Each firefly represents a type of hyper parameter configuration. Root mean square of the model training result is chosen for representing the fitness of each firefly and the loss function for the LSTM model.

$$RMSE = \sqrt{MSE} = \sqrt{\frac{SSE}{n}} = \sqrt{\frac{1}{n}\sum_{i=1}^{n} w_t(y_i - \widehat{y})^2} \qquad (3.7)$$

4 Experiments and Results Evaluations

4.1 IFA Performance Evaluation

In order to verify the actual performance for IFA (Improved Firefly Algorithm), we test algorithm with four functions. All functions are showed in Table 1.

Table 1. Test functions for IFA.

Function	Expression	Ranges
Sphere	$f(x) = \sum_{i=1}^{D} x_i^2$	$[-100, 100]$
Rosenbrock	$f(x) = \sum_{i=1}^{D-1} ((x_i - 1)^2 + 100(x_{i+1} - x_i^2)^2)$	$[-30, 30]$
Ackley	$f(x) = -20e^{-0.2\sqrt{\frac{1}{10}\sum_{i=1}^{D} x_i^2}} - e^{(\frac{1}{10}\sum_{i=1}^{D} \cos(2\pi x_i))} + 20 + e$	$[-100, 100]$
Rastrigin	$f(x) = \sum_{i=1}^{D} (x_i^2 - 10\cos(2\pi x_i) + 10)$	$[-10, 10]$

The simulation environment is python 3.6.4. And population size for firefly algorithm is set to 50. Dimension of the inputs is 20. The max iteration number is 200, and α_0 is 0.6. Each test function has been run 20 times to eliminate the random error. We analyse the fitness result for both IFA and basic FA. Test results are shown in Fig. 2 and Table 2.

Figure 2 and Table 2 prove that IFA achieves more satisfied results than basic FA, IFA converges faster and can repeatedly jump out of local optimal during whole process.

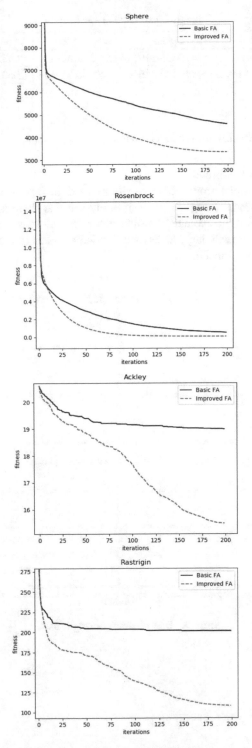

Fig. 2. Test results of IFA and FA.

Table 2. Test results of IFA and FA.

	Function	Sphere	Rosenbrock	Ackley	Rastrigin
D = 20	FA fitness	5530.119774	528292.30815	18.982729	204.693380
	Improved FA fitness	4142.266660	42230.312818	15.601827	96.525887

4.2   IFA-LSTM Performance Evaluation

Data for Model Development

We use real data samples from the university network in July, 2017. The box diagram of origin data is shown in Fig. 3. We can easily see that there are a lot of abnormal values, for that we use the average of values before and after abnormal point to smooth the data. After the normalizing process we divided samples into training part and test part, which are showed in Fig. 4.

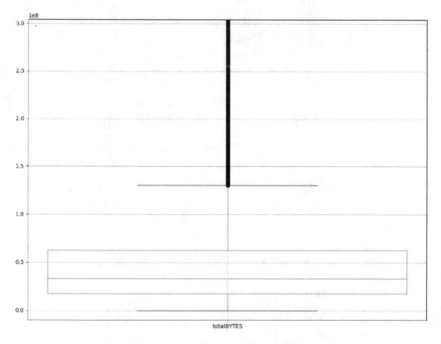

Fig. 3. Box diagram of origin data.

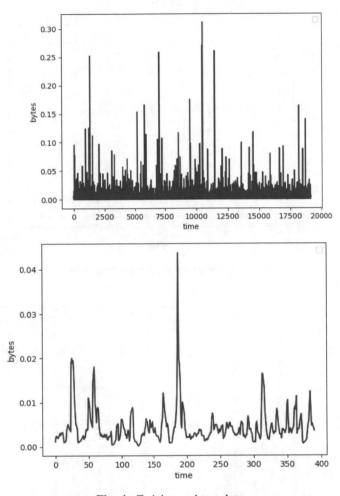

Fig. 4. Training and test data.

IFA-LSTM Training

We use both IFA and basic FA to optimize the hyper parameter configurations, the population size of fireflies is 50, the maximum number of iterations is 100, and the initial step factor α_0 is set to 0.6. The optimise result is showed in Fig. 5.

We can see that the IFA converges around 20th generation, with the best fitness of 0.0001, while the FA converges at 0.02 at 100th generation. Using IFALSTM, we can achieve faster and better result performance in training than FALSTM.

Model Performance Evaluation

We use same hyper parameter configuration from IFA calculating result for model training. With the training data from Fig. 4, we train the model 300 generations. Loss curves of the model are shown in Fig. 6. It can be seen that form the 100th generation, loss value of the model is stable near 0.00002.

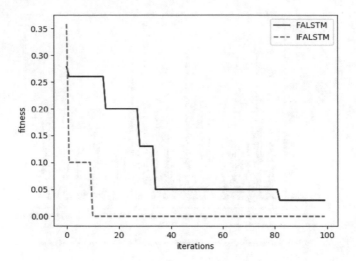

Fig. 5. Optimise result for hyper parameters.

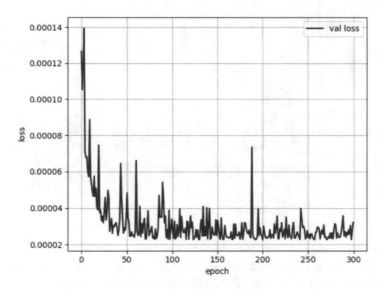

Fig. 6. Loss curves of the model.

In Fig. 7 we can see that IFALSTM achieves satisfactory prediction results, and we compare the prediction error results of IFALSTM, GRU and simple BP networks in Fig. 8, it turns out that IFALSTM can have more precise predicting results, even than GRU network.

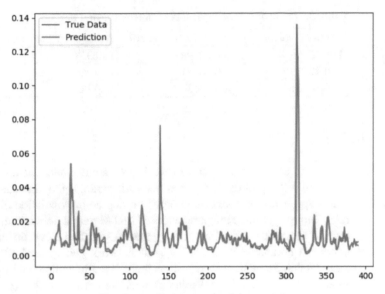

Fig. 7. Prediction result of the model.

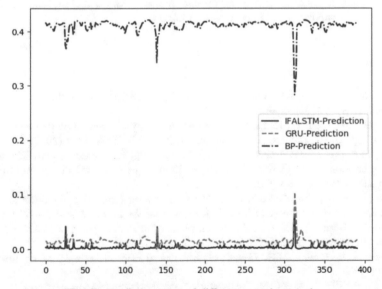

Fig. 8. Prediction error of different neural network.

Table 3 shows the performance states of different networks. We can see that IFALSTM can obtain more accurate prediction results on all indicators. Especially, for max prediction error, IFALSTM is 100 times less than BP network, and 30 times better than GRU. This tells us that when face with certain extreme conditions, IFALSTM can achieve better prediction results than other forecast methods.

Table 3. Prediction error indicators of different neural network.

	Max prediction error	Min prediction error	Average prediction error
IFALSTM	0.00356	0.00096	0.00333
BP	0.42128	0.28281	0.41066
GRU	0.10351	0.00034	0.01435

5 Conclusion

In order to improve the accuracy of neural network prediction model for time series prediction, and to solve the problem of neural network parameter optimization, this paper proposes an improved firefly algorithm based on degree of population diversity, then uses this algorithm to optimize the parameters of LSTM neural network and apply it. The prediction of network traffic time series generated on the university network management system is compared with GRU neural network and general BP neural network prediction model. Simulation results show that the optimized model has higher prediction accuracy and has a good application prospect in network traffic prediction.

Acknowledgment. This work was supported by State Grid Technology Project 'Edge Computing Research in Smart Grid Application and Security' (Grant. 52110118001H); National Natural Science Foundation of China (Grant. 61702048).

References

1. Qiu, J., Xia, J.B., Wu, J.X.: Research and development of network traffic prediction model. Comput. Eng. Des. **33**(3), 865–869 (2012)
2. Laner, M., Svoboda, P., Rupp, M.: Parsimonious fitting of long range dependent network traffic using ARMA models. IEEE Commun. Lett. **17**(12), 2368–2371 (2013)
3. Wang, J.: A process level network traffic prediction algorithm based on ARIMA model in smart substation. In: 2013 IEEE International Conference on Signal Processing, Communication and Computing, pp. 1–5. IEEE (2013)
4. Liu, X., Fang, X., Qin, Z., Ye, C., Xie, M.: A short-term forecasting algorithm for network traffic based on chaos theory and SVM. J. Netw. Syst. Manage. **19**(4), 427–447 (2011)
5. Dorigo, M., Birattari, M.: Ant colony optimization. In: Encyclopedia of Machine Learning, pp. 36–39. Springer, Boston (2011)
6. Goldberg, D.E., Holland, J.H.: Genetic algorithms and machine learning. Mach. Learn. **3**(2), 95–99 (1988)
7. Malhotra, P., Vig, L., Shroff, G., Agarwal, P.: Long short term memory networks for anomaly detection in time series. In: Proceedings of Presses universitaires de Louvain, p. 89 (2015)
8. Yang, X.S.: Firefly algorithms for multimodal optimization. In: International Symposium on Stochastic Algorithms, pp. 169–178. Springer, Heidelberg (2009)
9. Medsker, L.R., Jain, L.C.: Recurrent Neural Networks. Design and Applications 5. CRC Press, Boca Raton (2001)

Communication Analysis
and Application

This chapter aims to examine innovation in the fields of power electronics, power communication and network communication. This chapter contains 28 papers on emerging topics on communication analysis and application.

Unified Management Platform of Quantum and Classical Keys in Power Communication System

Jinsuo Liu[1,2(✉)], Gaofeng Zhao[1,2], Jiawei Wu[3], Wei Jia[2],
and Ying Zhang[2]

[1] State Grid Electric Power Research Institute,
NARI Group Corporation, Nanjing, China
liujinsuo2018@163.com
[2] NRGD Quantum CTek., Ltd., Nanjing, China
[3] State Grid Electric Power Company Information and Communication
Corporation, Shanghai, China

Abstract. With the advancement of quantum communication infrastructure, especially the completion of the "Quantum Beijing-Shanghai Trunk Line" and the launch of the Mozi quantum satellite, more and more industries have begun to incorporate quantum cryptography into business systems. However, it brought trouble to the key management of the original business system. In this paper, in order to solve problem of managing quantum and classical keys, an unified management architecture of quantum and classical keys is designed, which contains four key components: Power vertical encryption & authentication gateway, classic encryption device, QKD device, and unified key management center. All of these four components cooperate with each other to form a complete and efficient integrated key management system, which effectively solves the problem that the current power system cannot realize the quantum key management.

Keywords: Quantum key distribution · Key management · Unified platform · Power communication system

1 Introduction

Quantum communication has the characteristics of unconditional security and high efficiency, large channel capacity, ultra-high communication rate, long-distance transmission, and high information efficiency. Among these characteristics, the unconditional security is the most important feature which is unmatched over classic cryptographic communication. And its security is mainly based on the Heisenberg's uncertainty principle, quantum no-cloning theorem, etc.

There are two applications of quantum communication: quantum teleportation communication [1–3] and quantum cryptography communication [4], and both of them have unconditional security features. In quantum teleportation communication, since the two particles that transmit information are in the hands of A and B respectively, the information transmission is separated from the physical object and is not constrained by

© Springer Nature Switzerland AG 2020
Q. Liu et al. (Eds.): CENet 2018, AISC 905, pp. 695–705, 2020.
https://doi.org/10.1007/978-3-030-14680-1_76

the four-dimensional space. Both states change at the same time, so it is obviously impossible to be deciphered and absolutely safe. In quantum cryptography communication, it mainly transmits the encryption keys between two parties, says Alice and Bob, so quantum cryptography communication is also called quantum key distribution (QKD) [4–7]. In QKD, Alice sends photons of a certain state, Bob receives them using a transceiver, and then Alice and Bob correspond to photon states. Once a photon state is found, it is confirmed that the photon is being tapped. The method is also absolutely safe and cannot be deciphered. As quantum cryptography communication has a better application scenario, scientists all over the world have conducted in-depth research on it (especially for long-distance communication).

In 2006, the Los Alamos National Laboratory of the United States [8] and the Munich University [9] in Europe independently implemented the QKD experiments based on the decoy state, over more than 100 km, which opened the door to quantum communication applications. In 2007, Pan's team from the University of Science and Technology of China (USTC) [10] also implemented the decoy-state quantum key distribution (QKD) with one-way quantum communication in polarization space over 102 km. In 2010, researchers from USTC and Tsinghua University completed a free-space quantum communication experiment crossed the communication distance from the previous few hundred meters to 16 km [11]. In 2012, Pan et al. [12] firstly realized the 100-km free-space quantum teleportation and entanglement distribution in Qinghai Lake. China has included quantum communication technology in the "Thirteenth Five-Year Plan" and the National Energy Technology Innovation Action Plan (2016–2030). In August 2016, the world's first quantum science experiment satellite, namely "Mozi", was launched successfully in China. In September 2017, the "Quantum Beijing-Shanghai Trunk Line" [13], a quantum communications backbone network connecting Beijing and Shanghai with a total length of more than 2,000 km, was established. The quantum communication backbone network will promote the large-scale application of quantum security communication in the fields of finance, government affairs, national defense, and electronic information.

In the power industry, the security of both power grids and power communication systems is the focus for gird enterprises, especially the latter will directly affects the normal operation of the former. Since 2012, the State Grid Corporation of China has organized a number of special research projects such as Quantum Technology to enhance the feasibility of power system communication security, and the study of power multi-user quantum key distribution management systems. In 2013, the research team of China University of Science and Technology carried out related cooperation in the areas of quantum signal verification and quantum security communication technology, and conducted the test of the electric aerial cable environment quantum signal at the company's Changping, Bazhou UHV AC and DC bases, and obtained quantum communication. A large amount of test data applied in the power system preliminarily verifies the feasibility of transmitting quantum signals in an overhead power environment. During the G20 summit in 2016, Zhejiang Electric Power Company adopted quantum secret communication technology to provide safe voice, video, and encryption services for power-saving command systems. In 2016, a number of scientific research and industrial units within the organization's system carried out research on the application of quantum communication in power systems, and compiled a "State Grid

Corporation Quantum Communication Technology Research and Application Work Program" to guide the development and application of quantum secure communication technology in the power industry. In 2017, Quantum Secure Communication Demonstration Network was established in State Grid ICT, Beijing, Shandong, Anhui, Jiangsu, Zhejiang and other cities, verifying 13 categories including dispatch automation, telephone and teleconferencing, silver power lines, and source and payload coordination control. The business has initially verified the feasibility of carrying quantum power in quantum secret communications.

Although the "specification for functions and interfaces and security policy of unified key management system of State Grid Corporation of China" [14] has already been issued in 2014, but the specification only considers the management of classical keys. With the rapid development of quantum cryptography in the power industry in recent years, this specification is not suitable for the current key management situation. How to design a unified platform that integrates quantum keys and classical key management becomes an urgent problem to be solved at present. This paper is to solve this problem, and propose a unified management platform that integrates quantum key distribution management and classic key management. And the remainder of this paper is organized as follows, the classic key management of power communication system, quantum computation is briefly introduced in Sect. 2, a unified management platform of quantum key and classic key is presented in Sect. 3, and the conclusion is drawn in last section.

2 Preliminaries

2.1 Quantum Computation

As we know, the classic bit is the smallest unit in information processing, and its value is either 0 or 1. In quantum computing, the qubit (quantum bit) is quantum analogue of the classical bit, but it has two possible values $|0\rangle$ and $|1\rangle$ with a certain probability,

$$|\varphi\rangle = \alpha|0\rangle + \beta|1\rangle, \tag{1}$$

where $|\alpha^2| + |\beta|^2 = 1$, $\alpha, \beta \in C$. Since the vectors $|0\rangle$ and $|1\rangle$ can be represented as follows,

$$|0\rangle = \begin{pmatrix} 1 \\ 0 \end{pmatrix} \quad \text{and} \quad |1\rangle = \begin{pmatrix} 0 \\ 1 \end{pmatrix} \tag{2}$$

the qubit $|\varphi\rangle$ can be expressed in vector form $|\varphi\rangle = \begin{pmatrix} \alpha \\ \beta \end{pmatrix}$.

Quantum operators over a qubit are represented by 2×2 unitary matrices. An $n \times n$ matrix U is unitary if $UU^\dagger = U^\dagger U = I$, where $U^\dagger$ is the transpose conjugate of U. For instance, Pauli X, Pauli Z, and the Hadamard H operators are important quantum operators over one qubit and they are described in Eq. (3).

$$X = \begin{pmatrix} 0 & 1 \\ 1 & 0 \end{pmatrix} \quad Z = \begin{pmatrix} 1 & 0 \\ 0 & -1 \end{pmatrix} \quad H = \frac{1}{\sqrt{2}} \begin{pmatrix} 1 & 1 \\ 1 & -1 \end{pmatrix} \tag{3}$$

The operator described in Eq. (4) is *CNOT* operator, that flips the second qubit if the first (the controlled qubit) is the state $|1\rangle$.

$$CNOT = \begin{bmatrix} 1 & 0 & 0 & 0 \\ 0 & 1 & 0 & 0 \\ 0 & 0 & 0 & 1 \\ 0 & 0 & 1 & 0 \end{bmatrix} \tag{4}$$

In Quantum Physics, if a system is in a state which is a superposition $|\varphi\rangle = \alpha_0|1\rangle + \alpha_1|1\rangle$, upon measurement the system collapses to one of its basis state $|i\rangle$, $i \in \{0, 1\}$ probabilistically:

$$p(|i\rangle) = \frac{|\alpha_i|^2}{\||\varphi\rangle\|^2} = \frac{|\alpha_i|^2}{\sum_j |\alpha_j|^2}, \tag{5}$$

which is the probability that the system will be found in the ground state $|i\rangle$ after a measurement. After the first measurement of a state $|\varphi\rangle$ if one performs another measurements will get the same result. The collapse of the state after measurement says that one cannot see all the results generated by the quantum parallelism. The challenge in quantum computation is how to take advantage of the quantum parallelism before performing a measurement.

2.2   Classic Key Management in Power Communication System

The key management is an essential research area in cryptography because the security of key guarantees the security of communication system. Principles of key management system include security, robustness and efficiency, and the minor are convenience, maintainability and expandability. In power communication system, there is a specification for classical key management, including the functions, interfaces and the security policy.

The functions of the classic key management system at least include the symmetric key management module, the asymmetric key management module, the key monitoring module, the key aggregation management module, and the unified key management display module, etc. (The Function Block Diagram of Classic Key Management System is given in Fig. 1).

(1) **Symmetric key management module:** This module counts and manages the keys and cryptographic devices of symmetric key management system.
(2) **Asymmetric key management module:** This module provides management services for the entire life cycle of an asymmetric key pair. To be specific, it manages the certificates of SGCC CA system, asymmetric key pairs, and key devices; and provide key template management, key policy management,

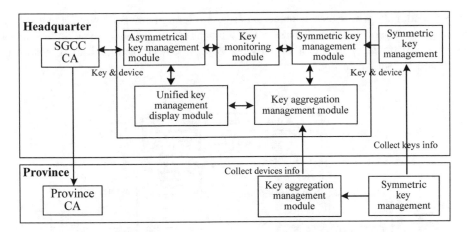

Fig. 1. The function block diagram of classic key management system.

application configuration management, key management, key store management, key library backup, key library restore and application service management functions.

(3) **Key monitoring management module:** It includes two parts, i.e. the compliance monitoring module, and the application effectiveness monitoring module.

(4) **Key aggregation management module:** It gathers provincial keys and cryptographic devices information, and sends to the headquarters through encryption authentication and data transmission protocols.

(5) **Unified key management display module:** It provides statistics and display of cryptographic devices and keys of SGCC CA system and symmetric key management systems, and provides the right management of operators, user policy configuration, device management, log management, and other functions.

The classic key management system adopts the main-tier deployment mode, and deploys all the modules in the headquarters, while in the province company it deploys provincial key aggregation management module, which implements the cryptographic devices and key management for the digital certificate authentication system (i.e. CA), and symmetric key management system.

3 The Unified Management Platform of Quantum and Classical Keys in Power Communication System

Key management focuses on managing all aspects of the secret key from its creation to its destruction. To be specific, it includes key generation, key distribution, key utilization, key replacement, key backup, and key destruction. For power communication system, due to its high security level, strong real-time performance, and complex network structure, the key management is very important and plays a major role in guaranteeing the security of power grid. In order to enhance the security of power communication system, quantum key distribution and it related technologies are introduced in the whole communication network. Through embedding the quantum key management into the original classical key management system, a new unified management platform of quantum & classical keys in power communication system is designed (as shown in Fig. 2).

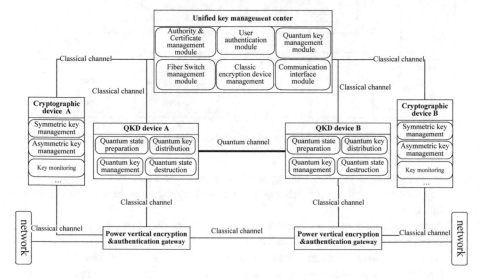

Fig. 2. Architecture of unified management platform for quantum & classical keys.

The entire platform is divided into four core components: the power vertical encryption & authentication gateways, the classic encryption devices, the quantum QKD devices, and the unified key management center. And the power vertical encryption & authentication gateways and the connected fibers are the physical layer foundation of the entire platform.

3.1   Power Vertical Encryption & Authentication Gateway

The power vertical encryption & authentication gateway is a unidirectional dedicated safety isolation device that is connected to the safety zone I, II and the safety zone III in the safety protection system of the secondary power system. In addition to using basic firewalls, proxy servers, and other security protection technologies, the device adopts a "dual-machine non-network" physical isolation technology, that is, the device can block the network directly and the two networks are not connected to the device at the same time; Blocking the network logical connection, that is, TCP/IP must be stripped, the original data is transmitted in a non-network manner; the isolated transmission mechanism is non-programmable; any data is completed through a two-stage proxy.

The power encryption and authentication gateway is located between the internal LAN of the power control system and the router of the power dispatch data network, and is used to guarantee the data confidentiality, integrity, and authenticity during the longitudinal data transmission of the power dispatch system. According to the requirements of "level management", the vertical encryption authentication gateway is deployed at all levels of the dispatch center and subordinate plant stations, and an encrypted tunnel is established according to the power dispatch communication relationship (in principle, an encrypted tunnel is only established between the upper and lower levels), and the encryption tunnel topology is configured.

3.2 Classic Encryption Device

In the classical encryption device, there are some foundational functions about Symmetric key and Asymmetric key management. So, this part can be viewed as part of the classic key management in power communication system described in Sect. 2.2. To be specific, there are some modules as follows,

(1) **Symmetric key management:** This module manages the whole six phases of the classical symmetric key (the shared session key K_{share}): key generation, key distribution, key utilization, key replacement, key backup, and key destruction.
(2) **Asymmetric key management:** This module provides management services for the entire life cycle (i.e. six phases) of an asymmetric key pair (public key K_{pub} and private key K_{pri}).
(3) **Key monitoring management:** It supervise the effectiveness of symmetric and asymmetric keys, such as, the integrity, security, availability, and availability of the keys.
(4) **The other functions:** There are some other functions to be included, such as key aggregation, statistical analysis, visual display, etc.

3.3 Quantum QKD Device

QKD is the foundation for our quantum application based on quantum communication technology, which is a communication method which implements a cryptographic protocol involving components of quantum mechanics. It enables two parties to produce a shared random secret key known only to them, which can then be used to encrypt and decrypt messages. It is often incorrectly called quantum cryptography, as it is the best-known example of a quantum cryptographic task. BB84 [4] is the most known QKD protocol, and its procedure can be illustrated in Fig. 3.

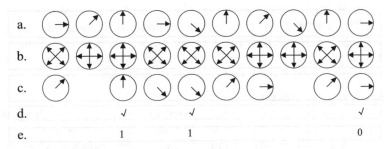

Fig. 3. Schematic diagram of the five-step BB84 protocol.

In these QKD devices, there are at least four module: Quantum state preparation, quantum key distribution, quantum key storage, and quantum state destruction. From the perspective of quantum key application, it can be further divided into two layers: key management layer and application layer (as shown in Fig. 4). In the application layer, all the quantum key is transferred into classical bits, so here we just discuss the four modules in the bottom layer.

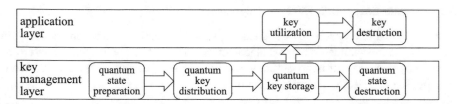

Fig. 4. Hierarchy diagram of main modules in QKD device.

(1) Quantum state preparation

This module or device unit is used to generate the quantum state, i.e. single photons with different polarization angles in our grid communication system. The ideal single-photon source is every hard to realize, and in experiment and practice, weak coherent light source is used to generate single-photon, but also contains elements of multi-photon. Single-photon can be created via post-selection for interference experiments and Hanbury-Brown and Twiss-style experiment [15, 16].

(2) Quantum key distribution

Quantum key distribution is key part which is used to transmit the secret key between two participants. During the distribution process, the eavesdropper Eve may capture some qubits, but some rules of quantum mechanism can either guarantee Eve cannot get any information about secret key or detect the cheating action when information is revealed. Most of QKD protocols are based on ideal single-photon resource, and they may be not secure in noisy environment. On the one hand, the post-processing operations, such as information reconciliation and privacy amplification [4, 6], are required, and some strategies for preventing attacks [17, 18] need to be taken into account.

(3) Quantum state storage

In quantum world, state storage is more difficult than the classical one, quantum memory [19] is used to store quantum state in short time. Quantum memory is instrumental for implementing long-distance quantum communication using quantum repeaters. There are several different quantum memory mechanisms, such as optical delay lines, cavities and electromagnetically induced transparency, as well as schemes that rely on photon echoes and the off-resonant Faraday interaction.

(4) Quantum state destruction

Quantum state destruction is a necessary process to ensure security. Compared with quantum state storage, quantum state destruction is easier to obtain. As we know, a quantum state will collapse into a certain state after being measured. If the malicious user measure these states in the correct base, he or she can get information about key. So, before abandoning these single photons, the measurement using random bases (Z-base $\{|0\rangle, |1\rangle\}$ or X-base $\{|+\rangle = \frac{1}{\sqrt{2}}(|0\rangle + |1\rangle), |-\rangle = \frac{1}{\sqrt{2}}(|0\rangle - |1\rangle)\}$) is suggested.

3.4 Unified Key Management Center

Unified key management center is the main body of the final integration of quantum key management and classical key management, and is also the center of the entire unified key management. The center implements the management and monitoring of the underlying communication equipment, middle-class classical encryption equipment, and quantum QKD equipment, realizes a quantum/classical integrated management platform, and also provides a unified quantum/classic key allocation interface. The purpose is to enhance the key management efficiency of power communication system, and enhances the external service quality.

As shown in Fig. 2, it is composed of six modules.

(1) **Fiber Switch management:** This module manages the device information, operating status, and configuration parameters of all high-speed fabric switches. It effectively manages the communication equipment at the bottom of the communication system, which provides the support for upward functionality.

(2) **Classic encryption device management:** The module implements the management of basic information, configuration parameters, and operating status of all classic encryption devices (including the symmetric key and Asymmetric key devices), and provides the necessary statistical analysis and visual display.

(3) **Communication interface:** It stipulates and manages all internal or external communication protocols, interfaces, data specifications, etc., and monitors the real-time status of all external communications.

(4) **Authority & Certificate management:** The module implements the monitoring of configuration parameters and operating status of all the power encryption and authentication gateway, and provides a unified communication interface for advanced applications of authority & certificate based on quantum or classical keys.

(5) **User authentication:** The module is based on the authority & certificate management module, and perform the authentication of all access users in the system.

(6) **Quantum key management:** This module manages the basic information, configuration parameters, and operating status of all QKD devices, and provides the necessary statistical analysis and visualization. In addition, it provides a unified call interface for advanced applications of quantum keys.

In the unified management platform, the four components cooperate with each other to form a complete and efficient integrated quantum and classical key management system, which effectively solves the problem that the current power system cannot realize the quantum key management.

4 Discussion and Conclusion

In this paper, we proposed an integrated architecture of unified quantum and classical keys management platform for power communication system, which contains four key components: Power vertical encryption & authentication gateway, classic encryption device, QKD device, and unified key management center. For each core component, we also conducted in-depth functional design and technical implementation discussions.

In the further work, there are three aspects to be studied in detail. Firstly, we will further refine and improve the functional details and technical solutions of each functional module. Secondly, we will carry out the regional experimental verification using existing SGCC quantum networks and classical networks. Finally, based on the above basic work, we hope to make some attempts in higher level quantum technology applications, such as quantum key agreement [20, 21], quantum sealed-bid auction [22, 23], etc.

Acknowledgement. This work is supported by 2017 Science and Technology Project of State Grid Corporation "Research and Development of Key Distribution Device in Dedicated Quantum Secure Communication for the Electrical Overhead Environment" (No. SGZJ0000KXJS1-700339).

References

1. Bouwmeester, D., Pan, J.W., Mattle, K., Eibl, M., Weinfurter, H., Zeilinger, A.: Experimental quantum teleportation. Nature **390**, 575–579 (1997)
2. Furusawa, A., Sorensen, J.L., Braunstein, S.L., Fuchs, C.A., Kimble, H.J., Polzik, E.S.: Unconditional quantum teleportation. Science **282**(5389), 706–709 (1998)
3. Wang, M.Y., Yan, F.L.: Perfect quantum teleportation and dense coding protocols via the 2N-qubit W state. Chin. Phys. B **20**(12), 120309 (2011)
4. Bennett, C.H., Brassard, G.: Quantum cryptography: public key distribution and coin tossing. Theoret. Comput. Sci. **560**, 7–11 (2014)
5. Ekert, A.K.: Quantum cryptography based on Bell's theorem. Phys. Rev. Lett. **67**(6), 661–663 (1991)
6. Bennett, C.H.: Quantum Cryptography using any 2 Nonorthogonal States. Phys. Rev. Lett. **68**(21), 3121–3124 (1992)
7. Shaari, J.S., Bahari, A.A.: Improved two-way six-state protocol for quantum key distribution. Phys. Lett. A **376**(45), 2962–2966 (2012)
8. Hiskett, P.A., Rosenberg, D., Peterson, C.G., Rice, P.: Long-distance quantum key distribution in optical fibre. New J. Phys. **8**(9), 193 (2006)
9. Schmitt-Manderbach, T., Weier, H., Fürst, M., Ursin, R., Tiefenbacher, F., Scheidl, T., Perdigues, J., Sodnik, Z., Kurtsiefer, Ch., Rarity, J., Zeilinger, A., Weinfurter, H.: Experimental demonstration of free-space decoy-state quantum key distribution over 144 km. Phys. Rev. Lett. **98**(1), 010504 (2007)
10. Peng, C.Z., Zhang, J., Yang, D., Gao, W.B., Ma, H.X., Yin, H., Zeng, H.P., Yang, T., Wang, X.B., Pan, J.W.: Experimental long-distance decoy-state quantum key distribution based on polarization encoding. Phys. Rev. Lett. **98**(1), 010505 (2007)
11. Jin, X.M., Ren, J.G., Yang, B., Yi, Z.H., Zhou, F., Xu, X.F., Wang, S.K., Yang, D., Hu, Y. F., Jiang, S., Yang, T., Yin, H., Chen, K., Peng, C.Z., Pan, J.W.: Experimental free-space quantum teleportation. Nat. Photonics **4**, 376 (2010)

12. Yin, J., Ren, J.G., Lu, H., et al.: Quantum teleportation and entanglement distribution over 100-kilometre free-space channels. Nature **488**, 185 (2012)
13. Xinhua: China Opens 2,000-km Quantum Communication Line. http://english.gov.cn/news/photos/2017/09/30/content_281475894651400.htm
14. SGCC: Specification for Functions and Interfaces and Security Policy of Unified Key Management System of State Grid Corporation of China. http://www.doc88.com/p-9761827189805.html
15. Resch, K.J., Lundeen, J.S., Steinberg, A.M.: Quantum state preparation and conditional coherence. Phys. Rev. Lett. **88**(11), 113601 (2002)
16. Agarwal, K., Bhatt, R.N., Sondhi, S.L.: Fast preparation of critical ground states using superluminal fronts. Phys. Rev. Lett. **120**(21), 210604 (2018)
17. Jiao, R.Z., Feng, C.X., Ma, H.Q.: Analysis of the differential-phase-shift-keying protocol in the quantum-key-distribution system. Chin. Phys. B **18**(3), 915–917 (2009)
18. Liu, W.J., Wang, F., Ji, S., et al.: Attacks and improvement of quantum sealed-bid auction with EPR Pairs. Commun. Theor. Phys. **61**(6), 686–690 (2014)
19. Lvovsky, A.I., Sanders, B.C., Tittel, W.: Optical quantum memory. Nat. Photonics **3**(12), 706–714 (2009)
20. Liu, W.J., Chen, Z.Y., Ji, S., Wang, H.B., Zhang, J.: Multi-party semi-quantum key agreement with delegating quantum computation. Int. J. Theor. Phys. **56**(10), 3164–3174 (2017)
21. Liu, W.J., Xu, Y., Yang, C.N., Gao, P.P., Yu, W.B.: An Efficient and secure arbitrary n-party quantum key agreement protocol using bell states. Int. J. Theor. Phys. **57**(1), 195–207 (2018)
22. Naseri, M.: Secure quantum sealed-bid auction. Optics Commun. **282**(9), 1939–1943 (2009)
23. Liu, W.J., Wang, H.B., Yuan, G.L., Xu, Y., Chen, Z.Y., An, X.X., Ji, F.G., Gnitou, G.T.: Multiparty quantum sealed-bid auction using single photons as message carrier. Quantum Inf. Process. **15**(2), 869–879 (2016)

Design of Signal-Oriented Automatic Test System Software Based on Flowchart

Luqing Zhu[(⊠)] and Heming Zhang

School of Instrumentation Science and Opto-Electronics Engineering,
Beihang University, Beijing 100191, China
syl617344@buaa.edu.cn

Abstract. In order to solve the low efficiency and high complexity of developing test program set (TPS) in automatic testing currently, an approach to developing TPS based on flowchart rather purely text code is proposed in this paper. By introducing test sequence module (TRM), TPS flowchart can be developed by combining and configuring the graphic test units logically according to the test requirements. Besides, the run-time module (RTS) is constructed to parse and execute the TPS flowchart developed upon. In process of executing, the RTS realizes resource mapping, signal path routing and finally implements the automatic test task. The fact has demonstrated that the method of developing TPS in this paper can reduce the complexity of developing TPS, improve efficiency and achieve the rapid development of TPS dramatically.

Keywords: Test program set · Test sequence module · Run-time module · Resource mapping · Signal path routing

1   Introduction

ATS refers to the system that drives the test equipment, completes the measurement and diagnosis of the unit under test (UUT), and displays or outputs the test results under the control of TPS developed by tester [1], therefore, TPS is an important part for ATS. However, the development of TPS is rather complicated and time-consuming currently, especially as the complexity and integration of UUTs is increasing rapidly.

In order to simplify the developing process and improve developing efficiency of TPS, much work has been done in this field. However, most of them adopt text language to describe TPS, focusing on simplifying test language grammar or optimizing language structure to achieve rapid development. Larry et al. studied a lot in the current problems of TPS development as well as what excellent properties a good TPS should have, but it didn't give any effective solutions [2]. Yu et al. proposed an architecture of testing system based on workflow technology, in which the testing process definition language STPDL based on XPDL was defined to describe TPS. Due to the hierarchical structure of STPDL, it could make the test program description clearer and easier [3]. Another prominent test description language, the ATLAS, realized the separation between TPS development and specific instruments, which enhanced the portability of TPS. Meanwhile, the ATLAS grammar lied on English, which made it easy to learn and read for testers [4]. Additionally, the ATML language,

© Springer Nature Switzerland AG 2020
Q. Liu et al. (Eds.): CENet 2018, AISC 905, pp. 706–717, 2020.
https://doi.org/10.1007/978-3-030-14680-1_77

based on XML, is popular in recent years for its excellent cross-platform and data-sharing ability. University of Electronic Science and Technology takes ATML as standard to describe TPS [5, 6]. However, the preparation of ATML files containing test information is so complex and arduous that the further use of this platform is seriously blocked.

Despite the fact that the text-based TPS developing language is rather mature now, testers are still unable to get rid of the problem of mastering lots of test grammars. Comparatively, if we apply visual technology into TPS development, the development will get much easier. The test platform–Virtual Instruments Test Environments (VITE), has provided a graphic means to develop the TPS, but its oriented-instrument design makes it weak to realize TPS portability and instrument interchangeability [7]; Luo et al. presented a visual approach to describe the TPS and converted the visual flowchart into C++ program. But the converter is difficult to make and it can not convert the complicated visual test flow very well [8].

Considering the defects of the researches upon, a signal-oriented ATS software based on flowchart is proposed in this paper. Firstly, the software provides a graphic method to develop TPS by assembling graphic units logically, which make it much easier for TPS development. Secondly, the TPS flowchart is constructed by signal units and has nothing to do with specific instrument, so it supports the instrument inter-changeability very well.

The paper is organized as follows. Firstly, the overall structure of the ATS software is introduced in Sect. 2. Then, in Sect. 3, we present the test resource models in detail. In Sect. 4, test sequence module is proposed which is used to develop TPS flowchart. Section 5 would explain how the TPS flowchart is parsed and executed. In the end, we will evaluate the ATS software to show its advantage and get the conclusion.

2 Structure of ATS Software

The ATS software typically comprises test sequence module, RTS module and test resource model, the detailed structure of which is shown in Fig. 1.

Test resource model: Test resources are hardware and software resources required by automatic testing, mainly including instruments, programmable switches, ports, UUTs, linking between instruments, and instrument drivers. Section 3 in this paper focuses on device model (DM), adaptor model (AM), configuration model (CM) and instrument driver model (IDM).

Test sequence module: The test sequence module (discussed in Sect. 4 in detail) is responsible for providing interface which allows tester to develop TPS flowchart rapidly by dragging graphic units.

Run-time service module: The run-time service module (discussed in Sect. 5 in detail), key part of ATS software, is in charge of executing TPS flowchart and realizing resource mapping signal path routing.

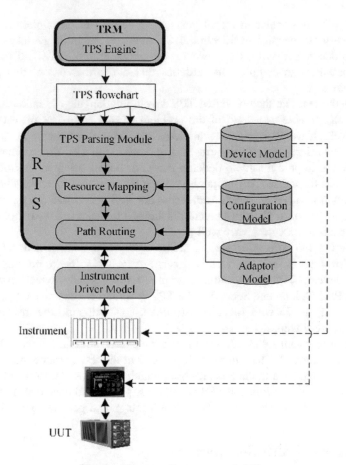

Fig. 1. The overall structure of ATS.

3 Test Resource Modeling

The transportability of ATS is built on the foundation of the standardization for interface of software and hardware as well as re-configurability of test resources information. Test resource model (TRM) includes device model (DM), adaptor model (AM), Configuration Model (CM), instrument driver model (IDM) [9]. The first three models are realized by database, which provide resource retrieval and matching services for ATS and enable resource sharing with other system, while the IDM is realized by COM component which provides method to access physical instruments.

3.1 Device Model

The device model mainly describes information relating to instrument, which is necessary for the initialization of instrument and matching between test signal and instrument. The instrument model is shown in Fig. 2. The basic properties mainly describe the instrument name, ID, address, instrument drive path and other basic attributes. Channel information is used to establish connection relationship with external devices, mainly including information about connectors and ports. Instrument capability information describes the instrument's ability of generating various signals.

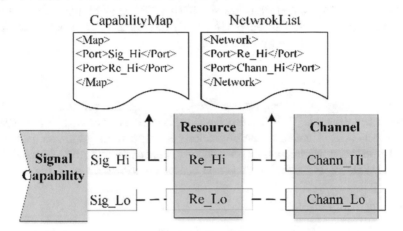

Fig. 2. Device model.

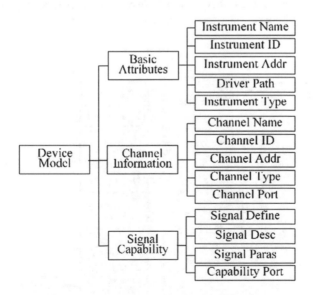

Fig. 3. Mapping between channel and signal.

Actually, signal is generated through channel. A single channel can produce multiple signals, and a single signal can also be generated by efforts of multiple channels. As we know the signal we need according to TPS, how to locate the designated physical channel that generates the signal becomes a problem that needs to solve. For this reason, we introduce the conception "resource", which is a intermediary, to establish the mapping relationship between channel and signal capability. As shown in Fig. 3, we superinduce logical Ports information for signal and resource and establish a CapabilityMap file to combinate these two ports together. By this way, we realize the linking between signal and resource. Similarly, if certain channel participate the generating of this signal, then we establish a NetworkList file to describe their relationship. Thanks to the "resource", it is easy to establish the mapping relationship between multiple signals with multiple channels.

3.2   Adaptor Model

The adaptor model describes the connection relationship between automatic test equipment (ATE) and UUT. The role of adaptor is to implement signal transformation, signal distribution, attenuation, amplification, etc. The most important part of AM should be the node information of adaptor, the connection path of each node to UUT nodes and ATE nodes. Therefore, the AM can be described below in Fig. 4.

3.3   Configuration Model

The configuration model describes the mutual connection relationship among hardware in order to implement reconfiguration of ATE. It typically includes: The connection among test instruments, the connection between test instruments and switches, as well as the connection between test instruments and ATE interface connector. The CM is shown in Fig. 5.

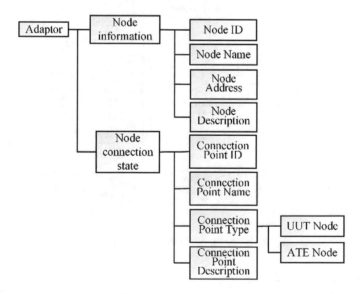

Fig. 4. Adaptor model.

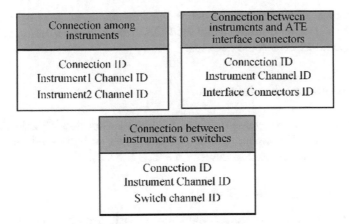

Fig. 5. Configuration model.

3.4 Instrument Driver Model

The instrument driver, whose internal implementation follows the design model of the VPP instrument driver, is implemented by COM component. The COM encapsulates secondarily the instrument's original driver interfaces according to functions or signal types and exposes unified interfaces for ATS, making it much easier to be called externally. When new instrument driver is required, the work is just to rewrite the content within the interface rather than the whole instrument driver. The IDM shows as follows in Fig. 6.

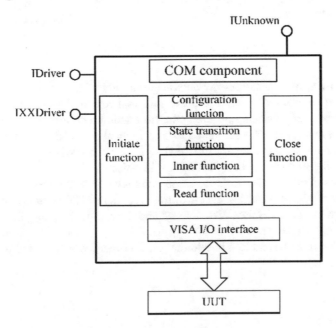

Fig. 6. Instrument driver model.

The interface IDriver is used to define public methods, such as the initialization, resetting, closing of the Instrument and the IXXDriver, named according to instrument itself, is used to define specific methods.

4 Test Sequence Module

Test sequence module provides platform for tester to develop TPS. In this section, firstly the look of TPS flowchart will be presented, then we will introduce the structure of TPS Engine which is used to develop the TPS flowchart.

4.1 Paradigm of TPS Flowchart

Each TPS flowchart is a logical combination of basic test elements, which include: process control units, signal units, data display units, and connection units. Each unit is represented by a bitmap, and the properties of the unit can be set or modified through property dialog provided by application. For example, tester can proceed parameter setup, such as name, signal value, range, and accuracy for the signal unit. Figure 7 shows an example of TPS flowchart and its main task is to apply a voltage excitation signal and a waveform signal to the UUT, then measure and display the frequency value. The figure is composed of signal units, flow control units, and data display units.

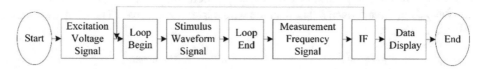

Fig. 7. Paradigm of TPS flowchart.

The following part describes the detailed definition of the basic test elements:

Flowchart control units: Refer to these units that control the execution direction of the test process, mainly including loop control, condition control and GOTO control.

Signal unit: Represents a reference to a real signal required by test requirements. Testers can configure parameters of the real signal by properties setting of signal unit. Actually, each real signal is designed by COM object.

Test result display units: Refer to these units which are used to show test results. Due to the diversification of test result forms, test result display units can be put into three categories: Text display, list display and waveform display. Each test result display unit is also designed by COM object.

Connection unit: Refers to a line in the test flowchart, which is used to express the execution order.

4.2   Design of TPS Engine

TPS Engine, which is the core part of test sequence module, allows testers to define a complete TPS flowchart by combining test graphics elements. TPS Engine is composed of three layers, namely user layer, execution layer and component layer, which is shown in Fig. 8.

The user layer is an interface for the user to interact with the software and it provides the user with the environment for drawing the TPS flowchart by dragging graphic units and combining them logically.

The execution layer is in charge of performing tasks coming from the user layer, including the selection, dragging and dropping of graphic units, drawing of TPS flowchart, and displaying of test results. It is mainly composed of four parts: flowchart control manager, configuration engine, drawing engine and display manager. The flowchart control manager is used to manage flowchart control units in order to control the direction of the test program. The drawing engine is used to draw TPS flowchart. The two configuration engine are responsible for configuring UUT pin parameters and signal parameters of signals respectively. The signal parameters include channel parameters, excitation parameters and measurement parameters. The display management module is used to manage the way of showing test results.

The component layer provides functional support for GTPE, which consists of COM components with specific functions. Each component corresponds to a graphic unit in the user layer. For example, a flow control unit corresponds to a flow control COM component, a signal unit corresponds to a signal COM component, and a test result display unit corresponds to a display COM component.

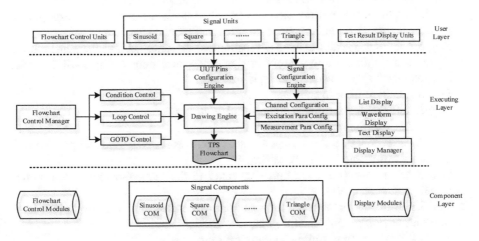

Fig. 8. Structure of TPS engine.

These basic graphic units that constitute the TPS flowchart are actually visual substitutions of the components from component layer. When the test process is executed, the COM components will be called in sequence. In this way, the transition from the TPS flowchart to the callable test program can be realized. The parsing and execution of TPS flowchart will be introduced in Sect. 5 in details.

5   Run-Time Service Module

The role of RTS is to parse and execute TPS flowchart designed in Sect. 4. In the process of execution, the two important parts-mapping of virtual resources and selection of routing channel, will be implemented.

5.1   Parsing and Execution of TPS Flowchart

In this section, the process that how the TPS flowchart is parsed and executed will be explained:

Step 1: The TPS flowchart is submitted to the test program parsing module entry after completed;

Step 2: The test program parsing module searches for the initial unit of the flow-chart, and sets the unit connected to the initial one as the current unit;

Step 3: Distinguish the type of current unit and receive the test data from the previous unit (if no data received, ignores it):

Step 3.1: If the current unit is a signal unit:

Step 3.1.1: Firstly, instantiate the COM component object mapped by the signal primitive;

Step 3.1.2: Then, pass the signal parameters and the device pin information contained in COM object to the resource mapping module (described in Sect. 5.2 in detail) and the signal routing module (described in Sect. 5.3 in detail) respectively, to achieve virtual resource matching and signal path search;

Step 3.1.3: Finally, call the matched instrument driver to achieve the generation of signal. If the signal is a measurement signal, the measurement value will be saved and returned to the flowchart analysis module;

Step 3.1.4: Step back to Step 3.

Step 3.2: If the current unit is a loop-start unit:

Step 3.2.1: Get the number of loops: N

Step 3.2.2: If N < 0, end the loop, and set the unit connected to the loop-end one as the current unit, and step back to Step 3;

Step 3.2.3: If N > 0, set the unit connected to the loop-start one as the current primitive, and step back to Step 3.

Step 3.3: If the current unit is a loop-end one, execute N = N − 1, and set the loop-start unit as current one, then step back to Step 3.

Step 3.4: If the current unit is a condition unit:

Step 3.4.1: Judge the condition result: true or false;

Step 3.4.2: If it is true, search the connection unit marked with true and set the unit connected to the connection unit as current unit. Then step back to Step 3;

Step 3.4.3: If it is false, search the connection unit marked with false and set the unit connected to the connection unit as current unit. Then step back to Step 3;

Step 3.5: If the current unit is a test result display unit, determine the type of the unit, then instantiate the COM component mapped by the unit according to the type of the unit, finally display the received test data in the screen. Step back to Step 3;
Step 3.6: If the current unit is the end unit, then stop and exit.
Step 4: If a failure occurs during execution, stop and exit.

5.2 Design of Resource Mapping Module

Resource mapping is in charge of finding the physical instrument that can generate the specific signal defined in the TPS.

In signal-oriented ATS, test requirement is described by the test signals required by the UUT, similarly, the capability of the instruments in the DM is also characterized by the ability of generating excitation or measurement signals. Therefore, it is viable to match test signals with instrument signal capability to find the available physical instrument resource that can generate the required signals. We define this process as resources mapping. Firstly, resource mapping module acquires the signal parameters information (e.g. current value, voltage value, frequency, etc.) from the signal unit in test program, and compares these parameters with the instrument's signal capability to determine which instrument has the ability to generate this signal.

Step 1: Get signal COM component interface pointer. Because signal parameters are saved in signal object, we need get the interface pointer of the COM object first;
Step 2: Extract parameter information of test signal, which includes signal type, signal values, range, accuracy and resolution, etc.;
Step 3: Load DM description information;
Step 4: Extract instrument capability information according to DM;
Step 5: Matching signal parameters to get the proper instrument. Compare the required signal parameters with the instrument capability information to get the instrument that can generate the signal. If the match succeeds, add the instrument to the alternative instrument library;
Step 6: Select the best instrument from the alternatives according to the preferred algorithm (e.g. highest accuracy, fastest, etc.);
Step 7: If the match fails, it needs to check the next instrument and execute Step 4 until all instruments have been traversed. If no instrument matches the test signal, a warning will be reported.

5.3 Design of Signal Path Routing Module

Path routing module is used to search the best signal path from the instrument port to the UUT port. In the DM, channel information of each hardware is stored in element <Channel Information>. Physical connection information among hardware (mainly refers to the relationship between test station and the adapter, as well as relationship between adapters and UUTs) are described by CM. Therefore, it is possible to unify these ports and connection information to establish a physically transmittable signal routing channel list. When the signal is generated and transmitted, the system searches for a suitable channel path from the signal routing channel list and selects the optimal path according to the actual situation.

The path of the excitation signal and the measurement signal can be shown as follows:

Excitation signal path: test instrument port→switch matrix→adapter→UUT test point.

Measurement Signal Path: UUT test point→adapter→switch matrix→test instrument port.

Considering that the connection of hardware resource is similar to the topology structure in the network. The signal transmission path can be abstracted as shown in Fig. 9:

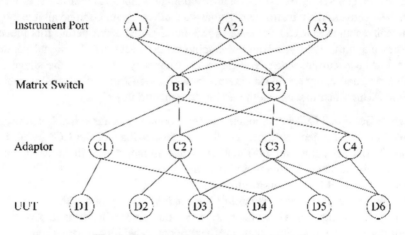

Fig. 9. Diagram of signal transmission path.

Therefore, the signal path routing process is shown as follow:

Step 1: Iterate through each DM, and extract the instrument port information;
Step 2: Analyze the CM, and take the test instrument port as the starting point, the UUT test port as the ending point, find the available physical path between the starting point and the ending point recursively;
Step 3: Save all available physical paths found in Step 2 in the database;
Sep 4: Query the reachable physical path in the path list according to the signal port and signal destination test point information;
Step 5: Determine whether the physical path is available currently (not being occupied or the matrix switch is open), if yes, add the path to the list of available logical paths, otherwise return to Step 4;
Step 6: Traverse the list of available logical paths, find the optimal path.

6 Evaluation

The ATS software scheme proposed in the paper provides a platform for tester to develop TPS in graphic method instead of text language. The tester can rapidly develop TPS by dragging graphic test units and assembling them together according to test requirement. Compared with the complex grammars, low efficiency of the text language, the graphic language is simple and easy for testers. In practical automatic testing, the proposed ATS software shows its advantage of high efficiency and simpleness.

7 Conclusion

The development of TPS is the foundation of automatic testing. However, the traditional way relies on test description language to model TPS, which is complicated and inefficient. This paper proposes a design scheme of signal-oriented ATS software that tester can develop TPS rapidly by assembling graphic units into TPS flowchart. The flowchart would be parsed and executed by RTS in order to implement automatic testing. Practice has proved that the ATS proposed in this paper has the characteristics of good versatility, ease of use, and high execution efficiency.

References

1. Guo, R.B., Zhao, X.C.: The trend of automatic test system. Foreign Electron. Meas. Technol. **33**(6) (2014). (in Chinese)
2. Kirkland, L.V., McConkey, C.N.: Recurrent TPS development issues or ascertaining the excellence of an automated unit. In: IEEE AUTOTESTCON, Institute of Electrical and Electronics Engineers Inc., USA, pp. 1–7 (2016)
3. Yu, D., Ye, G., Ma, S.L., Xiong, N.X., Yang, L.T.: The spacecraft automatic testing system based on workflow. In: 2008 IEEE Asia-Pacific Services Computing Conference, Yilan, pp. 890–895 (2008)
4. IEEE STD 771-1998: IEEE Guide to the Use of the ATLAS Specification (1998)
5. IEEE Trial-Use Standard for Automatic Test Markup Language (ATML) for Exchanging Automatic Test Equipment and Test Information via XML: Exchanging Test Descriptions. In: IEEE Standards Coordinating Committee 20, IEEE 1671.1 (2009)
6. Zhang, L.: Research and realization on automatic generation of TPS codes in signal-oriented ATS. University of Electronic Science and Technology of China, Sichuan (2014)
7. Zhu, F.X., Yu, Y.L., Li, X.X., Yuan, P.: Simulation of missile test signal based on VITE. Mod. Def. Technol. **45**(6) (2017). (in Chinese)
8. Luo, Y.M., Zhou, Y.O.: A program implementation of visual test flow. Meas. Control Technol. **5**, 67–69 (2005). (in Chinese)
9. Zhao, Y., Meng, X.F., Wang, G.H.: Method to design software architecture of automatic test system based-on component object model. Comput. Eng. Des. (4) (2008). (in Chinese)

A Low Voltage Power Line Model
for Broadband Communication

Yihe Guo$^{(\boxtimes)}$, Ran Huo, and Zhiyuan Xie

North China Electric Power University, Baoding 071003, China
yihe_guo@163.com

Abstract. The influence of skin effect and proximity effect on resistance and inductance of per-unit-length is analyzed. Based on the finite element method, the resistance and inductance are solved by electromagnetic simulation software, and a method combining with open circuit impedance measurement to solve capacitance and conductance is proposed. The accuracy of the cable model is verified by testing, and the transmission characteristics of T network are emphatically analyzed. The modeling method has the advantages of accuracy and simplicity.

Keywords: Power line broadband communication · Cable model ·
Transmission line · Distribution parameters · Scattering parameters

1   Introduction

In 1995, it was demonstrated that Low Voltage (LV) power lines could be used to carry high frequency (>1 MHz) communication signals. This brought about the concept of broadband Power Line Communication (PLC). It is of great theoretical significance and engineering application value to further study the channel model of broadband.

At present, a series of research achievements have been made on the broadband characteristics of low voltage cables at home and outdoors. The transmission line theory is used to simulate the low-voltage PLC channel, and the influence of different network topology on the channel characteristics of the frequency range from 2 to 30 MHz is analyzed [1]. The attenuation characteristics of T power line network are studied [2]. The propagation constant and characteristic impedance of the cable are approximated by the approximate formula, and the parameters are estimated by the open circuit and short circuit impedance test [3]. The method is influenced by the measurement accuracy, and the model error is difficult to eliminate when the line is long or the frequency is high. The classical approximate formula is used to calculate the main parameters of the power line [4]. The actual cable is often unable to meet the requirements for the uniform medium and the wide spacing formula. The insertion loss of two kinds of cables is analyzed [5]. The effect of proximity effect and insulation medium cannot be accurately calculated, and the error between the calculated result and the measurement result is large.

The difficulties in solving the distribution parameters and the effect of skin effect and proximity effect on the cable are analyzed in the case of relatively small distance between wires in this paper. At the same time, the influence of the branch of T network on the whole is also analyzed. Because the dielectric constant of insulating materials cannot be obtained and difficult to solve, a simple and easy method is proposed. Based on the theory of transmission lines and scattering parameters, a method for solving the cable model parameters is proposed in this paper, which combines electromagnetic field simulation and open circuit impedance measurement.

2 Description of Cable Model

2.1 Analysis of Model Parameters Based on Transmission Line Theory

It is appropriate to use transmission line equation to describe the cable model in low-voltage power line broadband communication.

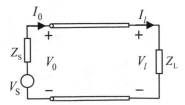

Fig. 1. Transmission model of power line.

Figure 1 is the transmission model of single phase power line carrier communication. The length of the power line is l, the per-unit-length parameters includes resistance R, inductance L, capacitance C and conductance G.

According to the per-unit-length parameters, the secondary parameters of the transmission line can be obtained. The propagation constant is [6]

$$\gamma = \alpha + j\beta = \sqrt{(R+j\omega L)(G+j\omega C)} \tag{1}$$

Where α is the attenuation constant and β is the phase constant. Characteristic impedance is

$$Z_C = \sqrt{\frac{(R+j\omega L)}{(G+j\omega C)}} \tag{2}$$

The source reflection coefficient Γ_S and load reflection coefficient Γ_L can be expressed as:

$$\Gamma_S = \frac{Z_S - Z_C}{Z_S + Z_C} \tag{3}$$

$$\Gamma_L = \frac{Z_L - Z_C}{Z_L + Z_C} \tag{4}$$

The source input resistance can be obtained as:

$$Z_{in} = Z_C \frac{1 + \Gamma_L e^{-2\gamma l}}{1 - \Gamma_L e^{-2\gamma l}} \tag{5}$$

The voltage and current at both ends of a uniform power line can be described by the transmission parameter matrix.

$$\begin{bmatrix} V_l \\ I_l \end{bmatrix} = \begin{bmatrix} A_{11} & A_{12} \\ A_{21} & A_{22} \end{bmatrix} \begin{bmatrix} V_0 \\ I_0 \end{bmatrix} \tag{6}$$

while the voltage at both ends is given by [7]

$$V_0 = \frac{A_{12} - Z_L A_{22}}{Z_L Z_S A_{21} - Z_L A_{22} - Z_S A_{11} + A_{12}} V_S \tag{7}$$

$$V_l = A_{12} \frac{V_S}{Z_S} + (Z_S A_{11} - A_{12}) \frac{V_0}{Z_S} \tag{8}$$

Further, the input impedance and voltage transmission coefficient can be solved.

2.2   Model Parameter Measurement Based on S Parameters

By measuring the S parameters of a network, the impedance and attenuation characteristics of the uniform power line can be obtained by the vector network analyzer [8].

S_{21} is the voltage transfer coefficient from port 1 to port 2. When using network analyzer, Z_S, Z_L and Z_0 are 50 Ω. Thus

$$S_{21} = \frac{V_l}{0.5 V_S} = \frac{(1 + \Gamma_L)(1 - \Gamma_S) e^{-\gamma l}}{1 - \Gamma_S \Gamma_L e^{-2\gamma l}} \tag{9}$$

A power line for an open circuit or short circuit can be regarded as a single port network. S_{11} is measured by a network analyzer, and the input impedance is obtained [9].

$$Z_{in} = 50 \frac{1 + S_{11}}{1 - S_{11}} \tag{10}$$

3 Characteristic of Distribution Parameter

3.1 Per-Unit-Length Resistance and Inductance

In low voltage PLC, the high frequency current is influenced by the eddy current, rather than uniformly distributed on the cross section. The skin effect leads to a reduction of the effective area of the conductor and a large increase in electrical resistance [10].

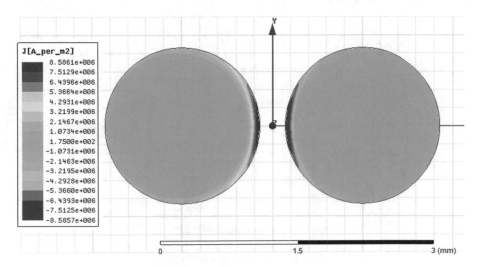

Fig. 2. Current density distribution.

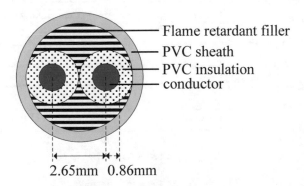

Fig. 3. Cross sections of the two cables.

Figure 2 shows the distribution of current densities of a typical two-conductor line based on the ANSYS software. It can be seen that the current distribution on the cross-section of the conductor is concentrated at the edge of the conductor.

When the distance between the conductors is small, the current will be concentrated on the inner side of the conductor. The proximity effect makes the R larger and L is also affected [11].

The cross-sectional schematic diagram of the cable used in this paper is shows in Fig. 3. It is a ZR-YJV cable with a cross-sectional area of 2.5 mm^2.

Based on the two-dimensional electromagnetic field simulation tool of ANSYS, the per-unit-length parameters of resistance and inductance are extracted, which are shown in Figs. 4 and 5 respectively.

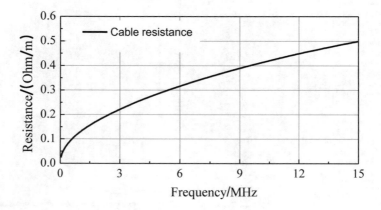

Fig. 4. Per-unit-length resistance.

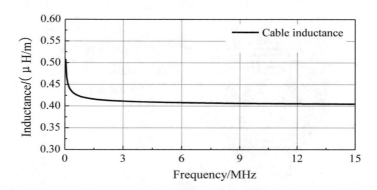

Fig. 5. Per-unit-length inductance.

3.2   Per-Unit-Length Capacitance and Conductance

In the transmission line model, the capacitance and conductance are all related to insulating materials. In order to better describe the two electrical characteristics of the insulation material, the complex permittivity is introduced.

$$\varepsilon_r = \varepsilon'_r - i\varepsilon''_r \tag{11}$$

where ε'_r is dielectric constant which is related to capacitance, and ε''_r represents loss dependent part. In practice, the loss angle tangent $\tan(\theta)$ is usually used to describe G.

$$\tan(\theta) = \frac{\varepsilon''_r}{\varepsilon'_r} \tag{12}$$

$$G = \omega \tan(\theta) C \tag{13}$$

Based on the measurement results, ε'_r and $\tan(\theta)$ are segmented and modeled respectively. Both of them decrease with the increase of frequency [12]. The ε'_r and ε''_r of rubber and PVC have been measured, which decrease as frequency increase [13].

The cable manufacturers cannot provide the ε_r value of the PVC material at high frequency. In many cases, more than one kind of insulating mediums is used around the copper wire, which causes the accurate calculations of ε_r cannot be completed. In this paper, the C and G are obtained indirectly by measuring the open circuit impedance.

4   Solution of Cable Model

4.1   Calculation of Conductance and Capacitance in Combination with Open Impedance

When the terminal is open, according to the transmission line theory, the input impedance of a transmission line of length l is

$$Z_{\text{ino}} = Z_C \text{atanh}(\gamma l) = Z_C \frac{1 + e^{-2\gamma l}}{1 - e^{-2\gamma l}} \tag{14}$$

In general, the line meets the low-loss conditions: $R \ll \omega L$ and $G \ll \omega C$. According to Eqs. (1) and (2), the corresponding approximate results are

$$Z_C \approx \sqrt{\frac{L}{C}} \tag{15}$$

$$\gamma \approx j\omega\sqrt{LC}[1 - j(\frac{R}{2\omega L} + \frac{G}{2\omega C})] \tag{16}$$

Then formula (14) can be simplified as

$$Z_{\text{ino}} \approx Z_C \frac{1 + e^{-(R/Z_C + G \cdot Z_C)l}e^{-2j\omega\sqrt{LC}l}}{1 - e^{-(R/Z_C + G \cdot Z_C)l}e^{-2j\omega\sqrt{LC}l}} \tag{17}$$

In order to determine the specific value of C and G, the nonlinear optimization algorithm is used in this paper. According to the values of L, R and l, C and $\tan(\theta)$ are used as the optimization variables, which minimize the mean square error of the measurement results and formula (17). The initial value of C' is based on the capacitance simulation results of ANSYS, where ε'_r of PVC is 3 and $\tan(\theta)$ is 0.08. The optimization results of $\tan(\theta)$ is 0.00448. The value of C' and C are shown in Fig. 6.

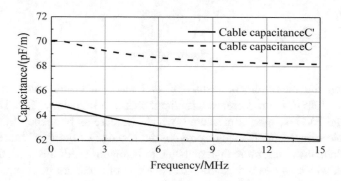

Fig. 6. Per-unit-length capacitance.

4.2   Open-Circuit Impedance Calculation and Analysis

Using the per-unit-length parameters that have been solved above, Z_{ino} can be calculated by (17) where the length of cable is 30 m. The magnitude and phase of the open circuit impedance are shown in Figs. 7 and 8 respectively. The results of theory and experiment have good coherence by comparing.

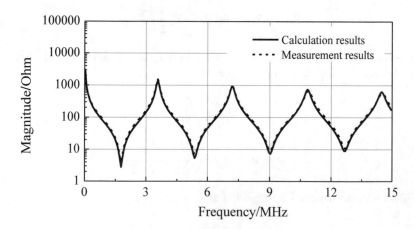

Fig. 7. Magnitude of the open-circuit impedance.

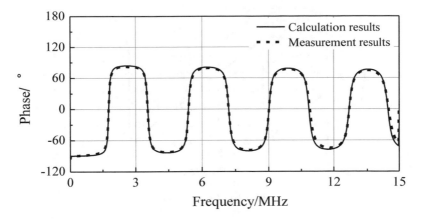

Fig. 8. Phase of the open-circuit impedance.

5 Verification of the Model

To verify the accuracy of the model, the transmission characteristics of the power line are calculated based on the model parameters, and the vector network analyzer is used for measurement. The magnitude and phase of S_{21} are shown in Figs. 9 and 10 for a 30 m cable, which shows that the calculation and measurement are in good agreement.

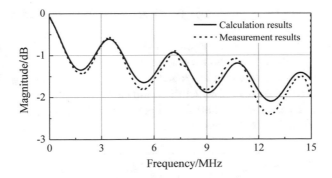

Fig. 9. Magnitude of S_{21}.

In order to further verify the accuracy of the model, a T-type network is built which is shown in Fig. 11. The link has one branch of 2–4 and port 4 is left open. The input impedance of the branch is Z_{in}, and the reverse transmission parameter matrix is

$$\begin{bmatrix} E & 0 \\ -Z_{in}^{-1} & E \end{bmatrix} \tag{18}$$

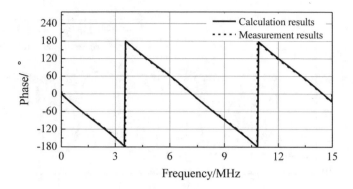

Fig. 10. Phase of S_{21}.

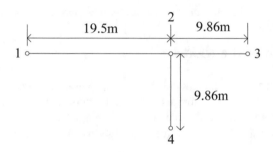

Fig. 11. T-type network.

The transmission parameter matrix of whole network is obtained by matrix multiplication of each section [14].

The S_{21} value between port 1 and 3 are shown in Figs. 12 and 13. The calculation results agree well with the measurement results.

As shown in Fig. 12, when the frequency is 5.3 MHz, there is a large fading point, which is caused by the branching structure in the cable network. The input impedance of the branch cable has a valley value at the frequency of 5.3 MHz, so the transmission characteristics are seriously affected.

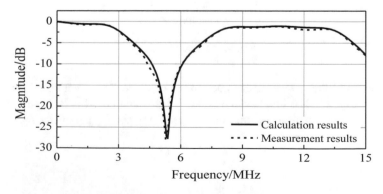

Fig. 12. Magnitude of S_{21} of the T-type network.

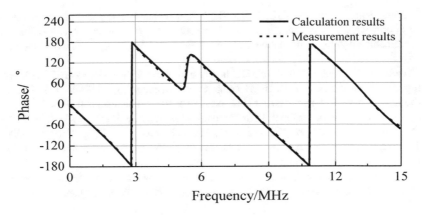

Fig. 13. Phase of S_{21} of the T-type network.

6   Conclusion

The broadband impedance and transmission model of common low-voltage cables are studied in this paper. For the case of relatively small conductor spacing, the difficulties in solving distributed parameters are analyzed. Based on the finite element method, the resistance and inductance were solved by electromagnetic simulation software. Since the dielectric constant of insulating materials cannot be solved directly, based on the measurement of the open-circuit impedance of a certain length of cable, the capacitance and conductance are solved by combining the simulation results. The measurements of the transmission characteristics verify the accuracy of the cable model. Some difficulties are encountered in the research process of this paper. In the actual measurement of S parameters, the data will also change due to the influence of space and magnetic field on the cable. More accurate data can be obtained only after repeated measurements and comparisons. This work has laid a good foundation for further study of multi-core cable channel models in real scenarios.

Acknowledgement. This study is supported by The Science and Technology Program of Hebei Province (No. 17211704D), the Fundamental Research Funds for the Central Universities (No. 2015MS97).

References

1. Lu, W.B., Zhang, H., Zhao, X.W.: The effect of network parameters for low-voltage broadband power line channels. Trans. China Electrotechnical Soc. **31**(a01), 221–229 (2016)
2. Wang, Y., Deng, Z.Q., W, H.A.: Study on multi-node broadband power line channel modeling. Power Syst. Prot. Control **46**(3), 18–25 (2018)
3. Zimmermann, M., Dostert, K.: A multipath model for the powerline channel. IEEE Trans. Commun. **50**(4), 553–559 (2002)

4. Thomas, M.S., Chandna, V.K., Arora, S.: Indoor power line channel characterization for data transfer and frequency response measurements. In: International Conference on Sustainable Mobility Applications, Renewables and Technology, Kuwait City, pp. 1–8 (2015)
5. Versolatto, F., Tonello, A.M.: An MTL theory approach for the simulation of MIMO power-line communication channels. IEEE Trans. Power Delivery **26**(3), 1710–1717 (2011)
6. Meng, H., Chen, S., Guan, Y.L., et al.: Modeling of transfer characteristics for the broadband power communication channel. IEEE Trans. Power Delivery **19**(3), 1057–1064 (2004)
7. Clayton, R.P.: Analysis of Multiconductor Transmission Lines. Wiley, Hoboken (2007)
8. Takmaz, E.: Impedance, attenuation and noise measurements for power line communication. In: Smart Grid Congress and Fair, Istanbul, pp. 1–4 (2016)
9. Lovric, D., Boras, V., Vujevic, S.: Accuracy of approximate formulas for internal impedance of tubular cylindrical conductors for large parameters. Prog. Electromagnet. Res. **16**(1), 171–184 (2011)
10. Cirino, A.W., De Paula, H., Mesquita, R.C., et al.: Cable parameter variation due to skin and proximity effects: determination by means of finite element analysis. In: 2009 35th Annual Conference of IEEE Industrial Electronics, Porto, pp. 4073–4079 (2009)
11. Lu, G.Z., Guo, Q.X., Zeng, D.D.: Proximity effect and the distribution parameters of multi-conductor transmission line. Chin. J. Radio Sci. **31**(3), 611–615 (2016)
12. Tsuzuki, S., Takamatsu, T., Nishio, H., et al.: An estimation method of the transfer function of indoor power-line channels for Japanese houses. In: International Symposium on Power Line Communications and Its Applications, Athens, pp. 1–5 (2002)
13. Kruizinga, B., Wouters, P.A.A.F., Steennis, E.F.: High frequency modeling of a shielded four-core low voltage underground power cable. IEEE Trans. Dielectr. Electr. Insul. **22**(2), 649–656 (2015)
14. Yarman, B.S.: Design of Ultra Wideband Power Transfer Network. Wiley, Hoboken (2010)

Research on the MAC Layer Performance of Wireless and Power Line Parallel Communication

Ran Liu, Jinsha Yuan, Zhixiong Chen[✉], Yingchu Liu,
Dongsheng Han, and Yihe Guo

North China Electric Power University, Baoding 071003, Hebei Province, China
chenzxl983@sohu.com

Abstract. Wireless technology is dominant in residential and enterprise networks and it offers mobility and attractive data-rates. Power Line Communication (PLC) is becoming popular in home networks. The main advantages of PLC is the no-new-wires connectivity and the high density of electrical plugs in any residential or enterprise environment. The MAC layer, as one of the keep foundations of networks, is important for improving network performance. However, previous research mainly focused on power line or wireless MAC layer and few studies MAC layer for power line and wireless parallel communication. Therefore, this paper proposes a hybrid PLC-Wireless network MAC protocol based on CSMA/CA (carrier sense multiple access with collision avoidance) and we verify its performance via simulation.

Keywords: Parallel communication · MAC layer · CSMA/CA

1 Introduction

With the development of home networks, power line and wireless hybrid communications can provide high-speed data rates and a wide range of coverage. The power line has the advantages of convenient use, plug and play, and wide network coverage. Although wireless can provide high data rate, but coverage is limited and there are network blind spots. Therefore, power line and wireless hybrid communications has a broad application prospect in smart home and Internet of things.

Many papers have studied the MAC layer of 802.11. Bianchi proposed a Markov model to describe the 802.11 back off process in Bianchi (2000) had been widely used. To apply Bianchi's analysis, Patras et al. (2011) estimated the number of stations by calculating the collision probability or channel state. The CSMA/CA algorithm of the 802.11 MAC layer was studied through the decoupling assumption in Kumar (2007). Mudriievskyi and Lehnert (2017) introduced a MAC mechanism for adaptive layer transformation between TDMA and CSMA/CA. Liu et al. (2017) proposed a new MAC protocol based on the time period. Huo et al. (2017) proposed a MAC mechanism based on clustering and probability competition. Nardelli and Knightly (2012) focused on performance evaluation of IEEE 802.11 EDCA. Vlachou et al. (2014)

© Springer Nature Switzerland AG 2020
Q. Liu et al. (Eds.): CENet 2018, AISC 905, pp. 729–737, 2020.
https://doi.org/10.1007/978-3-030-14680-1_79

analyzed the performance of IEEE 1901 by decoupling assumption model. Vlachou et al. (2016) proposed a coupling model and evaluated throughput. Fernandes et al. (2018) introduced the MAC-PHY cross-layer analysis of narrow-band power line communication systems in the presence of impulse noise. The additive noise in the PHY layer of narrowband power line communication is modeled by the Bernoulli-Gaussian process. Fernandes et al. (2018) focused on achievable data rate analysis of the incomplete HSRC (Hybrid power line-wireless Single Rrelay Channel) model.

Therefore, current researches mainly focused on power line or wireless MAC layer and few studies MAC layer for power line and wireless parallel communication. So this paper investigates a parallel MAC protocol for the hybrid PLC-Wireless network. In this paper, we evaluate the performance of Power line and Wireless Parallel Communication MAC systems, then analysis the performance via simulations. The paper includes the following parts: Part II describes the CSMA/CA algorithm of IEEE 1901. Part III describes an modified CSMA/CA mechanism for Power line and Wireless Parallel Communication. Part IV evaluates throughput, delay and collision probability of the MAC algorithm of Parallel Communication and PLC. Finally, conclude in Part V.

2   The 1901 MAC Mechanism

The IEEE 1901 backoff algorithm contains three counters in total: Back-off procedure counter (BPC), back-off counter (BC), and deferral counter (DC). The algorithm flow chart is as follows:

(1) When there is a packet to send, the BPC is initialized to 0 and BC randomly takes values in $\{0, \ldots, CW0 - 1\}$. Where CW0 represents the smallest contention window. The DC values corresponding to different back off stages are shown in Table 1 and enters Step (2).

(2) If the station detects the medium is idle, the BC value is decremented by 1, the DC value is unchanged, and then if BC = 0, the station transmit the packet immediately and enters Step (3), otherwise it waits for the arrival of the next time slot and return to Step (2). If the station detects that the medium is busy, the DC and BC values are simultaneously decremented by one. If DC is equal to 0 and the medium is detected busy the station enters the next back-off stage and BPC increases by one. When the BPC value is greater than MaxFrameRetries, where MaxFrameRetries denotes the maximum number of retransmissions, packet is discarded. Otherwise, return to Step (2).

(3) If the transmission is successful, it waits for the arrival of the next time slot to return to Step (1). If the transmission fails, the station will enter to the next back-off stage and BPC increases by one (if the station is already in the maximum back-off stage and reenter this back-off stage), the BC is newly set to any value in $\{0, \ldots, CWi - 1\}$, where CW_i is the contention window for back-off stage i. The values of CW and DC are shown in Table 1, and then enter to Step (2) (Fig. 1).

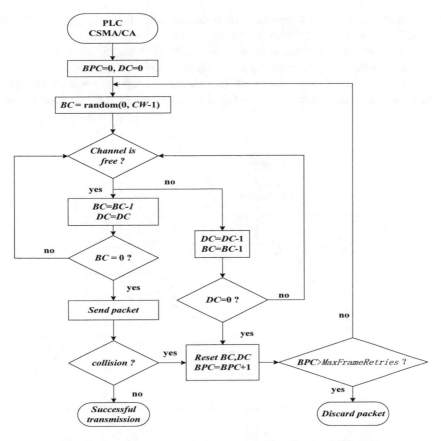

Fig. 1. Flowchart of the CSMA/CA algorithm for the IEEE 1901.

Table 1. CW and DC values based on BPC and priority.

	Priorities	CA0/CA1	Priorities	CA2/CA3
BPC	CW	DC	CW	DC
0	8	0	8	0
1	16	1	16	1
2	32	3	16	3
≥ 3	64	15	32	15

3 An Modified CSMA/CA Mechanism for Power Line and Wireless Parallel Communication

Assuming there are N stations in the network, the back-off process for each station uses two back-off counters, BC1 and BC2, which represent the back-off counter values for 802.11 and 1901, respectively. The 1901 is more complicated than 802.11. Due to the

DC, the station can enter the next back-off stage before sending data, and 802.11 enters the next back-off stage only when a collision occurs.

The back-off counter whose value reaches zero firstly waits the value of other back-off counter reaches zero. Then the station sends frame. If a collision occurs, the station enters the next back-off stage and resets the value of the deferral counter and back-off counter. If the transmission is successful, the station enters the first back-off stage (Fig. 2).

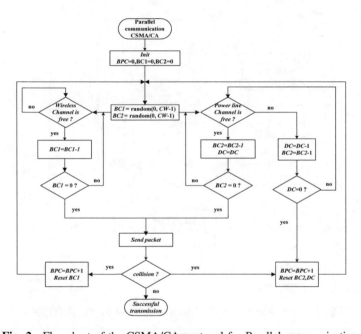

Fig. 2. Flowchart of the CSMA/CA protocol for Parallel communication.

The back-off process of parallel communication is described as follows:

The back-off algorithm of Parallel communication uses four counters, back-off procedure counter (BPC), deferral counter (DC), and two back-off counters (BC1 and BC2). BPC and BC represent the number of re-transmissions (back-off stage) and the random back-off time respectively.

(1) the value of BPC is initialized to zero when there is a data frame to send and BC1 and BC2 randomly select a number in $\{0, \ CW0 - 1\}$ as its value, where CW0 represents the smallest contention window.

(2) Station senses the channel in each slot. If the wireless medium is idle, BC1 is decreased by one. If the wireless channel is sensed busy, BC1 are frozen. For the power line, if station detects channel is idle, BC2 is decreased by one and DC is unchanged. BC2 and DC are decreased by one when the power line channel is detected busy.

(3) Whether BC1 or BC2 arrives zero first, it should wait the value of the other counter also reaches zero. Stations can now transmit their frames. If a packet is successfully transmitted, all the counter values of the station are initialized and the above steps are repeated; the station enters the next back-off stage if the transmission fails.

4 Performance Evaluations

The paper discusses the performance of Parallel communication in this part and compare throughput, delay and collision probability of Parallel communication against PLC.

The simulation parameters are shown in Table 2. Power line and wireless settings the same parameters. The throughput of a parallel communication network is expressed as follows:

$$S = \frac{p_s t_L}{p_s T_s + p_c T_c + p_e \sigma} \tag{1}$$

Where Ps1 indicates the probability that the station successfully sends data frames. P_c indicates the probability of a collision. The P_e indicates the probability that the channel is idle. t_L is time of transmission frame. Ts is the time for successful transmission of a frame. Tc is the time of collision and σ is the duration of a slot time.

$$Ts = PRS0 + PRS1 + P + t_L + RIFS + ACK + CIFS. \tag{2}$$

$$Tc = EIFS \tag{3}$$

Where PRS0, PRS1, P, RIFS, CIFS, ACK denote the duration of priority tone slots, a preamble, a response inter-frame space, contention inter-frame space, ACK.

EIFS denotes the extended inter-frame space. When a collision occurs, the collision time is equal to EIFS. After the EIFS time, it indicates that the other stations have finished sending data and the channel becomes idle.

Table 2. Simulation parameters.

Parameter	Time (us)
Slot σ	35.84
Priority slot PRS	35.84
Preamble P	110.48
CIFS	100
RIFS	140
Frame duration D	2050
ACK	110.48
EIFS	2920.64

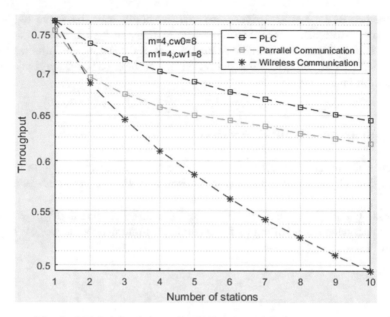

Fig. 3. MAC throughput under PLC and Parallel communication.

Figure 3 shows the throughput curve with different number of stations. From the results, when the number of stations N increases, the throughput gradually decreases, and the parallel communication throughput is lower than the PLC throughput because it requires a process of sending a wait, that is, one of the back-off counters needs to wait for another counter after reaching zero, which reduces the transmission probability so that the throughput decreases. Since wireless communication has only one back-off counter and has no delay counter, the probability of channel congestion increases when the number of stations increases. The probability of collision is greater, so the throughput is lower.

Figure 4 shows that the probability of collision increases when the number of stations N increases. For parallel communication, since the power line and the wireless are mutually restricted, only when value of two back-off counters reach zero can the station send frame, then many collisions due to a single channel congestion can be avoided. So the collision probability is very low and the reliability is high. Compared with power line communication, wireless communication has only one back-off counter, and the delay counter enables the station to enter the next back-off stage when it does not attempt transmitting, thereby avoiding collisions. Therefore, when the number of stations increases, the probability of collision of wireless communication is larger, which leads to lower throughput probability increases resulting in an increase in delay. For parallel communication, because sending packets is affected by the power line and the wireless back-off counters, Therefore, the average time spent on successfully sending a frame increases. When the number of stations increases in wireless communication, the greater the probability of channel congestion which leads to the greater probability of collision, so the throughput is low.

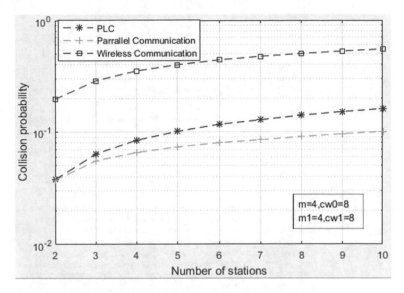

Fig. 4. MAC collision probability under PLC and Parallel communication.

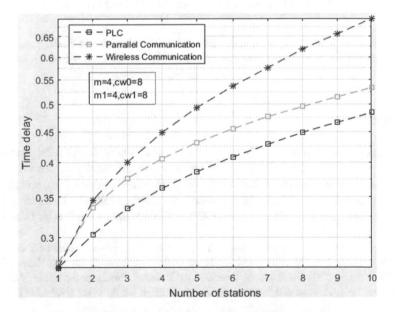

Fig. 5. MAC delay under PLC and Parallel communication.

Figure 6 shows the throughput for different values for CWmin, where CWmin Fig. 5 shows that when the number of stations increases, the collision represents the minimum contention window. When CWmin changes within a small range, the throughput decreases when the number of stations increases. When the CWmin increases to a relatively large value, the station randomly selects the value of the

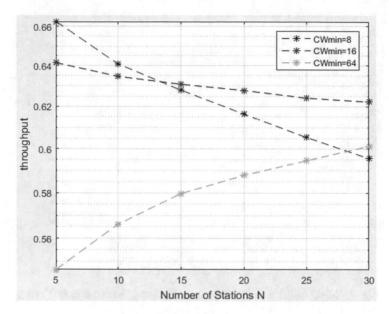

Fig. 6. MAC throughput for various values of CWmin.

back-off counter widely. The smaller the probability that the back-off counter value reaches zero simultaneously, and the smaller the probability of collision, so the throughput increases when the number of stations increases.

5 Conclusion

We discuss the performance analysis of wireless and power line parallel communication and verified via simulations in this paper. From the results, it can be seen that although the poor performance of parallel communication for throughput and delay, the collision probability is lower than power line communication. Therefore, parallel communication can be used to transmit QOS-ensured frames or higher-priority frames.

Acknowledgement. This study is supported by The National Natural Science Foundation of China (No. 61601182, No. 61771195), Natural Science Foundation of Hebei Province (No. F2017502059, F2018502047), the Fundamental Research Funds for the Central Universities (No. 2017MS109), Science and Technology Program of Hebei Province (No. 17211704D).

References

Bianchi, G.: Performance analysis of the IEEE 802.11 distributed coordination function. IEEE J. Sel. Areas Commun. **18**(3), 535–547 (2000)

Patras, P., Banchs, A., Serrano, P., Azcorra, A.: A control-theoretic approach to distributed optimal configuration of 802.11 WLANs. IEEE Trans. Mob. Comput. **10**(6), 897–910 (2011)

Kumar, A., Altman, E., Miorandi, D., Goyal, M.: New insights from a fixed-point analysis of single cell IEEE 802.11 WLANs. IEEE/ACM Trans. Netw. **15**(3), 588–601 (2007)

Mudriievskyi, S., Lehnert, R.: Adaptive layer switching for PLC networks with repeaters. In: IEEE International Symposium on Power Line Communications and its Applications, Madrid, pp. 1–6 (2017)

Liu, X., Liu, H., Cui, Y., Xu, D.: A new MAC protocol design based on time period for PLC network. In: IEEE 9th International Conference on Communication Software and Networks, Guangzhou, pp. 10–14 (2017)

Huo, C., Wang, L., Zhang, L.: Cluster and probability competition based MAC scheme in power line communications. In: 7th IEEE International Conference on Electronics Information and Emergency Communication, Macau, pp. 288–291 (2017)

Nardelli, B, Knightly, E.W.: Closed-form throughput expressions for CSMA networks with collisions and hidden terminals. In: Proceedings IEEE INFOCOM, pp. 2309–2317. IEEE (2012)

Vlachou, C., Banchs, A., Herzen, J., Thiran, P.: On the MAC for power-line communications: Modeling assumptions and performance tradeoffs. In: International Conference on Network Protocols, pp. 456–467. IEEE (2014)

Vlachou, C., Banchs, A., Herzen, J., Thiran, P.: How CSMA/CA with deferral affects performance and dynamics in power-line communications. IEEE/ACM Trans. Netw. **25**(1), 250–263 (2016)

Fernandes, V., Poor, H.V., Ribeiro, M.V.: Analyses of the incomplete low-bit-rate hybrid PLC-wireless single-relay channel. IEEE Internet Things J. **5**(2), 917–929 (2018)

An Automatic Discovery Method Based on SNMP and C4.5 in Cross-Regional Interconnected Power System

Zhiming Zhong[1]([✉]), Jie Wang[1], Chao Hu[2], Yuc Chen[2], and Tianming Huang[2]

[1] CSG Power Dispatching Control Center, Guangzhou 510670, China
zmizho@126.com
[2] NARI Information & Communication Technology Co., Ltd., Nanjing 210033, China

Abstract. Cross-regional interconnection connects the power system with the management information system, or with the Internet, and undermines the properties of the power monitoring system and the external network. Exposing the core assets of the power system originally deployed in the isolated environment to the external network brings great risks to the safe and stable operation of the system. In view of the seriousness of hidden information security risks in cross-region interconnections, this paper proposes an automatic discovery method based on SNMP and C4.5 algorithm for cross-regional interconnection of power systems. It mainly studies system network connection conditions. Based on the system's all equipment, it fully perceives the network connection of each device in the system, and finally realizes the automatic discovery and early warning of the inter-area interconnection of the system.

Keywords: Automatic discovery · Cross-regional interconnect · SNMP · C4.5 · Power network

1 Introduction

The electric power business is related to the national economy and livelihood of people, and it is also the basic guarantee for the sustained growth of the social economy [1]. Under the background of the continuous progress of the socialist modernization process, the electric power industry presents a good development prospect. But at the same time, due to the development characteristics of the power system itself, the security issue is always a key issue. It is easily affected by objective factors, threatens the operational safety of the power system, and affects the quality of power supply services. It has the potential for power supply companies to stand out in the fierce market competition [2]. Very far-reaching influence. Therefore, through the establishment of a power system, it provides a solid security protection for the safe and stable operation of the power system and reduces the hidden dangers of system operation. Once an abnormal situation is discovered, it can be handled in a timely and effective manner.

© Springer Nature Switzerland AG 2020
Q. Liu et al. (Eds.): CENet 2018, AISC 905, pp. 738–748, 2020.
https://doi.org/10.1007/978-3-030-14680-1_80

Under the background of the rapid development of computer technology, network technology and communication technology, a growing number of monitoring systems have been established in power production, fully combining the advantages of advanced technologies, enabling real-time monitoring of power system operating conditions [3]. Once an abnormal problem arises, with the help of communication technology and network technology, information is transmitted to the control center to realize the management and control of the power equipment and realize the exchange of data information. However, under the background of the rapid development of the Internet, due to the open nature of the Internet itself, the potential security problems in network applications have begun to highlight, which to a certain extent restricts the safe and stable operation of information systems, and even brings serious losses. Therefore, through the establishment of a power system to provide more reliable security protection, it will help provide the society with more secure and quality power supply services, improve the market competitiveness of power supply companies, and seek long-term survival and development [4].

From this point of view, it is critical to strengthen the research on the protection of power system security. To analyze the problems that exist, there are three main areas [5], (1) Partitioning errors; (2) Inter-regional interconnection; (3) The network security. With the development of power networks, cross-regional interconnection has become a problem that cannot be bypassed. Cross-regional interconnection may have a great impact on power monitoring networks. Therefore, research in this direction has attracted more and more attention from scholars, such as the research of transfer capability [6], management of transmission congestion [7], and joint dispatch approach [8]. This paper mainly addresses the issue of cross-regional interconnection of the second power system. In view of the seriousness of hidden information security risks in cross-regional interconnections, this paper proposes an automatic discovery method for cross-regional interconnection of power systems based on C4.5 algorithm. Based on the system's all equipment, it fully perceives the network connection of each device in the system, and finally realizes the automatic discovery and early warning of Cross-regional interconnection of the system.

Our paper is organized as follows. The basic knowledge of SNMP and C4.5 algorithm are briefly reviewed in Sect. 2, the method proposed in this paper is described in detail in Sect. 3. Subsequently, we compare and analyze the time consumption and accuracy of other algorithms with our method in Sect. 4, and made a conclusion in the last section.

2   Preliminary Knowledge

Before the formal introduction of our method, we first introduce the preliminary knowledge and the main techniques.

2.1   SNMP

The SNMP protocol was developed based on the Simple Network Monitoring Protocol (SGMP) released in 1987. Due to the urgent requirements of the industry for the standardization of network management protocols, the Internet Engineering Task Force (IETF) issued the official RFC document for SNMPv1 in 1990. Its design idea is to guarantee the simplicity, flexibility and expansibility of the protocol. It also hopes that the SNMP protocol will be a transitional network management protocol and realize the standardization of protocol when managing interconnected network devices. The SNMP management framework is shown in Fig. 1.

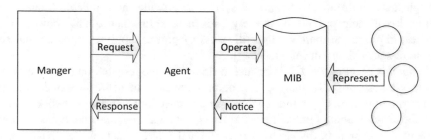

Fig. 1.   Management framework of SNMP.

The SNMP network management model adopts Manager and Agent modes and is mainly composed of (1) SNMP manger; (2) SNMP agent; (3) Management Information Base (MIB); and (4) SNMP protocol.

(1)   SNMP manger

SNMP manager is not a specific entity. It is essentially a set of procedures for conducting management activities through the issuance of orders, acceptance of responses, and active requests. The mode of operation of the manager is the polling mode. The request packet is issued to the managed device at intervals of a certain period of time, and the agent executes the command after receiving the request packet, and write the manager's information in the response data package and feed back to the manager.

(2)   SNMP agent

SNMP agent may be a hardware device such as a host, a switch, a bridge, a router, and a hub, or it may be a software service such as net-snmp. The SNMP agent on these devices or services must be able to receive command information from the management terminal, and the status of these agents must also be monitored by the management side. The SNMP agent responds to the command request of the management terminal to perform corresponding operations, and may also send information to the management terminal without request. The main functions of the SNMP agent are the interactive response to the SNMP Manger and the reporting of device emergency status based on the event trigger mode.

(3) MIB

The MIB is a management object database accessed by the network management protocol. It is essentially a tree-structured database, which is a collection of managed objects. The MIB includes two aspects:

(1) Object identifier (OID): The MIB's storage structure is tree-shaped. The nodes of the tree represent managed objects. Each node can be uniquely identified by a path from the root, and this path is called OID.
(2) MIB tree: MIB is constructed according to a tree structure. Each managed object is called a leaf node in the MIB structure, and there is only one unique parent node for the nodes in the next layer structure. A parent node can have more than one child node. Each node in the MIB tree network can be uniquely identified by an identifier.
(4) SNMP protocol

The SNMP protocol works on the fifth layer of the TCP/IP network structure and is an important part of the TCP/IP protocol suite. The transport layer adopts the UDP protocol. The manager accesses the agent's MIB to obtain information, and the manager and the agent provide corresponding interfaces to implement and manage the information interaction between the application and the managed device.

As we know, in order to achieve automatic monitoring, we must monitor the equipment. There are two general methods for monitoring: (1) Agent approach: Typically, the plug-in is installed on the monitored equipment or software to collect the operating data as an agent, and (2) Agent-less approach: Communication of monitored equipment or software through standard protocols enables the collection of operational data. The monitoring system monitors the operation of equipment and software using an agentless method based on a standard protocol. We take an agentless way to collect SNMP data, send notifications, and so on.

2.2 C4.5

There have been many variations for decision tree algorithms. C4.5 is one of the well-known decision tree induction algorithms [9]. In 1993, Ross Quinlin proposed the C4.5 algorithm which extents the ID3 algorithm [10]. Using information gain ratio to select the best attribute, C4.5 avoids ID3's bias toward features with many values that occurs [11, 12]. C4.5 has the ability to handle continuous attributes by proposing two different tests in function of each attribute values type.

At the training stage, the C4.5 uses the top down strategy based on the divide and conquer approach to construct the decision tree [13]. It maps the training set and uses the information gain ratio as a measurement to select splitting attributes and generates nodes from the root to the leaves. Every illustrating path from the root node to the leaf node forms a decision rule to determine which the class of a new instance [14]. The root node contains the whole training set, with all training case weights equal to 1.0, to take into account unknown attribute values. If all training cases of the current node belong to one single class, the algorithm terminates. Otherwise, if all training cases belong to more than one class, the algorithm calculates the information gain ratio for each attribute A_j. The attribute with the highest information gain ratio is selected to split information at the node [15]. For a discrete attribute A_j, the information gain ratio is computed by splitting training cases of the current node in function of each value of A_j [16]. If A_j is a continuous attribute, a threshold value must to be found for the splitting [17].

The C4.5 algorithm is an improvement of the ID3 algorithm. The basic workflow of this algorithm is the same as the ID3 algorithm. The difference is that the C4.5 algorithm uses the information gain rate to select attributes, which overcomes the disadvantage of the ID3 algorithm that uses the information gain to select attributes with more biased selection values [14]. The basis for the selection of attributes of the C4.5 algorithm is based on minimizing the entropy of the information contained in the nodes in the generated decision tree. The entropy of the set S is calculated as follows:

$$I(S) = -\sum_{i=1}^{k} ((freq(C_i, S)/|S|) * (\log_2(freq(C_i, S)/|S|))), \tag{1}$$

where $freq(C_i, S)$ represents the number of samples belonging to class C_i (one of the k possible classes) in set S, and $|S|$ represents the number of samples in set S. Equation (1) only gives the calculation of the entropy of a subset. If we partition after a certain attribute involves several subsets, we need to perform weighted sum calculations on these subsets. The formula is as follows:

$$E(A) = \sum_{i=1}^{n} ((|S_i|/|S|) * I(S_i)), \tag{2}$$

where E(A) represents the weighted sum of entropy, $|S_i|$ is the number of samples per class in the subtree based on a value of attribute A, and $I(S_i)$ is the entropy of S_i. In order to more clearly compare the entropy of different sets, we calculate the difference between the entropy of the pre-partition set and the entropy of the partition (this difference is called the gain). The node with the greatest gain is the node we want to select. The formula is as follows:

$$G(A) = I(S) - E(A). \tag{3}$$

GainRatio is defined as:

$$Gr(A) = G(S)/Sp(A), \tag{4}$$

where split information $split(A) = \sum_{i=1}^{k} \frac{|S_i|}{|S|} * \log_2 \frac{|S_i|}{|S|}$, S_1 to S_k are k sample subsets formed by splitting S of attribute A with k different values.

The C4.5 algorithm calculates the information gain rate for each attribute. The attribute with the highest information gain rate is selected as the test attribute of the given set S. Create a node and use this attribute as a tag attribute to create a branch for each value of the attribute and divide the sample accordingly.

3   Automatic Discovery Method Based on SNMP and C4.5 in Cross-Regional Interconnected Power System

The processes of our method is shown in Fig. 2. First, use the SNMP data acquisition module to complete the data collection of the device; then preprocess the collected SNMP data, wash out the redundant data and collect abnormal data, followed by the display of the collected data to facilitate the analysis and viewing of the inspectors; the next step is to discover regional interconnections, analyze and predict the collected data through the data security library and the trained C4.5 algorithm, and see if there is an abnormal cross-regional interconnection, if this situation is predicted, the SNMP Agent triggers the alert module and displays the alert event.

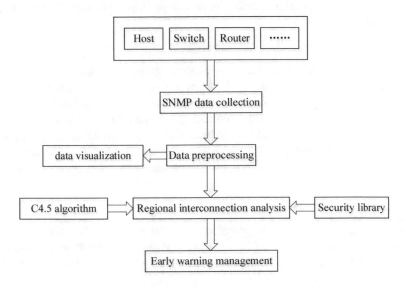

Fig. 2. The processes of our method.

3.1   SNMP Data Acquisition

SNMP data collection mainly includes data acquisition, data visualization and data preprocessing.

(1) Data collection: The bottom layer is an SNMP data acquisition engine. The network security equipment is the boundary of the system, and the system equipment assets are quickly scanned and accurately discovered, all-round discovery of the equipment connected to the system to fully understand the situation of asset interconnection. Implement the collection of security situation data in the MIB of the network device and the acquisition of Trap information of the core device;

(2) Data visualization: Visualize these data on the basis of data acquisition to help users intuitively understand the information in the data;

(3) Data preprocessing: Because the collected data may have problems of data redundancy and bad data, the preliminary analysis of the collected raw data is performed to achieve the compression of the data volume, which will pave the way for future data analysis.

3.2   Inter-regional Interconnection Prediction

The discovery of cross-area interconnection anomalies in the power monitoring system is mainly based on the combination of security libraries and C4.5 algorithms to predict and judge processed SNMP data. It is mainly divided into three parts: training of automatic discovery algorithm, prediction of inter-regional interconnection and warning management.

(1) Training of automatic discovery algorithm

The C4.5 decision tree is a kind of prediction model, which consists of three parts: Decision node, branch and leaf node. A decision node represents a sample test, usually representing an attribute of a sample to be classified, and different test results on the attribute represent a branch; the branch represents a different value of a decision node. Each leaf node represents a possible classification result. First, the training sample data with tags is collected, and the C4.5 algorithm is trained and tested. After the accuracy of the algorithm reaches the threshold, cross-area interconnection exception prediction can be performed.

(2) Prediction of inter-regional interconnection

The collected SNMP data is processed and put into the C4.5 decision tree for prediction, and a prediction result is obtained. If there is an abnormal situation in the inter-area interconnection, an exception report will be generated and sent by the SNMP agent to the SNMP manager.

(3) Warning management

Once an SNMP manager receives an SNMP agent to send an exception report, it directly uses the device alert to perform early warning management and displays the record storage to related devices. This feature greatly facilitates use. There are two main ways to collect warning information: The Trap accepts and actively polls the discovery. When an abnormal situation occurs, the device sends an exception report through the port. The exception report is implemented in the form of Trap; Regular polling means that the system sends an exception query command to devices and servers in the network within a certain period. This can help the device to send an exception to the exception. The combination of these two methods enables automatic discovery of cross-regional interconnections.

4 Experiment and Analysis

This experiment will reduce the power monitoring system network between regions and use multiple LANs to simulate the power monitoring system network in different regions. Each LAN includes 20 devices. First, the asset matching is performed by checking the system IP address, MAC address, host name, switch name, and roster name to find out the actual situation of the network connection of the system power monitoring system. Then, the network-related feature data of the power monitoring system of the device is automatically detected through SNMP. Naïve Bayes algorithm [18] and KNN algorithm [19] are used as comparison algorithms to test the efficiency and accuracy of our algorithm.

4.1 Efficiency Analysis

The computational complexity of the algorithm is related to the number of times the entropy and the information gain rate are calculated, and the efficiency of the algorithm is reflected in the length of time spent running the algorithm. The running times of the three algorithms are shown in Table 1 and Fig. 3 below.

Table 1. Time consumption of three methods (s).

	1	2	3	4	5	6	7	8
KNN	17	20	16	19	17	18	20	16
Naïve Bayes	13	15	13	10	13	10	15	14
Our method	11	11	9	11	12	10	9	7

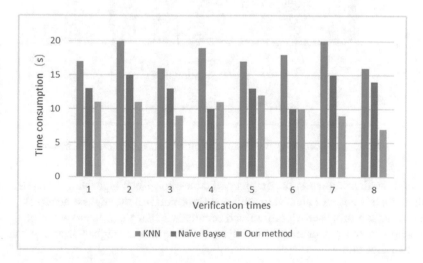

Fig. 3. Comparison of time consumption of Bayes, KNN, and our method.

From the table and the figure, we can clearly see that in the 8 experiments, the KNN algorithm takes the longest time, and Bayes takes less time than KNN, and the time spent by our method is obviously shorter than the other two.

4.2 Accuracy Analysis

We also analyzed the accuracy of the algorithm. We use accuracy figure to represent our comparative test results. Table 2 and Fig. 4 show the accuracy of Bayes, KNN, and our algorithm.

Table 2. Accuracy of four methods (%).

	1	2	3	4	5	6	7	8
KNN	88.1	87.4	82.9	85.9	82.2	87.3	84.2	82.4
Naïve Bayes	83.1	74.4	76.6	83.6	79.0	72.8	76.4	73.3
Our method	91.7	93.9	93.5	93.8	89.3	96.1	91.8	92.5

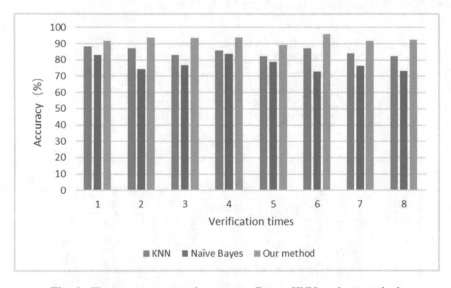

Fig. 4. The accuracy comparison among Bayes, KNN, and our method.

From the data shown in the graphs and tables, the KNN algorithm is slightly more accurate than the Bayes algorithm, and our algorithm has the highest accuracy. In this experiment, the proposed algorithm can predict whether there is cross-regional interconnection in the power monitoring system network with higher accuracy than the other two algorithms.

5 Conclusion

With the continuous expansion and development of power networks, power monitoring system networks have become an unavoidable reality across regions, and there are certain security issues in their cross-regional interconnections. How to quickly and accurately identify the interconnection between regions has attracted the attention of researchers. We proposed an auto-discovery algorithm for inter-regional interconnection problems. The SNMP is used to monitor the relevant data of the power monitoring system network in real time and C4.5 is used to predict whether there is interconnection between areas. In this paper, the efficiency and accuracy of the method are tested by comparison experiments with KNN and Naive Bayes. The experimental results show that the proposed method is superior to the other two algorithms both in efficiency and in accuracy.

Acknowledgement. This work is supported by Science and Technology Project of China Southern Power Grid Co., Ltd. "Research and Demonstration of Key Technologies of Network Security Situational Awareness in Power Monitoring System" (No. ZDKJXM20170002).

References

1. Ouyang, S., Shi, Y., Liu, Y.: Dynamic comprehensive evaluation of power quality for regional grid based on double inspiriting control lines. Power Syst. Technol. **36**(12), 205–210 (2012)
2. Sidorenko, A., Xu, Y.C.: Powering China: reforming the electric power industry in China. China J. **120**(50), 205 (2002)
3. Roy, L., Rao, N.D.: Real-time monitoring of power systems using fast second-order method. In: IEE Proceedings C-Generation, Transmission and Distribution, vol. 130(3), pp. 103–110 (2008)
4. Karaulova, I.V., Markova, E.V.: Optimal control problem of development of an electric power system. Autom. Remote Control **69**(4), 637–644 (2008)
5. Liu, H., Andresen, G.B., Greiner, M.: Cost-optimal design of a simplified highly renewable Chinese electricity network. Energy **147**(15), 534–546 (2018)
6. Shao, X.U., Miao, S., Zhou, L.: Calculation method of available transfer capability for cross-regional interconnected power grid. Autom. Electr. Power Syst. **41**(15), 74–80 (2017)
7. Liu, Y., Qiong, W.U., Jiang, Z., Qian, A.I.: Management of transmission congestion of cross-region interconnected electricity market based on demand response. Southern Power System Technology (2017)
8. Zhou, M., Wang, M., Li, J., Li, G.: Multi-area generation-reserve joint dispatch approach considering wind power cross-regional accommodation. CSEE J. Power Energy Syst. **3**(1), 74–83 (2017)
9. Quinlan, J.R.: Combining instance-based and model-based learning. In: Tenth International Conference on Machine Learning, pp. 236–243. Morgan Kaufmann Publishers Inc. (1993)
10. Quinlan, J.R.: Induction of decision trees. Mach. Learn. **1**(1), 81–106 (1986)
11. Ooi, S.Y., Tan, S.C., Cheah, W.P.: Temporal sampling forest: an ensemble temporal learner. Soft. Comput. **21**(23), 1–14 (2016)
12. Zhu, H., Zhai, J., Wang, S., Wang, X.: Monotonic decision tree for interval valued data. Commun. Comput. Inf. Sci. **481**, 231–240 (2014)

13. Liu, H., Gegov, A.: Induction of modular classification rules by information entropy based rule generation. In: Sgurev, V., Yager, R., Kacprzyk, J., Jotsov, V. (eds.) Innovative Issues in Intelligent Systems. Springer, Cham (2016)
14. Dai, W., Ji, W.: A mapreduce implementation of C4.5 decision tree algorithm. Int. J. Database Theory Appl. **7**(1), 49–60 (2014)
15. Hssina, B., Merbouha, A., Ezzikouri, H., Erritali, M.: A comparative study of decision tree ID3 and C4.5. Int. J. Adv. Comput. Sci. Appl. **4**(2), 13–19 (2014)
16. Ibarguren, I., Pérez, J.M., Muguerza, J.: CTCHAID: extending the application of the consolidation methodology. In: Pereira, F., Machado, P., Costa, E., Cardoso, A. (eds.) Progress in Artificial Intelligence, pp. 572–577. Springer, Cham (2015)
17. Perner, P.: Decision tree induction methods and their application to big data. In: Xhafa, F., Barolli, L., Barolli, A., Papajorgji, P. (eds.) Modeling and Processing for Next-Generation Big-Data Technologies, pp. 57–88. Springer, Cham (2015)
18. Ligeza, A.: Artificial intelligence: a modern approach. Neurocomputing **9**, 215–218 (1995)
19. Cover, T., Hart, P.: Nearest neighbor pattern classification. IEEE Trans. Inf. Theory **13**(1), 21–27 (2002)

End-to-End Network Fault Recovery Mechanism for Power IoT

ZanHong Wu[✉], Zhan Shi, Ying Wang, and Zhuo Su

Electric Power Dispatch & Control Center, Guangdong Power Grid Co., Ltd.,
Guangzhou, Guangdong, China
13602768748@139.com

Abstract. With the rapid development of power internet of things (PIoT) and increasing demands of power services, it is difficult for traditional network structure to fully differentiate QoS and reliability requirements of services. Software Defined Network (SDN) is an important virtualization technology. It separates network control plane from data forwarding infrastructure plane, simplifying network management and control. In this paper, a network resource allocation and fault recovery mechanism for PIoT is proposed. Firstly, it builds an SDN-based virtualized system model to abstract the physical network resources. Then, considering the factors such as network operation cost, revenue, service rate, QoS requirements, network load balancing, and network stability, this paper proposes an operating profit maximization model for multimedia services and uses dynamic resource load balancing (DRLB). The simulation results show that proposed mechanism can improve fault recovery ability of network on the basis of ensuring efficiency of resource utilization.

Keywords: Software Defined Network (SDN) · QoS · IoT · Fault recovery

1 Introduction

As gradually shifts to the Web 2.0 and Web 3.0 era, the internet is playing an increasingly important role in all walks of life. As one of components of the next generation network, IoT realizes a wide range of perceptual recognition, information transmission and intelligent processing from network equipment to any object. The smart grid is an important application area of IoT. IoT technology supports the development of the smart grid, improves the degree of information perception in all aspects of grid, and improves the power system's analysis, early warning, fault recovery and prevention capabilities. Because the application of power IoT includes different kinds of services such as visual monitoring, state detection, mapping and metering, etc. they have different requirements on bandwidth, delay and reliability of the power communication network, the network quality of service (QoS) requirements are also higher.

In order to satisfy these requirements, a network resource allocation and fault recovery mechanism for power IoT is proposed. Firstly, the SDN-based virtualized system model is built based on the characteristics of the general network to abstract physical network resources, and the SDN controller realizes perception of the physical

© Springer Nature Switzerland AG 2020
Q. Liu et al. (Eds.): CENet 2018, AISC 905, pp. 749–759, 2020.
https://doi.org/10.1007/978-3-030-14680-1_81

network resource status and differentiated demands of services. Then on basis of considering cost of network operations, earnings, services rate and QoS demand, network load balancing, network elements such as stability of multimedia services oriented operating profit maximization model is put forward and solved by dynamic resource load balancing, (DRLB) algorithm. At the same time, considering network's ability to recover after a failure, consider providing a failure recovery mechanism for network services, so that after physical network fails, service is still likely to resume normal operation.

2   Related Work

At present, there are some studies on improvement of services reliability in the context of SDN. Subbiah et al. [1] proposed an energy-aware network resource allocation in SDN. The proposed energy-aware particle swarm optimization algorithm defines network traffic forwarding policies in all switches in the network topology. Dutra et al. [2] proposed an end-to-end quality of service (QoS) solution based on queue support in OpenFlow, which enables operators with SDN networks to effectively allocate network resources according to user requirements, reducing or even eliminating over-provisioning. Gomes et al. [3] introduced two algorithms that implement virtual software to define network resource allocation. Some other people have also proposed SDN-based virtual resource allocation algorithms [4–8]. In addition, there is some other work focused on the virtualization of certain specific wireless networks, such as LTE, Wi-Fi, and WiMAX. However, the above solutions lack the following considerations: (1) Satisfaction of differentiated QoS requirements for different types of services; (2) Low utilization of network resources; (3) Unbalanced network load; (4) Poor recovery from physical failure.

3   Resource Allocation Model for Fault Recovery

The SDN-based virtualization system model is shown in Fig. 1. It is divided into physical layer, control layer and services layer from the bottom up. The services layer addresses various service requests and sends them through the northbound interface to the control layer. The SDN controller in the control layer is the core of virtualization. It is responsible for the centralized control of network resources and realizes the functions of unified scheduling and allocation of resources, abstraction of physical resources, and path planning. In resource management and scheduling, the SDN controller dynamically allocates node and link resources for the service according to the differentiated QoS requirements of different service requests. The physical layer is the underlying physical resources, which mainly include gateways, routers, and physical resources such as bandwidth and memory.

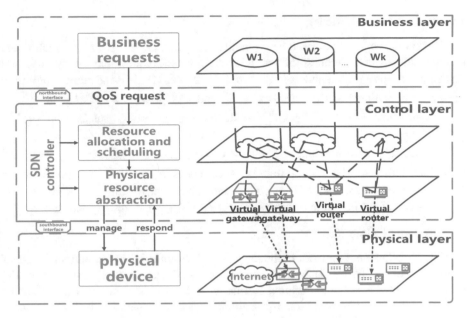

Fig. 1. SDN-based virtualization system model.

(1) Service Model

Services $W(C, T_W, T_d)$ has three attributes: priority, tolerance delay and processing delay. C indicates the priority of this service, when multiple services arrive at the same time, Services with large C values will be prioritized; T_W is for tolerating delay, indicating the time that services can wait; T_d indicates time services was processed in network.

(2) Physical Network Model

Physical network is represented by weighted undirected graph $G = (N, L)$, N is set of nodes in physical network. Physical node $n_i \in N | i \in \{1, 2, \ldots, N\}$, N indicates number of nodes. The type of node n_p is represented as $Type(n_p)$, including forwarding, encoding and other different node types. The capacity of node n_p to host service is $Cap(n_p)$. Service capacity includes CPU, memory and other aspects, expressed as $Cap(n_p) = \{CPU, Memory, \ldots\}$. Service capacity is quantified as capacity value: $C_p = \gamma_p \times CPU(n_p) + \delta_p \times Memory(n_p) + \ldots$, which can be used to measure capacity of node. CPU and Memory coefficients - γ_p and δ_p depends on type of node. Setting Boolean variables e to represent connectivity between nodes. If physical node n_i is connected to n_j, then e_{ij} is equal to 1, otherwise 0. L is collection of links in physical network. The physical link of any connection node n_i and n_i is represented as $l(i, j) \in L | i \neq j, i, j \in \{1, 2, \cdots, N\}$, its bandwidth is $B(l)$, and link delay is $T(l)$. $p(i, j) \in P$ represents set of all paths from node n_i to node n_i, P represents a collection of all physical paths.

(3) Virtual Network Model

Virtual network is expressed as an undirected graph $G^V = (N^V, L^V)$, which refers to physical network resources that are allocated for task request that satisfy node and link requirements. Virtual node $n_q^V \in N^V, q \in \{1, 2, \cdots, M\}$, M is the number of virtual nodes. The starting point of network is n_1^V and end point is n_M^V. For node n_q^V, its node type requirement is $Type(n_q^V)$, node capability requirement is $Cap^V(n_q^V) = \{CPU^V, Memory^V, \ldots\}$.

Quantified as demand value: $C_q^V = \gamma_q \times CPU^V(n_q) + \delta_q \times Memory^V(n_q) + \ldots$. L^V is virtual link set. The virtual link between nodes n_u^V and n_v^V is represented as $l^V(u, v) \in L^V | u \neq v, u, v \in \{1, 2, \ldots, M\}$, and link bandwidth requirement is $B(l^V(u, v))$. Let f_{ij}^{uv} be traffic occupied by virtual link $l^V(u, v)$ on physical link $l(i, j)$.

(4) Network Failure Recovery

Figure 2 illustrates network traffic from A to B, assume that probability of failure of original path (A, B, C, D) is p_1, then before the failure recovery mechanism, the probability of the network running normally is $1 - p_1$. The failure probability of backup path is p_2. The conditional probability that backup path (A, E, D) fails under condition of original path failure is:

$$P(\text{Backup path failure} \mid \text{Original path fault}) = p, \ (p_1, p_2, p \ll 1)$$

Then, the probability that task can run normally is $(1 - p \cdot p_1)$. Therefore, compared with the situation without backup, the probability of running normally $(1 - p \cdot p_1) > 1 - p_1$. It can be seen that the backup allows the network service to be quickly restored and the probability of normal operation of network is improved. Meanwhile, due to $p \geq p_2$, The equals sign can be established when primary and standby networks fail independently. Therefore, in order to maximize probability of normal operation, when mapping the backup path for the task, the mapped backup path and active path must be completely non-coincident. The probability of faults on lines is independent of each other, which maximizes probability that network will return to normal operation. At this point, probability of normal operation reaches $1 - p_2 \cdot p_1$.

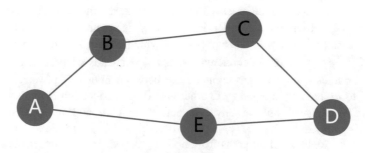

Fig. 2. Path diagram.

(5) Network Resource Status

At a certain moment, physical network resource status includes node resource status and link resource status.

(1) Node resource status. The node resources are divided into used capacity and remaining available capacity. Used capacity can be calculated by $Cap_{used}(n_i) = \sum_{\lambda} x_i^{\lambda} Cap^V(n_{\lambda})$, quantitative value is $C_{iused} = \sum_{\lambda} x_i^{\lambda} \times C_{\lambda}^V$. The available capacity is difference between total capacity of the node and the used capacity. Available capacity $Cap_{avi}(n_i) = Cap(n_i) - Cap_{used}(n_i)$, the value is $C_{iavl}(n_i) = C_i - C_{iused}$. And for the balanced allocation of resources, set resource perception coefficient $\alpha_i = \frac{\varphi_1}{C_{iavl}(n_i) + \mu_1} + \phi_1$ of node n_i, where φ_1 and μ_1 are constants used to adjust numerical range. ϕ_1 is set to a small value to avoid situation where overall value is 0.

(2) Link resource status. Like node resources, link bandwidth can also be divided into used bandwidth and available bandwidth. Used bandwidth $B_{used}(l) = \sum_{uv} B^V(l^V(u,v)) \times y_{ij}^{uv}$, available bandwidth $B_{avl}(l) = B(l) - B_{used}(l)$.

Assume $\beta_{ij} = \frac{\varphi_2}{B_{avl}(l) + \mu_2} + \phi_2$, among them φ_2 and μ_2 are constants used to adjust value range. ϕ_2 is set to a very small value to avoid a situation where overall value is 0. Among them x_i^{λ}, y_{ij}^{uv} are Boolean variable:

$$x_i^{\lambda} = \begin{cases} 1, \text{Physical node } n_i \text{ is assigned to the virtual node } n_{\lambda}^V \\ \quad 0, else \end{cases} \quad (1)$$

$$y_{ij}^{uv} = \begin{cases} 1, \text{Virtual link } l(i,j) \text{ is assigned to virtual link } l(u,v) \\ \quad 0, else \end{cases} \quad (2)$$

4 Problem Model

A. Service Costs

The cost of a service mainly includes the consumption of nodes and links to complete service. And set proportional coefficient related to current resource status of nodes and links and to adjust balance of use of physical network. The cost of using path with less remaining resources is higher, so that nodes and links with more remaining resources can be used preferentially.

The available cost function is as follows:

$$Co = a_1 \times \sum_i C_q^V + a_2 \times \sum_{ij} B^V(l^V) \quad (3)$$

Where a_1 a constant coefficient that represents the cost of unit node capacity value; a_2 indicates cost of unit link bandwidth.

B. Service Incomes

The revenue of service is related to characteristics of completed service W.

$$r(W) = \varepsilon \times C + \frac{\zeta_1}{T_d + \eta_1} + \frac{\zeta_2}{T_W + \eta_2} + \theta \qquad (4)$$

ε, ζ_1, ζ_2 as well as η_1, η_2 are used to adjust proportion of different attributes in benefits, which is related to QoS requirements of tasks and reflects weight of requirements in all aspects. θ is basic benefits of completing task.

The total revenue for entire network is sum of all revenues for completing service:

$$\text{Rev} = \sum_W r(W) \qquad (5)$$

C. Objective Function

The overall goal of network is to maximize overall profit:

$$f : Max\{\text{Rev} - Co\}$$
$$= Max\{\sum_W r(W) - [a_1 \times \sum_i C_q^V + a_2 \times \sum_{ij} B^V(l^V)]\} \qquad (6)$$

D. Constraints

Constraints of node capacity, node allocation, connectivity and bandwidth are described as (7)–(10).

$$\text{Cap}_{avl}(n_p) \geq \text{Cap}^V(n_q^V) \qquad (7)$$

$$\sum_p x_p^q = 1 \qquad (8)$$

$$e_{p,p+1} = \begin{cases} 1, x_p^q = 1 \,\&\, x_{p+1}^{q+1} = 1 \\ 0, else \end{cases} \qquad (9)$$

$$B_{avl}(l) \geq B^V(l^V) \qquad (10)$$

5 Algorithm Design

The main steps of resource-aware dynamic resource allocation algorithm are as follows:

(1) Initialization. Set network parameters.

(2) Determine service process order. For incoming service, first understand its properties. C Indicates priority of service, T_w is tolerate delays, it indicates time service can wait. B indicates bandwidth required by service. In the case of multiple services arriving at the same time, service with a large value of C is preferred, and service with a small value of T_w is preferred when value of C is same. Both of the first two give priority to service with large B value.

(3) Node resource allocation. In nodes satisfying above capacity constraints, allocation constraints, and connectivity constraints, The nodes with the lowest value of nodal resource perception coefficient were selected as assigned nodes.

(4) Determine link weight. The weight of link is determined by transmission delay and link load. It can be expressed as $W_{ij} = a_1 \times T(l)/T_W + b_1 \times \beta_{ij}$, I.e. The weighted sum of link delay and bandwidth perception coefficient. $a_1 + b_1 = 1$, and at beginning of algorithm initialization $a_1 = 0, b_1 = 1$. That is, priority is given to link load balancing.

(5) Link resource allocation. Use Dijkstra's algorithm to find shortest path and allocate link resources according to shortest path.

(6) Check whether obtained path meets delay requirement of service. If link delay is less than tolerable delay of service, link is allocated to service and Step 8 is entered.

(7) If delay requirement is not met, determine whether service wait time is greater than tolerance time or weight of a_1 reaches 1. If waiting time is not greater than tolerance time and weight is less than 1, go to Step 8.

(8) Verify that two completely different feasible paths have been obtained - one for primary path and one for backup path. If resulting number of paths is less than 2, increase value of a_1 and re-enter Step 4 for link assignment.

(9) If $a_1 = 1$ or time has been exceeded, allocation fails and there is no feasible solution to meet service requirements.

(10) After all service processes are finished, algorithm ends.

6 Simulation and Result Analysis

First, compare variance of resource utilization of physical nodes under different routing algorithms. This indicator reflects degree of balance of network load. As can be seen from Fig. 4, load variance of DRLB algorithm is significantly less than the other two algorithms, indicating that load on network nodes is more balanced and utilization of resources is more efficient.

Figure 5 shows that network operating profit of this algorithm is superior to the other two in both experimental time and overall trend. The DRLB algorithm can effectively reduce operating costs of network by increasing utilization of network resources (Fig. 3).

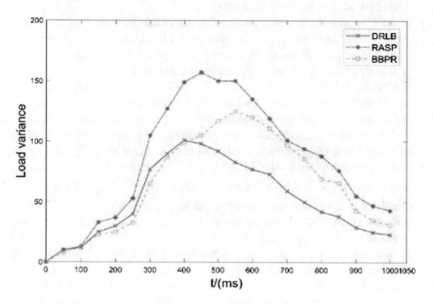

Fig. 3. Comparison of node load variances for three algorithms.

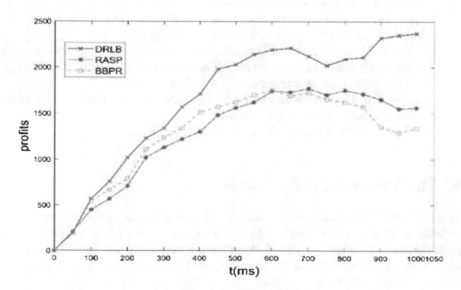

Fig. 4. Comparison of network operation profits of three algorithms.

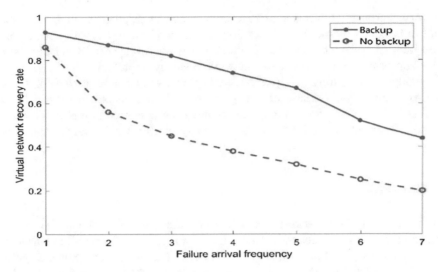

Fig. 5. Comparison of network recovery rates in case of backup.

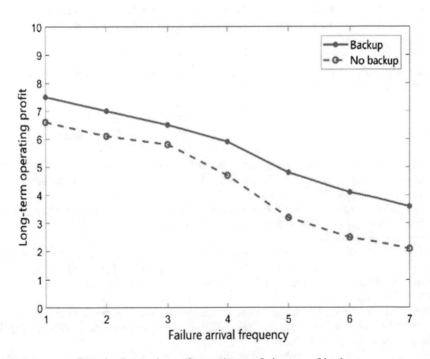

Fig. 6. Comparison of operating profit in case of backup.

In case of a physical network failure, consider probability that virtual network running on it can resume operation. From Fig. 6, we can see that in face of same network failure arrival frequency, in case of backup, probability of virtual network recovery is greater. At the same time, when failure rate rises, there is a relatively stable trend in backup network. This shows that since network map provides a backup path, when fault arrives and current path is affected, network service can be quickly restored to backup path, and survivability of network is greatly improved.

Finally, compare long-term profitability of network in absence of backup. It can be seen from Fig. 7 that because of strong survivability of network in case of backup, completion rate of service is high, and income from operation is high, so long-term operating profit of network with failure recovery mechanism is higher when failure arrives.

7   Conclusion

This paper proposes a network resource awareness algorithm for network resource allocation in context of network virtualization. It makes load of network as balanced as possible, improves resource utilization efficiency of network, and prevents network from being overstressed due to excessive local pressure. At the same time, in order to ensure reliability of network, a network mapping with a fault recovery mechanism is proposed. When resource is allocated, virtual network is provided with two master/slave network mapping results, making network traffic can be recovered quickly if original path fails.

Acknowledgment. This work was supported by the science and technology project of Guangdong power grid (036000KK52160025).

References

1. Subbiah, S., Perumal, V.: Energy-aware network resource allocation in SDN. In: 2016 International Conference on Wireless Communications, Signal Processing and Networking, Chennai, pp. 2071–2075 (2016)
2. Dutra, D.L.C., Bagaa, M., Taleb, T., Samdanis, K.: Ensuring end-to-end QoS based on multi-paths routing using SDN technology. In: IEEE Global Communications Conference, Singapore, pp. 1–6 (2017)
3. Gomes, R.L., Bittencourt, L.F., Madeira, E.R.M., Cerqueira, E., Gerla, M.: State-aware allocation of reliable virtual software defined networks based on bandwidth and energy. In: 2016 13th IEEE Annual Consumer Communications & Networking Conference, Las Vegas, NV, pp. 411–416 (2016)
4. Cao, B., Lang, W.Q., Chen, Z.: Power allocation in wireless network virtualization with buyer/seller and auction game. In: 2015 IEEE Global Communications Conference, San Diego, CA, USA, pp. 1–6 (2015)
5. Esposito, F., Chiti, F.: Distributed consensus-based auctions for wireless virtual network embedding. In: 2015 IEEE International Conference on Communications, London, UK, pp. 472–477 (2015)
6. Rahman, M.M., Despins, C., Affes, S.: HetNet cloud: leveraging SDN & cloud computing for wireless access virtualization. In: 2015 IEEE International Conference on Ubiquitous Wireless Broadband, Montreal, QC, Canada, pp. 1–5 (2015)

7. Zhou, B.Y., Cao, W., Zhao, S.S., Lu, X, Du, Z.: Virtual network mapping for multi-domain data plane in software-defined networks. In: 2014 4th International Conference on Wireless Communications, Vehicular Technology, Information Theory and Aerospace & Electronic Systems, Aalborg, Denmark, pp. 1–5 (2014)
8. Zubow, A., Doring, M., Chwalisz, M., Wolisz, A.: A SDN approach to spectrum brokerage in infrastructure-based cognitive radio networks. In: 2015 IEEE International Symposium on Dynamic Spectrum Access Networks, Stockholm, Sweden, pp. 375–384 (2015)

Cross-Domain Virtual Network Resource Allocation Mechanism for IoT in Smart Grid

Zhan Shi[1($\boxtimes$)], Zhou Su[1], and Peng Lin[2]

[1] Electric Power Dispatch & Control Center, Guangdong Power Grid Co., Ltd., No. 75 Meihua Road, Guangzhou, Guangdong Province, China
rhett.shi@139.com
[2] Beijing Vectinfo Technologies Co., Ltd., No. 15, Xueyuan South Road, Haidian District, Beijing, China

Abstract. In order to solve the problem of cross-domain resource allocation in IoT in Smart Grid, this paper proposes a virtual network resource allocation mechanism based on particle swarm algorithm. Its goal is to minimize mapping overhead under planning request. In this paper, VN request is divided into multiple virtual subnets according to matching set of virtual resource matching phase, virtual library resource type price information and border node information. We also propose a cross-domain virtual network mapping algorithm based on particle swarm optimization. It can be used to improve the efficiency of cross-domain virtual network mapping. Finally, the execution time, mapping cost and performance of the algorithm in different environments are tested by simulation experiment, which verifies its efficiency and stability performance in virtual network partitioning.

Keywords: Cross-domain virtual network resource allocation ·
Particle swarm algorithm · Virtual network mapping ·
Power communication network

1 Introduction

With the development of power system in digital direction, current scheduling data network provides support for multi-service bearer and data interaction of dispatching. With the increase of coverage area, network complexity, service type of power dispatching data network, its management of, traffic flow business priority and network security has been given higher requirements. SDN application can solve the problems that scheduling data network may face in the future, improving network utilization as well as ensuring business security and reliability [1].

Network virtualization under SDN environment has become a hotspot [2]. At present, there are many ways to achieve that.

Implementation of network virtualization through SDN demands physical network management, network resource virtualization [3] and network isolation. And these three parts are often completed through the special middle layer software, which is the

© Springer Nature Switzerland AG 2020
Q. Liu et al. (Eds.): CENet 2018, AISC 905, pp. 760–768, 2020.
https://doi.org/10.1007/978-3-030-14680-1_82

so-called network virtualization platform [4]. The virtualization platform is the middle layer between data network topology and tenant controller. Tenants can see their own virtual network through the controller merely and do not have knowledge of the real physical network. At the same time, in the data level point of view, virtualization platform is a controller, and switch does not know the existence of virtual plane. So the existence of a virtualization platform enables transparent virtualization for tenants and underlying networks that manage all physical network topologies and provide isolated virtual networks to tenants [5].

FlowVisor is proposed in [6] to achieve network virtualization under SDN platform. On this basis, the problem of virtual network mapping is studied. FlowVisor can divide physical network at different levels, such as topology, bandwidth, CPU and flow space. It provides experimental tools for network virtualization based on SDN, but lacks support for power communication services.

In this paper, a virtual network allocation mechanism of power communication network is proposed to realize cross-domain resource allocation of power communication network, which is combined with power communication services demands. Firstly, the problem model is constructed for network virtualization needs in power communication network. Secondly, we introduce a particle swarm algorithm to propose a virtual network mapping mechanism for service priority, and improve the reliability of the service. Finally, validity of the algorithm is verified by simulation.

2 Problem Model

A. System Description

SDN devices can be used as a feature of common network devices, the characteristics of distributed implementation ways, dispatching data network can smoothly transit to light domain SDN dispatching data network as shown in Fig. 1, and finally realizes data forwarding, network control double flat structure form visualization of the entire network physical and virtual logical topology, and provide entire network oriented ability for different topology control strategy adjustment. It can also realize flexible adjustment of logical topology, tech-oriented end-to-end business access, channel configuration, routing policy configuration as well as SLA management.

SDN network can better coordinate data forwarding strategy and cache mechanism of each node in the whole network. The network mechanism of SDN controller can improve network stability and security in the case of large network and relatively large nodes. At the same time, as the power communication network is divided into access side, backbone side and data center, a cross-domain SDN network is formatted as shown in Fig. 2. There is an urgent need for a cross-domain virtual resource allocation method for power communication services.

B. Problem Description

In order to solve the problem of cross-domain virtual resource allocation for power communication services, this paper takes business virtual network partitioning request as the goal to minimize overhead of VN mapping across domains. VN request is divided into multiple virtual subnets according to matching set obtained in virtual resource matching phase and price information of virtual resource type in the resource description knowledge base and the related information of boundary nodes [7].

The cost of cross-domain VN mapping consists of three parts: node mapping overhead (NODEcost), inter-domain link mapping overhead (LINKcost) and intra-domain link mapping overhead. Among them, the intra-domain link mapping overhead depends on InP internal specific network information. As InP does not fully expose its internal network information in the actual network environment, and the mapping overhead of intra-domain links is usually much smaller than the mapping overhead of inter-domain links, this paper ignores intra-domain link overhead when calculating cross-domain VN mapping overhead. In a word, cost is only determined by node mapping overhead and inter-domain link mapping overhead decision, namely: $\text{Cos}t = NODE_{\cos t} + LINK_{\cos t}$.

When a virtual link between a virtual node in an autonomous domain and a virtual node in another autonomous domain needs to be created, selecting different boundary nodes sThus, the VN partitioning scheme not only indicates InP for which each virtual node assumes its mapping, but also indicates which boundary node in the corresponding InP will complete the interdomain connection. In this paper, matching set MatchSet (v) of virtual node v is expanded to a collection MatchSet (v) consisting of all the boundary nodes contained by InP in the set. It should be noted that the boundary node does not assume a specific virtual node mapping. Mentioned will be mapped to a virtual node is presented in this paper. A boundary node, VN refers to is divided into phases, the virtual node into the boundary node of InP, subsequent virtual node mapping will be completed by InP, mapping and related virtual link to use the boundary nodes and other InP connection is established.

VN partitioning is understood to aim at minimizing cross-domain VN mapping overhead. There are two mappings:

$$f_n : n^v \to MatchSet'(n^v), n^v \in N^V \tag{1}$$

$$
\begin{aligned}
f_l &: l^v(i,j) \to Path(i',j') \to Path(i',j') \\
l^v(i,j) &\in E^V, i' = f_n(i), j' = f_n(j)
\end{aligned}
\tag{2}
$$

Among them, $Path(i',j')$ represents collection composed of acyclic path between physical boundary nodes i' and j' in network G^S, f_n represents a virtual node mapping that maps virtual node to a boundary node in its matching set; f_l represents virtual link mapping.

It means when two endpoints i and j of virtual link $l^v(i,j)$ in VN mapped to boundary nodes i' and j' respectively, $l^v(i,j)$ must be mapped to the $Path(i',j')$ of an acyclic Path for completing VN division.

On the basis of the above-mentioned problem model and problem description, virtual network partitioning problem can be solved. However, the problem of virtual network partitioning is a NP-C, and the current research on these large-scale virtual network division is generally inefficient. Aiming at these problems, this paper proposes a cross-domain virtual network mapping mechanism based on particle swarm optimization.

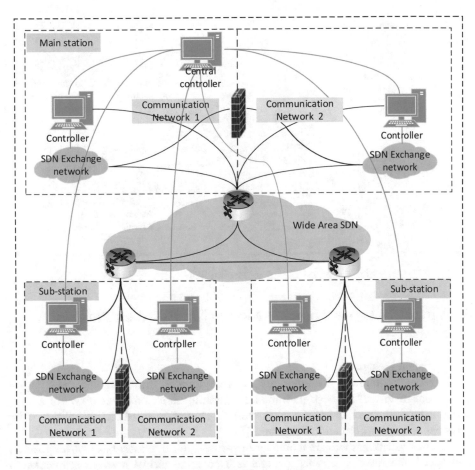

Fig. 1. Optical domain SDN scheduling data network.

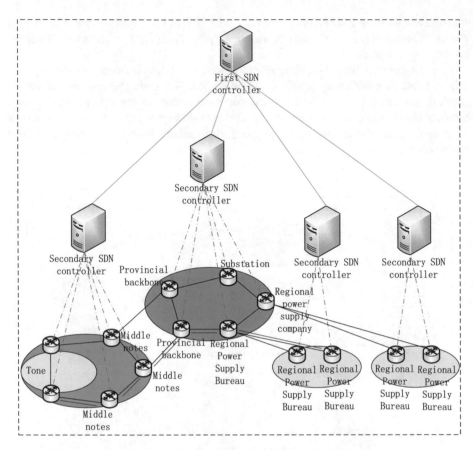

Fig. 2. Network edge application SDN typical scenario.

3   Virtual Network Mapping Mechanism Based on Particle Swarm

A. Algorithm Design

In order to improve the efficiency of cross-domain virtual network mapping, a cross-domain virtual network partitioning algorithm based on particle swarm is proposed to solve the problem of virtual network segmentation based on the model of finite physical network information disclosure [8].

In this algorithm, there are a number of particles, each particle has its own position and speed, the location of each particle is a potential virtual network segmentation program. The velocity of particles guides them to a new position. The position of particle i is expressed as $X_i = (x_{i1}, x_{i1}, \ldots, x_{iD})$, its velocity is expressed as $V_i = (v_{i1}, v_{i1}, \ldots, v_{iD})$. Where X_{id} represents mapping number of virtual node d when mapped to boundary. X_{id} represents the speed of the dth virtual node of particle i. If $V_{id} = 0$, V_{id} remains unchanged. If $V_{id} = 1$, X_{id} is inappropriate and should be replaced as thought.

In particular, it should be pointed out that a virtual node is not actually mapped to a boundary node, but communicates with other border nodes through the boundary node. Initialize optimal position of particle and population. Define the optimal position as initial position of the particle $P\_Best_i = (p_{i1}, p_{i2}, \ldots, p_{iD})$. Calculate the fitness of all particles f, finding the particle with greatest fitness and assign the position of the particle to the population optimal position $G\_Best = (g_1, g_1, \ldots, g_1)$. The fitness function f is:

$$f : \frac{1}{\sum_{n_v \in N_V, n_s \in N_p} C(n_v, n_s) + \sum_{i,j \in N_V, m,n \in N_p} C(l_{ij}, p_{mn})} \tag{3}$$

Where N_v is a set of virtual nodes and N_p is a set of boundary nodes. $C(n_v, n_s)$ represents the cost of mapping the virtual node N_v to the boundary node n_s. $C(l_{ij}, p_{mn})$ represents the cost of mapping the virtual link l_{ij} to the path p_{mn}. The denominator of the fitness function f is the sum of node cost and link cost of mapping virtual network. So the smaller the mapping cost of particles, the greater the fitness and vice versa.

Each particle also holds the best position that has ever reached. Simultaneously, the population has an optimal position, which is the optimal position of all particles'. Particle's speed and position are updated as follows:

$$v_{id} = \omega v_{id} + p_1(P\_Best_{id} - x_{id}) + p_2(G\_Best_d - x_{id}) \tag{4}$$

$$x_{id} = x_{id} + v_{id} \tag{5}$$

Where ω represents inertia weight, p_1 is the cognitive weight of individual, p_2 is the global weight, they are constant value, and $\omega + p_1 + p_1 = 1$. Particle swarms fly in the problem space. During the flight of the particle, a near virtual network segmentation scheme is generated.

B. Algorithm Flow

The algorithm is shown as follows:

Step 1: Set particle size N and iteration upper limit M of the algorithm, and randomly generate the particle initial position parameter X_i as well as initial velocity parameter V_i.

Step 2: Define the optimal position of the particle as its initial position P_Best_{id}. Calculate the fitness of all particles f. Finding the particles with greatest fitness and assign the position of the particle to the optimal position of the population G_Best.

Step 3: For the particles satisfying constraint condition, their velocity is updated according to Eq. (4), and position is updated according to Eq. (5). The position as well as velocity parameters are regenerated.

Step 4: The current number of iterations plus 1, recalculate the objective function of each particle.

Step 5: For each particle in the particle swarm, if $f(X_i) < f(X_{pb})$, $X_{pb} < X_i$; if $f(X_{pb}) < f(X_{gb})$, then $X_{gb} < X_{pb}$.

Step 6: If the current iteration times less than or equal to M, then go to Step 3, otherwise output.

4 Simulation Results and Analysis

In the simulation experiment, this study compares PSO-VNP algorithm with another exact method (LID-Exact) [9]. The experiment first compares the efficiency of the two methods, that is, the execution time of the two methods. The results are shown in Fig. 3(a).

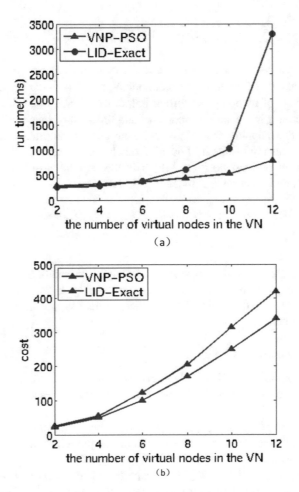

Fig. 3. The efficiency and the mapping overhead of two kinds of virtual network partitioning methods.

It can be seen from the figure that the execution time of PSO-VNP algorithm and LID-Exact algorithm is increasing with the increase of the number of virtual nodes in virtual network. Among them, the execution time of LID-Exact algorithm increases exponentially, and the execution time of PSO-VNP algorithm increases linearly.

It can be concluded that the PSO-VNP algorithm is more efficient than LID-Exact when the number of virtual nodes increases.

Experiment compares mapping overhead of the two methods further, results are shown in Fig. 3(b). Computing overhead of PSO-VNP algorithm and LID-Exact algorithm increases linearly with the number of virtual nodes increases. The cost of PSO-VNP algorithm increases faster, but the trend of the two algorithms are similar. Therefore, it can be concluded that the PSO-VNP algorithm can improve the efficiency of virtual network partitioning and ensure that the mapping overhead is not too large.

The experiment further validates the stability of the method in different problem scenarios. The experimental results are shown in Fig. 4.

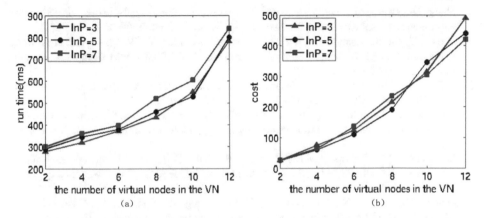

Fig. 4. The stability of PSO-VNP algorithm in different network environment.

It can be seen from Fig. 4(a) that, with the increase of the number of virtual nodes, in the three network topologies (InP = 3, InP = 5, InP = 7), PSO-VNP algorithm has a little difference in running time. And from Fig. 4(b), as the number of virtual nodes increases, in the three network topologies (InP = 3, InP = 5, InP = 7), PSO-VNP algorithm calculates that the mapping cost is not much different. Furthermore, the PSO-VNP algorithm is stable in different network environments.

5 Summary

In this paper, VN request is divided into multiple virtual subnets. It aims to minimize cross-domain VN mapping overhead on resource allocation and virtual network partitioning request for power communication services. Particle swarm cross-domain virtual network resource partitioning algorithm is proposed to solve large-scale virtual network partition in ubiquitous inefficient problem. The algorithm redefines position, velocity and update operation of particles according to problem model. Simulation results show that the proposed algorithm improves partition efficiency under the premise of guaranteeing mapping cost, and verifies its stability under different problem scenarios.

Acknowledgment. This work was supported by the science and technology project of Guangdong power grid (036000KK52160025).

References

1. Gopal, R.L.: Separation of control and forwarding inside a network element. In: IEEE International Conference on High Speed Networks and Multimedia Communications, pp. 161–166. IEEE, Piscataway (2002)
2. Blenk, A., Basta, A., Reisslein, M., Kellerer, W., et al.: Survey on network virtualization hypervisors for software defined networking. IEEE Commun. Surv. Tutor. **18**(1), 655–685 (2016)
3. Wang, A., Iyer, M., Dutta, R., Rouskas, G.N., Baldine, I.: Network virtualization: technologies, perspectives, and frontiers. J. Lightwave Technol. **31**(4), 523–537 (2013)
4. Cheng, X., Su, S., Zhang, Z., Wang, H.C., Yang, F.C., Luo, Y., Wang, J.: Virtual network embedding through topology-aware node rangking. ACM SIGCOMM Comput. Commun. Rev. **41**(2), 38–47 (2011)
5. Heller, B., Sherwood, R., Mckeown, N.: The controller placement problem. ACM SIGCOMM Comput. Commun. Rev. **42**(4), 7–12 (2012)
6. Sherwood, R., Gibb, G., Yap, K.K., Appenzeller, G., Casado, M., McKeown, N., Parulkar, G.: FlowVisor: a network virtualization layer. OpenFlow Consortium, Technical reports (2009)
7. Chowdhury, M., Rahman, M.R., Boutaba, R.: ViNEYard: virtual network embedding algorithms with coordinated node and link mapping. IEEE/ACM Trans. Networking **20**(1), 206–219 (2012)
8. Zhang, Z.B., Cheng, X., Su, S., Wang, Y.W., Shuang, K., Luo, Y.: A unified enhanced particle swarm optimization-based virtual network embedding algorithm. Int. J. Commun. Syst. **26**(8), 1054–1073 (2012)
9. Yuy, M., Yiz, Y., Rexfordy, J., Chiang, M.: Rethinking virtual network embedding: substrate support for path splitting and migration. ACM SIGCOMM Comput. Commun. Rev. **38**(2), 17–29 (2008)

Maintenance Scheduling Algorithm Based on Big Data for Power Communication Transmission Network

Wu Dong$^{(\boxtimes)}$, Qing Liu, Xu Liu, Cheng Cai, Dengchi Chen, Dili Peng, and Qi Tang

Power Dispatching Control Center, Guizhou Power Grid
Limited Liability Company, No. 17 Binhe Road, Nanming District,
Guiyang, Guizhou, China
4536298@qq.com

Abstract. With the development of the construction of power grid information, the scale of power communication transmission network is rapidly expanding, which greatly increases the difficulty of maintenance scheduling and requires the relevant intelligent algorithm to assist the operation and maintenance department to formulate an efficient maintenance plan. Aiming at solve maintenance problem of power communication transmission network, we proposed a maintenance task scheduling model which balancing the waiting time of maintenance task, considering the scheduling requirements of maintenance task of power communication network with non-interruptible conditions of service. Experiment results show that the feasibility and effectiveness of the algorithm are verified. A maintenance scheduling algorithm for power communication transmission network based on big data is useful to solve the problem which using heuristic method to generate the scheduling scheme for maintenance task of the power communication network.

Keywords: Power communication transmission network ·
Maintenance task scheduling · Ant colony algorithm

1 Background

The power communication transmission network is an important infrastructure for the power system. It is responsible for the transmission of important production control commands and services between power systems, including power system scheduling commands and information transmission services. It is an important guarantee for large scale data transmission. Therefore, regular maintenance is necessary for devices and lines that make up the power communication transmission network [1].

With the development of the construction of power grid information, the scale of power communication transmission network is rapidly expanding, which greatly increases the difficulty of maintenance power communication transmission network. However, because a large number of services are staggered and deployed on it, which is difficult to find an efficient solution to maintenance it. For this reason, there is an urgent need to design an intelligent algorithm to solving the maintenance problem [2–6].

© Springer Nature Switzerland AG 2020
Q. Liu et al. (Eds.): CENet 2018, AISC 905, pp. 769–776, 2020.
https://doi.org/10.1007/978-3-030-14680-1_83

To solve the above problems, we design an ant colony algorithm-based maintenance scheduling algorithm for power communication transmission network from the perspective of guaranteeing the normal operation of the service on the transmission network.

This paper is organized as follows: In Sect. 2 we described the problem mentioned above; in Sect. 3 a heuristic method based on big data is proposed to solve problem mentioned above; experiment results is analysed in Sect. 4; the next section is conclusion.

2 Problem Description

In existing work of the actual power transmission network, each maintenance unit submits the maintenance work including the equipment or multiplex section that needs to be overhauled, the expected start time of the repair, the latest start time of the repair, and the duration of the repair. Subsequently, the dispatching center organized professional personnel to conduct man-made scheduling of each service order and found a feasible solution through extensive analysis and comparison. Due to the huge arrangement of business paths, it is difficult to artificially find out more efficient solutions.

In addition, the actual repair time is not later than the latest start time, and during the inspection and repair, the qualified conditions such as the backup route must be arranged for the power communication service, which further enhances the complex line of formulating an efficient maintenance plan.

Optimization goal: Balance the wait time of maintenance tasks to avoid some services from being serviced for too long.

Restrictions:

(1) During the inspection and repair, spare routes must be arranged for the power communication service;
(2) The actual repair time is not later than the expected start time.

Establish a maintenance task scheduling model for power communication transmission network:

$$Minimize : Z = \sum_{i=1}^{n} \left(t_i - t_{i-early} - \mu \right)^2 \tag{1}$$

$$\text{S.T} \qquad t_{i\_early} \leq t_i \leq t_{i-late} - t_{i_r} \tag{2}$$

$$t_j - (t_i + t_{i_r}) \geq 0, j > i, R_j \cap R_i \neq \emptyset \tag{3}$$

$$t_{s_{i\_B}} - \left(t_{s_{i_w}} + t_{s_{ik_r}} \right) > 0 \tag{4}$$

Among them, t_i is the starting time for performing the inspection task i, $t_{i-early}$ is the expected starting time for performing the inspection task, μ is the average waiting time for the inspection task, t_{i-late} is the latest closing time for performing the inspection task, t_{i_r} is the maintenance time required for the inspection task i, R_i and R_j are the overhaul resources used for maintenance task i and task j, S_i is the service start time of the service work path $t_{S_{iw}}$, and the overhaul time of the kth resource on the overhaul work path $t_{S_{ikr}}$, $t_{S_{i\_B}}$ is the overhaul start times on the standby path; Eq. (3) means that if the path of the overhaul task node to the overhaul task node j has a common part, the time cannot be overlapped, that is, the task j cannot be started until the end of task i; otherwise, the service will be interrupted.

Mapping the overhaul task scheduling problem of the power communication transmission network into a directed graph G = (V, A). The node set V represents the maintenance task, and the edge set A represents that the maintenance tasks that can be carried out at the same time are connected by an edge. The feasible solution of the problem is obtained through the walking of the artificial ant from the starting point to the end point in the directed graph, for example the maintenance scheduling algorithm for Power Communication Network based on Ant Colony Optimization (MACO). Let $b_i(t)$ denotes the number of ants at time t at the overhaul task node i, $\pi_{ij}(t)$ denotes the amount of information on the path at time t, n denotes the number of overhaul task nodes, m denotes the total number of ants in the ant colony, and then the maintenance tasks in set C at time t can be denoted by $m = \sum_{i=1}^{n} b_i(t)$. $\tau = \tau_{ij}(t)/C_i, C_j \in C$ denotes the set of C residual information on the nodes connections l_{ij}. The amount of information on each path is equal at the initial time, and it is set that $\tau_{ij}(0) = \tau_0$ and then the ant searches the optimal path on the directed graph $g = (C, L, \tau)$ to achieve the optimization goal of the waiting time for the equalization maintenance task.

During the movement, ants k (k = 1, 2, ..., m) determine their direction of transfer based on the amount of information on each path. The tabulation table $tabu_k$ (k = 1, 2, ..., m) is used here to record the patrol task node that the ant k is currently walking through, and the collection set $tabu_k$ is dynamically adjusted along with the evolution process. During the search process, ants calculate the state transition probability based on the amount of information on each path and the heuristic information of the path. p_{ij}^k denotes the state transition probability that ant k is transferred from overhaul task node i to overhaul task node j at time t:

$$
p_{ij}^k = \begin{cases} \dfrac{[\tau_{ij}(t)]^{\alpha} * [\varphi_{ik}(t)]^{\beta}}{\sum_{s \in allowedk}([\tau_{is}(t)]^{\alpha} * [\varphi_{is}(t)]^{\beta})}, & if \ j \in allowedk \\ 0, & otherwise \end{cases}
\tag{5}
$$

In the formula, $allowedk = C - tabu_k$ denotes the next maintenance task node allowed by the ant k; α is the information heuristic factor, which indicates the relative importance of the trajectory, and the larger the value, the more the ant tends to choose the path that other ants pass through; β is the expected heuristic factor. The relative importance of visibility, the larger the value, the closer the state transition probability is to the greedy rule; the heuristic function is:

$$
\varphi_{ij}(t) = \frac{1}{d_{ij}} = \frac{1}{\left(t_i - t_{i-early} - \mu\right)^2 * \left(t_j - t_{j-early} - \mu\right)^2}
\tag{6}
$$

In the formula, d_{ij} represents the distance between two adjacent maintenance task nodes. For ant k, the smaller d_{ij}, the bigger $\mu_{ij}(t)$ and the greater $P_{ij}^k(t)$. This heuristic function represents the degree of expectation that the ant will move from the overhaul task node i to the overhaul task node j. In order to avoid the residual information pheromone from causing the residual information to overwhelm the heuristic information, after each ant has completed one step or completed the traversal of all n maintenance task nodes (i.e., one cycle is over), the residual information is updated. Thus the amount of information on path (i, j), at time t + n can be adjusted as follows:

$$\tau_{ij}(t+n) = (1 - \rho) * \tau_{ij}(t) + \Delta\tau_{ij}(t) \tag{7}$$

$$\Delta\tau_{ij}(t) = \sum_{k=1}^{m} \Delta\tau_{ij}^k(t) \tag{8}$$

In the formula, ρ means that the pheromone plays a coefficient, then $1 - \rho$ represents the information residual factor. In order to prevent the unlimited accumulation of information, the range of values ρ is $\rho \in [0, 1)$; $\Delta\tau_{ij}(t)$ represents the increment of pheromone on the path (i, j) in the current cycle. The initial value is $\Delta\tau_{ij}(0) = 0$. The moment indicates the amount of information that the k-th ant left on the path (i, j) in this cycle.

$$\Delta\tau_{ij}^k(t) = \begin{cases} \frac{Q}{L_k}, & \text{if ant k is on } (i,j) \\ 0, & \text{otherwise} \end{cases} \tag{9}$$

In the formula, Q represents the strength of pheromone, which affects the convergence speed of the algorithm to a certain extent; L_k represents the total length of the path taken by k ants in this cycle.

3   Algorithm

In this paper, we proposed maintenance scheduling algorithm for Power Communication Network based on Ant Colony Optimization (MACO) to solve problem described above. The steps of algorithm we proposed is shown as below.

Input: Maintenance plan, business path collection;

Output: Maintenance scheduling plan for power communication transmission network.

Step 1: Initialize. make time t = 0 and number of loops $N_c = 0$ set the maximum number of loops N_{Cmax}, place the ant on n maintenance task nodes, and initialize $\tau_{ij}(t) = \tau_0$ the amount of information on each side of the directed graph (i, j) where τ_0 is constant number, and initial time $\Delta\tau_{ij}(0) = 0$;

Step 2: Let loop iteration count $N_c = N_c + 1$;

Step 3: ant taboo index number of ants k = 1;

Step 4: Number of ants k = k + 1;

Step 5: The individual ant calculates the probability according to the state transition probability (see Eq. (5)) and selects the maintenance task node and proceeds $j \in \{C - tabu_k\}$.

Step 6: Modify the taboo table pointer, i.e. select the ant to move to the new overhaul task node, and move the node to the taboo table of the ant individual;

Step 7: If the repair task node in set C has not traversed, i.e., skip to Step 4; otherwise, go to Step 8;

Step 8: Update the amount of information on each path according to Eqs. (7) and (8);

Step 9: If the end condition is satisfied, i.e. if the number of cycles $N_c \geq N_{Cmax}$, the cycle ends and the maintenance task node sequence is output, otherwise the tabu list is cleared and the process jumps to Step 2;

4   Experimental Results

This article takes a monthly maintenance plan of the power communication transmission network as an example for simulation analysis. Our experiment is running in hadoop, with the data scale of 5 GB from 2010 to 2016 collected in actual network.

The number of iterations is 200; for simplicity, the resources required to set each maintenance task are the same, and the total maintenance resources remain constant. This is set to 3, which means that up to 3 maintenance tasks can be performed at the same time. Set the maximum number of services that can be carried on a device to 20. The topology of the power communication transmission network is as follows Fig. 1.

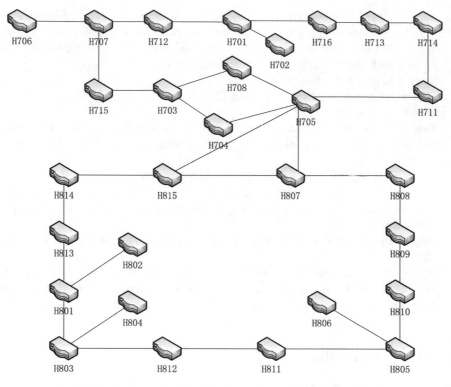

Fig. 1. Power communication transmission network topology.

The specific information of some services carried on the power communication transmission network is shown in the following Table 1.

Table 1. Power communication transmission network service information table.

Service name	Main route	Backup route
H708-H705. ODU9772	H708-H705	H708-H703-H715-H707-H712-H701- H716-H713-H714-H711-H705
H708-H706. ODU7715	H708-H703-H715-H707- H706	H708-H705-H711-H714-H713-H716- H701-H712-H707-H706
H705-H702. Client8090	H705-H711-H714-H713- H716-H701-H702	H705-H708-H703-H715-H707-H712- H701-H702
H706-H705. ODU5834	H706-H707-H715-H703- H708-H705	H706-H707-H712-H701-H716-H713- H714-H711-H705
H705-H704. Client5026	H705-H704	H705-H708-H703-H704
H708-H702. Client8092	H708-H703-H715-H707- H712-H701	H708-H705-H711-H714-H713-H716- H701-H702

(*continued*)

Table 1. (*continued*)

Service name	Main route	Backup route
H704-H708. Client4251	H705-H708	H705-H704-H703-H708
H807-H804. Client5617	H807-H815-H814-H813-H801-H803-H804	H807-H808-H809-H810-H805-H811-H812-H803-H804
H815-H804. Client5620	H815-H814-H813-H801-H803-H804	H815-H807-H808-H809-H810-H805-H811-H812-H803-H804

The specific information of a monthly maintenance plan for the grid system is shown in the following Table 2.

Table 2. Monthly maintenance work order.

Worker number	Device	Expected start time (day)	Latest start time (day)	Duration time (day)
J001	H712	1	1	2
J002	H715	1	15	3
J003	H713	4	13	1
J004	H714	3	15	5
J005	H711	1	15	5
J006	H703	2	16	2
J007	H716	2	15	1
J008	H808	10	30	5

The maintenance plans obtained by scheduling the above-mentioned maintenance tasks using the MACO algorithm proposed are as follows Tables 3 and 4. Table 3 shows the results using our proposed method, Table 4 shows the traditional method results. In these two figures we can see the method we proposed can improve the deviation.

Through Tables 3 and 4, the deviation in the scheduling scheme obtained by the MACO algorithm is significantly lower than the scheduling scheme that does not consider the waiting time of the equalization maintenance task, and the effectiveness of the algorithm can be verified. Our method is proved be efficient for solving maintenance scheduling problem of power communication transmission network. The traditional algorithm is a chaotic ant colony optimization algorithm based on the spark platform.

<table>
<tr><td colspan="3" align="center">**Table 3.** MACO.</td></tr>
</table>

Worker number	Device	Deviation
J001	H712	9
J002	H715	1
J003	H713	1
J004	H714	0
J005	H711	9
J006	H703	4
J007	H716	1
J008	H808	4

Table 4. Traditional methods.

Worker number	Device	Deviation
J001	H712	11.55612
J002	H715	16.98469
J003	H713	19.25715
J004	H714	2.69898
J005	H711	0.413265
J006	H703	16.32456
J007	H716	4.512639
J008	H808	0.215689

5 Conclusion

In this paper, the maintenance scheduling problem of power communication transmission network is researched in detail. With the goal of equalizing the maintenance task waiting time, a mathematical model for the maintenance scheduling problem of power communication transmission network is proposed. Then, we proposed mathematically modeled the problem and find new method to solve it. Then an ant colony algorithm based on big data for power communication transmission network is designed. Finally, experiments results show that the effectiveness of the algorithm is verified.

Acknowledgement. This paper is supported by China Southern Power Grid Co., Ltd. Science and Technology Project (GZKJXM20170077).

References

1. Feng, B., Fan, Q., Li, Y., Ding, D.Q.: Research on framework of the next generation power communication transmission network. In: 2012 International Conference on Computer Science and Electronic Engineering, pp. 414–417 (2012)
2. Liu, C., Luo, J., Zhang, Y., Zhang, J.: The electric power communication transmission network risk assessment index system research based on AHP. In: International Conference on Automation, Mechanical Control and Computational Engineering (2015)
3. Guo, R.: Application of SDH networking mode in electric power communication transmission network. Jiangxi Electric Power (2009)
4. Yao, W.U.: Application of OTN technology in power communication transmission network. Telecom Power Technology (2018)
5. Ginot, N., Mannah, M.A., Batard, C., Machmoum, M.: Application of power line communication for data transmission over PWM network. IEEE Trans. Smart Grid **1**(2), 178–185 (2012)
6. Mengxuan, M.A., Kang, X.: The necessary discussion of net bridge being used in Ningxia electric power communication transmission network. Low Voltage Apparatus (2009)

Reliability Oriented Probe Deployment Strategy for Power Cellular Internet of Things

Lei Wei[1], Daohua Zhu[2]([⊠]), Wenwei Chen[3], Yukun Zhu[3],
and Lin Peng[4]

[1] State Grid Jiangsu Electric Power Company, Ltd., Nanjing, China
[2] Electric Power Research Institute, State Grid Jiangsu Electric
Power Company, Ltd., Nanjing, China
daohua_zhu@aliyun.com
[3] State Grid Information and Telecommunication Group, Ltd., Beijing, China
[4] Beijing Vectinfo Technologies Company, Ltd., Beijing, China

Abstract. With the power cellular Internet of Things (IoT) communication system playing a more and more important role for information transmission, exchange, and processing in the smart grid, higher requirements for the maintenance management and monitoring technologies need to be satisfied. However, at present, the monitoring capability of the power cellular IoT communication network is relatively scarce, resulting in grid data transmission has to rely on the public network. The security, reliability, and real-time performance cannot be guaranteed. Therefore, this paper focuses on the characteristics of power wireless networks, studies the technology of distributed service quality monitoring, and designs a deployment strategy for Power Cellular Internet of Things to monitor performance nodes. We first introduce the loop rate of deployment probes as the evaluation index of network reliability, and model the minimum weighted vertex coverage problem of probe deployment as the largest weighted independent set problem. Then we propose a fast heuristic algorithm to find its minimum deployment overhead with satisfying reliability requirement. The simulation results show that the proposed algorithm can improve the ability of the smart grid for locating faults quickly and effectively, as well as the comprehensive monitoring and management capability, improving network performance.

Keywords: Probe deployment · Power Cellular IoT · Heuristic algorithm · Reliability

1 Introduction

With the continuous improvement of informatization, the Power Cellular Internet of Things (IoT) has developed rapidly, faces challenges. Its communication system is mainly used for intelligent transmission of information, which needs to meet the performance requirements of different power services such as QoS, timeliness, and reliability. However, with the continuous expansion of the network size, the structure has become increasingly complex. Over-relying communication technologies such as fiber-optic communications can no longer meet the development. The Power Cellular IoT

© Springer Nature Switzerland AG 2020
Q. Liu et al. (Eds.): CENet 2018, AISC 905, pp. 777–784, 2020.
https://doi.org/10.1007/978-3-030-14680-1_84

adopts new wireless technologies and is combined with power wireless resources. It aims to improve the access capability of power communication access networks and meet the requirements. As an important communication method for terminal access, it provides a new communication option for the smart grid service of electricity.

Businesses of the Power Cellular IoT is closely related to the production of electricity, requires high real-time and reliability of communications. The production scheduling of services has strict requirements on fault recovery time and communication failure rate, which requires a high requirement for operation and maintenance management. However, the current transmission capacity of the power communication network is poor, whose security, reliability and real-time performance cannot be guaranteed. Due to the lack of effective detection methods, the real-time monitoring and the amount of information in remote areas is limited. It seriously affects the flexibility of the network and the level of automation operations. Therefore, the research on the technical measures for distributed service quality monitoring is the basis for improving the ability of the smart grid to quickly and accurately locate faults.

To solve the above problems, the active probing method is active, efficient, and self-adaptive. The first is to determine the deployment of the monitoring probe. The deployment of monitoring probes will directly affect the ability to locate faults and the effectiveness of detection. Therefore, how to reduce the configuration overhead of the probe under the condition of ensuring the monitoring performance is a crucial link in network fault management.

When the network size is small and the network structure is simple, probes can be deployed on all network nodes to monitor the entire network. With the expansion of network size, deploying probes on all nodes will not only generate huge probe traffic, but also cause hardware and software consumption on network devices at other layers of the network [1].

In recent years, the selection of monitoring strategies has become a challenging topic for researchers. Zhang et al. [2] uses the method of clustering measurement nodes to achieve distributed measurement of the performance. It can reduce the cross-measurement, but the cluster head node is also more arbitrary. Gao et al. [3] proposes a method of measuring network performance by merging neighbor node measurement data. However, this algorithm has a high degree of dependence on the measurement nodes of the surrounding nodes. Pan et al. [4] proposes a method for calculating the performance of the path to be measured through the link relationship matrix. This algorithm can reduce the number of paths to be measured, but for a network with a complex topology, the optimization efficiency will decrease. Ge et al. [5] proposes a hybrid optimization algorithm based on tabu search and heuristic algorithm to select the measurement nodes. This algorithm can effectively reduce the redundancy overhead caused by flooding without reducing the coverage of the algorithm. However, this algorithm is computationally complex and time-intensive.

Unlike the above algorithms, our research is based on the Power Cellular IoT. We first consider the minimum weighted vertex coverage problem of probe deployment as the maximum weighted independent set problem. To consider the reliability of the network, we introduced the loop rate of probe deployment as an evaluation index of network reliability. Then, on the premise of satisfying the reliability, this paper proposes a fast heuristic algorithm to solve its minimum deployment cost. The simulation results show that the proposed algorithm can improve the smart grid's ability to comprehensively monitor and manage the operating status, better guaranteeing the network performance.

2   Probe Deployment Model for Power Cellular IoT Networks

The scale and structure of the LTE Power Cellular IoT are increasingly complex. In the past, deployment methods based on experience and local inference have not been adapted to the current status and needs. Without considering the noise of the network, it is considered that when one probe is successful, the nodes through which the probe passes are all working properly. Conversely, when one probe fails, all the nodes that the probe passes through may be malfunction nodes. In order to accurately locate the malfunction in the network, the optimization problem of power network probe deployment can be regarded as the minimum vertex coverage problem of undirected graphs.

There are many aspects of network reliability in the power network probe deployment model. This article only describes the probe loop rate.

The rate of probe loop formation is one of the important guarantees of network reliability [7]. It refers to the ratio of the number of probes forming the loop in the network to the total number of probes. The looping rate of the probes can reflect whether the deployment relationship of the probes is tight enough. The greater the looping rate of the probes, the higher the network reliability. Because in this way, the network can still be fully detected even if a probe fails, and it also represents a higher detection efficiency. However, it is negatively related to probe deployment costs, which will also be an important aspect of our need for trade-offs. The probe looping formula is

$$R = \frac{k}{m} \tag{2.1}$$

Where m represents the total number of probes, i.e., the number of x_i equal to 1, and k represents the number of looped probes. Then, the LTE power private network probe deployment model is:

$$\min_{x \in v}(cx) \tag{2.2}$$

S.t.

$$\sum_{j=1}^{n} a_{ij}(1 - x_i)(1 - x_j) = 0$$

$$R = \frac{k}{m} > \theta \tag{2.3}$$

$$x_i = \begin{cases} 1 \ v_i \in S \\ 0 \ v_i \notin S \end{cases}$$

The first constraint ensures that the vertex covers the network, and the second constraint ensures that the network reliability function is greater than the set threshold. Then the problem turns into the process of solving the minimum value of the probe deployment overhead economic function under the condition that the vertex coverage of the model undirected graph is satisfied and the reliability is guaranteed.

3   Solutions for the Model

3.1   Problem Translation

Before the above model is solved, the definition of maximum independent set is given first: Let simple graph $G = (V, E)$ be an undirected graph, where V is the set of vertices in the graph and E is the set of edges. Let S be a subset of V, $S \subseteq V$. If any two nodes in S are not adjacent in G, then S is called an independent set of G. For an independent set S of arbitrary G, if G does not contain any independent set S' of $|S'| > |S|$ ($|S|$ denotes the number of elements of the set, i.e. the potential of the set), S is said to be G The largest independent set.

Assume that a vertex cover set of the graph $G = (V, E)$ is C, then $S = V \backslash C$ is an independent set of graph G [8]. Therefore, the minimum weighted vertex cover problem can be transformed into the maximum weighted independent set problem.

3.2   Maximum Weighted Vertex Set Solving

This paper proposes a heuristic algorithm for fast execution. First, we construct an ordered set $\prod$ according to the vertex weight value (i.e., the probe deployment node overhead). With recursive calls, we find the solution of multiple sets of weighted independent sets. Finally, we compare multiple sets of feasible solutions to select the optimal one, the maximum weighted independent set solution S, then the solution of the minimum weighted vertex covering set is $V \backslash S$.

Algorithm 1 shows the process of solving the maximum weighted vertex independent set. The function EXPAND creates 2^{limit} independent sets recursively. The three parameters S, P and depth represent the current independent collection, the given set of nodes, and the depth of the recursive call respectively. $w(P)$ denotes the weighted sum of the vertex sets P, i.e. the total cost of probe deployment in the current network, where the limit is the depth of the recursive call, which is equivalent to the depth of the binary heap obtained by binarizing P. $R(S)$ denotes the looping rate of this set of probe deployment schemes for calculating reliability.

Algorithm 1 Multiple independents set construction

INPUT: $G = (V, E), \mathbf{w}[v_1, v_2, ..., v_n], \Pi = [v_1, v_2, ..., v_n]$, limit

OUTPUT: independent set S_{best}

1 $S_{best} \leftarrow \emptyset$
2 EXPAND$(\emptyset, \Pi, 0)$
3 **return** S_{best}
4 **function** EXPAND$(S, P, depth)$
5   **if** P is empty **then**
6   **if** $R(S) > \theta$ // Guaranteed reliability constraints
7      $S_{best} \leftarrow S$ if $w(S_{best}) < w(S)$
8   **return**
9   **end if**
10   **return** if $w(S) + w(P) \leq w(S_{best})$ // Pruning
11 **if** depth < limit **then**
12     $v_a \leftarrow$ the first vertex in P
13   let P_2 be a sequence made by removing v_a and elements of $N(v_a)$ from P
14   EXPAND$(S \cup \{v_a\}, P_2, depth + 1)$
15   rotate P (move v_a to the last)
16   EXPAND$(S, P, depth + 1)$
17 **else**
18   MAXIMALIZE(S, P)
19 **end if**
20 **end function**
21 **function** MAXIMALIZE(S, P)
22 **for all** v_i in P **do**
23    $B[v_i] \leftarrow 0$
24 **end for**
25 **for all** v_i in P **do**
26   **if** $B[v_i] = 0$ **then**
27      $S \leftarrow S \cup \{v_i\}$
28      $B[j] \leftarrow 1 \forall v_j \in N(v_i)$
29   **end if**
30   **return** if $w(S) + w(P) \leq w(S_{best})$
31 **end for**
32   **if** $R(S) > \theta$ // Guaranteed reliability constraints
33     $S_{best} \leftarrow S$ if $w(S_{best}) < w(S)$ // Update the optimal solution
34 **return**
35 **end function**

4   Simulation and Evaluation

4.1   Simulation Parameter Settings

In order to facilitate the analysis, considering that the total number of nodes in the actual business scenario is about dozens, the total number of nodes in the simulation scenario is set to 10–50, and the network structure diagram formed by the nodes is a connected graph. Set the looping rate of the probe to 30%–70% [10].

4.2   Analysis of Simulation Results

In order to evaluate the performance of the algorithm, we first generate a wireless network topology diagram composed of several bearer network routing nodes, and establishes a network monitoring model. Then we use an improved greedy hybrid algorithm to calculate the number of nodes needed to completely cover the graph. After averaging multiple measurements, the minimum number of covered nodes is obtained. Next we compare this algorithm with a random traversal algorithm. The random traversal algorithm randomly selects available nodes according to preset requirements in the graph, which is a fast but inaccurate algorithm. In contrast, the algorithm proposed in this paper will be more purposeful and selective. Calculate the number of selected nodes obtained by random traversal and compare them together with the total number of nodes as shown in Fig. 1:

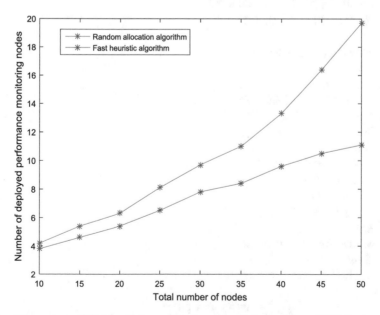

Fig. 1. Comparison of fast heuristic and random traversal algorithms (40% ringing rate).

Figure 1 shows the comparison between the algorithm used in this paper and the random traversal algorithm when setting the looping rate of the probe to 40%. As can be seen from the figure, after adopting the deployment algorithm of this paper, compared to the random traversal algorithm, the number of nodes to be selected is relatively small. Especially when the scale of the network map is enlarged, the advantages of the algorithm will become more apparent. Therefore, the fast heuristic algorithm can effectively reduce the number of monitoring nodes deployed, thus reducing the deployment cost.

We continue to explore the effect of probe looping rate on the number of deployed nodes. In the simulation process, we adjust the probe ring rate range of 30%–70%, the number of nodes is set to 20, 30, 40 and 50, the simulation results are shown below.

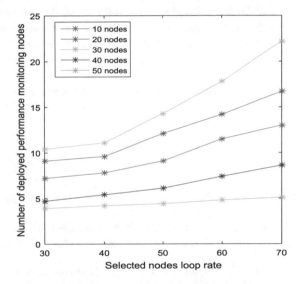

Fig. 2. Comparison of the effect of different loop rates on fast heuristic algorithm.

From Fig. 2, when the looping rate of the probe is required to be low, the difficulty of completely covering the entire network diagram is low, and the number of deployed nodes is less. However, as the probe loop rate requirement gradually increases, the difficulty of overlaying the network diagram gradually increases, and the number of required nodes also gradually increases. The looping rate of the probe is negatively related to the number of nodes to be selected. If the looping rate of the probe is required to be high, more nodes need to be selected for deployment, and vice versa. Therefore, the deployment of this strategy requires a compromise between the loop formation rate and the deployment cost based on the level of dependence of the service on reliability.

5 Conclusion

In order to ensure the quality of Power Cellular IoT communication services, we study the problems of fault detection. We choose the active detection method, and study the deployment strategy of its monitoring probe. We first consider the problem of minimum weighted vertex coverage for probe deployments in LTE Power Cellular IoTs as the largest weighted independent set problem. Then we introduce the deployment probe looping rate as an evaluation index of network reliability to monitor the network reliability. Finally, we propose a fast heuristic algorithm to solve the minimum deployment overhead of the model. By comparing the random traversal algorithm, the proposed algorithm selects fewer nodes and has lower deployment cost. It can improve the ability of the smart grid to quickly and accurately locate faults, as well as the comprehensive monitoring and management capabilities of the operating state, and is more suitable for end-to-end performance monitoring of Power Cellular IoTs. In addition, the deployment of this strategy needs to balance the ringing rate and the deployment cost according to the actual situation in the actual network.

Acknowledgement. This study is supported by 2017 state grid science and technology project "Research and Application for Adaption of Power Business based on LTE Wireless Private Network".

References

1. Yin, H., Li, F.: Research on the development of the internet performance measurement technologies. J. Comput. Res. Dev. **53**(1), 3–14 (2016)
2. Zhang, S., Lin, S., He, Z., Lee, W.J.: Ground fault location in radial distribution networks involving distributed voltage measurement. IET Gener. Transm. Distrib. **12**(4), 987–996 (2018)
3. Gao, J.L., Xu, Y.J., Li, X.W.: Weighted-median based distributed fault detection for wireless sensor networks. J. Softw. **5**, 1208–1217 (2007)
4. Pan, S., Zhang, Z., Yu, F., Hu, G.: End-to-end measurements for network tomography under multipath routing. IEEE Commun. Lett. **18**(5), 881–884 (2014)
5. Ge, H.W., Peng, Z.Y., Yue, H.B.: Hybrid optimization algorithm for efficient monitor-nodes selection in network traffic. Appl. Res. Comput. **4**, 1480–1483 (2009). 1486
6. McGregor, T., Braun, H.W., Brown, J.: The NLANR network analysis infrastructure. IEEE Commun. Mag. **38**(5), 122–128 (2000)
7. Khan, U.A., Kar, S., Moura, J.M.F.: Distributed sensor localization in random environments using minimal number of anchor nodes. IEEE Trans. Signal Process. **57**(5), 2000–2016 (2009)
8. Wu, Q., Hao, J.K.: A review on algorithms for maximum clique problems. Eur. J. Oper. Res. **242**(3), 693–709 (2015)
9. Avis, D., Imamura, T.: A list heuristic for vertex cover. Oper. Res. Lett. **35**(2), 201–204 (2007)
10. Battiti, R., Passerini, A.: Brain-computer evolutionary multiobjective optimization: a genetic algorithm adapting to the decision maker. IEEE Trans. Evol. Comput. **14**(5), 671–687 (2010)

Quality Evaluation System of Electric Power Backbone Optical Transport Network

Ying Wang[1(✉)], Zijian Liu[1], Zhongmiao Kang[1], Weijian Li[1], and Hui Yang[2]

[1] Power Grid Dispatching Control Center, Guangdong Power Grid Co., Ltd., Guangzhou, Guangdong Province, China
wangying@gddd.csg.cn
[2] Beijing University of Posts and Telecommunications, Beijing, China

Abstract. Optical transport network (OTN) provides strong support for the safe and stable operation of large capacity optical network, but the rapid development of large capacity optical network also puts forward higher requirements for the quality of the communication network. The existing evaluation system cannot effectively evaluate the operation quality of the OTN. In order to achieve the effective evaluation of the survivability of the backbone network, this paper puts forward a quality evaluation system based on the multi index comprehensive evaluation method.

Keywords: Quality evaluation · Power backbone network · OTN

1 Introduction

Electric power optical transport network (OTN) is an indispensable part of power system, which is an important foundation for the automation and management modernization of power grid dispatching. Electric power backbone OTN is one of an important part to ensure the safety and stability of power grid [1–3]. With the burst growth of large bandwidth, low latency service, which poses a great challenge to the safe and stable operation of OTN [4]. The increase of the probability of concurrent multiple failures will increase the complexity of the protection and recovery resource allocation scheduling. However, the existing power OTN evaluation system cannot evaluate the operational quality status and health status of the power communication network, and there is no quantitative and early warning mechanism of network risk.

Although the index system of multiple indicators can fully reflect the performance status of the power backbone network survivability, but it is difficult to evaluate the survivability performance [5, 6]. Because the various indicators cannot be unified comparison, while the indicators are used simultaneously. Therefore, it cannot be evaluated on the object for time and space on the overall comparison, and cannot judge the performance of the backbone power OTN. In order to solve the above problems, this paper puts forward a quality evaluation system based on the multi index comprehensive evaluation method. Numerical results show that the proposed quality evaluation system can get better performance in OTN and the process is simple and comprehensive.

Q. Liu et al. (Eds.): CENet 2018, AISC 905, pp. 785–793, 2020.
https://doi.org/10.1007/978-3-030-14680-1_85

2   The Evaluation Model of Quality System

In order to facilitate the operation, the index system is symbolized as shown in Fig. 1. The target index is the comprehensive performance of the operation quality of OTN. There are three primary index, which are Fault Location A1, Fault Protection A2 and Fault Recovery A3. In addition, each primary index has a number of secondary index. The index is shown in Fig. 1.

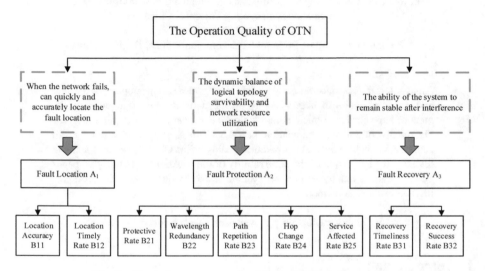

Fig. 1. The index system evaluation model of quality system.

Reasonable and effective evaluation and evaluation of survivability of large-capacity backbone network can not only play an important role in ensuring the normal demand of users and the development of enterprise information in the process of ensuring the large-capacity backbone network in the actual continuous operation. To reduce the risk of failure, improve network reliability, enhance the overall performance of the network to improve the important support. The framework of survivability evaluation index of large capacity backbone network can appear in the form of multi-layer structure: the first layer is the total target layer, the second layer is the target index layer and the third layer is the key factor layer. Corresponding to the survival of the three indicators of the system.

The following is the introduction of primary index.

(1) Location Accuracy B11: According to the collected alarm information, through a certain method to determine the network fault location and the number of fault accuracy.

$$B_{11} = \frac{AoNF}{ToNF} \times 100\% \qquad (1)$$

In which, AoNF represents the number of faults that are accurately positioned and ToNF represents all the number of failures.

(2) Location Timely Rate B12: In the Fault location process, the pretreatment time, optical path settling time and optical transmission delay and the relationship between the standard positioning times.

$$B_{12} = \begin{cases} 1 & , Tofl > Toat \\ \frac{Tofl}{Toat} & , Tofl < Toat \end{cases} \tag{2}$$

In which, Tofl is the actual time of the fault location, and Toat is the standard time for fault location.

(3) Protective rate B21: After the fault occurs in the power communication network, check whether the fault protection is successful.

$$B_{21} = \frac{NoPS}{NoF} \times 100\% \tag{3}$$

In which, NoF indicates the number of failures, and NoPS indicates the number of failure protection successes.

(4) Path Repetition rate B22: Detect whether the protection line and the working line are in the same line.

$$B_{22} = 0.95 \times \frac{NoDLT}{NoLT} + 0.05 \times \frac{MoD}{M} \tag{4}$$

In which, NoDLT represents the average number of heavy light transmission equipment, NoLT represents the number of optical transmission equipment, MoD represents the average number of heavy cable, M represents the total number of cables.

(5) Hop Change Rate B23: After the failure of the test, how many hops can be used to achieve fault protection.

$$B_{23} = \frac{|NoPJ - NoBJ|}{NoBJ} \times 100\% \tag{5}$$

In which, NoPJ indicates the average number of hops before the failure, and NoBJ indicates the average number of hops after the failure.

(6) Wavelength Redundancy Rate B24: This indicator analyzes the cost of the wavelength resource for fault protection.

$$B_{24} = \frac{NoPW}{NoTW} \times 100\% \tag{6}$$

In which, NoPW represents the sum of the protective wavelengths consumed, and NoTW represents the sum of the total occupied wavelengths.

(7) Service Affected Rate B25: Resource cost of fault protection for integrated response service.

$$B_{25} = \frac{NoRW}{NoRN} \times 100\% \tag{7}$$

In which, NoRW means the average number of working channels of production control business, NoRN said the total number of production control business network channels.

(8) Recovery Timeliness Rate B31: It can reflect the time of data recovery after the fault restoration of the power communication network.

$$B_{31} = \frac{ToRS}{ToRP} \times 100\% \tag{8}$$

In which, ToRP represents the standard time of failure recovery, and ToRS indicates the actual time of failure recovery.

(9) Recovery Success Rate B32: This indicator is to show the success rate of data recovery after the failure.

$$B_{32} = \frac{NoSR}{NoFA} \times 100\% \tag{9}$$

In which, NoSR indicates the number of times the fault recovery, NoFA said the total number of failures.

3 The Evaluation Algorithm of Quality System

The index system of electric power transmission network has the characteristics of multi-level and multi-angle, and the relationship is complex, so the distance evaluation method is used to classify the index [7–10].

3.1 Determine the Evaluation Matrix and the Index Weight Vector

Distance comprehensive evaluation method is used to compare the characteristics of the same type of company or company in different periods of development of the characteristics and balance. Assuming that P indicators are used to sort N companies or periods, the original evaluation matrix is $X = (x_{ij})_{n \times p}$. The weights of each index are determined by analytic hierarchy process. The weight vector of each index is recorded as: $W = (w_1, w_2, \ldots, w_p)$.

3.2 Constructing Weighted Normalized Evaluation Matrix Z

If the p indicators have inverse indicators, then take the countdown to deal with, that is:

$$x_{ij} = \frac{1}{x_{ij}} \tag{10}$$

Then constructs the weighted normalization evaluation matrix:

$$Z = (x_{ij})_{n \times p} = w_j * \frac{x_{ij}}{\sqrt{\sum_{i=1}^{n} x_{ij}^2}} \tag{11}$$

3.3 Determine Ideal Samples and Negative Ideal Samples

The maximum value of each index in the sample constitutes the ideal sample, the minimum value of each indicator constitutes a negative ideal sample, respectively, with Z^+ and Z^-.

$$Z^+ = (Z_1^+, Z_2^+, \ldots Z_m^+), \ Z^- = (Z_1^-, Z_2^-, \ldots Z_m^-) \tag{12}$$

In which, $Z_j^+ = \max\{z_{ij}\}$, $Z_j^- = \min\{z_{ij}\}$.

3.4 Calculate the Relative Proximity

Calculate the distance between each evaluation object, the ideal sample and the negative ideal sample Di^+ and Di^-, in which:

$$D_i^+ = \sqrt{(Z_i - Z^+)(Z_i - Z^+)^T}, \ D_i^- = \sqrt{(Z_i - Z^-)(Z_i - Z^-)^T} \tag{13}$$

Then the relative proximity of the evaluation object to the optimal sample is:

$$C_i = \frac{D_i^-}{D_i^+ + D_i^-}, (i = 1, 2, \ldots, n) \tag{14}$$

3.5 The Accuracy of Evaluation Model

To validate the performance of the proposed method, in addition to the above case study, a set of experiments are conducted. Precision, which is the standard measures that have been used in information retrieval for measuring the accuracy of an evaluation method, are selected as the criteria to test the accuracy of the proposed method. The formula for calculating precision is shown below:

$$P_{recision} = 1 - \frac{\left|N_{Qualified} - N_{Candidate}\right|}{N_{all}} \tag{15}$$

In which, $N_{Qualified}$ ($N_{Qualified}$ = 1, 2, 3... and $N_{Qualified} \leq N_{all}$) be the number of the qualified network selected by proposed method. Let $N_{Candidate}$ be the number of actual qualified network. N_{all} is the number of all network

4   Numerical Results and Analysis

The main idea of the evaluation system of the index system is to evaluate the things from multiple perspectives, in order to reflect the development of the evaluation object in many aspects. In this previous section, this paper has established the establishment of the index system and measurement methods after research and analysis. In addition, the evaluation model of the proposed index system and the measurement method is used to verify the effect of the evaluation.

There are 11 network statistics currently reported by a company, the parameters of the model are shown in Table 1.

Table 1. Index model parameters.

Network	Fault location	Fault protection	Fault recovery
1	0.5808	0.7660	0.8002
2	0.5802	0.7723	0.8504
3	0.6102	0.7823	0.8822
4	0.4431	0.7002	0.8642
5	0.5012	0.7302	0.8352
6	0.5433	0.7632	0.8432
7	0.5342	0.7529	0.8672
8	0.5565	0.7238	0.8522
9	0.5854	0.7108	0.8476
10	0.4908	0.7849	0.8752
11	0.9817	0.9458	0.9460

Assume that the weights of the indicators are 0.24, 0.52 and 0.24. Therefore, the ideal sample and the negative ideal sample are as follows.

$$Z^+ = (0.1178, 0.3783, 0.0798, 0.1135)$$
$$Z^- = (0.0369, 0.0238, 0.0358, 0.0757)$$

The relative proximity analysis of 11 reported data according to the approach described above is shown in Table 2.

Table 2. The data proximity analysis.

Network	Ideal sample distance D_i^+	Negative ideal sample distance D_i^-	Relative proximity
1	0.3606	0.0445	0.1099
2	0.3591	0.0371	0.0936
3	0.3588	0.0410	0.1026
4	0.3647	0.0391	0.0968
5	0.3627	0.0444	0.1091
6	0.3610	0.0370	0.0929
7	0.3612	0.0373	0.0936
8	0.3619	0.0389	0.0971
9	0.3628	0.0389	0.0969
10	0.3608	0.0411	0.1022
11	0.0440	0.3656	0.8926

In the above example, the eigenvalues of network 1 and network 11 are 0.1099 and 0.8926 respectively and the index values are N1 = 10.99 and N2 = 89.26 respectively. As can be seen from the data, the running status of network 1 has been very poor compared to the 11 network.

In order to evaluate the correlation between the proposed model and algorithm and the actual network. In our experiments, the number of network is set to 900, i.e. N_{all} = 600. The qualified rate varies from 10% to 70% with an increment of 20%. In order to illustrate the superiority of the proposed scheme, we compared it with the state-of-art quality evaluation method (such as The Grey Relational Analysis method [11] and The Rank-sum ratio method [12]) on the accuracy of model evaluation.

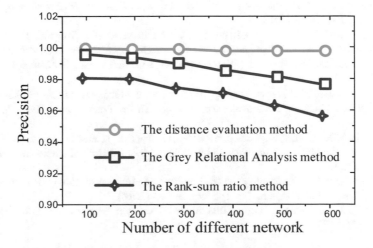

Fig. 2. The quality assessment results in actual network.

It can be concluded from Fig. 2 that, when the number of network is less than 200, the precision of the proposed method almost close to 100%. Moreover, with the network number increases, the precision does not decrease much. It means that, the quality evaluation results by using the proposed method are qualified for use's requirements and this scheme has high stability. The other two schemes are not satisfactory in terms of evaluation precision. Hence, the effectiveness of the proposed approach is apparent.

5   Conclusion

This paper puts forward the comprehensive quality evaluation system of the power backbone OTN, from the actual measurement point of view. Managers can evaluate the quality of the network from several aspects. This system can combine a variety of factors that affect the quality of the network together, using a more comprehensive method. To provide a more reasonable evaluation index for managers of power grid. Based on the actual measurement point of view, this paper puts forward the comprehensive evaluation system of survivability of the power backbone network. This paper chooses such an index system based on the hierarchical model, which can be more comprehensive and reasonable.

Acknowledgement. This study is supported by the Project on Demonstration of Communication Service Resource Management and Control and Research on Exploration and Expansion of Communication Service (Research on Theory and Market) under Grant No. GDKJXM20160901.

References

1. Morea, A., Charlet, G., Verchere, D.: Elasticity for dynamic recovery in OTN networks. In: Asia Communications and Photonics Conference. Optical Society of America (2014)
2. Wu, M., Guo, S., Chen, X. et al.: LM-BP based operation quality assessment method for OTN in smart grid. In: 18th Asia-Pacific Network Operations and Management Symposium, 2016, pp. 1–4. IEEE (2016)
3. Maier, G., Colombo, L., Costantino, D.G., Pattavina, A., Szegedi, P.: Quality of provisioning as an OPEX-related issue in research networks. In: 2008 4th International Telecommunication Networking Workshop on QoS in Multiservice IP Networks, pp. 33–39. IEEE (2008)
4. Zhou, S., Zhuang, C.: Measure and appraisal the national happiness index of china based on the distance comprehensive evaluation method. Theor. Pract. Finan. Econ. **5**, 112–115 (2008)
5. Palattella, M.R., Dohler, M., Grieco, A., Rizzo, G., Torsner, J., Engel, T.: Internet of Things in the 5G era: Enablers, architecture, and business models. IEEE J. Sel. Areas Commun. **34**(3), 510–527 (2016)
6. Mavromoustakis, C.X., Mastorakis, G., Batalla, J.M.: Internet of Things (IoT) in 5G mobile technologies. Model. Optim. Sci. Technol. **56**, 93 (2016)
7. Rader, T.: The existence of a utility function to represent preferences. Rev. Econ. Stud. **30**(3), 229–232 (1963)

8. Weiss, M., Cosart, L., Hanssen, J., Yao, J.: Precision time transfer using IEEE 1588 over OTN through a commercial optical telecommunications network. In: 2016 IEEE International Symposium on Precision Clock Synchronization for Measurement, Control, and Communication, pp. 1–5. IEEE (2016)
9. Ali, Z., Bonfanti, A., Hartley, M.: IANA allocation procedures for the GMPLS OTN signal type registry. Internet Engineering Task Force (2016)
10. Lee, W., Ryoo, J., Joo, B.S., Cheung, T.: Effective fault management for hierarchical OTN multiplexing using tandem connection monitoring coordinates. In: 2016 International Conference on Information and Communication Technology Convergence, pp. 254–257. IEEE (2016)
11. Shi, J., Ding, Z., Lee, W.J., Yang, Y.P., Liu, Y.Q., Zhang, M.M.: Hybrid forecasting model for very-short term wind power forecasting based on grey relational analysis and wind speed distribution features. IEEE Trans. Smart Grid 5(1), 521–526 (2014)
12. Xue, Q.F., Hao, Y.C.: Comprehensive assessment of power quality based on Rank-sum ratio method. Electr. Power Autom. Equipment 1, 1006–6047 (2015). (in Chinese)

Indoor-Outdoor Wireless Caching Relay System for Power Transformation Station in Smart Grid

Jingya Ma[(✉)], Dongchang Li, and Jiao Wang

Nari Group Corporation, Nanjing, China
majingya6613@163.com

Abstract. In power transformer station of smart grid, a large number of IoT terminal devices need to collect hotspot data in real time to realize monitoring and managing current operating environment of the grid. Since the electricity business requires high read-time performance and reliability, so how to reduce the communication delay between these terminals and increase the transmission efficiency of information has become a challenge for smart grid. Performance of cellular signals received by terminals in transformer station is severely impacted by participant terminals with poor channel conditions, which is usually the cases in the indoor environment due to strong penetration loss. To strengthen the indoor cellular signals, operators deploy the indoor-outdoor relay system (RS), which amplifies wireless signals and improves indoor throughput. Based on the present indoor-outdoor RS, this article proposes to augment it by adopting caching entities and developing caching mechanism that further improves the utilization of wireless resources. The resulting indoor-outdoor caching relay system (CRS) operates in two phases periodically under the control of management agent at the Macro Base Station (MBS). In Phase I, spectrum resources of the established links between MBS and terminals are extracted to support data caching in the system. In Phase II, CRS directly serves indoor terminals and the fronthaul resources are released to serve other terminals. Simulation results verify that the great throughput improvement achieved by CRS with our proposed caching-relay mechanism, only at the cost of temporarily suppressing the data rate to establish caching links.

Keywords: Smart grid · IoT · Caching relay system ·
Power transformation station

1 Introduction

In the automated system in transformer station, the historical data of a large number of terminals such as voltage values, frequency values of busbars, end points of AC line segments, transformer winding, circuit breakers, active power values, reactive power values, current values etc. are provided by the historical data service of the smart grid dispatch system. Then, the change of historical data is displayed in the form of a curve or a table to facilitate the dispatcher to monitor the operating status of the grid visually. Due to the real-time and reliability requirements of electricity business, how to reduce

© Springer Nature Switzerland AG 2020
Q. Liu et al. (Eds.): CENet 2018, AISC 905, pp. 794–801, 2020.
https://doi.org/10.1007/978-3-030-14680-1_86

the access delay between terminals in transformer station and scheduling systems and improve the access efficiency of historical data services has become a current challenge.

In practical applications, wireless rate usually depends on the worst terminal rate in power transformer station. In this way, weak wireless coverage area of the terminal will pull down the wireless business rate. Indoor terminals is such a typical scene, where the signals have to go through building walls, and this causes very high penetration loss, which significantly damages the data rate, spectral efficiency, and energy efficiency of wireless transmissions.

1.1   Indoor-Outdoor Relay System

In [1], an indoor-outdoor relay network structure is proposed by introducing the indoor-outdoor relay system (RS) with large antenna arrays outside each building to connect with MBS. Assisted by MIMO technology, wireless backhaul can achieve larger capacity gains. Meanwhile, the large antenna arrays are connected via cables with indoor wireless access points (APs) which directly serve indoor terminals and make emerging short-range communication technologies available, such as WiFi, etc. for better system throughput. The cellular architecture can significantly enhance the indoor coverage and improve traffic throughput, spectrum utility and energy efficiency of the wireless communication system.

1.2   Related Work

Caching technology means that caching popular content into data nodes in advance, which makes it possible for users to get the desired data from the nearest node [1]. Caching technology is applicable to various networks. For instance, [2] and [3] focus on caching mechanism used in wireless relay networks. In [2], it designed a caching relay mechanism concentrating on when and what to cache by relay stations, aiming at degrading energy consumption of data transmission. In [5], it emphasized on the topology of wireless relay caching network. These works uses stochastic process to model user requests and mainly focuses on the selection of locations and content of cache. But they does not pay attention to the process of caching data which can offer further improvement to transmission efficiency. It can alleviate network congestion and decrease transmission delay significantly, as well as efficiently utilize resource and improve network performance [4]. Existing works, however, mainly focuses on the selection of locations and content of cache. But they does not pay attention to the process of caching data which can offer further improvement to transmission efficiency.

2   Indoor-Outdoor Caching Relay System in Transformer Station

As shown in Fig. 1. There are relays to connect with MBS only, links are built between the MBS and the indoor APs which serve indoor terminals. Caching entities are added to the relays to store data.

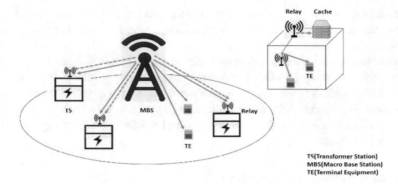

Fig. 1. Illustration of indoor-outdoor caching relay system.

The workflow of caching mechanism can be described as follows. Each work period T is divided into two phases, data caching phase t_1 and data retrieving phase t_2 as shown in Fig. 2, where solid lines represent physical links and dotted lines represent data flows. In Phase I, indoor terminals communicate with MBS via relay, at the same time relays cache data for Phase II. Therefore, the two data flows are from MBS via relay to indoor terminals and from MBS via relay to cache, respectively. In Phase II, indoor terminals get historical data from caches which are written during t_1. Thus, the data flows go from caches via relays to indoor terminals.

2.1   System Model

As is shown in Fig. 1, an MBS is located in the cell center with N outdoor relays connected to through fixed wireless backhaul links. Each relay has a functional caching entity attached. In order to simplify the analysis, we do not pay attention to the capacity of cache. The outdoor terminals are assumed to be uniformly distributed with the number of T_i^{out}. Each relay serves T_j indoor terminals, thus the total terminal number $T_i = T_i^{out} + \sum_{j=1}^{N} T_j$. We denote by T the set of all terminals t, T^{out} the set of outdoor terminals, and T^{in} the set of indoor terminals, where $T^{in} = \bigcup_{j=1}^{N} T_j$, $\mathrm{T}^{out} \cup \mathrm{T}^{in} = \mathrm{T}$.

During data caching phase, indoor terminals communicate with MBS via relays. Assumed that the data rate from MBS to relay is R_{ij}, and the data rate from relays to indoor terminals is R_{ju}. The received data rate of indoor terminals is $\min(R_{ij}, R_{ju})$. In this article, it is able to assume that $R_{ju} > R_{ij}$. Therefore, the received data rate of indoor terminals only relate to the data rate between MBS and relays, namely R_{ij}.

During data caching phase, the achievable rate can be denoted as c_{iu}. According to Shannon formula, c_{iu} can be represented by a logarithmic function of SINR. In which it is proportional to the transmit power of MBS i and the channel gain between terminal u to MBS i, and is inversely proportional to the co-channel interference from neighbor cells (which is assumed as a constant in this article) and the noise power level.

In OFDMA system, spectrum is usually divided into resource blocks (RBs). Thus, we assume that the number of available RBs is R. The bandwidth of each RB is B.

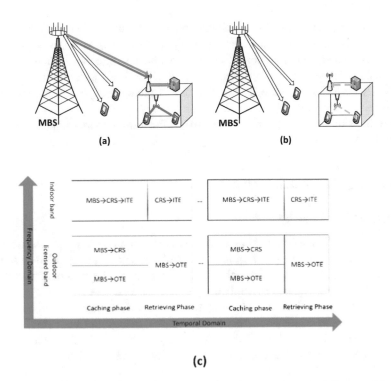

Fig. 2. Data flow and resource occupied by each link in two phases: (a) Data caching phase (Phase I); (b) Data retrieving phase (Phase II); (c) Resource occupation in two phases.

During data caching phase, MBS allocates $\alpha_j R$ RBs to each relay j to cache. Based on [6], the optimal bandwidth allocation is equal allocation, that is, the achievable bandwidth of each terminal is equal to the total bandwidth multiplied by the ratio assigned to the terminal divided by the total number of terminals.

During data retrieving phase, as shown in Fig. 2(b), cache serves indoor terminals individually, and the data rate R_{ju} is constant. Similarly, the bandwidth of each outdoor terminal can be written as the total bandwidth divided by the total number of outdoor terminals.

2.2 Problem Description

We assume that each relay in a cell has equal data caching duration, aiming at calculating achievable bandwidth per relay. Based on our analysis, this factor is proportional to indoor terminal number T_j and data rate between relay and terminal R_{ju}, and is inversely proportional to distance between MBS and relay d_{ij}. Thus the ratio of α_j among different relays is equal to the data rate from relays to indoor terminals multiplied the number of indoor terminals that each relay serves.

According to the ratio of α_j and the consensus that the received data volume in the first phase is equal to the total data volume transmitted to indoor terminals in the second phase, we can get reasonable t_1 and t_2.

In this article, our target is to maximize the throughput S_i in one period. we need to set a minimum data rate threshold R^{th} in the system which is served as a protection to the weakness. The throughout consists of three parts. The first part is the sum of rate of all terminals in Phase I. The second part is the sum of rate of outdoor terminals which is connected with MBS in Phase II. The last part is the sum of rate of indoor terminals which is served by cache in Phase II.

Note that situations mentioned above are all for terminals first requests, that is, the requested data are not in cache.

3   Caching Relay Mechanism

3.1   Applicability Analysis

In conventional RS, since relay is used to broadcast and amplify signals, we can ignore its impact on data rate. Then all terminals in a cell can simply be regarded as connected with MBS i. Thus, the system throughput can be written as the total bandwidth divided by the total indoor terminals multiplied by the achievable rate.

It is clear that the effect of caching relay mechanism is related to number of indoor and outdoor terminals, distance between MBS and relay, and etc. Thus, applicable scenarios need to be specified. This article analyses the mechanism applicability by comparing throughput of the caching relay and the conventional RS.

3.2   Optimal Bandwidth Allocation

In this article, during data caching phase, bandwidth of MBS can be divided into two parts, namely bandwidth to cache data and bandwidth to serve terminals. Obviously, S_i is an inverse proportional function with respect to α. Furthermore, the expression of S_i can be simplified into normal form as $A + E \cdot \frac{1}{Cx+D}$ in which $\left(-\frac{D}{C}, A\right)$ is its symmetry point and E determines its monotonicity. From the above we know, $S_i(\alpha)$ monotonically increases on $[0,1]$. Therefore, on the premise that terminal service quality is ensured, when α reaches its maximum, the overall system throughput would reach the maximum value.

Since we set R^{th} as minimum data rate threshold, each terminals can get $\frac{R^{th}}{\min(c_{iu})}$ bandwidth. Thus the maximum of α can be written as: 1 minus the ratio of the minimum of the resources allocated to the terminals, which is equal to the minimum bandwidth of each terminal multiplied by the total terminals and divided by the total bandwidth.

3.3   Detailed Caching Relay Mechanism

The detailed procedures of the caching relay mechanism can be summarized as follows.

Step 1: Management entity (ME) at MBS detects all terminal requests in the whole cell. If it detects that any relay receives requests for historical data, then enter Step 2.

Step 2: ME takes scenario parameters, including distance between MBS and relay d_{ij}, average distance between MBS and outdoor terminals, number of indoor terminal $\sum_{j=1}^{N} T_j$, and number of outdoor terminals T_i.

Step 3: ME at MBS judges whether caching mechanism is applicable. If our throughput is greater than the traditional throughput enter Step 4; if not, go back to Step 1. That is, ME continues to detect whether there are multimedia requests under relays.

Step 4: ME calculates the optimal bandwidth allocation. Use the maximum of α we have obtained to determine the ratio of bandwidth allocation. Then it sends control signals to MBS and relays in the cell.

Step 5: MBS and relays execute the caching mechanism, including caching data and retrieving data, as described in Sect. 2.

4 Numerical Results

In this part, we present numerical results based on simulation scenario we set. Each cell consists one MBS and three fixed relays which are 150 m, 100 m and 50 m away from MBS. According to our survey, in a transformer station, the number of outdoor terminals is generally 30–50, and the number of indoor terminals is generally 5–20, and in order to highlight the effect of the caching mechanism, we place more indoor terminals in the middle transformer station. So at last, we set the number of outdoor terminals is 30, with a mean distance of 200 m apart from MBS. The number of indoor terminals in each building is 8, 14, and 8, respectively. Thus, the total number of terminals in the system is 60. In modelling the propagation environment, path loss is set as $34 + 40\log10 \, (d[m])$ in dB, and the penetration loss for indoor terminals to MBS is set as 10 dB.

Since the effect of this system is related to scenarios parameters, including the number of terminals, distance between relay and MBS, repeated multimedia requests, etc. The analysis consists two parts. First, we compare the throughput situations among different ratio of indoor and outdoor terminals and number of terminals. Then, since further advantage of the system performance stands out in occasions of repeated multimedia requests, this article also discusses the relationship between percentage of terminals requesting the cached data and system throughput. The results are as follows.

In the simulation, the baseline system is the network in which only MBS serves terminals. Figure 3 illustrates the relationship of throughput and the ratio of indoor and outdoor terminals in two systems. First, with the ratio of indoor terminals increasing, the system throughput increases. That is because relays can amplify signals and enhance coverage, then average SINR of relay terminals is a bit larger than that of MBS terminals. Second, Fig. 3 shows that when the ratio of relay terminals is relatively large, the throughput of no caching relay system is a little larger, which leads to inapplicability of caching mechanism. The reason is that caching consumes too much bandwidth resources, thereby greatly reduce data rate of Phase I. In addition, by comparison, Fig. 3 suggests that the conventional indoor-outdoor RS can improve throughput and further it can achieve better efficiency with caching mechanism.

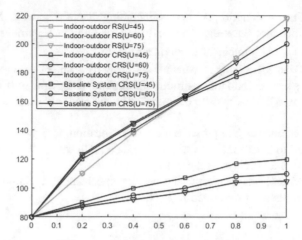

Fig. 3. Average system throughput for different indoor to outdoor terminals ratio (black lines mean not applicable).

Due to the repeatability of historical data requests, caching mechanism can save backhaul bandwidth. As shown in Fig. 4 under the conventional RS, the overall system throughput is constant. However, under indoor-outdoor CRS, caching function can decrease the waste of bandwidth to transmit repeated data though as ratio of repeated requested increases, system throughput increases obviously (nearly 142%), from 156 Mbps to 377 Mbps. Thus it can be seen that caching mechanism can improve performance of system.

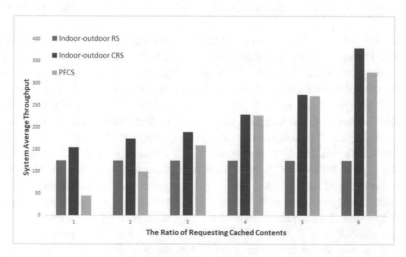

Fig. 4. Comparison of system average throughput related to the ratio of requesting cached contents.

5 Conclusion

This article innovatively proposes an indoor-outdoor CRS, which adds caching entities to indoor-outdoor RS, thereby substantially promoting the utilization of bandwidth and system throughput. This article mainly puts forward a caching mechanism which is executed periodically in two phases and analyses the optimal allocation of bandwidth and applicable scenarios. In simulation part, we confirm the efficacy of this system in consideration of effect of parameters varying with comparison to other systems. The numerical results suggest under the best situation, the performance of proposed CRS can be improved approximately 142% if indoor terminals reuse the caching data. There are still some details to be discussed, such as the limitation of cache capacity and cleaning cache content. These problems will be considered in our further study.

Acknowledgement. This study is supported by 2017 state grid science and technology project "Research and Application for Adaption of Power Business based on LTE Wireless Private Network".

References

1. Wang, C.X., Haider, F., Gao, X., You, X.H., Yang, Y., Yuan, D.: Cellular architecture and key technologies for 5G wireless communication networks. Commun. Mag. **52**(2), 122–130 (2014)
2. Wang, X., Bao, Y., Liu, X., Niu, Z.: On the design of relay caching in cellular networks for energy efficiency. In: 2011 IEEE Conference on Computer Communications Workshops (INFOCOM WKSHPS), pp. 259–264 (2011)
3. Xiaofei, W., Min, C., Taleb, T., Ksentini, A., Leung, V.: Cache in the air: exploiting content caching and delivery techniques for 5G systems. Commun. Mag. **52**(2), 131–139 (2014)
4. Cáceres, R., Douglis, F., Feldmann, A., Glass, G., Rabinovich, M.: Web proxy caching: the devil is in the details. ACM Sigmetrics Perform. Eval. Rev. **26**(3), 11–15 (1998)
5. Xie, F., Hua, K.A.: A caching-based video-on-demand service in wireless relay networks. In: International Conference on Wireless Communications & Signal Processing, pp. 1–5 (2009)
6. Ye, Q., Rong, B., Chen, Y., Al-Shalash, M., Caramanis, C., Andrews, J.G.: User association for load balancing in heterogeneous cellular networks. IEEE Trans. Wirel. Commun. **12**(6), 2706–2716 (2013)

An Improved Propagation Prediction Model Based on Okumura-Hata Model

Rui Gao[1,2], Yanan Zhao[1], Yixuan Wang[1(✉)], and Tianfeng Yan[1]

[1] School of Electronic and Information Engineering, Lanzhou Jiaotong University, Lanzhou 730070, China
roy17007@126.com
[2] Key Laboratory of Opto-Technology and Intelligent Control, Ministry of Education, Lanzhou Jiaotong University, Lanzhou 730070, China

Abstract. The explosive development of wireless communications has brought severe challenges to radio spectrum strategy and planning. In the complex electromagnetic environment, accurate spectrum availability estimation and simulation are playing an increasingly important role in economic development. In recent decades, many researchers and engineers presented a large number of mathematical models to attempt to solve this problem. Okumura-Hata model is one of the most popular models for predicting macrocell path loss in a general flat terrain, but is poor in the mountainous terrain. This paper proposes an improved propagation prediction model based on the Okumura-Hata model, considering the effect of ductile diffraction in the hilly terrain environment. Furthermore, we designed a comparative experiment to verify the effectiveness of improved propagation prediction model. Finally, the results of simulation analysis show that this improved propagation prediction model has high accuracy result in mountains terrain environment.

Keywords: Wireless communications · Propagation prediction model · Okumura-Hata

1 Introduction

The radio spectrum resources are important strategic resources for technological innovation and economic development in a country. The demand for wireless communications service will increase substantially, especially as it stimulates the emergence of new applications and pushes towards the growth of the Internet of Things and 5G mobile communications [1, 2]. The higher numbers of users and the less spectrum resources contributes to the increased interest in spectrum evaluation and simulation. Therefore, the study of radio propagation prediction model is critical to the radio frequency spectrum network specification.

In the complex electromagnetic environment, there are numerous differences from free-space propagation. For instance, it may cause reflection, diffraction and scattering. The propagation characteristics of mobile radio are closely related to its surrounding environment. The irregular terrain, the various shapes of architectural structures, the changeable climate characteristics and other factors make it difficult to predict the

© Springer Nature Switzerland AG 2020
Q. Liu et al. (Eds.): CENet 2018, AISC 905, pp. 802–809, 2020.
https://doi.org/10.1007/978-3-030-14680-1_87

accuracy of path loss. Propagation prediction model is a good way for simulating and calculating the wireless propagation in the complex electromagnetic environment. Therefore, choosing an appropriate propagation prediction model is directly related to the reasonability and accuracy of spectrum planning. Okumura model is a radio propagation model for predicting macrocell path loss in the exterior environment, which has been as one of the most popular macroscopic propagation models and used all over the world [3, 4]. For the purpose of facilitating computer implementation and reducing the amount of computation, Hata presented a set of equations for path loss prediction instead of Okumura's curves, and Okumura-Hata model was further encouraged with its formal recognition by ITU [5]. It is more suitable to predict the path loss over flat terrain, but is poor in the mountainous terrain. For these reasons, this paper proposes an improved propagation prediction model based on the Okumura-Hata model, considering the effect of ductile diffraction in Lanzhou city, China.

2 Theoretical Propagation Model

Path loss is the reduction in power density of an electromagnetic wave as it propagates through space, which is caused by the radiation spread of the transmit power and the characteristics of propagation of channel. First, one calculation method assumes that the transmit and receive antennas is to be located in an otherwise empty environment, also called free space propagation model. Friis [6] presented an equation to explain the relationship between the receive signal power $P_r(d)$ and transmission distance d, as follows Eq. (2.1):

$$P_r(d) = g_b g_m \left(\frac{\lambda}{4\pi d}\right)^2 P_t \tag{2.1}$$

In this equation, P_t is the transmit power, g_b and g_m are the transmitter and receiver antenna gain respectively. According to Eq. (2.1), we can get the path loss $Loss(dB)$ in free space, as shown in Eq. (2.2):

$$Loss(dB) = 32.4 + 20\log(d_{km}) + 20\log(f_{MHz}) \tag{2.2}$$

In free space propagation, the causes of path loss are merely distance d_{km} and frequency f_{MHz}. But in fact that path-loss may be due to many effects, such as free space path loss (FSPL), refraction, diffraction, reflection, absorption and other factors.

3 Improved Propagation Prediction Model

In this section, the empirical propagation model Okumura-Hata and our improved propagation prediction model will be introduced.

3.1 Okumura-Hata Model

The Okumura model is a radio propagation model, which is one of the most widely used models for signal prediction in urban areas built using the data collected in the city of Tokyo, Japan.

$$Loss(dB) = 69.55 + 26.16 \log f_{MHz} - 13.82 \log h_1 - \alpha(h_2)$$
$$+ (44.9 - 6.55 \log h_1) \log d_{km} - C_{type} \tag{3.1}$$

Where h_1 (30–200 m) is base station antenna height; h_2(1–10 m) is mobile antenna height. f_{MHz}(150–1500 MHz) is frequency and d_{km} (1–20 km) is link distance. Besides, $\alpha(h_2)$ and C_{type} are functions that depend on frequency and antenna height, as shown in Eq. (3.2):

$$\alpha(h_2) = \begin{cases} (1.1 \log f_{MHz} - 0.7)h_2 - (1.56 \log f_{MHz} - 0.8) & \textit{small or medium-size city, urban environment} \\ 8.29(\log 1.54h_2)^2 - 1.1 & \textit{metropolitan city, } f_{MHz} \le 300MHz, \textit{ urban environment} \\ 3.2(\log 11.75h)^2 - 4.97 & \textit{metropolitan city, } f_{MHz} \ge 300MHz, \textit{ urban environment} \end{cases}$$
$$\tag{3.2}$$

$$C_{type} = \begin{cases} 0 & \textit{urban environment} \\ 2\left[\log\left(\frac{f_{MHz}}{28}\right)\right]^2 + 5.4 & \textit{suburban environment} \\ 4.78(\log f_{MHz})^2 - 18.33 \log f_{MHz} + 40.9 & \textit{open area} \end{cases}$$

3.2 An Improved Propagation Prediction Model

However, there are some aspects of the application of the Okumura-Hata model that we should be pay attention to: (1) The data of this model was defined as "urban environment", "suburban environment" and "rural environment" on the basis of Tokyo area's data, which may be not match the Lanzhou area very well. (2) This model tends to average the extreme variation of the signal due to sudden changes in terrain elevation, which is more suitable to predict path loss over a general flat terrain without knowing the particular terrain configuration.

For above reasons, an improved propagation prediction model was proposed, as shown in Fig. 1. In our study, a set of measure data from Lanzhou city were used to train the Okumura-Hata model in order to match the best Correction Factors $\alpha(h_2)$ and C_{type} for local area. In addition, Okumura-Hata model has been proven to be more suitable for a relatively flat or smooth ground surface environment. In this paper, our research area is located in Lanzhou, a narrow and curved river valley with surrounding mountains city. Therefore, diffraction effect must be taken into account in our improved propagation prediction model.

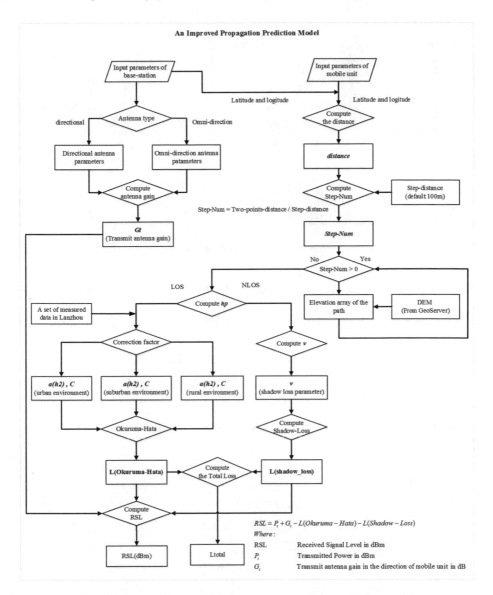

Fig. 1. The workflow of the improved propagation prediction model.

In mobile radio environment, any obstructed objects are much larger than the wavelength of its frequency that the influence of knife-edge diffraction should be considered. According to the Fresnel-Kirchhoff diffraction theory, Anderson presented a simplified method of calculating the knife edge diffraction in the shadow region, which is suitable for most obstacle gain paths [7, 8]. In our study, this method is integrated into our improved propagation prediction model, considering the effect of

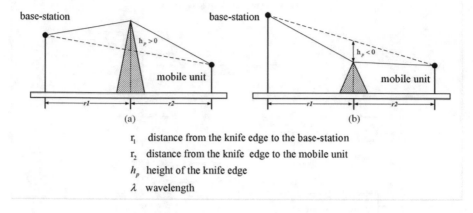

r_1 distance from the knife edge to the base-station

r_2 distance from the knife edge to the mobile unit

h_p height of the knife edge

λ wavelength

Fig. 2. Diffraction of knife peak.

ductile diffraction in Lanzhou city, China. In Fig. 2. r_1, r_2, h_p and λ four parameters are required:

We use the v to present the parameter of shadow loss in knife edge diffraction, as shown in Eq. (3.3):

$$v = (-h_p)\sqrt{\left(\frac{2}{\lambda}\right)\left(\frac{1}{r_1} + \frac{1}{r_2}\right)} \tag{3.3}$$

Once the parameter v is obtained, the shadow loss L_{shadow_loss} can be calculated according to the Eq. (3.4).

$$L_{shadow\_loss} = \begin{cases} 0 & v \geq 1 \\ 20\log(0.5 + 0.62v) & 0 \leq v < 1 \\ 20\log(0.5e^{0.95v}) & -1 \leq v < 0 \\ 20\log(0.4\sqrt{0.1184 - (0.1v + 0.38)^2}) & 2.4 \leq v < -1 \\ 20\log\left(-\frac{0.255}{v}\right) & v < 2.4 \end{cases} \tag{3.4}$$

4 Results and Discussions

As discussed in Sect. 3, when employing this improved propagation prediction model for path loss, the results are more suitable and accurate for complex electromagnetic environment in comparison with the Okumura-Hata model. Thus, a comparative experiment is designed to verify the accuracy of the two models. As can be seen from Fig. 3, it shows schematically the distribution and distance of mobile-unit. Data is collected from different sites in Lanzhou city. According to the roughness of terrain, it can be divided into three different types: (1) Test 1 is a relatively flat terrain environment. (2) Test 2, Test 3 and Test 5 are a relatively rough surface environment.

(3) Test 4 is a hilly terrain environment. Table 1 shows the specific configuration of this experiment, such as Frequency point, Base-station parameters, Transmit antenna parameters and other relevant parameters.

Fig. 3. The distribution site diagram of base-station and mobile-unit.

Table 1. Input parameters of experiment.

Parameter	Value	Parameter	Value
Frequency point	430 MHz	Base-station location	103.721146, 36.10360737
Power	25 W	Feeder loss	5 dB
Base-station antenna height	58 m	Transmit antenna gain	5 dBi
Base-station elevation	1550 m	Mobile-unit antenna height	2 m
Mobile-unit sensitivity	120 dB	Antenna type	Omni-direction

The results of this experiment are shown in Table 2. In a flat terrain environment, the results of two models are similar. In a relatively rough surface environment, the improved model gets smaller error than the Okumura-Hata model. Especially in the Test 4, the mobile unit can't receive the signal from base-station, which means that path-loss is larger than 164 dB. In this case, Okumura-Hata model gets a bad result that will be inclined to misjudge communication status from base-station to mobile unit. Therefore, we get a conclusion that this improved propagation prediction model can provide a more accurate result in complex electromagnetic environment.

Table 2. The results of path-loss prediction.

	Measured (dB)	Okumura-Hata (dB)	Error (dB)	Improved model (dB)	Error (dB)
Test 1	−149.59	−145.59	4.21	−145.998	3.8
Test 2	−156.5	−145.45	11.06	−157.752	1.252
Test 3	−151.9	−145.42	6.48	−154.651	2.75
Test 4	Not received (>164)	−145.43	-	−171.98	-
Test 5	−155.3	−144.68	10.62	−152.37	2.923

5 Conclusions

This paper proposes an improved propagation prediction model based on the Okumura-Hata model. And a comparative experiment is designed to evaluate the effectiveness of this model. The results of simulation analysis suggest that this model has better accuracy results. This model is suitable for Lanzhou city, but it might not be fully adapted in all regions. It is only a simplified and approximate method of computing knife edge diffraction under the multiple-edges diffraction. Therefore, further improvement of our study has been suggested. [8–10] approaches will be considered to integrate into this improved model in order to improve the impact of multiple-edges.

Acknowledgments. Thanks to the experimental data provided by the Institute of Digital Signal Processing and Software-Defined Radio, Lanzhou Jiaotong University. In addition, the authors gratefully acknowledge the financial support provided by Opening Foundation of Key Laboratory of Opto-technology and Intelligent Control (Lanzhou Jiaotong University), Ministry of Education (KFKT2018-16), Youth Science Foundation of Lanzhou Jiaotong University under Grant No. 2018003, Innovation Fund Project of Lanzhou Jiaotong University and Tianjin University under Grant No. 2018062, Scientific Research plan projects of Gansu Education Department under Grant No. 2017C-09, Lanzhou Science and Technology Bureau under Grant No. 2018-1-51.

References

1. Dao, N.N., Park, M., Kim, J., Paek, J., Cho, S.: Resource-aware relay selection for inter-cell interference avoidance in 5G heterogeneous network for Internet of Things systems. Futur. Gener. Comput. Syst. **93**, 877–887 (2018)
2. Gavrilovska, L., Latkoski, P., Atanasovski, V., Prasad, R., Mihovska, A., Fratu, O., Lazaridis, P.: Radio spectrum: evaluation approaches, coexistence issues and monitoring. Comput. Netw. **121**, 1–12 (2017)
3. Garg, V.: Wireless Communications & Networking. Elsevier, Amsterdam (2010)
4. Hata, M.: Empirical formula for propagation loss in land mobile radio services. IEEE Trans. Veh. Technol. **29**(3), 317–325 (1980)
5. Nadir, Z., Ahmad, M.I.: Pathloss determination using Okumura-Hata model and cubic regression for missing data for Oman. In: Proceedings of the International Multi-Conference of Engineering and Computer scientist, p. 2 (2010)

6. Lee, W.C.Y.: Mobile Communications Design Fundamentals. Wiley, Hoboken (2010)
7. Anderson, L., Trolese, L.: Simplified method for computing knife edge diffraction in the shadow region. IEEE Trans. Antennas Propag. **6**(3), 281–286 (1958)
8. Lee, W.C.Y.: Integrated Wireless Propagation Models. McGraw-Hill Education, New York (2015)
9. Epstein, J., Peterson, D.W.: An experimental study of wave propagation at 850 MC. Proc. IRE **41**(5), 595–611 (1953)
10. Giovaneli, C.L.: An analysis of simplified solutions for multiple knife-edge diffraction. IEEE Trans. Antennas Propag. **32**(3), 297–301 (1984)

A Improved Algorithm of MAC Layer for Wireless and Power Line Cooperative Communication

Yingchu Liu[1], Zhixiong Chen[1($\boxtimes$)], Jinsha Yuan[1], Ran Liu[1], and Xincheng Tian[2]

[1] North China Electric Power University, Baoding 071003, China
chenzxl983@sohu.com
[2] State Grid Tangshan Power Corporation, Tangshan 063000, China

Abstract. Power Line Communication (PLC) and Wireless Communication (WLC) are the popular research directions in the field of communication in the future. The concept of cooperative communication between the power line and wireless has been proposed. However, the current researches focus mainly on the physical layer, and the MAC layer is rarely researched. Therefore, this paper proposes a new CSMA/CA algorithm-W/P-CSMA (Wireless/Power Line alternative communication of CSMA/CA). It combines the advantages of PLC and WLC and formulates a new algorithm to make frames well transmitted in power line and wireless. Then through the MATLAB simulation, we study the key indicators. From the analysis of results, alternative communications have significantly improved time delay and throughput compared with PLC.

Keywords: PLC · Wireless communication · CSMA/CA · Throughput · Time delay

1   Introduction

Power line communication and wireless communication technologies have a wide range of application prospects in smart electricity and Internet of Things (IOT) [1]. Power line communication has the advantages of convenience, low cost, and wide coverage, but communication condition is harsh and needs improving. The advantage of wireless communication is that the communication channel is relatively good and the transmission rate is high, but it is difficult to achieve full coverage because of the usage of 2.4 GHz frequency resource. Therefore, the cooperation of power line and wireless communication can complement each other and improve the overall performance of the communication system.

The bit error ratio of a cooperation communication based on power line and wireless has been analyzed in [2]. The paper [3–5] proposed a model for coupling and decoupling between stations. Bianchi studied the performance of 802.11 using a discrete-time Markov chain model [6]. [7] proposed a Markov model with a deferral counter similar to [6]. The paper of [8] discussed the effect of wireless channel fading on the CSMA performance of the 802.11n standard. For cross-layer cooperative

© Springer Nature Switzerland AG 2020
Q. Liu et al. (Eds.): CENet 2018, AISC 905, pp. 810–817, 2020.
https://doi.org/10.1007/978-3-030-14680-1_88

control, the paper [9] proposed a cooperative control strategy for smart distribution network and wireless network. Regarding key issues such as time synchronization and energy consumption, Kong Qingping et al. proposed an improved scheme based on the original strategy [10].

At present, power line and wireless hybrid communication mainly focus on technologies such as physical layer cooperation, hardware design, and hybrid networking. It rarely studies the performance analysis and optimization mechanism of MAC layer protocol in the process of power line and wireless hybrid communication. So, the work presented here offers contributions in this area and introduces some original results. For the wireless 802.11 protocol and the IEEE 1901 CSMA/CA algorithm, the paper proposes an improved mechanism of MAC layer for hybrid cooperative communication based on Wireless/Power Line alternative communication of CSMA/CA both in 802.11 and 1901, named as W/P-CSMA. By introducing this new communication method, the power line communication can be greatly improved. Because the impedance in the power grid greatly changes with the change of the load, a strong power noise is generated, and the power line channel has a strong attenuation characteristic. These factors will cause the communication quality to be seriously affected. The article introduces the wireless channel, using its good information transmission environment, can fundamentally weaken the influence of the power line channel. In addition, in the peak period of power usage, the power line channel can easily cause congestion. Applying this cooperative communication method at this time can effectively improve network performance and improve communication service quality. From the perspective of difficulty of implementation, due to the perfection and full coverage of the power line infrastructure, it is more convenient to install a wireless module with a higher degree of integration on the power facilities, and it can cover a large range of users.

The main performance indexes such as MAC layer throughput and time delay are simulated and analyzed.

2 CSMA Algorithm and Its Throughput in Power Line and Wireless Communication

The 802.11 of wireless communication adopts the CSMA protocol. The channel throughput S can be expressed as:

$$S = \frac{P_S P_{tr} E[P]}{(1 - P_{tr})\alpha + P_{tr} P_s T_s + P_{tr}(1 - P_s)T_c} \tag{1}$$

In Eq. (1), P_s is the probability of successful transmission and P_{tr} is the probability of at least one station sending. The $E[P]$ is the frame length and α is the length of the time slot. T_s and T_c are the time spent on successful transmission and the time spent on collision. As follows:

$$T_S = H + E[P] + SIFS + \delta + ACK + DIFS + \delta \tag{2}$$

$$T_c = H + E[P] + DIFS + \delta \tag{3}$$

Where H denotes the frame header and δ denotes the propagation delay. Through the calculation of the above parameters, we can finally get the throughput.

The latency of wireless MAC is expressed as the total time it takes to contend for the channel from the station to successfully send the data frame. The expression is as follows:

$$T_{sy} = T_{tb} + T_{tr} \tag{4}$$

Where T_{rb} is the time spent for back-off and T_{tr} is the time spent for transmitting data frames.

In the CSMA/CA algorithm of the IEEE1901 of power line communication standard, m is defined as the number of back-off steps, and n_i is the number of stations in the back-off phase i, and τ_i is one station's transmission probability of the back-off stage i in an arbitrary slot. p_i indicates the probability of transmission from at least one other station. p_s indicates the probability of the station successfully sending a data frame.

$$p_i = 1 - \frac{1}{1 - \tau_i} \prod_{k=0}^{m-1} (1 - \tau_k)^{n_k} \tag{5}$$

Since the PLC has a DC counter, x_k^i is introduced. Let T be the number of time slots detected when the channel is busy:

$$x_k^i = P(T > d) = \sum_{j=d_i+1}^{k} \binom{k}{j} p_i^j (1 - p_i)^{k-j} \tag{6}$$

Let bc_i be the expected number of time slots that the station spends in back-off stage i.

$$bc_i = \frac{1}{CW_i} \sum_{k=d_i+1}^{CW_i-1} \left[(k+1)(1 - x_k^i) + \sum_{j=d_i+1}^{k} j(x_j^i - x_{j-1}^i) \right] + \frac{(d_i+1)(d_i+2)}{2CW_i} \tag{7}$$

And τ_i can be represented by bc_i and x_k^i:

$$\tau_i = \frac{\sum_{k=d_i+1}^{CW_i-1} \frac{1}{CW_i}(1 - x_k^i) + \frac{d_i+1}{CW_i}}{bc_i} \tag{8}$$

Through the above parameters, throughput can be expressed as:

$$S = \frac{p_i p_s E[N_{payload}]}{(1 - p_i)\alpha + p_i(p_s T_s + (1 - p_s)T_c)} \tag{9}$$

Where $E[N_{\text{paylaod}}]$ is the average effective frame length. Tc and Ts are respectively the time taken by collision and the time taken by successful transmission. α is the length of the idle time slot. Above all, this section makes theoretical analysis on the key indicators of cooperative communication, and gives detailed explanation and formula derivation of each parameter, which lays a theoretical foundation for the following simulation.

3 An Improved CSMA Algorithm for Power Line and Wireless Hybrid Collaboration

Though the wireless communication use IEEE 802.11 standard and power line communications (PLC) comply with IEEE 1901 standard respectively, both systems use the CSMA/CA (Carrier Sense Multiple Access/Collision Avoidance) mechanism. The main difference between 1901 and 802.11 is the introduction of DC in 1901. In order to make a full use of both Wireless and Power line channel, the W/P-CSMA proposed in this paper detects the channels simultaneously and choose the idle channel to transmit packet whose back-off counter reaches 0 firstly.

The back-off process of 1901 uses two counters: the back-off counter (BC) and the deferral counter (DC), and the 802.11 uses only one back-off counter (BC_0). The contention window (CW) is doubled between two successive back-off stages. DC was introduced as a countermeasure in the CSMA/CA process of 1901. BC_0 is decreased by 1 in each time slot if the station senses the medium to be idle and it is frozen when the medium is sensed busy; for 1901, whether busy or not, the BC counter continuously decrements by 1 in each time slot, which is the difference from the wireless. The DC counter is decreased by 1 when the power line channel is busy and is frozen when the medium is idle. When BC or BC_0 reaches to 0, the stations attempts to transmit the packet. If the transmission is successful, the station returns to initial back-off stage and re-draws BC, BC_0 and DC. If a collision occurs, the corresponding transmission module enters the next level of back-off and the other one is not affected. To sum up, the flowchart and steps of the alternative communication are described as follows:

(1) When there is packet to be transmitted, the station's wireless and power line counters are initialized.

(2) The wireless back-off counter is frozen when the channel is busy, and decreases by 1 when the channel is idle; when the power line channel is busy, the delay counter decreases by 1 and it is frozen when the channel is idle. For 1901, the back-off counter continuously decreases by 1 whether the channel is busy or not.

(3) When one of the back-off counters reaches 0, the station attempts to transmit packet. What needs to be emphasized is that, when the value of DC reaches 0, the power line will directly enter the next back-off stage at this time, and re-draws BC.

(4) If a packet is successfully transmitted, all the counter values of the station are initialized and the above steps are repeated; the station enters the next back-off stage if the transmission fails, and the other back-off counter is not affected, then repeats (2)–(4).

Alternative communication methods have two more counters than wireless communication, and one more counter than power line communication, which leads to a higher complexity of the algorithm. However, since the rules of the wireless and power line MAC layers have not changed, it is relatively easy to implement (Fig. 1).

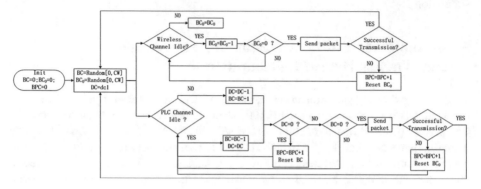

Fig. 1. Alternative communication basic flow.

4   Performance and Simulation Analysis

This paper uses the MATLAB simulation platform for the simulation of power line and wireless cooperation communication, and focuses on the throughput, time delay and reliability of power line communication and cooperative communication. The simulation is based on the following assumptions.

(1)   All station data is saturated, that is, the station always has data to send.
(2)   The data frame is lost or error only due to channel collision without considering the external environment; if the data frame is not sent successfully, it will not be discarded.

In this paper, the throughput, delay and reliability of the optional communication in cooperation are simulated and compared with PLC. The application parameters are shown in Table 1.

Table 1. Simulation parameters.

Mode	Number of sites (N)	Wireless back-off phase (m)	Power-line back-off phase (m1)	Wireless minimum back-off window (cw0)	Power line minimum back-off window (cw1)	Frame transmission time (Ts)	Frame collision duration (Tc)	Frame length
PLC	1–10		4		8	2542.64	2920.64	2050
Alternative communication	1–10	4	4	8	8	2542.64	2920.64	2050

It should be emphasized that the parameter T_s is the total time elapsed since the data frame was sent to the sender to confirm successful reception; T_c is the total time since the data frame was sent to the sender to confirm the collision, and the unit is microseconds. cw0 and cw1 are the window sizes for the initial back-off phases of the wireless and power lines, respectively; both m and m1 indicate the total number of back-off phases.

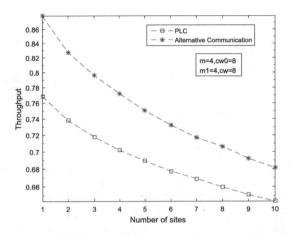

Fig. 2. Throughput comparison.

As can be seen from Fig. 2, with fixed m and cw0, as the number of stations N increases, the competition between system stations intensifies, which in turn leads to decreased throughput. For alternative communication, it is equivalent to using two channels to transmit data frames. If one of the channel environments is poor, then another channel can be used to send data, which reduces the probability of collision, so the throughput of alternative communication is higher than that of power line communication.

From Fig. 3 we can see that with fixed m and cw0, as the number of stations increases, the collision probability increases and the time delay increases. For alternative communications, the opportunity to send data frames at each station is increased by following the principle of who goes to zero and who sends it. When one of the channel conditions is not good, another channel can be used to send, greatly increasing the probability of successful transmission. In other words, the waiting time for the successful transmission of data is reduced, that is, the latency of the alternative communication is lower than the power line communication.

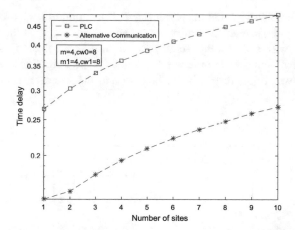

Fig. 3. Comparison of time delays.

As shown in Fig. 4, when N increases, the competition increases, causing the collision probability to increase. For the alternative communication, although its throughput is high and the time delay is low, but due to the collision should not only consider the wireless, also should to consider the collision of power lines, if one side is heavily collided, even if the other channel is in good condition, it will affect the overall frame error rate. So the reliability of the alternative communication is lower than the power line communication.

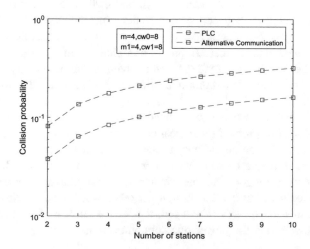

Fig. 4. Comparison of the reliability.

5   Conclusion

At present, there is very little research based on the cooperation of the MAC layer, and there is still a huge gap in the field. In this context, the paper proposes the strategy of implementing wireless and power line cooperative communication on MAC, clarifies the algorithm and flow in detail, and promotes the field of MAC-based cooperative communication. Through the analysis of the results, we can see that in terms of throughput and time delay, the alternative communication mode is obviously better than power line communication, but its reliability is slightly insufficient. In the actual application process, various requirements should be comprehensively considered and two communication devices should be flexibly configured so that all indicators of the entire network can reach expectations.

Acknowledgement. This study is supported by The National Natural Science Foundation of China (No. 61601182, No. 61771195), Natural Science Foundation of Hebei Province (No. F2017502059, F2018502047), the Fundamental Research Funds for the Central Universities (No. 2017MS109).

References

1. Vlachou, C., Henri, S., Thiran, P.: Electri-Fi your data: measuring and combining power-line communications with Wifi. In: Internet Measurement Conference, pp. 325–338. ACM (2015)
2. Chen, Z.X., Han, D.S., Qiu, L.J.: Research on performance of indoor wireless and power line dual-media cooperative communication system. Proc. CSEE **37**(9), 2589–2598 (2017)
3. Vlachou, C., Banchs, A., Herzen, J., Thiran, P.: How CSMA/CA with deferral affects performance and dynamics in power-line communications. IEEE/ACM Trans. Netw. **25**(1), 250–263 (2017)
4. Vlachou, C., Banchs, A., Herzen, J., Thiran, P.: On the MAC for power-line communications: modeling assumptions and performance tradeoffs. In: International Conference on Network Protocols, pp. 456–467. IEEE (2014)
5. Vlachou, C., Banchs, A., Herzen, J., Thiran, P.: Analyzing and boosting the performance of power-line communication networks. In: Proceedings of the 10th ACM International on Conference on emerging Networking Experiments and Technologies, pp. 1–12 (2014)
6. Min, Y.C., Jung, M.H., Lee, T.J., Lee, Y.: Performance analysis of HomePlug 1.0 MAC with CSMA/CA. IEEE J. Sel. Areas Commun. **24**(7), 1411–1420 (2006)
7. Bianchi, G.: Performance analysis of the IEEE 802.11 distributed coordination function. IEEE J. Sel. Areas Commun. Arch. **18**(3), 535–547 (2000)
8. Xu, X.H., Lin, X.K.: Improved algorithm of IEEE 802.11 DCF in fading channels. J. Tsinghua Univ. (Sci. Technol.) **47**(1), 57–60 (2007)
9. Fang, R.J., Wang, J.P., Sun, W.: Cross-layer cooperative control of wireless sensor communication network in smart distribution network. J. Electron. Meas. Instrum. (2) (2018)
10. Kong, Q.P.: Research on time synchronization protocol and algorithm improvement of high accuracy and low power wireless sensor network. Kunming University of Science and Technology (2017)

Coverage Enhancement for Cellular Internet of Things Based on D2D Multicasting

Jinsuo Liu[1], Daohua Zhu[2(✉)], Jingzhi Xue[1], and Tao Ma[1]

[1] Nari Group Corporation, Nanjing, China
[2] Electric Power Research Institute,
State Grid Jiangsu Electric Power Company, Ltd., Nanjing, China
sgdaohuazhu@163.com

Abstract. Cellular Internet of Things (cIoT), which will create a huge network of billions or trillions of Things communicating with one another, are facing many technical and application challenges. In particular, in this paper, we have proposed a mechanism based on D2D multicast group communication. The goal of us is to enhance the coverage ability of the cellular network system which is used in IoT. Firstly, we propose a D2D clustering algorithm with the consideration of energy consumption and the outage of the system. Secondly, we want to improve the throughput of the whole system. We decomposed this optimization problem into two sub problems: power control and channel assignment. Referring to the Genetic Algorithm, we get the optimal transmit power of the whole devices. Then, we use the Greedy Algorithm to obtain the perfect channel allocation. Finally, the simulation results demonstrated the efficiency of the communication mechanism proposed by us.

Keywords: Coverage enhancement · Cellular Internet of Things ·
D2D multicasting · Genetic algorithm

1 Introduction

With the rapid development of the IoT, more and more devices exist in the same network. This will reduce the overall system throughput and affect the functionality of the device. Cellular network is considered one of the most promising approaches for supporting diverse IoT services. In a traditional cellular IoT network, data are transmitted between base station (BS) and IoT devices. However, many IoT devices in today communication are limited by energy and coverage ability. The emergence of Device-to-Device (D2D) communications can decrease energy consumption of IoT service.

By using direct communication (D2D), terminals transmit data directly to each other without circumventing it through a base station (BS). The concept of network controlled direct communication in a cellular network not only offers seamless operation and connection setup, but is also able to guarantee QoS. Also, the regular cellular connection will always be available as a backup and ensure the service continuity in case the D2D link between devices deteriorates. Compared to traditional communication, D2D both reduces interference to other users, releases resources and increases

© Springer Nature Switzerland AG 2020
Q. Liu et al. (Eds.): CENet 2018, AISC 905, pp. 818–827, 2020.
https://doi.org/10.1007/978-3-030-14680-1_89

the spatial reuse of radio resources, benefiting the network as a whole: The BS doesn't have to process the data D2D terminals are transmitting, thus allowing it to handle more connections. Moreover, direct communication has potential to offer smaller delays, thus better user experience, than its cellular counterpart, since the data doesn't have to traverse first to the BS and from there to the receiver.

When the number of devices in the vicinity of a base station increases to a certain amount, the traffic between the devices rapidly increases, resulting in some devices failing to work properly. By using D2D communication, the consumption of IoT devices in the process of communication can be reduced, which can increase the number of devices that can be accommodated in a certain network area. But traditional D2D technology is not perfect. There are lots of possible practical limitations for it such as: (1) The excessive interference and poor propagation channel will reduce the advantages of D2D communication. (2) Propagation channel between devices is worse than propagation channel between a BS and devices. (3) Only supporting one-to-one direct communications between IoT devices. There are already some researchers on these issues.

Although D2D communication has high probability to improve performance of overall cellular systems, it can cause interference between UEs due to the use of the same wireless resources. A researcher on dynamic power control for interference coordination of Device-to-Device communication in IoT Cellular Networks has been done in [1]. Dynamic power control mechanism is proposed in [1] to reduce interference and improve performance of cellular systems. In the context of D2D communication, it is crucial to set up reliable direct links between the IoT devices and D2D users in the network. Direct communication with clustering is analyzed in [2] using tele traffic approach to calculate blocking probabilities. In that study the devices were able communicate directly while being members of different clusters. This research proves that the clustering algorithm can achieve one-to-many interconnection of IoT devices in D2D communication. As a state-of-the-art method in D2D, the clustering method is expected to reduce the devices' energy consumption [3]. The benefits of this method had been proofed in several works [4–6]. A new method in D2D for energy efficiency is proposed in [7] called Cluster Head Rotation method. The method does not only reduce energy consumption but also balance the energy consumption of Cluster Head (CH) and Cluster Member (CM). Our cluster model is cited and inspired from cluster head rotation method. Although these studies have reduced the interference in the network to a certain extent, they have not completed a deeper optimization, and the mechanism we propose will optimize the system throughput and coverage in detail by optimizing the energy control and appropriate channel allocation to enhance the overall performance of the system to allow more IoT devices to work properly.

In this paper, we see the optimizing problem as tow sub-problems: power control and channel assignment. We propose two algorithms for them, one is genetic algorithm for power control, and another is greedy algorithm for channel assignment. In computer science and operations research, a genetic algorithm (GA) is a meta-heuristic inspired by the process of natural selection that belongs to the larger class of evolutionary algorithms (EA) [8].

2 System Model and Problem Formulation

2.1 System Model

Figure 1 shows the scenario of D2D multicast cellular network. Each cellular network cell contains one macro base station and multiple user devices. The number of clusters in the system is M, which means there are M D2D multicast groups. We use set $A^H = \{A_m^H | m = 1, 2, 3, \ldots, M\}$ to represent the cluster head devices, where A_m^H represents the receiver in the mth D2D multicast group. The number of the transmitters within the D2D multicast group is Z, we use set $D^T = \{D_{m,i}^T | i = 1, 2, 3, \ldots, Z\}$ to represent it. Except the D2D devices, there are some cellular devices which are not divided into clusters. The number of them is K and we use set $C = \{C_j | j = 1, 2, \ldots, K\}$ to represent it. So the total number of the devices in the edge zone of the system is $S = M + Z + K$. We use set $A = \{A_s | s = 1, 2, 3, \ldots, S\}$ to represents it and we use B^s to represent the base station.

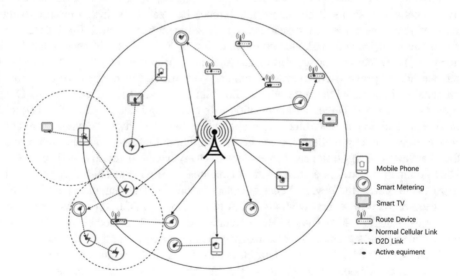

Fig. 1. The scenario of D2D multicast cellular network.

2.2 Problem Formulation

According to Shanna's theorem, when active device A_i is communicating with A_j, the channel transmission rate can be expressed as follow:

$$R_{ij} = B_{ij} log_2 (1 + g_{ij}) \tag{2.1}$$

Where B_{ij} represents the bandwidth of the channel between them, and g_{ij} represents the signal interference and noise ration between them.

When $D_{m,i}^T$ communicates with cluster head A_m^H or C_j communicates with base station, since $D_{m,i}^T$ has reused channel resource of C_i, so there are interference between them and the SINR of them can be represented as below:

$$R_{D_{m,i}^T,A_m^H} = B_{ij}log_2\left(1 + \frac{P_{D_{m,i}^T,A_m^H}G_{D_{m,i}^T,A_m^H}}{P_{C_jB^s}G_{C_jA_m^H} + N_0}\right) \tag{2.2}$$

$$R_{C_jB^s} = B_{ij}log_2\left(1 + \frac{P_{C_jB^s}G_{C_jB^s}}{P_{D_{m,i}^T,A_m^H}G_{D_{m,i}^T,B^s} + N_0}\right) \tag{2.3}$$

where $G_{D_{m,i}^T,A_m^H}$, $G_{C_jA_m^H}$, $G_{C_jB^s}$, $G_{D_{m,i}^T,B^s}$ represent the channel gain and N_0 is the noise.

In this manuscript, by the way of clustering to form D2D multicast group to improve the coverage ability of the cellular system. At the same time, since the limitation of the channel resources in cellular system, we assume that the channel resources can be reused by $D_{m,i}^T$ and C_j. In order to guarantee the communication quality and minimize the interrupt probability, our goal is to maximum the transmit rate of the system. We define a matrix $\Pi = \left[\phi_{i,j}\right]_{M \times K}$ to represent the channel resource allocation situation, where $\phi_{i,j}$ is 0 or 1. If $\phi_{i,j} = 1$, it means $D_{m,i}^T$, can reuse the channel resource of C_j, otherwise $\phi_{i,j} = 0$. Therefore, the optimization problem in this paper can be expressed by the following formula:

$$P: max \sum_{i=1}^M \sum_{j=1}^K \phi_{i,j}\left(R_{D_{m,i}^T,A_m^H} + R_{C_jB^s}\right) \tag{2.4}$$

$$s.t \ 0 \le P_{D_{m,i}^T,A_m^H} \le P^{max} \quad 0 \le P_{C_jB^s} \le P^{max} \tag{2.5}$$

$$\gamma_{D_{m,i}^T,A_m^H} \ge \gamma_{th} \quad \gamma_{C_jB^s} \ge \gamma_{th} \tag{2.6}$$

$$\sum_{i=1}^M \emptyset_{ij} \le 1 \quad j \in [1,2,3,\ldots,K] \tag{2.7}$$

$$\sum_{i=1}^K \emptyset_{ij} \le 1 \quad i \in [1,2,3,\ldots,M] \tag{2.8}$$

Constrain (2.5) indicate that transmit power levels are within maximum limit. (2.6) is the constraint of the SINR, while the minimum SINR threshold is $SINR_{min}$, which ensure the communication quality of each IoT devices. (2.7) is used to constrain each transmitter device in D2D group can only reuse one cellular device's channel resource and (2.8) ensures that each cellular device's resource can only be reused by one transmitter device D2D group.

3   Resources Allocation

3.1   Power Control

In order to maximize the channel rate after channel multiplexing, the power optimization problem of the multiplexed channel transmission rate can be modeled as:

$$P1 : max\ R_{D_{m,i}^T,A_m^H} + R_{C_jB^s} \tag{3.1}$$

The power optimization problem in P1 can be viewed as an optimization problem in a multivariable dynamic system. Each IoT device wants to obtain higher transmission power to improve efficiency, however, any increase in IoT device power will increase the interference to other IoT devices and reduce the effectiveness of other IoT devices, so that affected device also resist by increasing the transmission power. Its interference, where $P_{D_{m,j}^T,A_m^H}$ and $P_{C_jB^s}$ are the key variable elements. For this problem can be solved using genetic algorithm (GA).

The structure of a GA is composed by an iterative procedure through the following five main Steps: 1. Creating an initial population P0, 2. Evaluation of the performance of each individual pi of the population, by means of a fitness function, 3. Selection of individuals and reproduction of a new population, 4. Application of genetic operators: crossover and mutation, 5. Iteration of Steps 2–4 until a termination criterion is fulfilled.

To apply GA to the optimization of our strategy, a fitness function is required in order to evaluate the status of each solution. In this study, the fitness function is considered to be the inverse of the objective function described by P1:

$$J(x) = max\left(R_{D_{m,i}^T,A_m^H} + R_{C_jB^s}\right) \tag{3.2}$$

Using this approach, the performance requirements are then considered as constrained. However, as GA is directly applicable only to unconstrained optimization problem, the constraints are handled by using penalty functions that penalize the infeasible solutions by reducing their fitness values. Here a penalty value is added into the violating solutions taking the number of violated constraints and their distance from feasibility into account. In this case, the fitness function will take the following form:

$$F(x) = J(x) + \sum_{i=1}^{n_{con}} \alpha_i \times P_i(x) \tag{3.3}$$

where $F(x)$ is the fitness function, $J(x)$ is the objective function, $P_i(x)$ is the penalty function related to the ith constraint and α_i is a positive constant value that determines the degree to which the ith constraint is penalized, normally called penalty factor. These factors are treated as constant here and their values are obtained by trial and error.

The dimension of the solution space is five. Therefore, the following five variables are coded in a chromosome using a binary coding scheme consisting of the following genes:

$$x = \left\{ \Delta J(x), P_{D_{m,i}^T A_m^H}, P_{C_j B^s}, G_{D_{m,i} A_m^H}, G_{C_j A_m^H} \right\} \quad (3.4)$$

To start the algorithm, an initial population of individuals (chromosomes) is defined.

A fitness value is associated with each individual, expressing the performance of the related solution with respect to a fixed objective function to be minimized.

Reproduction is the process of generating a new population from the current population. Selection is the mechanism for selecting the individuals with high fitness over low-fitted ones to produce the new individuals for the next population. The variant used here is the roulette wheel method in which the probability to choose a certain individual is proportional to its fitness:

$$Prob[p_i] = \frac{f(p_i)}{\sum_{k=1}^{n} f(p_k)} \quad (3.5)$$

3.2   Channel Assignment

Since the matrix $\Pi = [\phi_{i,j}]_{M \times K}$ represents the channel resource allocation situation. The value of $\phi_{i,j}$ is 0 or 1. This can be seen as an assignment problem. In this section, we will use the bipartite maxima matching method to optimize the channel allocation. The Greedy Algorithm can be used to solve the problem.

Given a graph and weights $w_e \geq 0$ for the edges, the goal is to find a matching of large weight. The greedy algorithm starts by sorting the edges by weight, and then adds edges to the matching in this order as long as the set of a matching.

Greedy Algorithms for Matching: $M = \emptyset$. For all $e \in E$ in decreasing order of w_e add e to M if it forms a matching. The Greedy algorithm is detailed as follows.

Step 1: Since the number of cellular devices and the number of ordinary cluster devices are not necessarily equal. When the relationship between the number of D^T and the number of C is N = M + K, we add vertices to the dichotomous vertex set D^T. When the relationship is $N = M - K$, we add $\hbar$ vertices to the dichotomous vertex set C. The weight of the add vertices'edges with other vertices is zero. Sort all edges in descending order according to weight size, get the results:

$$\left\{ \left[D_{m,2}^T, C_3 \right] = 7, \left[D_{m,4}^T, C_2 \right] = 5, \left[D_{m,2}^T, C_1 \right] = 4, \left[D_{m,3}^T, C_3 \right] = 4, \right.$$
$$\left. \left[D_{m,1}^T, C_1 \right] = 3, \left[D_{m,1}^T, C_3 \right] = 2, \left[D_{m,3}^T, C_2 \right] = 2, \left[D_{m,4}^T, C_3 \right] = 2 \right\}$$

Step 2: Find the max value edge: $\left[D_{m,2}^T, C_3 \right] = 7$, add it to the M. Step three: Find the max value edge: $\left[D_{m,4}^T, C_2 \right] = 5$, add it to the M. Step four: Check the edges

$\left[D_{m,2}^T, C_1\right] = 4$, $\left[D_{m,3}^T, C_3\right] = 4$, both of them don't match, so add the next max value edge $\left[D_{m,1}^T, C_1\right] = 3$ to the M. and there is not more edges to match, we get the maxima matching set M.

4 Simulation Analysis

In this section, we will present our simulation results and corresponding simulation analysis. Our simulation is for the uplink data transmission process of a single IoT cell scenario. Our simulation is based on the MATLAB simulation platform. In this paper, because we want to improve the edge coverage of the cell, we randomly distribute 1000 users evenly within a radius of 500 m and within 100 m of the coverage edge. In each simulation process, we randomly wake up some user equipments near the edge of the cell for data transmission.

In the specific simulation, we mainly focus on the change of the outage probability of the system and the change of the throughput of the system, and verify the performance of the proposed IoT network uplink communication algorithm based on D2D communication.

The Fig. 2 compares the system's outage probability with the increase of SNR based on the D2D group communication and normal communication. As can be seen from the Fig. 2, as the system SNR increases, the system interruption probability under the ordinary IoT communication mode and the D2D communication mechanism is constantly rising. Compared with the ordinary IoT system, we propose The algorithm significantly reduces the outage probability of the system.

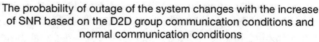

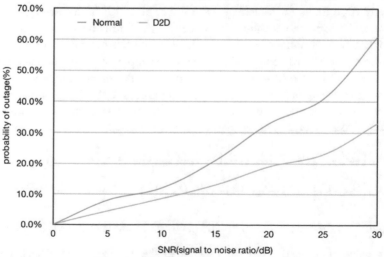

Fig. 2. Probability of outage of the system changes with the increase of SNR.

The Fig. 3 compares the system throughput under different user sizes as the number of channel resources in the system changes. As seen from the Fig. 3, with the increase in the number of resources allocated by the system, as more devices can access the system, the throughput of the system continues to increase. At the same time, the optimal channel allocation strategy adopted by the greedy algorithm based on D2D communication proposed by us has significantly improved the system throughput compared with the ordinary IoT system.

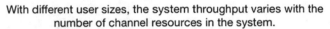

With different user sizes, the system throughput varies with the number of channel resources in the system.

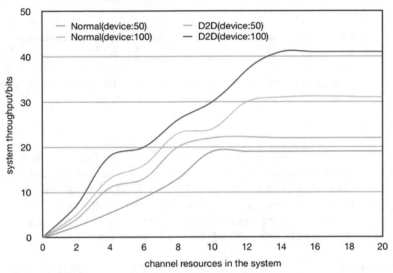

Fig. 3. System throughput varies with the number of channel resources in the system.

The Fig. 4 compares the throughput of D2D multicast group devices with the number of system device resources at different scales. When Devices = 50, when the number of system resources is less than 10, the D2D communication equipment throughput continuously increases with the increase of system resources. When the number of system resources is greater than 10, the throughput will remain unchanged. The same is true for the number of devices with 100, but due to the larger number of devices, there are many D2D communication pairs, so when the system resources become 12, the throughput does not begin to equalize. But overall, system throughput has been greatly improved through our D2D clustering algorithm.

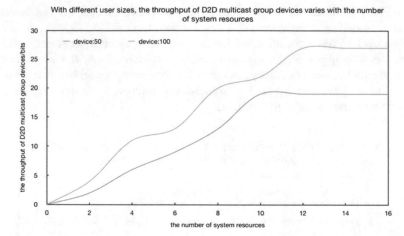

Fig. 4. Throughput of D2D multicast group devices varies with the number of system resources.

5 Conclusion

This paper proposes a coverage enhancement method for cellular IoT based on D2D multicasting communications. The work principle and resources allocation model of the IoT D2D clustering is proposed. And the greedy-based algorithms are given to solve power control and channel assignment problem respectively. The simulation results verify the model and algorithm proposed in this work. The coverage ability and performance of the cellular IoT network system is enhanced.

Acknowledgement. This study is supported by 2017 state grid science and technology project "Research and Application for Adaption of Power Business based on LTE Wireless Private Network".

References

1. Gu, J., Bae, S.J., Min, Y.C.: Dynamic power control mechanism for interference coordination of device-to-device communication in cellular networks. In: Third International Conference on Ubiquitous and Future Network, pp. 71–75 (2011)
2. Lehtomaki, J., Suliman, I., Umebayashi, K., Suzuki, Y.: Teletraffic analysis of direct communication with clustering. IEICE Trans. Fundam. Electron. Commun. Comput. Sci. **E92-A**(5), 1356–1362 (2009)
3. Narottama, B., Fahmi, A., Syihabuddin, B., Isa, A.J.: Cluster head rotation: a proposed method for energy efficiency in D2D communication. In: 2015 IEEE International Conference on Communication, Networks and Satellite, pp. 89–90. IEEE (2015)
4. Yaacoub, E., Ghazzai, H.: Achieving energy efficiency in LTE with joint D2D communications and green networking techniques. In: Wireless Communications and Mobile Computing Conference. IEEE (2013)

5. Yaacoub, E., Kubbar, O.: Energy-efficient device-to device communications in LTE public safety networks. In: International Workshop on Green Internet of Things, Anaheim, CA, USA (2012)
6. Narottama, B., Fahmi, A., Syihabuddin, B.: Impact of number of devices and data rate variation in clustering method on device-to-device communication. In: IEEE Asia Pacific Conference on Wireless and Mobile, Bandung (2015)
7. Koskela, T., Hakola, S., Chen, T., LehtomäKi, J.: Clustering concept using Device-to-Device communication in cellular system. In: Wireless Communications and Networking Conference, Sydney, NSW Australia, vol. 29, no. 16, pp. 1–6 (2010)
8. Eddine, K.T.: A new multilevel inverter with genetic algorithm optimization for hybrid power station application. Optimization and Applications, IEEE (2018)

Storage Optimization Algorithm
for Publication Blockchain

Qingtao Zeng[1,2(✉)], Kai Xie[1], Yeli Li[1], Xinxin Guan[1],
Chufeng Zhou[1], and Shaoping Ma[2]

[1] Beijing Institute of Graphic Communication, Beijing 102600, China
zengqingtao@bigc.edu.cn
[2] Postdoctoral Research Station in Computer Science and Technology
of Tsinghua University, Beijing 100084, China

Abstract. "We Media" is developing rapidly and there is a sharp increase in the number of various electronic publications. Meanwhile, copyright issues between author and publisher are becoming increasingly prominent. To solve this problem, storage optimization algorithm for publication blockchain is based on Pearson similarity algorithm and K-means algorithm. First, the author and publisher association table is built by cooperation record between author and publisher and familiarity calculation method is designed. Subsequently, clustering algorithm is established by using the maximum and minimum principle. On this basis, the prediction algorithm is established. These algorithms are used to adjust Merkle tree structure. Finally, effectiveness of the algorithm is verified by experiments.

Keywords: Publication · Blockchain · Optimization algorithm

1 Introduction

With development of Internet technology, the speed of knowledge renewal and dissemination is becoming faster. In this context, "We Media" is developing rapidly and there is a sharp increase in the number of various electronic publications. Meanwhile, copyright issues between author and publisher are becoming increasingly prominent. In order to solve this problem, blockchain technology is introduced into publishing field. The block chain technology can solve the trust problem between author and publisher without relying on the third party. In publication blockchain, if a node deceives, it will be rejected and suppressed by other nodes, so that there is no need to rely on authority. All transaction records in publishing process are stored in blockchain in the form of trading tickets.

In recent years, with continuous coverage of mobile networks, more and more people are transacting and transmitting electronic publications through mobile networks. However, it is difficult to build trust between author and publisher, and author fears that publisher will not pay royalty after submitting manuscript, or modify the work and publish it in a magazine without the consent of author. Therefore, author and publisher need to guarantee by third parties, but this will generate additional costs. Blockchain technology can gradually replace the third party transactions before,

Q. Liu et al. (Eds.): CENet 2018, AISC 905, pp. 828–835, 2020.
https://doi.org/10.1007/978-3-030-14680-1_90

making transactions simpler and less costly. At the same time, rights and interests of author can be better protected [1–3].

However, the transaction process of block chain needs several stages, broadcasts a lot of information and performs complex validation algorithm. Consequently, a large number of communication bandwidth and computing resources are consumed. And the time of transaction is too long. These factors restrict the application of block chain in publishing field. Therefore, we need to design storage optimization algorithm for publication blockchain (SOPB) to improve trading time.

2 Publication Blockchain

Blockchain uses digital signature to safeguard the rights and interests of author and publisher. The sender in blockchain uses its own private key and public key to verify the following contents: one is to verify integrity of the message through private key and public key. The other is to verify that the message is sent by sender's signature. Data and information in all processes of blockchain is open and transparent. Each transaction can be seen at each node and do not need mutual trust each other. Therefore, there is no identity between these nodes and all the information is anonymous. When author sends an article to publisher, he can use his private key to generate and hash the message. The encrypted digest is sent to publisher as a digital signature together with a written article. Then, publisher first uses the same hash function as the author to calculate message digest from original message received. If two abstracts are the same, the author's public key is used to decrypt digital signature attached to message. Publisher can confirm that digital signature is the sender. Vice versa, for publishers the alliance chain between public chains and private chains can be established. Author can use public chain to intervene freely in blockchain without any control and work freely in publishing.

Blockchain uses a classic hash algorithm in cryptography, which converts input content into a fixed length of digital or alphanumeric output. At the same time, block chains are irreversible, and new blocks can be sorted in chronological order. If a node is used by one person to modify data, it can't affect database of other nodes, unless it can control more than half of the nodes to modify it at the same time. This situation is almost impossible, and the modified node will lead to other nodes. Therefore, if someone wants to tamper with block chain information, it is easy to track, thereby inhibiting the generation and enforcement of related illegal activities. In blockchain, transactions that have been generated cannot be modified and will be recorded in chronological order. Each transaction in blockchain can use cryptography to make two adjacent blocks form serial mode. Therefore, transaction data in blockchain can be found in past and present through cryptographic knowledge. At the same time, hash tree can be used to quickly verify the integrity of large-scale data. In bitcoin or ether network, Merkle tree is used to summarize all transaction information in block, and ultimately generate a uniform hash of all transaction information in the block [4, 5].

3   Storage Optimization Algorithm

For a transaction, it occurs in two accounts. At the same time, mutual transfer between acquaintances is more than stranger. So based on the above two points, author and publisher can be modified in the storage position of Merkle tree. Each account operation in block chain is divided into two accounts for a transaction separately. An account often transfers transactions to a regular account. Such an account is called an associated account. So if leaves of two accounts are in a parent node, then a transaction only needs to locate the location of a parent node and modify leaf nodes under the node, and the cost of changing a data and changing multiple data for each hash operation is the same. This process reduces hash process of one parent to root node in the computing root hash operation. For two accounts under the same parent, the calculation root hash overhead of a transaction is reduced by half (Fig. 1).

Get the cooperation record between author and publisher from the website, and build the association matrix $R_{n \times n}$. We use Pearson similarity algorithm to calculate familiarity between author and publisher.

$$Familiarity_{i,j}^{A} = \frac{\sum\limits_{k \leq n} \left(r_{k,i} - \overline{r_i}\right) \cdot \left(r_{k,j} - \overline{r_j}\right)}{\sqrt{\sum\limits_{k \leq n} \left(r_{k,i} - \overline{r_i}\right)^2} \cdot \sqrt{\sum\limits_{k \leq n} \left(r_{k,j} - \overline{r_j}\right)^2}} \tag{1}$$

In formula (1), $r_{k,i}, r_{k,j} \in R_{n \times n}$ represent the number of cooperation between publisher k and author i and author j. $\overline{r_i}, \overline{r_j}$ indicating average value of $r_{k,i}, r_{k,j}$.

$$Familiarity_{i,j}^{P} = \frac{\sum\limits_{k \leq n} \left(r_{i,k} - \overline{r_i}\right) \cdot \left(r_{j,k} - \overline{r_j}\right)}{\sqrt{\sum\limits_{k \leq n} \left(r_{i,k} - \overline{r_i}\right)^2} \cdot \sqrt{\sum\limits_{k \leq n} \left(r_{j,k} - \overline{r_j}\right)^2}} \tag{2}$$

In formula (2), $r_{i,k}, r_{j,k} \in R_{n \times n}$ represent the number of cooperation between author k and publisher i and publisher j. $\overline{r_i}, \overline{r_j}$ indicating average value of $r_{i,k}, r_{j,k}$.

In order to optimize the storage structure, authors and publishers need to be clustered according to familiarity. Although K-means clustering algorithm can be well classified, K-means algorithm needs input value and initial cluster center. In publication blockchain, selection of the input value and initial cluster center is based on structure of Merkle tree and number of accounts. Therefore, selection of the initial clustering center is particularly important. To solve this problem, the maximum minimum method is mainly used, as shown in formula (3) or formula (4). Firstly, this algorithm selects the farthest samples from adjacent distance that is calculated by formula (1) or formula (2) as initial two points, and the other points are selected according to recurrence. After classification is finished, nodes in the same class are assigned the same parent node.

$$Kernel_{k+1} = Max\{\min\{Familiarity_{i,k+1}\}\}$$
$$i \notin \{Kernel_1, \cdots, Kernel_k\} \tag{3}$$

$$Kernel'_{k+1} = Max\{\min\{Familiarity_{j,k+1}\}\}$$
$$j \notin \{Kernel_1, \cdots, Kernel_k\} \tag{4}$$

When new author or publisher insert into publication blockchain, storage optimization algorithm is based on familiarity and neighbor set N(i) or N(p) of author i or publisher p, which assesses the familiarity of author or publisher, as shown in formula (5).

$$Familiarity'_{p,i} = \lambda \left(\overline{r_i} + \frac{\sum\limits_{k \in N(i)} (r_{p,k} - \overline{r_k}) \cdot (Familiarity_{i,k})}{\sum\limits_{k \in N(i)} |Familiarity_{i,k}|} \right)$$
$$+ (1 - \lambda) \left(\overline{r_p} + \frac{\sum\limits_{j \in N(p)} (r_{p,j} - \overline{r_j}) \cdot (Familiarity_{p,j})}{\sum\limits_{j \in N(p)} |Familiarity_{p,j}|} \right) \tag{5}$$

In formula (5), $Familiarity'_{p,i}$ represents familiarity prediction for author i or publisher p. N(i) or N(p) shows neighborhoods of author i or publisher p, $\lambda \in [0, 1]$. When $Familiarity_{i,k}$ is zero, $Familiarity'_{p,i}$ can be calculate by formula (6). In addition, $Familiarity'_{p,i}$ can be calculate by formula (7), while $Familiarity_{p,j}$ is zero.

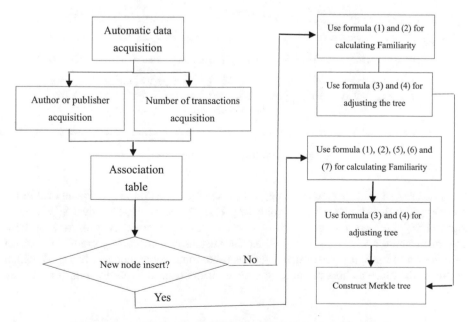

Fig. 1. Storage optimization algorithm for publication blockchain.

$$Familiarity'_{p,i} = \bar{r}_i + \frac{\sum\limits_{k \in N(i)} \left(r_{p,k} - \bar{r}_k\right) \cdot \left(Familiarity_{i,k}\right)}{\sum\limits_{k \in N(i)} \left|Familiarity_{i,k}\right|} \qquad (6)$$

$$Familiarity'_{p,i} = \bar{r}_p + \frac{\sum\limits_{j \in N(p)} \left(r_{p,j} - \bar{r}_j\right) \cdot \left(Familiarity_{p,j}\right)}{\sum\limits_{j \in N(p)} \left|Familiarity_{p,j}\right|} \qquad (7)$$

4 Analysis of Experimental Results

Comparison of RLP(Recursive Length Prefix) algorithm and the SOPB algorithm in this paper between 100 and 1000 readers and publishers is carried out, which shows that SOPB algorithm has shorter execution time (Figs. 2 and 3).

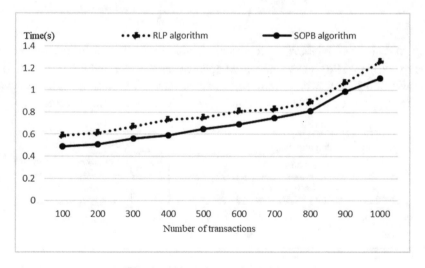

Fig. 2. 100 readers and publishers.

The error of RLP (Recursive Length Prefix) algorithm and the SOPB algorithm for forecasting transaction is seen in following Figs. 4 and 5, which shows SOPB algorithm has lower error rate. In Figs. 4 and 5, Y-axis represents error rate of RLP algorithm and the SOPB algorithm, X-axis represents the size of neighborhood node set of author. In Fig. 5, the performance gain is more remarkable than that in Fig. 4, which is due to the more authors are used, the fewer errors caused by transaction forecasting.

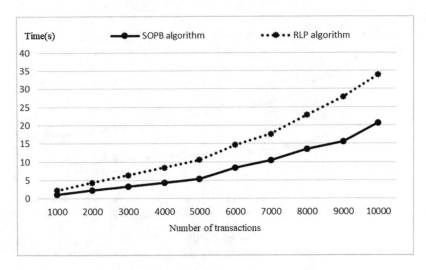

Fig. 3. 1000 readers and publishers.

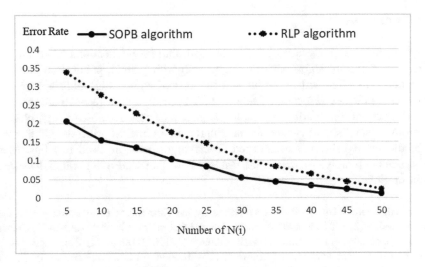

Fig. 4. 100 readers and publishers.

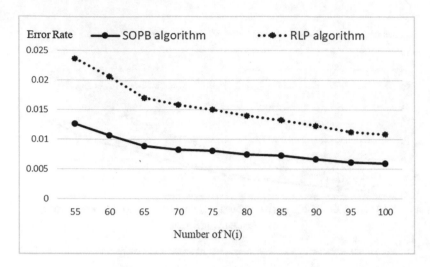

Fig. 5. 1000 readers and publishers.

5 Conclusion

This paper focuses on Storage optimization algorithm for publication blockchain based on Pearson similarity algorithm and *K*-means algorithm. First, publication blockchain is introduced to safeguard the rights and interests of author and publisher. Then, the author and publisher association table is built by cooperation record between author and publisher and familiarity calculation method is designed. Subsequently, clustering algorithm is established by using the maximum and minimum principle. On this basis, the prediction algorithm is established. These algorithms are used to adjust Merkle tree structure. Finally, experimental results show that SOPB algorithm has shorter execution time.

Acknowledgement. This work was supported by the Curriculum construction project-Linux Program design (22150118005/014), doctoral research funding (04190117003/044), Construction of school teachers-doctoral research funding (27170118003/007), Construction of the publication data asset evaluation platform (04190118002/039) and Construction of computer science and technology in predominant construction (22150118010/006).

References

1. Xi, X.B., Lu, C., Cao, W.Z.: A preliminary study of block chain based automated demand response system. In: Proceedings of the CSEE, vol. 7, pp. 1–13 (2017)
2. Wu, M.L., Chang, C.H., Liu, R.Z.: Integrating content-based filtering with collaborative filtering using co-clustering with augmented matrices. Expert Syst. Appl. **41**(6), 2754–2761 (2014)
3. Alain, R.L.: Factors influencing the usage of an electronic book collection: size of the e-book collection, the student population, and the faculty population. Coll. Res. Libr. **1**, 39–59 (2013)

4. Bobadilla, J., Ortega, F., Hernando, A., Gutiérrez, A.: Recommender systems survey. Knowl.-Based Syst. **46**(1), 109–132 (2013)
5. Merkle, R.C.: A digital signature based on a conventional encryption function. In: Conference on the Theory and Applications of Cryptographic Techniques on Advances in Cryptology, vol. 293, no. 1, pp. 369–378. Springer, London (1987)

Improved Hybrid Query Tree Anti-collision Algorithm

Hongwei Deng[1,2,3(✉)], Lang Li[1], and Xiaoman Liang[1]

[1] School of Computer Science and Technology, Hengyang Normal University,
Hengyang 421002, Hunan, China
dhwwhd@163.com
[2] School of Information Science and Engineering, Central South University,
Changsha 410083, Hunan, China
[3] Hunan Provincial Key Laboratory of Intelligent Information Processing
and Application, Hengyang 421002, China

Abstract. An improved algorithm based on the hybrid query tree (HQT) algorithm is proposed in this work. Tags are categorized according to the combined information of the highest bit of collision and second-highest bit of collision. Then, these tags are used to decide the postponement of several delayed-response time slots, to diminish the collision probability. The performance analysis results show that this algorithm is superior to the query tree (QT) and HQT algorithms. It reduces the query time and system communication traffic and improves the tag identification efficiency.

Keywords: Hybrid query tree algorithm · Highest bit of collision ·
Second highest bit of collision · Anti-collision algorithm

1 Introduction

Radio-frequency identification (RFID) technology is one of the key technologies in the perception layer of the RFID system architecture [1]. The RFID system mainly consists of four parts: the reader, electronic tags, RFID middleware, and application system software. The reader and electronic tags accomplish wireless data transmission through a shared communication channel. If two or more tags communicate with the reader simultaneously, and are in the same readable reader range, data collision problems can occur, making the reader unable to identify the tags. A robust RFID-tag anti-collision algorithm is an important part of the RFID system, and a solution for the tag collision problems is vital for improving the performance of the RFID system [2, 3].

Currently, there are two types of tag anti-collision algorithms: nondeterministic algorithms based on ALOHA and deterministic algorithms based on the binary-tree structure [4, 5]. Nondeterministic algorithms include the ALOHA algorithm, slotted ALOHA algorithm, framed slotted ALOHA algorithm, and dynamic framed slotted ALOHA algorithm. However, these algorithms exhibit large randomness and low channel utilization (the ideal state is 36.8%). Moreover, individual tags cannot be identified (starvation phenomenon) [6]. Deterministic algorithms include the binary-search algorithm, dynamic binary-search algorithm, backward binary algorithm, and

Q. Liu et al. (Eds.): CENet 2018, AISC 905, pp. 836–845, 2020.
https://doi.org/10.1007/978-3-030-14680-1_91

query tree (QT) algorithm. These algorithms provide high tag-recognition rates; however, they exhibit high identification delay rates [7].

In this paper, from the information on the highest bit of collision and second-highest bit of collision, we generate the query prefix dynamically, and use it in the hybrid query tree (HQT) algorithm, to reduce the query time and traffic for tag identification.

2 Hybrid Query Tree Algorithm

2.1 Algorithm Instruction Conventions

In order to make full use of the information of the collision bits, thereby reducing the query time and traffic, the following improvements are made to the original algorithm:

(1) Query instruction: REQUEST. There are three types of REQUEST commands, based on the number of parameters-single-parameter, two-parameter, and three-parameter commands. Accordingly, there are three REQUEST instructions: REQUEST(#), REQUEST(Ht, Hr), and REQUEST(p, Ht, Hr). REQUEST(#) is the initial query instruction to which all the tags associated with the reader must respond. When a tag responds to REQUEST(#), Ht and Hr of REQUEST(Ht, Hr) are obtained, which are the highest bit of collision and second-highest bit of collision, respectively. REQUEST(Ht, Hr) requests the tags to calculate a combined HtHr decimal value and store it in the accumulator C. p in REQUEST (p, Ht, Hr) is known as the prefix. The tags use p as the prefix and the highest bit of collision and second-highest bit of collision after the p position, in that order.

(2) Select instruction: SELECT. SELECT(ID) selects a number as the ID tag for the read or write data to prepare.

(3) Read instruction: READ. READ(ID) reads out numbers for ID tags in the data.

(4) Shield instruction: UNSELECT. UNSELECT(ID) puts the tag with that particular ID in an inactive state, so that it does not respond to the REQUEST command.

2.2 Algorithm Description

The proposed algorithm determines the tag whose response to the reader is to be postponed by C time slots, based on the value of XtXr, which is a combination of the highest collision bit Ht and second-highest collision bit Hr. The reader side is provided with two query queues Q_0 and Q_1. Q_0 stores the new query prefix p constituted by the Xt.•••Xr values between the highest collision bit Ht and the second-highest collision bit Hr. Q_1 stores a new combination (Ht, Hr) of the highest collision bit and second-highest collision bit, and its initial value is null. The initial value of the number of accumulator k requests that send the statistics about the used times of command REQUEST is 1. The algorithm process is shown in Fig. 1.

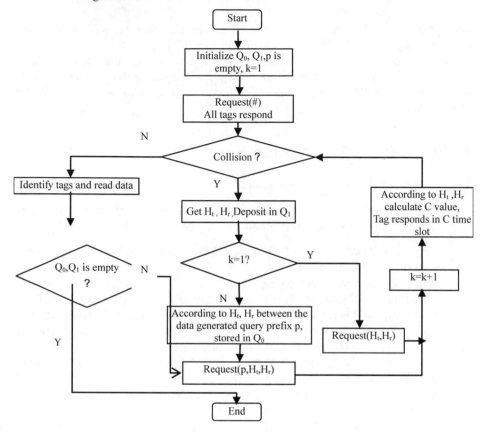

Fig. 1. Algorithm flowchart.

(1) Let us assume that there are N tags for identification in the reader recognition range. The reader first sends REQUEST(#), and N tags respond. The corresponding bit stream is generated according to the Manchester encoding principle. Then, the highest collision bit Ht and second-highest collision bit Hr are acquired, and a new query combination parameter (Ht, Hr) is inserted into the tail of the queue Q_1. The reader takes out (Ht, Hr) from Q_1 to generate the query instruction REQUEST (Ht, Hr) to send to the tag. Accordingly, the tag calculates the Ht, Hr collision bit combination XtXr, and stores its decimal value in the accumulator C. Then, according to the value C, a response is provided in the corresponding time slot.

(2) If only one tag responds in the corresponding time slot, the tag is identified directly. We use the READ instruction to read out the data in the tag, and the UNSELECT instruction shields this tag. In all other cases, the processing moves on to step (3).

(3) If more than one tag responds in a time slot, we acquire the Xt•••Xr value between the highest collision bit Ht and second-highest collision bit Hr, and use this value as a new query prefix p. It is inserted into the tail of the queue Q_0 in the time slot sequence. In this case, the responses of multiple tags in the same time slot

will collide. We can take out the highest collision bit Ht and second-highest collision bit Hr from the collision bit that is behind the original Hr to constitute a new query parameter (Ht, Hr). This is inserted into the tail of the queue Q_1 in the time slot sequence.

(4) The reader obtains the query prefix p and combination parameter (Ht, Hr) from the heads of Q_0 and Q_1, respectively, and generates a new query instruction REQUEST(p, Ht, Hr). This is transmitted to the tags, and the tag whose prefix is p responds. This tag calculates the Ht, Hr collision bit combination XtXr, stores its decimal value in its own accumulator C, and responds according to the value of C in the corresponding time slot.

(5) When the value of Q_0 and Q_1 in the queue is empty, it indicates that the algorithm has identified all the N tags and has finished execution. Otherwise, it returns to step (2).

2.3   Algorithm Example

Here, we explain the algorithm implementation process through an example. Let us assume that there are eight tags in the recognition range of the reader, waiting for identification. The ID numbers of the tags consist of eight digits, as shown in Table 1.

Table 1. Manchester encoding.

Tags	Manchester encoding
Tag1	1 0 1 0 1 1 0 1
Tag2	0 0 1 0 0 1 0 1
Tag3	1 0 1 1 0 1 1 0
Tag4	0 0 0 0 0 1 0 0
Tag5	1 0 0 1 0 0 0 0
Tag6	1 0 0 0 0 1 0 1
Tag7	1 0 0 1 0 1 0 0
Tag8	1 0 1 1 1 1 1 1

The realization of the algorithm is as follows:

(1) Reader initializes the query queues Q_0 and Q_1. Their initial values are null. Furthermore, initialize k = 1.

(2) Reader sends REQUEST(#). The reader identifies all the tags within the scope of the range, according to the principle of Manchester encoding, and obtains the decoding information X0XXXXXX, as shown in Table 2. The highest collision bit is Ht = 8 and the second-highest collision bit is Hr = 6. It stores (Ht, Hr) = (8, 6) in the tail of Q_1 and the number of accumulator k, which counts the number of command REQUESTs is 1. The reader reads (8, 6) from Q_1 and generates a new query instruction REQUEST (8, 6). Then, it sends the command to the tags. Meanwhile, the value of k is incremented by one. Each tag calculates the highest collision bit and second-highest collision bit (8th bit, 6th bit) combination XtXr,

and stores its decimal value in its own accumulator C. Then, according to the value of C in the corresponding time slot, it responds to the reader. For Tag4, XtXr = 00, corresponding to C = 0. For Tag2, XtXr = 01, corresponding to C = 1. These two tags respond and are identified separately in the time slots slot0 and slot1, respectively. For Tag5, Tag6, and Tag7, XtXr = 10, corresponding to C = 2. Therefore, they respond in time slot slot2. For Tag1, Tag3, and Tag8, XtXr = 11, corresponding to C = 3. Therefore, they respond in time slot slot3. Collisions occur in these two cases, as shown in Table 3.

Table 2. Collision bit information.

Position	8 7 6 5 4 3 2 1
Information	X 0 X X X X X X
Tag1	1 0 1 0 1 1 0 1
Tag2	0 0 1 0 0 1 0 1
Tag3	1 0 1 1 0 1 1 0
Tag4	0 0 0 0 0 1 0 0
Tag5	1 0 0 1 0 0 0 0
Tag6	1 0 0 0 0 1 0 1
Tag7	1 0 0 1 0 1 0 0
Tag8	1 0 1 1 1 1 1 1

Table 3. Response time slots of tags.

Ht	Hr	Collision	slot0	slot1	slot2	slot3
8	6	XtXr	00	01	10	11
		Xt0Xr	000	001	100	101
			Tag4	Tag2	Tag5	Tag1
					Tag6	Tag3
					Tag7	Tag8

(3) Table 3 shows that Tag5, Tag6, and Tag7 respond in time slot slot2. Their combination value Xt0Xr is 100 and they collide from the 5th position to the 1st position. The collision information is shown in Table 4. The highest collision bit is Ht = 5 and second-highest collision bit is Hr = 3. The reader generates a new query instruction REQUEST(100, 5, 3), sends it to the tags, and requests the top three (8th, 7th and 6th bit) tags, which are 100, to respond. Tag5, Tag6, and Tag7 calculate the 5th and 3rd collision bit combination XtXr, and store its decimal value in their accumulators. Then, according to the value of C, they respond in the corresponding time slot. For Tag6, XtXr = 01, corresponding to C = 1. For Tag 5, XtXr = 10, corresponding to C = 2. For Tag7, XtXr = 11, corresponding to C = 3. All these tags respond separately and are identified in the time slots slot1, slot2, and slot3, respectively. The identification process is shown in Table 5.

Table 4. Collision bit information

Position	8 7 6 5 4 3 2 1
Information	1 0 0 X 0 X 0 X
Tag5	1 0 0 1 0 0 0 0
Tag6	1 0 0 0 0 1 0 1
Tag7	1 0 0 1 0 1 0 0

Table 5. Tag response time slots.

Ht	Hr	Collisions	slot0	slot1	slot2	slot3
5	3	XtXr	00	01	10	11
		Xt0Xr	000	001	100	101
				Tag6	Tag5	Tag7

(4) Table 3 shows that Tag1, Tag3, and Tag8 respond in time slot slot3. Their combination value Xt0Xr is 101 and they collide from the 5th position to the 1st position. The collision information is shown in Table 6. The highest collision bit is Ht = 5 and second-highest collision bit is Hr = 4. The reader generates a new query instruction REQUEST(101, 5, 4), sends it to the tags, and requests the top three (8th, 7th, and 6th bit) tags, which are 101, to respond. Tag1, Tag3, and Tag8 each calculate the 5th and 4th collision bit combination XtXr, and store the corresponding decimal value in their individual accumulators. Then, according to the value of C, they respond in the corresponding time slot. For Tag1, XtXr = 01, corresponding to C = 1. For Tag3, XtXr = 10, corresponding to C = 2. For Tag8, XtXr = 11, corresponding to C = 3. All the tags respond separately and are identified in the time slots slot1, slot2, and slot3, respectively. The identification process is shown in Table 7.

Table 6. Collision bit information

Position	8 7 6 5 4 3 2 1
Information	1 0 1 X X 1 X X
Tag1	1 0 1 0 1 1 0 1
Tag3	1 0 1 1 0 1 1 0
Tag8	1 0 1 1 1 1 1 1

Table 7. Tags response time slots.

Ht	Hr	Collisions	slot0	slot1	slot2	slot3
5	4	XtXr	00	01	10	11
				Tag1	Tag3	Tag8

(5) The above-mentioned identification process for eight tags, in a tree structure, is shown in Fig. 2. By using the highest collision bit and second-highest collision bit to build a quad-tree with a tree height of two layers, the collision time slots has 3, and free time slot has 2. Compared to the QT and HQT algorithms, it reduces the query time and system communication traffic and improves the algorithm identification efficiency.

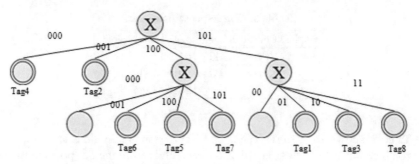

Fig. 2. NHQT algorithm.

3   Performance Analysis

The parameters used in the proposed algorithm are defined as follows:

K: Tag length
N: Number of tags to be queried
Q: Number of query instructions that the reader must transmit
T: Transmission delay
S: Throughput rate

3.1   Analysis of the Number of Query Instructions

In practical applications, the number of tags to be identified is massive. That is, the value of N is very large. When the reader sends the initial command REQUEST(#), the tags can have a K bit collision. The query process is similar to querying a quad tree.

While obtaining the collision bit information, we need to know the collision bit information of at least two bits. If we assume that the highest bit and second-highest bit all have L bits of information, each collision involves $N/2^L$ tags. Let us use one of the collisions as an example. The query process for each collision can be expressed through binary tree traversal operations, where the number of query instructions is the binary tree node and the degree is two. Then, we inquire each collision time slot and assume the number of query instructions sent by the reader, Q_e, to be

$$Q_e = 2 * 2^{K-L} \tag{1}$$

There are four search trees in the corresponding collision time slots, and every tree has 2^{K-L} branches. Q_e represents the search times in each branch. Therefore, the number of query instructions is:

$$Q = 4 * Q_e = 4 * 2 * 2^{K-L} = 2^{K+3-L}$$ (2)

3.2 Transmission Delay Analysis

(1) When starting a query, the reader first broadcasts a K-bit collision query instruction, the tag receives the command, and responds with the K-bit collision bit information. The bit for transmission over a channel is:

$$L_1 = K$$ (3)

(2) The summation of the length of the query instruction, which the reader sends every time, and the length of the tag response is equal to K, that is the tag ID sequence. Therefore, the transmitted data bit length in each collision time slot is equal to the multiplication of the number of the query instructions sent by the reader and the length of the tag ID number.

$$L_d = KQ_e = K * 2 * 2^{K-L}$$ (4)

We can obtain the total number of bits needed to transmit in the channel about search four collision time slots, L_e, as:

$$L_2 = 4 * L_d = 4 * K * 2 * 2^{K-L}$$ (5)

Therefore, the total number of bits required to transmit in the channel, L, is:

$$L = L_1 + L_2 = K + K * 2^{K+3-L}$$ (6)

Let us assume that the data transmission rate in the channel is V bit/s. The formula for the transmission delay is T = L/V. Therefore, the transmission delay will be:

$$T = K * (1 + 2^{K+3-L})/V$$ (7)

3.3 Throughput Rate Analysis

The formula for the throughput rate is:

$$S = (N/Q) * 100\% = N/2^{K+3-L} * 100\%$$ (8)

4   Algorithm Emulation

This work uses MATLAB simulation tools to verify the algorithm through simulations. The simulation results show that, when comparing the NHQT algorithm with the QT and HQT algorithms under the same experimental conditions, the NHQT algorithm significantly improves the performance in terms of the identification delay, recognition time, size of data transferred, and so on, as shown in Fig. 3.

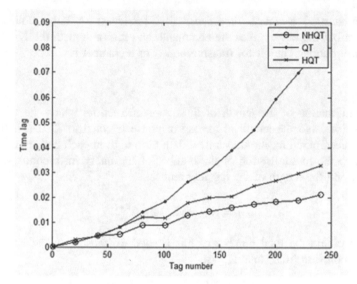

Fig. 3.   Identification delays of three algorithms.

5   Conclusion

The proposed algorithm, based on the HQT algorithm, could generate the query prefix dynamically, from the details of the highest collision bit and second-highest collision bit. It made full use of the known bit information and used a quad-tree, thereby increasing the convergence of judgment and greatly reducing the query time for tag identification and traffic volume. The simulation results showed that this algorithm exhibited significant improvement in the delay recognition performance, when compared with the QT and HQT algorithms.

Acknowledgement.   This research is supported by the National Natural Science Foundation of China (No: 61572174), the Science and Technology Plan Project of the Hunan Province (No: 2016TP1020), the Hunan Province Special Funds of Central Government for Guiding Local Science and Technology Development (No. 2018CT5001) and Subject group construction project of Hengyang Normal University (No. 18XKQ02).

References

1. Song, J.H., Guo, Y.J., Han, L.S., Wang, Z.H.: An adjustive hybrid tree anti-collision algorithm for RFID multi-tag identification. ACTA Electronica Sin. **42**, 685–689 (2014)
2. Zhou, Y.C., Sun, X.C., Gu, J.H.: Research on improved binary anti-collision algorithm. Appl. Res. Comput. **29**, 256–262 (2012)
3. Long, Z.H., Gong, T.F.: Estimation algorithm of radio frequency identification tags based on non-empty slot number. J. Comput. Appl. **36**(1), 101–106 (2016)
4. Zhou, Q., Cai, M.: Improved hybrid query tree anti-collision algorithm in RFID system. Comput. Eng. Des. **33**(1), 209–213 (2012)
5. Nan, J.C., Shan, X.Y., Gao, M.M.: Improved hybrid query tree anti-collision algorithm in RFID system. Comput. Eng. **38**(23), 291–293 (2012)
6. Jiang, W., Yang, H.X., Zhang, Y.: An improved query tree anti-collision algorithm in RFID systems. Comput. Technol. Dev. **25**(2), 86–89 (2015)
7. Yan, S., Shi, C.Q., Chen, R., Chen, Y.H., Zou, Q.: Multi-cycle anti-collision algorithm for RFID tag based on parity packet. Comput. Eng. **42**(2), 312–315 (2016)

Research on GNSS Positioning Aided by SVR

Zhifei Yang[1(⊠)], Tianfeng Yan[1], Yifei Yang[2], and Jinping Qi[1]

[1] School of Electronic and Information Engineering,
Lanzhou Jiaotong University, 730070 Lanzhou, China
yzf@mail.lzjtu.cn
[2] Taiyuan Works Section, Daqin Railway Co., Ltd., 030000 Taiyuan, China

Abstract. In order to solve the problem of the influence of the surrounding environment on GNSS signal leads to the loss of GNSS measurement. Introducing the least squares support vector regression LS-SVR and the monotonicity of geological short-term subsidence deformation, the model of LS-SVR assisted GNSS positioning is established. To explain the model, a GNSS settlement monitoring experiment was adopted. Example analysis shows when GNSS monitoring points are seriously affected by the surrounding area, the LS-GNSS model can obtain more stable positioning results. It eliminates the serious positioning error under the GNSS unlock, achieves the positioning ability under the ideal environment, expands the use range of the GNSS positioning, and improves the positioning accuracy of the GNSS.

Keywords: Measurement · Monitor · SVR · GNSS

1 Introduction

In the monitoring of railway subgrade settlement, the main monitoring methods are traditional manual method, displacement meter method, total station monitoring, GNSS monitoring method and so on. Traditional manual method requires a large amount of manpower and material resources. Displacement meter measurement is characterized by simple structure, convenient installation and low cost. However, the method can only be used for small-scale monitoring and can not meet the large-scale measurement. Although total station monitoring has a high measurement accuracy, it cannot eliminate the problem of high dispersion under manual measurement. GNSS monitoring method can meet the wide range of regional measurement and satisfy the high precision. PPP is adopted to improve the positioning accuracy of GNSS [1]. However, the method is enslaved to the topography of the monitoring site, and there is the risk of satellite unlock. The positioning problem of GPS unlock is compensated by established an algorithm combined with inertial navigation [2, 3]. However, the model is applied to high-speed motion vector, and the positioning accuracy is the meter level, which can not meet the high precision monitoring requirements for subgrade settlement and deformation. This paper presents a new monitoring method with SVR aimed GNSS named LS-GNSS to solve the problems above. Using the geologic evolution characteristics that settlement or deformation is slow to serious, the method can effectively compensate for the monitoring of GNSS unlock by SVR regression.

© Springer Nature Switzerland AG 2020
Q. Liu et al. (Eds.): CENet 2018, AISC 905, pp. 846–854, 2020.
https://doi.org/10.1007/978-3-030-14680-1_92

2 Deformation and Settlement Based on SVR

SVM [4] is widely used in Classification and regression of time Series [4–7].

2.1 Basic SVR Regression

The deformation curves of the monitoring values usually are nonlinear in the GNSS monitoring. The function is summarized as:

$$f(x) = \omega\varphi(x) + b \tag{1}$$

The essential theory of SVR regression is mapping the input data X to the high dimensional feature space and using the nonlinearity function to do linear regression.

A sample set is given as $A = \{(x_i, y_i)|x_i \in R, y_i \in R\}$, where x_i is the variable value of the input and y_i is the corresponding output.

Using SVR to study sample data, the sample set is divided into a 0/1 model by Logistic regression function and a hyperplane is finded. The function of the hyperplane can be explained as:

$f(x) = \omega^T \varphi(x) + b$, where $\omega^T \varphi(x)$ is the inner product of vector quantity ω^T and $\varphi(x)$. The dimension of ω^T is the dimension of characteristic space.

According to the optimality condition of the dual problem.

$$\omega^T = \sum_{i=1}^{N} \alpha_i \cdot \varphi^T(x_i) \tag{2}$$

Introduce the kernel function $k(x, x_i) = \langle \varphi(x_i), \varphi(x) \rangle = \varphi^T(x_i) \cdot \varphi(x)$, the hyperplane function can be revised to the ultima regression model

$$f(x) = \sum_{i=1}^{N} \alpha_i \cdot k(x, x_i) + b \tag{3}$$

the gaussian kernel function [5, 6] $k_{rbf}(x, x_i) = exp\left[-\lambda|x - x_i|^2\right]$ is introduced, λ is the reciprocal value of nuclear radius.

According to statistical theory [5], the optimization problem is as follows:

$$\min_{\alpha, b} \frac{1}{2}\|\alpha\|^2 + C\sum_{i=1}^{N}(\xi_i + \xi_i^*) \tag{4}$$

$$\text{st.} \begin{cases} y_i - \alpha^T K(x, A) + b \le \varepsilon + \xi_i \\ \alpha^T K(x, A) + b - y_i \le \varepsilon + \xi_i \\ \xi_i, \xi_i^* \ge 0, i = 1, \dots, N \end{cases}$$

2.2   The Analysis of Deformation and Settlement Based on SVR

The geological deformation model is studied in three-dimensional direction. The monotonicity principle is met at each direction. Therefore, the model of monotonic knowledge SVM is established.

For one dimensional single variable regression. The first order nonnegative difference equation is constructed directly from the kernel regression function in matrix form [6] $f(x) = \alpha^T K(x, A) + b$. (Take the monotone decline as an example.)

$$f(v_m + \Delta h) - f(v_m) \leq 0 \Rightarrow \tag{5}$$

$$\alpha^T [K(v_m + \Delta h, A) - K(v_m, A)] \leq 0$$

f is the kernel regression function. $v_m, v_m + \Delta h$ is the vector quantity with m discrete points, which are the corresponding points of geological settlement. Δh is the increment value relative to v_m, which is the deformation of each position.

Therefore, the optimization problem is revised as follows:

$$\min_{\alpha, b} \|\alpha\|_1 + C \sum_{i=1}^{N} (\xi_i) \tag{6}$$

$$\begin{cases} y_i - \alpha^T K(x, A) + b \leq \varepsilon + \xi_i \\ \alpha^T K(x, A) + b - y_i \leq \varepsilon + \xi_i \\ \alpha^T [K(v_m + \Delta h, A) - K(v_m, A)] \leq 0 \\ \xi_i \geq 0, i = 1, \cdots, N \end{cases}$$

2.3   SVR Recursive Process for Deformation and Settlement

In the process of GNSS, the short-term sliding window regression of LS-SVM is proposed in order to estimate the truth-value of the monitoring point. As shown in the Fig. 1:

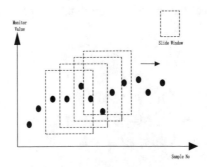

Fig. 1. Slide window processes.

A SVR supported short-time regression model is established for predicting. As shown in the Fig. 1, the slide window includes a series of resent values. Each time, an observation is obtained, the sliding window is postponed. A new observation enters into the window and the oldest observation gets out. In the way. The effective combination of sliding windows and SVR can capture most recent dynamic behavior and track short-term changes. Therefore, the accuracy of the overall forecast is significantly improved.

2.4   SVR Aided GNSS Positioning Model

In order to eliminate the influence of positioning data noise, it is assumed that the positioning data noise distribution is normal distribution. SVR is used for regression of short-term small sample historical data, then the velocity, displacement and other factors of the monitoring point are calculated. At the same time, according to the monotonicity of the deformation and displacement of the monitoring point, the trajectory calculation is carried out by inertial navigation position analysis theory. Finally, the deformation and settlement value of the monitoring point is solved. The model is as follows (Fig. 2):

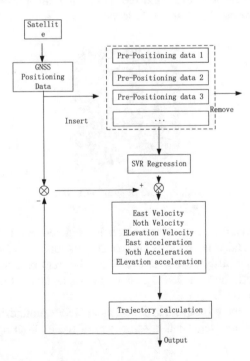

Fig. 2. LS-GNSS model.

In the model, GNSS positioning data is used not only for regression analysis of SVR, but also for feedback calibration of positioning results.

The model steps are as follows:

(1) Initialize data and build slide window.
(2) Use the improved SVR model to analysis of data source in the slide window.
(3) Calculate the track velocity, acceleration and location information based on regression curve.
(4) Do Trajectory calculation.
(5) Compare the output result with GNSS positioning value, then feedback the bias into the model.
(6) Slide window forwards a step.
(7) Repeat 2 to 6 until data is over.

3   Experiment

The pre-monitoring deformation data are used to verify the effectiveness of LS-SVR aided GNSS positioning model. The data sets are from March 2015 to November 2015 in Lanzhou north railway station. The project lasted for more than half a year, and the settlement monitor cycle was 24 h, and the signal sampling period was 1 s.

The track monitoring charts are as follows:

Fig. 3. The monitoring point.

Figure 3 shows that the monitoring point is located between two lanes, affected by the parking of train, and it is height is shorter than the train's. Thus, the signal of the monitoring point is seriously obscured, and When the train passes, the track vibration will cause severe interference to GNSS positioning, so, results of the GNSS positioning is very discrete or unfixed.

In the above case, there is severe gross errors in GNSS monitoring, which seriously affects the analysis of the deformation settlement of the monitoring points.

3.1 Data Processing

In the experiment, the monitoring data sets are analyzed by SVR. Set $\{x_0, x_1, \ldots, x_9\}$ is as a training sample of SVR, kernel function is selected as gaussian kernel function:

$$k_{rbf}(x, x_i) = exp\left[-\lambda|x - x_i|^2\right] \tag{7}$$

The SVR regression parameter $C = 0.92, \lambda = 10$. The experimental results are shown in Figs. 4 and 5:

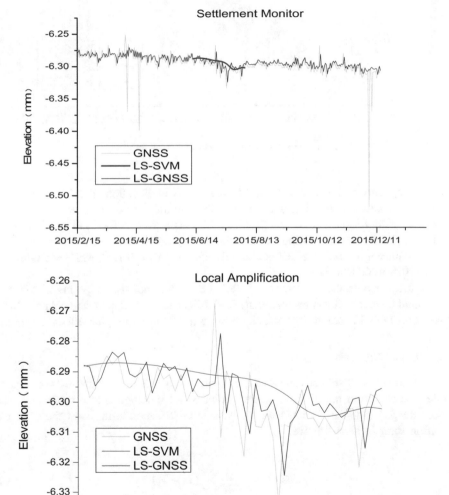

Fig. 4. Elevation monitor.

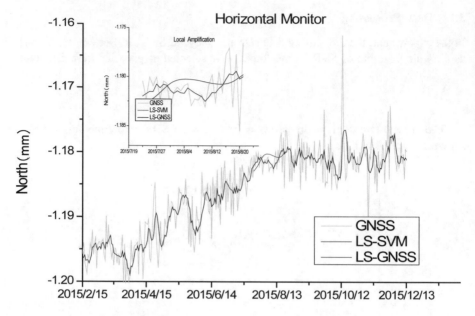

Fig. 5. Horizontal monitor-north and local amplification.

Figure 4 shows the curves of GNSS, LS-GNSS in elevation. The LS-SVM curve shows the smooth regression analysis of GNSS points, so as to obtain the current monitoring speed and other information. The data were processed by LS-GNSS without affecting the monitoring trend of monitoring data. It can be seen that LS-GNSS has a significant inhibitory effect compared with the previous two, while excluding the gross errors in GNSS monitoring.

Figure 5 shows the analysis of GNSS, LS-GNSS monitoring points in terms of horizontal direction. It can be seen that, LS-GNSS can well adapt to the changes in the direction of GNSS monitoring, which shows a good approximation effect.

3.2   Error Analysis

In order to verify the effectiveness of the algorithm, the three algorithms are compared under the condition that the monitoring point is not moving. The root means square error (RMSE) is generally used for analysis in GNSS monitoring, and the error calculation formula is as follows:

$$RMSE = \sqrt{\frac{1}{N}\sum_{i=1}^{N}(\hat{Y}_i - Y_i)^2} \qquad (3.2)$$

A total of 50 sets of data were formed, and the analysis results were shown in Table 1 below.

Table 1. Analysis of error data of monitoring point immovable

Data/RMSE	GNSS	LS-GNSS	Date time
1	−7.6258	−7.6291	2015/9/2
2	−7.6272	−7.6290	2015/9/3
3	−7.6263	−7.6275	2015/9/4
.	.	.	.
49	−7.6260	−7.6260	2015/10/20
50	−7.6274	−7.6264	2015/10/21
Average	−7.6277	−7.6276	
RMSE	0.0013	0.0008	

In table, the two models are in accordance on average. However, in terms of RMSE, LS-GNSS regression analysis is more effective than GNSS.

The simulation results show that LS-GNSS can reflect the deformation settlement of monitoring points more accurately than GNSS and improve the environmental adaptability of GNSS. Meanwhile, in the experiment, the GNSS Model needs about ten minutes, but LS-SVM only needs dozens of seconds, thus, time consuming of regression of LS-SVM is far less than that of GNSS, which can be negligible, the model of LS-GNSS time consumption is approximately equal to GNSS's.

4 Conclusion

In this paper, in the case of serious distortion of GNSS positioning in bad environment, a new model (LS-GNSS) combined SVR regression analysis by using the short-term monotonicity prior knowledge of deformation and settlement of monitoring points is put forward. The experiment shows that under the condition of not changing the GNSS measurement trend, the model (LS-GNSS) can well complete the high precision analysis of the short-term trajectory of the monitoring point and not impact on GNSS monitoring trends, then improve the interference ability and the applicability of GNSS post-processing. Error analysis shows that the measurement accuracy of the system is approximately doubled in the model.

Acknowledgement. This study is supported by Gansu science and Technology Department (17YF1FA122); Lanzhou Science and Technology Bureau (2018-1-51); School youth fund of Lanzhou Jiaotong University (2015008).

References

1. Alkan, R.M., İlçi, V., Murat Ozulu, İ., Saka, M.H.: A comparative study for accuracy assessment of PPP technique using GPS and GLONASS in urban areas. Measurement **65**, 1–8 (2015)
2. Li, J., Song, N.F., Yang, G.L., Li, M., Cai, Q.Z.: Improving positioning accuracy of vehicular navigation system during GPS outages utilizing ensemble learning algorithm. Inf. Fusion **35**, 1–10 (2017)
3. Zhou, J., Chen, M.H., Wu, H.J.: Research on plane dead reckoning based on inertial navigation system. Comput. Sci. **44**(6A), 582–586 (2017)
4. Wang, H.B., Wang, Y., Hu, Q.H.: Self-adaptive robust nonlinear regression for unknown noise via mixture of Gaussians. Neurocomputing **235**, 274–286 (2017)
5. Zhang, Q., Yan, X.F.: Improved support vector regression algorithm combining with probability distribution and monotone property. Control Theory Appl. **34**(5), 671–676 (2017)
6. Bao, C.R.: Applying nonlinear generalized autoregressive conditional heteroscedasticity to compensate ANFIS outputs tuned by adaptive support vector regression. Fuzzy Sets Syst. **157**(13), 1832–1850 (2006)
7. Bai, P.: Support Vector Machine and Its Application in Mixed Gas Infrared Spectrum Analysis. Xi'an Electronic Science and Technology University Press, Xi'an (2008)
8. Zhou, J.Z., Huang, J.: Multiple kernel linear programming support vector regression incorporating prior knowledge. J. Autom. **37**(3), 360–370 (2011)

Interference Emitter Localization Based on Hyperbolic Passive Location in Spectrum Monitoring

Yixuan Wang[1(✉)], Zhifei Yang[1], Rui Gao[1], Jianhui Yang[1],
Tianfeng Yan[1], and Qinghua Hu[2]

[1] School of Electronic and Information Engineering,
Lanzhou Jiaotong University, Lanzhou 730070, China
royl7007@126.com
[2] School of Computer Science and Technology,
Tianjin University, Tianjin 300072, China

Abstract. Interference signals can always be found during spectrum monitoring, which has a serious impact in the regular use of radio business [1]. Sometimes is difficult to shied it by suppress signal, so it is becoming increasingly important to find the location of the interference emitter [2]. This paper proposed an effective technique in interference emitter localization based on intersections of hyperbolic curves defined by the time differences of arrival of a signal received at three monitoring stations. The approach is noniterative and gives an explicit solution. In the end, this paper token a field test based on the hardware platform, the final result is given.

Keywords: TDOA · Spectrum monitoring · Hyperbolic · Passive location

1 Introduction

Frequency band of the radio business have continued to expand, and the radio spectrum resource becoming more and more valuable in recent years. However, the radio business often encountered an increasingly serious interference, which has natural interference and artificial interference, they all caused serious problems to radio business, especially the artificial interference, it has greatly influence to people's daily work routine.

Passive location technology has a very good development for its good concealment ability. They include time difference of arrival localization technology (TDOA), angle of arrival localization technology (AOA), frequency difference of arrival localization technology (FDOA) [3]. Multiple station case of TDOA localization technology has largely use because its high localization accuracy and low requirements to the receiving system [4]. Due to these advantages, TDOA playing an important role in interference signal localization.

© Springer Nature Switzerland AG 2020
Q. Liu et al. (Eds.): CENet 2018, AISC 905, pp. 855–865, 2020.
https://doi.org/10.1007/978-3-030-14680-1_93

2   Three Stations Localization Algorithm

2.1   Hyperbolic Passive Location Based on TDOA

Three stations are distributed as shown in Fig. 1. The coordinates of three receiving station is R1(x1, y1), R2(x2, y2) and R3(x3, y3), the coordinates of emitter is E(x, y), the set of TDOA measurements equations is [3]:

$$\begin{cases} c(t_2 - t_1) = \sqrt{(x_2 - x)^2 + (y_2 - y)^2} - \sqrt{(x_1 - x)^2 + (y_1 - y)^2} = d_{21} \\ c(t_3 - t_1) = \sqrt{(x_3 - x)^2 + (y_3 - y)^2} - \sqrt{(x_1 - x)^2 + (y_1 - y)^2} = d_{31} \end{cases} \qquad (2.1)$$

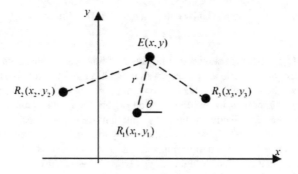

Fig. 1. Sketch map of receiving stations and signal emitter.

Where c be the signal propagation speed, is the time delay between station 2 and station 1, is the time delay between station 3 and station 1, and are the distance differences.

$$\begin{cases} x = r\cos\theta + x_1 \\ y = rsin\theta + y_1 \end{cases} \qquad (2.2)$$

From (2.2), so that (2.1) can be rewritten as

$$\begin{cases} -2x_{21}r\cos\theta - 2y_{21}r\sin\theta = d_{21} - x_{21}^2 - y_{21}^2 + 2d_{21}r \\ -2x_{31}r\cos\theta - 2y_{31}r\sin\theta = d_{31} - x_{31}^2 - y_{31}^2 + 2d_{31}r \end{cases} \qquad (2.3)$$

$$\begin{cases} x_{21} = x_2 - x_1 \\ x_{31} = x_3 - x_1 \\ y_{21} = y_2 - y_1 \\ y_{31} = y_3 - y_1 \end{cases} \qquad (2.4)$$

So,

$$r = \frac{x_{21}^2 + y_{21}^2 - d_{21}^2}{2(d_{21} + x_{21}\cos\theta + y_{21}\sin\theta)} = \frac{x_{31}^2 + y_{31}^2 - d_{31}^2}{2(d_{31} + x_{31}\cos\theta + y_{31}\sin\theta)} \tag{2.5}$$

From (2.5), we obtain

$$a\cos\theta + b\sin\theta = c \tag{2.6}$$

Where

$$\begin{cases} a = (d_{31}^2 - x_{31}^2 - y_{31}^2)x_{21} - (d_{21}^2 - x_{21}^2 - y_{21}^2)x_{31} \\ b = (d_{31}^2 - x_{31}^2 - y_{31}^2)y_{21} - (d_{21}^2 - x_{21}^2 - y_{21}^2)y_{31} \\ a = -(d_{31}^2 - x_{31}^2 - y_{31}^2)d_{21} - (d_{21}^2 - x_{21}^2 - y_{21}^2)d_{31} \end{cases} \tag{2.7}$$

The final solution is

$$\theta = \arcsin\frac{c}{\sqrt{a^2 + b^2}} - \Phi(a, b) \tag{2.8}$$

or

$$\theta = \pi - \arcsin\frac{c}{\sqrt{a^2 + b^2}} - \Phi(a, b) \tag{2.9}$$

$\Phi(y, x)$ is the phase of vector (x, y).

2.2   Method to Choose the Correct Solution

From conclusion above, we can indicate that the final result of the localization will have least two solutions, one of them must be correct, while the other one is wrong which needs to remove. In order to analysis the pattern of two points, do the following simulation [5]:

Fist assume that the three monitoring stations are fixed, and the interference emitter uniformly step through the whole area (interference emitter move with a fixed step), then use Eqs. (2.8) and (2.9), switch the resultand r to Cartesian coordinates, at last the result is two sets of localization point. As shown in Fig. 2:

Under ideal conditions, if the results of two equations are both correct, the distribution of interference emitter location calculated by equations should be as same as location of interference emitter, at every point of the area. But we can know from the Fig. 2A, when use Eq. (2.8), no matter how much time delay was given, it is impossible to locate the interference emitter in the area which enclosed by vertical angle and its corresponding extension lines of triangle sides. We name this area vertical angle area. However, in the non vertical angle area, the location calculated by equation is as same as interference emitter's location. In this scenario, we need special attention to the area which near the extension lines, there are a lot of error locations distribute, while the correct locations of those error ones should be in the vertical angle area. In Fig. 2B,

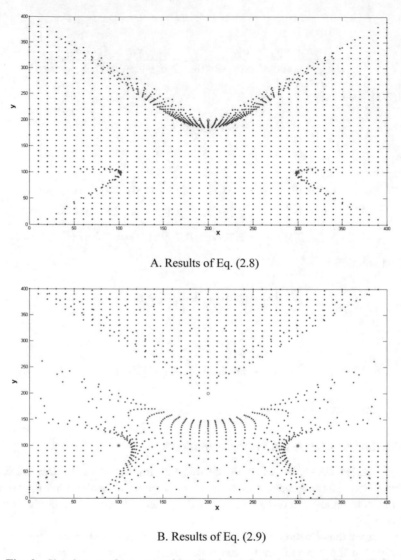

A. Results of Eq. (2.8)

B. Results of Eq. (2.9)

Fig. 2. Sketch map of two sets of localization point using Eqs. (2.8) and (2.9).

when we use Eq. (2.9), it has a good localization performance with only little disturb of error locations in the vertical angle area. However, the performance of localization in the non vertical angle area is not that good.

In order to verify this, following simulation has made:

Continue to assume that the three monitoring stations are fixed, for the sake of exclude the effect of different distribution of stations may bring, took 4 different distributions as shown in Fig. 3. The interference emitter also uniformly step through the whole area, and use Eqs. (2.8) and (2.9) at same time, then switch the result of and r to Cartesian coordinates, and gives a distance threshold Δd. If the distance between

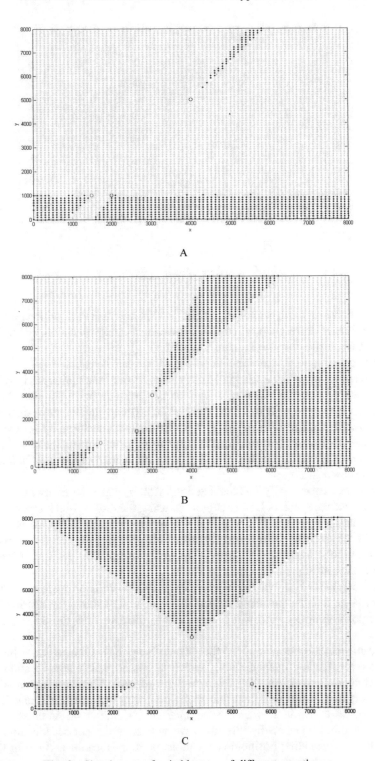

Fig. 3. Sketch map of suitable area of different equations.

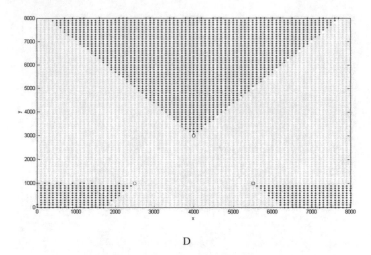

D

Fig. 3. (*continued*)

correct location and result calculated by Eq. (2.8) is less than Δd, we draw a light point in the location of corresponding correct interference emitter location. Instead, draw a dark point when the distance is more than Δd.

The simulation results are shown in Fig. 3, as can be seen, even in the case of distribution is extreme acute triangle, final results fit our assumption very well. So we can propose a hyperbolic localization method that exclude fuzzy solutions based on a three station localization: In the most general case (three stations have distribution of triangle), switch three stations into a triangle, the area that enclosed by vertical angle of the arbitrary angle and corresponding two extension lines are suitable for Eq. (2.9) to calculate, the other part are suitable for Eq. (2.8). Flow chart of the specific localization method is shown in Fig. 4.

In order to further verify this localization method, make the following simulation:

Under the assumption of former condition, using the localization method mentioned in Fig. 4, and also gives a threshold that distance between real location and calculate location, if the distance is less than threshold we consider the result is correct, or is incorrect. In the simulation, the value of threshold is 1×10^{-3} m. Localization effect diagram shows in Fig. 5.

Figure 5(A) shows the distribution of all correct positioning locations which have shorter distance than the threshold we just given, Fig. 5(B) shows the distribution of all incorrect positioning location which have longer distance than the threshold we just given.

We can conclude from Fig. 5(A) that, there always have an area that is blank when we use the localization method above, which means no matter how location of inter-ference emitter changes, we will never have our estimated location in this area. After further analyzing, it can be found that the reason why those area is blank is that the incorrect equation was used when the real interference emitter was in those blank area. If we use another equation by this time, we will have the correct location. What is worse, the corresponding positioning locations of those locations which in the blank

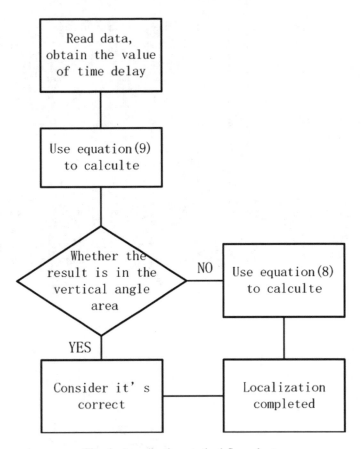

Fig. 4. Localization method flow chart.

area are in the vertical angle area, that's a pollution to the vertical angle area. And we found that when the positioning locations obtained by the method above are in the the vertical angle area, the closer the location to the extension lines of triangle, the greater probability of location are incorrect.

For these problems, we can use the following method to improve it:

(1) Evaluate the final localization result, if the result is in the vertical angle area and close to the extension lines of station triangle, we use another equation to calculate it, if the result is in the blank area, we consider the first result is wrong, use the second to replace it as the final result.

(2) Design a grid model of station distribution, we could use the difference assemble of station to avoid the having incorrect location. At the same time, we could also avoid the Influence of non line of sight (NLOS) may bring.

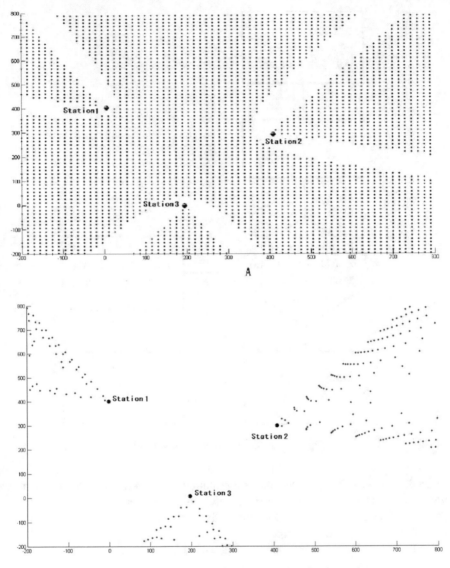

Fig. 5. Diagram of correct and incorrect localization point.

3   The Field Test

3.1   The Hardware Environment

In the field test, we use the new generation of broadband radio receiver based on software defined radio (SDR) WDDC, PFB architecture of digital down convert and Multi-channel broadband signal processing unit developed by digital signal processing and software defined radio research institute of LanZhou JiaoTong University. Based on this, we added time synchronization module, guarantee the synchronization of each receiver.

3.2   Data Acquisition

Data acquisition software developed based on the LabWindows/CVI. The type of transmitting signal is FSK, sampling rate is 3.125 MHz. Each data package has the size of 8250 bits, with 58 bits GPS data and 8192 bits I/Q data [7].

The latitude and longitude of three receivers are: Receiver 1: N36.0759293, E103.4325505; Receiver 2: N36.0639057, E103.4357563; Receiver 3: N36.0689753, E103.4378592.

3.3   The Localization Results of Field Test

In Here we analysis one set of data, and take receiver 1 as the main receiver, use the generalized cross correlation (GCC) to acquire the time delay [8, 9], then obtained differences of distance of receiver 1 and receiver 2, receiver 3:

The points in Figs. 6 and 7 represents difference of distance, Y-axis represents the ordinal numbers of package, and has collected for about 5 min, that has about 300 packages.

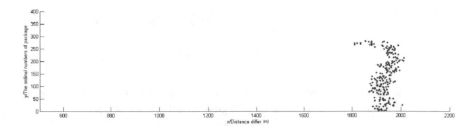

Fig. 6. Distance differ of receiver 1 and receiver 2 from TDE.

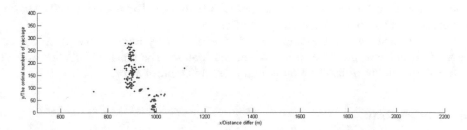

Fig. 7. Distance differ of receiver 1 and receiver 3 from TDE.

The average distance difference from the one set of data is: d12 = 1930 m, d13 = 920 m. While the true distance difference is: d12 = 2019 m, d13 = 879 m. As can be seen, the average and the true distance difference is less than 5%.

We use every result from each package to calculate the location with the method we proposed. Figure 8 shows the results of all three minutes (every second have one result).

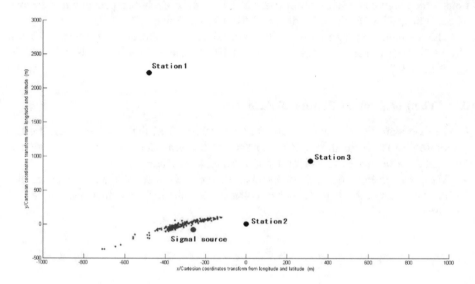

Fig. 8. Diagram of field test result.

We can obtain that the location calculated by the equation and the true location is quite close, and the average distance can be guaranteed within 70 m.

3.4 Error Analysis

Errors caused by hardware are basically constant, we can correct those errors though lots of tests with the guarantee of stability with hardware. The fluctuation of distance difference in Figs. 6 and 7 are this kind of errors.

Error caused by the algorithm can only be correct by improve the algorithm or analyzing and screening the results of the calculation. The zonal distribution which affect the accuracy of localization in Fig. 8 are this kind of error.

4   Conclusion

Hyperbolic passive location is a classic method, it has a quick calculation speed, without iterative computing, combined with the time delay estimation of generalized cross correlation could ignore the type of signal that the emitter send. In this paper, we consider the characteristics of the interference that we meet in the spectrum monitoring, and proposed a method to choose the correct solution among the multiple solutions based on the distribution feature of different solutions of hyperbolic localization method in passive location, and have a satisfied result in both the simulation and field test.

Acknowledgement. Thanks to the experimental data provided by the Institute of Digital Signal Processing and Software-Defined Radio, Lanzhou Jiaotong University. In addition, this work was supported by Youth Science Foundation of Lanzhou Jiaotong University under Grant No. 2018003, No. 2018028, No. 2015009, Opening Foundation of Key Laboratory of Opto-technology and Intelligent Control (Lanzhou Jiaotong University) Ministry of Education (KFKT2018-16), Innovation Fond Project of Lanzhou Jiaotong University and Tianjin University under Grant No. 2018062, Scientific Research plan projects of Gansu Education Department under Grant No. 2017C-09, Lanzhou Science and Technology Bureau under Grant No. 2018-1-51.

References

1. Li, J.S., Shao, Y.B., Long, H.: Research on wireless interference source localization based on grid spectrum monitoring. Computer Science (2017)
2. Bull, J.F., Ward, M.L.: Interference detection, characterization and location in a wireless communications or broadcast system. In: IEEE International Conference on Communications, pp. 2979–2984 (2015)
3. Hu, L.Z.: Passive Location. National Defense Industry Press, Beijing (2004)
4. Chang, Y.T.: Simulation and implementation of an integrated TDOA/AOA monitoring system for preventing broadcast interference. J. Appl. Res. Technol. **12**(6), 1051–1062 (2014)
5. Zhang, Z.M., Li, H.W.: TDOA localization method and analysis. Electron. Warfare **5**, 19–24 (2000)
6. Chan, Y.T., Ho, K.C.: A simple and efficient estimator for hyperbolic location. IEEE Trans. Signal Process. **42**(8), 1905–1915 (2002)
7. Xu, B.X., Xiao, Y.: The Mathematical Transformation and Estimation Method in Signal Processing. Tsinghua University Press Ltd., China (2004)
8. Jiang, X.: Research of Passive Location Technology and Application. University of Electronic Science and Technology (2008)
9. Knapp, C., Carter, G.: The generalized correlation method for estimation of time delay. IEEE Trans. Acoust. Speech Signal Process. **24**(4), 320–327 (2003)

Research and Application of Improved Genetic Algorithm in Lanzhou Self-service Terminal Patrol System

Jiangwei Bai[1], Yi Yang[1,2(✉)], and Lian Li[1]

[1] School of Information Science and Engineering, Lanzhou University,
Lanzhou 730000, China
yy@lzu.edu.cn
[2] Silk Road Economic Belt Research Center, Lanzhou University,
Lanzhou 730000, China

Abstract. During the 2016–2017 period, Lanzhou government deployed 5000 self-service terminals throughout the city [1]. In order to ensure the normal operation of these devices, a patrol team of about 20 people was organized to check the operating status of the devices and repair the faulty every day. However, due to the wide distribution, and large quantity of devices and the frequent drainage of patrol personnel, the patrol task cannot be completed scientifically and efficiently. Most employees arrange the patrol sequence of devices on rules of thumb so that the efficiency of the patrol work cannot be further improved. In this paper, we made three new improvements to the genetic algorithm, such as using Greedy Ideas to generate initial population, combinating of superior group retains and roulette strategy and superior offspring to stop mutation. And we use the genetic algorithm to design the daily patrol path, ensuring that the patrol work of the devices can be conducted scientifically and efficiently.

Keywords: Genetic algorithm · Path planing · Patrol system

1 Introduction

Nowadays, self-service terminals can be seen everywhere in today's financial industry, hospitals, retail shop, and etc., but their maintenances are particularly challenging. And many people tried to solve this kind of problem by using genetic algorithm. Li used Greedy Ideas to generate initial population to reduce the time for better solutions [2]. To speed up the convergence of results, a team researching robot path planing used dynamic mutation rate and crossover rate [3]. Using greedy ideas to generate initial population can not speed up the convergence rate. Using dynamic crossover probability and mutation probability can easily lead to a local optimal solution when the population is large.

The issue is clearly exposed in the project of Lanzhou self-service terminals, and this paper tries to solve this problem by using improved genetic algorithm. First, we further improve the initial population generation algorithm based on the research results of others to make it adapt to current issues. Second, we regard combination of superior

© Springer Nature Switzerland AG 2020
Q. Liu et al. (Eds.): CENet 2018, AISC 905, pp. 866–875, 2020.
https://doi.org/10.1007/978-3-030-14680-1_94

group retains and roulette strategy as select operator. In this way, we ensure that superior group can be reserved regardless of the mutation rate and crossover rate. Third, we will stop mutation when superior offsprings appear. So the intermediate results are preserved and accelerate the speed of convergence.

The content of this paper is organized as follows. Firstly, this paper analyzes the patrol problem in a mathematical way, then improve the genetic algorithm according to the actual scene. Finally, we use the improved genetic algorithm to design the daily patrol path, ensuring that the patrol work can be conducted scientifically and efficiently, thereby improving patrol efficiency and reducing terminal maintenance costs.

2 Mathematical Model

2.1 Problem Description

The number of self-service terminals in Lanzhou is approximately 5,000. A patrol team of about 20 people is responsible for inspections, one inspector for one region. The deployment of self-service terminals in the city is shown as Fig. 1:

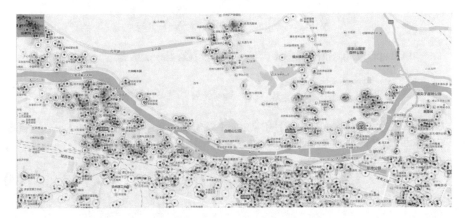

Fig. 1. The deployment of self-service terminals in Lanzhou.

During work hours, the inspector starts from his residence and inspects the device in the area under his jurisdiction. After completing one-day task, he can go home. On the next day, he continues to inspect the devices that has not been patrolled. After a few days, all devices will be inspected to once, i.e. a complete parol process ends. This paper names this issue as the patrolling traversal problem (PTP).

2.2   Mathematical Analysis

In this paper, the inspector's residence is marked as h, the number of devices that the inspector is responsible to as n, the device itself as m, and the inspector's devices as m1, m2, m3 … mn.

So, the entire patrol process can be described as: in a complete graph, there are n points, m1, m2, m3, …, mn. Starting from point h outside the graph, traversing Ni (Ni ≪ n) points and returning to point h, the traversal process is recorded as (t1, t2, t3 … tNi). Assuming the looping reaches k (0 < k < n) times, then all points are traversed to once, and i (1 ≤ i ≤ k) is the number of times starting from point h. As each traversal starts from point h, so h repeatedly traversed k times. The formula is shown as:

$$N1 + N2 + \ldots + Nk = n + k \tag{2.1}$$

During the patrol process, patrolling more device with the least cost is most important. Here we use the cost function V to describe the price that inspectors have to pay, which in this paper only refers to the walking distance between nodes. Then the conclusion is:

$$V(t1, t2, t3) = V(t1, t2) + V(t2, t3) \tag{2.2}$$

The kth inspection, the number of devices is Nk, the cost is:

$$Vk = \sum_{j=N_1+N_2+\cdots+N_{k-1}+1}^{N_1+N_2+\cdots+N_{k-1}+N_k} V\left(t_j, t_{j+1}\right) + V\left(t_1, t_{N1+N2+\cdots+Nk}\right) \tag{2.3}$$

Obtained by the formulas (2.1), (2.2), and (2.3), the cost which is paid to complete the inspection of all devices is [3]:

$$V = \sum_{j=1}^{n+k} V\left(t_j, t_{j+1}\right) + \sum_{i=1}^{k} V\left(t_1, t_{N1+N2+\cdots+Ni}\right) \tag{2.4}$$

2.3   Mathematical Description

There are n nodes of m1, m2, m3 … mn, a node h outside the graph, and the weight between nodes in a complete graph as shown in Fig. 2. Starting from point h, after traversing a certain number of nodes, we return to point h and loop through k times to

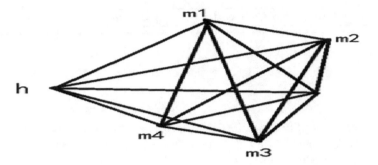

Fig. 2. The relations between the inspector and the devices.

complete traversal of all nodes. So we should find the traversal order (t1, t2, t3, t4, t5…
tn + k) to make the traversal cost meet our expectations [4]:

$$\sum_{j=1}^{n+k} V(t_j, t_{j+1}) + \sum_{i=1}^{k} V(t_1, t_{N1+N2+\cdots+Ni}) \leq \delta \qquad (2.5)$$

3 Solutions

The PTP in this paper is a typical combinational optimization problem. And the number
of devices that each inspector is responsible for is about 200. So all the permutations
and combinations results will have 200!. So deterministic algorithms is very unrealistic.
But non-deterministic algorithms have a great advantage in this type of problem [5].
Genetic algorithms have been widely used to figure out approximate optimal solutions
for combinational optimization problems [6]. In this paper, we will use genetic algorithms to design path for the patrol process, and regard the cost of inspecting as the
individual's fitness. Combined with the actual situation, three improvements are made
to the genetic algorithm to improve the convergence speed and the final result performance of the algorithm.

3.1   Using Greedy Ideas to Generate Initial Population

Individual Coding
In this paper, the device number is used to code the chromosome and the length of the
chromosome is n (the total number of devices that are not inspected). The number
consists of 7 characters, which can not be duplicated in a chromosome. A valid
chromosome is as Fig. 3 [7]:

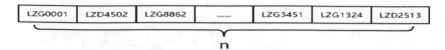

Fig. 3. A valid chromosome.

Step

(a) Initialize origin[i] (the initial value is the inspector's residence h) and subRoute[i] [avgCheck], length[i] (initial value is 0); i = ceil(n/avgCheck)

(b) Generate a random number r (0 ≤ r ≤ i)

(c) If length[r] == avgCheck, skip to step (b), else perform step (d)

(d) Finds the nearest device which is not inspected so far from origin[r]. The device number is represented by devNo, origin[r] ← devNo, subRoute[r][length[r]] ← devNo, length[r] ← length[r] + 1, and mark this device as "visited"

(e) Skip to step (b) until all points have been visited.

Complexity Analysis

The basic operation of the above algorithm is step (d). Generating a chromosome requires this step n (n is the number of devices which the inspector responsible for) times. If the scale of the population is "s", then we need to execute the algorithm "s" times. So the time complexity of this algorithm is o(sn).

The length of an individual chromosome is n. But the length which is actually calculated is n + k (k = ceil(n/avgCheck)) when calculating the individual fitness. An individual fitness is the distance to sequentially traverse devices in the chromosome shown in Fig. 4. The location data of the devices are from the Lanzhou self-service terminal database which uses latitude and longitude to determine the device location. Before calculating the fitness, we need to calculate the walking distance between the two devices based on latitude and longitude. Here we use the Baidu map API to perform the conversion. The request method is as follows: http://api.map.baidu.com/routematrix/v2/walking?output=json&origins=40.45,116.34&destinations=40.34,116.45.

Fitness Calculation

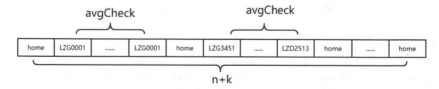

Fig. 4. The chromosome used to calculate fitness.

3.2   Combination of Superior Group Retains and Roulette Strategy

Description

The elite group retains 1/10 of the size of the population. This 1/10 is the competition between the offspring and the father, and the remaining 9 out of 10 come into being in the offspring by roulette strategy. The advantage is that when the variation is very large, the algorithm also has a good stability. In this way, we not only retains quality groups, but also ensures that the diversity of the groups will not be greatly reduced.

Fitness Function

The level of fitness is expressed in the order of 105. So using 1/fitness as the fitness function directly will reduce the precision and get a poor result. Here we use relative values. The lowest inspection cost is recorded as "best". Then the fitness function of this generation is [8]:

$$1/(\text{fitness - best} + 1) \tag{3.1}$$

3.3   Superior Offspring to Stop Mutation

Description

(a) Use dynamic mutation times. Generating random numbers of mutations makes that it have stronger neighborhood search capabilities
(b) Calculate the fitness of the intermediate results after a mutation. If it is better than the optimal individuals in this population, the mutation is stopped immediately

Steps

(a) Randomly generate mutation times r (0 < r < avgCheck/2), and calculate the optimal individual fitness in the current population as parameter best.
(b) Randomly generate swaps for k1 and k2, swapping the k1 and k2 positions in the chromosome.
(c) Calculate the fitness of the new individual and compare it with parameter best. If fitness < best, the mutation is stopped, then perform step (d), otherwise repeat steps (a), (b), (c) r times
(d) Return the new individuals generated to the population.

3.4   Experiment Results

In this paper, we use Greedy Ideas to generate initial population, use "Order Crossover (OX)" as crossover operator, "Superior Offspring to stop Mutation" as mutation operator, "Combination of superior group retains and roulette strategy" as select operator.

In the experiment, we choose a inspector responsible for 200 devices (Chromosome length), and he can inspect 20 devices a day on average (avgCheck). At the same time, we set the probability of mutation and crossover in each individual to 0.3 (Mutation rate) and 0.8 (Crossover rate) respectively. The number of individuals in offspring population is 100 (Population size). Then we take the best individual after 1000 generations as the final result (Genetic algebra).

The parameters used during the test are shown in Fig. 5:

Parameter	Value
Population size	100
Crossover rate	0.8
Mutation rate	0.3
Genetic algebra	1000
Chromosome length	200
avgCheck	20

Fig. 5. Parameters used in the test.

The results of the algorithm before and after optimization are shown in Fig. 6:

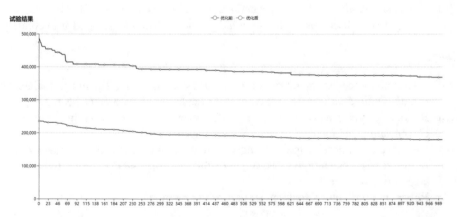

Fig. 6. Experiment results.

From the perspective of the algorithm alone, the performance of the optimization was improved by 50% compared with that before the optimization. However, in the actual patrol process, the inspection order arranged by the inspector with his/her own experience is actually based on greedy thoughts before using this system. The next device which the inspector will inspects is the nearest from the current position, but this will make the cost of inspections higher day by day. According to the experimental results, the inspectors will save about 40 km, and the time to complete each round of inspection will be shortened by about 1.5 days by using this system.

4   Related Functions Pages

Related functional pages are shown in Figs. 7, 8, 9, and 10. Figure 7 shows the patrol order. The blue dot indicates that the device has been inspected, and the red dot indicates that the device is not inspected. Figure 8 is the inspection form page. When a device is inspected, the inspector needs to fill in the form and the corresponding equipment status will change into "finished" after submission. Figures 9 and 10 is the process of calling the local map to navigate to the next device when starting the inspection [9].

Fig. 7. Patrol route.

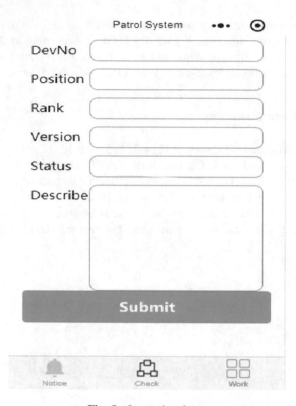

Fig. 8. Inspection form.

5   Conclusion

In this paper, an improved genetic algorithm is used to analyze, calculate, and plan the patrol process. Considering the operating efficiency and final results, the population scale is 100, the crossover rate is 0.8, the mutation rate is 0.3, and the genetic algebraic is 1000. At this time, the initial planning takes an average of 120 s. Completing each round of patrol according to experimental simulation data, the inspection staff can save about 40 km. If the population size and genetic algebra are further expanded, the performance of the final result can be further improved. However, as the performance period is limited, it cannot be put into use. Therefore, there is room of further optimization for the algorithm.

Acknowledgements. This study is supported by Science and Technology Innovation Project of Foshan City, China (Grant No. 2015IT100095), the Fundamental Research Funds for the Central Universities (Grant No. lzujbky-2016-br03), CERNET Innovation Project (Grant No. NGII20150603) and Science and Technology Planning Project of Guangdong Province, China (Grant No. 2016B010108002).

References

1. The new practice of governing the country - the Gansu chapter. Serving the last mile and more provinces and cities to praise the Lanzhou 3D digital platform. (in Chinese). http://news.cctv.com/2017/01/17/ARTIgIPwNiaEgyDb4hp5M1kb170117.shtml. Accessed 17 Jan 2017–7 May 2018
2. Li, J.W.: Research on Logistics Distribution TSP. Harbin Institute of Technology (2017). (in Chinese)
3. Wang, L., Li, M., Cai, J.C., Liu, Z.H.: Research on application of improved genetic algorithm in path planning of mobile robot. Mech. Sci. Technol. Aerosp. Eng. 36(05), 711–716 (2017). (in Chinese)
4. Liu, S.T., Ling, J., Xiao, P.: The computational complexity of TSP. Mod. Comput. 06, 6–9 (2001). (in Chinese)
5. Chen, L., Pan, D.Z.: Improved genetic algorithm to solve TSP problem. Intell. Comput. Appl. 6(05) (2016). (in Chinese)
6. Jiang, R.: Application of improved genetic algorithm in TSP. Softw. Guide 15(12), 127–129 (2016). (in Chinese)
7. Zhao, X.C., Guo, S.: Genetic algorithm for solving multiple traveling salesman problem with relative solution space. CAAI Trans. Intell. Syst. 1–9 (2018). (in Chinese)
8. Chen, G., Deng, Y.: Some new methods of constructing fitness function in feature selection of genetic algorithms and their applications. Mech. Sci. Technol. Aerosp. Eng. 30(01) (2011). (in Chinese)
9. Zhang, G.J., Wu, J.F., Liu, L.F., Peng, Z.F.: Digital community learning map development based on WeChat applet. J. Guangzhou Univ. (Soc. Sci. Ed.) 16(11), 57–63 (2017). (in Chinese)

An Optimized AES Masking Method for Resisting Side Channel Analysis

Ge Jiao[1,2(✉)], Lang Li[1], and Yi Zou[1]

[1] College of Computer Science and Technology, Hengyang Normal University,
Hengyang 421002, Hunan, China
jiaoge@126.com

[2] Hunan Provincial Key Laboratory of Intelligent Information Processing
and Application, Hengyang 421002, Hunan, China

Abstract. In order to against the side channel analysis attack such as power, electromagnetic waves, and time attack, an optimized masking method is proposed for Advanced Encryption Standard (AES) algorithm in this paper. This scheme adopts random hamming distance mask and offset randomization strategy in the first two rounds, the last round and two rounds randomly selected in the middle of the AES algorithm to ensure the security of each intermediate value. The scheme then adopts fixed-value mask strategy in the five rounds in the middle, which can reduce the time and space consumption to recalculate S box, improve the efficiency of the algorithm and reduce the production cost. To compare with other defence methods, we build a power analysis platform and conduct extensive experiments. The experimental results show that our optimized AES masking method is more secure than the compared methods, and it is able to resist second-order correlation power analysis (CPA) attacks.

Keywords: Masking · AES · Resisting side channel analysis

1 Introduction

In 1999, Kocher et al. proposed the differential power analysis (DPA) method at the CRYPTO conference [1, 2]. The proposed method has made the security of the crypto chip enormous challenged. Side Channel Attack has posed a serious threat to the implementation of the AES algorithm. Mask technology is the most representative method of algorithm layer protection. The core idea is that during the execution of the algorithm, all intermediate results are hidden by performing arithmetic operations on the intermediate results and random numbers, which make the attackers fail to select intermediate results that can be attacked. Before the end of the execution, remove the mask and restore the correct output value to achieve the purpose of anti-side channel attacks.

Kocher et al. first proposed using random mask techniques to defend against DPA attacks [1, 2]. In 2001, Itoh et al. first proposed a fixed valued masking algorithm. The

© Springer Nature Switzerland AG 2020
Q. Liu et al. (Eds.): CENet 2018, AISC 905, pp. 876–884, 2020.
https://doi.org/10.1007/978-3-030-14680-1_95

core idea of the algorithm is to pre-calculate multiple sets of masks and their corresponding modified S-boxes, then store them in ROM, and randomly select a group of masks during execution, which can reduce the processor load and RAM occupancy of the resource-constrained crypto chip [3]. However, fixed value mask scheme cannot resist second-order differential power attack. In 2007, Mangard et al. put forward a kind of common hardware mask scheme, which can protect all processing unit of AES encrypted circuit, including adding random mask to the input value and output value which possibly leak, so as to achieve the aim of resist the DPA attacks [4]. In 2012, Nassar et al. proposed a rotating S-box masking(RSM) algorithm for AES, which can effectively resist variance-based power attack (VPA) and second-order zero-offset correlation power analysis (CPA), but the method still has a first-order leakage [5]. In 2014, Ding et al. proposed a statistical model for high-order DPA attacks on masked devices, further demonstrating that high-order DPA is the best choice to deal with low signal-noise ratio (SNR) power curves [6]. In 2017, Zhang et al. improved the efficiency of the Coron's scheme [7] by decreasing the random generations according to modifying each pair of intermediate values and reusing some randomness. The application of the AES inversion circuit mask proves that the proposed scheme is significantly better than the original one [8].

2 Optimization and Implementation of Random Mask Based on Hamming Distance

2.1 Random Mask Optimization

The Rotating S-Box Masking scheme is one of the most popular mask schemes currently discussed in smart card protection applications. Although this algorithm has fixed the second-order leakage of the fixed-valued mask scheme, there is an obvious first-order leakage of the algorithm because of the loophole in the choice of the random mask and the logical design of the algorithm. This paper considers the security factor of the random mask algorithm and optimizes the random mask. The random mask is generated with the Hamming distance as the seed. In information theory, the Hamming Distance represents the number of different characters in the corresponding positions of two equal-length strings. We denote the Hamming distance between the strings x and y by $d(x, y)$.

In the random mask generation design, the Hamming distance is a random seed. For each 8-bit mask value, the possible Hamming distance range is [0, 8]. The random mask generation procedure is as follows:

(1) Determine the number of random masks n;
(2) Using [0, 8] as seeds, randomly generate n different Hamming distance values HD_i, $HD_i \in [0, 8]$, $i \in [0, n-1]$, and randomly generate n different values V_i, $V_i \in [0, 255]$, $i \in [0, n-1]$;
(3) According to $HD_i = d(V_i, m_i)$, a specific 8-bit mask value m_i is generated by HD_i and V_i;
(4) Repeat Step (3) until all 8-bit mask values are generated.

2.2 Random Mask Implementation

The specific implementation of random mask generation uses the ith encryption as an example to set the hamming distance as a random seed. Since the masking scheme is for a 128-bit AES algorithm, 16 random masks are needed. First, randomly generate 16 hamming distance values and integers, which are denoted as HD_i = {4, 8, 4, 0, 7, 0, 2, 2, 5, 3, 6, 6, 1, 4, 8, 1}, V_i = {124, 48, 51, 10, 160, 151, 53, 130, 150, 48, 254, 72, 223, 115, 200, 96} then call the ProduceMask() function to generate a mask, denoted as M_i = {0x8C, 0xCF, 0xC3, 0x0A, 0x5E, 0x97, 0xF5, 0x42, 0x6E, 0xD0, 0x02, 0xB4, 0x5F, 0x83, 0x37, 0xE0}, as shown in Fig. 1. For example: HD_0 = 4, V_0 = 01111100B, bitwise negation from the (8-HD_0)th bit of V_0 to get m_0 = 10001100B.

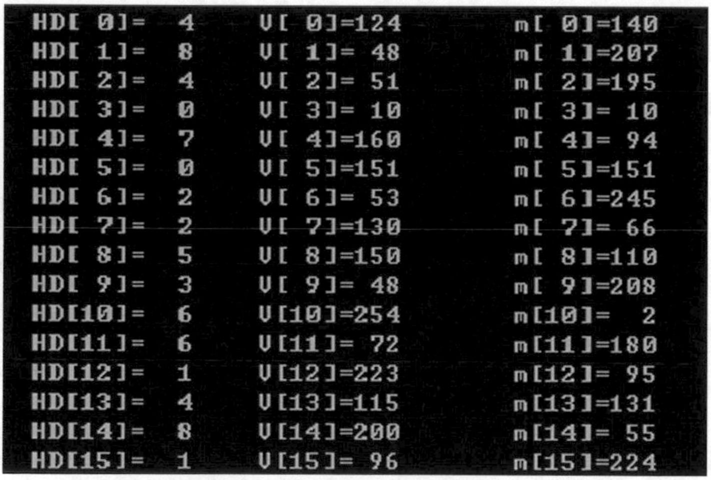

```
HD[ 0]=   4      V[ 0]=124      m[ 0]=140
HD[ 1]=   8      V[ 1]= 48      m[ 1]=207
HD[ 2]=   4      V[ 2]= 51      m[ 2]=195
HD[ 3]=   0      V[ 3]= 10      m[ 3]= 10
HD[ 4]=   7      V[ 4]=160      m[ 4]= 94
HD[ 5]=   0      V[ 5]=151      m[ 5]=151
HD[ 6]=   2      V[ 6]= 53      m[ 6]=245
HD[ 7]=   2      V[ 7]=130      m[ 7]= 66
HD[ 8]=   5      V[ 8]=150      m[ 8]=110
HD[ 9]=   3      V[ 9]= 48      m[ 9]=208
HD[10]=   6      V[10]=254      m[10]=  2
HD[11]=   6      V[11]= 72      m[11]=180
HD[12]=   1      V[12]=223      m[12]= 95
HD[13]=   4      V[13]=115      m[13]=131
HD[14]=   8      V[14]=200      m[14]= 55
HD[15]=   1      V[15]= 96      m[15]=224
```

Fig. 1. Random mask generation.

The C++ code for improved random masking is as follows:

```
#include <iostream.h>
#include <iomanip.h>
#include "stdlib.h"
#include  <time.h>
int ProduceMask(int hd,int v);
void main()
{
  srand((unsigned)time(NULL));
  int HD[16];
  int V[16],m[16];
  for(int i=0;i<=15;i++)
  {
    V[i]=rand() % 256;
    HD[i]=rand() % 9;
    m[i]=ProduceMask(HD[i],V[i]);
      cout<<"
HD["<<setw(2)<<i<<"]="<<setw(3)<<HD[i]<<"\t"<<"V["<<setw(2)<<i<<"]="<<set
w(3)<<V[i]<<"\t"<<"m["<<setw(2)<<i<<"]="<<setw(3)<<m[i]<<endl;
  }
}
int ProduceMask(int hd,int v)
{
int xorExp=0;
int m;
int pos=8-hd;
for(int i=pos;i<8;i++)
{
    xorExp|=(1<<i);
}
m=v^xorExp;
return m;
}
```

3 S-Box Mask Optimization

The key of implementation of AES algorithm based on random mask is to mask the only non-linear transformation, subByets. During the byte replacement operation, the original S-box transformation cannot guarantee the correct implementation of the mask protection, and the new S-box needs to be recalculated according to the mask. Normally, the new S-box S_{new} and the original S-box need to satisfy the Eq. 1:

$$S_{new}(x \oplus m) = S(x) \oplus m \tag{1}$$

For a 128-bit AES algorithm, each round of cryptographic operations requires the calculation of 16 new S-boxes. The same as the normal mask operation, the first step is to preprocess the input of the S box, that is, remove the mask added in the previous round key XOR operation; then perform the byte replacement operation and add a new mask. The specific steps are as follows:

(1) According to the design of Chapter 2, 16 random masking constants m_{0-15} are generated, as shown in Eq. (2). Using these constants as mask seeds, the mask values needed for each round of encryption are obtained by converting these 16 constants.

$$Mask_0 = \{m_0, m_1, m_2, \ldots, m_{15}\} \tag{2}$$

(2) For each round of SubBytes operations, 16 new S-boxes need to be recalculated based on the value of the subkey and mask.

$$S_{new\_j}(P\prime) = S(P\prime \oplus Mask_{Offset\_j}) \oplus Mask_{(Offset\_j+1 \bmod 16)}$$
$$P\prime = PlainText \oplus (RoundKey \oplus Mask_j) \tag{3}$$

In Eq. (3), $Offset\_j$ ($j \in [0, 15]$) represents the offset of the S-box, which is random and satisfies a uniform distribution and is generated randomly before the algorithm is run. The randomization of the S-box offset guarantees the unpredictability of the offset and can successfully resist the offset-based first-order CPA attack.

(3) Each round of encryption operations requires a different random mask. As described in the first step, the specific mask value will be obtained by transforming 16 mask seeds. The transformation rules are as follows:

$$Mask_j = \left\{ m_j, m_{(j+1) \bmod 16}, m_{(j+2) \bmod 16}, \ldots, m_{(j+15) \bmod 16} \right\} j \in [1, 15] \tag{4}$$

The 16 sub-masks in the random mask sequence $Mask_{0-15}$ can be mapped one by one according to the value of $Offset\_j$. The random mask selection order is indirectly generated by the mask seed when it is encrypted, which is determined by $Mask_j$. When the encryption round iterates, the mask elimination operation is $Mask_{(Offset\_j+1 \bmod 16)}$ and the added mask is $Mask_{(Offset\_j+2 \bmod 16)}$.

4 Mask Scheme Optimization

The existing AES mask protection mainly focuses on the first round and the last round, so it cannot resist high-order DPA attacks. In order to resist high-order DPA attacks, all potential leaks need to be masked, but this will make the algorithm run less efficiently. Therefore, from the point of view of algorithm operation efficiency, the overall mask protection scheme is improved. This scheme adopts random hamming distance mask and offset randomization strategy in the first two rounds, the last round and two rounds randomly selected in the middle of the AES algorithm to ensure the security of each intermediate value. The scheme then adopts fixed-value mask strategy in the five rounds in the middle, which can reduce the time and space consumption to recalculate S box, improve the efficiency of the algorithm; the correct cipher text can be output after the last round of mask compensation operations. The overall process of the improved design plan is shown in Fig. 2. In this way, the overall mask protection scheme can guarantee the security of the AES algorithm on the smart card, and it can also reduce the protection cost to some extent.

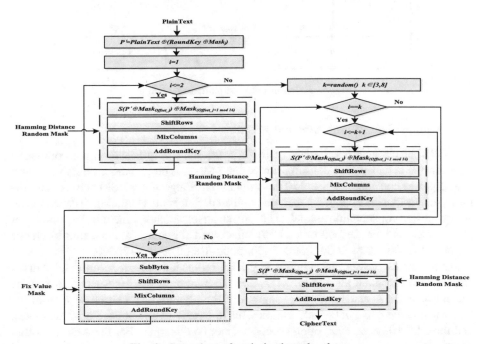

Fig. 2. Procedure of optimized mask scheme.

5 Experimental Results and Analysis

5.1 Experiment Platform

The hardware components of the experimental platform include the SAKURA-G side-channel attack development board, PicoScope oscilloscopes and PC [9]. The SAKURA-G development board consists of two Xilinx Spartan-6 FPGA chips. One is the main FPGA and the other is the control FPGA. The ROM of the development board has been written into the officially provided AES circuit and control circuit, and will be automatically loaded after power-on FPGA chip. The relationship between the hardware modules of the experimental platform is shown in Fig. 3.

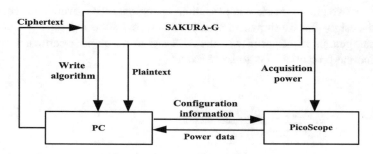

Fig. 3. Relation between hardware modules.

5.2 Comparison of Experimental Results

The CPA attack analyzes the key of the cryptographic chip based on the correlation between the actual power consumption value of the cryptographic chip and the simulated power consumption value. The ordinary AES algorithm, RSM scheme and the mask scheme designed in this paper are compared. Select the Hamming distance model as the power consumption model. The power attack methods include the first-order CPA attack and the second-order CPA attack and the attack sample number range includes 0-1000, 1000–2000, 2000–5000, 5000–10000, 10000–20000.

Experiments proves that without any AES protection measures, only 128 curves can be used to obtain all 128-bit keys, and the correct rate is 100%. Figure 4 is the second-order CPA attack analysis diagram of the algorithm. It can be seen from the figure that in the range of 0–5000 samples, all 16-byte keys do not have obvious spikes, and the AES 10-round encryption process cannot be discerned. For the AES algorithm of the mask protection scheme designed in this paper, no attack domain can be selected for any of the AES encryption keys to achieve a correct value. Using the mask protection scheme designed in this paper can effectively resist the second-order CPA attack and ensure the key security of the encryption algorithm in the smart card.

According to the small sample size (1000–5000) and the large sample size (10000–20000), the operating efficiency of each program is calculated several times and averaged. The experimental results are shown in Table 1. From the results of the encryption efficiency, it can be seen that when the number of collected samples is small

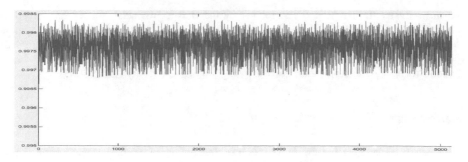

Fig. 4. Second-order CPA attack analysis.

Table 1. Efficiency of algorithms.

Sample size	AES (min)	RSM (min)	[10] (min)	Ours (min)
1000	2	42	45	44
2000	4	86	87	84
5000	8	210	200	181
10000	16	421	410	362
20000	32	867	839	757

(1000 to 5000), the mask scheme and the RSM scheme designed in this paper are close to each other in terms of the operating efficiency of the algorithm, which is far higher than that of ordinary AES. High security will inevitably lead to efficiency. When the number of collected samples is large (10000 to 20000), Compare the mask scheme designed in this paper with the RSM scheme and the literature [10], the time used is greatly reduced, and the efficiency is obviously improved. It is proved that the mask scheme designed in this paper can reduce the protection cost and improve encryption efficiency to some extent.

6 Conclusion

Taking the resource-constrained smart card device as the research object, based on the analysis and research of the existing mask algorithm, an improved mask defense algorithm is proposed. The algorithm has considered security and resource overhead comprehensively, and is a kind of high security and easy to implement masking defense algorithm, and can resist first-order and second-order CPA attacks. The proposed masking scheme is theoretically resistant to high-order CPA attacks, and subsequent work can be verified through experiments.

Acknowledgement. This study is supported by the Hunan Provincial Natural Science Foundation of China (2017JJ2010), the Scientific Research Fund of Hunan Provincial Education Department (16B039), the Science and Technology Plan Project of Hunan Province (2016TP1020), Open Fund Project of Hunan Provincial Key Laboratory of Intelligent Information Processing and Application for Hengyang Normal University (IIPA18K03).

References

1. Kocher, P., Jaffe, J., Jun, B.: Introduction to differential power analysis and related attacks (1998). www.cryptography.com/resources/whitepapers/DPATechInfo.pdf
2. Kocher, P., Jaffe, J., Jun, B.: Differential power analysis. In: Advances in Cryptology-CRYPTO 1999, pp. 388–397. Springer, Heidelberg (1999)
3. Itoh, K., Takenaka, M., Torii, N.: DPA countermeasure based on the "masking method". In: Information Security and Cryptology-ICISC 2001, pp. 440–456. Springer, Heidelberg (2002)
4. Mangard, S., Oswald, E., Popp, T.: Power Analysis Attacks: Revealing the Secrets of Smart Cards. Graz University of Technology, pp. 1–306, Springer, Austria (2007)
5. Nassar, M., Souissi, Y., Guilley, S., Danger, J.L.: RSM: a small and fast countermeasure for AES, secure against 1st and 2nd-order zero-offset SCAs. In: Design, Automation and Test in Europe Conference and Exhibition-DATE 2012, pp. 1173–1178. IEEE, Dresden (2012)
6. Ding, A.A., Zhang, L., Fei, Y., Luo, P.: A statistical model for higher order dpa on masked devices. In: Proceedings of CHES 2014, pp. 147–169. Springer, Berlin (2014)
7. Coron, J.S., Prouff, E., Rivain, M., Roche, T.: Higher-order side channel security and mask refreshing. In: International Workshop on Fast Software Encryption, pp. 410–424. Springer, Heidelberg (2013)
8. Zhang, R., Qiu, S., Zhou, Y.: Further improving efficiency of higher order masking schemes by decreasing randomness complexity. IEEE Trans. Inf. Forensics Secur. **12**(11), 2590–2598 (2017)
9. Jiao, G., Li, L., Zou, Y.: Research on power attack comprehensive experiment platform based on SAKURA-G hardware circuit. In: Proceedings of the 2017 The 7th International Conference on Computer Engineering and Networks, Shanghai, pp. 343–349 (2017)
10. Xu, P.: Research and Implementation with Mask Technology on AES Encryption Module of Smartcard against Side Channel Attack. Chongqing University (2015). (in Chinese)

Research on the Confirmation Mechanism of SCPS-TP Protocol in Satellite Network

Yueqiu Jiang[✉], Liyuan Yang, Qixue Guan, and Shijie Guan

School of Information Science and Engineering,
Shenyang Ligong University, Shenyang, China
missjiangyueqiu@sina.com

Abstract. In order to improve the bandwidth utilization of the satellite communication link and the performance of the SCPS-TP transmission protocol, an improvement of TCP Vegas algorithm is proposed in this paper. For the satellite link bandwidth asymmetry, the proposed algorithm distinguishes the relative queue delay congestion state on the satellite network reverse link. The forward and reverse link states are subdivided and correspondingly improved in the congestion avoidance phase. On this basis, the reverse link acknowledgement packet transmission frequency is dynamically adjusted. The OPNET simulation result shows that the algorithm improves the satellite link throughput when the bit error rate is high. The algorithm has low complexity and obvious improvement in effect.

Keywords: Satellite communication · SCPS-TP · Confirmation frequency · OPNET

1 Preface

With the continuous development of space network technology, satellite communication network has become one of the most popular communication methods in modern times with its unique advantages [1]. It has the advantages of wide coverage, real-time monitoring and strong scalability, especially in places where the environment is harsh, sparsely populated and the disaster areas after the earthquake are unreachable. However, satellite communication networks have drawbacks such as high bit error rate, asymmetric uplink and downlink bandwidth, extended round-trip time, and limited storage capacity [2].

In order to overcome the above shortcomings, some improvements must be made to the existing protocol, and so far, many solutions have been formed. In recent years, the better methods of improvement are TCP-Hybla, which separate the transmission rate of satellite network from the transmission delay, and TCP-Westwood can measure the network link bandwidth in real time and the SCPS protocol designed by NASA [3, 4]. The above method improves the satellite link throughput and provides a good idea and direction for further improvement research.

© Springer Nature Switzerland AG 2020
Q. Liu et al. (Eds.): CENet 2018, AISC 905, pp. 885–893, 2020.
https://doi.org/10.1007/978-3-030-14680-1_96

2   Satellite Communication Features

The satellite communication network consists of two parts, a satellite and a ground station, which use the satellite as a relay for information communication. The main features of satellite communication: (1) The data error rate is high. Compared with other communication methods, the natural environment such as weather has a great influence on the communication performance of satellite communication. The error rate of general transmission data is between 10^{-4} and 10^{-7}. (2) The higher transmission delay, the communication satellite is far away from the ground receiving station, and the data transmission delay is large. Generally, the one-way transmission delay is between 250 ms and 280 ms. (3) The forward and reverse link bandwidth is asymmetrical. Due to the complexity of the satellite's transmission and control technology and the cost of establishing a satellite communication network system, the satellite forward link transmission bandwidth is much larger than the reverse link transmission bandwidth.

3   Introduction to SCPS Protocol

In order to solve the problem of space communication series and ensure the reliability of spatial data transmission, NASA's Jet Power Laboratory and the International Spatial Data System Advisory Committee jointly designed and developed the SCPS protocol, which is now included in Military Specifications and ISO [5]. SCPS uses the existing Internet protocol as the standard. Although the original protocol hierarchy model is still used, the differences between space communication and terrestrial communication are based on the space network communication environment for IP, IP-SEC, TCP/UDP, and FTP protocols. The characteristics are also modified and extended at the same time of imitation, so that the series of problems encountered by the Internet protocol in space communication are solved, such as large channel noise, large round-trip delay, large Doppler shift, frequent air-to-ground communication interruption and so on. To develop SCPS-NP, SCPS-TP, SCPS-FP and SCPS-SP protocols adapted to space communication for related problems, and SCPS-TP must be used whenever SCPS is used, while the other three protocols can be replaced with the corresponding protocols in the Internet TCP/IP protocol suite. The SCPS high-level protocol, with the support of the underlying protocol, forms a complete spatial communication network model. The air-to-ground and inter-satellite end-to-end data transmission becomes a reality. On the one hand, it can adapt to the characteristics of the space communication environment; on the other hand, it is also compatible with existing Internet protocols.

The SNACK option is an extension of SCPS-TP to improve bandwidth limited and lost recovery. It is a process of selectively sending negative acknowledgments, that is, the data receiver notifies the sender which data was not received by sending a SNACK option [6]. This option contains more than one data segment, so it is a good choice for satellite networks with long delays. The traditionally used ACK confirmation information can only contain one transmission data information. If the data packet has multiple errors in one window, it must send a corresponding multiple ACK information to notify the sender, which will be a link with asymmetric bandwidth in the forward and reverse directions. It has a great impact. The uplink-downlink bandwidth ratio is between 1:10 and 1:1000. When there is no congestion in the forward direction, the reverse link is congested and the data throughput is reduced [7].

4 Algorithm Improvement Strategy

4.1 Traditional Vegas Algorithm

The traditional TCP Vegas protocol is a congestion control algorithm based on round-trip delay (RTT) measurement proposed by Brakmo and Peterson. The network's state is determined by calculating the expected transmission rate of the link and the actual transmission rate through the network round-trip delay RTT value, thereby adjusting the size of the congestion window at the transmitting end.

After the data connection is established, Vegas will obtain a minimum round-trip delay BaseRTT [8], and calculate the expected throughput:

Expected = cwnd/BaseRTT

The actual throughput is calculated from the actual measured round trip delay RTT:

Actual = cwnd/RTT

TCP Vegas judges the state of the network based on the difference Diff between the expected transmission rate and the actual transmission rate:

Diff = (Expected - Actual)/BaseRTT

And adjust the size of the congestion window:

$$cwnd = \begin{cases} diff < \alpha, \text{ Linear increase of congestion window} \\ \alpha < diff < \beta, \text{ Congestion window unchanged} \\ diff > \beta, \text{ Linearly reduce the congestion window} \end{cases}$$

α and β are two thresholds based on experience, 1 and 3 [9]. When the value of Diff is less than α, the network bandwidth is considered to be surplus, and the congestion window is increased to increase the transmission rate. When the value of Diff is greater than β, the link is considered to be congested, and the congestion window is reduced to alleviate congestion; when the value of Diff is between α and β, the congestion window remains unchanged.

4.2 Reverse Link Confirmation Frequency Adjustment

Through the above analysis, it can be seen that the Vegas algorithm in the SCPS-TP protocol is based on the link round-trip delay RTT, and detects the forward link state in real time to control the congestion window. However, due to the limitations of the satellite network itself, the bandwidth of the forward and reverse link is extremely asymmetric, and the round-trip delay fluctuates greatly. When the reverse link is congested, the value of the RTT is increased; or when the reverse acknowledgement data frame is lost, the forward timer expires to make the forward link think that congestion has occurred.

Since the Vegas algorithm cannot distinguish the cause of congestion very well, the increase of the round-trip delay RTT defaults to congestion on the forward link, which reduces the congestion window, but actually reduces the congestion window without congestion. Transmission efficiency is difficult to improve.

In view of the above situation, based on the original Vegas algorithm, this paper adds real-time detection of the reverse link state, adjusts the transmission frequency of the acknowledgement packet in time through different states of the link, and reduces the impact of the reverse link on the RTT. Improve forward link bandwidth utilization.

4.3 Congestion Control on the Reverse Link

Based on the Vegas algorithm, combined with the congestion state discrimination method based on relative queue delay, an improved Vegas congestion control algorithm for satellite network environment is proposed. The algorithm is mainly used to improve the congestion avoidance phase of Vegas. Considering the asymmetry of the bandwidth of the forward and reverse links of the satellite network, the communication status of the satellite network is further subdivided. First, the current throughput of the network and the throughput in the previous RTT are compared. If the current throughput is greater than the previous throughput, it indicates that the network is not saturated; otherwise, it indicates that the network is close to saturation state. At this time, the reverse link is used to confirm the packet relative queue delay to further judge the network state, and corresponding measures are taken according to the result of the judgment.

It is assumed that the receiving end transmits two consecutive acknowledgment packets at the time of S_n and S_{n+1}, and the transmitting end is received at the time of R_n and R_{n+1}. Defining the difference between the ACK reception time interval of the two consecutive packets and the transmission time interval of the two packets, the relative queue delay can be expressed as follows:

$$\Delta = (R_{n+1} - R_n) - (S_{n+1} - S_n) \tag{1}$$

Judging the congestion state of the reverse link network through the Delta. If $\Delta \leq 0$, the ACK reception time interval of two consecutive packets is less than or equal to the transmission time interval, indicating that the packet does not pass the queue delay, and the reverse link is in a stable state; If $\Delta > 0$, the ACK reception time interval of two consecutive packets is greater than the transmission time interval, it indicates that the queue delay of the packet increases and the reverse link is in a congestion state. The specific implementation process is as follows: Where $Th(t)$ represents the current throughput, $Th(t-RTT)$ represents the throughput of the previous RTT, and the initial values of α and β are taken as 1 and 3.

```
If (α≤δ≤β) {
  If (Th(t) > Th(t-RTT)) {

        cwnd=cwnd+1;α=α+1;β=β+1;}
    Else If (Th(t)≤Th(t-RTT)) {
      If (Δ≤0) {
            no update of cwnd,α,β;}
      Else If (Δ>0) {
            cwnd=cwnd+1;α=α+1;β=β+1;
            }
        }
}
Else If (δ < α) {

  If (α≥1&&Th(t) > Th(t-RTT)) {

    cwnd=cwnd+2;}
Else If (α > 1&&Th(t)≤Th(t-RTT)) {

  If (Δ≤0) {
    cwnd=cwnd+2;α=α;β=β;}
Else If (Δ > 0) {

    cwnd=cwnd+2;α=α+1;β=β+1;}
    }
    Else If (δ > β) {

If (Th(t) > Th(t-RTT)) {

    cwnd=cwnd+1;α=α;β=β;}
Else If (Th(t)≤Th(t-RTT)) {
  If (Δ≤0){
    cwnd=cwnd-1;α=α-1;β=β-1;
    If (cwnd < 2) {
```

```
    cwnd=2;}
    }
Else If (Δ > 0){
    cwnd=cwnd+1;α=α;β=β;}
    }
    }
```

When $\alpha \leq \delta \leq \beta$, the congestion window will increase or not change. Because when $\alpha \leq \delta \leq \beta$, the expected throughput is close to the actual throughput, and the actual communication network state needs further prediction. When the new throughput is greater than the previous throughput, it indicates that the communication network is in a clear state, the congestion window is increased by one within a RTT, and the alpha and beta values are also increased by one, respectively. Conversely, when the new throughput is less than or equal to the previous throughput, it indicates that the communication network is close to saturation, and then the network status is further subdivided by using two consecutive acknowledgment packets relative to the queue delay Δ. If $\Delta \leq 0$, it indicates that the reverse link of the communication network is in a clear state, and the forward link is close to saturation, so the congestion window remains unchanged. If $\Delta > 0$, it indicates that the network reverse link is congested, and the forward link is unblocked. At this time, the congestion window, α and β values are increased by one. At this time, the congestion window is increased to make the algorithm have a stronger ability to utilize bandwidth and achieve higher TCP throughput.

If $\delta < \alpha$, the congestion window will increase. Because when $\delta < \alpha$, it indicates that the expected throughput is too small, and the actual communication network state is relatively smooth. When $\alpha \geq 1$ and the new throughput is greater than the previous throughput, the congestion window is increased by a large margin, so that the network is fully utilized, so the congestion window is increased by two data segments in each RTT. When $\alpha \geq 1$ and the new throughput is less than or equal to the previous throughput, further consider the two consecutive acknowledgment packet relative queue delays Δ. If $\Delta \leq 0$, it means that the reverse link is smooth at this time, the forward link is a bit congested, and all congestion windows increase by one in one RTT, but the values of α and β are unchanged; If $\Delta > 0$, it indicates that the reverse link is congested and the forward link is unobstructed. Therefore, the congestion window increases by two in one RTT, and the values of α and β are increased by one and adjust the reverse link confirmation frequency.

When $\delta > \beta$, the congestion window will increase, remain the same or decrease. Because if $\delta > \beta$, it means that the expected throughput is much higher than the actual throughput, and the communication network at this time is not smooth. When the new throughput is greater than the previous throughput, the congestion window is increased by one, and the alpha and beta values are also increased by one. When the new throughput is less than or equal to the previous throughput, if $\Delta \leq 0$, the congestion window, α and β values are all decreased by one. At this time, if the congestion window is less than two and the value is two; If Δ > 0, the congestion window is increased by one and the reverse link acknowledge frequency is adjusted.

5   Algorithm Simulation

5.1   Simulation Environment

This experiment uses opnet14.5 network simulation software to simulate the simulation environment and verify the performance of the algorithm. The simulation model library is the CCSDS protocol model library, which adopts the standard SCPS-TP communication protocol. The simulation parameters are given in Table 1. In the simulation, the satellite is used as the information source to send information to the ground and receive the return confirmation. The network topology of the simulation experiment is shown in Fig. 1.

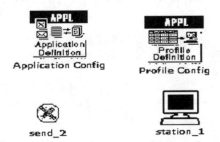

Fig. 1. Topology diagram.

Table 1. Values of simulation experiment parameters.

Parameter item	Parameter value
Uplink	1 Mb
Downlink	100 Mb
Satellite	LEO
Time	500 s
Ber	$10^{-5}, 10^{-6}$

5.2   Impact of Different Bit Error Rates on Throughput

This simulation experiment uses the bit error rate as a variable parameter to obtain a comparison of link throughput. As shown in Figs. 2 and 3 below, the comparison of the congestion windows of the two algorithms is for LEO satellites.

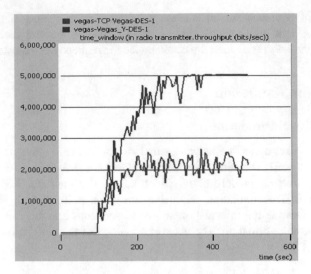

Fig. 2. BER is 10-5 the throughput in the environment.

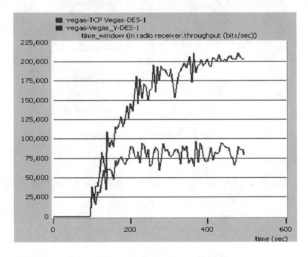

Fig. 3. BER is 10-6 the throughput in the environment.

By comparing the simulation results in the following figure, it can be seen that the new improved algorithm throughput is better than the original algorithm when the bit error rate is not the same, and the new algorithm throughput size fluctuation is smaller than the original algorithm. By analyzing the two simulation results, the new algorithm has better improvement in throughput and stability, and verifies the correctness of the new algorithm.

6 Conclusion

By comparing with the original algorithm, the new algorithm can predict the communication link state more accurately, the congestion window adjustment is more rapid and accurate, the reverse link data packet confirmation frequency is adjusted in time, reduce the impact of reverse link on network throughput, greatly improved satellite network bandwidth utilization. The correctness of the algorithm is verified by OPNET simulation. The algorithm has low complexity and obvious improvement effect.

Acknowledgements. This work was supported by the National Natural Science Foundation of China (6150130), Postdoctoral Science Foundation of Shenyang Ligong University (2016), Leading Academic Discipline Open-ended Foundation of Shenyang Ligong University (4771004kfx24), Middle-aged and Young Science and Technology Innovation Talents Support Program Project of Shenyang (2017).

References

1. Li, L.Q., Zhu, J., Yang, Y.T., Hu, Z.: Performance analysis of TCP congestion control algorithm for satellite IP network. Shanghai Aerosp. **33**(6), 109 (2016)
2. Dai, S., Xiao, N., Liang, J., Yuan, T.: TCP congestion control algorithm for satellite networks based on processing delay. Mod. Def. Technol. **42**(3), 128 (2014)
3. Du, L.H., Wu, X.J.: Research and implementation of TCP protocol accelerator based on SCPS-TP. Radio Eng. **47**(9), 13 (2017)
4. Li, L.Q., Zhu, J., Yang, Y.T., Hu, Z.: Performance analysis of TCP congestion control algorithm for satellite IP network. Shanghai Aerosp. **6**, 110 (2016)
5. Jiang, H., Huang, J.Q., Wang, X.: Simulation of TCP congestion control based on bandwidth estimation of MANET. Comput. Eng. Des. **37**(1), 201 (2016)
6. Hu, J.L., Wang, R.H., Sun, X., Yu, Q.: Memory dynamics for DTN protocol in deep-space communications. Aerosp. Electron. Syst. Mag. **29**(2), 22 (2014)
7. Cheng, R.S., Deng, D.J.: Congestion control with dynamic threshold adaptation and cross-layer response for TCP vegas over IEEE802.11 wireless networks. Int. J. Commun Syst **27**, 2919 (2014)
8. Sreekumari, P., Chung, S.H., Lee, M., Kim, W.S.: Detection of fast retransmission losses using TCP timestamp for improving the end-to-end performance of TCP over wireless networks. Int. J. Distrib. Sens. Netw. **11**(9), 1 (2013)
9. Jamali, S., Alipasandi, N., Alipasandi, B.: TCP pegas: a PSO-based improvement over TCP vegas. Appl. Soft Comput. **32**, 165–166 (2015)

A Power Network Illegal Access Monitoring Method Based on Random Forests

Wenzhe Zhang[1(✉)], Wenwei Tao[1], Song Liu[1], Yang Su[1],
Zhaohui Hu[2], and Chao Hu[3]

[1] CSG Power Dispatching Control Center, Guangzhou 510670, China
zhangwenzhe@csg.cn
[2] Dingxin Information Technology Co., Ltd., Guangzhou 510627, China
[3] NARI Information and Communication Technology Co., Ltd.,
Nanjing 210033, China

Abstract. With the development of network technology, people have begun to pay more and more attention to the impact of illegal access on the power network, and have tried to take measures to monitor whether the power network is illegally accessed. People try to use machine learning to monitor the power network. Because of the high classification accuracy and high efficiency of the random forest algorithm, we proposed a random forest-based power network monitoring algorithm. Comparison experiments with other algorithms have proved that our algorithm is more superior both in terms of time consumption and accuracy.

Kerwords: Network illegal access · Power network monitoring · Random forests

1 Introduction

With the development of power network technology, the scale of power networks is increasing, and their complexity is also increasing. However, some hackers exploit some of the loopholes in the power network to attack the power network and obtain information in an illegal manner. This can have a huge impact on the power network and cause significant losses. The security of the power network is closely related to people's lives. Therefore, the research in this area is of great significance. The power network's requirements for data confidentiality, timeliness, integrity, and reliability are very high. In general, there are mainly the following aspects.

(1) High robustness. That is, in the case of complex and harsh environments or where the equipment is destroyed by force majeure, it is ensured that the power network service can still work uninterruptedly.
(2) High security. Security and confidentiality are important principles in the construction of power networks. Adopting necessary security measures to prevent loss of confidentiality and confidentiality is an important goal of wireless network construction.

Q. Liu et al. (Eds.): CENet 2018, AISC 905, pp. 894–903, 2020.
https://doi.org/10.1007/978-3-030-14680-1_97

(3) High reliability. The data transmitted on the confidential network must be reliable, complete, and real-time. If there is any false information or illegal information, it will have a serious impact and cause significant losses.

In order to solve the problem of illegal access, people have already proposed some methods, such as the well-known network mapping tool "nmap" implements various techniques to find services and vulnerabilities in a network, SSID hiding technology, the technology of binding household authentication and IP-Mac-Port, wireless encryption technology, frame detection method. As an example of a cooperated SCAN activity, distributed traceroutes may be carried out to find out the entry point (router) of an Intranet [1]. and people are exploring the use of better methods to monitor illegal access under power networks. The prediction of data by machine learning has made great progress in the past two years. People have also begun to use machine learning methods to deal with power network monitoring problems.

Random forest [2] refers to a classifier that uses multiple trees to train and predict samples. The classifier was originally proposed by Leo Breiman and Adele Cutler. As the name suggests, Random Forests uses a random approach to establish a forest. There are many decision trees in the forest. Each decision tree in a random forest is not related. After getting the forest, when a new input sample comes in, let each decision tree in the forest make a separate judgment to see which sample the sample should belong to, and then see which one is selected the most, to predict this sample is that type. Random forests are widely used, such as image Classification [3], variable selection [4], real time 3D face analysis [5], exploratory data analysis [6], profiling computer network users [7], location recognition [8] and so on. This paper uses random forest method to analyze data and judge whether there is illegal access.

The outline of this paper is as follows. The basic knowledge of illegal intervention and random forest are briefly reviewed in Sect. 2, the method proposed in this paper is described in detail in Sect. 3. Subsequently, we compare and analyze the time consumption and accuracy of other algorithms with our method in Sect. 4, and made a conclusion in the last section.

2 Preliminary Knowledge

2.1 Illegal Access in Power Network

Illegal access refers to the act of directly connecting various types of computers and mobile devices to the intranet without permission. Infrequent use of IP, IP, and MAC mapping tables and inconsistencies on the network can all be considered as illegal access. The illegal access generally includes two aspects. First, illegal users access the device by cracking the identity authentication of the access device. Second, the illegal access device tricks users into connecting by means of camouflage, collects user information without the user's knowledge, and performs phishing attacks. Although this type of illegal incident is not a violent invasion, it may have serious consequences such as spreading viruses on the Intranet, porting Trojan horses, and leaking confidential information of intranet services. The problem of illegal access to the power network will have a certain impact on the important attributes of these data. Therefore, the monitoring of the illegal access of the power network has become a problem that must be faced.

General illegal access methods include: 1. Proxy-based proxy server is illegally accessed; 2. Nat's proxy server is illegally accessed; 3. IP address theft: It infringes the rights of normal users in the Internet network, and brings huge negative impact on network charging, network security and network operations. Therefore, the solution to IP address theft becomes a pressing problem at present. Therefore, for illegal access, there are some solutions as follows:

(1) SSID hiding technology. An SSID (Service Set Identifier) is a unique identifier of a wireless local area network and is used to distinguish different networks. With SSID hiding technology, SSID broadcasts can be cancelled, and attackers cannot directly obtain the necessary information about this network. However, the prior art can find the hidden SSID through a simple scan, and the effectiveness of this technique has been greatly reduced.

(2) The technology of binding household authentication and IP-MAC-PORT. First of all, ensure that legitimate users pass the authentication, then write relevant network management software through SNMP protocol, poll network devices such as switches, and obtain information such as the IP address and MAC address of hosts in the current network, and compare them with the original IP-MAC comparison table. Then find illegal IP addresses and access points.

(3) Wireless encryption technology. The 802.11 standard describes the specifications of the MAC layer and the physical layer of WLAN and WMAN, and introduces the WEP (Wired Equivalent Privacy) encryption technology. However, cryptanalysis experts have discovered that WEP has serious weaknesses. WPA (Wi-Fi Protected Access) technology is an improvement of WEP. It is an improvement of the security of the interim solution before the completion of 802.11i. WPA2 is an upgraded version of WPA. It implements all the specifications of 802.11i, and its security is greatly enhanced. Since March 2006, WPA2 has become a mandatory standard for companies.

(4) Frame detection method: Send some forged data frames through the network and check the response of the corresponding client to determine whether the client is illegally outbound.

(5) IPsec technology. IPsec (Internet Protocol Security) is the long-term direction of network security. End-to-end security is used to prevent attacks from inside and outside networks. IPsec has security features such as non-repudiation, integrity, and authentication.

2.2   Random Forest

Applying the ensemble learning algorithm to the data classification field of data mining, the earliest is the Boosting method [9] and Bagging method [10]. These two methods are based on the central content of the ensemble learning algorithm, aiming at the practical problem that the single classifier performance cannot be improved, and propose the idea of reducing the generalization error of the algorithm by generating multiple classifiers, thereby improving the performance of the algorithm. And this idea is applied to the field of data classification.

With the development of ensemble learning methods, Ho [11] proposed the concept of Random Decision Forests in 1995. Three years later, he also proposed an integration method for stochastic subspaces [12]. In 2001 Breiman [2] systematically described the random forest method. The random forest algorithm has become an important part of the data mining classification algorithm. Random forests improve prediction accuracy without significant increase in computation. Random forests are not sensitive to multivariate collinearity. The results are robust to missing data and non-equilibrium data. They can well predict the effects of up to several thousand explanatory variables and are hailed as one of the best algorithms currently available.

Random forest algorithm is an extension of Bagging. Random forest is based on the bagging integration of decision tree-based learner, and further introduces random property selection in the process of decision tree training. The algorithm flow is shown in the Algorithm 1.

Algorithm 1. Framework of Random forest algorithm

Input: Training data D ; Base learner h_t ; Training rounds T
Process:

For $t = 1, 2, \text{L}, T$ do

training h_t
End for

Majority voting to obtain $H(x)$

Output: $H(x)$

The schematic diagram of the random forest is as follows in Fig. 1.

Random forest algorithm has some advantages, such as (1) It is unexcelled in accuracy among current algorithms; (2) It runs efficiently on large data bases; (3) It can handle thousands of input variables without variable deletion; (4) It gives estimates of what variables are important in the classification; (5) It generates an internal unbiased estimate of the generalization error as the forest building progresses; (6) And it has an effective method for estimating missing data and maintains accuracy when a large proportion of the data are missing.

Random forests are mainly used for regression and classification. This article mainly discusses the classification problem based on random forests. Random forests are somewhat similar to bagging using a decision tree as a basic classifier. The bagging based on the decision tree is used to generate a decision tree after each bootstrap sampling, and how many trees are generated by how many samples are taken. No more interventions are made when these trees are generated. The random forest also performs bootstrap sampling, but it differs from bagging in that each node variable is generated only in a randomly selected minority of variables when each tree is generated. Therefore, not only are the samples random, but even the production of each node

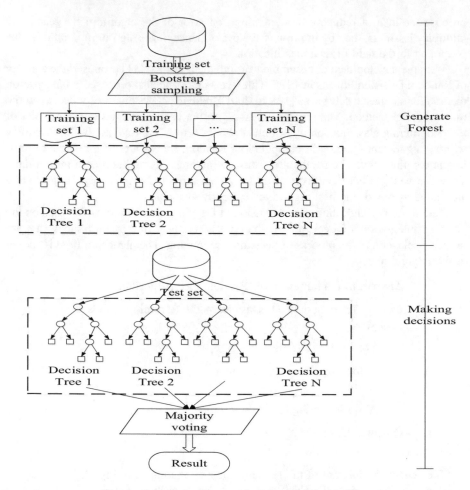

Fig. 1. Schematic diagram of random forest algorithm.

variable is random. Many studies have shown that the combined classifier performs better than a single classifier. Random forest is a method that uses multiple classification trees to distinguish and classify data. While classifying the data, it can also give the importance score of each variable and evaluate the role of each variable in the classification.

3 Power Network Illegal Access Monitoring Method Based on Random Forests

Random forest is an algorithm that integrates multiple trees through the idea of ensemble learning. Its basic unit is a decision tree, and its essence belongs to a major branch of machine learning-integrated learning methods. Integrated learning solves single prediction problems by establishing several model combinations. Its working principle is to generate multiple classifiers/models and learn and make predictions independently of each other. These predictions are finally combined into single predictions and are therefore superior to any single classification in making predictions. Random forest is a subclass of ensemble learning. It depends on the voting choice of the decision tree to determine the final classification result. This article selects the C4.5 decision tree.

The C4.5 decision tree was proposed by Quinlan [13] in 1995 and is an improvement based on the ID3 decision tree. C4.5 overcomes the disadvantages of ID3's use of information gain to select attributes that are biased towards selecting more attribute values of the branch, i.e., attributes with more values, and the inability to handle coherent attributes. The classification rules produced by the C4.5 algorithms are easy to understand and have a high accuracy, so the algorithm is used as a base classifier for random forests to construct algorithms. And then we proposed a power network illegal access monitoring method based on random forests.

The random forest construction steps in this algorithm are as follows:

(1) Data preparation: Given N tagged samples $\{(x_1, y_1), (x_2, y_2), \cdots, (x_N, y_N)\}$, where $x_i \in X$ (X is called the sample space of the power network), $y_i \in Y$ (Y is called the illegal access class space of the power network) where $i = 1, 2, \cdots, N$, there are M features per sample The number of trainings T is known.

(2) Data sampling: There are N samples randomly selected for return. The selected samples of N are used to train a decision tree. As a sample at the root node of the decision tree, the decision tree of the t-th round h_t is constructed according to the $C4.5$ algorithm.

(3) Forming decision trees: When each sample has M attributes, when each node of the decision tree needs to be split, m attributes are randomly selected from the M attributes, satisfying the condition $m \ll M$. Then use the information gain from the m attributes to select an attribute as the split attribute of the node.

$$\Delta = I(parent) - \sum_{g=1}^{2} \frac{N(v_g)}{N} * I(v_g), \tag{1}$$

where, v_g represents the number of records in the child node, and $N(v_g)$ represents the number of child node v_g. I represents the impurity, and $I(v_g)$ is the impurity of v_g, they are calculated by the following entropy formula,

$$-\sum_{i=1}^{n} P(i) * \log_2^{p(i)}, \tag{2}$$

and $P(i)$ represents the probability of i.

During the formation of the decision tree, each node must split in accordance with this step until it can no longer be split.

(4) Build forests: repeat steps (2)–(3) T times to build a large number of decision trees to form a random forest.

(5) Majority voting: Each decision tree constructed from random forests is very weak, so we need to combine the resulting learners h_t. For the task of unauthorized

access to the power network, learner h_t predicts a tag from class tag set $\{c_1, c_2, \cdots, c_N\}$, and uses a combination of majority voting to monitor the state of the power network.

$$H(x) = \begin{cases} c_j, & if \ \sum_{t=1}^{T} h_t^j(x) > 0.5 \sum_{k=1}^{N} \sum_{t=1}^{T} h_t^k(x) \\ reject, & otherwise \end{cases}. \tag{3}$$

That is, if the number of votes for a mark is more than half, the prediction is the mark, otherwise the prediction is rejected.

(6) Use the trained random forest to predict and classify the input test data to see if there is any illegal access, and obtain the correct rate of the algorithm.

4 Experiment and Analysis

In order to prove the usability, validity and accuracy of the algorithm, an experimental environment was set up in a non-cluster network environment. 1000 host computers in the Windows system were selected as the check target set, and 10 of them were illegally accessed. The relevant data of the power network are collected, and the SVM algorithm [14], the Adaboost algorithm [15] and the KNN algorithm [16] are used as comparison test algorithms of the algorithm. The measurement methods include efficiency analysis and accuracy analysis.

4.1 Analysis of Time Consumption

Assuming that the computational complexity of the base learner is $O(m)$, the complexity of the algorithm is roughly $T(O(m) + O(s))$, where $O(s)$ is the complexity of sampling and voting and can be ignored, and T is usually a not too large constant. Therefore, this algorithm is a very efficient integration algorithm. First compare the efficiency of this algorithm with the SVM algorithm, Adaboost algorithm and KNN algorithm. The Table 1 and Fig. 2 below show the running time of the four algorithms.

Table 1. Time consumption of four methods (second).

	1	2	3	4	5	6	7	8	9	10
KNN	60	55	49	49	58	60	52	58	56	57
Adaboost	43	50	46	41	47	45	45	51	53	44
SVM	42	35	33	32	40	30	31	37	33	36
Ours	35	25	31	33	29	27	24	21	21	23

From the comparision of time consumption, we can easliy know that KNN takes the longest time, Adaboost algorithm takes less time than KNN, but more than SVM, our method can spend less time than other methods.

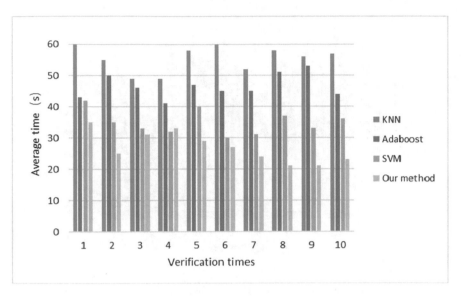

Fig. 2. Comparison of time consumption of KNN, Adaboost, SVM algorithms and our method.

4.2 Analysis of Accuracy

We also analyze the accuracy of these four algorithms, the Table 2 gives a comparison of the generalization performance of the four methods and compares the correctness of the three methods (as shown in Fig. 3).

Table 2. Accuracy of four methods (%).

	1	2	3	4	5	6	7	8	9	10
KNN	85.1	78.5	89.4	87.5	86.3	79.8	82.7	88.4	86.2	89.8
Adaboost	78.8	82.1	78.8	72.7	73.9	73.0	80.2	82.3	77.6	79.7
SVM	91.0	90.4	91.5	88.0	91.1	88.2	89.2	90.1	87.9	89.3
Ours	98.1	96.3	93.5	98.2	97.8	94.7	97	92.8	91.5	96.2

From the accuracy comparison above, in terms of accuracy of these algorithms, Adaboost has the lowest accuracy, KNN algorithm and SVM algorithm are slightly higher than Adaboost, and our method has the highest accuracy. Therefore, in terms of algorithm efficiency and accuracy, our algorithm is superior to KNN, Adaboost and SVM algorithms.

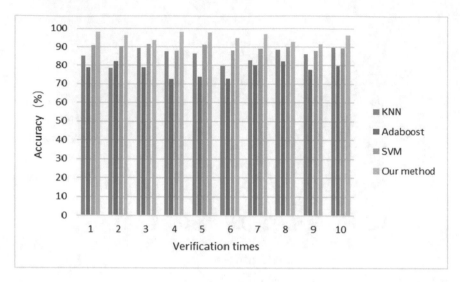

Fig. 3. The accuracy comparison among KNN, Adaboost, SVM algorithms and our method.

5 Conclusion

With the continuous development of the electric power network, people are increasingly demanding safety. The problem of illegal access to the network has attracted people's attention. This paper proposes a power network monitoring method based on random forest, uses random forest algorithm to classify data, and identifies devices that are illegally accessed. This paper compares the experiment with KNN, Adaboost and SVM algorithms, and verifies the efficiency and accuracy of this method from two aspects of running time and accuracy. Experiments have shown that this method is superior to the other three algorithms in terms of operating efficiency and accuracy.

Acknowledgement. This work is supported by Science and Technology Project of China Southern Power Grid Co., Ltd. "Research and Demonstration of Key Technologies of Network Security Situational Awareness in Power Monitoring System" (No. ZDKJXM20170002).

References

1. Stephen, N., Novak, J.: Network Intrusion Detection, An Analyst's Handbook, pp. 207–219. Peoples Posts & Telecommunications Publishing House, Beijing (2000)
2. Breiman, L.: Random forests. Mach. Learn. **45**(1), 5–32 (2001)
3. Bosch, A., Zisserman, A., Munoz, X.: Image classification using random forests and ferns. In: IEEE 11th International Conference on Computer Vision (2007)
4. Genuer, R., Poggi, J.M., Tuleau-Malot, C.: Variable selection using random forests. Pattern Recognit. Lett. **31**(14), 2225–2236 (2010)
5. Fanelli, G., Dantone, M., Gall, J., Fossati, A., Gool, L.V.: Random forests for real time 3D face analysis. Int. J. Comput. Vis. **101**(3), 437–458 (2013)

6. Jones, Z., Linder, F.: Exploratory data analysis using random forests. J. Open Source Softw. **1**(6) (2016)
7. Nowak, J., Korytkowski, M., Nowicki, R., Scherer, R., Siwocha, A.: Random forests for profiling computer network users. In: Rutkowski, L., Scherer, R., Korytkowski, M., Pedrycz, W., Tadeusiewicz, R., Zurada, J.M. (eds.) ICAISC 2018. LNCS (LNAI), vol. 10842, pp. 734–739. Springer, Cham (2018). https://doi.org/10.1007/978-3-319-91262-2_64
8. Lee, S., Moon, N.: Location recognition system using random forest. J. Ambient. Intell. Hum. Comput. **11**, 1–6 (2018)
9. Sehapire, R.E.: The strength of weak learnability. Mach. Learn. **5**(2), 197–227 (1990)
10. Breiman, L.: Bagging predicators. Mach. Learn. **24**(2), 123–140 (1996)
11. Ho, T.K.: Random decision forest. In: Proceedings of the 3rd International Conference on Document Analysis and Recognition. Montreal, Canada, pp. 278–282 (1995)
12. Ho, T.K.: The random subspace method for constructing decision forests. IEEE Trans. Pattern Anal. Mach. Intell. **20**(8), 832–844 (1998)
13. Quinlan, J.R.: C4.5: Programs for Machine Learning. Morgan Kaufmann Publishers Inc., Burlington (1993)
14. Cortes, C., Vapnik, V.: Support-vector networks. Mach. Learn. **20**(3), 273–297 (1995)
15. Freund, Y., Schapire, R.E.: A desicion-theoretic generalization of on-line learning and an application to boosting. In: Vitányi, P. (ed.) EuroCOLT 1995. LNCS, vol. 904, pp. 23–37. Springer, Heidelberg (1995). https://doi.org/10.1007/3-540-59119-2_166
16. Cover, T., Hart, P.: Nearest neighbor pattern classification. IEEE Trans. Inf. Theory **13**(1), 21–27 (2002)

Time-Series Based Ensemble Forecasting Algorithm for Out-Limit Detection on Stable Section of Power Network

Haizhu Wang[1(✉)], Chao Hu[2], Yue Chen[2], Bo Zhou[2],
and Zhangguo Chen[1]

[1] Electric Power Dispatching and Control Center of Guangdong
Power Grid Co., Ltd., Guangzhou 510600, China
20801244@qq.com
[2] NARI Information & Communication Technology Co., Ltd.,
Nanjing 210033, China

Abstract. With the development of power company network technology, out-limit forecast of grid stable section is an important point of grid operation and control. However, due to the large amount of grid stable section data in power grid, traditional single classic forecasting algorithms are difficult to predict efficiently and accurately. In order to solve this problem, we proposed a time-series based ensemble forecasting algorithm (TSEFA) for out-limit detection on grid stable section which integrates multiple classification forecasting algorithms to classify and predict the collected grid stable section data, and then to realize the forecasting of the out-limit quantity with comprehensive optimal accuracy. Compared with the other four single-model algorithms (i.e. SWAF, RA, ANN, SVM), our TSEFA algorithm achieves the effect of efficient and accurate forecasting, and enhances the security and stability of the grid stable section analysis platform.

Keywords: Time-series · Out-limit detection · Ensemble forecasting ·
Grid stable section

1 Introduction

Grid stability of the limit of the prediction is a challenging problem and it is impossible to ignore the problem. It is the key point of monitoring and controlling by the power grid operators. Based on the mass storage section history state data, we can count and analyze the stable section history overload information of the equipment customized by the user freely to carry out stability section early warning by scheduling [1]. Relying on the cross section analysis to study the weak link and making rapid identification of important transmission sections are of great significance to ensure safety and prevent the occurrence of large area of cascading failure.

In order to ensure the safety of power grid section operation, it is necessary to analyze and predict the operation data. At present, there are many methods available for stability cross - sectional analysis, such as sliding window average forecasting algorithm, regression analysis model, artificial neural network, time series analysis model

Q. Liu et al. (Eds.): CENet 2018, AISC 905, pp. 904–916, 2020.
https://doi.org/10.1007/978-3-030-14680-1_98

etc. But most of the algorithms have some limitations: Kapoor et al. [2] studied weather forecasting using sliding window algorithm, but inappropriate selection of sliding window data and time size has a big effect on the results. Regression analysis algorithm which used by Punzo et al. [3] requires high sample size, and too little observation data will seriously influence the results. Park et al. [4] utilized artificial neural network to forecast the electric load, however their method expose a shortcoming that it is difficult to select the training set from the problem, which directly affects the approximation and promotion capabilities of the network model. Percival et al. [5] proposed time series analysis model which has poor model adaptability and inappropriate time series interval. Therefore, traditional single classic forecasting algorithms are difficult to predict efficiently and accurately. In order to solve this problem, we proposed a time-series based ensemble forecasting algorithm (TSEFA) for out-limit detection on grid stable section. This algorithm integrates multiple classification forecasting algorithms to classify and predict the collected grid stable section data, and then to realize the forecasting of the out-limit quantity with comprehensive optimal accuracy.

The remaining part of the paper is organized as follows: before the text, we made a few reviews of knowledge points in Sect. 2. In Sect. 3, we described the time-series based ensemble forecasting algorithm elaborately. Besides, we give the corresponding procedures of this algorithm. And then we show asymmetric error cost function to deal with the application scenario of abnormal quantity forecasting in power industrial control system. In Sect. 4, compared with the other four single-model algorithms (i.e. SWAF, RA, ANN, SVM) and the relevant evaluation indicators, we conducted specific experiments through MATLAB, and then assessed and analyzed the results of the experiments based on the evaluation criteria. Finally, Sect. 5 is dedicated for conclusion.

2   Preliminaries

2.1   Definition of the Time Window

The time-series based integration algorithms are usually applied to establish a forecasting model and a verification model: one model is to split the data set into two parts of the training set and the test set. The second verification model is to implement forecasting which use the statistics of various alarm events to describe the characteristics of network operations as a decision feature for fault forecasting [6]. The time axis is divided into time windows of a certain size. The objective of failure forecasting is to determine whether there is a failure event within the forecasting time window. The division of the time axis is shown in Fig. 1.

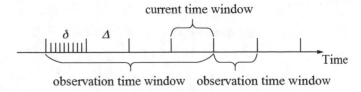

Fig. 1. The division of the time axis.

As shown in Fig. 1, we define the time window as follows: the size of the unit time window is resident Δ, the forecasting time window is the unit time window for predicting whether a fault event occurs, and the window size is resident Δ. The current time window is the forecast time window's previous unit time window, window size is resident Δ. The observation time window is n unit time window before the forecast time window, including the current time window, the window size is stationed in $\Delta * n$. The sample window is more unit time window for small partitions, the window size is δ, and the number of sample windows per unit time window is Δ/δ.

2.2 Sliding Window Average Forecasting

Sliding window average forecasting [7] is the simplest way to predict time series. Essentially, it is a finite-impulse-response filter that performs analysis by computing the average of a subset of the entire data set. When using sliding window average forecasting, you need to set a sliding window with a fixed length of n.

Let the forecasting time point be t, then the forecasting value of time point t is

$$\hat{v}^{(t)} = MA_{t-1} = \frac{v^{(t-1)} + v^{(t-2)} + \cdots + v^{(t-n)}}{n}, \tag{1}$$

Here, MA_{t-1} represents the moving average at time t. If incremental calculations are used, the average number of sliding windows can be calculated

$$\hat{v}_{(t)} = MA_{t-1} \frac{v^{(t-n)}}{n} + \frac{v^{(t-1)}}{n}. \tag{2}$$

The calculation is simple the sliding window average forecasting, but if the time series change is unstable, its forecasting efficiency will be greatly reduced. So how to correctly choose the length of the sliding window is a problem. If n is chosen too large, the change in the forecast value will be underestimated; if the n choice is too small, past history will not be fully utilized.

2.3 Regression Analysis

Regression analysis [8] is a statistical data analysis method. The main purpose is to understand whether two or more variables are related, related directions and intensity, and to establish a mathematical model to achieve optimal data fitting. RA is divided into two main categories: linear regression analysis and nonlinear regression analysis. Linear regression need to satisfy a given sample and satisfy the linear condition, the formula is as follows

$$y = \beta_0 + \beta_1 x_1 + \beta_2 x_2 + \cdots + \beta_i x_i + \varepsilon. \tag{3}$$

that is

$$y = X\beta + \varepsilon. \tag{4}$$

In order to estimate β, an appropriate evaluation criterion is needed to find the optimal estimate of β. The least square method provides a standard. The basic principle is to minimize the mean square error of the sample data, and the cost function is

$$J(\beta) = \frac{1}{2k} \sum_{i-1}^{k} \left(X^{(i)}\beta - y^{(i)} \right)^2. \tag{5}$$

where k is the number of samples and its vector form can be expressed as

$$\beta = \left(X^T X \right)^{-1} X^T y. \tag{6}$$

this formula needs to satisfy the linearly independent column vectors of X.

The regression model is divided into linear regression and nonlinear regression. Quadratic polynomial regression model and exponential regression model belong to linear regression. Polynomial regression is the generalization of linear regression, its formula is

$$y = \beta_0 + \beta_1 x + \beta_2 x^2 + \varepsilon. \tag{7}$$

let $x_1 = x$, $x_2 = x^2$, quadratic polynomials can be converted into binary linear regression and the value of β and estimation methods are not changed in parameter estimation.

In mathematics, the exponent represents the power, that is, the general formula of exponential regression is an operational form of rational number multiplication square. The formula can be written as

$$y = \alpha \beta_1^{x_1} \beta_2^{x_2} (\alpha > 0, \ \alpha \neq 1), \tag{8}$$

where α is the exponential regression coefficient, and β_1, β_2 are the exponential partial regression coefficients. In this paper, there is only one independent variable, which only needs the regression of the unary index, whose formula is

$$y = \alpha \beta^x. \tag{9}$$

2.4 Artificial Neural Network

As mentioned before, if there is a complex nonlinear relationship in the data, it is difficult for the regressive forecasting to correctly fit the relationship between the data. In order to more effectively predict data with nonlinear relationships, artificial neural network [9] is the suitable model. Artificial neural network is a mathematical model that imitates biological neural network. It is a data mining algorithm that can find the complex relationship between data input and output. A typical artificial neural network generally has 3 levels: Input layer, Hidden layer, and Output layer, and the hidden layer can be 0 to multiple layers.

Figure 2 shows the topology of a 3-layer artificial neural network, including a hidden layer.

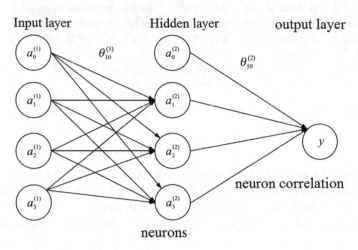

Fig. 2. The topological diagram of artificial neural network.

As shown in Fig. 2, the number of neurons in each layer is 4, 4 and 1, respectively, in which the input layer and hidden layer each contain a deviating neuron (the value of the deviating neuron is fixed at 1). θ_{ij}^k is the weight of the value output from the j-th neuron in the k-th layer to the i-th neuron in the $k + 1$-th layer. $a_i^{(k)}$ denotes the value of the i-th neuron in the k-th layer. If it is the input layer, the value is directly obtained from the data, otherwise the value is calculated by the neuron in the upper layer (except for the deviating neuron). Let the $k + 1$ layer calculate the neuron, then the value can be obtained by

$$a^{k+1} = g\left(\theta^{(k)}a^{(k)}\right), \tag{10}$$

where $\theta^{(k)}$ is the neuron weight matrix for the kth to $k + 1$th layers, $a^{(k)} = [a_0^{(K)}, a_1^{(K)}, a_2^{(K)}]^T$ is the vector of all neurons in the kth layer, and $a^k = [a_1^{k+1}, \cdots, a_m^{k+1}]^T$, $\theta^{(k)} = [\theta_0]^T$.

In view of the artificial neural network updating the weights after learning the whole training sample set, and the complex topological structure of the neural network, when the data size of the training set is too large, the general gradient descent algorithm has low efficiency and long training time. In order to speed up the training, a more practical method is necessary which allows each training sample to be updated with a gradient descent.

2.5   Support Vector Machine

Support vector machine [10] is a high-precision supervised learning algorithm for solving classification problems. In a binary-class (training set with only positive and negative classes) classification problem, it obtains decision-making hyperplanes with well-divided functions by analyzing training samples. Since the algorithm always tends to find the decision hyperplane with the largest margin, it has a good generalization of the test sample.

A support vector regression machine is a variant of support vector machine. Its basic principle is to map the features of the trainer data from the n-dimensional space to the higher dimensions through a nonlinear transformation vertebra. Do a linear regression in the m-dimensional space, and then the basic idea can be expressed as

$$f(x) = \omega\Phi(x) + b \quad \Phi: R \to \Gamma, \ \omega \in \Gamma. \tag{11}$$

where $\omega = [\omega_1, \omega_2, \cdots, \omega_m]^T$ is the weight of each feature in m-dimensional $(m > n)$ space after the original sample mapping, and b is a threshold. If the mapping function vertebra in the formula is removed, Eq. (11) is a general linear regression expression. Therefore, the mapping relationship between the linear regression in the high-dimensional space and the nonlinear regression in the low-dimensional space is shown in Fig. 3.

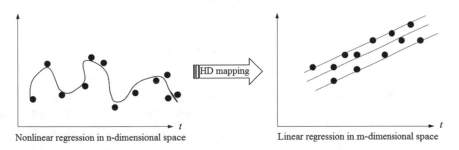

Nonlinear regression in n-dimensional space   Linear regression in m-dimensional space

Fig. 3. The mapping relationship between nonlinear regression and linear regression in support vector machine.

3   Time-Series Based Ensemble Forecasting Algorithm for Out-Limit Detection

Since single forecasting model is difficult to predict accurately, we adopt ensemble learning strategy and combines the forecasting ability of different forecasting models to propose a TSEFA algorithm for out-limit detection on stable section of power network. The ensemble learning strategy has been widely used in data mining classification problems. In the classification problem, the data samples are considered as independent and identically distributed, i.e. the samples are considered to be extracted independently from the same distribution. However, for the time series analysis problem, there is a strong association of time dimension between the samples.

In addition, because the index of the time series analysis problem is a continuous value, a voting mechanism similar to the classification cannot be used to obtain the final result. For time series analysis ensemble learning, ensemble learning strategies update the weight of each predictor primarily through predictive evaluation criteria.

3.1   The Procedures of Our TSBFA Algorithm

In order to analyze the large amount of data grid section to guarantee the safe operation of power grid section, we proposed a TSEFA algorithm for out-limit detection on grid stable section. The algorithm integrates multiple classification forecasting algorithms to classify and predict the collected grid stable section data, and then to realize the forecasting of the out-limit quantity with comprehensive optimal accuracy. For the ensemble learning of time series analysis, the ensemble learning strategy mainly updates the weight of each predictor through the prediction evaluation criteria [11].

In our ensemble forecasting algorithm, four different time series based forecasting methods (algorithms) are used. All the algorithms and their related descriptions are shown in Table 1.

Table 1. All the techniques which are used in our ensemble forecasting algorithm

Method's name	Its related description
Sliding window average forecasting	Naive forecast
Regression analysis	Linear regression
Artificial neural network	Nonlinear regression
Support vector machine	Nonlinear kernel learning algorithm

In order to integrate the prediction results of the above methods, a weighted linear combination strategy is proposed in this paper. Assume that the prediction result of the forecasting algorithm $p \in P$ at time t is $\hat{v}_p^{(t)}$, and its corresponding weight at time t is $w_p^{(t)}$, then the predicted value for a certain out-limit data at time t is

$$\hat{v}^{(t)} = \sum_p \omega_p^{(t)} \hat{v}_p^{(t)} \left(\sum_p \omega_p^{(t)} = 1 \right). \tag{12}$$

In the initial state, $t = 0$, all prediction algorithms contribute to the same prediction result, such as $w_p^{(t)} = 1/\|P\|$. Weighted update strategy based on ensemble algorithm is different from traditional classification basic strategy. In the classified predicted scenario, the results can only be expressed as "correctly smelting" or "incorrectly smelting", and the purpose of updating the weights of the ensemble algorithm is precisely to increase the weight of those classifiers whose classification results are correct. The prediction result is a continuous value and the weights of the prediction algorithm directly affect the result of the integration algorithm.

In order to update the weight of the ensemble predictive algorithm, the difference between each predictor $\hat{v}_p^{(t)}$ and the real value $v^{(t)}$ is used. The relative error $e_i^{(t)}$ of the prediction algorithm i at time t can be expressed as

$$e_i^{(t)} = \frac{c_i^{(t)}}{\sum_p c_p^{(t)}} \omega_i^{(t)}. \tag{13}$$

where $c_i^{(t)}$ represents the predicted cost of the prediction algorithm and i or p can be calculated from the predictive evaluation cost function.

The relative error is not normalized and therefore cannot be used in the update weight of the prediction algorithm. Since the final prediction result is a linear combination of multiple prediction algorithms, the weights are normalized as

$$\omega_i^{(t+1)} = \frac{e_i^{(t)}}{\sum_p e_p^{(t)}}. \tag{14}$$

According to this weight update strategy, the weight of the optimal prediction algorithm at each time point can be increased.

3.2   Asymmetric Error Cost Function

In order to deal with the application scenario of abnormal quantity forecasting in power industrial control system, this paper proposes an Asymmetric Error Cost (AEC) to evaluate the prediction error. AEC is an asymmetric heterogeneous error cost function, because over-prediction and under-prediction are two different cost components, respectively expressed as $R(v^{(s)}, \hat{v}^{(s)})$ and $P(v^{(s)}, \hat{v}^{(s)})$. Here, represents the number of abnormalities at future time points s, while $\hat{v}^{(s)}$ represents the number of abnormal predictions at future time points s. Therefore, the total cost function can be expressed as

$$A = \beta P\left(v^{(s)}, \hat{v}^{(s)}\right) + (1 - \beta) R\left(v^{(s)}, \hat{v}^{(s)}\right). \tag{15}$$

In the formula, β is a parameter used to adjust the two kinds of cost weights, and the weight of the over-prediction and the under-prediction can be artificially adjusted by changing the β value.

AEC is a generalized error cost function that can meet the needs of the application scenario of this article. The expression of AEC changes according to $P(v^{(s)}, \hat{v}^{(s)}) \geq 0$ the $v^{(s)}$ specific application. The defined P function and R function satisfy both non-negative and consistent properties:

(1) Non-negative. For any non-negative $v^{(s)}$ and $\widehat{v}^{(s)}$, and $R(v^{(s)}, \widehat{v}^{(s)}) \geq 0$.

(2) Consistency. If $v_1^{(s)} - \widehat{v}_1^{(s)} \geq v_2^{(s)} - \widehat{v}_2^{(s)}$, then $P(v_1^{(s)}, \widehat{v}_1^{(s)}) - P(v_2^{(s)}, \widehat{v}_2^{(s)})$ should maintain positive and negative consistency; Similarly, if $v_1^{(s)} - \widehat{v}_1^{(s)} \geq v_2^{(s)} - \widehat{v}_2^{(s)}$, then $R(v_1^{(s)}, \widehat{v}_1^{(s)}) - R(v_2^{(s)}, \widehat{v}_2^{(s)})$ should also maintain positive and negative consistency.

4 Algorithm Evaluation

4.1 Experimental Data Processing

In forecasting algorithm, we select a stable sectional area between January 2015 and June 2017 and predict the limit of the data using MATLAB preparation of variable sliding window method to simulate calculation for the selected data.

The selected data mainly include: The equipment which is out-limit or the section name, effective quota, the limit of the single frequency, the number of out-limit condition, voltage and daily maximum and minimum voltage, rate of capacitance which is put into operation, accumulated the limit time. Through the analysis and calculation of these data, it is possible to better predict the over-limit of the stable section of power network by mastering the operation section of power network.

4.2 Algorithm Evaluation Criteria

In addition to using the AEC function proposed in this paper as an evaluation criterion, in order to verify the effectiveness of the ensemble prediction algorithm and the accuracy of the evaluation criteria, other traditional evaluation criteria have also been used. The first prediction model uses a common error cost function as a predictive evaluation criterion. The error cost function used in the traditional time series analysis method is usually mean square error (MSE), mean absolute error (MAE), mean absolute error ratio (MAER), and the like. Essentially speaking, these error cost functions are symmetric cost functions, i.e., the over-estimation and under-estimation of the error cost function are treated equally. The starting point of these functions is to measure the geometric error of the time series analysis method and guide the method to approximate the true value in the form of minimal geometric error during training.

These functions are generally suitable for most time series prediction problems, and these cost functions are defined as follows,

MSE function

$$MSE = \frac{1}{n} \sum_{t=1}^{n} (o_t - p_t)^2 \tag{16}$$

The mean absolute error (MAE) can evaluate the degree of change of the data. The smaller the value of MSE, the better the accuracy of the prediction model to describe the experimental data. And o_t represents the measured value and p_t represents predicted value.

MAE function

$$MAE = \frac{1}{n} \sum_{t=1}^{n} |(f_i - y_i)| \tag{17}$$

The average absolute error (MAE) is the average of the absolute error, it can better reflect the actual situation of the prediction error. In Eq. (17), f_i represents the predicted value and y_i represents the true value.

MAER function

$$MAPE = \sum_{t=1}^{n} \left| \frac{o_t - p_t}{o_t} \right| \times \frac{100}{n} \tag{18}$$

The larger the MAER (mean absolute error ratio) value, the greater the difference between the predicted value and the original value.

4.3   Time Granularity Selection

The time series obtained by different time granularity clustering can be seen in Fig. 4. The X axis represents the time granularity and the Y axis represents the anomaly. We can see that the larger the time granularity, the greater the anomaly at each time point.

For large time granularity, due to the large number of bases, even small prediction deviations may make the absolute error of prediction large, resulting in a large prediction error. Selecting too small time granularity is also not suitable for time series analysis. Too small time granularity can make the data at each time point statistically insignificant. In addition, the time series thus clustered shows irregularities in a number of indicators.

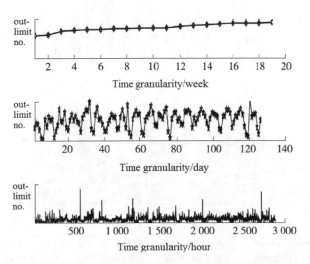

Fig. 4. Time series with different time granularity.

4.4   Results Comparison Among TSBFA and Single-Model Forecasting Algorithms

In order to evaluate our TSBFA algorithm in this paper, we select four representative single-model algorithms, i.e. sliding window average forecasting (SWAF for short) [12], regression analysis (RA) [13], artificial neural network (ANN) [14], support vector machine (SVM) [15] as comparison references. According the running results, we can calculate their evaluation indicators, MAE, MSE, MAER, respectively. In order to more intuitively compare the difference between our algorithm and other algorithms, we give its histogram form as shown in Fig. 5.

From Fig. 5, we can find that ensemble learning algorithm is always the most optimal method, or with an optimal method of effect on the whole. In total, ensemble learning algorithm integrates the advantages of all which can be used as the most accurate method to use.

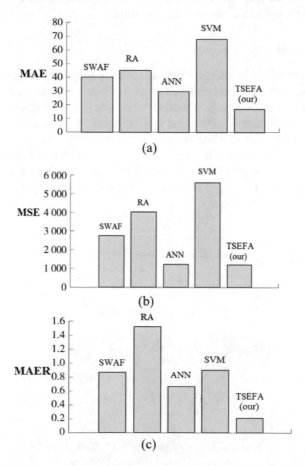

Fig. 5. Result comparison of five algorithms under three evaluation criteria. Here, (a), (b), (c) represent the comparison results of mean absolute error (MAE), average absolute error (MSE) mean absolute error ratio (MAER), respectively.

In order to better evaluate the superiority of our algorithm, we also compare the performance of the algorithm by selecting different time windows. In our experiments, we divided the Unit time window as three different division, i.e. 1 h (window 1), 8 h (window 2), 16 h (window 3). Under different time window, we can analyze the prediction performance results of different forecasting models (shown in Fig. 6).

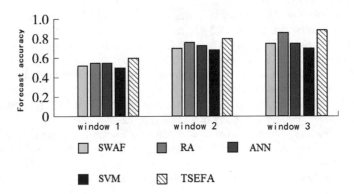

Fig. 6. Average absolute ratio error evaluation results under different time window

As shown in Fig. 6, when the time window increases (window 1 < window 2 < window 3), the prediction accuracy of each algorithm tends to increase. And At the same time window, the prediction accuracy of TSEFA maintains a higher level than the other four algorithms.

5 Conclusion

As we all know, out-limit forecast of grid stable section is an important point of grid operation and control. However, due to the large amount of grid stable section data in power grid, traditional single classification forecasting algorithms are difficult to predict efficiently and accurately. In order to solve this problem, we proposed TSEFA for out-limit detection on grid stable section. This algorithm integrates multiple classification prediction algorithms, which classifies and predicts the collecting grid stable section data and realizes the output prediction, and predicts the comprehensive optimization precision. Through the comparison of some other algorithms and the relevant evaluation indexes, the efficient and accurate prediction results are achieved, and the security and stability of the grid stability section analysis platform are improved. Through experiment, ensemble prediction algorithm obtains the better prediction accuracy and effect, overall. The accuracy of the ensemble algorithm has advantages in various evaluation index, and reaches a similar level with the algorithm with the best effect.

Acknowledgement. This work is supported by Science and Technology Project of Guangdong Power Grid Co., Ltd. "New Generation Grid Dispatch Operation Data Storage and Analysis Service Based on Big Data" (No. 036000KK52160030).

References

1. Dong, Y., Zhang, J., Garibaldi, J.M.: Neural networks and AdaBoost algorithm based ensemble models for enhanced forecasting of nonlinear time series. In: International Joint Conference on Neural Networks, pp. 149–156. IEEE (2014)
2. Kapoor, P., Bedi, S.S.: Weather forecasting using sliding window algorithm. ISRN Sig. Process. **3**(1), 1–5 (2013)
3. Punzo, P.D., Robust, M.: Clustering in regression analysis via the contaminated gaussian cluster-weighted model. J. Classif. **34**(1), 1–45 (2017)
4. Park, D.C., El-Sharkawi, M.A., Marks, R.J.I., Atlas, L.E., Damborg, M.J.: Electric load forecasting using an artificial neural network. IEEE Trans. Power Syst. **6**(2), 442–449 (1991)
5. Percival, D.B., Walden, A.T.: Wavelet Methods for Time Series Analysis, vol. 6, no. 1, pp. 13–20. China Machine Press, Beijing (2004)
6. Wan, C.H., Yang-Ping, O.U.: Deformation data of dam monitoring based on wavelet analysis method. Beijing Surveying & Mapping, pp. 2–5 (2010)
7. Song, C., Zhang, Q.: Sliding-window algorithm for asynchronous cooperative sensing in wireless cognitive networks. In: IEEE International Conference on Communications, pp. 3432–3436. IEEE (2008)
8. Walker, E.: Applied regression analysis and other multivariable methods. Technometrics **31**(1), 117–118 (2008)
9. Ostad-Ali-Askari, K., Shayannejad, M., Ghorbanizadeh-Kharazi, H.: Artificial neural network for modeling nitrate pollution of groundwater in marginal area of Zayandeh-rood River, Isfahan, Iran. KSCE J. Civil Eng. **21**(1), 1–7 (2016)
10. Chen, N., Lu, W., Yang, J.: Support Vector Machine, pp. 24–52. Springer, London (2016)
11. Janacek, G.: Time series analysis forecasting and control. J. Time **31**(4), 303 (2012)
12. Alberg, D., Last, M.: Short-term load forecasting in smart meters with sliding window-based ARIMA algorithms. In: Nguyen, N.T., Tojo, S., Nguyen, L.M., Trawiński, B. (eds.) ACIIDS 2017. LNCS (LNAI), vol. 10192, pp. 299–307. Springer, Cham (2017). https://doi.org/10.1007/978-3-319-54430-4_29
13. Zhang, H., Wang, P., Sun, J.: Regression analysis of interval-censored failure time data with possibly crossing hazards. Stat. Med. **37**(5), 497 (2018)
14. Folkes, S.R., Lahav, O., Maddox, S.J.: An artificial neural network approach to the classification of galaxy spectra. Mon. Not. R. Astron. Soc. **283**(2), 651–665 (2018)
15. Yuan, Y.: Canonical duality solution for alternating support vector machine. J. Ind. Manag. Optim. **8**(3), 611–621 (2017)

Improvement of Wireless Positioning Algorithm Based on TOA

Shengyu You[1(✉)], Shengqing Liu[2], and Hongling Wang[1]

[1] East China University of Technology, Nanchang 330013, China
youshengyu@ecit.cn
[2] Jiangxi Expressway Connection Management Centre,
Nanchang 330011, China

Abstract. With the development of Internet of Things (IoT) and Internet of Everything (IoE) technology and continuous promotion of industrialization, wireless positioning technology is playing an increasingly important role in the intelligent field. The location services of the high precision, high coverage and low energy overhead is a hot research issues. This paper studies the time of arrival (TOA) positioning algorithm for wireless positioning technology. To deal with the problem of the positioning accuracy decreased while the of TOA algorithm is reduced in NLOS measurement, an improved TOA algorithm is proposed. Through MatLab simulation experiment, the simulation results confirm the feasibility and effectiveness of the improved algorithm.

Keywords: Wireless location technology · Localization algorithm · Time of arrival

1 Introduction

Wireless location technology analyzes received radio wave signals. Techniques for calculating goals based on a specific algorithm, to get information such as transmission time, phase, and angle of arrival [1, 2]. Wireless location technology is now widely used in communications, logistics, navigation, transportation and military and other fields [3].

This paper studies the wireless location algorithm based on time of arrival (TOA). For TOA algorithm, the problem of declining positioning accuracy in non-line-of-sight measurement, to improved TOA algorithm. The simulation verified the effectiveness of the improved algorithm.

2 The Principle of the TOA Algorithm

Time of Arrival (TOA) technology measures distance by measuring signal propagation time. In the TOA method, if the radio wave travels from the unknown node to the reference node, the propagation time is t. The distance from the unknown node to the reference node is R = c × t, In the formula, c is the propagation speed of wireless electromagnetic waves in the air [4].

© Springer Nature Switzerland AG 2020
Q. Liu et al. (Eds.): CENet 2018, AISC 905, pp. 917–923, 2020.
https://doi.org/10.1007/978-3-030-14680-1_99

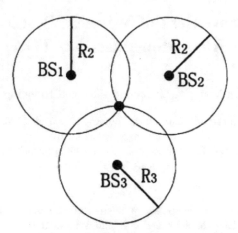

Fig. 1. TOA positioning technology.

Assume that the coordinates of the unknown node are (x, y), the coordinates of the three reference nodes BS_1, BS_2, BS_3 are (x1, y1), (x2, y2), (x3, y3). After the distances from the unknown node MS to the three reference nodes BS1, BS2, and BS3 are measured as R1, R2, and R3. As shown in Fig. 1, three reference nodes can be used as the center. The distance from the unknown node MS to the three reference nodes makes three circles for the radius, and lists the system of equations for the circle [5, 6]:

$$(x1 - x)^2 + (y1 - y)^2 = R12$$

$$(x2 - x)^2 + (y2 - y)^2 = R22$$

$$(x3 - x)^2 + (y3 - y)^2 = R32$$

Solving the equation can get the coordinate of the intersection of the three circles, which is the coordinates of the unknown node.

The arrival time TOA positioning algorithm requires that the measurement signal arrives at the reference node after it is sent from the unknown node. Therefore, it is required that the nodes have very accurate clock synchronization, and therefore the cost is also increased. Because the calculated distance completely depends on the measurement time, the TOA algorithm requires a very high time synchronization of the system. All existing time errors are magnified many times, even if they are very small. In the non-line-of-sight environment, due to the influence of multipath propagation will bring greater errors, so the application of simple TOA positioning in practice is rarely.

3 Improved TOA Algorithm

For the TOA algorithm in the non-line-of-sight propagation environment, due to the multipath effect caused by the increase of the distance of the radio wave from the unknown node to the reference base station, an improved optimization algorithm is proposed. The main idea of the algorithm is to use the three base station circumferential intersection line positioning method to measure the three points intersected by the three intersection lines to form a triangle. The coordinates of the unknown node are the coordinates of the center of mass of the triangle.

3.1 The Algorithm Flow of TOA Algorithm in NLOS

The idea of the algorithm is to first calculate the intersection of two circles. and determine the distance between the two intersection points to the center of the third circle, and take the point where the distance is small. By analogy, the three points from each of the three groups of intersection points to the third circle center point are taken out to form a triangle. The center of mass of the triangle is the coordinates of the unknown node, as shown in Fig. 2.

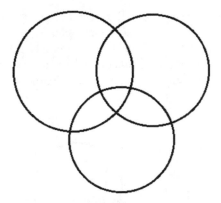

Fig. 2. TOA algorithm under NLOS.

In the actual environment, especially the urban environment, signal propagation paths between base stations and mobile nodes are often obstructed by obstacles. Radio wave signals can only be propagated by non-line-of-sight propagation (NLOS) methods such as refraction and reflection. At this point, the signal path distance from the mobile node is definitely longer than the path distance under the LOS propagation LOS. In the NLOS channel, the distance from the mobile station to the base station is measured to be greater than or equal to the actual distance from the mobile station to the base station:

$$D > R$$

Where D is the distance measured in the NLOS channel, R is the distance measured in the LOS channel. That is, the actual distance from the mobile station to the base station. Since D is always not less than the true value of the mobile station to the base station, it can be considered that the distance from the coordinates of the intersection of the two or two circles to the base station under NLOS channel transmission is always not less than the distance from the actual mobile station coordinates to the base station.

Since the distribution of NLOS in the TOA measurement is related to the distribution of obstacles on the radio wave propagation path, So the NLOS error is random [7, 8], that is:

$$D = R + N * rand(1)$$

In the formula, D is a measure containing NLOS error. Relative to NLOS error, and measurement noise is zero. The Gaussian stochastic process with small mean standard deviation can be temporarily ignored. Rand (1) is a random number from 0 to 1, N is the maximum non-line-of-sight propagation error value, as show in Fig. 4 value setting. At this point, three reference base stations are used as the center of gravity. The three circles made by the unknown distance from the unknown node to the base station no longer intersect at a single point, but to form a region.

3.2   Improved TOA Algorithm Flow in NLOS

The idea of the algorithm is to transform three circles into two intersecting lines. Solved by three measurements, to get three measurement intersections, and find their centroids to determine the position of the mobile station. This method requires at least 3 base stations. The algorithm focuses on the situation where the base station is in non-line-of-sight, simple structure, and small amount of calculation and great improvement in positioning accuracy, as shown in Fig. 3.

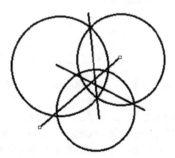

Fig. 3.   TOA improved algorithm schematic.

The improved intersecting line algorithm model first obtains TOA circumferential measurement equations for 3 base stations, then the equations of the three intersecting lines are obtained by subtracting the two equations of the circumferential measurement equation. For base station 1 and base station 2, the two measurement equations are subtracted, the resulting intersection line equation is:

$$2(x1 - x2)x + x2^2 - x1^2 + 2(y1 - y2)y + y2^2 - y1^2 = d2^2 - d1^2$$

Similarly, the equation of the intersection line passing through the intersection point of the base station 2 and the base station 3 is:

$$2(x2 - x3)x + x3^2 - x2^2 + 2(y2 - y3)y + y3^2 - y2^2 = d3^2 - d2^2$$

The equation of the intersection line passing through the intersection of base station 1 and base station 3 is:

$$2(x1 - x3)x + x3^2 - x1^2 + 2(y1 - y3)y + y3^2 - y1^2 = d3^2 - d1^2$$

Combine the three intersecting lines to get three points, finding the coordinates of the center of mass is ((x1 + x2 + x3)/3, (y1 + y2 + y3)/3), which is the coordinates of the unknown node.

For improved positioning algorithm, there are two points that require special attention: The first, in a two-dimensional plane positioning system, improved positioning algorithm requires a minimum of 3 reference nodes. When the number of reference nodes is more than 3, the least square method can be used to solve the system of equations. The second, analysis of the improved positioning algorithm from the equation requires that the base station can not be on the same vertical or horizontal line. It is possible to get the coordinates of multiple unknown nodes. If the base station is on the same vertical or horizontal line, you need to reselect the reference base station for positioning.

4   Algorithm Simulation Analysis

Three reference nodes and one unknown node coordinate are randomly generated in the simulation experiment. The coordinates of the unknown node are calculated by the design position function [9], and compare it with the set coordinates. Calculate the mean square error.

Set the maximum non-line-of-sight propagation error value N = 1, perform 20 consecutive simulations. Set N = 5, perform 20 consecutive simulations. Set N = 10, perform 20 consecutive simulations. As shown in Fig. 4 the comparison data between the original algorithm and the improved algorithm.

According to the experimental data can be seen, when the non-line-of-sight propagation error N = 1, the average error of the improved TOA algorithm is 3.4776385, and the average error of the improved algorithm is 1.00110585. It can be seen that when the non-line-of-sight propagation error is small, the improved TOA algorithm results in more accurate positioning results. Simultaneously, at N = 5, the average error of the improved TOA algorithm measured is 11.123775, and the average error of the improved algorithm is 4.6470885, at N = 10, the average error of the pre-improvement TOA algorithm measured is 21.03031, the average error of the improved algorithm is 14.47108. At this time, the non-line-of-sight propagation error is large. So improved positioning accuracy of the algorithm has been greatly improved.

	N=1			N=5			N=10	
	Before improving	After improving		Before improving	After improving		Before improving	After improving
	2.6862	1.2636		14.3544	3.8928		25.7854	4.7844
	3.5765	0.919		9.4687	2.1707		18.5328	7.7999
	1.2563	0.3635		3.7897	12.2124		15.4854	23.7624
	1.8624	0.3219		16.4856	1.9564		19.7892	12.3463
	0.8424	1.7919		13.4646	6.7764		34.8792	37.4968
	9.4631	2.3649		11.4726	2.2397		29.4452	7.3139
	4.5423	2.0589		9.4268	11.0324		11.8782	18.0896
	2.7521	0.8871		8.7413	7.1047		19.4652	9.031
	4.4566	0.1412		16.4832	4.0679		38.4564	4.3987
	2.4562	0.4747		15.1495	5.5568		19.7265	1.2772
	3.4209	0.4678		7.183	1.8786		22.4791	5.2911
	0.6575	0.6833		9.4287	2.3139		17.7264	20.499
	2.4657	0.6826		20.9274	1.582		11.8356	9.7813
	6.6876	0.5108		9.4383	3.0401		33.7162	21.5798
	1.4567	3.7666		15.1795	3.3523		28.4019	11.7496
	3.7698	0.3254		7.9681	3.2814		9.4209	46.2582
	5.4564	0.6301		14.7872	2.1918		7.0652	16.7746
	8.4564	0.8449		7.1158	2.1781		19.4068	13.3963
	0.4892	0.6599		3.4992	13.8351		22.9271	9.2132
	2.7985	0.903		8.1257	2.2765		14.1836	8.5783
Average error	3.4776	1.0011		11.1238	4.6471		21.0303	14.4711

Fig. 4. Experimental data results.

5   Conclusion

This paper analyzes the TOA algorithm in the non-line-of-sight propagation environment and brings a huge error due to multipath effect, an improved TOA algorithm is proposed. And through MatLab simulation software for simulation experiments, experimental results show that the improved algorithm has greatly improved positioning accuracy in non-line-of-sight environments. Although the improved positioning algorithm also has a certain degree of improvement, however, the average error of the algorithm is still large, therefore, this TOA-based improved algorithm still has deficiencies that need to be improved.

Acknowledgement. This study is supported by Open Fund of Jiangxi Engineering Laboratory on Radioactive Geoscience and Big data Technology (No. JELRGBDT201708).

References

1. Liu, Z., Wu, Y.P., Fan, X.: Improvement of wireless positioning algorithm based on TDOA. Fire Control Command Control **39**, 52–54 (2014)
2. Liang, J.Z.: Wireless Positioning System, pp. 36–37. Electronincs and Industry Press, Beijing (2013)
3. Abouzar, P., Michelson, D.G., Hamdi, M.: RSSI-based distributed self-localization for wireless sensor networks used in precision agriculture. IEEE Trans. Wirel. Commun. **15**(10), 6638–6650 (2016)

4. Zhang, H.T.: Research on the Localization Technology of Wireless Sensor Networks. Beijing Jiaotong University, Beijing (2017)
5. Jamali, R.H., Leus, G.: Sparsity-aware multi-source TDOA localization. IEEE Trans. Sig. Process. **61**(19), 4874–4887 (2013)
6. Jin, C., Ye, C., Han, Z.B.: DV-hop localization algorithm improvements in wireless sensor network. Comput. Eng. Des. **34**(2), 45–54 (2013)
7. Zhang, D., Fang, Z.Y., Sun, H.Y.: A range-free location algorithm based on homothetic triangle cyclic refinement in wireless sensor network. Emerg. Sources Citation Index **8**(2), 1–15 (2017)
8. Chee-Yee, C., Srikanta, P.: Sensor networks evolution opportunities and challenges. Proc. IEEE **91**(8), 1247–1256 (2008)
9. Chen, X.Y., Ke, W., Du, L.: Simulation of target location algorithm for internet of things in matlab. J. Wuhan Inst. Technol. **37**(3), 69–73 (2015)

Method of Complex Internet of Things Simulation Attack and Reliability Analysis

Bingwei Chen[✉], Zidi Chen, Hecun Yuan, and Lanlan Rui

State Key Laboratory of Networking and Switching Technology,
Beijing University of Posts and Telecommunications, No. 10 Xitucheng Road,
Haidian District, Beijing, China
584490885@qq.com

Abstract. In recent years, with the rapid development of Internet of Things, the scale of networks has gradually increased, link capacity has become higher and higher. In this context, the research on the reliability of complex networks has been paid more and more attention. The network reliability analysis and evaluation technology need to be implemented by a corresponding network simulation and reliability comprehensive test analysis system, and presented to the user in a variety of visual forms more intuitively, so as to be further applied to the actual network environment. Simulate the network attack mode under complex network environment and get the reliability evaluation results under different attack strategies. Establishing complex network simulation attacks and analyzing them is of great significance in evaluating the reliability of complex IoT topologies. In this paper, a complex network evaluation model was further established. At the same time, three kinds of strategies are designed to simulate attacks on complex networks. Based on Django and the smart home network, the method is implemented in a more appropriate form of visualization. Finally, the method is tested and analyzed. The results show that the method can better represent the process of complex network simulation attacks and the changes of network reliability indicators.

Keywords: IoT · Simulation attack · Reliability analysis

1 Introduction

In 2005, the International Telecommunication Union released the "ITU Internet Report 2005: Internet of Things" at the Information Summit held in Tunis, formally proposed the concept of "IoT", and pointed out that the era of "IoT" communication in the smart world is approaching [1]. With the realization of the IoT from theory to practical applications, ensuring the efficient and reliable work of the system has become an issue that the IoT must consider. Whether the IoT is working properly and is reliable has become a major obstacle to the successful promotion of IoT technology, so it is necessary to assess the reliability of the IoT network system.

The network reliability analysis and assessment technology need to be implemented by a corresponding network simulation and reliability comprehensive test analysis system, and presented to the user in a variety of visual forms for further application to

Q. Liu et al. (Eds.): CENet 2018, AISC 905, pp. 924–933, 2020.
https://doi.org/10.1007/978-3-030-14680-1_100

the actual network environment. Since the risk factors that affect reliability such as faults or damages in a real complex network may have negative effects, it is imperative to develop a network simulation and reliability comprehensive analysis system to simulate the status and behavior of real networks, and to risk factors to understand the reliability of complex IoT under various test scenarios. In modern complex network scenarios, such as the smart home which is a key part of the smart grid is an important means for realizing real-time interaction between users and grid, enhancing the capacity of integrated services of the grid, meeting the demand of interactive marketing and improving the quality of service [2].

The innovations in this paper are: (1) Selecting appropriate indicators and cut-in dimensions to establish a reliability assessment model for complex networks, which can effectively evaluate the reliability of the network; (2) Three selective attack strategies and methods for constructing initial solutions are proposed. The complex network simulation attack model can reproduce the reliability change process of the attacked network.

2 Related Work

The study of network reliability problems can be traced back to Mr. Lee's research on telecom switching networks, which mainly focused on the field of communication networks [3]. It mainly discusses the reliability of IoT in terms of wireless link layer, transport layer and routing, application layer, and system architecture. Research on network reliability at home and abroad has had a lot of work. For the issue of simulated attacks on complex networks [4], many people have also conducted research in different directions. In Wu Ting's research, a weighted network is used to analyze the reliability of nodes, and different types of complex networks are simulated and analyzed under certain metrics [5].

However, there are still many deficiencies and pertinences, such as visual simulation level, easy operation and maintenance personnel operation; For example, the tacticality of simulated attacks and the diversity of coverage attacks; For example, the reliability assessment model can closely describe the reliability of the network. Therefore, for the evaluation of complex IoT, a suitable method is needed to establish the evaluation criteria and present it through appropriate means. With the idea of data visualization, the reliability indicators can be intuitively presented, and at the same time, the changes in network performance can be intuitively presented.

3 Model Establish

3.1 Complex Network Reliability Evaluation Model

Reliability Evaluation Index
When a node attack sequence K is constructed in advance and the nodes in the network G are attacked one by one according to the sequence of nodes in K, the size of the

largest connected component of the network G will continue to decrease as the attack progresses. The reliability measure $R(G, K)$ of network G under attack sequence K can be defined as:

$$R(G,K) = \frac{1}{n} \sum_{i=1}^{n} S(i)$$

(3.1)

Where n is the number of nodes originally included in the network G, and $S(i)$ is the size of the largest connected component after the i nodes in the network G are removed.

However, the drawback of the above standard is that it can only explain the degree of network connectivity at a relatively coarse granularity. If the index is used to measure the overall performance of the network, it is not accurate enough. Therefore, the indicators need to be improved. Definition: In network S, there are a total of m connected components. The number of nodes contained in the $i(0 \leq i \leq m)$-th connected component is n_i. Then the availability index U of the network S is:

$$U = \frac{\sum_{i=1}^{m} n_i(n_i - 1)/2}{n(n-1)/2} = \frac{\sum_{i=1}^{m} n_i(n_i - 1)}{n(n-1)}$$

(3.2)

When attacking network G with an attack sequence K, the availability-based measure of network invulnerability is defined as:

$$R(G,K) = \frac{1}{n} \sum_{i}^{n} U(i)$$

(3.3)

Smart Home Network

There are a lot of complex IoT and their comprehensive reliability needs further analysis. For example, the smart home based on IoT is an important means for realizing real-time interaction between users and the grid. The architecture of IoT consists of three layers: the perception layer, the network layer and the application layer. The perception layer of smart home is mainly composed of sensors, machines and terminals, which is responsible for the perception and collection of data in a way similar to how human sensory organs function. Telecommunication network transmission channel constitutes the network layer, including public 4G mobile network, WiFi, ZigBee, etc. which passes information to the processing and application layer in the way like nervous system does to the brain. The application layer is the brain of the system, in charge of data processing [2].

Based on these three parts of the network architecture, three systems are designed, namely a series system, a parallel system, and a series-parallel hybrid system. For a simple series system, the reliability R is defined as the product of the reliability Rr of each subsystem, which is:

$$R = \prod_{r=1}^{n} R_r$$

(3.4)

For a parallel system, if all subsystems fail, the system fails. Similarly, the definition of the parallel system reliability F is:

$$F = \prod_{r=1}^{n} F_r = 1 - \prod_{r=1}^{n} (1 - R_r) \tag{3.5}$$

3.2   Complex Network Simulation Attack Model

In the complex network simulation attack model, three attack strategies are proposed. In order to control the cost, a node attack sequence and an attack initial solution are proposed.

Selective Attack Strategy
Assuming that the network G has V nodes and E edges, denoted as G = (V, E). This article proposes three selective attack strategies defined as follows:

(1) Degree-first [6] attack strategy: the algorithm attacks the node according to the initial topology graph in descending order of node degree size.
(2) Betweenness-first [7] attack strategy: according to the initial topology graph, the algorithm attacks the nodes according to the descending order of the node intervening numbers.
(3) Random attack strategy: According to the initial network topology diagram, this attack strategy does not consider the centrality and attacks the nodes in a random order.

Initial Solution
In order to improve the performance of the algorithm, a method for constructing an initial solution is proposed:

(1) For the betweenness-first attack strategy, if there are n nodes in the network, the labels are $1, 2, \ldots, n$. The betweenness of node i is defined as

$$BT(i) = \sum_{1 \leq i,j \leq n; i \neq p \neq q} c_{pq}^{i} / c_{pq} \tag{3.6}$$

Where c_{pq} is the number of shortest paths between nodes p and q, c_{pq}^{i} represents the number of nodes i that have the shortest path between nodes p and q.

(2) For the degree-first attack strategy, if there are n nodes in the network, the labels are $1, 2, \ldots, n$ respectively, and the degree-centrality of node i is defined as

$$DG(i) = hi \tag{3.7}$$

Where hi is the number of edges directly connected to node i.

Simulation Attack Algorithm
Based on the above considerations, the simulation attack algorithm is designed as Fig. 1.

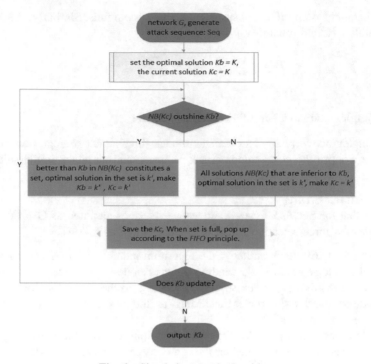

Fig. 1. Simulation attack algorithm.

Detailed steps are as follows:

Step 1: For a given network G, randomly generate an attack sequence Seq, initialize an empty cost bad attack set, set the optimal solution to $Kb = K$, and the current solution to $Kc = K$;

Step 2: Determine whether the neighborhood $NB(Kc)$ of Kc has an attack sequence superior to Kb. If yes, go to Step 3, otherwise go to Step 4;

Step 3: All solutions better than Kb in $NB(Kc)$ form a set, the optimal solution in the set is K', let $Kb = K'$, $Kc = K'$, perform Step 5;

Step 4: All solutions in $NB(Kc)$ that are inferior to Kb and are not stored in the bad attack set constitute a set. The optimal solution in the set is K', $Kc = K'$, and Step 5 is performed;

Step 5: Save Kc in the bad attack set. When the bad attack set is full, the attack sequence pops up according to the *FIFO* principle;

Step 6: Whether Kb is updated within a certain number of iteration steps, and if so, returns to Step 2, otherwise the search ends and Kb is output as the optimal solution.

4 Simulation

4.1 Simulation Topology

Design the smart home network topology as shown in the Fig. 2 to land. It includes three parts: The perception layer, the network layer and the application layer.

Fig. 2. Smart home network.

4.2 Simulation Results

Dimension One: Node Degree
Take the degrees of node in strategy one as the first dimension of comparison. The result is shown in the Fig. 3. As can be seen from the figure, after the nodes are continuously attacked, the sum of the node degrees of the three attack strategies is decreasing. However, strategy 1 is to attack based on the degree of node, so the rate of decline is the fastest among the three. The second is strategy 2, because strategy 2 is based on the betweenness, and its principle has a certain relationship with the importance of the node. As for strategy 3, there is no regular pattern, it is a random attack, and its rate of decline is irregular.

Dimension Two: Betweenness
Take the betweenness in strategy 2 as the second dimension of comparison as Fig. 4. It can be seen from the figure that when the number of attacked nodes increases, the sum of betweenness of the nodes of the three strategies is all reduced, but strategy 2 is based

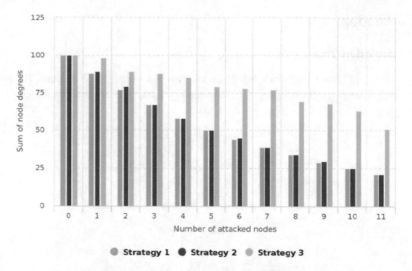

Fig. 3. First dimension of comparsion.

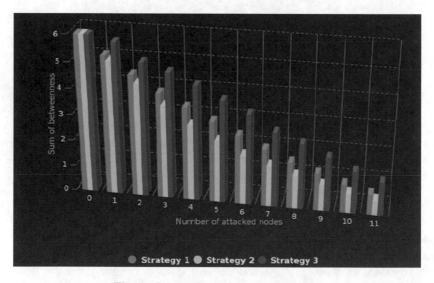

Fig. 4. Second dimension of comparsion.

on betweenness, so the rate of decline is the fastest among the three. Followed by strategy 1, because the strategy is based on degree, there is a great degree of correlation between the degree of the node and the importance of the node. The strategy 3 is a random attack strategy, and its decline has no rules, and the degree of damage to the network performance is minimal.

Dimension Three: Overall Reliability Measurement

The overall reliability measurement in the aforementioned reliability assessment model is taken as the third dimension of comparison as Fig. 5. From the comparison chart of the above three strategies, we can see that the network performance under the random attack strategy is relatively even, the overall reliability index of the network showed irregular fluctuations. After experiencing certain attacks, the reliability index of the entire network gradually dropped to zero. For Strategy 1 and Strategy 2, the attack effects of the two are all significant. Among them, attacks against the betweenness cause most of the middle performance fluctuations. The reasons for this jitter are as follows: After several rounds of attacks, there may be points in the network where the degree is not high and the betweenness is relatively large. After attacking this type of node, the reliability index may instead rise for the entire network. Therefore, when there are many such nodes in the network, the reliability index of the network will fluctuate greatly.

Fig. 5. Third dimension of comparsion.

Algorithm Comparison

For a network G with n nodes, if the attack sequence is simply constructed, the problem size will reach n!, even if it is a network with only 20 nodes, the solution space scale will reach 10^{18}. Therefore, it is necessary to consider the node attack sequence under the construction cost, that is, an initial solution is beneficial to improve the performance of algorithm.

It can be seen from Fig. 6 that the performance of the Algorithm 2 based on the initial solution and the bad attack set is significantly better than the Algorithm 1 without the initial solution. At the beginning, the difference between the two algorithms is not large; By 60 s, the reliability of the network topology in Algorithm 2 has dropped by half, while Algorithm 1 has just dropped a little bit; By 120 s, the network topology reliability in Algorithm 2 is almost zero, that is, the network has been completely attacked, while Algorithm 1 only drops by half; until 180 s, the network in Algorithm 1 is completely paralyzed. Based on the previous description, the conclusion is that the Algorithm 2 based on the initial solution and the bad attack set can obviously reduce the performance loss of the attack.

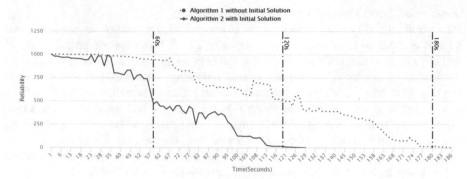

Fig. 6. Algorithm comparsion.

5 Conclusions

At present, with the development of complex IoT networks in various fields, network reliability has received increasing attention as an important aspect of ensuring network security. Based on the structure and properties of complex networks, this paper selects suitable indicators as evaluation criteria for complex networks and establishes a reliability assessment model. Based on the indicators, a simulation attack model is established. This paper designs three attack strategies which define the attack sequences based on different central indicators. Then based on Django and smart home network this method is implemented with visual modeling. Finally, the simulation results show that this method can well present the complex network simulation attack process and the changes of the network reliability index. In the future, how to make the process of simulated attacks more relevant to the actual situation is a direction worthy of in-depth research.

Acknowledgments. The work presented in this paper was supported by the National Natural Science Foundation of China (61302078, 61702048), 863 Program (2011AA01A102).

References

1. Islam, K., Shen, W., Wang, X.: Wireless sensor network reliability and security in factory automation: a survey. IEEE Trans. Syst. Man Cybern. Part C **42**(6), 1243–1256 (2012)
2. Chen, X.Z., Zhen, J.M., Liu, J., Zhen, Y.: Applications of IoT in smart grid. Electricity **33**(2), 15–19 (2012)
3. Ni, X.P., Mei, S.W., Zhang, X.M.: Transmission lines' vulnerability assessment based on complex network theory. Autom. Electr. Power Syst. **32**, 1–5 (2008)
4. Wang, J., Sun, E., Xu, B., Li, P., Ni, C.: Abnormal cascading failure spreading on complex networks. Chaos Solitons Fractals Interdisc. J. Nonlinear Sci. Nonequilibrium Complex Phenomena **91**, 695–701 (2016)
5. Opsahl, T., Agneessens, F., Skvoretz, J.: Node centrality in weighted networks: generalizing degree and shortest paths. Soc. Netw. **32**(3), 245–251 (2010)

6. Meghanathan, N.: Spectral radius as a measure of variation in node degree for complex network graphs. In: International Conference on u- and e-Service, Science and Technology, pp. 30–33. IEEE (2015)
7. Nie, T., Guo, Z., Zhao, K., Lu, Z.M.: The dynamic correlation between degree and betweenness of complex network under attack. Phys. A Stat. Mech. Appl. **457**, 129–137 (2016)

Distributed Anomaly Detection Method in Wireless Sensor Networks Based on Temporal-Spatial QSSVM

Zhili Chen[1,2], Huarui Wu[3(✉)], Huaji Zhu[3], and Yisheng Miao[3]

[1] National Engineering Research Center for Information
Technology in Agriculture, Beijing 100097, China
[2] College of Information and Electrical Engineering,
China Agricultural University, Beijing 100083, China
[3] Beijing Research Center for Information Technology in Agriculture,
Beijing Academy of Agriculture and Forestry Sciences,
Beijing 100097, China
wuhuarui1975@163.com

Abstract. In Wireless Sensor Networks (WSNs), abnormal sensing data is easily generated due to factors such as the harsh working environment, sensor faults and external events. In order to enhance the detection rate of abnormal data and reduce the false positive rate, we propose a distributed anomaly detection method using one-class quarter-sphere support vector machine (QSSVM) based on temporal-spatial fusion in WSNs. Firstly, according to the synthetic data, the temporal-spatial QSSVM model is trained to determine the relevant parameters. Secondly, the trained QSSVM model is used to classify the streaming data in WSNs, and the abnormal data types are classified into noise, faults and events. Finally, the method decides whether to update the classification model based on whether the new sample has an effect on the boundary of the hypersphere. The experimental results illustrate that the proposed method has a detection rate of 96% compared with other three methods, and the false positive rate is only 14%.

Keywords: Anomaly detection · QSSVM · Temporal-spatial ·
Wireless sensor networks

1 Introduction

As the main sensory network of Internet of Things (IoT) technology, wireless sensor network (WSN) is particularly important to enhance the accuracy and reliability of sensing data, providing strong support for subsequent data analysis, decision-making model and intelligent control [1]. Because of the particularity and complexity of the agricultural environment, WSNs in the agricultural Internet of Things are mainly for the fusion and analysis of multi-source data rather than just analyzing a single type of data. Therefore, it has higher requirements for the accuracy and reliability of the sensing data [2].

© Springer Nature Switzerland AG 2020
Q. Liu et al. (Eds.): CENet 2018, AISC 905, pp. 934–943, 2020.
https://doi.org/10.1007/978-3-030-14680-1_101

The reasons for the abnormal data generated by the agricultural scene mainly include: (1) Noise and faults mainly caused by the hardware and software faults of the sensor itself. (2) Specific events occur in the area where the sensor nodes are distributed. The abnormal data generated by the specific events requires timely warning and urgent measures [3]. Due to the above reasons, the abnormal sensing data seriously reduces the quality of the data, and wastes a lot of valuable and limited network resources. This paper mainly studies the sensor faults and external event detection scheme in WSNs. The successful implementation of the scheme has important theoretical significance and application value for ensuring the reliability of WSNs.

The remainder of this paper is organized as follows. Section 2 presented the related work on anomaly detection technologies. The proposed distributed TSAD method in WSNs based on Temporal-spatial QSSVM is explained in Sect. 3. Experimental results and analysis are informed in Sect. 4. The conclusion and future study are described in Sect. 5.

2 Related Work

Anomaly detection is an important application and a research hotspot in WSNs. The latest definition of anomalous data is given by Sadik et al.: An anomaly is a data point that is very different from other data or conforms to the defined anomalous behavior [4]. Related scholars have proposed a series of anomaly detection methods applied in WSNs. There are a statistical-based, nearest neighbor-based, artificial Intelligence-based, classification-based approaches [2]. Gil et al. studied SVM-based techniques for Anomaly detection in wireless sensor network environments. The machine learning technology is based on LS-SVM and Sliding window learning [5]. Yao et al. proposed an anomaly detection method based on QSSVM algorithm and applied it [6]. Cheng et al. proposed two lightweight anomaly detection algorithms LADS and LADQA for WSNs. The results illustrate that compared with QSSVM, the algorithm can maintain low time complexity and maintain the detection accuracy [7]. Rajasegarar et al. found that QSSVM has less network communication overhead than CESVM in distributed anomaly detection of WSNs. Therefore, we propose a distributed anomaly detection method using QSSVM to detect anomalous nodes effectively [8, 9].

Based on the existing work, aiming at the shortcomings of previous research results, we fully exploits the inherent temporal-spatial correlation between network nodes to achieve distributed detection. We proposes a distributed anomaly detection method in WSNs based on temporal-spatial QSSVM (TSAD). The method implements detection and recognition of abnormal data and identifies types of abnormalities such as noise, faults, and external events.

3   Distributed Anomaly Detection Based on Temporal-Spatial QSSVM in WSNs

3.1   Fundamentals of the QSSVM

In this paper, we use the QSSVM to train the classifier and find the anomalies in the classification data set and it can solve the anomaly detection problem in high-dimensional data well [10]. The optimization problem of QSSVM is expressed as follows:

$$\min_{R \in \mathbb{R}, \xi \in \mathbb{R}^m} R^2 + \frac{1}{vm} \sum_{i=1}^{m} \xi_i$$

$$subject\ to: \|\phi(x_i)\|^2 \leq R^2 + \xi_i,\ \xi_i \geq 0,\ i = 1, 2, \ldots, m \tag{3.1}$$

where the normalization parameter $v \in (0, 1)$ represents the penalty factor of the abnormal data vector and it reflects the tradeoff of the minimum abnormal data and the smallest possible supersphere radius R. Parameter m represents the number of data vectors in the training set, and ξ_i is a slack variable that allows the data vector to be located outside the hypersphere to control the proportion of data that falls outside the hypersphere. The dual form is expressed as follows:

$$\min_{\alpha \in \mathbb{R}^m} - \sum_{i=1}^{m} \alpha_i k(x_i, x_i)$$

$$subject\ to: \sum_{i=1}^{m} \alpha_i = 1,\ 0 \leq \alpha_i \leq \frac{1}{vm},\ i = 1, 2, \ldots, m \tag{3.2}$$

where α_i is the Lagrangian multiplier, and some effective linear optimization techniques can be used to obtain the value of α_i.

For any boundary support vector x_i, the hypersphere radius R can be determined as $R^2 = k(x_i, x_i)$. The distance-based kernel problem can be explained by feature space center kernel matrix as Eq. (3.3):

$$\phi(x_i)_c = \phi(x_i) - \frac{1}{m} \sum_{i=1}^{m} \phi(x_i) \tag{3.3}$$

The inner product of the central image vector $K_c = \left(\phi(x_i)_c \cdot \phi(x_j)_c \right)$ can be obtained by using $K_c = K - 1_m K - K 1_m + 1_m K 1_m$ and kernel matrix $K = k(x_i, x_i) = (\phi(x_i) \cdot \phi(x_i))$, where 1_m is an $m \times m$ matrix with all values equal to $1/m$.

The main parameters of the QSSVM model include a kernel function and an abnormal penalty parameter v [11]. (1) The value of different parameters v has a significant impact on the solution of QSSVM, directly affecting the decision boundary.

(2) Different kernel functions represent machine learning algorithms that select different data characteristics. Commonly used kernel functions are as follows [12]:

$$K(x,y) = \begin{cases} (x \cdot y), \ Linear \ kernel \ function \\ (x \cdot y + 1)^p, \ Polynomial \ kernel \ function \\ \exp\left(-\|x - y\|^2 / (2\sigma^2)\right), Gaussian \ radial \ basis \ function \ (RBF) \\ \tanh(kx \cdot y - \delta), \ Sigmoid \ kernel \ function \end{cases} \quad (3.4)$$

3.2 Temporal-Spatial Correlation Analysis

The information collected by each sensor node is highly correlated with its own historical state information and its spatial location, so the sensing data has a high temporal-spatial correlation. Using the temporal-spatial correlation of sensor nodes, the collected data is analyzed to detect abnormal behaviors such as noise, faults and events in a timely and accurate manner [13].

Definition 1. Temporal correlation: Temporal correlation refers to the fact that the data collected by the sensor nodes before and after the time illustrates a certain continuity. Each node needs to save its own m measurement readings and radius information R_i. The temporal correlation formula for defining the data x_i is as follows:

$$d_i = \frac{1}{2} \left(\sqrt{\|\phi(x_i) - \phi(x_{i-1})\|^2} + \sqrt{\|\phi(x_i) - \mu\|^2} \right) \quad (3.5)$$

where $\mu = \frac{1}{m} \sum_{j=1}^{m} \phi(x_j)$ represents the average within the feature space. The first part calculates the distance between the current reading and the previous reading, and the latter part calculates the distance between the current reading and the mean value in the feature space, thus taking into account both the temporal correlation and the interference of the abnormal point.

Definition 2. Sliding window model: If the data loss exception is not considered, the sliding window size is assumed to contain b small blocks of equal size, each of which is n in length. When the data of the next sampling time enters the sliding window, the data value of the last sampling time is replaced with the data value of the new entering window. It is represented as follows:

$$\text{mod}(t_{next}, |W|) = \text{mod}(t_{before}, |W|) \quad (3.6)$$

where $\text{mod}(a, b)$ represents the remainder function.

Definition 3. Spatial correlation: Spatial correlation refers to the existence of a quantitative functional relationship between sensor node data within a certain spatial range. The spatial correlation can effectively filter out the noise data. First define the inverse distance weighting (IDW) to represent the weight of the neighbor, which is inversely proportional to the distance of the neighbor node relative to its own position, namely:

$$IDW\left(s_i, s_j\right) = \frac{1}{d\left(s_i, s_j\right)} \tag{3.7}$$

where $d(s_i, s_j)$ represents the distance between two nodes. The node obtains its own location and broadcasts its own location information according to a certain time interval, so every node can receive the location of all neighbor nodes. Considering the influence of neighbor node readings at the same time, the spatial correlation formula of node s_i is calculated as follows:

$$R_{im} = \alpha \left(\frac{\sum\limits_{j=1}^{k} IDW\left(s_i, s_j\right) R_j}{\sum\limits_{j=1}^{k} IDW\left(s_i, s_j\right)} \right) + (1 - \alpha) \frac{1}{k} \sum\limits_{j=1}^{k} \sqrt{\left\| \phi(x_i) - \phi\left(x_j\right) \right\|^2} \tag{3.8}$$

where $IDW\left(s_i, s_j\right)$ is the inverse distance weighting to represent the weight of the neighbor, $\phi(x_i)$ is the value of the new data, $\phi\left(x_j\right)$ is the neighboring value of the node s_i. The first part is the radius information of the node s_i neighbor, and the second part $\frac{1}{k} \sum\limits_{j=1}^{k} \sqrt{\left\| \phi(x_i) - \phi\left(x_j\right) \right\|^2}$ is the average value of the difference between the node s_i and the neighbor node reading, which $\alpha \in (0, 1)$ is a coefficient. We are taking into account both the spatial correlation and the neighbor's sensing data value.

When it is judged based on the temporal correlation that the data x_i is a suspected abnormal data, the node s_i compares d_i and R_{im}, and if $d_i > R_{im}$, the data x_i is abnormal data. So define the decision function as follows:

$$f(x) = \text{sgn}(R_i - d_i) \oplus \text{sgn}(R_{im} - d_i) \tag{3.9}$$

when $f(x) = -1$, the data x_i is an abnormal data, and the radius information R_i and R_{im} are important decision criteria.

3.3   Temporal-Spatial QSSVM Algorithm

This paper proposed a distributed anomaly detection method based on temporal-spatial QSSVM (TSAD). First, the TSAD algorithm uses training data to train the model, determines basic parameters, and then performs real-time distributed anomaly detection on real-time streaming data. The pseudo code of the TSAD algorithm is described as follows (Table 1):

Table 1. The pseudo code of the TSAD algorithm.

---

1 **TrainingSVM()**
2   every node counts R'_{im} and R_{im};
3   initialization **AnomalyDetectionProcess**(R'_{im}, R_{im});
4 **AnomalyDetectionProcess**(R'_{im}, R_{im})
5   when x_i is arriving at S_i
6   compute d_i;
7   if ($d_i > R_i$ and $d_i > R'_{im}$)
8   x_i is an anomaly;
9   **SourceOfAnomalyProcess**(R_i, R'_{im}, R_{im}, d_i);
10   else
11   x_i is a normal measurement;
12   if (x_i is an anomaly or border support vector)
13   initialization **UpdatingProcess**($x_{i-n'+1}$, ..., x_i);
14 **SourceOfAnomalyProcess**(R_i, R'_{im}, R_{im}, d_i)
15   compute d'_{im} using the distance of x_1^i, x_2^i, ..., x_k^i from their own origin collected by S_i;
16   if ($d_i > R_i$ and $d'_{im} > R'_{im}$)
17   if ($d_i > R_{im}$ and $d'_{im} > R_{im}$)
18   x_i is an anomaly caused by event;
19   else
20   x_i is an anomaly caused by sensor faults;
21   else
22   x_i is an anomaly caused by noise;
23 **UpdatingProcess**($x_{i-n'+1}$, ..., x_i)
24   update the training data set;
25   recompute R_i, R'_{im} and R_{im};
26   return;

---

4 Simulation Experiment and Result Analysis

4.1 Experimental Environment and Data Set

In this paper, we use the MATLAB R2014a software for simulation experiments, and 200 wireless sensor nodes with a wireless communication radius of 20 m are set in a plane area of 200×200 m. The simulated data set is injected into 200 nodes. Manually set 15 fault nodes, 40 event nodes, 120 normal nodes and 25 noise nodes. The Z-score standardization method is used to standardize the sensing data. The original value x of A is normalized to x' using Z-score. The formula is:

$$x' = (x - \mu)/\sigma \tag{10}$$

where μ and σ are the mean and standard deviation of the original data, respectively.

In this paper, Detection Rate (DR), False Positive Rate (FPR), and Receiver Operating Characteristic (ROC) curve are used to be the evaluation criteria for anomaly detection algorithms.

4.2 Simulation Experiment

We compare the four algorithms of the proposed TSAD, LAD [7], COD [6], CHD [8]. The experimental results are illustrated in the Figs. 1, 2, 3, 4, 5. Figure 1 illustrates the variation of the FPR when the TSAD uses different kernel functions. As can be seen from the Fig. 1, the parameter v is set from 0.01 to 0.25 with an interval of 0.01. Meanwhile, the FPR reaches converged. When the algorithm uses the RBF kernel function, it can maintain a low false positive rate. Therefore, the subsequent experiments in this paper we use the RBF kernel function. Figure 2 illustrates the effect of parameter v on the DR and FPR of the TSAD. The DR of the TSAD is gradually increased from the lowest 30%. When v approaches 0.25, the detection rate remains stable at about 96%. When $0 \leq v \leq 8$, the FPR tends to 0. As v increases, the FPR increases fluctuated. When v approaches 0.25, the FPR remains stable at approximately 14%.

Figure 3 illustrates a comparison of the DR of the four algorithms with the RBF kernel function. Compared with other three algorithms, the TSAD algorithm proposed in this paper gradually stabilizes with the increase of v, and maintains a relatively higher detection rate. Figure 4 illustrates a comparison of the FPR of the four algorithms with the RBF kernel function. Compared with other three algorithms, the TSAD algorithm increases with v and maintains a relatively lower false positive rate. Figure 5 demonstrates the comparison of ROC curves with RBF kernel. Compared with other algorithms, the AUC of TSAD is larger, so the TSAD algorithm performs better in distributed anomaly detection.

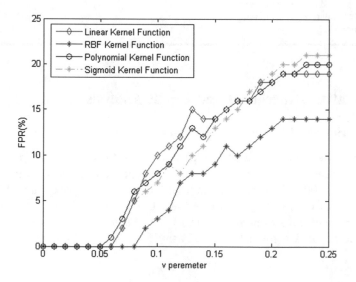

Fig. 1. Comparison of FPR under different kernel.

Time complexity analysis of the four algorithms is illustrated in Table 2, $O(L)$ explains the time complexity of solving a linear optimization problem. The simulation results illustrate that the proposed TSAD has lower time complexity and maintains the higher detection accuracy compared with other three techniques.

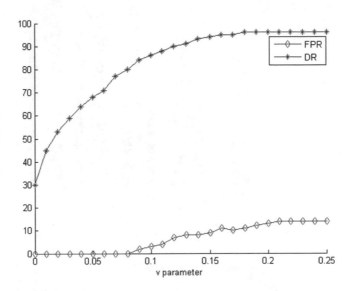

Fig. 2. Effect of v on the TSAD under the RBF kernel

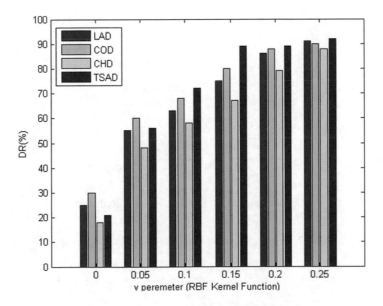

Fig. 3. Comparison of DR with RBF kernel.

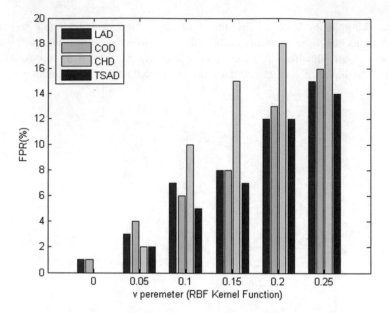

Fig. 4. Comparison of FPR with RBF kernel.

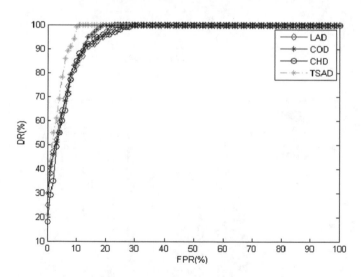

Fig. 5. Comparison of ROC curves with RBF kernel.

Table 2. Time complexity analysis of the four algorithms.

Techniques	Time complexity
LAD	$O(n \log n)$
COD	$O(n)$
CHD	$O(L)$
TSAD	$O(n \log n)$

5 Conclusion

In this paper, we propose a distributed anomaly detection method based on temporal-spatial QSSVM in WSNs. We performed simulation experiments using synthetic datasets and compared performance with the other three algorithms. Experimental results illustrate that our method can achieve higher DR and lower FPR while maintaining lower communication overhead. Our future research involves studying the effects of kernel parameters and the applying implementation at sensor nodes in practice.

Acknowledgement. This work was supported by Natural Science Foundation of China (61471067, 61571051) and Beijing Natural Science Foundation (4172024, 4172026).

References

1. Zhang, Y., Meratnia, N., Havinga, P.: Outlier detection techniques for wireless sensor networks: a survey. IEEE Commun. Surv. Tutor. **12**(2), 159–170 (2010)
2. Ayadi, A., Ghorbel, O., Obeid, A.M., Abid, M.: Outlier detection approaches for wireless sensor networks: a survey. Comput. Netw. **129**, 319–333 (2017)
3. Fei, H., Xiao, F., Li, G.H., Sun, L.J.: An anomaly detection method of wireless sensor network based on multi-modals data stream. Chin. J. Comput. **40**(8), 1829–1842 (2017). (in Chinese)
4. Sadik, S., Gruenwald, L.: Online outlier detection for data streams. In: IDEAS 2011, Proceedings of the 15th Symposium on International Database Engineering & Applications, pp. 88–96, 2011
5. Gil, P., Martins, H., Januário, F.: Detection and accommodation of outliers in wireless sensor networks within a multi-agent framework. Appl. Soft Comput. **42**, 204–214 (2016)
6. Yao, H.Q., Cao, H., Li, J.: Comprehensive outlier detection in wireless sensor network with fast optimization algorithm of classification model. Int. J. Distrib. Sens. Netw. **11**(7), 1–20 (2015)
7. Cheng, P., Zhu, M.H.: Lightweight anomaly detection for wireless sensor networks. Int. J. Distrib. Sens. Netw. **11**(8), 1–8 (2015)
8. Rajasegarar, S., Leckie, C., Bezdek, J.C., Palaniswami, M.: Centered hyperspherical and hyperellipsoidal one-class support vector machines for anomaly detection in sensor networks. IEEE Trans. Inf. Forensics Secur. **5**(3), 518–533 (2010)
9. Zhang, Y., Meratnia, N., Havinga, P.: An online outlier detection technique for wireless sensor networks using unsupervised quarter-sphere support vector machine. In: 2008 International Conference on Intelligent Sensors, Sensor Networks and Information Processing, pp. 151–156 (2008)
10. Zhang, Y., Meratnia, N., Havinga, P.: Adaptive and online one-class support vector machine-based outlier detection techniques for wireless sensor networks. In: International Conference on Advanced Information Networking and Applications Workshops, pp. 990–995 (2009)
11. Xie, M., Hu, J.K., Guo, S.: Segment-based anomaly detection with approximated sample covariance matrix in wireless sensor networks. IEEE Trans. Parallel Distrib. Syst. **26**(2), 574–583 (2015)
12. Garg, S., Batra, S.: A novel ensembled technique for anomaly detection. Int. J. Commun. Syst. **30**(11), 1–16 (2017)
13. Li, M.: Study on algorithms for outlier detection in wireless sensor networks. Jiangsu University (2010)

Automatic Security Baseline Verification Method Based on SCAP and Cloud Scanning

Yanzhou Chen[1]([⊠]), Qi Wang[1], Meng Sun[1], Peng Chen[1],
Zhizhong Qiao[2], and Zhangguo Chen[2]

[1] CSG Power Dispatching Control Center, Guangzhou 510670, China
yanzhou.chen@foxmail.com
[2] NARI Information & Communication Technology Co., Ltd.,
Nanjing 210033, China

Abstract. With the development of power networks, automated verification of security baselines has become increasingly important. Traditional verification methods have disadvantages such as low efficiency, inability to centralize management, and difficulty in maintaining and upgrading. In this paper, we proposed a method of automatically checking the security baseline based on the security baseline model and using the SCAP standard combined with efficient Cloud scanning technology. Our method not only improves efficiency, but also facilitates centralized management and maintenance of upgrades.

Keywords: Automatic security baseline verification · SCAP ·
Cloud scanning · Power networks

1 Introduction

With the increase in the degree of informatization demand of power companies, the dependence of business logic and process systems on information systems will continue to increase, and the value of information will continue to be promoted. The information security protection work of power companies is a key component of information construction, which directly restricts the production process, the stable development of the industry, and the operational effectiveness of the core business. In recent years, the information security protection facilities of power companies are gradually becoming more and more perfect. Through the unified safety planning of enterprises, it has been possible to successfully implement the unique security protection mechanism and supervision model of the power industry.

The security baseline is the minimum security control for maintaining the confidentiality, integrity, and reliability of information systems, and is the minimum security guarantee for information systems [1]. The construction of a safety baseline is the first step in the safety engineering of a system. The basic components of the security baseline are as shown as Fig. 1.

As shown in the figure, there is a wide range of security baselines, including security vulnerability, security configuration, and system status [2].

© Springer Nature Switzerland AG 2020
Q. Liu et al. (Eds.): CENet 2018, AISC 905, pp. 944–954, 2020.
https://doi.org/10.1007/978-3-030-14680-1_102

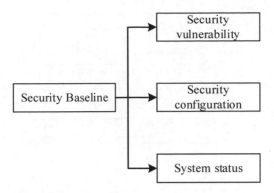

Fig. 1. Basic components of the security baseline.

(1) Security vulnerability: Refers to the defects of the information system itself, which enables an attacker to access or destroy the system without authorization, and achieve the purpose of illegal use or malicious destruction.

(2) Security configuration: Including the system's account password, authorization, auditing and other aspects of security management configuration. Security configuration problems reflect the security vulnerabilities of system and should be actively avoided.

(3) System status: The network port status, process status, service operation status, account status, and monitoring of changes in important documents during system operation. These contents reflect the current security environment of the power network system and the dynamics of system operation.

Security baselines are now widely used, such as security event advance alarm analysis [3], enterprise application [4], power information system [5], inspection of big data component [6], and so on. At present, the main methods and existing problems in the security baseline verification of power networks are shown in Table 1.

Taking full consideration of the status quo of the industry and industry best practices, and referring to the relevant security policy documents, and inheriting and absorbing the experience of national grade protection and risk assessment, we have established a security baseline model based on the power network system [7], as shown in the Fig. 2.

Table 1. Main methods and existing problems in the security baseline verification

Security baseline verification method	Existing problems
Manual verification	Low efficiency Difficult to guarantee the verification quality
Script/specific tool verification	Difficult to maintain Unable to centralize management
Software interface extension customization	Need secondary development Difficult to maintain and upgrade

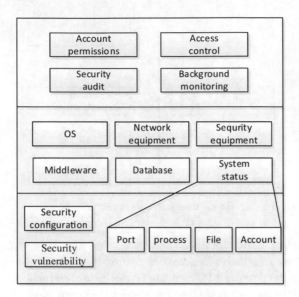

Fig. 2. Security baseline model.

Our goal is to make security measurable and manageable [8]. By using this baseline model, this paper proposes an automated security baseline verification method based on SCAP and Cloud scan to achieve two goals for the security network of power networks.

(1) Standardized verification: A security baseline (i.e., checklist) for power monitoring systems is constructed. These checklists consist of security vulnerability, security configuration and other related inspection contents, providing a framework for standardized security operations;

(2) Automated verification: Based on the characteristics of the power monitoring system, standardized check contents and methods are adopted and executed through automated tools to provide basis for the power monitoring system and equipment configuration compliance.

The outline of this paper is as follows. The basic knowledge of SCAP and Cloud scanning technology are briefly reviewed in Sect. 2, the method proposed in this paper is described in detail in Sect. 3. Subsequently, we perform an comparison experiment and analyze the time consumption and accuracy in Sect. 4, and finally, we made a brief conclusion.

2 Preliminary Knowledge

Before the formal introduction of the algorithm, we first introduce the preliminary knowledge of the algorithm and the main techniques.

2.1   SCAP

With the rapid development of computer technology, the misconfiguration of some computer systems has led to more and more security vulnerabilities [9]. SCAP (Security Content Automation Protocol) [10] is proposed by the NIST (National Institute of Standards and Technology), and NIST expects to use SCAP to solve three difficult problems: The first is to achieve the implementation of high-level policies and regulations (such as FISMA, ISO27000 series, etc.) to the bottom of the implementation. The second is to standardize the various elements involved in information security (such as the naming and severity measurement of uniform vulnerabilities). The third is to automate complex system configuration verification. SCAP provides a standardized, automated approach to maintaining the security of enterprise systems through standardized expressions, the use of secure data, and the assessment of security issues. For example, by implementing a security configuration baseline, verifying a patch, performing continuous monitoring of system security configuration settings, checking the system's compliance flags and giving the system's security status at any given time. Its standardized, automated thinking has had a profound impact on the information security industry [10].

Because of the large number and diversity of systems that require protection, the constant emergence of new threats, and the lack of interoperability, SACP has evolved further, extending from the original only to the Windows development Checklist to Unix, Series systems, web browsers, firewalls, and anti-virus software. The formulation of the SCAP standard provides an important guarantee for the realization of a unified automated safety baseline inspection system and the ability to give the safety status of the system in real time [11].

NIST explains SCAP in two ways: Protocol and Content. The protocol means that SCAP consists of a series of existing public standards, which are called SCAP Elements. The Protocol specifies how these Elements interact with each other. Content refers to the data generated by the Element description and applied to the actual inspection work according to the protocol. For example, official checklist data formats such as FDCC (Federal Desktop Core-Configuration) and USGCB (United States Government Configuration Baseline) are SCAP Content.

SCAP version 1.0 contains the following six SCAP Elements: XCCDF, OVAL, CVE, CCE, CPE and CVSS. These standards existed before SCAP was created and played an important role in their respective fields. SCAP provides solutions for the standardization of security tools: standard input data formats, standard processing methods and standard output data formats, which is very conducive to the exchange of data between security tools. SCAP elements can be divided into the following three types:

(1) Language class, the standard used to describe the evaluation content and evaluation method, including XCCDF and OVAL (1.2 SCAP [12] adds OCIL);
(2) Enumeration class, describing the naming format of the assessment object or configuration item, and providing a library that follows these naming, including CVE, CCE, CPE;

(3) Metrics class, which provide a metric for quantifying the evaluation results. The corresponding element is CVSS (1.2 CCAP added CCSS). The relationship between the SCAP elements is illustrated in the Fig. 3.

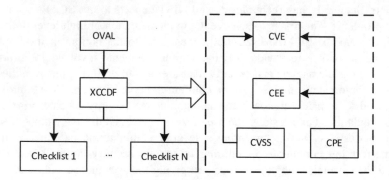

Fig. 3. Relationship between SCAP elements.

2.2 Cloud Scanning Technology

Cloud scanning technology is an instantiation of cloud computing technology which have attracted more and more attentions of scholars [13]. Currently, information scans and killings against Trojans and viruses are common in the field of information security. Such as 360 cloud scanning, trend officescan 10, Norton 360 and other domestic security vendors virus killing products [14]. In the cloud scanning process, each host is a cloud front end (i.e. cloud), and each cloud can acquire the capability support from the cloud backend to scan spontaneously, and feed the front end scan results back to the centralized processing and analysis, and finally feedback the results to the user. The core idea of the cloud-based security configuration verification mechanism is as follows: A verification cloud is composed of all network hosts to be checked and the baseline security check master device, and each host is a cloud. The cloud has the capability of spontaneous verification and can initiate verification. The results of the verification are fed back to the main equipment through the network for unified processing and analysis, and the main equipment gives the relevant verification guidance and feedback to the user.

Combined with the following figure, this mechanism is analyzed by the Fig. 4.

(1) Cloud verification: Could End has self-checking capabilities that can be supported by the Scan Server, such as upgrading the cloud;
(2) Back-end processing: processing and analysis work is delivered to the back-end, and the back-end's powerful processing and analysis capabilities are fully utilized;
(3) Two-way interaction: The cloud and the back end have two-way interaction and clear responsibilities.

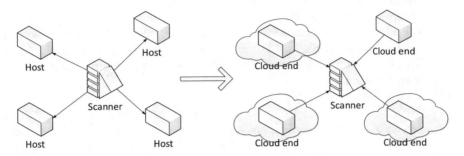

Fig. 4. Comparison between existing scanning mechanism and cloud scanning mechanism.

3   Automatic Security Baseline Verification Method Based on SCAP and Cloud Scan

The architectural design of this method is shown in the Fig. 5.

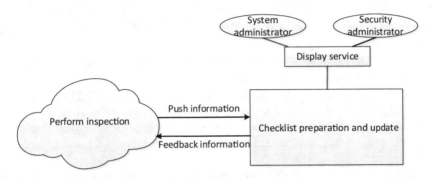

Fig. 5. Architectural design of our method.

3.1   Establishing the Security Baseline Checklist

According to the power network security baseline model, checklists for detailed vulnerability requirements, configuration requirements, and status requirements for power network systems can be formed, which provide operational and executable standards for standardized and automated technical security operations [15]. The security baseline checklist mainly includes three aspects:

(1) Security vulnerability checklist
When the security vulnerability baseline is established, it is first necessary to meet and comply with the security specification requirements, and consider the technical reasons for the patching of defects and limit the range of vulnerability of the assets. This scope is used as the minimum standard for assets in terms of security

breaches, that is, the baseline of the vulnerability. Standard implementation of vulnerability baselines is generally the white list.

(2) Security configuration checklist

The security configuration checklist refers to the standards that must be met to meet the requirements of the security specification and asset security configuration. It is measured by checking whether the security configuration parameters meet the standards.

(3) System status checklist

The system status checklist is the most important and most critical content. It refers to a set of process whitelists, port whitelists, account whitelists, and important file lists that are included on a specific asset, and is used as a measure of security.

The specific content of security verification will be carried out around these three levels, and a specific security checklist will be established. The verification items are formulated according to each product model and usage. The specific contents can be divided into accounts, passwords, authorizations, logs, IP protocols, etc., and mainly relate to items that have a large impact on the safety of equipment or systems.

3.2   Automatic Verification of Security Baseline

After the Checklist is determined, the security baseline is checked through the cloud scanning technology. The verification process is divided into the following three stages.

(1) Information push phase

The corresponding security configuration information is collected on the target device, and then the obtained information is transmitted to the baseline inspection system and stored in the database. The collected vulnerability configuration information of the detection target includes account management, password management, authorization management, log configuration, communication protocol, screen protection, shared folder and access rights, patch management, and startup items.

According to the scheduled task time, the cloud device logs in to the main equipment through a connection protocol such as Telnet or SSH. The cloud first verifies the user name, password, and other information of the host. After the verification succeeds, the security baseline verification starts. The security baseline checklist, data collection template, and security knowledge base developed in the first step are pushed to the cloud to initialize the cloud environment.

(2) Inspection and feedback phase

After the cloud receives the data collection template, it performs data collection on the database. After this task is completed, the automatic verification of the security baseline is performed through cloud scanning. The fragile configuration information and collected data information stored in the database are extracted and compared with the security baseline checklist, the security knowledge base, and the vulnerability database, and the non-conformances of the target device security configuration information are formed, and the result is fed back to the target device.

(3) Forming a reporting phase

After the main equipment receives the result automatically generated by the cloud, it will analyze and process the result and generate an analysis report according to the regulations and requirements.

3.3 Update of Security Baseline

Security baselines are not static. When the security baseline is used for verification work, different security baselines may exist due to different devices or different functions, at the same time, new vulnerabilities and virus databases may also be generated due to new viruses and vulnerabilities. Therefore, the security baseline should also be adjusted according to different requirements and situations so as to avoid some new risks affecting the power network. Security baselines are mainly updated based on international/national standards, corporate standards, security regulations, and vulnerability vaults. The checklist of the security baseline is adjusted by the management personnel or security personnel in accordance with changes in the security baseline, and is saved as a new baseline.

4   Experiment and Analysis

In order to fully verify the availability, effectiveness, and accuracy of the cloud scanning method, an experimental environment was set up in a non-clustered network reachable environment. A total of 15 hosts equipped with Windows system, were selected as the check target set. We performed security baseline-related configurations on these 15 computers, and then examined these computers in different ways. Five groups of comparison experiments were conducted. The results of the inspections were analyzed for efficiency and accuracy.

First compare the efficiencies of the three methods (Manual verification, Script verification, and our method). The security baselines of 15 computers were checked using three methods respectively. The Table 2 shows the average time required by the three methods (as shown in Fig. 6).

Table 2. Time consumption of three methods (second).

	1	2	3	4	5	6	7	8	9	10
Manual verification	273	305	292	287	281	282	277	293	275	294
Script verification	59	54	57	55	57	52	55	51	58	53
Our method	36	37	43	35	33	40	39	38	31	43

In addition, the Table 3 gives a comparison of the generalization performance of the three methods and compares the correctness of the three methods (as shown in Fig. 7).

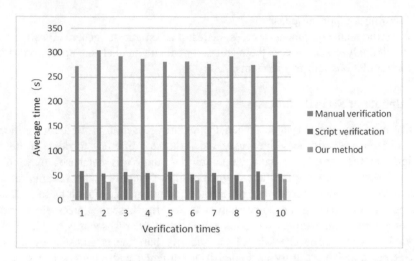

Fig. 6. Comparison of time consumption of Manual verification, Script verification and our method.

Table 3. Accuracy of three methods (%).

	1	2	3	4	5	6	7	8	9	10
Manual verification	87.7	79.3	82.9	89.7	83.5	82.8	78.5	86.8	83.3	78.8
Script verification	93	95.7	94.2	94.1	94.2	94.9	95.6	93.7	94	93.4
Our method	98.6	99.1	98.3	98.7	99.2	97.1	99.6	99.8	99.5	98.6

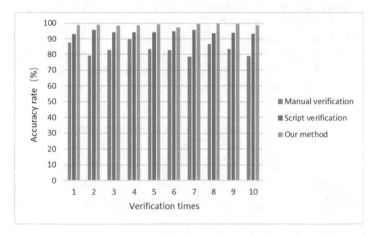

Fig. 7. The accuracy comparison among Manual verification, Script verification and our method.

From the comparison of the experimental results, it can be seen intuitively that manual verification takes much longer than the script verification and the method of this paper, and the accuracy is also lower than other methods and is not stable. The script method is somewhat better than the manual verification method, but its efficiency and accuracy are also not as good as this method. Therefore, the method proposed in this paper is better than the other two methods in terms of efficiency and generalization performance. This method has certain advantages for automated verification of security baselines.

5 Conclusion

This article proposes a method of automated verification of a security baseline. This method follows the security baseline framework and SCAP standards, combined with cloud scanning, and uses its efficiency to verify security baselines. From the experimental results, compared to other security baseline verification methods, this method has better results both in terms of efficiency and accuracy. This method can perform safety baseline verification with relatively high accuracy in a relatively short period of time.

Acknowledgement. This work is supported by Science and Technology Project of China Southern Power Grid Co., Ltd. "Research and Demonstration of Key Technologies of Network Security Situational Awareness in Power Monitoring System" (No. ZDKJXM20170002).

References

1. Zhang, X., Chen, X.H., Liu, X.: Construction of information security baseline standardization system for power systems. Electric Power Inf. Commun. Technol. **11**(11), 110–114 (2013). (in Chinese)
2. Gui, Y.H.: Study and applications of operation system security baseline. Comput. Secur. **10**, 11–15 (2011)
3. Kozlovszky, M.: Cloud security monitoring and vulnerability management. Crit. Infrastruct. Protect. Res. **12**, 265–269 (2016)
4. Chen, Z. H.: Security baseline management in the enterprise application. Computer Security, 2013
5. Gao, S., Wang, Q.Q.: A new security baseline reinforcement method for the power information system. Appl. Mech. Mater. 2407–2411 (2013)
6. Wang, K., Lu, Y.J.: Automated baseline inspection of big data components. Secur. Informatization **11**, 108–110 (2017). (in Chinese)
7. Shen, Z.H.: Application of security baseline management in enterprises. Comput. Secur. **3**, 19–21 (2013). (in Chinese)
8. Martin, R.A.: Making security measurable and manageable. In: IEEE Military Communications Conference, pp. 1–9 (2008)
9. Na, S., Kim, T., Kim, H.: A study on the classification of common vulnerabilities and exposures using Naïve Bayes. In: International Conference on Broadband and Wireless Computing, Communication and Applications, pp. 657–662. Springer International Publishing, Heidelberg (2016)

10. Radack, S., Kuhn, R.: Managing security: the security content automation protocol. IEEE Educational Activities Department (2011)
11. Li, C., Wang, W.: Application of safety baseline control in risk management process. Netw. Secur. Technol. Appl. **9**, 4–7 (2009). (in Chinese)
12. Waltermire, D., Quinn, S., Scarfone, K., Halbardier, A.: The technical specification for the security content automation protocol-SCAP: SCAP version 1.2 recommendations of the national institute of special publication 800–126 revision 2. Acta Obstetrica Et Gynaecologica Japonica **37**(5), 608–609 (2012)
13. Shi, W., Zhang, L., Wu, C., Li, Z., Laue, F.C.M.: An online auction framework for dynamic resource provisioning in cloud computing. IEEE/ACM Trans. Netw. **24**(4), 2060–2073 (2016)
14. Zhang, Z., Feng, W., Yan, J.T.: A security configuration baseline verification system and method based on the cloud scanning system. Telecommun. Eng. Technol. Stand. **5**(12), 20–23 (2012). (in Chinese)
15. Warrenl, W.J.B.M., Hutchinson, W.: A security evaluation criteria for baseline security standards. In: IFIP TC11 International Conference on Information Security: Visions and Perspectives, pp. 79–90. Kluwer (2002)

Improved Multi-swarm PSO Based Maintenance Schedule of Power Communication Network

Minchao Zhang[1($\boxtimes$)], Xingyu Chen[1], Yue Hou[2], and Guiping Zhou[3]

[1] Institute of Network Technology, Beijing University of Posts
and Telecommunications, No. 10, Xitucheng Road, Haidian District,
Beijing 100876, China
n0001n0600@163.com
[2] Beijing Guodiantong Network Technical Co. Ltd., The 28th Floor of Fortune
World in No. 1 Hangfeng Road, Fengtai District, Beijing, China
[3] State Grid Liaoning Electric Power Co., Ltd., No. 18, Ningbo Road,
Heping District, Shenyang 110006, Liaoning, China

Abstract. Maintenance schedule is an important and complex task in power communication system. This paper builds a maintenance schedule model that considers decreasing average waiting time of maintenance as well as some constraints. This paper uses Hadoop and MapReduce to handle huge amount of information in power communication network. An improved multi-swarm PSO (Particle Swarm Optimization) algorithm is proposed to schedule maintenance. This algorithm combines the MPSO algorithm with the bacterial chemotaxis. Experiment demonstrates accuracy and efficiency of the improved MPSO algorithm.

Keywords: Power communication network · Particle swarm optimization ·
Maintenance schedule · Dynamic learning factor · Population diversity

1 Introduction

The power communication network is another physical network of the power system besides the power grid. This network is an import guarantee for safe production of power grid. The power communication network brings a variety of communication businesses [1]. The states of these communication businesses have a great influence on power grid.

To ensure the normal operation of the power communication network, maintenance is of great significance. With the development of communication technology and the extension of the power communication network, it has been more and more impossible to schedule maintenance manually [2]. So designing an algorithm which can schedule maintenance properly is important.

In the era of big data, the information about equipment, business and their relationships of power communication network explodes. As a result, the amount of

© Springer Nature Switzerland AG 2020
Q. Liu et al. (Eds.): CENet 2018, AISC 905, pp. 955–963, 2020.
https://doi.org/10.1007/978-3-030-14680-1_103

maintenance works needed to be scheduled is huge. This paper uses Hadoop and MapReduce to analyze the influence caused by maintenance on the power communication network.

This paper proposes an improved MPSO (Multi-swarm Particle Swarm Optimization) algorithm. This algorithm combines bacterial chemotaxis with MPSO algorithm, enhancing the algorithm's global search performance further. Besides, this algorithm adjusts the learning factors dynamically to speed up algorithm's convergence.

2 Related Works

Because maintenance schedule is an important job for power communication network and power grid network, many studies have been conducted to solve the problem. Xie et al. propose an auto-generated method for power grid maintenance scheduling to improve the accuracy of security correction work of power grid maintenance [3]. Samuel et al. use a hybrid particle swarm optimization algorithm to schedule maintenance for generation units aiming at enhancing the reliability of the units [4]. Ming et al. propose a risk-based maintenance model to find the optimal maintenance schedule through the assessment of the risk rate of transmission equipment and the equilibrium of system security, risk and maintenance cost [5]. Those above papers research the problem of scheduling maintenance from different perspectives. This paper focuses on reducing the average waiting time of all maintenance works. Too long waiting time will reduce the maintenance work efficiency, waste maintenance resources and then cause a certain economic loss.

Maintenance scheduling is a multi-constrained nonlinear mixed integer programming problem, which belongs to non-polynomial hard problems [6]. This paper proposes an improved MPSO algorithm to solve the problem. PSO is a kind of swarm intelligence search algorithm which has the advantages of fewer parameters to be adjusted and lower difficulty in implementation. In addition, its convergence speed can meet the actual engineering needs [7]. But original PSO has the disadvantage of being trapped easily in local best solution, MPSO algorithm is an improved PSO algorithm, which divides the whole swarm into several sub-swarms, every sub-swarm searches best solution independently.

3 Power Communication Network Maintenance Schedule Model

3.1 Background Information About Communication Business

A communication business of power communication network has two routes: One is primary route and the other is alternate route. Normally a business runs in its primary route, but if primary route is interrupted, the business will switch to its alternate route.

3.2   Target Function

The target function is defined as follow:

$$Minimize\ f = \sum_{i=1}^{n} (T_i - E_i)/n \qquad (3.1)$$

Where n represents the number of maintenance works, T_i is the actual start time of maintenance i, E_i is the expected start time of maintenance i.

3.3   Constraints

1. Start time constraint

$$Early_i \leq T_i \leq Late_i \qquad (3.2)$$

Where $Early_i$ is the earliest start time of maintenance i, $Late_i$ is the latest start time of maintenance i.

2. Maintenance resource constraint

$$\sum_{i=1}^{n} R_t^i \leq R_t, t = 1, \ldots, T \qquad (3.3)$$

Where R_t^i is the resource needed by maintenance i at time t, R_t is the total resource available at time t.

3. Mutual constraint

$$[T_i T_i + L_i] \cap [T_j, T_j + L_j] = \emptyset \qquad (3.4)$$

Where L_i is the cost time of work i. This constraint means if maintenance i will break a business's primary route, and maintenance j will break the same business's alternate route, then maintenance i and maintenance j cannot be executed at same time.

4   Big Data Analysis Technology Usages on Maintenance Schedule

Nowadays, the 4Vs data (data with features of volume, variety, velocity and veracity) in power communication network is hardly handled by traditional tools. This paper uses Hadoop and MapReduce to analyze the influence caused by a maintenance work on business of power communication network.

Hadoop is a distributed framework that processes large data sets, which composed of MapReduce, a distributed data process model, and HDFS, a distributed file system [8].

Figure 1 shows the process of analyzing influence caused by a maintenance work using MapReduce.

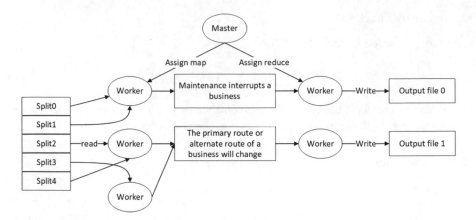

Fig. 1. The analyze procedure of maintenance work using MapReduce.

5 Improved Multi-swarm Particle Swarm Optimization

5.1 Multi-swarm Particle Swarm Optimization

In the original PSO algorithm, each particle in swarm only has velocity and position, without volume and mass, moving within the N-dimensional search spaces [9]. One particle represents a solution to the problem. In this paper, a particle represents a schedule plan. The quality of a particle is determined by the fitness.

As for the multi-swarm PSO algorithm, a swarm is divided into k sub-swarms $S_1, S_2, \ldots, S_k$ by choosing seed particles. Each seed particle has range γ. In this paper, euclidean distance is employed to represent the distance between two particles.

The strategy of dividing sub-swarms is: Firstly all particles are sorted according to the fitness, the first particle is chosen as seed particle, if the distance between the second particle and the first particle is less than γ, the second particle is allocated to sub-swarm whose seed is the first particle, else the second particle is chosen as another seed particle. Then we traverse all particles in swarm. If a particle does not belong to any sub-swarm or it belongs to several sub-swarm, it will be allocated to the nearest sub-swarm.

The velocity and position of a particle are updated by the following formulas:

$$\overrightarrow{V_i^{t+1}} = W\overrightarrow{V_i^t} + C_1R_1\left(\overrightarrow{P_i} - \overrightarrow{X_i^t}\right) + C_2R_2\left(\overrightarrow{Seed_s^t} - \overrightarrow{X_i^t}\right) + C_3R_3\left(\overrightarrow{G_{good}^t} - \overrightarrow{X_i^t}\right) \quad (5.1)$$

$$\overrightarrow{X_i^{t+1}} = \overrightarrow{X_i^t} + \overrightarrow{V_i^{t+1}} \quad (5.2)$$

Where $\overrightarrow{V_i^t}$ is the velocity of particle i at step t, $\overrightarrow{X_i^t}$ is the position of particle i at step t, C_1, C_2, C_3 are learning factors, W is the inertia weight, $R_1, R_2, R_3 \sim U(0,1)$, $\overrightarrow{P_i}$ is the personal best position of particle i at step t, $\overrightarrow{Seed_s^t}$ is the position of sub-swarm s's seed particle (particle i belongs to sub-swarm s) and $\overrightarrow{G_{good}^t}$ is the global best position of all particles at step t.

5.2 Improved Multi-swarm Particle Swarm Optimization

Adjustment of Learning Factor

Learning factor C_1 represents the personal affects on the position of particle, learning factor C_2 represents the sub-swarm affects on the position of particle, while learning factor C_3 represents the global affects on position of particle. At the beginning of algorithm, C_1 should be big while C_2 and C_3 should be small, making particles search widly. With the interaction increasing, C_1 should decrease while C_2 and C_3 should increase. So at the later period, particles can assemble around the best solution position quickly. The updates of C_1 and C_2 are as follows:

$$C_1 = 1 + cos^2 \left[\pi \left(1 - \frac{t}{t_{max}} \right) \right] \tag{5.3}$$

$$C_2 = 1 + cos^2 \left[\pi \left(\frac{t}{t_{max}} \right) \right] \tag{5.4}$$

$$C_3 = \ln \left(\frac{t}{t_{max}} + 1 \right) \tag{5.5}$$

Where t is the current interation, t_{max} is the maximum iteration.

Addition of Repulsion Operation

Bacteria tends to approach to a favorable environment when foraging, while escaping from the unfavorable environment. This paper combines the multi-swarm PSO algorithm with the behaviors of bacteria. The MPSO algorithm strengthens algorithm's global search performance through dividing particles into several sub-swarms. However, in a sub-swarm, the problem that diversity decreases quickly still exists. This paper adds the behavior of forcing particles away from bad positions in sub-swarm, making further efforts on algorithm's global search performance.

In this paper, "particles tend to the optimal area" is defined as attraction operation, while "particles tend to escape from the bad area" is defined as repulsion operation. The update rules of position in MPSO algorithm is attraction operation. As for the repulsion operation, the update rules are as follow:

$$\overrightarrow{V_i^{t+1}} = W\overrightarrow{V_i^t} - C_1 R_1 \left(\overrightarrow{W_i} - \overrightarrow{X_i^t} \right) - C_2 R_2 \left(\left(\overrightarrow{Rseed_s^t} - \overrightarrow{X_i^t} \right) \right) - C_3 R_3 \left(\overrightarrow{G_{bad}^t} - \overrightarrow{X_i^t} \right) \tag{5.6}$$

Where $\overrightarrow{W_i}$ is the personal worst position of particle i, $\overrightarrow{Rseed_s^t}$ is the worst position in sub-swarm s at step t, and $\overrightarrow{G_{bad}^t}$ is the global worst position of all particles at interation t.

Whether attraction operation or repulsion operation is adopted depends on the value of population diversity. It is calculated as follow:

$$Diversity = \frac{1}{|L||P|} * \sum_{i=1}^{|P|} \sqrt{\sum_{j=1}^{N} (X_{id}^t - \overline{S_d^t})^2} \qquad (5.7)$$

Where P is the number of particles in sub-swarm, L is the length of the longest diagonal in the search space, t is the current iteration, X_{id}^t is the value in the dimension d of the particle i at iteration t, $\overline{S_d^t}$ is the average value of particles in dimension d. The larger the value of diversity is, the more sparse the swarm is, the smaller the value of diversity is, the more dense the swarm is.

Two thresholds D_h and D_l are defined. When population diversity is greater than D_h, the algorithm performs attraction operation. When population diversity is less than D_l, the algorithm performs replusion operation. Otherwise, the algorithm keeps particles' velocity unchaned.

Flows of Improved MPSO Algorithm

Step 1: Initialize the particle swarm, generate the initial position and velocity of each particle randomly.

Step 2: Generate seed particles, then divide all particles into several sub-swarms.

Step 3: Calculate the fitness of each particle, update the individual best and worst solution for each particle, update the worst position in a sub-swarm, update the global best and worst position for all particles.

Step 4: Calculate the population diversity of all sub-swarms.

Step 5: For each sub-swarms, compare diversity with the two threholds: D_h and D_l. If diversity is greater than D_h, then perform the attraction operation; if diversity is less than D_l, perform the replusion operation; otherwise keep the velocity of particles unchanged.

Step 6: Repeat Steps 2–5 until the iteration reaches the maximum number.

6   Experimentation

The improved MPSO algorithm was tested using two test cases: monthly maintenance data from two area's power communication network of China. The scale of test data is shown in Table 1.

Table 1. Scale of test data.

Case	Area	Month	Number of equipment	Number of business	Number of maintenance
Case 1	JB	3	18400	16100	3200
Case 2	GZ	4	12700	14200	2500

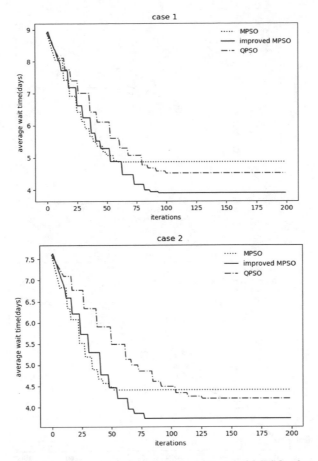

Fig. 2. Convergence process of MPSO, QPSO and improved MPSO of two test cases.

The improved MPSO algorithm was compared with MPSO algorithm and QPSO (quantum particle swarm optimization) algorithm [10]. The basic parameters of algorithm are set as p(number of particles) = 40, $t_{max} = 200$, $W_{max} = 0.9$, $W_{min} = 0.5$, $D_h = 0.5$, $D_l = 0.2$, $R_t = 480$, $\gamma = 5$. In order to reduce the amount of error, each algorithm was executed 10 times and the average values were adopted.

Figure 1 displays the convergence process of the three algorithms using test case 1 and test case 2. It can be seen from Fig. 2 that in the two test cases, the improved MPSO algorithm's final solution is better than the other two algorithms. We can draw the conclusion that the improved MPSO has the better global search performance than MPSO and QPSO. Figure 3 shows how average wait time changes when number of maintenance increases in power communication network of two test cases. It can be seen from Fig. 3 that when maintenance works increases, the improved MPSO is always better than the other two algorithms. Besides, the improved MPSO is more steady than MPSO and QPSO.

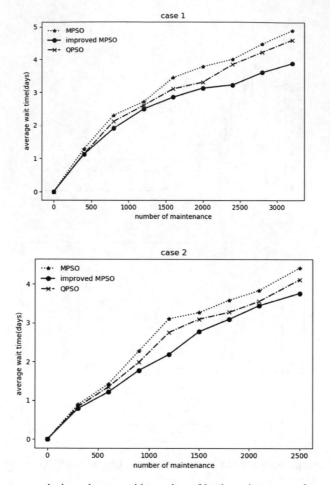

Fig. 3. Average wait time changes with number of business increases of two test cases.

7 Conclusion

Targeting at the problem of maintenance schedule in power communication network, this paper builds the maintenance schedule model to minimize the average wait time of all maintenance works with the help of big data analysis technology. An improved MPSO algorithm is proposed to solve the problem. The proposed algorithm adjusts the learning factors dynamically. Moreover, this algorithm adds repulsion operations to MPSO algorithm, which strengthens the algorithm's global search performance further. Finally the example demonstrated the effectiveness of this improved MPSO algorithm.

Acknowledgement. This work is supported by National Key R&D Program of China (2016YFB0901200).

References

1. Jiang, D., Tian, D., Liu, X., Shen, Z.: Security analysis of topology structure of electric power communication network. In: 2016 First IEEE International Conference on Computer Communication and the Internet, Wuhan, China, pp. 76–79. IEEE (2016)
2. Zhu, Q., Peng, H., Timmermans, B., van Houtum, G.J.: A condition-based maintenance model for a single component in a system with scheduled and unscheduled downs. Int. J. Prod. Econ. **193**, 365–380 (2017)
3. Xie, C., Liu, W., Wen, J., Wang, J.: An auto-generated method of day-ahead forecast powerflow for security correction of power grid maintenance scheduling. In: 2012 Asia-Pacific Power and Energy Engineering Conference, Shanghai, China, pp. 1–4. IEEE (2012)
4. Samuel, G.G., Rajan, C.C.A.: Hybrid particle swarm optimization – genetic algorithm and particle swarm optimization – evolutionary programming for long-term generation maintenance scheduling. In: 2013 International Conference on Renewable Energy and Sustainable Energy, Coimbatore, India, pp. 237–232. IEEE (2014)
5. Zeng, M., Huang, L., Qiu, L., Tian, K., The risk-based optimal maintenance scheduling for transmission system in smart grid. In: 2010 International Conference on Electrical and Control Engineering, Wuhan, China, pp. 4446–4449. IEEE (2010)
6. Suresh, K., Kumarappan, N.: Coordination mechanism of maintenance scheduling using modified PSO in a restructured power market. In: 2013 IEEE Symposium on Computational Intelligence in Scheduling, Singapore, Singapore, pp. 36–43. IEEE (2013)
7. Wilhelm, P.A.: Pheromone Particle Swarm Optimization of Stochastic Systems. Iowa State University, Ames (2008)
8. Thuraisingham, B., Khan, L.R., Husain, M.F.: Data intensive query processing for semantic web data using Hadoop and MapReduce. The University of Texas at Dallas, Richardson (2011)
9. Gaonkar, V., Nanannavar, R.B., Manjunatha: Power system congestion management using sensitivity analysis and particle swarm optimization. In: 2017 International Conference on Energy, Communication, Data Analytics and Soft Computing, Chennai, India, pp. 1268–1271 (2017)
10. Yang, L.H., Wang, Y.J., Zhu, C.M.: Study on fuzzy energy management strategy of parallel hybrid vehicle based on quantum PSO algorithm. Int. J. Multimed. Ubiquit. Eng. **11**(05), 147–158 (2016)

Author Index

ted States